Study Guide to Accompany
Human Anatomy and Physiology

Third Edition

Elaine N. Marieb, R.N., Ph.D.
Holyoke Community College

with Contributions from
Marie Fitzgerald
University of Massachusetts

and
Margaret G. Ott
Tyler Junior College

The Benjamin/Cummings Publishing Company, Inc.

Redwood City, California • Menlo Park, California
Reading, Massachusetts • New York • Don Mills, Ontario
Wokingham, U.K. • Amsterdam • Bonn • Sydney
Singapore • Tokyo • Madrid • San Juan

Science Executive Editor: Johanna Schmid

Assistant Editor: Natasha Banta

Production Supervisor: Larry Olsen

Production Editor: Brian Jones

Copyeditor: Betsy Dilernia

Text Designer: Gary Head

Artists: Barbara Cousins, Sibyl Graber-Gerig, Pauline Phung, and Ben Turner Graphics

Compositor: Fog Press

Cover Designer: Yvo Riezebos

Cover Photo: Copyright © Annie Leibovitz/Contact Press Images

Figure 3.10 courtesy of Carolina Biological Supply

Library of Congress Cataloging-in-Publication Data
Study guide to accompany human anatomy and physiology
         p.   cm.

   1. Human physiology—Problems, exercises, etc.   2. Human anatomy—Problems, exercises, etc.
QP31.2.M36 1991 Suppl.
612'.0076—dc20   94-39756
                                                                     CIP

       4 5 6 7 8 9 10—CRK—98 97

ISBN 0-8053-4283-4

The Benjamin/Cummings Publishing Company, Inc.
390 Bridge Parkway
Redwood City, California 94065

Preface

This *Study Guide* is intended to help health professional and life science students master the basic concepts of human anatomy and physiology through review and reinforcement exercises. The order of topics reflects the organization of *Human Anatomy and Physiology,* Third Edition, by Elaine N. Marieb, and each chapter begins with a listing of *Student Objectives* taken from that textbook. However, this *Study Guide* may be used with other human anatomy and physiology textbooks as well.

Scope

Exercises in this *Study Guide* review human anatomy from microscopic to macroscopic levels. Topics range from simple chemistry to entire body systems. Pathophysiology is briefly introduced with each system so that students can apply their learning. Wherever relevant, clinical aspects of study are covered—for example, muscles used as injection sites, the role of ciliated cells in protecting the respiratory tract, and reasons for skin ulcer formation. Developmental stages of youth, adulthood, and old age are emphasized to encourage conceptualization of the human body as a dynamic and continually changing organism.

Learning Aids

Each chapter is divided into three major sections. The first section, called **Building the Framework,** walks the student through the chapter material more or less in the order in which it is presented in the corresponding chapter of the textbook. This section is intentionally inviting and uses a variety of techniques to involve the student in the learning process. The use of coloring exercises helps promote visualization of key structures and processes. Coloring exercises have been shown to be a unique motivating approach for learning and reinforcement. Each illustration has been carefully prepared to show sufficient detail for learning key concepts without overwhelming the student with complexity or tedious repetition of coloring. There are over 120 coloring exercises in this *Study Guide*. When completed, the colored diagrams provide an ideal reference and review tool.

Other question formats in this section include selecting from key choices, matching terms with appropriate descriptions, defining important terms, and labeling diagrams. Elimination questions require the student to discover similarities and dissimilarities among a number of structures or processes. Correcting true/false questions adds a new dimension to this traditional exercise format. In addition, students are asked to construct graphs and complete tables, exercises that not only reinforce learning but also provide a handy study aid. If applicable, each **Building the Framework** section ends with a visualization exercise, another unique feature of this *Study Guide.* These *Incredible Journey* exercises ask students to imagine themselves in miniature traveling through various organs and systems within the body. Each organ system (as well as the topic of chemistry) is surveyed by a visualization exercise that summarizes the content of the chapter. These exercises allow the student to consider what has been learned from another point of view in an unusual and dynamic way.

The second major section of each chapter, called **Challenging Yourself,** typically consists of two groups of questions: Questions in *At the Clinic* focus on applying knowledge to clinical situations, whereas those in *Stop and Think* stress comprehension of principles pertaining to nonclinical situations. For the most part, the clinical questions are written to approximate real situations and require short answers. The *Stop and Think* questions require critical thinking. They cross the lines between topics and prod the student to put two and two together, to synthesize old and new information, and to think logically. Occasionally, a third group of questions appears in the **Challenging Yourself** section. Called *Closer Connections: Checking the Systems,* these questions prod the student to see the connections between body systems and to describe the interrelationships considered in previous chapters.

The final section of each chapter is called **Covering All Your Bases.** This section reintroduces concepts already covered but tests them via different formats. It includes multiple-choice questions and a section called *Word Dissection* that asks the student to define word roots encountered in the chapter and to come up with an example. Although this section is perhaps less interesting than the first two, knowing word roots is an advantage in studying A&P, and the multiple-choice questions are much more challenging than they might at first appear. Most have more than one correct answer, and the student is asked to choose *all* the correct responses. When this third and final section has been completed, the student will have been exposed to essentially all the basic and necessary information of the related textbook chapter.

Answers for all the exercises are provided at the back of the book in Appendix A.

Throughout this *Study Guide,* selected illustrations include a "walking man" icon to indicate where learning can be enhanced through the use of A.D.A.M. software. Appendix B, at the back of this *Study Guide,* correlates these illustrations (as well as the corresponding figures in the textbook) to key structures, views, and layers found in the Standard version of A.D.A.M.

Acknowledgments

I would like to thank Rose Leigh Vines, at California State University, Sacramento, for providing a strong and helpful link (Appendix B) between this *Study Guide* and A.D.A.M. software. I also thank Marie Fitzgerald, at the University of Massachusetts, and Betsy Ott, at Tyler Junior College, both of whom contributed questions to this or previous editions. I am also grateful to those educators, colleagues, and students who provided suggestions during preparation of this *Study Guide,* and to the staff at Benjamin/Cummings, who continually supported my efforts to create a *Study Guide* that will benefit both educators and students. I particularly want to thank Thor Ekstrom for his diligent manuscript preparation. Special kudos to Brian Jones, the production editor on this edition. His "can-do" attitude made the revision as easy a process as possible. My sincere thanks go to Larry Olsen, Production Supervisor, who assigned Brian to this "job." I also appreciate the diligence of Betsy Dilernia, the copyeditor, and the support provided by my editor, Johanna Schmid. I also wish to acknowledge with gratitude the excellent drawings by Barbara Cousins, whose work so beautifully complements that of Sibyl Graber-Gerig, Pauline Phung, and Ben Turner Graphics.

Instructions for the Student: How to Use This Book

This *Study Guide to Accompany Human Anatomy and Physiology* is the outcome of years of personal attempts to find and create exercises most helpful to my own students when they study and review for a lecture test or laboratory quiz.

Although I never cease to be amazed at how remarkable the human body is, I would never try to convince you that studying it is easy. The study of human anatomy and physiology has its own special terminology. It requires that you become familiar with the basic concepts of chemistry to understand physiology, and often it requires rote memorization of facts. It is my hope that this *Study Guide* will help simplify your task. To make the most of the exercises, carefully read these descriptions of what to expect before starting work.

This *Study Guide* has three sections—each with a special focus and purpose—as explained here.

1. Building the Framework

This first section of each chapter is the most varied. It begins with a list of *Student Objectives* (keyed to the text) and covers virtually all of the major points you are expected to know about the information in the corresponding chapter. The types of exercises you can expect to see are described below.

• **Labeling and Coloring.** Some of these questions ask you only to label a diagram, but most also ask that you do some coloring of the figure. You can usually choose whichever colors you prefer. Soft colored pencils are recommended so that the underlying diagram shows through. Since most figures have multiple parts to color, you will need a variety of colors—18 should suffice. In the coloring exercises, you are asked to choose a particular color for each structure to be colored. That color is then used for both a color-coding circle found next to the name of the structure or organ, and the structure or organ on the figure. This allows you to identify the colored structure quickly and by name in cases where the diagram is not labeled. In a few cases, you are given specific coloring instructions to follow.

• **Matching.** Here you are asked to match a term denoting a structure or physiological process with a descriptive phrase or sentence. In many exercises, some terms are used more than once, and others are not used at all.

• **Completion.** You are asked to select the correct term or phrase to answer a specific question, or to fill in blanks to complete a sentence or a table or chart.

• **Definitions.** You are asked to provide a brief definition of a particular structure or process.

• **True/False.** One word or a phrase is underlined in a sentence. You decide if the sentence is true as it is written. If if is not, you are asked to correct the underlined word or phrase.

• **Elimination.** Here you are asked to find the term that does not belong in a particular grouping of related terms. In this type of exercise, you must analyze how each term is like or different from the others.

• **Graphs.** You are given data and asked to plot them on a graph. Graphs are important because they enable you to pinpoint important variations in some physiological factor.

• **Visualization.** *The Incredible Journey* is a special type of completion exercise. For these exercises, you are asked to imagine that you have been miniaturized and injected into the body of a human being (your host). Anatomical landmarks and physiological events are described from your miniaturized viewpoint, and you are then asked to identify your observations. Although these exercises are optional, my students have found them fun to complete, and I hope you will, too.

Each exercise has complete instructions. Read through them before beginning. When there are multiple instructions, complete them in the order given.

2. Challenging Yourself

This second section of each chapter consists of short-answer essay questions. Don't worry; as long as you've studied, you do have the information you need in your memory bank to cope with this section. Its two major parts, *At the Clinic* and *Stop and Think,* ask you to apply your new learning to both clinical and nonclinical situations. At first, you may have to stretch a little to answer some of these questions, but stick with it—success builds. In cases where you have been unable to come up with the correct answer, pay particular attention to just what the right answer is, and try to analyze just how that answer was arrived at (that is, the problem-solving route). This approach will help you develop those needed problem-solving skills. Occasionally, this section will contain a third part called *Closer Connections: Checking the Systems,* which asks you to figure out how the system currently being studied relates to or interacts with other body systems.

3. Covering All Your Bases

This third section of each chapter is for the student "darned and determined" to do well in the course. Using multiple-choice questions, this section retests information already tested in the first two sections. But watch out, because the multiple-choice questions are special. You are not just looking for that one right answer. While there may be only one right answer, it is more likely that there will be two, three, or even four right answers to a given question. If you can "catch" them all, you really have this chapter in your pocket! The other part of this section is the *Word Dissection* exercise, which tests your understanding of word roots used in the chapter. Do not ignore this section. Once you learn word roots, your (A&P) life becomes a lot easier.

At times, it may appear that information is being duplicated in the different types of exercises. Although there *is* some duplication, the concepts being tested reflect different vantage points or approaches in the different exercises. Remember, when you understand a concept from several different perspectives, you have mastered that concept.

Using A.D.A.M.

You will notice that some illustrations include a "walking man" icon. When you see this icon, turn to Appendix B in the back of the *Study Guide* to find out how you can learn more anatomy by using A.D.A.M. software.

I sincerely hope that this *Study Guide* challenges you to increase your knowledge, comprehension, retention, and appreciation of the structure and function of the human body. Please write and let me know what you like or dislike about it (and why). This will help me tailor the next edition even more to the needs of "my" A&P students.

Good luck!

Elaine Marieb

Department of Science, Engineering, and Math
Holyoke Community College
303 Homestead Avenue
Holyoke, Massachusetts 01030

Contents

1

The Human Body: An Orientation   1

BUILDING THE FRAMEWORK   2
An Overview of Anatomy and
Physiology   2
The Hierarchy of Structural Organization   3
Maintaining Life   8
Homeostasis   9
The Language of Anatomy   11
CHALLENGING YOURSELF   17
At the Clinic   17
Stop and Think   18
COVERING ALL YOUR BASES   20
Multiple Choice   20
Word Dissection   23

2

Chemistry Comes Alive   24

BUILDING THE FRAMEWORK   25
Part I: Basic Chemistry   25
Definition of Concepts: Matter and
Energy   25
Composition of Matter: Atoms and
Elements   26
How Matter Is Combined: Molecules and
Mixtures   27
Chemical Bonds   29
Chemical Reactions   32
**Part II: Biochemistry: The Composition
and Reactions of Living Matter   33**
Inorganic Compounds   33
Organic Compounds   35
The Incredible Journey: A Visualization
Exercise for Biochemistry   42

CHALLENGING YOURSELF   44
At the Clinic   44
Stop and Think   45
COVERING ALL YOUR BASES   48
Multiple Choice   48
Word Dissection   51

3

Cells: The Living Units   52

BUILDING THE FRAMEWORK   53
Overview of the Cellular Basis of Life   53
The Plasma Membrane: Structure and
Functions   54
The Cytoplasm   60
The Nucleus   64
Cell Growth and Reproduction   65
Extracellular Materials   70
Developmental Aspects of Cells   70
The Incredible Journey: A Visualization
Exercise for the Cell   71
CHALLENGING YOURSELF   72
At the Clinic   72
Stop and Think   74
COVERING ALL YOUR BASES   76
Multiple Choice   76
Word Dissection   78

4

Tissues: The Living Fabric   80

BUILDING THE FRAMEWORK   81
Overview of Body Tissues   81
Epithelial Tissue   84
Connective Tissue   88
Epithelial Membranes   91
Muscle Tissue   92
Nervous Tissue   93
Tissue Repair   93
Developmental Aspects of Tissues   94
CHALLENGING YOURSELF   95
At the Clinic   95
Stop and Think   96
COVERING ALL YOUR BASES   97
Multiple Choice   97
Word Dissection   100

5

The Integumentary System   101

BUILDING THE FRAMEWORK   102
The Skin   102
Appendages of the Skin   106
Functions of the Integumentary System   109
Homeostatic Imbalances of Skin   110
Developmental Aspects of the Integumentary
    System   111
The Incredible Journey: A Visualization
    Exercise for the Skin   111
CHALLENGING YOURSELF   113
At the Clinic   113
Stop and Think   114
COVERING ALL YOUR BASES   116
Multiple Choice   116
Word Dissection   118

6

Bones and Bone Tissue   120

BUILDING THE FRAMEWORK   121
Skeletal Cartilages   121
Functions of the Bones   122
Classification of Bones   122
Bone Structure   123
Bone Development (Osteogenesis)   126
Bone Homeostasis: Remodeling and
    Repair   129
Homeostatic Imbalances of Bone   134
Developmental Aspects of Bones: Timing
    of Events   134
The Incredible Journey: A Visualization
    Exercise for the Skeletal System   136
CHALLENGING YOURSELF   137
At the Clinic   137
Stop and Think   138
COVERING ALL YOUR BASES   140
Multiple Choice   140
Word Dissection   142

7

The Skeleton   143

BUILDING THE FRAMEWORK   144
Part I: The Axial Skeleton   146
The Skull   146
The Vertebral Column   150
The Bony Thorax   154
**Part II: The Appendicular
    Skeleton   155**
The Pectoral (Shoulder) Girdle and
    the Upper Limb   155
The Pelvic (Hip) Girdle and the
    Lower Limb   160
Developmental Aspects of the
    Skeleton   163
CHALLENGING YOURSELF   164
At the Clinic   164
Stop and Think   166
COVERING ALL YOUR BASES   168
Multiple Choice   168
Word Dissection   170

⑧

Joints   172

BUILDING THE FRAMEWORK   173
Classification of Joints   173
Fibrous, Cartilaginous, and
   Synovial Joints   174
Homeostatic Imbalances of Joints   180
Developmental Aspects of Joints   181
The Incredible Journey: A Visualization
   Exercise for the Knee Joint   181
CHALLENGING YOURSELF   182
At the Clinic   182
Stop and Think   183
COVERING ALL YOUR BASES   185
Multiple Choice   185
Word Dissection   187

⑨

Muscles and Muscle Tissue   188

BUILDING THE FRAMEWORK   189
Overview of Muscle Tissues   189
Skeletal Muscle   191
Smooth Muscle   202
Developmental Aspects of Muscles   204
The Incredible Journey: A Visualization
   Exercise for Skeletal Muscle
   Tissue   205
CHALLENGING YOURSELF   206
At the Clinic   206
Stop and Think   207
COVERING ALL YOUR BASES   210
Multiple Choice   210
Word Dissection   213

10

The Muscular System   214

BUILDING THE FRAMEWORK   215
Muscle Mechanics: The Importance of
   Leverage and Fascicle Arrangement   215
Interactions of Skeletal Muscles in
   the Body   218
Naming Skeletal Muscles   218
Major Skeletal Muscles of the Body   219

CHALLENGING YOURSELF   241
At the Clinic   241
Stop and Think   243
Closer Connections: Checking the Systems—
   Covering, Support, and Movement   243
COVERING ALL YOUR BASES   244
Multiple Choice   244
Word Dissection   246

11

Fundamentals of the Nervous System and Nervous Tissue   247

BUILDING THE FRAMEWORK   248
Organization of the Nervous System   248
Histology of Nervous Tissue   250
Neurophysiology   256
Basic Concepts of Neural Integration   263
Developmental Aspects of Neurons   265
CHALLENGING YOURSELF   266
At the Clinic   266
Stop and Think   267
COVERING ALL YOUR BASES   269
Multiple Choice   269
Word Dissection   272

12

The Central Nervous System   273

BUILDING THE FRAMEWORK   274
The Brain   274
The Spinal Cord   289
Diagnostic Procedures for Assessing CNS
   Dysfunction   292
Developmental Aspects of the Central
   Nervous System   292
The Incredible Journey: A Visualization
   Exercise for the Nervous System   293
CHALLENGING YOURSELF   294
At the Clinic   294
Stop and Think   297
COVERING ALL YOUR BASES   299
Multiple Choice   299
Word Dissection   302

The Peripheral Nervous System and Reflex Activity 303

BUILDING THE FRAMEWORK 304
Overview of the Peripheral
Nervous System 304
Cranial Nerves 309
Spinal Nerves 312
Reflex Activity 316
Developmental Aspects of the Peripheral
Nervous System 320
CHALLENGING YOURSELF 320
At the Clinic 320
Stop and Think 322
COVERING ALL YOUR BASES 324
Multiple Choice 324
Word Dissection 326

The Autonomic Nervous System 327

BUILDING THE FRAMEWORK 328
Overview of the Autonomic Nervous
System 328
Anatomy of the Autonomic Nervous
System 330
Physiology of the Autonomic Nervous
System 333
Homeostatic Imbalances of the Autonomic
Nervous System 335
Developmental Aspects of the Autonomic
Nervous System 335
CHALLENGING YOURSELF 336
At the Clinic 336
Stop and Think 338
COVERING ALL YOUR BASES 339
Multiple Choice 339
Word Dissection 340

Neural Integration 341

BUILDING THE FRAMEWORK 342
Sensory Integration: From Reception to
Perception 342
Motor Integration: From Intention
to Effect 346
Higher Mental Functions 350
CHALLENGING YOURSELF 354
At the Clinic 354
Stop and Think 356
COVERING ALL YOUR BASES 357
Multiple Choice 357
Word Dissection 359

The Special Senses 361

BUILDING THE FRAMEWORK 362
The Chemical Senses: Taste and Smell 362
The Eye and Vision 365
The Ear: Hearing and Balance 374
Developmental Aspects of the
Special Senses 380
The Incredible Journey: A Visualization
Exercise for the Special Senses 381
CHALLENGING YOURSELF 382
At the Clinic 382
Stop and Think 384
COVERING ALL YOUR BASES 386
Multiple Choice 386
Word Dissection 389

The Endocrine System 391

BUILDING THE FRAMEWORK 392
The Endocrine System and Hormone
Function: An Overview 392
Hormones 394
Major Endocrine Organs of the Body 396
Other Endocrine Structures 404
Developmental Aspects of the Endocrine
System 404

The Incredible Journey: A Visualization
  Exercise for the Endocrine System   405
CHALLENGING YOURSELF   406
  At the Clinic   406
  Stop and Think   408
  Closer Connections: Checking the
  Systems—Regulation and Integration of
  the Body   409
COVERING ALL YOUR BASES   410
  Multiple Choice   410
  Word Dissection   412

18

Blood   413

BUILDING THE FRAMEWORK   414
  Overview: Composition and Functions of
  Blood   414
  Blood Plasma   415
  Formed Elements   416
  Hemostasis   422
  Transfusion and Blood Replacement   424
  Diagnostic Blood Tests   425
  Developmental Aspects of Blood   426
  The Incredible Journey: A Visualization
  Exercise for the Blood   426
CHALLENGING YOURSELF   428
  At the Clinic   428
  Stop and Think   431
COVERING ALL YOUR BASES   432
  Multiple Choice   432
  Word Dissection   435

19

The Cardiovascular System:
The Heart   436

BUILDING THE FRAMEWORK   437
  Heart Anatomy   437
  Blood Supply to the Heart: Coronary
  Circulation   441
  Properties of Cardiac Muscle Fibers   442
  Heart Physiology   444
  Developmental Aspects of the Heart   452
CHALLENGING YOURSELF   453
  At the Clinic   453
  Stop and Think   455

COVERING ALL YOUR BASES   456
  Multiple Choice   456
  Word Dissection   459

20

The Cardiovascular System:
Blood Vessels   460

BUILDING THE FRAMEWORK   461
  Overview of Blood Vessel Structure and
  Function   461
  Physiology of Circulation   466
  Circulatory Pathways: Blood Vessels of the
  Body   473
  Developmental Aspects of the
  Blood Vessels   486
  The Incredible Journey: A Visualization
  Exercise for the Circulatory System   487
CHALLENGING YOURSELF   488
  At the Clinic   488
  Stop and Think   491
COVERING ALL YOUR BASES   493
  Multiple Choice   493
  Word Dissection   495

21

The Lymphatic System   497

BUILDING THE FRAMEWORK   498
  Lymphatic Vessels   498
  Lymphoid Cells, Tissues, and Organs: An
  Overview   501
  Lymph Nodes   502
  Other Lymphoid Organs   504
  Developmental Aspects of the Lymphatic
  System   506
CHALLENGING YOURSELF   506
  At the Clinic   506
  Stop and Think   508
  Closer Connections: Checking the
  Systems—Circulation and Protection   508
COVERING ALL YOUR BASES   509
  Multiple Choice   509
  Word Dissection   511

22

Nonspecific Body Defenses and Immunity   512

BUILDING THE FRAMEWORK   513
  Part I: Nonspecific Body Defenses   513
  Surface Membrane Barriers   513
  Nonspecific Cellular and Chemical
    Defenses   514
  **Part II: Specific Body Defenses:
    Immunity   519**
  Antigens   519
  Cells of the Immune System:
    An Overview   519
  Humoral Immune Response   522
  Cell-Mediated Immune Response   524
  Homeostatic Imbalances of Immunity   528
  Developmental Aspects of the Immune
    System   530
  The Incredible Journey: A Visualization
    Exercise for the Immune System   531
CHALLENGING YOURSELF   532
  At the Clinic   532
  Stop and Think   534
COVERING ALL YOUR BASES   535
  Multiple Choice   535
  Word Dissection   537

23

The Respiratory System   538

BUILDING THE FRAMEWORK   539
  Functional Anatomy of the Respiratory
    System   539
  Mechanics of Breathing   549
  Gas Exchanges in the Body   553
  Transport of Respiratory Gases
    by Blood   555
  Control of Respiration   557
  Respiratory Adjustments During Exercise and
    at High Altitudes   558
  Homeostatic Imbalances of the Respiratory
    System   559
  Developmental Aspects of the Respiratory
    System   560
  The Incredible Journey: A Visualization
    Exercise for the Respiratory System   560

CHALLENGING YOURSELF   562
  At the Clinic   562
  Stop and Think   564
COVERING ALL YOUR BASES   566
  Multiple Choice   566
  Word Dissection   568

24

The Digestive System   570

BUILDING THE FRAMEWORK   571
  Overview of the Digestive System   571
  Functional Anatomy of the
    Digestive System   577
  Physiology of Chemical Digestion and
    Absorption   588
  Developmental Aspects of the Digestive
    System   590
  The Incredible Journey: A Visualization
    Exercise for the Digestive System   591
CHALLENGING YOURSELF   593
  At the Clinic   593
  Stop and Think   595
COVERING ALL YOUR BASES   596
  Multiple Choice   596
  Word Dissection   600

25

Nutrition, Metabolism, and Body Temperature Regulation   602

BUILDING THE FRAMEWORK   603
  Nutrition   603
  Metabolism   607
  Body Energy Balance   619
  Developmental Aspects of Nutrition
    and Metabolism   621
CHALLENGING YOURSELF   622
  At the Clinic   622
  Stop and Think   624
COVERING ALL YOUR BASES   626
  Multiple Choice   626
  Word Dissection   628

The Urinary System   629

BUILDING THE FRAMEWORK   630
Kidney Anatomy   632
Kidney Physiology: Mechanisms of Urine
    Formation   636
Ureters, Urinary Bladder, and Urethra   643
Micturition   643
Developmental Aspects of the Urinary
    System   644
The Incredible Journey: A Visualization
    Exercise for the Urinary System   646
CHALLENGING YOURSELF   647
At the Clinic   647
Stop and Think   649
COVERING ALL YOUR BASES   650
Multiple Choice   650
Word Dissection   653

Fluid, Electrolyte, and Acid-Base Balance   654

BUILDING THE FRAMEWORK   655
Body Fluids   655
Water Balance   658
Electrolyte Balance   660
Acid-Base Balance   662
Developmental Aspects of Fluid, Electrolyte,
    and Acid-Base Balance   667
CHALLENGING YOURSELF   668
At the Clinic   668
Stop and Think   669
COVERING ALL YOUR BASES   671
Multiple Choice   671
Word Dissection   673

The Reproductive System   674

BUILDING THE FRAMEWORK   675
Anatomy of the Male Reproductive
    System   675
Physiology of the Male Reproductive
    System   678
Anatomy of the Female Reproductive
    System   682
Physiology of the Female Reproductive
    System   685
Sexually Transmitted Diseases   688
Developmental Aspects of the Reproductive
    System: Chronology of Sexual
    Development   689
The Incredible Journey: A Visualization
    Exercise for the Reproductive
    System   690
CHALLENGING YOURSELF   692
At the Clinic   692
Stop and Think   693
COVERING ALL YOUR BASES   694
Multiple Choice   694
Word Dissection   697

Pregnancy and Human Development   699

BUILDING THE FRAMEWORK   700
From Egg to Embryo   700
Events of Embryonic Development   705
Events of Fetal Development   708
Effects of Pregnancy on the Mother   711
Parturition (Birth)   712
Adjustments of the Infant to Extrauterine
    Life   713
Lactation   713
CHALLENGING YOURSELF   714
At the Clinic   714
Stop and Think   716
COVERING ALL YOUR BASES   718
Multiple Choice   718
Word Dissection   720

30

Heredity   722

BUILDING THE FRAMEWORK   723
The Vocabulary of Genetics   723
Sexual Sources of Genetic Variation   724
Types of Inheritance   725
The Influence of Environmental Factors
   on Gene Expression   728
Nontraditional Inheritance   729
Genetic Screening, Counseling, and
   Therapy   729
CHALLENGING YOURSELF   731
At the Clinic   731
Stop and Think   733
COVERING ALL YOUR BASES   734
Multiple Choice   734
Word Dissection   735

Epilogue: A Day in the Life   737

Appendix A   Answers   739

**Appendix B   How to Use A.D.A.M.
with This *Study Guide*   869**

1 The Human Body: An Orientation

Student Objectives

When you have completed the exercises in this chapter, you will have accomplished the following objectives:

An Overview of Anatomy and Physiology

1. Define *anatomy* and *physiology* and describe their subdivisions.
2. Explain the principle of complementarity.

The Hierarchy of Structural Organization

3. Name (in order of increasing complexity) the different levels of structural organization that make up the human body and explain their relationships.
4. List the 11 organ systems of the body and briefly explain the major function(s) of each system.

Maintaining Life

5. List the functional characteristics common to humans (and other organisms) and explain the importance of each to maintaining life.
6. List the survival needs of the body.

Homeostasis

7. Define *homeostasis* and explain its importance.
8. Define *negative feedback* and describe its role in maintaining body homeostasis.
9. Define *positive feedback* and explain why it usually causes homeostatic imbalance. Also note specific situations in which it contributes to homeostasis, or normal body function.
10. Describe the relationship between homeostatic imbalance and disease.

The Language of Anatomy

11. Describe the anatomical position.
12. Use correct anatomical terminology to describe body directions, regions, and body planes or sections.
13. Locate and name the major body cavities and their subdivisions and list the major organs in each cavity or subdivision.
14. Name the specific serous membranes and note their common function.
15. Name the nine regions or four quadrants of the abdominopelvic cavity and list the organs they contain.

Ready, set, go!

1

Most of us have a natural curiosity about our bodies, and the study of anatomy and physiology elaborates on this interest. Anatomists have developed a universally acceptable set of reference terms that allows body structures to be located and identified with a high degree of clarity. Initially, students might have difficulties with the language used to describe anatomy and physiology, but without such a special vocabulary, confusion is inevitable.

The topics in Chapter 1 enable students to test their mastery of terminology commonly used to describe the body and the relationships among its various parts. Concepts concerning the functions vital for life and homeostasis are also reviewed. Additional topics include body organization from simple to complex levels and an introduction to organ systems.

BUILDING THE FRAMEWORK

An Overview of Anatomy and Physiology

1. Contrast briefly the following pairs of anatomical terms.

1. Regional anatomy versus systemic anatomy _____

_____

_____

_____

2. Gross anatomy versus microscopic anatomy _____

_____

_____

_____

3. Developmental anatomy versus embryology _____

_____

_____

_____

4. Histology versus cytology _____

_____

_____

_____

2. Circle all the terms or phrases that correctly relate to the study of *physiology;* use a highlighter to identify those terms or phrases that pertain to the study of *anatomy.*

   A. Measuring an organ's size, shape, and weight

   B. Can be studied in dead specimens

   C. Often studied in living subjects

   D. Chemistry principles

   E. Measuring the acid content of the stomach

   F. Principles of physics

   G. Observing a heart in action

   H. Dynamic

   I. Dissection

   J. Experimentation

   K. Observation

   L. Directional terms

   M. Static

3. "Function always follows structure." This statement is referred to as the principle of ____Complemtarity____.

The Hierarchy of Structural Organization

1. The structures of the body are organized in successively larger and more complex structures. Fill in the answer blanks with the correct terms for these increasingly larger structures.

   Chemicals ➔ ___Cellular___ ➔ ___tissue___ ➔ ___Organs___ ➔ ___Systems___ ➔ Organism

2. Circle the term that does not belong in each of the following groupings.

   1. Atom   Cell   Tissue   Alive   Organ

   2. Brain   Stomach   Heart   Liver   Epithelium

   3. Epithelium   Heart   Muscle tissue   Nervous tissue   Connective tissue

   4. Human   Digestive system   Horse   Pine tree   Amoeba

3. Figures 1.1–1.6 on pp. 4–6 represent some of the various body organ systems. First identify and name each organ system illustrated by filling in the blank directly under the illustration. Then select a different color for each organ and use it to color the coding circles and corresponding structures in the illustrations.

○ Blood vessels

○ Heart

○ Nasal cavity

○ Lungs

○ Trachea

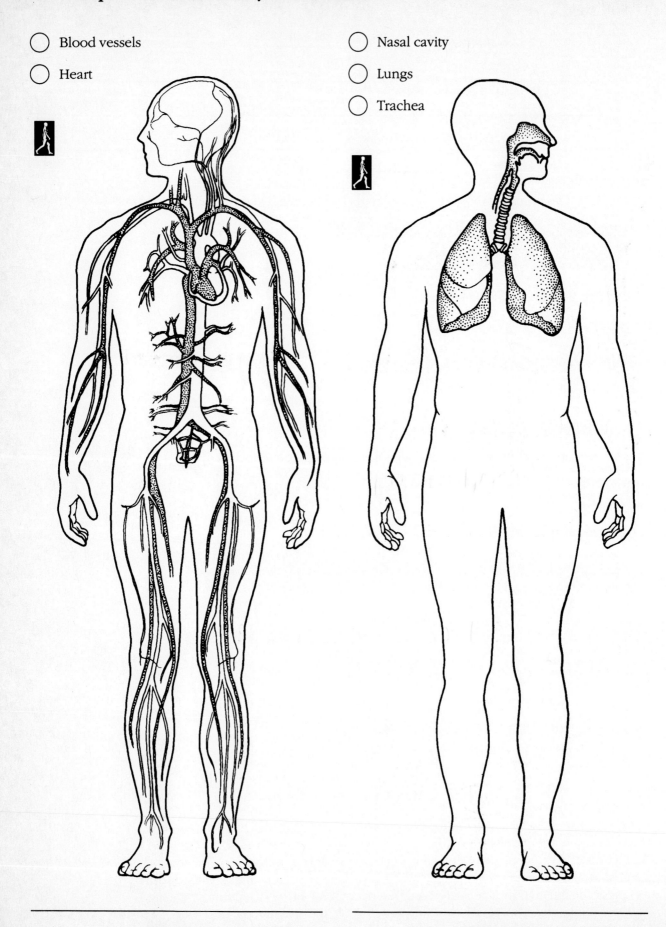

Figure 1.1

Figure 1.2

○ Brain

○ Spinal cord

○ Nerves

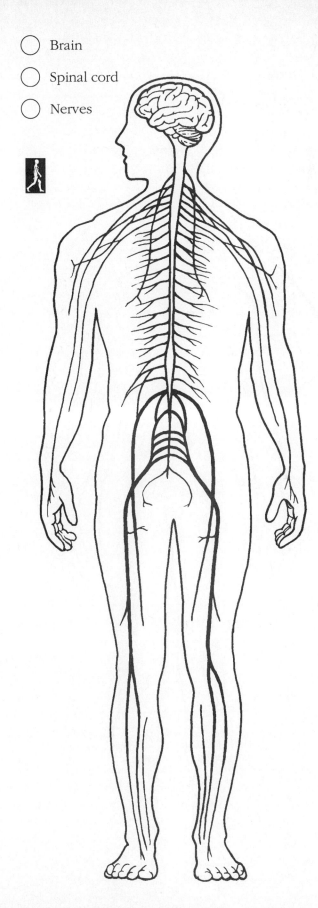

Figure 1.3

○ Kidneys   ○ Urethra

○ Ureters

○ Bladder

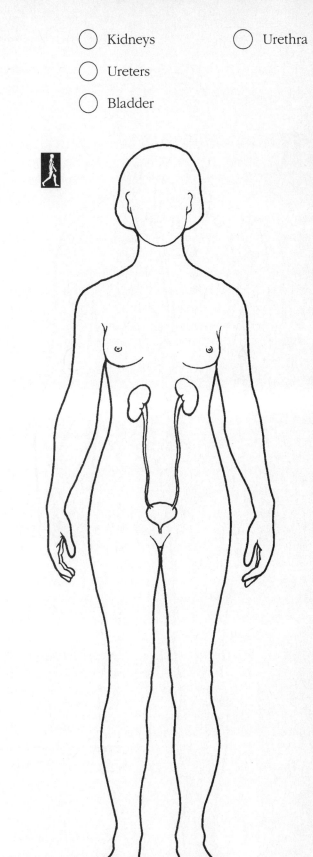

Figure 1.4

○ Oral cavity   ○ Esophagus   ○ Ovaries

○ Stomach   ○ Rectum   ○ Uterus

○ Intestines

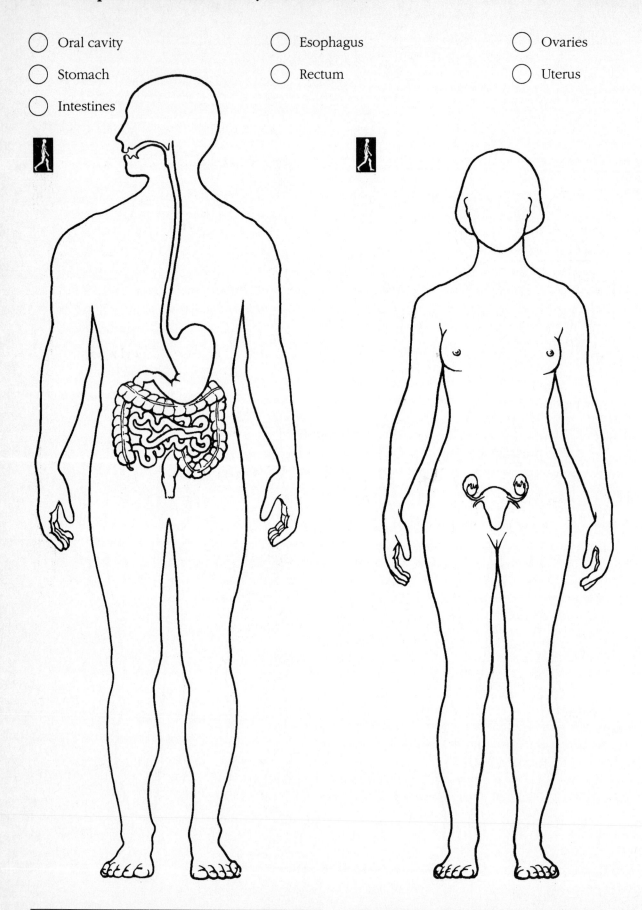

Figure 1.5   **Figure 1.6**

4. Using the key choices, identify the body systems to which the following organs or functions belong. Insert the correct answers in the answer blanks.

KEY CHOICES

A. Cardiovascular   **D.** Integumentary   **G.** Nervous   **J.** Skeletal

B. Digestive   **E.** Lymphatic/Immune   **H.** Reproductive   **K.** Urinary

C. Endocrine   **F.** Muscular   **I.** Respiratory

Urinary   **1.** Rids the body of nitrogen-containing wastes

Endocrine   **2.** Is affected by the removal of the thyroid gland

Skeletal   **3.** Provides support and levers on which the muscular system can act

Cardiovascular   **4.** Includes the heart

Integumentary   **5.** Protects underlying organs from drying out and mechanical damage

Lymphatic   **6.** Protects the body; destroys bacteria and tumor cells

Digestive   **7.** Breaks down foodstuffs into small particles that can be absorbed

Respiratory   **8.** Removes carbon dioxide from the blood

Cardiovascular   **9.** Delivers oxygen and nutrients to the body tissues

Muscular   **10.** Moves the limbs; allows facial expression

Urinary   **11.** Conserves body water or eliminates excesses

Reproductive   **12.** Allows conception and childbearing

Endocrine   **13.** Controls the body with chemicals called hormones

Integumentary   **14.** Is damaged when you cut your finger or get a severe sunburn

5. Using the key choices from Exercise 4, choose the organ system to which each of the following sets of organs belongs. Enter the correct letters in the answer blanks.

_A_   **1.** Blood vessels, heart

_C_   **2.** Pancreas, pituitary, adrenal glands

_K_   **3.** Kidneys, bladder, ureters

_H_   **4.** Testis, vas deferens, urethra

_B_   **5.** Esophagus, large intestine, stomach

_J_   **6.** Breastbone, vertebral column, skull

_G_   **7.** Brain, nerves, sensory receptors

Maintaining Life

1. Match the terms pertaining to functional characteristics of organisms in Column B with the appropriate descriptions in Column A. Fill in the blanks with the appropriate answers.

Column A

Maintenance of Boundaries **1.** Keeps the body's internal environment distinct from the external environment

Reproduction **2.** Provides new cells for growth and repair

Growth **3.** Occurs when constructive activities occur at a faster rate than destructive activities

Digestion **4.** The tuna sandwich you have just eaten is broken down to its chemical building blocks

Excretion **5.** Elimination of carbon dioxide by the lungs and of nitrogenous wastes by the kidneys

Responsiveness **6.** Ability to react to stimuli; a major role of the nervous system

Movement **7.** Walking, throwing a ball, riding a bicycle

Metabolism **8.** All chemical reactions occurring in the body

_D_ **9.** At the cellular level, membranes; for the whole organism, the skin

Column B

A. Digestion

B. Excretion

C. Growth

D. Maintenance of boundaries

E. Metabolism

F. Movement

G. Responsiveness

H. Reproduction

2. Using the key choices, correctly identify the survival needs that correspond to the following descriptions. Insert the correct answers in the answer blanks.

KEY CHOICES

A. Appropriate body temperature   **C.** Nutrients   **E.** Water

B. Atmospheric pressure   **D.** Oxygen

_C_ **1.** Includes carbohydrates, proteins, fats, and minerals

_B_ **2.** Essential for normal operation of the respiratory system and breathing

_E_ **3.** Single substance accounting for over 60 percent of body weight

_B_ **4.** Required for the release of energy from foodstuffs

_E_ **5.** Provides the basis for body fluids of all types

_A_ **6.** When too high or too low, physiological activities cease, primarily because molecules are destroyed or become nonfunctional

Homeostasis

1. Define *homeostasis.* _____

   _____

2. For each of the following sets of data, graph the provided data as shown in the example just below. Then, determine whether the data show negative or positive feedback.

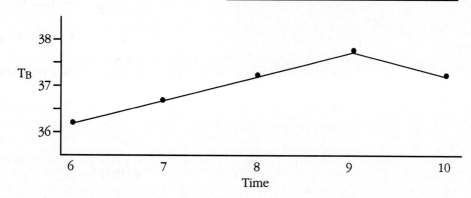

EXAMPLE:

Body temperature (T_B), °C:	36.5	37.0	37.5	38.0	37.5
Time, hours, A.M.:	6	7	8	9	10

Type of feedback: _____negative_____

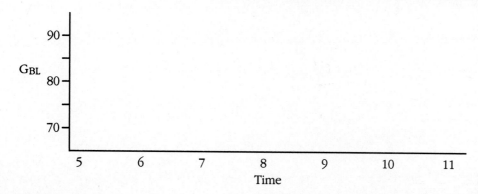

1.

Blood glucose (G_{BL}), mg/100 ml:	70	73	78	87	90	88	76
Time, hours, A.M.:	5	6	7	8	9	10	11

Type of feedback: _____

2. Blood pH:

7.36	7.38	7.42	7.43	7.41	7.39	7.37	7.40
1 P.M.	2 P.M.	3 P.M.	4 P.M.	5 P.M.	6 P.M.	7 P.M.	8 P.M.

Time:

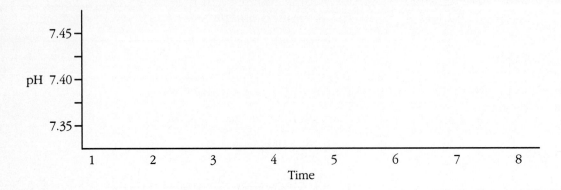

Type of feedback: _____

3. Blood pressure (BP), mm Hg:

95	89	86	80	73	66	58
0	10	15	20	25	30	35

Time, minutes:

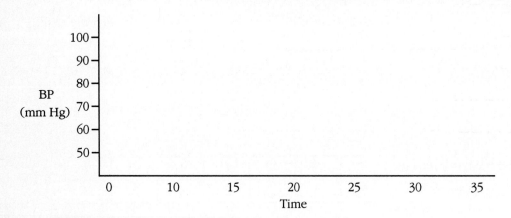

Type of feedback: _____

The Language of Anatomy

1. Complete the following statements by choosing an anatomical term from the key choices. Enter the appropriate answers in the answer blanks.

KEY CHOICES

A. Anterior   **D.** Inferior   **G.** Posterior   **J.** Superior

B. Distal   **E.** Lateral   **H.** Proximal   **K.** Transverse

C. Frontal   **F.** Medial   **I.** Sagittal

_____ 1.

_____ 2.

_____ 3.

_____ 4.

_____ 5.

_____ 6.

_____ 7.

_____ 8.

_____ 9.

_____ 10.

_____ 11.

_____ 12.

_____ 13.

_____ 14.

In the anatomical position, the nose and belly button are on the __(1)__ body surface, the calves and shoulder blades are on the __(2)__ body surface, and the soles of the feet are the most __(3)__ part of the body. The nipples are __(4)__ to the shoulders and __(5)__ to the armpits. The heart is __(6)__ to the spine and __(7)__ to the lungs. The knee is __(8)__ to the toes but __(9)__ to the thigh. In humans, the ventral surface can also be called the __(10)__ surface; however, in four-legged animals, the ventral surface is the __(11)__ surface.

If an incision cuts the brain into right and left parts, the section is a __(12)__ section, but if the brain is cut so that superior and inferior parts result, the section is a __(13)__ section. You are told to cut an animal along two planes so that the paired lungs are observable in both sections. The two sections that meet this requirement are the __(14)__ and __(15)__ sections.

The thoracic cavity is __(16)__ to the abdominopelvic cavity and __(17)__ to the spinal cavity.

_____ 15.

_____ 16.

_____ 17.

2. In 1–6 below, a directional term (e.g., distal in 1) is followed by terms indicating different body structures or locations (e.g., the elbow/the shoulder). In each case, choose the structure or location that best matches the directional term given.

1. Distal—the elbow/the shoulder

2. Lateral—the shoulder/the breastbone

3. Superior—the forehead/the chin

4. Superficial—the skeleton/the muscles

5. Proximal—the knee/the foot

6. Inferior—the liver/the small intestine

3. Circle the term or phrase that does not belong in each of the following groupings.

   1. Transverse   Distal   Frontal   Sagittal

   2. Lumbar   Flank   Antecubital   Abdominal   Scapular

   3. Calf   Brachial   Femoral   Popliteal

   4. Epigastric   Hypogastric   Lumbar   Left upper quadrant

   5. Orbital cavity   Nasal cavity   Ventral cavity   Oral cavity

4. Select the key choices that identify the following body parts or areas. Enter the appropriate answers in the blanks.

   KEY CHOICES

A. Abdominal	**E.** Calf	**I.** Gluteal	**M.** Popliteal
B. Axillary	**F.** Cervical	**J.** Groin	**N.** Pubic
C. Brachial	**G.** Digital	**K.** Lumbar	**O.** Scapular
D. Buccal	**H.** Femoral	**L.** Occipital	**P.** Umbilical

   ___Axillary___   1. Armpit

   ___Femoral___   2. Thigh region

   ___Gluteal___   3. Buttock area

   ___Cervical___   4. Neck region

   ___oa Umbilical___ 5. "Belly button" area

   ___Digital___   6. Pertaining to the toes

   ___Scapular___   7. Posterior aspect of head

   ___Inguinal___   8. Area where trunk meets thigh

   ___Lumbar___   9. Back area from ribs to hips

   ___Buccal___   10. Pertaining to the cheek

5. Correctly label all body areas indicated with leader lines on Figure 1.7. Some of the terms you will need are in the key of Exercise 4; others will have to be provided from memory. In addition, identify the sections labeled A and B in the figure.

   Section A: _____   Section B: _____

The body position indicated by the diagrams in Figure 1.7 are fairly close to the anatomical position. What body part(s) are incorrectly depicted? _____

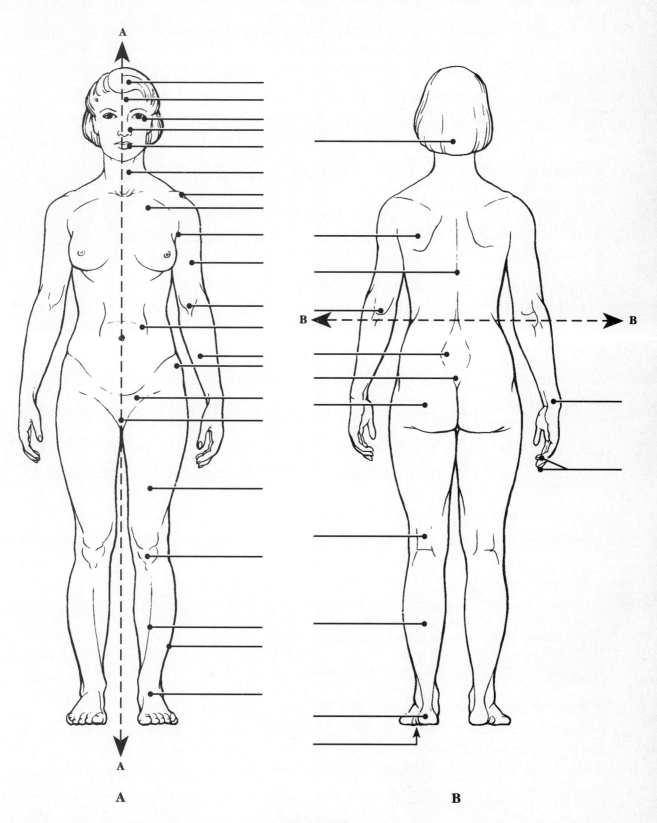

Figure 1.7

6. Complete the following statements by filling in the answer blanks with the correct terms.

_____ **1.**   The abdominopelvic and thoracic cavities are subdivisions of the __(1)__ body cavity; the cranial and spinal cavities are parts

_____ **2.**   of the __(2)__ body cavity. The __(3)__ body cavity is totally surrounded by bone and provides very good protection to the

_____ **3.**   structures it contains.

7. Identify the following serous membranes. Insert your answers in the answer blanks.

   1. Covers the lungs and cavity of the thorax _____

   2. Forms a slippery sac around the heart _____

   3. Surrounds the abdominopelvic cavity organs _____

8. Using the key choices, identify the body cavities where the following body organs are located. Enter the appropriate answers in the answer blanks.

KEY CHOICES

   A. Abdominopelvic   **B.** Cranial   **C.** Spinal   **D.** Thoracic

_____ **1.** Stomach   _____ **7.** Bladder

_____ **2.** Small intestine _____ **8.** Heart

_____ **3.** Large intestine _____ **9.** Lungs

_____ **4.** Spleen   _____ **10.** Brain

_____ **5.** Liver   _____ **11.** Rectum

_____ **6.** Spinal cord   _____ **12.** Ovaries

9. Refer to the organs listed in Exercise 8. In the answer blanks below, record the numbers of those that would be found in each abdominal region. Some organs may be found in more than one abdominal region.

_____ **1.** Hypogastric region

_____ **2.** Right lumbar region

_____ **3.** Umbilical region

_____ **4.** Epigastric region

_____ **5.** Left iliac region

10. Select different colors for the *dorsal* and *ventral* body cavities. Color the coding circles below and the corresponding cavities in part A of Figure 1.8. Complete the figure by labeling those body cavity subdivisions that have a leader line. Complete part B by labeling each of the abdominal regions indicated by a leader line.

◯ Dorsal body cavity   ◯ Ventral body cavity

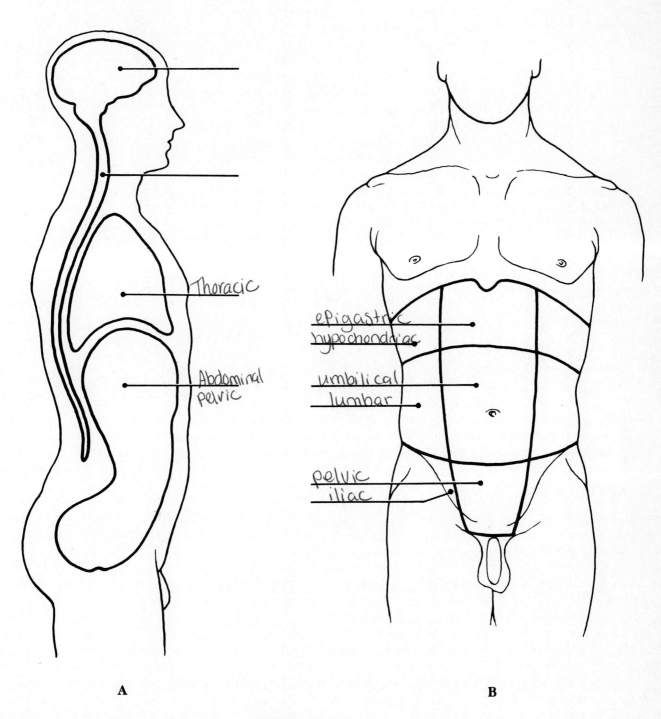

Thoracic

Abdominal
pelvic

epigastric
hypochondriac

umbilical
lumbar

pelvic
iliac

A

B

Figure 1.8

11. Choose different colors for the cavities listed with color coding circles. Color the coding circles and the corresponding cavities in Figure 1.9. Complete this exercise by identifying the structure provided with a leader line.

○ Pleural cavities

○ Pericardial cavity

○ Abdominal cavity

○ Pelvic cavity

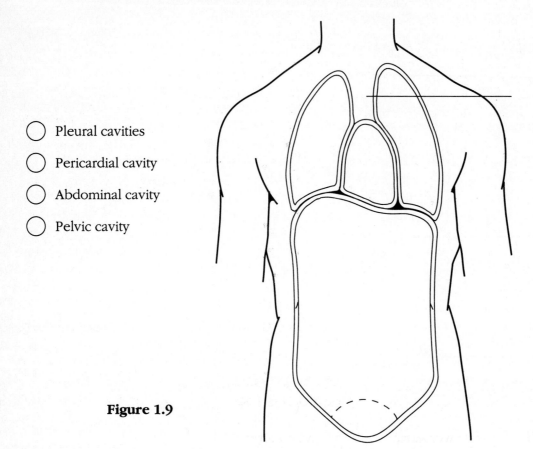

Figure 1.9

12. From the key choices, select the body cavities where the following surgical procedures would occur. Insert the correct answers in the answer blanks. Be precise; select the name of the cavity subdivision if appropriate.

KEY CHOICES

A. Abdominal	**C.** Dorsal	**E.** Thoracic
B. Cranial	**D.** Pelvic	**F.** Ventral

_____ **1.** Surgery to remove a cancerous prostate gland

_____ **2.** Coronary bypass surgery (heart surgery)

_____ **3.** Removal of a brain tumor

_____ **4.** Surgery to remove the distal part of the colon and establish a colostomy

_____ **5.** Gallbladder removal

CHALLENGING YOURSELF

At the Clinic

1. A jogger has stepped in a pothole and sprained his ankle. What organ systems have suffered damage?

2. A newborn baby is unable to hold down any milk. Examination reveals a developmental disorder in which the esophagus fails to connect to the stomach. Which survival needs are most immediately threatened?

3. In congestive heart failure, the weakened heart is unable to pump with sufficient strength to empty its own chambers. As a result, blood backs up in the veins, blood pressure rises, and circulation is impaired. Describe what will happen as this situation worsens owing to positive feedback. Then, predict how a heart-strengthening medication will reverse the positive feedback.

4. The control center for regulating body temperature is in a region of the brain called the hypothalamus. When the body is fighting an infection, the hypothalamus allows body temperature to climb higher than the normal value of about 38°C. Explain how this can be considered a negative feedback mechanism.

5. A patient reports stabbing pains in the right hypochondriac region. The medical staff suspects gallstones. What region of the body will be examined?

6. Number the following structures, from darkest (black) to lightest (white), as they would appear on an X ray. Number the darkest one 1, the next darkest 2, etc.

_____ **(a)** Soft tissue

_____ **(b)** Femur (bone of the thigh)

_____ **(c)** Air in lungs

_____ **(d)** Gold (metal) filling in a tooth

7. Mrs. Bigda is 3 months pregnant, and her obstetrician wants to measure the head size of the fetus in her womb to make sure development is progressing normally. Which medical imaging technique will the physician probably use and why?

8. Mr. Petros is behaving abnormally, and doctors strongly suspect he has a brain tumor. Which medical imaging device—conventional X ray, DSA, PET, ultrasound, or MRI—would be best for precisely locating a tumor within the brain? Explain your choice.

9. The Chan family was traveling in their van and had a minor accident. The children in the backseat were wearing seat belts but they still sustained bruises around the abdomen and had some internal organ injuries. Why is this area more vulnerable to damage than others?

Stop and Think

1. Circle the *most likely* structure or function of the items listed below:

    1. Thick outer layer of skin: *protection* or *sensation*

    2. Pigment of eyes: *absorb light* or *attract a mate*

    3. Thin lining of intestine: *protection* or *absorption*

    4. Pumping blood: *thick, hollow, muscular organ* or *thin, narrow vessels*

    5. Breathing: *thin, elastic tissue* or *dense, tough tissue*

2. Name the organ systems that come in contact with the external environment and then list some common infections of these systems.

3. List the functional characteristics of life that are *essential* to maintaining life.

4. Which of the survival needs must be renewed routinely? Of these, which can be stored for later use?

5. Why is the type of feedback that maintains homeostasis referred to as *negative* feedback?

6. Explain how the two-layered condition of serous membranes comes about.

7. Why do anatomists prefer abdominopelvic regions, while medical personnel tend to favor quadrants?

8. Make a simple line drawing of a chair. Then, add three different (common) planes to divide the chair structure. Identify each plane by name. Which of these planes could yield two parts, one consisting of the back of the chair and the other consisting mostly of the seat? Which would give you two mirror images of chair parts? Which could provide sections passing only through the legs, seat, or back of the chair?

9. The following advanced imaging techniques are discussed in the text: CT, DSA, PET, ultrasound, and MRI. Which of these techniques uses X rays? Which uses radio waves and magnetic fields? Which uses radioisotopes? Which displays body regions in sections? (You may have more than one answer for each question.)

COVERING ALL YOUR BASES

Multiple Choice

Select the best answer or answers from the choices given.

1. Which of the following is not a discipline of anatomy?
   - **A.** Developmental
   - **B.** Regional
   - **C.** Microscopic
   - **D.** Homeostatic
   - **E.** Systemic

2. Histology is the same as:
   - **A.** light microscopy
   - **B.** ultrastructure
   - **C.** functional morphology
   - **D.** surface anatomy
   - **E.** microscopic anatomy

3. Which of the following activities would *not* represent an anatomical study?
   - **A.** Making a section through the heart to observe its interior
   - **B.** Drawing blood from recently fed laboratory animals at timed intervals to determine their blood sugar levels
   - **C.** Examining the surface of a bone
   - **D.** Viewing muscle tissue through a microscope

4. Which subdivision of anatomy involves the study of organs that function together?
   - **A.** Regional
   - **B.** Developmental
   - **C.** Systemic
   - **D.** Histology
   - **E.** Radiographic

5. Consider the following listed levels: (1) chemical; (2) tissue; (3) organ; (4) cellular; (5) organismal; (6) systemic. Which of the following choices has the levels listed in order of increasing complexity?
   - **A.** 1, 2, 3, 4, 5, 6
   - **B.** 1, 4, 2, 5, 3, 6
   - **C.** 3, 1, 2, 4, 6, 5
   - **D.** 1, 4, 2, 3, 6, 5
   - **E.** 4, 1, 3, 2, 6, 5

6. Which of the following is (are) characteristic of a living organism?
   - **A.** Maintenance of boundaries
   - **B.** Metabolism
   - **C.** Waste disposal
   - **D.** Reproduction
   - **E.** Response to environment

7. Which is not essential to survival?
   A. Water   D. Atmospheric pressure
   B. Oxygen   E. Nutrients
   C. Gravity

8. Two of the fingers on Sally's left hand are fused together by a web of skin. Twenty-year-old Joey has a small streak of white hair on the top of his head, although the rest of his hair is black. Both Sally and Joey exhibit:
   A. rare but minor diseases
   B. serious birth defects
   C. examples of normal anatomical variation

9. Which of the following is (are) involved in maintaining homeostasis?
   A. Effector   D. Feedback
   B. Control center   E. Lack of change
   C. Receptor

10. In a negative feedback mechanism:
    A. the response of the effector is to depress or end the original stimulus
    B. the response of the effector is to enhance the original stimulus
    C. the effect is usually damaging to the body
    D. the physiological function is maintained within a narrow range

11. When a capillary is damaged, a platelet plug is formed. The process involves platelets sticking to each other. The more platelets that stick together, the more the plug attracts additional platelets. This is an example of:
    A. negative feedback
    B. positive feedback

12. In the preceding exercise, the overall effect is to maintain total blood volume within the normal range of 4.5 to 5.5 liters. Thus, the occurrence of a platelet plug preserves homeostasis and is in fact:
    A. negative feedback
    B. positive feedback

13. A coronal plane through the head:
    A. could pass through both the nose and the occiput
    B. could pass through both ears
    C. must pass through the mouth
    D. could lie in a horizontal plane

14. Which of these organs would not be cut by a section through the midsagittal plane of the body?
    A. Urinary bladder   C. Small intestine
    B. Gallbladder   D. Heart

15. The dorsal aspect of the human body is also its:
    A. anterior surface
    B. posterior surface
    C. lateral aspect
    D. superior aspect

16. A neurosurgeon orders a spinal tap for a patient. Into what body cavity will the needle be inserted?
    A. Ventral   D. Cranial
    B. Thoracic   E. Pelvic
    C. Dorsal

17. An accident victim has a collapsed lung. Which cavity has been entered?
    A. Mediastinal   D. Vertebral
    B. Pericardial   E. Ventral
    C. Pleural

18. A patient has a ruptured appendix. What condition does the medical staff watch for?
    A. Peritonitis   B. Pleurisy

19. The membrane that forms the outermost layer of the heart is the:
    A. visceral pleura
    B. visceral peritoneum
    C. parietal pericardium
    D. visceral pericardium
    E. parietal pleura

20. A patient complains of pain in the lower right quadrant. Which system is most likely to be involved?
   A. Respiratory   **D.** Skeletal
   B. Digestive   **E.** Muscular
   C. Urinary

21. Which of the following groupings of the abdominopelvic regions is medial?
   A. Hypochondriac, hypogastric, umbilical
   B. Hypochondriac, lumbar, inguinal
   C. Hypogastric, umbilical, epigastric
   D. Lumbar, umbilical, iliac
   E. Iliac, umbilical, hypochondriac

22. The radiographic technique used to provide information about brain metabolism is:
   A. DSR   **D.** ultrasonography
   B. CT   **E.** any X ray technique
   C. PET

23. Which microscopic technique provides sharp pictures of three-dimensional structures at high magnification?
   A. Scanning electron microscopy
   B. Transmission electron microscopy
   C. X ray microscopy
   D. Light microscopy

24. Which body system would be affected by degenerative cartilage?
   A. Muscular   **D.** Skeletal
   B. Nervous   **E.** Lymphatic
   C. Cardiovascular

25. The position of the heart relative to the structures around it would be described accurately as:
   A. deep to the sternum (breastbone)
   B. lateral to the lungs
   C. superior to the diaphragm
   D. inferior to the ribs
   E. anterior to the vertebral column

26. What term(s) could be used to describe the position of the nose?
   A. Intermediate to the eyes
   B. Inferior to the brain
   C. Superior to the mouth
   D. Medial to the ears
   E. Anterior to the ears

27. Which of the following statements is correct?
   A. The brachium is proximal to the antebrachium.
   B. The femoral region is superior to the tarsal region.
   C. The orbital region is inferior to the buccal region.
   D. The axillary region is lateral to the sternal region.
   E. The crural region is posterior to the sural region.

28. Which of the following body regions is (are) found on the torso?
   A. Gluteal   **D.** Acromial
   B. Inguinal   **E.** Olecranal
   C. Popliteal

29. Which of the following body regions would be moved by a muscle called the abductor pollicis longus?
   A. Arm   **D.** Thumb
   B. Forearm   **E.** Fingers
   C. Wrist

30. The *CT* in CT scanning stands for:
   A. cut transversely
   B. Charles Thorgaard, the inventor's name
   C. contrast medium of thin sections
   D. computed tomography

Word Dissection

For each of the following word roots, fill in the literal meaning and give an example, using a word found in this chapter.

Word root	Translation	Example
1. ana	_____	_____
2. chondro	_____	_____
3. corona	_____	_____
4. cyto	_____	_____
5. epi	_____	_____
6. gastr	_____	_____
7. histo	_____	_____
8. homeo	_____	_____
9. hypo	_____	_____
10. lumbus	_____	_____
11. meta	_____	_____
12. ology	_____	_____
13. org	_____	_____
14. para	_____	_____
15. parie	_____	_____
16. pathy	_____	_____
17. peri	_____	_____
18. stasis	_____	_____
19. tomy	_____	_____
20. venter	_____	_____
21. viscus	_____	_____

2 Chemistry Comes Alive

Student Objectives

When you have completed the exercises in this chapter, you will have accomplished the following objectives:

PART I: BASIC CHEMISTRY

Definition of Concepts: Matter and Energy

1. Differentiate clearly between matter and energy and between potential energy and kinetic energy.
2. Describe the major energy forms.

Composition of Matter: Atoms and Elements

3. Define *chemical element* and list the four elements that form the bulk of body matter.
4. List the subatomic particles; describe their relative masses, charges, and positions in the atom.
5. Define *atom, atomic weight, isotope,* and *radioisotope.*

How Matter Is Combined: Molecules and Mixtures

6. Distinguish between a compound and a mixture. Define molecule.
7. Compare solutions, colloids, and suspensions.

Chemical Bonds

8. Differentiate between ionic and covalent bonds. Contrast these bonds with hydrogen bonds.
9. Compare and contrast polar and nonpolar compounds.

Chemical Reactions

10. Contrast synthesis, decomposition, and exchange reactions. Comment briefly on the nature and importance of oxidation-reduction reactions.
11. Explain why chemical reactions in the body are often irreversible.
12. Describe factors that affect chemical reaction rates.

PART II: BIOCHEMISTRY: THE COMPOSITION AND REACTIONS OF LIVING MATTER

Inorganic Compounds

13. Explain the importance of water and salts to body homeostasis.
14. Define *acid* and *base,* and explain the concept of pH.

Organic Compounds

15. Compare the building blocks, general structures, and biological functions of organic molecules in the body.
16. Explain the role of dehydration synthesis and hydrolysis in the formation and breakdown of organic molecules.
17. Compare the functional roles of neutral fats, phospholipids, and steroids in the body.
18. Describe the four levels of protein structure.
19. Describe the general mechanism of enzyme activity.
20. Describe the function of stress proteins.
21. Compare and contrast DNA and RNA.
22. Explain the role of ATP in cell metabolism.

Ready, set, go!

Everything in the universe is composed of one or more elements, the unique building blocks of all matter. Although over 100 elemental substances exist, only four of these (carbon, hydrogen, oxygen, and nitrogen) make up over 96 percent of all living material.

Student activities in Chapter 2 test mastery of the basic concepts of both inorganic and organic chemistry. Chemistry is the science of the composition, properties, and reactions of matter. Inorganic chemistry is the science of nonliving substances such as salts and minerals, which generally do not contain carbon. Organic chemistry is carbon-based chemistry. Biochemistry is the science of living organisms, whether they are plants, animals, or microorganisms.

Student understanding of atomic structure, the bonding behavior of elements, and the structure and activities of the most abundant biochemical molecules (protein, fats, carbohydrates, and nucleic acids) is tested in various ways. Mastering these concepts is necessary to understanding the functions of the body.

BUILDING THE FRAMEWORK

PART I: BASIC CHEMISTRY

Definition of Concepts: Matter and Energy

1. Select *all* terms that apply to each of the following statements and insert the letters in the answer blanks.

_____ **1.** The energy located in the bonds of food molecules:
     A. is called thermal energy   **C.** causes molecular movement
     B. is a form of potential energy   **D.** can be transformed to the bonds of ATP

_____ **2.** Heat is:
     A. thermal energy   **C.** kinetic energy
     B. infrared radiation   **D.** molecular movement

_____ **3.** Whenever energy is transformed:
     A. the amount of useful energy decreases
     B. some energy is lost as heat
     C. some energy is created
     D. some energy is destroyed

2. Use choices from the key to identify the energy *form* in use in each of the following examples.

KEY CHOICES

A. Chemical   **B.** Electrical   **C.** Mechanical   **D.** Radiant

_____ **1.** Chewing food

_____ **2.** Vision (two types, please—think!)

_____ **3.** Bending your fingers to make a fist

_____ **4.** Breaking the bonds of ATP molecules to energize your muscle cells to make that fist

_____ **5.** Lying under a sunlamp

Composition of Matter: Atoms and Elements

1. Complete the following table by inserting the missing words.

Particle	Location	Electrical charge	Mass
		+ 1	
Neutron			
	Orbitals		

2. Insert the *chemical symbol* (the chemist's shorthand) in the answer blanks for each of the following elements.

____ **1.** Oxygen   ____ **4.** Iodine   ____ **7.** Calcium   ____ **10.** Magnesium

____ **2.** Carbon   ____ **5.** Hydrogen   ____ **8.** Sodium   ____ **11.** Chloride

____ **3.** Potassium ____ **6.** Nitrogen   ____ **9.** Phosphorus ____ **12.** Iron

3. Using the key choices, select the correct responses to the following descriptive statements. Insert the appropriate answers in the answer blanks.

KEY CHOICES

A. Atom   **C.** Element   **E.** Ion   **G.** Molecule   **I.** Protons

B. Electrons   **D.** Energy   **F.** Matter   **H.** Neutrons   **J.** Valence

_____ **1.** An electrically charged atom or group of atoms

_____ **2.** Anything that takes up space and has mass (weight)

_____ **3.** A unique substance composed of atoms having the same atomic number

_____ **4.** Negatively charged particles, forming part of an atom

_____ **5.** Subatomic particles that determine an atom's chemical behavior, or bonding ability

_____ **6.** The ability to do work

_____ **7.** The smallest particle of an element that retains the properties of the element

_____ **8.** The smallest particle of a compound, formed when atoms combine chemically

_____ **9.** Positively charged particles forming part of an atom

_____ **10.** The combining power of an atom

_____ **11.** _____ **12.** Subatomic particles responsible for most of an atom's mass

4. From the list of elements below, select the element or elements that match the descriptions. Insert their chemical symbols in the answer blanks.

Oxygen	Iodine	Calcium	Magnesium
Carbon	Hydrogen	Sodium	Chloride
Potassium	Nitrogen	Phosphorus	Iron

_____ **1.** Found as a salt in bones and teeth

_____ **2.** Make up more than 96% of the mass of a living cell

_____ **3.** Essential for transport of oxygen in red blood cells

_____ **4.** Essential cations in muscle contraction

_____ **5.** Essential for production of thyroid hormones

_____ **6.** Present in nucleic acids (in addition to C, H, O, and N)

_____ **7.** The most abundant negative ion in extracellular fluids

How Matter Is Combined: Molecules and Mixtures

1. Match the following terms to the appropriate descriptions below:

Mixture   Solution   Molarity   Colloid   Suspension

_____ **1.** Concentration expressed in moles

_____ **2.** Components are physically, not chemically, combined

_____ **3.** Homogeneous combination of solvent and solute(s)

_____ **4.** Large particles can settle out unless constantly mixed

_____ **5.** Large particles will not settle out

_____ **6.** Exhibits the sol-gel phenomenon

_____ **7.** An example is sand and water

2. Briefly explain how the following pairs of chemical species differ. Identify each substance as a molecule of an element, a molecule of a compound, an atom, or an ion.

1. H_2O_2 and $2OH^-$ _____

_____

2. $2O^{2-}$ and O_2 _____

_____

3. $2H^+$ and H_2 _____

_____

3. Using the periodic table in the appendix of your textbook, determine the molecular weight of each of the following:

____ **1.** Water (H_2O)   ____ **2.** Urea (N_2H_4)   ____ **3.** Carbonic acid (H_2CO_3)

4. Circle the term that does not belong in each of the following groupings.

1. Cholesterol $C_6H_{12}O_6$ Perspiration H_2O NH_3 Compound

2. Urine $Ca_3(PO_4)_2$ Inhaled air   Exhaled air   Plasma

3. Scatters light   Transparent   True solution   Saline

4. Blood cells   Water   Solute   Salt (NaCl)   Glucose molecules

5. Weight/volume method   5 g NaCl and 100 ml H_2O 5% NaCl   One-molar NaCl

6. Heterogeneous mixture   Scattering of light   Salt water   Milky   Colloid

Chemical Bonds

1. Figure 2.1 is a diagram of an atom. Select two different colors and use them to color the coding circles and the subatomic structures on the diagram. Identify each valence electron by inserting an X in the correct location on the diagram. Then respond to the questions that follow, referring to this diagram. Insert your answers in the answer blanks.

◯ Nucleus   ◯ Electrons

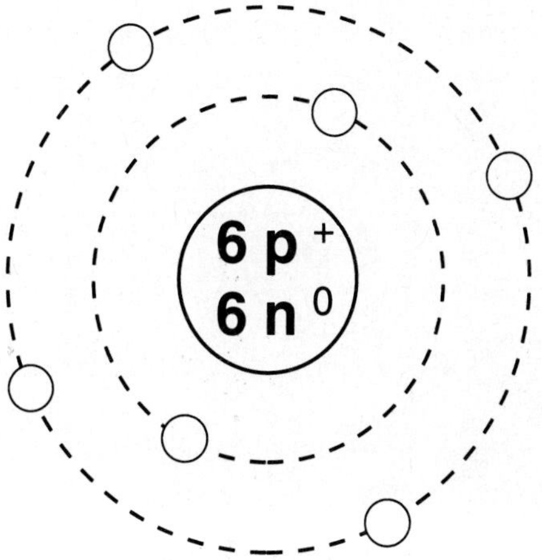

_____ **1.** What is the atomic number?

_____ **2.** Give the atomic mass of this atom.

_____ **3.** Name the atom represented here.

_____ **4.** How many electrons are needed to fill its outer shell?

_____ **5.** What is the valence of this atom?

Figure 2.1

_____ **6.** Is the atom chemically active *or* inert?

_____ **7.** Imagine this atom with one additional neutron. What is the name given to this slightly different form of the same element?

_____ **8.** Imagine this atom with two additional neutrons and a nucleus that can eject beta particles. What is the name given to this type of atom?

_____ **9.** The electrons in the second electron shell are more energetic. True or false?

2. Two types of chemical bonding are shown in Figure 2.2. Select three different colors. Use two colors to color the electrons named at each coding circle. Use the third color to add an arrow to show the direction of transfer of any electron. Then insert the name of the type of bond and the name of the compound in the blanks provided below the figure.

◯ Electrons shared   ◯ Electron(s) transferred

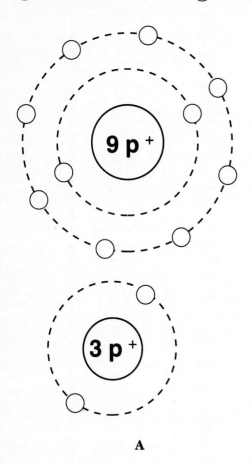

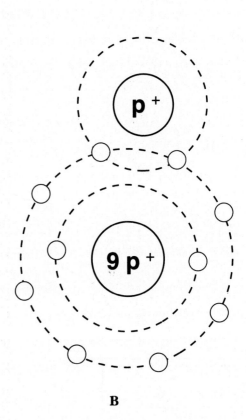

A   B

Figure 2.2

A. Type of bond _____   **B.** Type of bond _____

Name of compound _____   Name of compound _____

3. Complete the following table by filling in the missing parts.

Chemical symbol	Atomic number	Atomic mass	Electron distribution
	1		
		12	
N			
			2, 6, 0
		23	
	12		
			2, 8, 5
S			
	17		
		39	
Ca			

4. Using the information in the table you constructed in Exercise 3, complete the following table, which concerns atoms that form ionic bonds.

Chemical symbol	Loss/gain of electrons	Electrical charge
H		
Na		
Mg		
Cl		
K		
Ca		

5. Complete the following table relating to atoms that form covalent bonds.

Chemical symbol	Number of electrons shared	Number of bonds made
H		
C		
N		
O		

6. Figure 2.3 illustrates five water molecules held together by hydrogen bonds. First, correctly identify the oxygen and hydrogen atoms both by color and by inserting their atomic symbols on the appropriate circles (atoms). Then label the following structures in the figure:

○ Oxygen

○ Hydrogen

○ Positive pole

○ Negative pole

○ Hydrogen bonds

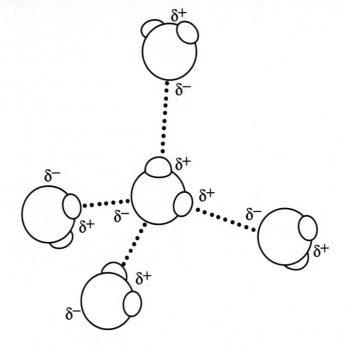

Figure 2.3

7. Circle each structural formula that is *likely* to be a polar covalent compound.

A. Cl — C — Cl
with Cl above and Cl below

B. H — Cl

C. with H, H above N and H below

D. Cl — Cl

E. H, O, H

Chemical Reactions

1. Match the terms in Column B with the equations listed in Column A. Enter the letters of all correct answers in the answer blanks.

Column A

_____ **1.** A + B → AB

_____ **2.** AB + CD → AD + CB

_____ **3.** XY → X + Y

_____ **4.** AB + CD ⇄ AD + CB

Column B

A. Synthesis (combination)

B. Decomposition (dissociation)

C. Displacement (exchange)

D. Designates chemical equilibrium

2. Circle the term or phrase that does not belong with the grouping in each of the following cases.

   1. High reaction rate   Low product concentration   High reactant concentration

      Low reactant concentration

   2. High reaction rate   Large particles   Small particles   Many forceful collisions

   3. Catalyst   Enzyme   Reactant   Increase reaction rate

   4. Oxidized substrate   Electron donor   Electron acceptor   Loss of hydrogen

   5. Catabolic   Simple → complex   Complex → simple   Degradation

PART II: BIOCHEMISTRY: THE COMPOSITION AND REACTIONS OF LIVING MATTER

1. Use an X to designate which of the following are organic compounds.

   ____ 1. Carbon dioxide   ____ 3. Fats   ____ 5. Proteins   ____ 7. H_2O

   ____ 2. Oxygen   ____ 4. KCl   ____ 6. Glucose   ____ 8. DNA

Inorganic Compounds

1. Complete the following statements concerning the properties and biological importance of water.

   _____ 1.

   _____ 2.

   _____ 3.

   _____ 4.

   _____ 5.

   _____ 6.

   _____ 7.

   The ability of water to maintain a relatively constant temperature and thus prevent sudden changes is because of its high __(1)__. Biochemical reactions in the body must occur in __(2)__. About __(3)__ % of the volume of a living cell is water. Water molecules are bonded to other water molecules because of the presence of __(4)__ bonds. The cooling effect of perspiration as it evaporates from the skin is the result of the high __(5)__ of water. Water, as H^+ and OH^- ions, is essential in biochemical reactions such as __(6)__ and __(7)__ reactions.

2. Circle the term that does not belong in each of the following groupings.

   1. HCl H_2SO_4 Vinegar   Milk of magnesia

   2. NaOH   Blood (pH 7.4) $Al(OH)_2$ Urine (pH 6.2)

   3. KCl   NaCl $NaHCO_3$ $MgSO_4$

    4. Ionic   Organic   Salt   Inorganic

    5. pH 7   pH 4   Neutrality $OH^- = H^+$

    6. Basic   pH 8   Alkaline   pH 2

3. Use the key choices to identify the substances described in the following statements. Insert the appropriate answers in the answer blanks.

KEY CHOICES

    A. Acids   **B.** Bases   **C.** Buffers   **D.** Salts

_____ **1.** _____ **2.** _____ **3.** Substances that ionize in water; good electrolytes

_____ **4.** Proton (H^+) acceptors

_____ **5.** Substances that dissociate in water to release hydrogen ions and a negative ion other than hydroxide (OH^-)

_____ **6.** Substances that dissociate in water to release ions other than H^+ and OH^-

_____ **7.** Substances formed when an acid and a base are combined

_____ **8.** Substances such as lemon juice and vinegar

_____ **9.** Substances that prevent rapid or large swings in pH

_____ **10.** Substances such as ammonia and milk of magnesia

4. Define *pH*. _____

5. Using the key choices, fully characterize weak and strong acids.

KEY CHOICES

    A. Ionize completely in water   E. Ionize at high pH

    B. Ionize incompletely in water   F. Ionize at low pH

    C. Act as part of a buffer system   G. Ionize at pH 7

    D. When placed in water, always act to change the pH

Weak acid: _____   **Strong acid:** _____

Organic Compounds

1. Match the terms in Column B with the descriptions in Column A. Enter the correct letters in the answer blanks.

	Column A	Column B
_____	1. Building blocks of carbohydrates	**A.** Amino acids
_____	2. Building blocks of fat	**B.** Carbohydrates
_____	3. Building blocks of protein	**C.** Lipids (fats)
_____	4. Building blocks of nucleic acids	**D.** Fatty acids
_____	5. Cellular cytoplasm is primarily composed of this substance	**E.** Glycerol
		F. Nucleotides
_____	6. The single most important fuel source for body cells	**G.** Monosaccharides
_____	7. Not soluble in water	**H.** Proteins
_____	8. Contains C, H, and O in the ratio CH_2O	
_____	9. Contain C, H, and O, but have relatively small amounts of oxygen	

_____ 10. _____ 11. These building blocks contain N in addition to C, H, and O

_____ 12. Contain P in addition to C, H, O, and N

_____ 13. Used to insulate the body and found in all cell membranes

_____ 14. Primary components of meat and cheese

_____ 15. Primary components of bread and lollipops

_____ 16. Primary components of egg yolk and peanut oil

_____ 17. Includes collagen and hemoglobin

_____ 18. Class that usually includes cholesterol

2. For each of the following statements that is true, insert T in the answer blank. For each false statement, correct the underlined word(s) and insert your correction in the answer blank.

_____ **1.** Phospholipids are <u>polarized</u> molecules.

_____ **2.** Steroids are the major form in which body fat is stored.

_____ **3.** Water is the most abundant compound in the body.

_____ **4.** Nonpolar molecules are generally soluble in water.

_____ **5.** The bases of RNA are A, G, C, and <u>U</u>.

_____ **6.** The universal energy currency of living cells is <u>RNA</u>.

_____ **7.** RNA is <u>single-stranded</u>.

_____ **8.** The bond type linking the subunits of proteins together is commonly called the <u>hydrogen</u> bond.

_____ **9.** The external fuel of choice used by cells as a ready source of energy is <u>starch</u>.

_____ **10.** The nucleotide base complementary to G is <u>C</u>.

_____ **11.** The backbone of a nucleic acid molecule consists of sugar and <u>base</u> units.

3. If any of the following statements about enzymes is true, write T in the answer blank. If a statement is false, correct the underlined word and write the correction in the answer blank.

_____ **1.** All enzymes are <u>proteins</u>.

_____ **2.** The substances on which enzymes act are called <u>cofactors</u>.

_____ **3.** The name of an enzyme usually ends in the suffix <u>-ide</u>.

_____ **4.** Coenzymes are vitamins that assist the chemical action of enzymes.

_____ **5.** Enzymes <u>increase</u> the activation energy of biochemical reactions.

_____ **6.** The active site of an enzyme is the location of <u>substrate</u> attachment.

_____ **7.** Changes in pH or temperature decrease enzyme activity because bonds break and the enzyme returns to its <u>tertiary</u> structure.

4. Five simplified diagrams of the structures of five biochemical molecules are shown in Figure 2.4.

First, identify the molecules and insert their correct names in the answer blanks on the figure (A through E).

Second, select a different color for each type of molecule listed below and use it to color the coding circle and the corresponding molecular structure in the illustration.

○ Fat

○ Nucleotide

○ Monosaccharide

○ Functional protein

○ Polysaccharide

Third, answer the following questions relating to these diagrams by inserting your answers in the answer blanks.

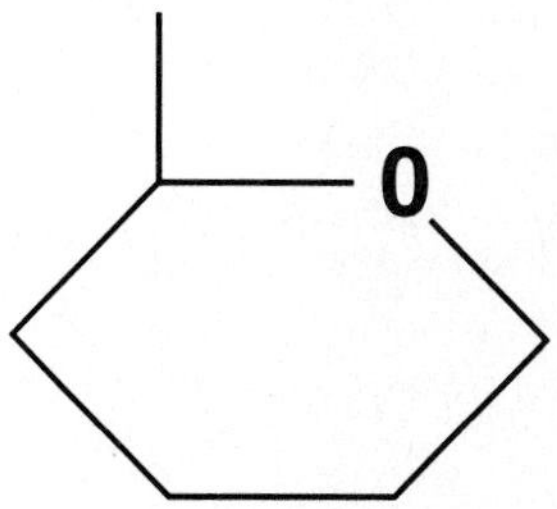

A. _____

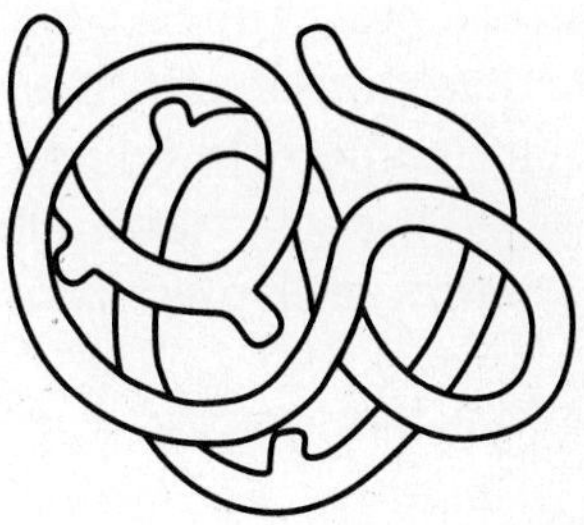

B. _____

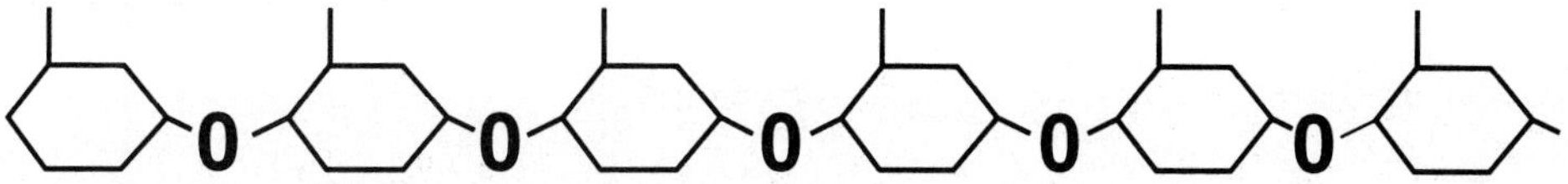

C. _____

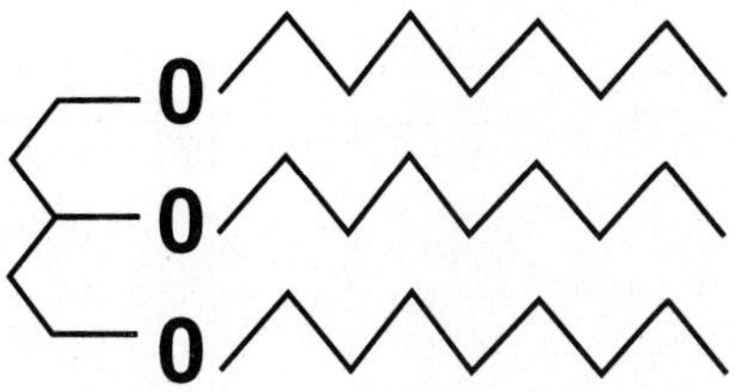

D. _____

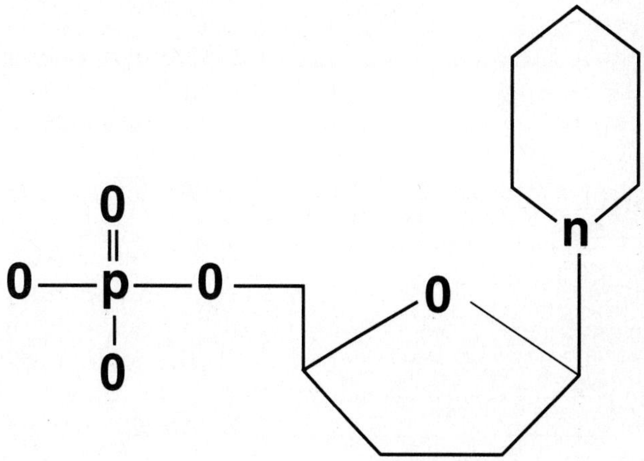

E. _____

Figure 2.4

_____ **1.** Give an example of a biochemical having a structure like diagram C.

_____ **2.** Which two diagrams illustrate structures of monomers (building blocks)?

_____ **3.** Name the level of structure illustrated in diagram B.

_____ **4.** Which two diagrams illustrate molecules used for energy storage?

_____ **5.** Which diagram shows a structure that most looks like a molecule of ATP?

5. The biochemical reaction shown in Figure 2.5 represents the complete digestion of a polymer (as consumed in food) down to its constituent monomers (building blocks). Select two colors and color the coding circles and the structures. Then, select the one correct answer for each statement below and insert your answer in the answer blank.

◯ Monomer   ◯ Polymer

Figure 2.5

_____ **1.** If starch is the polymer, the monomer is:

       A. glycogen   **B.** amino acid   **C.** glucose   **D.** maltose

_____ **2.** During polymer digestion, water as H^+ and OH^- ions would:

       A. be a product of the reaction.

       B. act as a catalyst.

       C. enter between monomers, bond to them, and keep them separated.

       D. not be involved in this reaction.

_____ **3.** Another name for the chemical digestion of polymers is:

       A. dehydration   **B.** hydrolysis **C.** synthesis   **D.** displacement

_____ **4.** If the monomers are amino acids, they may differ from each other by their:

       A. R group   **B.** amino group   **C.** acid group   **D.** peptide bond

6. Place an X in the blank before each phrase that accurately describes stress proteins.

_____ **1.** Crucial to protein structure

_____ **2.** Numbers increase during fever

_____ **3.** Numbers decline when proteins are denatured

_____ **4.** Originally called heat shock proteins

_____ **5.** Defensive molecules

7. This exercise concerns the way polar molecules—phospholipids, in this example—interact with water. As you can see, the phospholipid molecules have arranged themselves to form a hollow sphere within the aqueous environment of the beaker.

First, color all the water-containing areas blue, all areas that exclude water yellow, and the polar heads of the phospholipids gray-black.

Second, draw in a row of phospholipids on the surface of the water in the beaker, showing how they would orient themselves (heads vs. tails) to the water in the beaker.

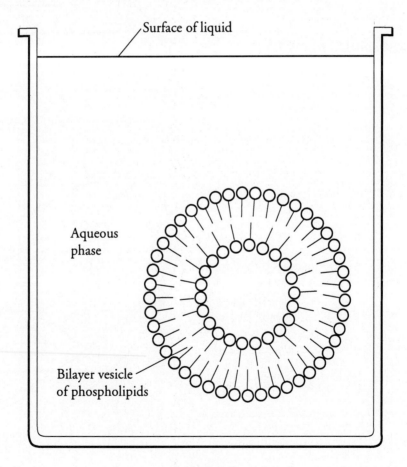

Figure 2.6

8. Figure 2.7 shows the molecular structure of DNA, a nucleic acid.

A. Identify the two unnamed nitrogen bases and insert their correct names and symbols in the two blanks beside the color coding circles.

B. Complete the identification of the bases on the diagram by inserting the correct symbols in the appropriate spaces on the right side of the diagram.

C. Select different colors and color the coding circles and the corresponding parts of the diagram.

D. Label one d-R sugar unit and one P unit of the "backbones" of the DNA structure by inserting leader lines and labels on the diagram.

E. Circle the associated nucleotide.

○ Deoxyribose sugar (d-R)

○ Phosphate unit (P)

○ Adenine (A)

○ Cytosine (C)

○ _____ ( )

○ _____ ( )

F. Answer the following questions by writing your answers in the answer blanks.

1. Name the bonds that help hold the two DNA strands together. _____

2. Name the three-dimensional shape of the DNA molecule. _____

3. How many base pairs are present in this segment of a DNA model? _____

4. What is the term that means "base pairing"?

_____

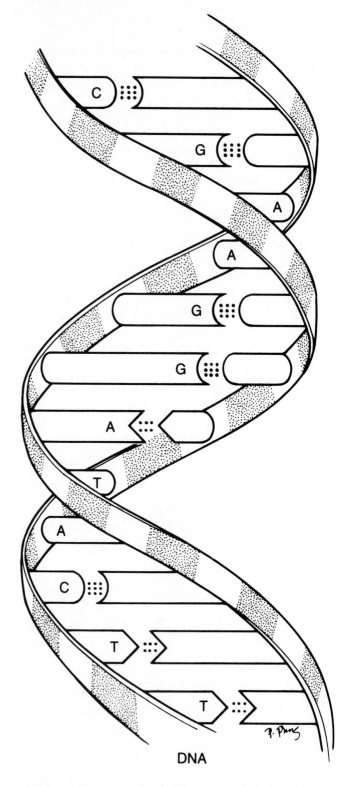

DNA

Figure 2.7 Part of a DNA molecule (coiled)

9. Circle the term that does not belong in each of the following groupings.

   1. Adenine   Guanine   Glucose   Thymine

   2. DNA   Ribose   Phosphate   Deoxyribose

   3. Galactose   Glycogen   Fructose   Glucose

   4. Amino acid   Polypeptide   Glycerol   Protein

   5. Glucose   Sucrose   Lactose   Maltose

   6. Synthesis   Dehydration   Hydrolysis   Water removal

   7. Amino acid   Compound   Fatty acid   Carbon

   8. Enzymes   98°F   Active site   Denaturation   Favorable pH

10. Two diagrams, A and B, are shown in Figure 2.8. These represent the stylized structures of ATP and cyclic AMP. Select a different color for each term with a color-coding circle and color the structures. Identify each biochemical by inserting its name in blanks A and B below. Label all high-energy bonds with corresponding colored arrows on the diagrams.

   ◯ Phosphate   ◯ Ribose   ◯ Adenine   ◯ High-energy bond

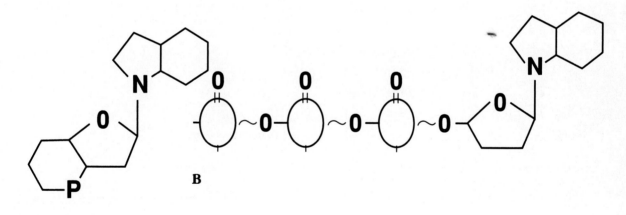

A.

A. _____   B. _____

Figure 2.8

11. Complete the following chart on proteins by responding to the questions that are printed in the headings of the chart. Note the clues that are given! Insert your answers in the spaces provided.

Name of protein?	Secondary or tertiary structure?	Molecular shape: fibrous or globular?	Water-soluble or insoluble?	Function of protein?	Location in body?
Collagen					Cartilage and bone
Proteinase					Stomach
	Quaternary			Carries O_2	Red blood cells
Keratin	Secondary		Insoluble		

The Incredible Journey: A Visualization Exercise for Biochemistry

. . . you are suddenly up-ended and are carried along in a sea of water molecules at almost unbelievable speed.

1. Complete the narrative by inserting the missing words in the answer blanks.

_____ **1.**

_____ **2.**

_____ **3.**

For this journey, you are miniaturized to the size of a very small molecule by colleagues who will remain in contact with you by radio. Your instructions are to play the role of a water molecule and to record any reactions that involve water molecules. Since water molecules are dipoles, you are outfitted with an insulated rubber wetsuit with one __(1)__ charge at your helmet and two __(2)__ charges, one at the end of each leg.

As soon as you are injected into your host's bloodstream, you feel as though you are being pulled apart. Some large, attractive forces are pulling at your legs from different directions! You look about but can see only water molecules. After a moment's thought, you remember the dipole construction of your wetsuit. You record that these forces must be the __(3)__ in water that are easily formed and easily broken.

_____ **4.**

_____ **5.**

_____ **6.**

_____ **7.**

_____ **8.**

_____ **9.**

_____ **10.**

_____ **11.**

_____ **12.**

_____ **13.**

_____ **14.**

_____ **15.**

_____ **16.**

_____ **17.**

_____ **18.**

_____ **19.**

_____ **20.**

After this initial surprise, you are suddenly up-ended and carried along in a sea of water molecules at almost unbelievable speed. You measure and record this speed as 1000 miles/sec. You are about to watch some huge, red, disk-shaped structures (probably __(4)__) taking up O_2 molecules, when you are swept into a very turbulent environment. Your colleagues radio that you are in the small intestine. With difficulty, because of numerous collisions with other molecules, you begin to record the various types of molecules you see.

In particular, you notice a very long helical molecule made of units with distinctive R groups. You identify and record this type of molecule as a __(5)__ , made of units called __(6)__ that are joined together by __(7)__ bonds. As you move too close to the helix during your observations, you are nearly pulled apart to form two ions, __(8)__ , but you breathe a sigh of relief as two ions of another water molecule take your place. You watch as these two ions move between two units of the long helical molecule. Then, in a fraction of a second, the bond between the two units is broken. As you record the occurrence of this chemical reaction, called __(9)__ , you are jolted into another direction by an enormous globular protein, the very same __(10)__ that controls and speeds up this chemical reaction.

Once again you find yourself in the bloodstream, heading into an organ identified by your colleagues as the liver. Inside a liver cell, you observe many small monomers, made up only of C, H, and O atoms. You identify these units as __(11)__ molecules because they are being bonded to form very long, branched polymers called __(12)__ by the liver cells. You record that this type of chemical reaction is called __(13)__ , and you happily note that this reaction also produces __(14)__ molecules like you!

Once more in the bloodstream, you are ferried, according to your colleagues, into the cartilage of the knee. What an environment you have found! Long, glistening, parallel white fibrous structures indicate that this protein must be __(15)__ . Before you fully appreciate and record your observations, you are immobilized along with thousands of other water molecules by this long twisting protein. Like them, you are trapped, all dipoles in the same orientation, in layers of __(16)__ around this protein. Your colleagues are quick to advise that movement of your host's knee will press water out of the cartilage and you will be released. And so you are with a great "woosh."

Via another speedy journey through the bloodstream, you next reach the skin. You move deep into the skin and finally gain access to a sweat gland. In the sweat gland, you collide with millions of water molecules and some ionized salt molecules that are continually attracted to your positive and negative charges. Suddenly, the internal temperature rises, and molecular collisions __(17)__ at an alarming rate, propelling you through the pore of the sweat gland onto the surface of the skin. As you watch, other water molecules on the surface begin to __(18)__ , using __(19)__ energy from the body. You record that the cooling effect just observed is the result of a property of water called high heat of __(20)__ . Saved from the fate of evaporating into thin air, you are speedily rescued by your colleagues.

CHALLENGING YOURSELF

At the Clinic

1. It is determined that a patient is in acidosis. What does this mean, and would you treat the condition with a chemical that would *raise* or *lower* the pH?

2. Carbon dioxide concentration influences blood pH in the following manner: High levels of CO_2 increase the rate of formation of carbonic acid. If a patient has difficulty ventilating the lungs (particularly in exhaling, as in emphysema), would you expect the patient to be in acidosis or alkalosis?

3. Hugo, a patient with kidney disease, is unable to excrete sufficient amounts of hydrogen ions. Would you expect Hugo to hyperventilate or hypoventilate?

4. Vitamin D is essential for absorption of calcium, which contributes to the hardness of bones. What would most likely correlate with a softening of the bones: an inability to absorb fat, or excess fat absorption?

5. A newborn is diagnosed with sickle-cell anemia, a genetic disease in which an amino acid substitution results in abnormal hemoglobin. Explain to the parents how the substitution can have such a drastic effect on the structure of the protein.

6. Explain the effect of mineral and vitamin deficiency on enzyme activity.

7. Johnny's body temperature is spiking upward. When it reaches 104°F, his mother puts in a call to the pediatrician. She is advised to give Johnny children's aspirin and sponge his body with cool to tepid water to prevent a further rise in temperature. How might a fever (excessively high body temperature) be detrimental to Johnny's welfare?

Stop and Think

1. Explain why the formation of ATP from ADP and P_i requires more energy than the amount released for cellular use when ATP is broken down.

2. Can all molecular structures be explained by the concepts of chemical bonding presented in this chapter? What about ozone (O_3) and carbon monoxide (CO)?

3. Using the octet rule, predict the atomic number of the next two inert gases after helium (atomic number 2).

4. Proteins within a cell commonly have negative surface charges. Indicate whether aqueous mixtures containing proteins form colloids or suspensions, and explain why.

5. Explain why a one-molar solution of sodium chloride has twice as many solute particles as a one-molar solution of glucose. How many particles would a one-molar solution of *calcium chloride* have compared to a one-molar glucose solution?

6. Combining carbonic acid, H_2CO_3, and calcium hydroxide, $Ca(OH)_2$, results in the formation of two water molecules and what salt?

7. Draw the following molecules, showing the bonds as lines between the chemical symbols: N_2H_4, H_2CO_3, CH_3COOH.

8. Hydrogen bonding determines the spacing between water molecules in ice crystals. What would happen to lakes in the winter if water molecules were packed more tightly in the solid form than in the liquid form?

9. Explain, in terms of chemical bonding, how ammonia functions as a base.

10. What happens to the pH of water when its temperature is raised to 38°C? What does this imply about our definition of neutral pH?

11. What is "special" about valence shell electrons?

12. Neutral fats float on water. How does this explain the use of neutral fats as the main form of reserve body fuel?

13. Dissect the word "atom."

14. Can chlorine make covalent bonds? If so, how many will each chlorine atom make?

15. What can a cell do to actively control the direction of a reversible reaction?

16. Justify classifying water as both an acid and a base.

17. Given a hydrogen ion concentration [H$^+$] of 0.000001 mol/liter, what is the pH? How does the relative acidity of this solution compare to that of a solution of pH 7?

18. Graph the following data relating to an enzyme-catalyzed reaction. Then answer the associated questions.

Substrate concentration (mol)	0.1	0.2	0.3	0.4	0.5
Reaction rate (mol product/min)	0.05	0.06	0.068	0.069	0.069

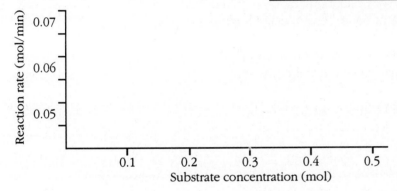

1. Why is there a plateau in the graph?

2. How would an alteration of the enzyme's active site, such as might occur in competitive or noncompetitive inhibition, affect the reaction rate?

19. Matter *occupies space* and *has mass*. Describe how energy *must be* described in terms of these two criteria. Then list the criteria usually used to define or describe energy.

20. How do saturated and unsaturated fats differ in molecular structure and gross appearance?

21. Relative to the chemical equation X + YC → XC + Y, the reactants are

_____; the products are _____. If interpreting

this reaction in terms of mole quantities, how many moles of each

(reactants and products) are indicated by the equation? _____

22. Describe several uses of radioisotopes in biological research or medicine.

23. Evelyn is quite proud of her slender, model-like figure and boasts that she doesn't have an "ounce of excess body fat." Barbara, on the other hand, is grossly overweight. She complains of being "hot" most of the time, and on a hot day she is miserable. Evelyn generally feels chilled except on very hot days. Explain the relative sensitivity to environmental temperature of these two women on the basis of information you have been given.

COVERING ALL YOUR BASES

Multiple Choice

Select the best answer or answers from the choices given.

1. Energy:
   A. has mass
   B. does work
   C. has weight
   D. takes up space
   E. causes motion

2. Energy contained in molecular bonds is:
   A. electrical
   B. mechanical
   C. chemical
   D. radiant
   E. kinetic

3. Which of the following are among the four major elements in the body?
   - **A.** Iron
   - **B.** Carbon
   - **C.** Oxygen
   - **D.** Copper
   - **E.** Sulfur

4. Which of the following is (are) true concerning the atomic nucleus?
   - **A.** Contains the mass of the atom
   - **B.** The negatively charged particles are here
   - **C.** Particles can be ejected
   - **D.** Contains particles that determine atomic number
   - **E.** Contains particles that interact with other atoms

5. Which of the following would indicate an isotope of oxygen?
   - **A.** Atomic number 7
   - **B.** Atomic number 8
   - **C.** Atomic mass 16
   - **D.** Atomic mass 18
   - **E.** Gain of 2 electrons

6. An example of a suspension is:
   - **A.** sugar in tea
   - **B.** cornstarch in water
   - **C.** proteins in plasma
   - **D.** carbon dioxide in soda water
   - **E.** oil in water

7. Pick out the correct match(es) of element and valence number.
   - **A.** Oxygen-6
   - **B.** Chlorine-8
   - **C.** Calcium-2
   - **D.** Nitrogen-3
   - **E.** Carbon-4

8. In which mixtures would you expect to find hydrogen bonds?
   - **A.** Sugar in water
   - **B.** Oil in water
   - **C.** Oil in gasoline
   - **D.** Carbonic acid in water
   - **E.** Salt in water

9. Which of the following is (are) a synthesis reaction?
   - **A.** Glucose to glycogen
   - **B.** Glucose + fructose to sucrose
   - **C.** Starch to glucose
   - **D.** Polypeptide to dipeptides
   - **E.** Amino acids to dipeptide

10. Which of the following will have the greatest reaction rate, assuming the same temperature, concentration, and catalyst activity?
    - **A.** Carbon dioxide and water
    - **B.** Glucose and fructose forming sucrose
    - **C.** Amino acids combining to dipeptides
    - **D.** Fatty acids combining with glycerol
    - **E.** All of these will have the same reaction rate with the listed factors being equal.

11. Organic compounds include:
    - **A.** water
    - **B.** carbon dioxide
    - **C.** oxygen
    - **D.** carbonic acid
    - **E.** glycerol

12. All of the following are important functions of water except:
    - **A.** cushioning
    - **B.** cooling the body
    - **C.** participation in chemical reactions
    - **D.** solvent for sugars, salts, and oils
    - **E.** reduces temperature fluctuations

13. Which of the following molecules does not have at least one double covalent bond?
    - **A.** Carbon dioxide
    - **B.** Formaldehyde (H_2CO)
    - **C.** Oxygen gas
    - **D.** Hydrogen cyanide (HCN)
    - **E.** Ethanol (CH_3CH_2OH)

14. Which of the elements listed is the most abundant extracellular anion?
    - **A.** Phosphorus
    - **B.** Sulfur
    - **C.** Potassium
    - **D.** Chloride
    - **E.** Calcium

15. The element essential for normal thyroid function is:
    - A. Iodine
    - B. Iron
    - C. Copper
    - D. Selenium
    - E. Zinc

16. To prepare a 15% salt solution such as potassium chloride (KCl) by the weight-volume method, you would:
    - A. weigh out 15 g of KCl and add water to obtain 100 g of solution
    - B. weigh out 15 g of KCl and add water to obtain 100 ml of solution
    - C. weigh out the number of grams equal to the molecular weight of KCl and add water to obtain 1 liter of solution
    - D. weigh out 15 g of KCl and add 15 ml of water
    - E. measure out 15 ml of KCl and add water to obtain 100 ml of solution

17. Alkaline substances include:
    - A. gastric juice
    - B. water
    - C. blood
    - D. orange juice
    - E. ammonia

18. Which of the following is (are) not a monosaccharide?
    - A. Glucose
    - B. Fructose
    - C. Sucrose
    - D. Glycogen
    - E. Deoxyribose

19. Which is a building block of neutral fats?
    - A. Ribose
    - B. Guanine
    - C. Glycerol
    - D. Glycine
    - E. Glucose

20. What lipid type is the foundation of the cell membrane?
    - A. Triglyceride
    - B. Steroid
    - C. Vitamin D
    - D. Phospholipid
    - E. Prostaglandin

21. Which of the following is (are) not derived from cholesterol?
    - A. Vitamin A
    - B. Vitamin D
    - C. Steroid hormones
    - D. Bile salts
    - E. Prostaglandins

22. Which of the following is primarily responsible for the helical structure of a polypeptide chain?
    - A. Hydrogen bonding
    - B. Tertiary folding
    - C. Peptide bonding
    - D. Quaternary associations
    - E. Complementary base pairing

23. Absence of which of the following nitrogen bases would prevent protein synthesis?
    - A. Adenine
    - B. Cytosine
    - C. Guanine
    - D. Thymine
    - E. Uracil

24. Which of the following is (are) not true of RNA?
    - A. Double-stranded
    - B. Contains cytosine
    - C. Directs protein synthesis
    - D. Found primarily in the nucleus
    - E. Can act as an enzyme

25. DNA:
    - A. contains uracil
    - B. is a helix
    - C. is the "genes"
    - D. contains ribose

26. ATP is not associated with:
    - A. a basic nucleotide structure
    - B. high-energy phosphate bonds
    - C. deoxyribose
    - D. inorganic phosphate
    - E. reversible reaction

Word Dissection

For each of the following word roots, fill in the literal meaning and give an example, using a word found in this chapter.

Word root	Translation	Example
1. ana	_____	_____
2. cata	_____	_____
3. di	_____	_____
4. en	_____	_____
5. ex	_____	_____
6. glyco	_____	_____
7. hydr	_____	_____
8. iso	_____	_____
9. kin	_____	_____
10. lysis	_____	_____
11. mono	_____	_____
12. poly	_____	_____
13. syn	_____	_____
14. tri	_____	_____

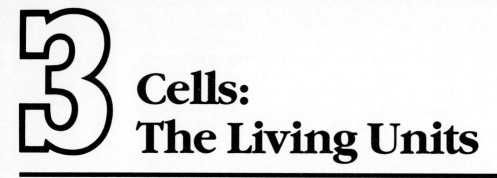

3 Cells: The Living Units

Student Objectives

When you have completed the exercises in this chapter, you will have accomplished the following objectives:

Overview of the Cellular Basis of Life

1. Define *cell*.

2. List the three major regions of a generalized cell and indicate the general function of each region.

The Plasma Membrane: Structure and Functions

3. Describe the chemical composition of the plasma membrane.

4. Compare the structure and function of tight junctions, desmosomes, and gap junctions.

5. Relate plasma membrane structure to active and passive transport mechanisms. Differentiate clearly between these transport processes relative to energy source, substances transported, direction, and mechanism.

6. Define *membrane potential* and explain how the resting membrane potential is maintained.

7. Describe the role of the plasma membrane in cells' interactions with their environment.

The Cytoplasm

8. Describe the composition of the cytosol; define *inclusions* and list several types.

9. Discuss the structure and function of mitochondria.

10. Discuss the structure and function of ribosomes, the endoplasmic reticulum, and the Golgi apparatus; note functional interrelationships among these organelles.

11. Compare the functions of lysosomes and peroxisomes.

12. Name and describe the structure and function of cytoskeletal elements.

13. Describe the roles of centrioles in mitosis and in the formation of cilia and flagella.

The Nucleus

14. Describe the chemical composition, structure, and function of the nuclear membrane, nucleolus, and chromatin.

Cell Growth and Reproduction

15. List the phases of the cell life cycle and describe the events of each phase.

16. Describe the process of DNA replication. Explain the importance of this process.

17. Define *gene* and explain the function of genes. Also explain the meaning of the genetic code.

18. Name the two phases of protein synthesis and describe the roles of DNA, mRNA, tRNA, and rRNA in each phase. Contrast triplets, codons, and anticodons.

Extracellular Materials

19. Name and describe the composition of extracellular materials.

Developmental Aspects of Cells

20. Discuss some of the theories of cellular aging.

Ready, set, go!

The basic unit of structure and function in the human body is the cell. Each of a cell's parts, or organelles, as well as the entire cell, is organized to perform a specific function. Cells have the ability to metabolize, grow, reproduce, move, and respond to stimuli. The cells of the body differ in shape, size, and specific roles in the body. Cells that are similar in structure and function form tissues, which, in turn, form the various body organs.

Student activities in this chapter include questions relating to the structure and functional abilities of the generalized animal cell.

BUILDING THE FRAMEWORK

Overview of the Cellular Basis of Life

1. Answer the following questions by inserting your responses in the answer blanks.

   1. List the four concepts of the cell theory. _____

      _____

      _____

      _____

   2. List three different cell shapes. _____

      _____

   3. Name the three major parts of any cell. _____

      _____

   4. Define *generalized* or *composite cell*. _____

      _____

      _____

      _____

The Plasma Membrane: Structure and Functions

1. Figure 3.1 is a diagram of a portion of a plasma membrane. Select four different colors and color the coding circles and the corresponding structures in the diagram. Then respond to the questions that follow, referring to Figure 3.1 and inserting your answers in the answer blanks.

   ○ Phospholipid molecules   ○ Carbohydrate molecules

   ○ Protein molecules   ○ Cholesterol molecules

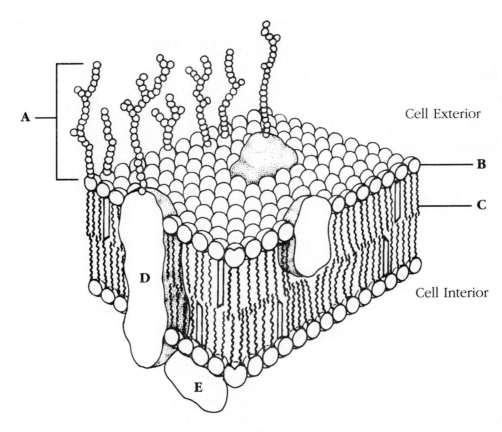

Figure 3.1

1. What name is given to this model of membrane structure? _____

2. What is the function of cholesterol molecules in the plasma membrane? _____

   _____

3. Name the carbohydrate-rich area at the cell surface (indicated by bracket A). _____

4. Which label, B or C, indicates the nonpolar region of a phospholipid molecule? _____

5. Does nonpolar mean hydrophobic or hydrophilic? _____

6. Which label, D or E, indicates an integral protein and which a peripheral protein? _____

   _____

2. Label the specializations of the plasma membrane, shown in Figure 3.2, and color the diagram as you wish. Then, answer the questions provided that refer to this figure.

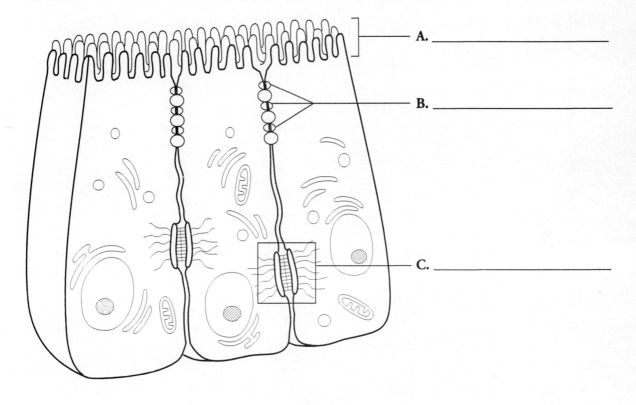

A. _____

B. _____

C. _____

Figure 3.2

1. What is the structural significance of microvilli? _____

_____

2. What type of cell function(s) does the presence of microvilli typically

indicate? _____

3. What protein acts as a microvilli "stiffener"? _____

4. Name two factors in addition to special membrane junctions that help

hold cells together. _____

5. Which cell junction forms an impermeable barrier? _____

6. Which cell junction is a buttonlike adhesion? _____

7. Which junction has linker proteins spanning the intercellular space? _____

8. Which cell junction (not shown) allows direct passage from one cell's

    cytoplasm to the next? _____

9. What name is given to the transmembrane proteins that allow this direct

    passage? _____

3. Figure 3.3 is a simplified diagram of the plasma membrane. Structure A
    represents channel proteins constructing a pore, structure B represents an
    ATP-energized solute pump, and structure C is a transport protein that does
    not depend on energy from ATP. Identify these structures and the membrane
    phospholipids by color before continuing.

    ◯ Pore   ◯ Solute pump   ◯ Passive transport pump   ◯ Phospholipids

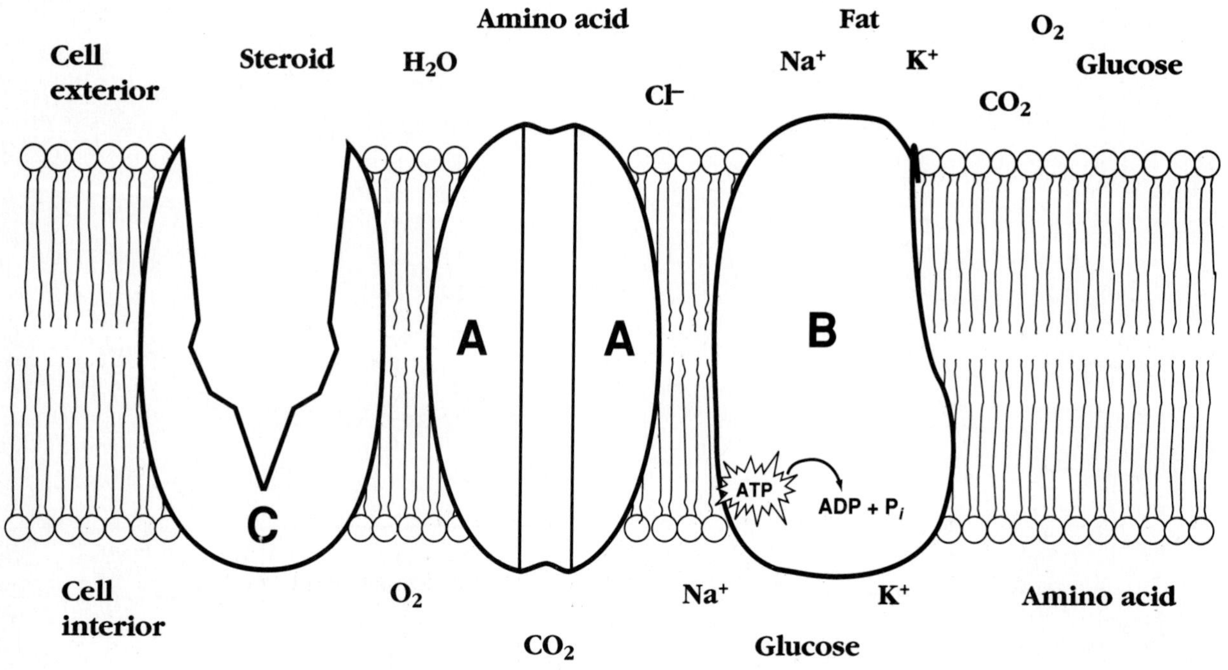

Figure 3.3

Now add arrows to Figure 3.3 as instructed next: For each substance that moves through the
plasma membrane, draw an arrow indicating its (most likely) direction of movement (into or out of
the cell). If it is moved actively, use a red arrow; if it is moved passively, use a blue arrow.

Finally, answer the following questions referring to Figure 3.3:

1. Which of the substances shown move passively *through the lipid* part of

   the membrane? _____

2. Which of the substances shown enter the cell by attachment to a passive-

   transport protein carrier? _____

3. Which of the substances shown moves passively through the membrane by moving through its

   pores? _____

4. Which of the substances shown would have to use a solute pump to be transported through

   the membrane? _____

4. Select the key choices that characterize each of the following statements.
   Insert the appropriate answers in the answer blanks.

KEY CHOICES

A. Diffusion, dialysis   C. Filtration   E. Phagocytosis

B. Diffusion, osmosis   D. Solute pumping   F. Pinocytosis

_____   1. Engulfment processes that require ATP

_____   2. Driven by molecular energy

_____   3. Driven by hydrostatic (fluid) pressure (typically blood pressure in
                      the body)

_____   4. Moves down (with) a concentration gradient

_____   5. Moves up (against) a concentration gradient; requires a carrier

5. Figure 3.4 shows three microscope fields containing red blood cells. Arrows indicate the direction of net osmosis. Select three different colors and use them to color the coding circles and the corresponding cells in the diagrams. Then, respond to the questions below, referring to Figure 3.4 and inserting your answers in the spaces provided.

◯ Water moves into the cells   ◯ Water enters and exits the cells at the same rate

◯ Water moves out of the cells

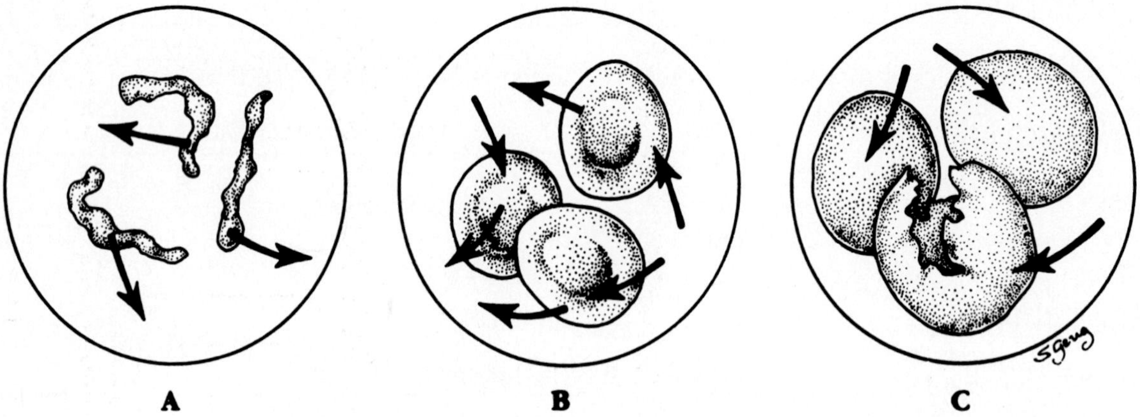

Figure 3.4

1. Name the type of tonicity illustrated in diagrams A, B, and C.

   A. _____   **B.** _____   **C.** _____

2. Name the terms that describe the cellular shapes in diagrams A, B, and C.

   A. _____   **B.** _____   **C.** _____

3. What does *isotonic* mean? _____

4. Why are the cells in diagram C bursting? _____

   _____

5. What is the difference between tonicity and osmolarity? _____

   _____

6. The differential permeability of the plasma membrane to sodium (Na⁺) and potassium (K⁺) ions results in the development of a voltage (resting membrane potential) of about –70 mV across the membrane as indicated in the simple diagram in Figure 3.5.

   First, draw in some Na⁺ and K⁺ ions in the cytoplasm and extracellular fluid, taking care to indicate their *relative* abundance in the two sites.

   Second, add positive and negative signs to the inner and outer surfaces of the "see-through" cell's plasma membrane to indicate its electrical polarity.

   Third, draw in arrows and color them to match each of the coding circles associated with the conditions noted just below.

   ◯ Potassium electrical gradient   ◯ Sodium electrical gradient

   ◯ Potassium concentration gradient   ◯ Sodium concentration gradient

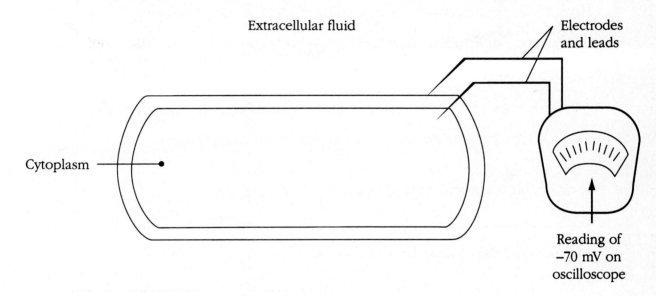

Figure 3.5

7. Referring to plasma membranes, circle the term or phrase that does not belong in each of the following groupings.

   1. Fused protein molecules of adjacent cells   Tight junction   Lining of digestive tract

   Communication between adjacent cells   No intercellular space

   2. Glycoprotein filaments   Binding of tissue layers   Heart muscle

   Impermeable junction   Desmosome

(continues on next page)

3. Impermeable intercellular space   Molecular communication   Embryonic cells

   Gap junction   Protein channel

4. Resting membrane potential   High extracellular potassium ion (K^+) concentration

   High extracellular sodium ion (Na^+) concentration   Nondiffusible protein anions

   Separation of cations from anions

5. −50 to −100 millivolts   Electrochemical gradient   Inside membrane negatively charged

   K^+ diffuses across membrane more rapidly than Na^+ Protein anions move out of cell

6. Active transport   Sodium-potassium pump   Polarized membrane

   More K^+ pumped out than Na^+ carried in   ATP required

7. Carbohydrate chains on cytoplasmic side of membrane   Cell adhesion

   Glycocalyx   Recognition sites   Antigen receptors

8. Facilitated diffusion   Nonselective   Glucose saturation   Carrier molecule

9. Clathrin-coated pit   Exocytosis   Receptor-mediated   High specificity

10. CAMs   Membrane receptors   G proteins   Channel proteins

11. Second messenger   NO Ca^{2+} Cyclic AMP

The Cytoplasm

1. Define *cytosol*. _____

   _____

2. Differentiate clearly between *organelles* and *inclusions*. _____

   _____

   _____

   _____

3. Using the following terms, correctly label all cell parts indicated by leader lines in Figure 3.6. Then select different colors for each structure and use them to color the coding circles and the corresponding structures in the illustration.

◯ Plasma membrane   ◯ Mitochondrion   ◯ Nuclear membrane   ◯ Centrioles

◯ Chromatin threads   ◯ Nucleolus   ◯ Golgi apparatus   ◯ Microvilli

◯ Rough endoplasmic reticulum (rough ER)   ◯ Smooth endoplasmic reticulum (smooth ER)

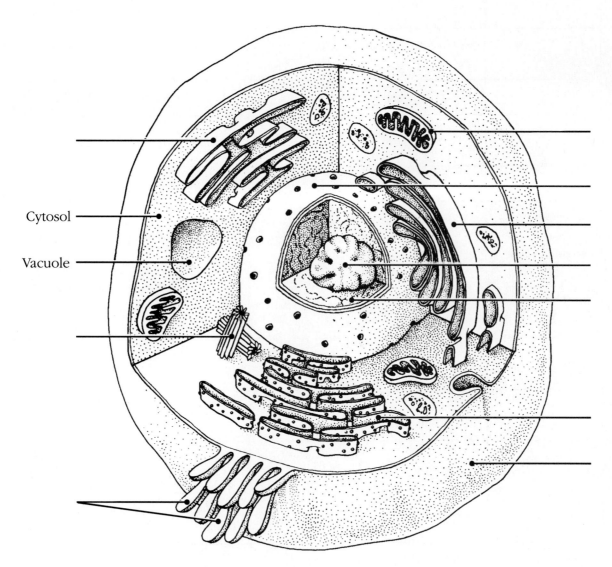

Figure 3.6

4. Complete the following table to fully describe the various cell parts. Insert your responses in the spaces provided under each heading.

Cell structure	Location	Function
	External boundary of the cell	Confines cell contents; regulates entry and exit of materials
Lysosome		
	Scattered throughout the cell	Controls release of energy from foods; forms ATP
	Projections of the plasma membrane	Increase the membrane surface area
Golgi apparatus		
	Two rod-shaped bodies near the nucleus	"Spin" the mitotic spindle
Smooth ER		
Rough ER		
	Attached to membranes or scattered in the cytoplasm	Synthesize proteins
		Act in unison to move substances across cell surface in one direction
	Internal structure of centrioles; part of the cytoskeleton	
Peroxisomes		
		Contractile protein (actin); moves cell or cell parts as microvilli
Intermediate filaments	Part of cytoskeleton	
Inclusions		

5. Relative to cellular organelles, circle the term or phrase that does not belong in each of the following groupings.

1. Peroxisomes   Enzymatic breakdown   Centrioles   Lysosomes

2. Microtubules   Intermediate filaments   Cytoskeleton   Cilia

3. Ribosomes   Smooth ER   Rough ER   Protein synthesis

4. Mitochondrion   Cristae   Self-replicating   Vitamin A storage

5. Centrioles   Basal bodies   Mitochondria   Cilia   Flagella

6. Name the cytoskeletal element (microtubules, microfilaments, or intermediate filaments) described by each of the following phrases.

_____   **1.** give the cell its shape

_____   **2.** resist tension placed on a cell

_____   **3.** radiate from the cell center

_____   **4.** interact with myosin to produce contractile force

_____   **5.** are the most stable

_____   **6.** have the thickest diameter

7. Different organelles are abundant in different cell types. Match the cell types with their abundant organelles by selecting a letter from the key choices.

____ **1.** cell lining the small intestine (assembles fats)

____ **2.** white blood cell; a phagocyte

____ **3.** liver cell that detoxifies carcinogens

____ **4.** muscle cell (contractile cell)

____ **5.** mucous cell (secretes a protein product)

____ **6.** cell at external skin surface (withstands friction and tension)

____ **7.** kidney tubule cells (makes and uses large amounts of ATP)

KEY CHOICES

A. mitochondria

B. smooth ER

C. rough ER

D. peroxisomes

E. microfilaments

F. lysosomes

G. intermediate filaments

H. Golgi apparatus

8. Describe the components and importance of the endomembrane system.

_____

_____

_____

_____

_____

The Nucleus

1. Complete the following brief table to describe the nucleus and its parts. Insert your responses in the spaces provided.

Nuclear structure	General location/appearance	Function
Nucleus		
Nucleolus		
Chromatin		
Nuclear membrane		

2. Figure 3.7 shows a portion of the proposed model of a chromatin fiber. Select two different colors for the coding circles and the corresponding structures on the figure. Then respond to the two questions that follow.

◯ DNA helix   ◯ Nucleosome

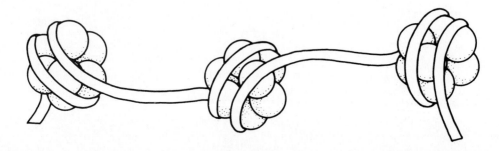

Figure 3.7

1. What is the chemical composition of a nucleosome? (Be specific.) _____

2. What is the function of the nucleosome components named above? _____

_____

Cell Growth and Reproduction

1. The cell life cycle consists of interphase and _____, during which the cell

_____. Name the three phases of interphase in order, and indicate the

important events of each phase.

Phase of interphase	Important events

2. Diagram (by making a pie-shaped graph) the relative lengths of time of the G_1,
    S, G_2, and M phases for the cell types listed:

    A. Rapidly dividing cell type

    B. Slowly dividing cell type

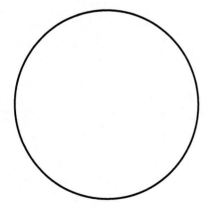
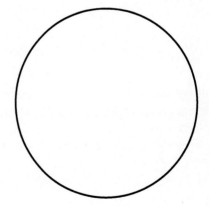

Figure 3.8

3. The following statements describe events that occur during the different phases of mitosis. Identify the phase by choosing the correct responses from the key choices and inserting the answers in the answer blanks.

KEY CHOICES

A. Anaphase   **B.** Metaphase   **C.** Prophase   **D.** Telophase   **E.** None of these

_____   **1.** Chromatin coils and condenses to form deeply staining bodies.

_____   **2.** Centromeres break, and chromosomes begin migration toward opposite poles of the cell.

_____   **3.** The nuclear membrane and nucleoli reappear.

_____   **4.** Chromosomes cease their poleward movement.

_____   **5.** Chromosomes align on the equator of the spindle.

_____   **6.** The nucleoli and nuclear membrane disappear.

_____   **7.** The spindle forms through the migration of the centrioles.

_____   **8.** DNA replication occurs.

_____   **9.** Chromosomes obviously are duplex structures.

_____   **10.** Chromosomes attach to the spindle fibers.

_____   **11.** Cytokinesis occurs.

_____   **12.** The nuclear membrane is absent during the entire phase.

_____   **13.** This is the period during which a cell is not in the M phase.

_____   **14.** Chromosomes (chromatids) are V-shaped.

4. Identify the phases of mitosis depicted in Figure 3.9 by inserting the correct terms in the blanks under each diagram. Then select different colors to represent the structures below, and use them to color the coding circles and the corresponding structures in the illustration. When you have completed the work on Figure 3.9, identify all of the mitotic stages provided with leader lines in the photomicrograph of an onion root tip in Figure 3.10.

◯ Nuclear membranes, if present   ◯ Centrioles   ◯ Chromosomes

◯ Nucleoli, if present   ◯ Spindle fibers, if present

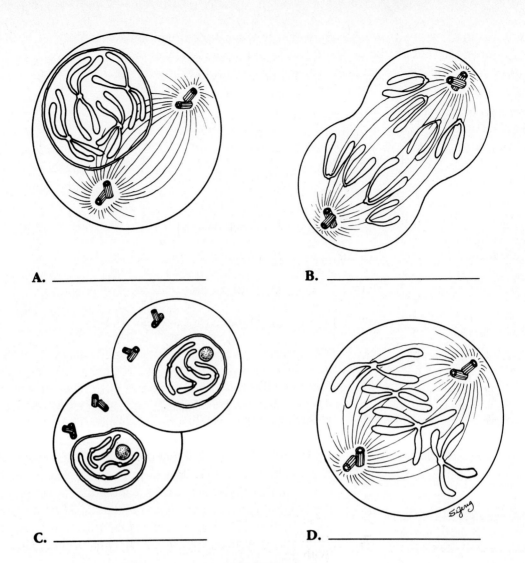

A. _____

B. _____

C. _____

D. _____

Figure 3.9

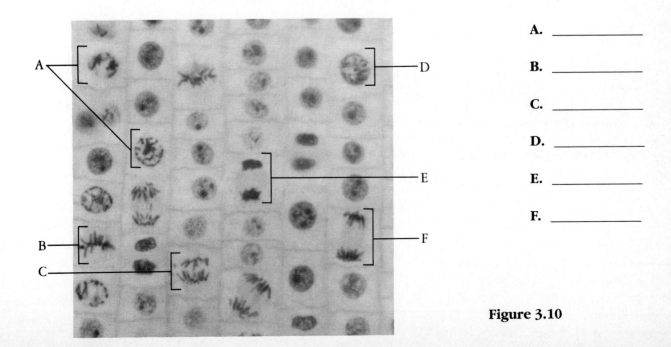

A. _____

B. _____

C. _____

D. _____

E. _____

F. _____

Figure 3.10

5. The following statements provide an overview of the structure of DNA (the genetic material) and its role in the body. Choose responses from the key choices that complete the statements. Insert the appropriate answers in the answer blanks.

KEY CHOICES

A. Adenine	**G.** Enzymes	**M.** Nucleotides	**S.** Ribosome
B. Amino acids	**H.** Genes	**N.** Old	**T.** Sugar (deoxyribose)
C. Bases	**I.** Growth	**O.** Phosphate	**U.** Template, or model
D. Codons	**J.** Guanine	**P.** Proteins	**V.** Thymine
E. Complementary	**K.** Helix	**Q.** Replication	**W.** Transcription
F. Cytosine	**L.** New	**R.** Repair	**X.** Uracil

_____ 1.

_____ 2.

_____ 3.

_____ 4.

_____ 5.

_____ 6.

_____ 7.

_____ 8.

_____ 9.

_____ 10.

_____ 11.

_____ 12.

_____ 13.

_____ 14.

_____ 15.

_____ 16.

_____ 17.

_____ 18.

DNA molecules contain information for building specific **(1)** . In a three-dimensional view, a DNA molecule looks like a spiral staircase; this is correctly called a **(2)** . The constant parts of DNA molecules are the **(3)** and **(4)** molecules, forming the DNA-ladder uprights, or backbones. The information of DNA is actually coded in the sequence of nitrogen-containing **(5)** , which are bound together to form the "rungs" of the DNA ladder. When the four DNA bases are combined in different three-base sequences, called triplets, different **(6)** of the protein are called for. It is said that the N-containing bases of DNA are **(7)** , which means that only certain bases can fit or interact together. Specifically this means that **(8)** can bind with guanine, and adenine binds with **(9)** .

The production of proteins involves the cooperation of DNA and RNA. RNA is another type of nucleic acid that serves as a "molecular slave" to DNA. That is, it leaves the nucleus and carries out the instructions of the DNA for the building of a protein on a cytoplasmic structure called a **(10)** . When a cell is preparing to divide, in order for its daughter cells to have all its information, it must oversee the **(11)** of its DNA so that a "double dose" of genes is present for a brief period. For DNA synthesis to occur, the DNA must uncoil, and the bonds between the N-bases must be broken. Then the two single strands of **(12)** each act as a **(13)** for the building of a whole DNA molecule. When completed, each DNA molecule formed is half **(14)** and half **(15)** . The fact that DNA replicates before a cell divides ensures that each daughter cell has a complete set of **(16)** . Cell division, which then follows, provides new cells so that **(17)** and **(18)** can occur.

6. Figure 3.11 is a diagram illustrating protein synthesis.

First, select four different colors and use them to color the coding circles and the corresponding structures in the diagram.

Second, using the letters of the genetic code, label the nitrogen bases on strand 2 of the DNA double helix, on the mRNA strands, and on the tRNA molecules.

Third, answer the questions that follow referring to Figure 3.11, inserting your answers in the answer blanks.

○ Backbones of the DNA double helix   ○ tRNA molecules

○ Backbone of the mRNA strands   ○ Amino acid molecules

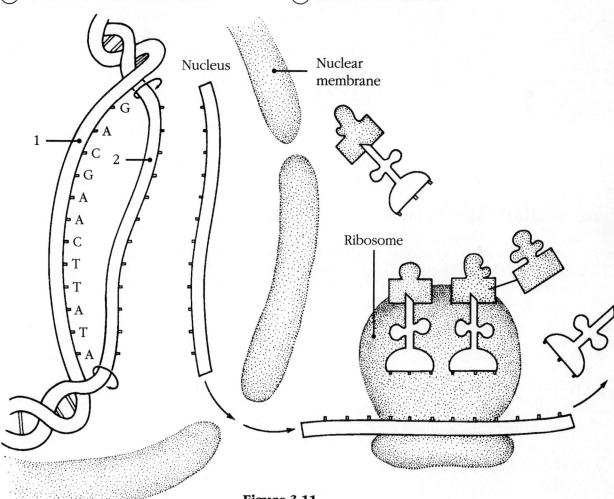

Figure 3.11

1. Transfer of the genetic message from DNA to mRNA is called _____.

2. Assembly of amino acids according to the genetic information carried by mRNA is called

_____.

3. All three types of RNA are made on the _____.

4. The set of three nitrogen bases on tRNA that is complementary to an mRNA codon is called

a _____. The complementary three-base sequence on DNA is called a

_____.

5. Define *gene*. _____

_____

7. Complete the following statements. Insert your answers in the answer blanks.

_____ **1.**

_____ **2.**

_____ **3.**

_____ **4.**

_____ **5.**

_____ **6.**

_____ **7.**

Division of the **(1)** is referred to as mitosis. Cytokinesis is division of the **(2)**. The major structural difference between chromatin and chromosomes is that the latter are **(3)**. Chromosomes attach to the spindle fibers by undivided structures called **(4)**. If a cell undergoes nuclear division but not cytoplasmic division, the product is a **(5)**. The structure that acts as a scaffolding for chromosomal attachment and movement is called the **(6)**. **(7)** is the period of cell life when the cell is not involved in division.

Extracellular Materials

1. Name the three major categories of extracellular materials in the body, provide examples of each class, and cite some of the important roles of these substances.

1. _____

2. _____

3. _____

Developmental Aspects of Cells

1. Correctly complete each statement by inserting your responses in the answer blanks.

_____ **1.**

_____ **2.**

_____ **3.**

_____ **4.**

The fertilized egg, or **(1)**, divides thousands of times to form the multicellular embryo. Although cells begin to specialize very early in development, all cells have identical numbers and kinds of **(2)**. Specialization of particular cells is fostered by **(3)** signals that activate or deactivate particular genes in neighboring cells. Consequently, the synthesis of only certain kinds of **(4)** is directed by the activated genes in specialized cells.

_____ **5.**

_____ **6.**

_____ **7.**

_____ **8.**

_____ **9.**

_____ **10.**

After birth, cell divisions continue throughout childhood and adolescence until the body reaches adult size. Thereafter, cell divisions occur only for **(5)** and **(6)**. Under certain circumstances, as in anemia, **(7)**, or accelerated cell division, occurs.

Cellular aging may be the result of continual insults over a long period of time by chemical toxins, such as **(8)** and **(9)**, and from high-energy radiation, such as **(10)**.

2. List three theories of cellular aging. _____

_____

_____

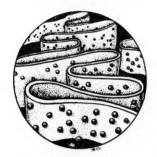

The Incredible Journey: A Visualization Exercise for the Cell

A long, meandering membrane with dark globules clinging to its outer surface now comes into sight.

1. Complete the narrative by inserting the missing words in the answer blanks.

_____ **1.**

_____ **2.**

_____ **3.**

For this journey, you will be miniaturized to the size of a small protein molecule and will travel in a microsubmarine, specially designed to enable you to pass easily through living membranes. You are injected into the intercellular space between two epithelial cells, and you are instructed to observe one cell firsthand and to identify as many of its structures as possible.

You struggle briefly with the controls and then maneuver your microsub into one of these cells. Once inside the cell, you find yourself in a kind of "sea." This salty fluid that surrounds you is the **(1)** of the cell.

Far below looms a large, dark oval structure, much larger than anything else. You conclude that it is the **(2)**. As you move downward, you pass a cigar-shaped structure with strange-looking folds on its inner surface. Although you have a pretty good idea that it must be a **(3)**, you decide to investigate more thoroughly. After passing through the membrane of the structure, you are confronted with yet another membrane. Once past this membrane, you are inside the strange-looking structure. You activate the analyzer switch in your microsub for a readout on which molecules are in your immediate vicinity.

_____ 4.

_____ 5.

_____ 6.

_____ 7.

_____ 8.

_____ 9.

_____ 10.

_____ 11.

_____ 12.

As suspected, there is an abundance of energy-rich __(4)__ molecules. Having satisfied your curiosity, you leave this structure to continue the investigation.

A long, meandering membrane with dark globules clinging to its outer surface now comes into sight. You maneuver closer and sit back to watch the activity. As you watch, amino acids are joined together and a long, threadlike protein molecule is built. The globules must be __(5)__, and the membrane, therefore, is the __(6)__. Once again you head toward the large dark structure seen and tentatively identified earlier. On approach, you observe that this huge structure has very large openings in its outer wall; these openings must be the __(7)__. Passing through one of these openings, you discover that from the inside the color of this structure is a result of dark, coiled, intertwined masses of __(8)__, which your analyzer confirms contain genetic materials, or __(9)__ molecules. Making your way through this tangled mass, you pass two round, dense structures that appear to be full of the same type of globules you saw outside. These two round structures are __(10)__. All this information confirms your earlier identification of this cellular structure, so now you move to its exterior to continue your observations.

Just ahead, you see what appears to be a mountain of flattened sacs with hundreds of small vesicles at its edges. The vesicles appear to be migrating away from this area and heading toward the outer edges of the cell. The mountain of sacs must be the __(11)__. Eventually you come upon a rather simple-looking membrane-bounded sac. Although it doesn't look too exciting and has few distinguishing marks, it does not resemble anything else you have seen so far. Deciding to obtain a chemical analysis before entering this sac, you activate the analyzer, and on the screen you see "Enzymes—Enzymes—Danger—Danger." There is little doubt that this apparently innocent structure is actually a __(12)__.

Completing your journey, you count the number of organelles identified so far. Satisfied that you have observed most of them, you request retrieval from the intercellular space.

CHALLENGING YOURSELF

At the Clinic

1. An infant is brought in with chronic diarrhea, which her mother says occurs whenever the baby drinks milk. The doctor diagnoses lactose intolerance. She explains to the parents that their baby is unable to digest milk sugar and suggests adding lactase to the baby's milk. How would lactose intolerance lead to diarrhea? How does adding lactase prevent diarrhea?

2. Anaphylaxis is a systemic (bodywide) allergic reaction in which capillaries become excessively permeable. This results in increased filtration and fluid accumulation in the tissues, leading to edema. Why is this condition life-threatening even if no frank bleeding occurs?

3. Some people have too few receptors for the cholesterol-carrying low-density lipoprotein (LDL). As a result, cholesterol builds up in blood vessel walls, restricting blood flow and leading to high blood pressure. By what cellular transport process is cholesterol taken up from the blood in a person with normal numbers of LDL receptors?

4. Sugar (glucose) can appear in the urine in nondiabetics if sugar intake is exceptionally high (when you pig out on sweets!). What functional aspect of carrier-mediated transport does this phenomenon demonstrate?

5. Plasma proteins such as albumin have an osmotic effect. In normally circulating blood, the proteins cannot leave the bloodstream easily and, thus, tend to remain in the blood. But, if stasis (blood flow stoppage) occurs, the proteins will begin to leak out into the interstitial fluid (IF). Explain why this leads to edema.

6. Hydrocortisone is an antiinflammatory drug that acts to stabilize lysosomal membranes. Explain how this effect reduces cell damage and inflammation. Why is this steroid hormone marketed in a cream (oil) base and used topically (applied to the skin)?

7. Streptomycin (an antibiotic) binds to the small ribosomal subunit of bacteria (but not to the ribosomes of the host cells infected by bacteria). The result is the misreading of bacterial mRNA and the breakup of polysomes. What process is being affected, and how does this kill the bacterial cells?

8. Kareem had a nervous habit of chewing on the inner lining of his lip and the lip grew thicker and thicker. Kareem's dentist noticed his thickened lip and suggested he have it checked. A biopsy revealed hyperplasia and scattered areas of dysplasia, but no evidence of neoplasia. What do these terms mean? Did Kareem have cancer of the mouth?

Stop and Think

1. Think *carefully* about the chemistry of the plasma membrane, then answer this question: Why is minor damage to the membrane usually not a problem?

2. Knowing that diffusion rate is inversely proportional to molecular weight, predict the results of the following experiment: Cotton balls are simultaneously inserted in opposite ends of a 1-meter-long glass tube. One cotton ball is saturated with ammonium hydroxide (NH_4OH), the other with hydrochloric acid (HCl). The two gases diffuse until they meet, at which point a white precipitate of ammonium chloride is formed. At what relative point along the tube does the precipitate form?

3. The upper layers of the skin constantly slough off. Predict the changes in the integrity of desmosomes as skin cells age (and move closer to the skin's surface).

4. Some cells produce lipid-soluble products. Can you deduce how such products are stored, that is, prevented from exiting the cell?

5. List three examples of folding of cellular membranes to increase membrane surface area.

6. Should the existence of mitochondrial ribosomes come as a complete surprise? Explain your response.

7. Examine the organelles of the cells depicted in Figure 3.12 A and B. Predict the product of each cell and state your reasons.

Figure 3.12

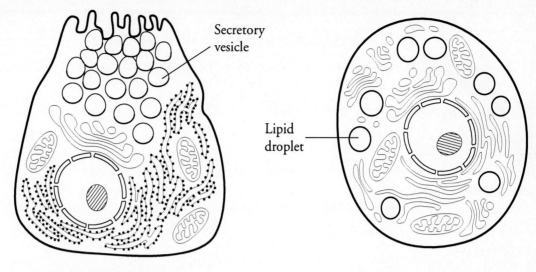

| A. Acinar cell from pancreas | B. Interstitial cell from testis |

8. If a structure (such as the lens and cornea of the eye) contains no blood vessels (that is, is *avascular*), is it likely to be very thick? Why or why not?

9. In Figure 3.13, an artificial cell with an aqueous solution enclosed in a selectively permeable membrane has just been immersed in a beaker containing a different solution. The membrane is permeable to water and to the simple sugars glucose and fructose, but is completely impermeable to the disaccharide sucrose. Answer the following questions pertaining to this situation.

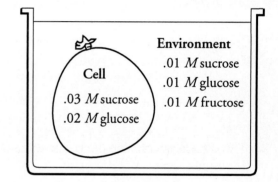

Figure 3.13

   1. Which solute(s) will exhibit net diffusion into the cell?

   2. Which solute(s) will exhibit net diffusion out of the cell?

   3. In which direction will there be net osmotic movement of water?

   4. Will the cell crenate or swell?

10. Using the information provided:

    mRNA codon: AUG   AAC   CGU   GAA   AGU   UAG

    amino acid: Met   Asn   Arg   Glu   Ser   Stop

    and given the DNA sequence   TA CGCATCACTTTT GATC:

    1. What is the amino acid sequence encoded?

    2. If the nucleotides that are underlined were deleted by mutation, what would the resulting amino acid sequence be?

COVERING ALL YOUR BASES

Multiple Choice

Select the best answer or answers from the choices given.

1. Which of the following is not a basic concept of the cell theory?
   A. All cells come from preexisting cells.
   B. Cellular properties define the properties of life.
   C. Organismic activity is based on individual and collective cellular activity.
   D. Cell structure determines and is determined by biochemical function.
   E. Organisms can exhibit properties that cannot be explained at the cellular level.

2. The dimension outside the observed range for human cells is:
   A. 10 micrometers
   B. 30 centimeters long
   C. 2 nanometers
   D. 1 meter long

3. A cell's plasma membrane would *not* contain:
   A. phospholipid
   B. nucleic acid
   C. protein
   D. cholesterol
   E. glycolipid

4. Which of the following would you expect to find in or on cells whose main function is absorption?
   A. Microvilli
   B. Cilia
   C. Desmosomes
   D. Gap junctions
   E. Secretory vesicles

5. Adult cell types expected to have gap junctions include:
   A. skeletal muscle
   B. bone
   C. heart muscle
   D. smooth muscle

6. For diffusion to occur, there must be:
   A. a selectively permeable membrane
   B. equal amounts of solute
   C. a concentration difference
   D. some sort of carrier system
   E. all of these

7. Fluid moves out of capillaries by filtration. If the plasma glucose concentration is 100 mg/dl, what will be the concentration of glucose in the filtered fluid?
   A. Less than 100 mg/dl
   B. 100 mg/dl
   C. More than 100 mg/dl
   D. The concentration cannot be determined.

8. Lack of cholesterol in the plasma membrane would result in the membrane's:
   A. excessive fluidity
   B. excessive rigidity
   C. increased protein content
   D. excessive fragility
   E. reduced protein content

9. Which of the following membrane components is involved in glucose transport?
   A. Phospholipid bilayer
   B. Transmembrane protein
   C. Cholesterol
   D. Peripheral protein
   E. Glycocalyx

10. If a 10% sucrose solution within a semi-permeable sac causes the fluid volume in the sac to increase a given amount when the sac is immersed in water, what would be the effect of replacing the sac solution with a 20% sucrose solution?
    A. The sac would lose fluid.
    B. The sac would gain the same amount of fluid.
    C. The sac would gain more fluid.
    D. There would be no effect.

11. Intestinal cells absorb glucose, galactose, and fructose by carrier-mediated transport. Poisoning the cells' mitochondria inhibits the absorption of glucose and galactose but not that of fructose. By what processes are these sugars absorbed?
    A. All are absorbed by facilitated diffusion.
    B. All are absorbed by active transport.
    C. Glucose and galactose are actively transported, but fructose is moved by facilitated diffusion.
    D. Fructose is actively transported, but glucose and galactose are moved by facilitated diffusion.

12. In a polarized cell:
    A. sodium is being pumped out of the cell
    B. potassium is being pumped out of the cell
    C. sodium is being pumped into the cell
    D. potassium is being pumped into the cell

13. Which of the following are possible functions of the glycocalyx?
    A. Determination of blood groups
    B. Binding sites for toxins
    C. Aiding the binding of sperm to egg
    D. Guiding embryonic development
    E. Increasing the efficiency of absorption

14. A cell stimulated to increase steroid production will have abundant:
    A. ribosomes   D. Golgi apparatus
    B. rough ER   E. secretory vesicles
    C. smooth ER

15. A cell's ability to replenish its ATP stores has been diminished by a metabolic poison. What organelle is most likely to be affected?
    A. Nucleus   D. Microtubule
    B. Plasma membrane E. Mitochondrion
    C. Centriole

16. Steroid hormones increase protein synthesis in their target cells. How would this stimulus be signified in a bone-forming cell, which secretes the protein collagen?
    A. Increase in heterochromatin
    B. Increase in endocytosis
    C. Increase in lysosome formation
    D. Increase in formation of secretory vesicles
    E. Increase in amount of rough ER

17. In certain nerve cells that sustain damage, the rough ER disbands and most ribosomes are free. What does this indicate?
    A. Decrease in protein synthesis
    B. Increase in protein synthesis
    C. Increase in synthesis of intracellular proteins
    D. Increase in synthesis of secreted proteins

18. What cellular inclusions increase in number in a light-skinned human after increased exposure to sunlight?
    A. Melanin granules
    B. Lipid droplets
    C. Glycogen granules
    D. Mucus
    E. Zymogen granules

19. A cell with abundant peroxisomes would most likely be involved in:
    A. secretion
    B. storage of glycogen
    C. ATP manufacture
    D. movement
    E. detoxification activities

20. Biochemical tests show a cell with replicated DNA, but incomplete synthesis of proteins needed for cell division. In what stage of the cell cycle is this cell most likely to be?
   A. M   **C.** S
   B. G_2 **D.** G_1

21.–23. Consider the following information for Questions 21–23:
A DNA segment has this nucleotide sequence:
   A A G C T C T T A C G A A T A T T C

21. Which mRNA matches or is complementary?
   A. A A G C T C T T A C G A A T A T T C
   B. T T C G A G A A T G C T T A T A A G
   C. A A G C U C U U A C G A A U A U U C
   D. U U C G A G A A U G C U U A U A A G

22. How many amino acids are coded in this segment?
   A. 18   **C.** 6
   B. 9   **D.** 3

23. What is the tRNA anticodon sequence for the fourth codon from the left?
   A. G   **C.** GCU
   B. GC   **D.** CGA

24. Which statement concerning lysosomes is false?
   A. They have the same structure and function as peroxisomes.
   B. They form by budding off the Golgi apparatus.
   C. They are abundant in phagocytes.
   D. They contain their digestive enzymes to prevent general cytoplasmic damage.

25. The fundamental structure of the plasma membrane is determined almost exclusively by:
   A. phospholipid molecules
   B. peripheral proteins
   C. cholesterol molecules
   D. integral proteins

26. Centrioles:
   A. start to duplicate in G_1
   B. reside in the centrosome
   C. are made of microtubules
   D. lie parallel to each other

27. The *trans* face of the Golgi apparatus:
   A. is where products are dispatched in vesicles
   B. is its convex face
   C. receives transport vesicles from the rough ER
   D. is in the center of the Golgi stack

Word Dissection

For each of the following word roots, fill in the literal meaning and give an example, using a word found in this chapter.

Word root	Translation	Example
1. chondri	_____	_____
2. chrom	_____	_____
3. crist	_____	_____
4. cyto	_____	_____

Word root	Translation	Example
5. desm	_____	_____
6. dia	_____	_____
7. dys	_____	_____
8. flagell	_____	_____
9. meta	_____	_____
10. mito	_____	_____
11. nucle	_____	_____
12. onco	_____	_____
13. osmo	_____	_____
14. permea	_____	_____
15. phag	_____	_____
16. philo	_____	_____
17. phobo	_____	_____
18. pin	_____	_____
19. plasm	_____	_____
20. telo	_____	_____
21. tono	_____	_____
22. troph	_____	_____
23. villus	_____	_____

# 4 Tissues: The Living Fabric

Student Objectives

When you have completed the exercises in this chapter, you will have accomplished the following objectives:

Epithelial Tissue

1. List several structural and functional characteristics of epithelial tissue.

2. Classify the epithelia.

3. Name and describe the various types of epithelia; also indicate their chief function(s) and location(s).

4. Compare endocrine and exocrine glands relative to their general structure, product(s), and mode of secretion.

5. Describe how multicellular exocrine glands are classified structurally and functionally.

Connective Tissue

6. Describe common characteristics of connective tissue.

7. List the structural elements of connective tissue and describe each element.

8. Describe the types of connective tissue found in the body and indicate their characteristic functions.

Epithelial Membranes

9. Describe the structure and function of the three varieties of epithelial membranes.

Muscle Tissue

10. Compare and contrast the structures and body locations of the three types of muscle tissue.

Nervous Tissue

11. Note the general characteristics of nervous tissue.

Tissue Repair

12. Describe the process of tissue repair involved in the normal healing of a superficial wound.

Developmental Aspects of Tissues

13. Indicate the embryonic derivation of each tissue class.

14. Briefly mention tissue changes that occur with age and indicate the possible causes of such changes.

Ready, set, go!

In the human body, as in any multicellular body, every cell has the ability to perform all activities necessary to remain healthy and alive. In the multicellular body, however, the individual cells are no longer independent. Instead, they aggregate, forming cell communities made of cells similar to one another in form and function. Once specialized, each community is committed to performing a specific activity that helps to maintain homeostasis and serves the body as a whole.

Fully differentiated cell communities are called tissue (from *tissu,* meaning "woven"). Tissues, in turn, are organized into functional units called organs (such as the heart and brain). Because individual tissues are unique in cellular shape and structure, they are readily recognized and are often named for the organ of origin—for example, muscle tissue and nervous tissue.

Student activities in Chapter 4 include questions relating to the structure and function of tissues, membranes, glands and glandular tissue, tissue repair, and the developmental aspects of tissues.

BUILDING THE FRAMEWORK

Overview of Body Tissues

1. Circle the term that does not belong in each of the following groupings.

   1. Columnar   Areolar   Cuboidal   Squamous

   2. Collagen   Cell   Matrix   Cell product

   3. Cilia   Flagellum   Microvilli   Elastic fibers

   4. Glands   Bones   Epidermis   Mucosae

   5. Adipose   Hyaline   Osseous   Nervous

   6. Blood   Smooth   Cardiac   Skeletal

   7. Polarity   Cell-to-cell junctions   Regeneration possible   Vascular

   8. Matrix   Connective tissue   Collagen   Keratin

   9. Cartilage   GAGs   Vascular   Water retention

2. Twelve tissue types are diagrammed in Figure 4.1. Identify each tissue type by inserting the correct name in the blank below each diagram. Select different colors for the following structures and use them to color the coding circles and corresponding structures in the diagrams.

◯ Epithelial cells ◯ Nerve cells

◯ Muscle cells ◯ Matrix (Where found, matrix should be colored differently from the living cells of that tissue type. Be careful. This may not be as easy as it seems!)

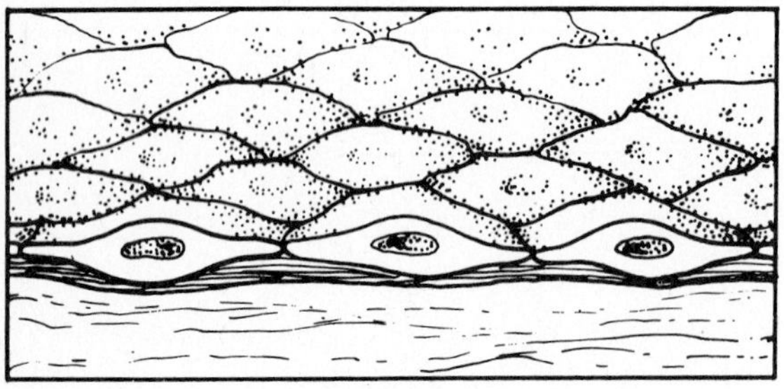

A. _____

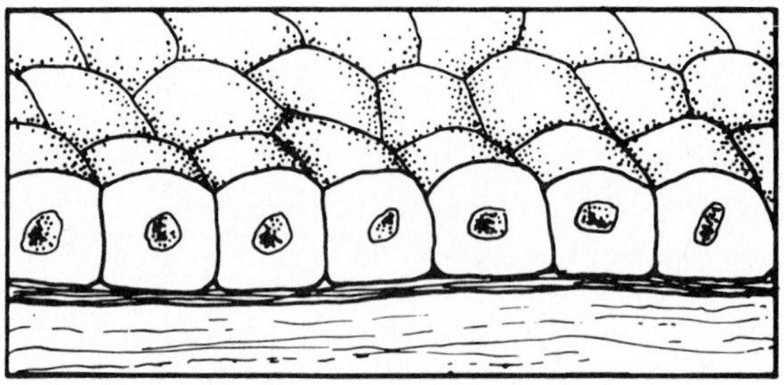

B. _____

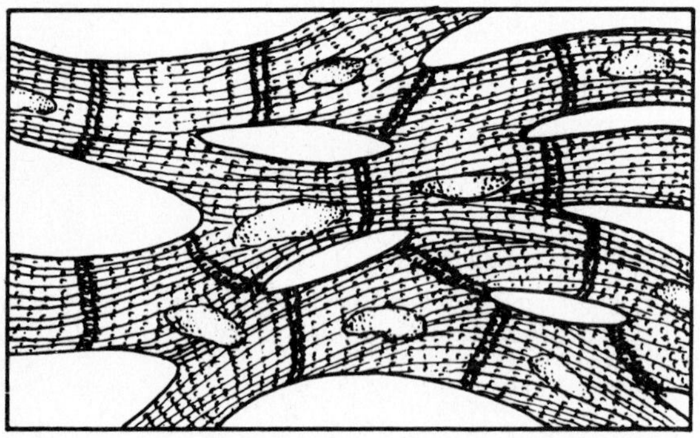

C. _____

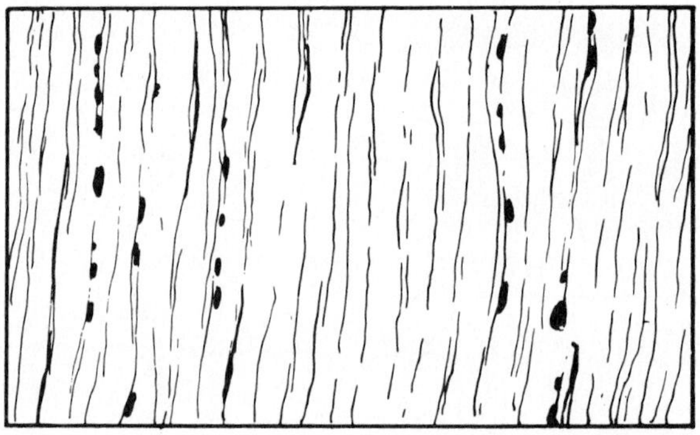

D. _____

E. _____

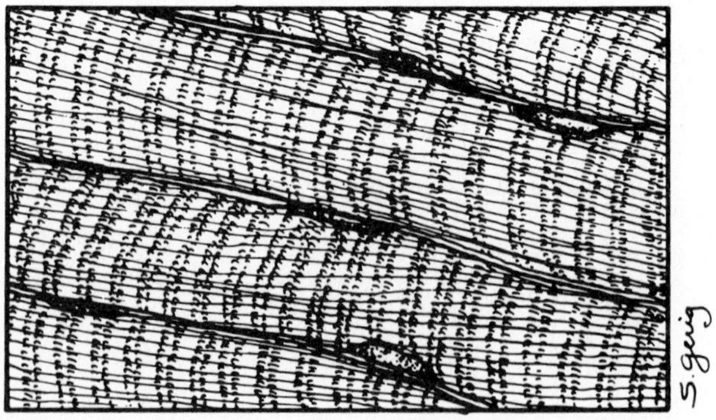

F. _____

Figure 4.1

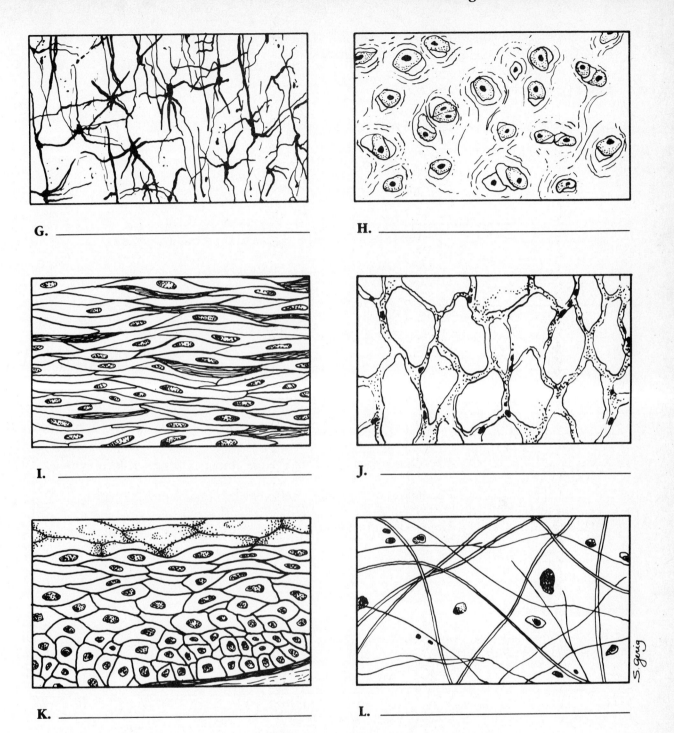

G. _____

H. _____

I. _____

J. _____

K. _____

L. _____

Figure 4.1 (continued)

3. Using the key choices, correctly identify the following *major* tissue types.
Enter the appropriate answer in the answer blanks.

KEY CHOICES

A. Connective   **B.** Epithelium   **C.** Muscle   **D.** Nervous

_____ **1.** Forms membranes

_____ **2.** Allows for movement of limbs and for organ movements within
                              the body

_____ **3.** Uses electrochemical signals to carry out its functions

_____ **4.** Supports and reinforces body organs

_____ **5.** Cells of this tissue may absorb and/or secrete substances

_____ **6.** Basis of the major controlling system of the body

_____ **7.** Its cells shorten to exert force

_____ **8.** Forms endocrine and exocrine glands

_____ **9.** Surrounds and cushions body organs

_____ **10.** Characterized by having large amounts of extracellular material

_____ **11.** Allows you to smile, grasp, swim, ski, and throw a ball

_____ **12.** Widely distributed; found in bones, cartilages, and fat depots

_____ **13.** Forms the brain and spinal cord

Epithelial Tissue

1.   List the six major functions of epithelium. _____

_____

2.   List six special characteristics of epithelium. _____

_____

_____

_____

_____

3. For 1–5, match the epithelial type named in Column B with the appropriate *location* in Column A.

Column A   Column B

_____ **1.** Lines the stomach and most   **A.** Pseudostratified ciliated
                              of the intestines   columnar

_____ **2.** Lines the inside of the mouth   **B.** Simple columnar

_____ **3.** Lines much of the respiratory tract **C.** Simple cuboidal

_____ **4.** Endothelium and mesothelium   **D.** Simple squamous

_____ **5.** Lines the inside of the urinary bladder **E.** Stratified columnar

                                                                         F. Stratified squamous

                                                                         G. Transitional

For 6–10, match the epithelium named in Column B with the appropriate *function* in Column A.

Column A   Column B

_____ **6.** Protection   **H.** Endothelium

_____ **7.** Small molecules pass through rapidly **I.** Simple columnar

_____ **8.** Propel sheets of mucus   **J.** A ciliated epithelium

_____ **9.** Absorption, secretion, or ion transport **K.** Stratified squamous

_____ **10.** Stretches   **L.** Transitional

4. Arrange the following types of epithelium from 1 to 5 in order of increasing *protectiveness*.

_____ **A.** Simple squamous   _____ **D.** Pseudostratified

_____ **B.** Stratified squamous   _____ **E.** Simple columnar

_____ **C.** Simple cuboidal

5. Arrange the following types of epithelium from 1 to 4 in order of increasing *absorptive* ability.

_____ **A.** Simple squamous   _____ **C.** Simple cuboidal

_____ **B.** Stratified squamous   _____ **D.** Simple columnar

6. Epithelium exhibits many of the plasma membrane modifications described in conjunction with our discussion of the composite cell in Chapter 3. Figure 4.2 depicts some of these modifications.

First: Choose a color for the coding circles and the corresponding structures in the figure.

◯ Epithelial cell cytoplasm   ◯ Connective tissue

◯ Epithelial cell nucleus   ◯ Blood vessel

◯ Nerve fibers

Second: Correctly identify the following structures or regions by labeling the appropriate leader lines using terms from the list below:

A. Epithelium   **E.** Connective tissue   **I.** Cilia

B. Basal region   **F.** Basement membrane   **J.** Reticular lamina

C. Apical region   **G.** Basal lamina   **K.** Tight junctions

D. Capillary   **H.** Microvilli   **L.** Desmosome

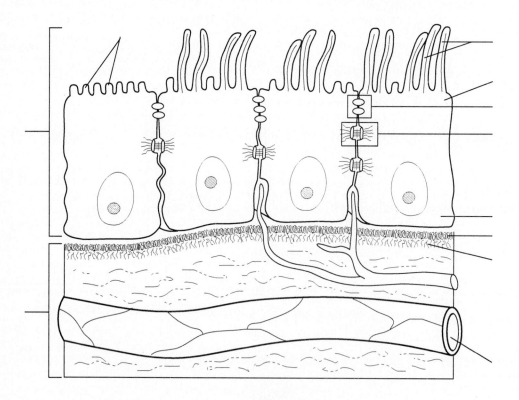

Figure 4.2

_____ **7.** Insulates the body

_____ **8.** Firm, slightly "rubbery" matrix; milky white and "glassy" in appearance

_____ **9.** Cells are arranged in concentric circles around a nutrient canal; matrix is hard due to calcium salts

_____ **10.** Contains collagenous fibers; found in intervertebral discs

_____ **11.** Makes supporting framework of lymphoid organs

_____ **12.** Found in umbilical cord

_____ **13.** Found in external ear and auditory tube

_____ **14.** Provides the medium for nutrient transport throughout the body

_____ **15.** Forms the "stretchy" ligaments of the vertebral column

2. Areolar connective tissue is often considered to be the prototype of connective tissue proper because of its variety of cell types and fibers. Figure 4.2 shows most of these elements. Identify all structures or cell types provided with leader lines. Color the diagram as your fancy strikes you.

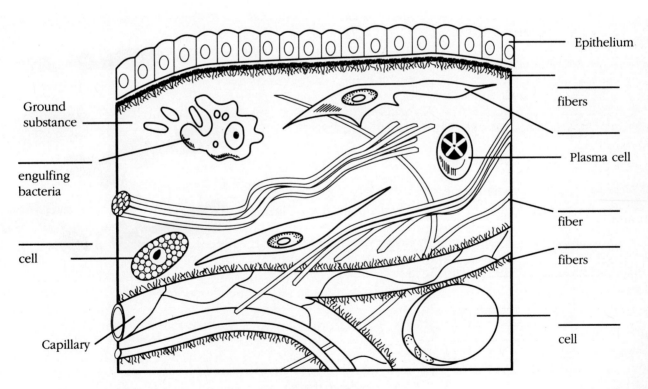

Figure 4.3

3. Arrange the following tissue types from 1 to 3 in order of *decreasing* vascularity.

____ **A.** Cartilage

____ **B.** Areolar connective

____ **C.** Dense connective

4. Using the key choices, select the structural or related elements of connective tissue (CT) types that permit specialized functions. Insert the appropriate answers in the answer blanks.

KEY CHOICES

A. Adipocytes	**D.** Elastic fibers	**G.** Macrophages	**J.** Osteocytes
B. Chondrocytes	**E.** Ground substance	**H.** Matrix	**K.** Osteoblasts
C. Collagen fibers	**F.** Hemocytoblast	**I.** Mesenchyme	**L.** Reticular fibers

_____ **1.** Composed of ground substance and structural protein fibers

_____ **2.** Composed of glycoproteins and water-binding glycosaminoglycans

_____ **3.** Tough protein fibers that resist stretching or longitudinal tearing

_____ **4.** Primary bone marrow cell type that remains actively mitotic

_____ **5.** Fine, branching protein fibers that construct a supportive network

_____ **6.** Large, irregularly shaped cells, widely distributed, often found in CT; they engulf cellular debris and foreign matter and are active in immunity

_____ **7.** The medium through which nutrients and other substances diffuse

_____ **8.** Living elements that maintain the firm, flexible matrix in cartilage

_____ **9.** Randomly coiled protein fibers that recoil after being stretched

_____ **10.** The structural element of areolar tissue that is fluid and provides a reservoir of water and salts for neighboring tissues

_____ **11.** In a loose CT, the nondividing cells that store nutrients

_____ **12.** The embryonic tissue that gives rise to all types of CT

_____ **13.** Cellular elements that produce the collagen fibers of bone matrix

Epithelial Membranes

1. Five simplified diagrams are shown in Figure 4.4. Select different colors for the membranes listed below and use them to color the coding circles and the corresponding structures.

○ Mucosae

○ Visceral pleura (serosa)

○ Parietal pleura (serosa)

○ Visceral pericardium (serosa)

○ Parietal peritoneum (serosa)

○ Visceral peritoneum (serosa)

○ Mesentery

○ Endothelium

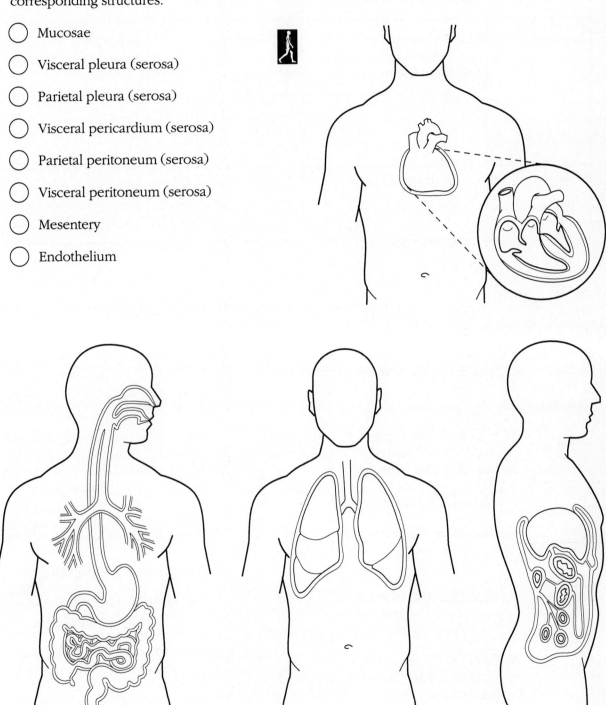

Figure 4.4

2. Complete the following table relating to epithelial membranes. Enter your responses in the areas left blank.

Membrane	Tissue type (epithelial/connective)	Common locations	Functions
Mucous	Epithelial sheet with underlying connective tissue (lamina propria)		Protection, lubrication, secretion, absorption
Serous		Lines internal ventral cavities and covers their organs	
Cutaneous			Protection from external insults; protection from water loss

Muscle Tissue

1. The three types of muscle tissue exhibit certain similarities and differences. Insert Sk (skeletal), C (cardiac), or Sm (smooth) into the appropriate blanks to indicate which muscle type exhibits each characteristic.

Characteristic

_____ **1.** Voluntarily controlled

_____ **2.** Involuntarily controlled

_____ **3.** Banded appearance

_____ **4.** Uninucleate

_____ **5.** Multinucleate

_____ **6.** Found attached to bones

_____ **7.** Enables you to swallow

_____ **8.** Found in the walls of the small intestine, uterus, bladder, and veins

_____ **9.** Contains spindle-shaped cells

_____ **10.** Contains cylindrical cells with branching ends

_____ **11.** Contains long, nonbranching cylindrical cells

_____ **12.** Displays intercalated disks

_____ **13.** Concerned with locomotion of the body as a whole

_____ **14.** Changes the internal volume of an organ as it contracts

_____ **15.** Tissue of the circulatory pump

Nervous Tissue

1. Describe briefly how the particular structure of a neuron relates to its function

in the body. _____

_____

2. Circle the word that does *not* apply to neuroglia:

Support   Insulate   Conduct   Protect

Tissue Repair

1. For each of the following statements about tissue repair that is true, enter T in
the answer blank. For each false statement, correct the underlined word(s) by
writing the correct word(s) in the answer blank.

_____   **1.** The nonspecific response of the body to injury is called
underline{regeneration}.

_____   **2.** Intact capillaries near an injury dilate, leaking plasma, blood cells,
and underline{antibodies}, which cause the blood to clot. The clot at the
surface dries to form a scab.

_____   **3.** During organization, the first phase of tissue repair, capillary buds
invade the clot, forming a delicate pink tissue called underline{endodermal}
tissue.

_____   **4.** Fibroblasts synthesize fibers across the gap.

_____   **5.** When damage is not too severe, the surface epithelium migrates
beneath the dry scab and across the surface of the granulation
tissue. This repair process is called underline{proliferation}.

_____   **6.** If tissue damage is very severe, tissue repair is more likely to occur
by underline{fibrosis}, or scarring.

_____   **7.** During fibrosis, fibroblasts in the granulation tissue lay down
underline{keratin} fibers, which form a strong, compact, but inflexible mass.

_____   **8.** The repair of cardiac muscle and nervous tissue occurs only by
underline{fibrosis}.

_____   **9.** Organization is replacement of a blood clot by granulation tissue.

_____   **10.** Granulation tissue resists infection by secreting underline{viral-inhibiting}
substances.

_____   **11.** Problems associated with underline{regeneration} include shrinking, loss of
elasticity, and formation of adhesions.

Developmental Aspects of Tissues

1. Complete the following statements by filling in the appropriate answer blanks.

_____ 1.   During embryonic development, cells specialize and aggregate
                                  to form **(1)** . Mitotic cell division is responsible for overall
_____ 2.   body **(2)** . Organization of the three primary germ layers,
                                   (3) , **(4)** , and **(5)** , occurs between 9 and 13 days after
_____ 3.   fertilization. After approximately 60 days, the four primary
                                  tissues have developed. All tissues, except **(6)** tissue,
_____ 4.   continue to undergo cell division until the end of adolescence.
                                  After this time, **(7)** tissue also becomes amitotic.
_____ 5.

_____ 6.   _____ 7.

2. Figure 4.5 represents the three embryonic germ layers that form the embryonic
   body and give rise to the four primary tissues.

   First, color each germ layer and the corresponding color-coding circle as
   follows: ectoderm, blue; mesoderm, red; and endoderm, yellow. Do not color
   the cells of the body stalk, yolk sac, or amnion.

   Second, in the answer blanks below, insert the names of the four primary
   tissues arising from each of the numbered regions (one or more of the germ
   layers), as indicated by the brackets on the diagram.

   ◯ Ectoderm (layer of columnar-shaped cells)   ◯ Endoderm (layer of cuboidal-shaped cells)

   ◯ Mesoderm (layer of loose, migrating cells)

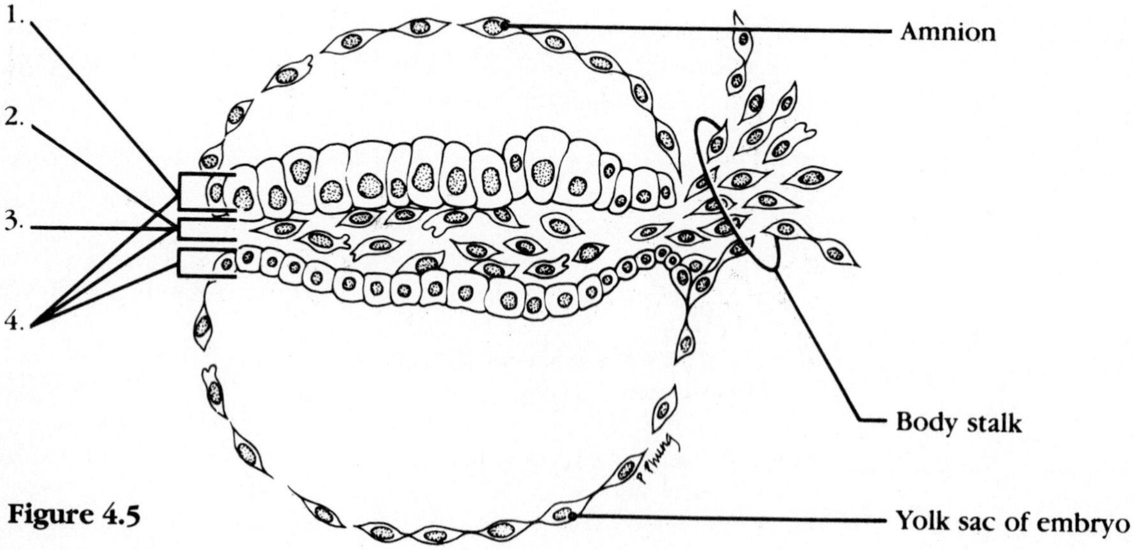

Figure 4.5

1. _____   3. _____

2. _____   4. _____

3. Complete the following table on embryological development by filling in the appropriate terms:

Germ layer	Tissue type
Endoderm	
	Muscle
	Nerve tissue
	Bone
	Epidermis
	Mesothelium

4. State the effect of aging on the following tissue types and components:

Epithelium **1.** _____

Collagen   **2.** _____

Muscle   **3.** _____

CHALLENGING YOURSELF

At the Clinic

1. Since adipocytes are incapable of cell division, how is weight gain accomplished?

2. A woman sees her gynecologist because she is unable to become pregnant. The doctor discovers granulation tissue in her vaginal canal and explains that sperm are susceptible to some of the same chemicals as bacteria. What is inhibiting the sperm?

3. A histological examination of the brain during an autopsy reveals an extremely high number of microglial cells in an area that had suffered recent trauma. What do these cells signify?

4. Why do cartilage and tendons take so long to heal?

5. In mesothelial cancer of the pleura, serous fluid is hypersecreted. How does this contribute to respiratory problems?

6. In cases of ruptured appendix, what serous membrane is likely to become infected? Why can this be life-threatening?

Stop and Think

1. Try to come up with some *advantages* of the avascularity of epithelium and cartilage. Also, try to deduce why tendons are poorly vascularized.

2. Structurally speaking, why is simple columnar epithelium more resistant to being torn apart than simple squamous epithelium?

3. What types of tissue can have microvilli? Cilia?

4. Why do endocrine glands start out having ducts?

5. On his anatomy test, Bruno answered two questions incorrectly. He confused a basal lamina with a basement membrane, and a mucous membrane with a sheet of mucus. What are the differences between each of these sound-alike pairs of structures?

6. Time for an educated guess. Do you think the elastic connective tissue layer in the large arteries is regularly *or* irregularly arranged? Explain your reasoning.

7. What kind of tissue surrounds a bone shaft?

8. Why are skeletal muscle cells multinucleate?

9. Explain how the nervous system (obviously an internal organ system) can be derived from the same primary germ layer as the outer covering of the body.

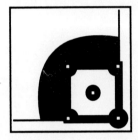

COVERING ALL YOUR BASES

Multiple Choice

Select the best answer or answers from the choices given.

1. Scar tissue is a type of:
   A. epithelium   C. muscle
   B. connective tissue   D. nervous tissue

2. Which of the following terms/phrases could *not* be applied to epithelium?
   A. Basement membrane   D. Strong matrix
   B. Free surface   E. Ciliated
   C. Desmosomes present

3. Which is not a type of epithelium?
   A. Reticular   D. Transitional
   B. Simple squamous   E. Stratified columnar
   C. Pseudostratified

4. In which of the following tissue types might you expect to find goblet cells?
   A. Simple cuboidal
   B. Simple columnar
   C. Simple squamous
   D. Stratified squamous
   E. Transitional

5. Mesothelium is found in:
   A. kidney tubules
   B. mucous membranes
   C. serosae
   D. the liver
   E. the lining of cardiovascular system organs

6. An epithelium "built" to withstand friction is:
   A. simple squamous
   B. stratified squamous
   C. simple cuboidal
   D. simple columnar
   E. pseudostratified

7. Functions of keratin include:
   A. absorbing sunlight   D. waterproofing
   B. energy storage   E. providing resilience
   C. directing protein synthesis

8. What cellular specialization causes fluid to flow over the epithelial surface?
   A. Centrioles   D. Microvilli
   B. Flagella   E. Myofilaments
   C. Cilia

9. The gland type that secretes its product continuously, by exocytosis, into a duct is:
   A. merocrine   C. endocrine
   B. holocrine   D. apocrine

10. Which of the following is not a function of some kind of connective tissue?
    A. Binding   D. Sensation
    B. Support   E. Repair
    C. Protection

11. The original embryonic connective tissue is:
    A. mucous connective
    B. mesenchyme
    C. vascular
    D. areolar connective
    E. connective tissue proper

12. Components of connective tissue matrix include:
    A. hyaluronic acid   D. glycocalyx
    B. basal lamina   E. glycosaminoglycans
    C. proteoglycans

13. Which of the following fibrous elements gives a connective tissue high tensile strength?
    A. Reticular fibers   C. Collagen fibers
    B. Elastic fibers   D. Myofilaments

14. The cell that forms bone is the:
    A. fibroblast   D. osteoblast
    B. chondroblast   E. reticular cell
    C. hemocytoblast

15. Which of the following cell types secretes histamine and perhaps heparin?
    A. Macrophage   D. Fibroblast
    B. Mast cell   E. Histiocyte
    C. Reticular cell

16. Resistance to stress applied in a longitudinal direction is provided best by:
    A. fibrocartilage
    B. elastic connective
    C. reticular connective
    D. dense regular connective
    E. areolar connective

17. What kind of connective tissue acts as a sponge, soaking up fluid when edema occurs?
    A. Areolar connective
    B. Adipose connective
    C. Dense irregular connective
    D. Reticular connective
    E. Vascular tissue

18. Viewed through the microscope, most cells in this type of tissue have an empty appearance.
    A. Reticular connective
    B. Adipose connective
    C. Areolar connective
    D. Osseous tissue
    E. Hyaline cartilage

19. The major function of reticular tissue is:
    A. nourishment   D. protection
    B. insulation   E. movement
    C. stroma formation

20. What type of connective tissue prevents muscles from pulling away from bones during contraction?
    A. Dense irregular connective
    B. Dense regular connective
    C. Areolar
    D. Elastic connective
    E. Hyaline cartilage

21. The type of connective tissue that provides flexibility to the vertebral column is:
    A. dense irregular connective
    B. dense regular connective
    C. reticular connective
    D. areolar
    E. fibrocartilage

22. Phrases that describe cartilage include:
    A. highly vascularized
    B. holds large volumes of water
    C. has no nerve endings
    D. grows both appositionally and interstitially
    E. can get quite thick

23. Which type of cartilage is most abundant throughout life?
    A. Elastic cartilage
    B. Fibrocartilage
    C. Hyaline cartilage

24. Serous membranes:
    A. line the mouth
    B. have parietal and visceral layers
    C. consist of epidermis and dermis
    D. have a connective tissue layer called the lamina propria
    E. secrete a lubricating fluid

25. Select the one false statement about mucous and serous membranes.
    A. The epithelial type is the same in all serous membranes, but there are different epithelial types in different mucous membranes.
    B. Serous membranes line closed body cavities, while mucous membranes line body cavities open to the outside.
    C. Serous membranes always produce serous fluid, and mucous membranes always secrete mucus.
    D. Both membranes contain an epithelium plus a layer of loose connective tissue.

26. Which of the following terms describe cardiac muscle?
    A. Striated
    B. Intercalated discs
    C. Multinucleated
    D. Involuntary
    E. Branching

27. Events of tissue repair include:
    A. regeneration
    B. organization
    C. granulation
    D. fibrosis
    E. inflammation

28. Which tissues are derived from ectoderm?
    A. Muscle
    B. Epidermis
    C. Nerve tissue
    D. Mucosae
    E. Cartilage

29. Which primary germ layer gives rise to the lining of the intestine?
    A. Endoderm
    B. Mesoderm
    C. Ectoderm

30. Tissues derived from mesoderm include:
    A. connective tissue
    B. nerve tissue
    C. muscle
    D. dermis
    E. epidermis

Word Dissection

For each of the following word roots, fill in the literal meaning and give an example, using a word found in this chapter.

Word root	Translation	Example
1. ap	_____	_____
2. areola	_____	_____
3. basal	_____	_____
4. blast	_____	_____
5. chyme	_____	_____
6. crine	_____	_____
7. endo	_____	_____
8. epi	_____	_____
9. glia	_____	_____
10. holo	_____	_____
11. hormon	_____	_____
12. hyal	_____	_____
13. lamina	_____	_____
14. mero	_____	_____
15. meso	_____	_____
16. retic	_____	_____
17. sero	_____	_____
18. squam	_____	_____
19. strat	_____	_____

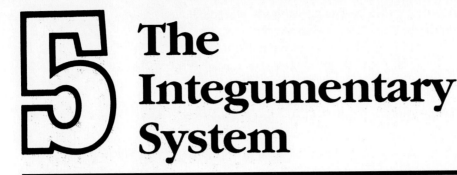 # The Integumentary System

Student Objectives

When you have completed the exercises in this chapter, you will have accomplished the following objectives:

The Skin

1. Name the specific tissue types composing the epidermis and dermis. List the major layers of each and describe the function of each layer.

2. Describe the factors that normally contribute to skin color. Briefly describe how changes in skin color may be used as clinical signs of certain disease states.

Appendages of the Skin

3. List the parts of a hair follicle and explain the function of each part. Describe the functional relationship between arrector pili muscles and the hair follicle.

4. Explain the basis of hair color. Describe the distribution, growth, and replacement of hairs and the changing nature of hair during the life span.

5. Compare the structure, distribution, and most common locations of sweat and oil glands. Compare the composition and functions of their secretions.

6. Compare and contrast eccrine and apocrine glands.

7. Describe the structure of nails.

Functions of the Integumentary System

8. Describe how the skin accomplishes at least five different functions.

9. Explain why serious burns are life-threatening. Describe how to determine the extent of a burn and differentiate between first-, second-, and third-degree burns.

Homeostatic Imbalances of Skin

10. Summarize the characteristics and warning signs of skin cancers.

Developmental Aspects of the Integumentary System

11. Briefly describe the changes that occur in the skin from birth to old age. Attempt to explain the causes of such changes.

Ready, set, go!

The integumentary system consists of the skin and its derivatives—glands, hairs, and nails. Although the skin is very thin, it provides a remarkably effective external shield that acts to protect our internal organs from what is outside the body.

This chapter reviews the anatomical characteristics of the skin (composed of the dermis and the epidermis) and its derivatives. It also reviews the manner in which the skin responds to both internal and external stimuli to protect the body.

BUILDING THE FRAMEWORK

The Skin

1. 1. Name the tissue type composing the epidermis. _stratified squamouse pith_

2. Name the tissue type composing the dermis. _dense irregular connective_

2. The more superficial cells of the epidermis become less viable and ultimately die. What two factors account for this natural demise of the epidermal cells?

1. _They are increasingly further from the blood supply_

2. _Surrounded by glyco lipids + lack nutrients_

3. Several types of skin markings may reveal structural characteristics of the dermis. Complete the following statements by inserting your responses in the answer blanks.

_____ 1. Skin cuts that run parallel to _(1)_ gape less than cuts running across these skin markings.

_____ 2. A more scientific term for "stretch marks" is _(2)_.

_____ 3. Skin markings that occur where the dermis is secured to deeper structures are called _(3)_.

_____ 4. _(4)_ appear when dermis elasticity declines from age or excessive sun exposure.

4. Figure 5.1 depicts a longitudinal section of the skin. Label the skin structures and areas indicated by leader lines and brackets on the figure. Select different colors for the structures below and color the coding circles and the corresponding structures on the figure.

- ◯ Arrector pili muscle
- ◯ Adipose tissue
- ◯ Hair follicle
- ◯ Nerve fibers
- ◯ Sweat (sudoriferous) gland
- ◯ Sebaceous gland

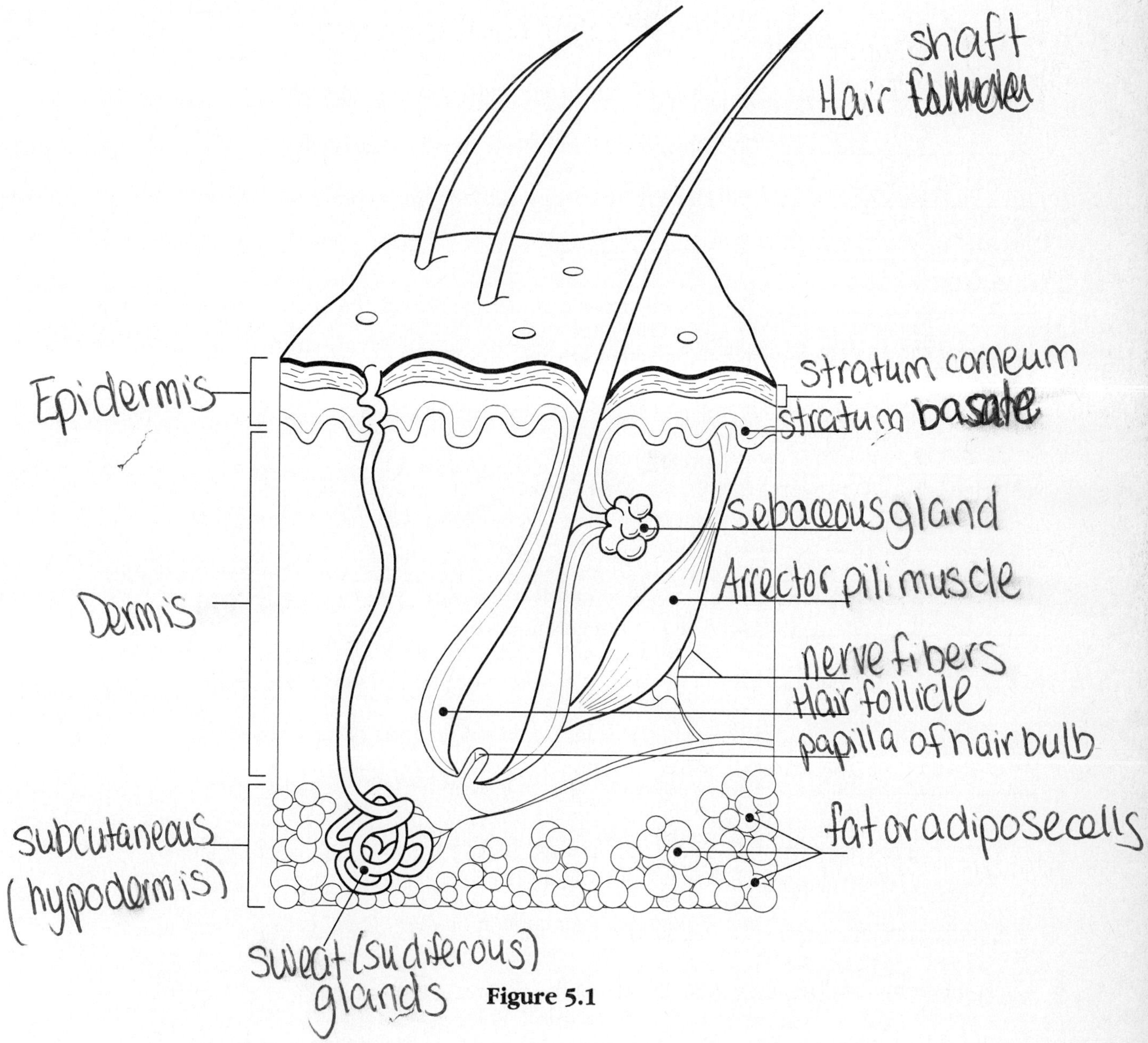

shaft
Hair follicle

Epidermis

stratum corneum
stratum basale

sebaceous gland
Arrector pili muscle

nerve fibers
Hair follicle
papilla of hair bulb

Dermis

subcutaneous
(hypodermis)

fat or adipose cells

sweat (sudiferous)
glands

Figure 5.1

5. Using the key choices, choose all responses that apply to the following descriptions. Enter the appropriate letters and/or terms in the answer blanks. (Note: S. = stratum)

KEY CHOICES

A. S. basale   **D.** S. lucidum   **G.** Reticular layer   **J.** Hypodermis

B. S. corneum   **E.** S. spinosum   **H.** Epidermis (as a whole)

C. S. granulosum   **F.** Papillary layer   **I.** Dermis (as a whole)

_____ **1.** Layer of translucent cells, absent in thin skin

_____ **2.** Strata containing all (or mostly) dead cells

_____ **3.** Dermal layer responsible for fingerprints

_____ **4.** Vascular region

_____ **5.** Actively mitotic epidermal region, the deepest epidermal layer

_____ **6.** Shinglelike cells that slough off; the basis of "dandruff"

_____ **7.** Site of elastic and collagen fibers

_____ **8.** General site of melanin formation

_____ **9.** Major skin area where derivatives (hair, nails) *reside*

_____ **10.** Largely adipose tissue; anchors the skin to underlying tissues

_____ **11.** The stratum germinativum

_____ **12.** Epidermal layer where most melanocytes are found

_____ **13.** Cells of this layer contain granules of keratohyalin

_____ **14.** Accounts for the bulk of epidermal thickness

_____ **15.** When tanned, becomes leather; provides mechanical strength to the skin

_____ **16.** Epidermal layer containing the "oldest" cells

6. Circle the term that does not belong in each of the following groupings.

1. Reticular layer   Keratin   Dermal papillae   Meissner's corpuscles

2. Melanin   Freckle   Wart   Malignant melanoma

3. Prickle cells   Stratum basale   Stratum spinosum   Cell shrinkage

4. Langerhans' cells   Phagocytes   Keratinocytes   Macrophages

5. Meissner's corpuscles   Pacinian corpuscles   Merkel cells   Arrector pili

6. Waterproof substance   Elastin   Keratin   Produced by keratinocytes

7. Mast cells   Macrophages   Fibroblasts   Melanocytes

8. Tonofilaments   Keratin fibrils   Keratohyaline   Laminated granules

9. Keratinocyte   Fibroblast   Merkel cell   Langerhans' cell

7. This exercise examines the relative importance of three pigments in determining skin color. Indicate which pigment is identified by the following descriptions by inserting the appropriate answer from the key choices in the answer blanks.

KEY CHOICES

A. Carotene   **B.** Hemoglobin   **C.** Melanin

Melanin **1.** Most responsible for the skin color of dark-skinned people

Carotene **2.** Provides an orange cast to the skin

melanin **3.** Provides a natural sunscreen

Hemoglobin **4.** Most responsible for the skin color of Caucasians

Melanin **5.** Phagocytized by keratinocytes

Carotene **6.** Found predominantly in the stratum corneum

Hemoglobin **7.** Found within red blood cells in the blood vessels

8. Abnormalities of skin color can be helpful in alerting a physician to certain pathologies. Match the clinical terms in Column B with the possible-cause descriptions in Column A. Place the correct letter in each answer blank.

Column A

_____ **1.** A bluish cast of the skin resulting from inadequate oxygenation of the blood

_____ **2.** Observation of this condition might lead to tests for anemia or low blood pressure

_____ **3.** Accumulation of bile pigments in the blood; may indicate liver disease

_____ **4.** Clotted mass of blood that may signify bleeder's disease

_____ **5.** A common result of inflammation, allergy, and fever

Column B

A. Cyanosis

B. Erythema

C. Hematoma

D. Jaundice

E. Pallor

Appendages of the Skin

1. Figure 5.2 shows longitudinal and cross-sectional views of a hair follicle.

Part A

1. Identify and label all structures provided with leader lines.

2. Select different colors to identify the structures described below and color both the coding circles and the corresponding structures on the diagram.

   ◯ Contains blood vessels that nourish the growth zone of the hair

   ◯ Secretes sebum into the hair follicle

   ◯ Pulls the hair follicle into an upright position during fright or exposure to cold

   ◯ The follicle sheath that consists of dermal tissue

   ◯ The follicle sheath that consists of epidermal tissue

   ◯ The actively growing region of the hair

3. Draw in the nerve fibers and blood vessels that supply the follicle, the hair, the hair root, and the arrector pili.

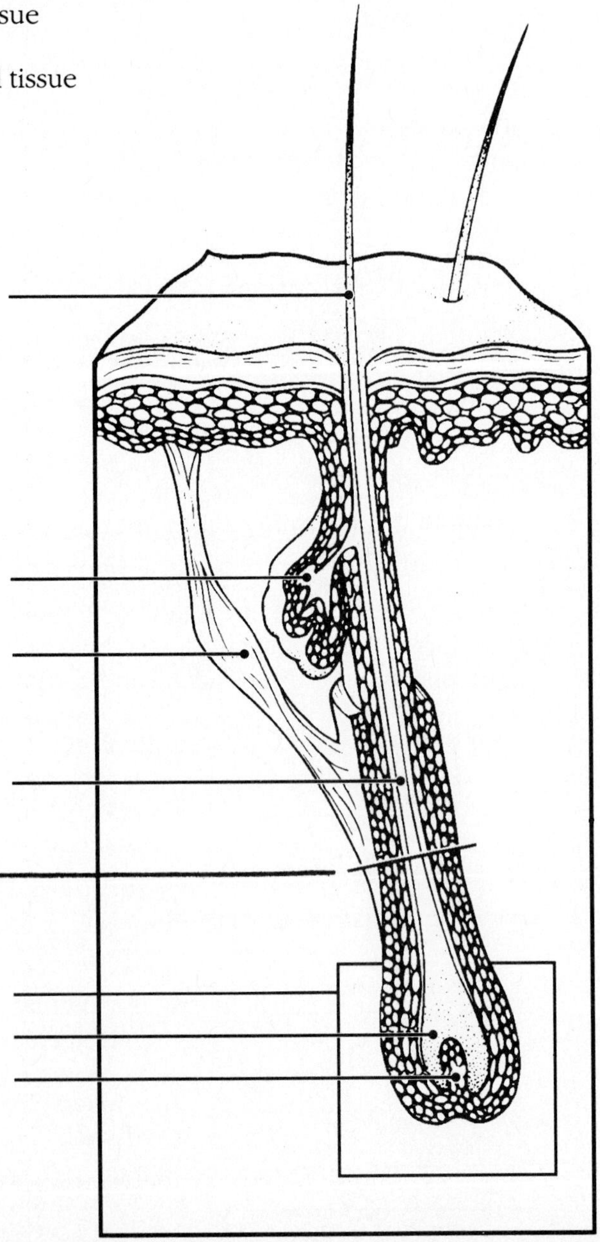

See Figure 5.2B for cross section

Figure 5.2A

Part B

4. Identify the two portions of the follicle wall by placing the correct name of the sheath at the end of the appropriate leader line.

5. Color these regions using the same colors used for the identical structures in Part A.

6. Label, color code, and color these three regions of the hair:

   ◯ Cortex　　　◯ Cuticle　　　◯ Medulla

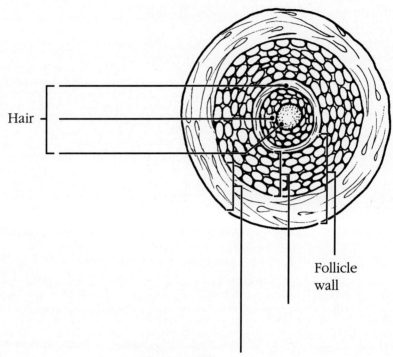

Hair

Follicle wall

Figure 5.2B

2. Circle the term that does not belong in each of the following groupings.

   1. Luxuriant hair growth　　　Testosterone　　　Poor nutrition　　　Good blood supply

   2. Vitamin D　　　Cholesterol　　　UV radiation　　　Keratin

   3. Stratum corneum　　　Nail matrix　　　Hair bulb　　　Stratum germinativum

   4. Scent glands　　　Eccrine glands　　　Apocrine glands　　　Axilla

   5. Terminal hair　　　Vellus hair　　　Dark, coarse hair　　　Eyebrow hair

   6. Hard keratin　　　Hair shaft　　　Desquamation　　　Durable

   7. Growth phase　　　Resting phase　　　Atrophy　　　Inactive

3. What is the scientific term for baldness? _____

4. Name four factors that can cause hair loss and hair thinning *other than* nutritional or

circulatory factors. _____

_____

5. Draw a simple diagram of a fingertip *bearing a fingernail* in the space at the right. Identify and label the following nail regions on your sketch: free edge, body, lunula, lateral nail folds, proximal nail fold (eponychium).

   1. What is the common name for the eponychium? _____

   2. Why does the lunula appear whiter than the rest of the nail? _____

         _____

         _____

6. Using the key choices, complete the following statements. Insert the appropriate letters in the answer blanks.

KEY CHOICES

A. Sebaceous glands   **B.** Sweat glands (apocrine)   **C.** Sweat glands (eccrine)

_____ **1.** Their products are an oily mixture of lipids, cholesterol, and cell fragments.

_____ **2.** Functionally, these are merocrine glands.

_____ **3.** The less numerous variety of perspiration gland, their secretion (often milky in appearance) contains proteins and other substances that favor bacterial growth.

_____ **4.** Their ducts open to the external environment via a pore.

_____ **5.** These glands are found everywhere on the body except the palms of the hands and soles of the feet.

_____ **6.** Their secretions contain bactericidal substances.

_____ **7.** They become more active at puberty under the influence of androgens.

_____ **8.** Their secretions, when oxidized, are seen on the skin surface as a blackhead.

_____ **9.** The ceruminous glands that produce ear wax are a modification of this gland variety.

_____ **10.** These glands are involved in thermoregulation.

Functions of the Integumentary System

1. The skin protects the body by providing three types of barriers. Classify each of the protective factors listed below as an example of a chemical barrier (C), a biological barrier (B), or a mechanical (physical) barrier (M).

   _____ **1.** Langerhans' cells and macrophages   _____ **4.** Keratin

   _____ **2.** Intact epidermis   _____ **5.** Melanin

   _____ **3.** Bactericidal secretions   _____ **6.** Acid mantle

2. Substances that can penetrate the skin in limited amounts include (circle all that apply):

   Fat-soluble vitamins   Steroid hormones   Water-soluble substances

   Organic solvents   Oxygen   Mercury, lead, and nickel

3. In what way does a sunburn impair the body's ability to defend itself? (Assume the sunburn is mild.) _____

   _____

4. Explain the role of sweat glands in maintaining body temperature homeostasis.

   In your explanation, indicate how their activity is regulated. _____

   _____

   _____

   _____

5. Complete the following statements. Insert your responses in the answer blanks.

   _____ **1.**

   _____ **2.**

   _____ **3.**

   _____ **4.**

   _____ **5.**

   _____ **6.**

   _____ **7.**

   _____ **8.**

   _____ **9.**

   The cutaneous sensory receptors that reside in the skin are actually part of the __(1)__ system. Four types of stimuli that can be detected by certain of the cutaneous receptors are __(2)__, __(3)__, __(4)__, and __(5)__.

   Vitamin D is synthesized when modified __(6)__ molecules in the __(7)__ of the skin are irradiated by __(8)__ light. Vitamin D is important in the absorption and metabolism of __(9)__ ions.

Homeostatic Imbalances of Skin

1. Overwhelming infection is one of the most important causes of death in burn patients. What is the other major problem they face, and what are its possible consequences?

   _____

   _____

2. This section reviews the severity of burns. Using the key choices, select the correct burn type for each of the following descriptions. Enter the correct answers in the answer blanks.

   KEY CHOICES

   A. First-degree burn   **B.** Second-degree burn   **C.** Third-degree burn

   _____ 1. Full-thickness burn; epidermal and dermal layers destroyed; skin is blanched

   _____ 2. Blisters form

   _____ 3. Epidermal damage, redness, and some pain (usually brief)

   _____ 4. Epidermal and some dermal damage; pain; regeneration is possible

   _____ 5. Regeneration impossible; requires grafting

   _____ 6. Pain is absent because nerve endings in the area are destroyed

3. What is the importance of the "rule of nines" in the treatment of burn patients?

   _____

4. Fill in the type of skin cancer that matches each of the following descriptions:

   _____ 1. Cells of the stratum spinosum develop lesions; metastasizes to lymph nodes.

   _____ 2. Cells of the lowest level of the epidermis invade the dermis and hypodermis; exposed areas develop ulcer; slow to metastasize.

   _____ 3. Rare but deadly cancer of pigment-producing cells.

5. What does ABCD mean in reference to examination of pigmented areas? _____

   _____

   _____

Developmental Aspects of the Integumentary System

1. Match the choices in Column B with the appropriate descriptions in Column A.

Column A

_____ 1. Skin inflammations that increase in frequency with age

_____ 2. Cause of graying hair

_____ 3. Small white bumps on the skin of newborn babies, resulting from accumulations of sebaceous gland material

_____ 4. Reflects the loss of insulating subcutaneous tissue with age

_____ 5. A common consequence of accelerated sebaceous gland activity during adolescence

_____ 6. Oily substance produced by the fetus's sebaceous glands

_____ 7. The hairy "cloak" of the fetus

Column B

A. Acne

B. Cold intolerance

C. Dermatitis

D. Delayed action gene

E. Lanugo

F. Milia

G. Vernix caseosa

The Incredible Journey: A Visualization Exercise for the Skin

Your immediate surroundings resemble huge, grotesquely twisted vines . . . you begin to climb upward.

1. Complete the narrative by inserting the missing words in the answer blanks.

_____ 1.

_____ 2.

For this trip, you are miniaturized for injection into your host's skin. Your journey begins when you are injected into a soft gel-like substance. Your immediate surroundings resemble huge, grotesquely twisted vines. But when you peer carefully at the closest "vine," you realize you are actually seeing connective tissue fibers. Most of the fibers are fairly straight, although tangled together, and look like strong cables. You identify these as the __(1)__ fibers. Here and there are fibers that resemble coiled springs. These must be the __(2)__ fibers

_____ 3.

_____ 4.

_____ 5.

_____ 6.

_____ 7.

_____ 8.

_____ 9.

_____ 10.

that help give skin its springiness. At this point, there is little question that you are in the __(3)__ region of the skin particularly since you can also see blood vessels and nerve fibers around you.

Carefully, using the fibers as steps, you begin to climb upward. After climbing for some time and finding that you still haven't reached the upper regions of the skin, you stop for a rest. As you sit, a strange-looking cell approaches, moving slowly with parts alternately flowing forward and then receding. Suddenly you realize that this must be a __(4)__ that is about to dispose of an intruder (you) unless you move in a hurry! You scramble to your feet and resume your upward climb. On your right is a large fibrous structure that looks like a tree trunk anchored in place by muscle fibers. By scurrying up this __(5)__ sheath, you are able to escape from the cell and again scan your surroundings. Directly overhead are tall cubelike cells, forming a continuous sheetlike membrane. In your rush to escape you reached the __(6)__ region of the skin. As you watch the activity of the cells in this layer, you notice that many of the cells are pinching in two and that the daughter cells are being forced upward. Obviously, this is the specific layer that continually replaces cells that rub off the skin surface, and these cells are the __(7)__ cells.

Looking through the transparent cell membrane of one of the basal cells, you see a dark mass hanging over the nucleus. You wonder if this cell could have a tumor; but then, looking through the membranes of the neighboring cells, you find that they also have dark umbrella-like masses hanging over their nuclei. As you consider this matter, a black cell with long tentacles begins to pick its way carefully between the other cells. As you watch, one of the transparent cells engulfs the end of a tentacle of the black cell, and within seconds contains some of its black substance. Suddenly, you remember that one of the skin's protective functions is to protect the deeper layers from sun damage; the black substance must be the protective pigment __(8)__.

Once again you begin your upward climb and notice that the cells are becoming shorter and harder and are full of a tough, waxy substance. This substance has to be __(9)__, which would account for the increasing hardness of the cells. Climbing still higher, the cells become flattened like huge shingles. The only material apparent in the cells is the waxy substance; there is no nucleus, and there appears to be no activity in these cells. Considering the clues—shinglelike cells, no nuclei, full of the waxy substance, no activity—these cells are obviously __(10)__ and therefore are very close to the skin surface.

Suddenly, you feel a strong agitation in your immediate area. The pressure is tremendous. Looking upward through the transparent cell layers, you see your host's fingertips vigorously scratching the area directly overhead. You wonder if you are causing his skin to sting or tickle. Then, within seconds, the cells around you begin to separate and fall apart, and you are catapulted out into the sunlight. Since the scratching fingers might descend once again, you quickly advise your host of your whereabouts.

CHALLENGING YOURSELF

At the Clinic

1. Xeroderma pigmentosum is a severe, genetically linked skin cancer in which DNA repair mechanisms are impaired. Why would sufferers of this condition need to stay out of the sun?

2. A new mother brings her infant to the clinic, worried about a yellowish, scummy deposit that has built up on the baby's scalp. What is this condition called, and is it serious?

3. During a diaper change, an alert day care worker notices a dark, bruised-looking area at the base of a baby's spine. Worried about possible child abuse, she reports the spot to her supervisor, who tells her not to worry because it is a Mongolian spot. What is a Mongolian spot?

4. Hives are welts, or reddened "bumps," that indicate sites of local inflammation. They are often a sign of an allergic reaction. Recall from Chapter 3 the role of capillary permeability and plasma loss in causing edema. Would systemic hives be cause for worry?

5. A worker in a furniture refinishing establishment fell into a vat of paint stripper, but quickly removed his clothes and rinsed off in the safety shower. Were his safety measures adequate? What vital organs might suffer early damage from poisoning through skin by organic solvents?

6. What two factors in the treatment of critical third-degree burn patients are absolutely essential?

7. Mr. Bellazono, a fisherman in his late 60s, comes to the clinic to complain of small ulcers on both forearms as well as on his face and ears. Although he has had them for several years, he has not had any other problems. What is the likely diagnosis, and what is the likely cause?

8. The hypodermis of the face is quite loose and has few connections to the deep fascia of the muscles. Explain how this relates to the greater need to suture cuts on the face compared to other body regions.

9. Martha, the mother of a 13-month-old infant, brings her child to the clinic because his skin has turned orange. Why does the pediatrician inquire about the child's diet?

10. Mrs. Ibañez volunteered to help at a hospital for children with cancer. When she first entered the cancer ward, she was upset by the fact that most of the children had no hair. What is the explanation for their baldness?

Stop and Think

1. How can the skin be both a membrane and an organ?

2. Why does the border between the epidermis and the dermis undulate?

3. The skin covering your shins is *not* freely movable. Palpate (feel) your shins and compare that region to the other regions of the body. Then try to deduce why there is little free movement of the skin of the shins.

4. In terms of both function and benefit, why are surface keratinocytes dead?

5. What nerve endings in the skin respond to the lightest touch?

6. How do basal keratinocytes and melanocytes avoid being pushed into higher (more superficial) epidermal layers?

7. Why does sunburned skin peel *in sheets?*

8. Explain each of these familiar phenomena in terms of what you learned in this chapter:
   - pimples and blackheads
   - goose bumps
   - greasy hair and shiny nose
   - stretch marks from gaining weight
   - turning blue from holding your breath
   - leaving fingerprints
   - pores on the face
   - the almost hairless body of humans
   - blisters
   - bruises

9. Would increasing protein intake (such as by taking gelatin supplements) increase hair and nail strength in an otherwise healthy individual?

10. Studies have shown that women who live or work together tend to develop synchronized monthly cycles. What aspect of the integumentary system might explain this sexual signaling?

11. If our cells and body fluids are hyperosmotic to the water of a swimming pool (and they *are*), then why do we not swell and pop when we go for a swim?

COVERING ALL YOUR BASES

Multiple Choice

Select the best answer or answers from the choices given.

1. Which is *not* part of the skin?
   A. Epidermis   C. Dermis
   B. Hypodermis   D. Superficial fascia

2. Which of the following is *not* a tissue type found in the skin?
   A. Stratified squamous epithelium
   B. Loose connective tissue
   C. Dense irregular connective tissue
   D. Ciliated columnar epithelium
   E. Vascular tissue

3. Epidermal cells that aid in the immune response include:
   A. Merkel cells   C. melanocytes
   B. Langerhans' cells D. spinosum cells

4. Which organelle is most prominent in cells manufacturing the protein keratin?
   A. Ribosomes
   B. Golgi apparatus
   C. Smooth endoplasmic reticulum
   D. Lysosomes

5. Which epidermal layer has the highest concentration of Langerhans' cells, and has numerous desmosomes and thick bundles of keratin tonofilaments?
   A. Stratum corneum
   B. Stratum lucidum
   C. Stratum granulosum
   D. Stratum spinosum

6. Fingerprints are caused by:
   A. the genetically determined arrangement of dermal papillae
   B. the conspicuous epidermal ridges
   C. the sweat pores
   D. all of these

7. Find the *false* statement concerning vitamin D.
   A. Dark-skinned people make no vitamin D.
   B. If vitamin D production is inadequate, one may develop weak bones.
   C. If the skin is not exposed to sunlight, one may develop weak bones.
   D. Vitamin D is needed for the uptake of calcium from food in the intestine.

8. Use logic to deduce the answer to this question. Given what you now know about skin color and skin cancer, which of the following groups would have the highest rate of skin cancer?
   A. Blacks in tropical Africa
   B. Scientists in research stations in Antarctica
   C. Whites in northern Australia
   D. Norwegians in the southern part of the U.S.
   E. Blacks in the U.S.

9. Which structure is not associated with a hair?
   A. Shaft   D. Matrix
   B. Cortex   E. Cuticle
   C. Eponychium

10. Which of the following hair colors is not produced by melanin?
    - **A.** Blonde
    - **B.** Brown
    - **C.** Black
    - **D.** Gray
    - **E.** White

11. Which is the origin site of cells that are directly responsible for growth of a hair?
    - **A.** Hair bulb
    - **B.** Hair follicle
    - **C.** Papilla
    - **D.** Hair bulge
    - **E.** Matrix

12. Concerning movement of hairs:
    - **A.** Movement is the function of the arrector pili.
    - **B.** Movement is sensed by the root hair plexus.
    - **C.** Muscle contraction flattens the hair against the skin.
    - **D.** Muscle contraction is prompted by cold or fright.

13. Some infants are born with a fuzzy skin; this is due to:
    - **A.** vellus hairs
    - **B.** terminal hairs
    - **C.** lanugo
    - **D.** hirsutism

14. A particular type of tumor of the adrenal gland causes excessive secretion of sex hormones. This condition expresses itself in females as:
    - **A.** male pattern baldness
    - **B.** hirsutism
    - **C.** increase in growth of vellus hairs over the whole body
    - **D.** increase in length of terminal hairs

15. At the end of a follicle's growth cycle:
    - **A.** the follicle atrophies
    - **B.** the hair falls out
    - **C.** the hair elongates
    - **D.** the hair turns white

16. In investigating the cause of thinning hair, which of the following questions needs to be asked?
    - **A.** Is the diet deficient in proteins?
    - **B.** Is the person taking megadoses of vitamin C?
    - **C.** Has the person been exposed to excessive radiation?
    - **D.** Has the person recently suffered severe emotional trauma?

17. Which structures are not associated with a nail?
    - **A.** Nail bed
    - **B.** Lunula
    - **C.** Nail folds
    - **D.** Nail follicle

18. One of the following is not associated with production of perspiration. Which one?
    - **A.** Sweat glands
    - **B.** Sweat pores
    - **C.** Holocrine gland
    - **D.** Eccrine gland
    - **E.** Apocrine gland

19. Components of sweat include:
    - **A.** water
    - **B.** sodium chloride
    - **C.** sebum
    - **D.** ammonia
    - **E.** vitamin D

20. Which of the following is *true* concerning oil production in the skin?
    - **A.** Oil is produced by sudoriferous glands.
    - **B.** Secretion of oil is via the holocrine mode.
    - **C.** The secretion is called sebum.
    - **D.** Ceruminous glands are a specialized type of oil gland.
    - **E.** Oil is usually secreted into hair follicles.

21. A disorder *not* associated with oil glands is:
    - **A.** seborrhea
    - **B.** cystic fibrosis
    - **C.** acne
    - **D.** whiteheads

22. Contributing to the chemical barrier of the skin is/are:
    - **A.** perspiration
    - **B.** stratified squamous epithelium
    - **C.** Langerhans' cells
    - **D.** sebum
    - **E.** melanin

23. Which of the following provide evidence of the skin's role in temperature regulation?
    A. Shivering
    B. Flushing
    C. Blue fingernails
    D. Sensible perspiration
    E. Acid mantle

24. Contraction of the arrector pili would be "sensed" by:
    A. Merkel discs
    B. Meissner's corpuscles
    C. root hair plexuses
    D. Pacinian corpuscles

25. A dermatologist examines a patient with lesions on the face. Some of the lesions appear as a shiny, raised spot; others are ulcerated with a beaded edge. What is the diagnosis?
    A. Melanoma
    B. Squamous cell carcinoma
    C. Basal cell carcinoma
    D. Either squamous or basal cell carcinoma

26. A burn patient reports that the burns on her hands and face are not painful, but she has blisters on her neck and forearms and the skin on her arms is very red. This burn would be classified as:
    A. first-degree only   C. third-degree only
    B. second-degree only   D. critical

27. Which of the following is associated with vitamin synthesis in the body?
    A. Cholesterol
    B. Calcium metabolism
    C. Carotene
    D. Ultraviolet radiation
    E. Melanin

28. A patient has a small (about 1 mm), regular, round, pale mole. Is this likely to be melanoma?
    A. Yes
    B. No, because it is small
    C. No, because it is pale
    D. No, because it is round
    E. No, because it is regular

Word Dissection

For each of the following word roots, fill in the literal meaning and give an example, using a word found in this chapter.

Word root	Translation	Example
1. arrect	_____	_____
2. carot	_____	_____
3. case	_____	_____
4. cere	_____	_____
5. corn	_____	_____
6. cort	_____	_____
7. cutic	_____	_____

Word root	Translation	Example
8. cyan	_____	_____
9. derm	_____	_____
10. folli	_____	_____
11. hemato	_____	_____
12. hirsut	_____	_____
13. jaune	_____	_____
14. kera	_____	_____
15. lanu	_____	_____
16. lunul	_____	_____
17. medull	_____	_____
18. melan	_____	_____
19. pall	_____	_____
20. papilla	_____	_____
21. pili	_____	_____
22. plex	_____	_____
23. rhea	_____	_____
24. seb	_____	_____
25. spin	_____	_____
26. sudor	_____	_____
27. tegm	_____	_____
28. vell	_____	_____

Bones and Bone Tissue

Student Objectives

When you have completed the exercises in this chapter, you will have accomplished the following objectives:

Skeletal Cartilages

1. Explain the functional properties of each of the three types of cartilage tissue.
2. Locate the major cartilages of the adult skeleton.
3. Explain how cartilage grows.

Functions of the Bones

4. List and describe five important functions of bones.

Classification of Bones

5. Compare and contrast the structure of the four bone classes and provide examples of each class.

Bone Structure

6. Describe the gross anatomy of a typical long bone and flat bone. Indicate the locations and functions of red and yellow marrow, articular cartilage, periosteum, and endosteum.
7. Distinguish clearly between the several types of bone markings and note their relative functions.
8. Describe the histology of compact and spongy bone.

9. Discuss the chemical composition of bone and the relative advantages conferred by its organic and its inorganic components.

Bone Development (Osteogenesis)

10. Compare and contrast the two types of bone formation: intramembranous and endochondral ossification.
11. Describe the process of long bone growth that occurs at the epiphyseal plates.

Bone Homeostasis: Remodeling and Repair

12. Compare the locations and functions of the osteoblasts, osteocytes, and osteoclasts in bone remodeling.
13. Explain how hormonal controls and physical stress regulate bone remodeling.
14. Describe the steps of fracture repair.

Homeostatic Imbalances of Bone

15. Contrast the disorders of bone remodeling seen in osteoporosis, osteomalacia, and Paget's disease.

Developmental Aspects of Bones: Timing of Events

16. Describe the timing and cause of changes in bone architecture and bone mass throughout life.

Ready, set, go!

Bone and cartilage are the principal tissues that provide support for the body; together they make up the skeletal system. The early skeleton is formed largely from cartilage. Later, the cartilage is replaced almost entirely by bone. Bones provide attachments for muscles and act as levers for the movements of body parts. Bones perform additional essential functions. They are the sites for blood cell formation and are storage depots for many substances, such as fat, calcium, and phosphorus.

Bone tissue has an intricate architecture that is both stable and dynamic. Although it has great strength and rigidity, bone tissue is able to change structurally in response to a variety of chemical and mechanical factors.

Chapter 6 topics for student review include an overview of skeletal cartilages, the structures of long and flat bones, the remodeling and repair of bone, and bone development and growth.

BUILDING THE FRAMEWORK

Skeletal Cartilages

1. Use the key choices to identify the type of cartilage tissue found in the following body locations:

		KEY CHOICES
_____ **1.** At the junction of a rib and the sternum		**A.** Elastic cartilage
_____ **2.** The skeleton of the ear		**B.** Fibrocartilage
_____ **3.** Supporting the trachea walls		**C.** Hyaline cartilage
_____ **4.** Forming the intervertebral discs		
_____ **5.** Forming the epiglottis		
_____ **6.** At the ends of long bones		
_____ **7.** Most of the fetal skeleton		

2. In comparing bone and cartilage tissue, indicate whether each of the following statements is true (T) or false (F).

____ **1.** Cartilage is more resilient than bone.

____ **2.** Cartilage is especially strong in resisting shear (bending and twisting) forces.

____ **3.** Cartilage can grow faster than bone in the growing skeleton.

____ **4.** In the adult skeleton, cartilage regenerates faster than bone when damaged.

____ **5.** Neither bone nor cartilage contains capillaries.

____ **6.** Bone tissue contains relatively little water compared to cartilage tissue, which contains a large amount of water.

____ **7.** Nurtients diffuse quickly through cartilage matrix but very poorly through solid bone matrix.

Functions of the Bones

1. List and explain five important functions of bones. Write your answers in the answer blanks below.

1. _____

2. _____

3. _____

4. _____

5. _____

Classification of Bones

1. Identify each of the following bones as a member of one of the four major bone categories. Use L for long bone, S for short bone, F for flat bone, and I for irregular bone. Enter the appropriate letters in the answer blanks.

____ **1.** Calcaneus   ____ **4.** Humerus   ____ **7.** Radius

____ **2.** Frontal   ____ **5.** Mandible   ____ **8.** Sternum

____ **3.** Femur   ____ **6.** Metacarpal   ____ **9.** Vertebra

Bone Structure

1. Figure 6.1A is a drawing of a sagittal section of the femur. Do not color the articular cartilage. Leave it white. Select different colors for the bone regions listed at the coding circles below. Color the coding circles and the corresponding regions on the drawing. Complete Figure 6.1A by labeling compact bone and spongy bone.

   Figure 6.1B is a mid-level, cross-sectional view of the diaphysis of the femur. Label the membrane that lines the cavity and the membrane that covers the outside surface. Indicate by an asterisk (*) the membrane that contains both osteoblasts and osteoclasts.

   ○ Diaphysis   ○ Area where red marrow is found

   ○ Epiphyseal plate   ○ Area where yellow marrow is found

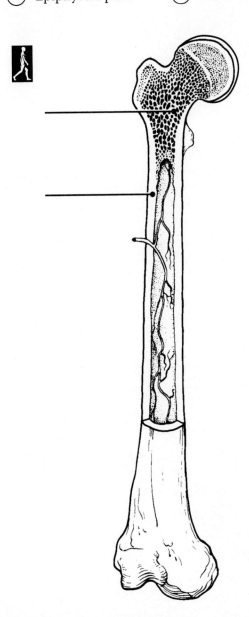

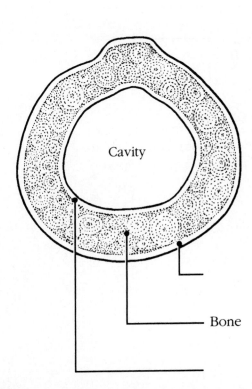

Cavity

Bone

A   **Figure 6.1**   B

2. Using the key choices, characterize the following statements relating to the structure of a long bone. Enter the appropriate answers in the answer blanks.

KEY CHOICES

A. Diaphysis   **C.** Epiphysis   **E.** Yellow marrow cavity

B. Epiphyseal plate   **D.** Red marrow

_____   **1.** Location of spongy bone in an adult's bone

_____   **2.** Location of compact bone in an adult's bone

_____   **3.** Site of hematopoiesis in an adult's bone

_____   **4.** Scientific name for bone shaft

_____   **5.** Site of fat storage

_____   **6.** Region of longitudinal growth in a child

_____   **7.** Composed of hyaline cartilage until the end of adolescence

3. Figure 6.2 is a sectional diagram showing the five-layered structure of a typical flat bone. Select different colors for the layers below. Add labels and leaders to identify Sharpey's fibers and trabeculae. Then answer the questions that follow, referring to Figure 6.2 and inserting your answers in the answer blanks.

◯ Spongy bone   ◯ Compact bone   ◯ Periosteum

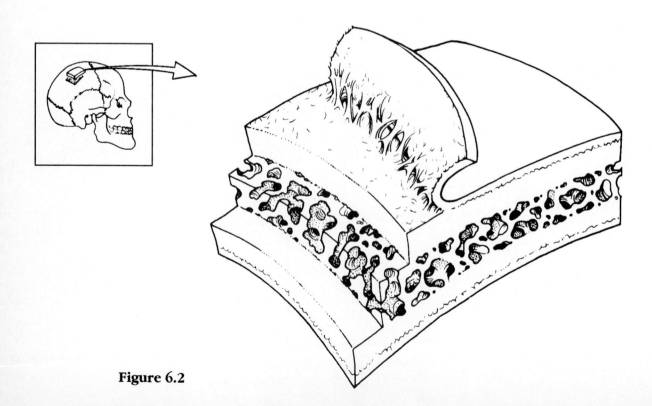

Figure 6.2

1. Which layer is called the diploë? _____

2. Name the membrane that lines internal bone cavities. _____

4. Five descriptions of bone structure are provided in Column A.

First, identify the structure by choosing the appropriate term from Column B and placing the corresponding answer in the answer blank.

Second, consider Figure 6.3A, a diagrammatic view of a cross section of bone, and Figure 6.3B, a higher-magnification view of compact bone tissue. Select different colors for the structures and bone areas in Column B and use them to color the coding circles and corresponding structures on the diagrams. Since concentric lamellae would be difficult to color without confusing other elements, identify one lamella by using a bracket and label.

Column A	Column B
_____ 1. Layers of calcified matrix	**A.** Central (Haversian) canal ○
_____ 2. "Residences" of osteocytes	**B.** Concentric lamellae ○
_____ 3. Longitudinal canal, carrying blood vessels and nerves	**C.** Lacunae ○
	D. Canaliculi ○
_____ 4. Nonliving, structural part of bone	**E.** Bone matrix ○
_____ 5. Tiny canals connecting lacunae	**F.** Osteocyte ○

A

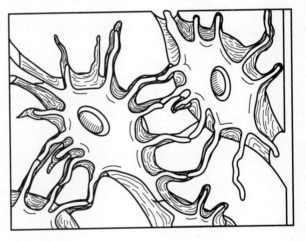

B

Figure 6.3

5. Classify each of the following terms as a projection (P), a depression (D), or an opening (O). Enter the appropriate letter in the answer blanks.

____ **1.** Condyle	____ **4.** Foramen	____ **7.** Ramus	____ **10.** Fossa
____ **2.** Crest	____ **5.** Head	____ **8.** Spine	____ **11.** Facet
____ **3.** Fissure	____ **6.** Meatus	____ **9.** Tuberosity	____ **12.** Sinus

6. Circle the term that does not belong in each of the following groupings.

1. Rigidity   Calcium salts   Hydroxyapatites   Collagen   Hardness

2. Hematopoiesis   Red marrow   Yellow marrow   Spongy bone   Diploë

3. Lamellae   Cellular extensions   Canaliculi   Circulation   Osteoblasts

4. Osteon   Marrow cavity   Volkmann's canals   Haversian canal   Canaliculi

5. Epiphysis   Articular cartilage   Periosteum   Hyaline cartilage

6. Perichondrium   Periosteum   Appositional growth   Osteoblasts

7. Spongy   Cancellous   Woven   Lamellar   Trabecular

Bone Development (Osteogenesis)

1. The following events apply to the endochondral ossification process as it occurs in the primary ossification center. Put these events in their proper order by assigning each a number (1–6).

____ Cavity formation occurs within the hyaline cartilage.

____ Collar of bone is laid down around the hyaline cartilage model just beneath the periosteum.

____ Periosteal bud invades the marrow cavity.

____ Perichondrium becomes vascularized to a greater degree and becomes a periosteum.

____ Osteoblasts lay down bone around the cartilage spicules in the bone's interior.

____ Osteoclasts remove the cancellous bone from the shaft interior, leaving a marrow cavity that then houses fat.

2. For each statement that is true, insert T in the answer blank. For false statements, correct the underlined words by inserting the correct words in the answer blanks.

_____ **1.** When a bone forms from a fibrous membrane, the process is called <u>endochondral</u> ossification.

_____ **2.** Membrane bones develop from <u>hyaline cartilage</u> structures.

_____ **3.** The organic bone matrix is called the <u>osteoid</u>.

_____ **4.** The enzyme alkaline phosphatase encourages the deposit of <u>collagen fibers</u> within the matrix of developing bone.

_____ **5.** When trapped in lacunae, osteoblasts change into <u>osteocytes</u>.

_____ **6.** Large numbers of <u>osteocytes</u> are found in the inner periosteum layer.

_____ **7.** During endochondral ossification, the <u>periosteal bud</u> invades the deteriorating hyaline cartilage shaft.

_____ **8.** Primary ossification centers appear in the <u>epiphyses</u>.

_____ **9.** Epiphyseal plates are made of <u>spongy</u> bone.

_____ **10.** In appositional growth, bone reabsorption occurs on the <u>periosteal</u> surface.

_____ **11.** "Maturation" of newly formed (noncalcified) bone matrix takes about <u>10 days</u>.

3. Figure 6.4 is a diagram representing the histological changes in the epiphyseal plate of a growing long bone.

First, select different colors for the types of cells named below. Color the coding circles and the corresponding cells in the diagram.

◯ Dead cells   ◯ Older, enlarging, vesiculating cells   ◯ Region of ossification

◯ Resting cells   ◯ Dividing cartilage cells

Second, complete the statements that follow, referring to Figure 6.4 and the labeled regions on the diagram. Insert the correct words in the spaces provided.

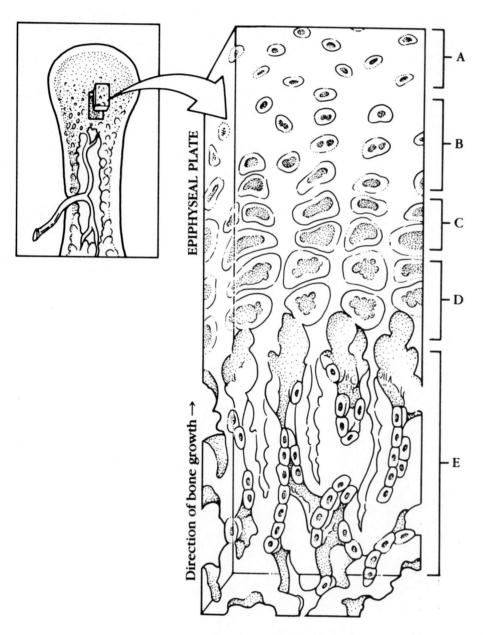

Figure 6.4

_____ 1. The type of cell in region A is __(1)__.

_____ 2. The type of cell in region B is __(2)__.

_____ 3. Calcification of the cartilage matrix begins in region __(3)__.

_____ 4. The cartilaginous matrix begins to deteriorate in region __(4)__.

_____ 5. The type of cell found in region E is __(5)__.

_____ 6. The matrix being deposited in region E is __(6)__ matrix.

Bone Homeostasis: Remodeling and Repair

1. Using the key choices, insert the correct answers in the answer blanks below.

KEY CHOICES

A. Atrophy   C. Gravity   E. Osteoclasts   G. Parathyroid hormone

B. Calcitonin D. Osteoblasts   F. Osteocytes   H. Stress and/or tension

_____ 1. When blood calcium levels begin to drop below homeostatic levels, __(1)__ is released, causing calcium to be released from bones.

_____ 2. Mature bone cells, called __(2)__, maintain bone in a viable state.

_____ 3. Disuse such as that caused by paralysis or severe lack of exercise results in muscle and bone __(3)__.

_____ 4. Large tubercles and/or increased deposit of bony matrix occur at sites of __(4)__.

_____ 5. Immature, or matrix-depositing, bone cells are referred to as __(5)__.

_____ 6. __(6)__ causes blood calcium to be deposited in bones as calcium salts.

_____ 7. Bone cells that liquefy bone matrix and release calcium to the blood are called __(7)__.

_____ 8. Astronauts must perform isometric exercises when in outer space because bones atrophy under conditions of weightlessness or lack of __(8)__.

2. Circle the term that does not belong in each of the following groupings.

1. Bone deposit   Injury sites   Growth zones   Repair sites   Bone resorption

2. Osteoid   Organic matrix   Calcium salts   Osteoblasts   10 μ wide

3. Hypercalcemia   Hypocalcemia   Calcium   Salt deposit in soft tissue

   Ca^{2+} over 11 mg/100 ml blood

4. Mechanical forces   Gravity   Muscle pull   Blood calcium levels

   Wolff's law

5. Growth in diameter   Growth in length   Appositional growth

   Increase in thickness

3. According to Wolff's law, bones form according to the stresses placed upon them. Figure 6.5 is a simple diagram of the proximal end of a femur (thighbone). There are two sets of double arrows and two single arrows. Your job is to decide which of the single or paired arrows represents each of the following conditions and indicate those conditions on the diagram. Color the arrows to agree with your coding circles.

◯ Site of maximal compression   ◯ Load (body weight) exertion site

◯ Site of maximal tension   ◯ Point of no stress

Explain why long bones can "hollow out" without jeopardy to their integrity (soundness of structure).

_____

_____

_____

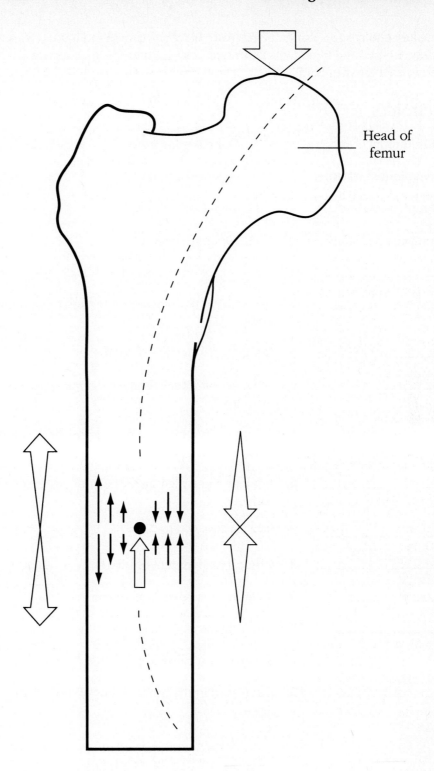

Head of femur

Figure 6.5

4. Use the key choices to identify the fracture (fx) types shown in Figure 6.6 and the fracture types and treatments described below. Enter the appropriate answer in each answer blank.

KEY CHOICES

A. Closed reduction	**E.** Depressed fracture	**I.** Simple fracture
B. Comminuted fracture	**F.** Greenstick fracture	**J.** Spiral fracture
C. Compression fracture	**G.** Impacted fracture	
D. Compound fracture	**H.** Open reduction	

_____   **1.** Bone is broken cleanly; the ends do not penetrate the skin

_____   **2.** Nonsurgical realignment of broken bone ends and splinting of bone

_____   **3.** Bone breaks from twisting forces

_____   **4.** A break common in children; bone splinters, but break is incomplete

_____   **5.** A fracture in which the bone is crushed; common in the vertebral column

_____   **6.** A fracture in which the bone ends penetrate through the skin surface

_____   **7.** Broken ends are pushed into each other

_____   **8.** Surgical realignment of broken bone ends

_____   **9.** A common type of skull fracture

_____   **10.** Also called a closed fracture

_____   **11.** A common sports fracture

_____   **12.** Often seen in the brittle bones of the elderly

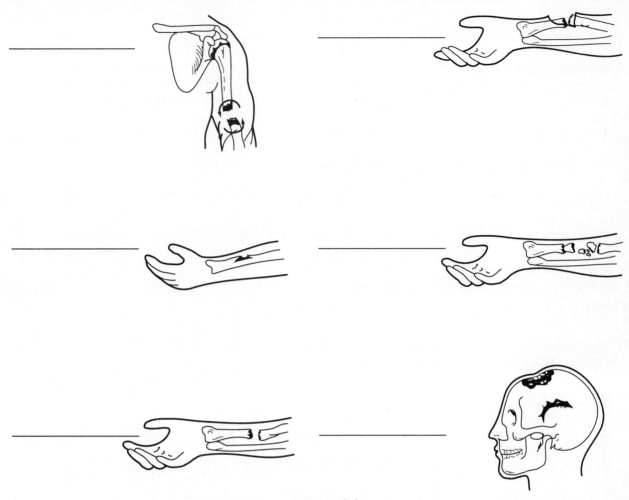

Figure 6.6

5. For each of the following statements about bone breakage and the repair process that is true, insert T in the answer blank. For false statements, correct the underlined words by inserting the correct words in the answer blanks.

_____ 1. A <u>hematoma</u> usually forms at a fracture site.

_____ 2. Deprived of nutrition, <u>osteocytes</u> at the fracture site die.

_____ 3. Nonbony debris at the fracture site is removed by <u>osteoclasts</u>.

_____ 4. Osteocytes produce collagen fibers that span the break.

_____ 5. Osteoblasts from the <u>medullary cavity</u> migrate to the fracture site.

_____ 6. The <u>fibrocartilaginous callus</u> is the first repair mass to splint the broken bone.

_____ 7. The bony callus is composed of <u>compact</u> bone.

Homeostatic Imbalances of Bone

1. Circle the term that does not belong in each of the following groupings.

    1. Bacterial infection   Osteoporosis   Inflammation   Osteomyelitis

    2. Osteomalacia   Elderly   Vertebral compression fractures   Osteoporosis

    3. Increased reabsorption   Decreased density   Paget's disease   Elderly

    4. Soft bones   Rickets   Osteomalacia   Porous bones

    5. Bone-filled marrow cavity   Pagetic bone   Osteomalacia   Bone thickenings

    6. Rickets   Calcified epiphyseal discs   Vitamin D deficiency   Children

Developmental Aspects of Bones: Timing of Events

1. Using the key choices, identify the body systems that relate to bone tissue viability. Enter the appropriate answers in the answer blanks.

    KEY CHOICES

    A. Endocrine   **C.** Muscular   **E.** Reproductive

    B. Integumentary   **D.** Nervous   **F.** Urinary

    _____   **1.** Conveys the sense of pain in bone and joints

    _____   **2.** Activates vitamin D for proper calcium usage

    _____   **3.** Regulates the uptake and release of calcium by bones

    _____   **4.** Increases bone strength and viability by pulling action

    _____   **5.** Influences skeleton proportions and adolescent growth of long bones

    _____   **6.** Provides vitamin D for proper calcium absorption

2. Select the one best answer to complete each of the following statements. Insert the appropriate letters and terms (if desired) in the answer blanks.

    _____   **1.** The embryonic skeleton arises from:

            A. ectoderm   **B.** mesoderm   **C.** endoderm

    _____   **2.** Most hyaline cartilage "long bones" show primary ossification centers by:

            A. 2 weeks   **B.** 8 weeks   **C.** 12 weeks   **D.** 6 months

_____ **3.** At birth, bones generally lack:

      A. ossification centers   **C.** medullary cavities
      B. trabeculae   **D.** bone markings

_____ **4.** Long bone growth during childhood and adolescence is provided by persistence of the:

      A. epiphyseal plates   **C.** articular cartilage
      B. tuberosities   **D.** Sharpey's fibers

_____ **5.** The longitudinal growth of long bones ceases when:

      A. chondroblasts are less active mitotically
      B. the epiphyseal plates become thinner
      C. the epiphyses and the diaphyses fuse
      D. all of these

_____ **6.** Appositional growth, which continues after the cessation of longitudinal long bone growth, means that:

      A. osteoblasts secrete bony matrix on the internal bone surface
      B. osteoclasts destroy bone in the periosteal surface
      C. long bones increase in diameter
      D. articulating cartilages become thicker

3. Arrange each of the following groups in consecutive order of development:

_____ **1.** Red blood cell production occurs in:

      A. red marrow of all bones
      B. liver
      C. red marrow of torso bones and femur and humerus heads

_____ **2.** Relationship between osteoblast and osteoclast activity:

      A. about equal
      B. osteoclast more than osteoblast
      C. osteoblast more than osteoclast

_____ **3.** Elements of matrix formation:

      A. osteocytes
      B. osteoblasts
      C. calcification front
      D. osteoid seam
      E. mesenchyme

_____   **4.** Events of ossification:

        A. hematopoietic stem cells invade medullary cavity
        B. endochondral ossification begins
        C. epiphyses begin to ossify
        D. intramembranous ossification begins

_____   **5.** Growth in long bones:

        A. long bones reach their peak density
        B. sex hormones induce a growth spurt
        C. drop in estrogen causes loss of bone mass
        D. most bones stop growing in length

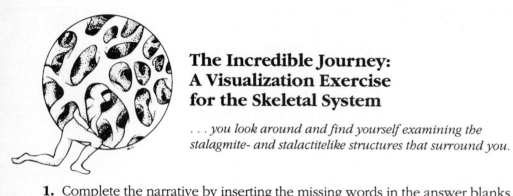

The Incredible Journey:
A Visualization Exercise
for the Skeletal System

. . . you look around and find yourself examining the stalagmite- and stalactitelike structures that surround you.

1. Complete the narrative by inserting the missing words in the answer blanks.

_____   **1.**

_____   **2.**

_____   **3.**

_____   **4.**

_____   **5.**

_____   **6.**

_____   **7.**

_____   **8.**

_____   **9.**

_____   **10.**

_____   **11.**

_____   **12.**

For this journey you are miniaturized and injected into the interior of the largest bone of your host's body, the __(1)__. Once inside this bone, you look around and find yourself examining the stalagmite- and stalactitelike structures that surround you. Although you feel as if you are in an underground cavern, you know that it has to be bone. Since the texture is so full of holes, it obviously is __(2)__ bone.

Although the arrangement of these bony spars seems to be haphazard, as if someone randomly had dropped straws, they are precisely arranged to resist points of __(3)__. All about you is frantic, hurried activity. Cells are dividing rapidly, nuclei are being ejected, and disclike cells are appearing. You decide that these disclike cells are __(4)__ and that this is the __(5)__ cavity. As you explore further, strolling along the edge of the cavity, you spot many tunnels leading into the solid bony area on which you are walking. Walking into one of these drainpipelike openings, you notice that it contains a glistening white ropelike structure, a __(6)__, and blood vessels running the length of the tube. You eventually come to a point in the channel where the horizontal passageway joins with a vertical passage that runs with the longitudinal axis of the bone. This is obviously a __(7)__

canal. Since you would like to see how nutrients are brought into __(8)__ bone, you decide to follow this channel. Reasoning that there is no way you can possibly scale the slick walls of the channel, you leap up and grab onto a white cord hanging down its length. Since it is easier to slide down than to try to climb up the cord, you begin to lower yourself, hand over hand. During your descent, you notice small openings in the wall, which are barely large enough for you to wriggle through. You conclude that these are the __(9)__ that connect all the __(10)__ to the nutrient supply in the central canal. You decide to investigate one of these tiny openings and begin to swing on your cord, trying to get a foothold on one of the openings. After managing to anchor yourself, and squeezing into an opening, you use a flashlight to illuminate the passageway in front of you. You are startled by a giant cell with many dark nuclei that appears to be plastered around the entire lumen directly ahead of you. As you watch this cell, the bony material beneath it, the __(11)__, begins to liquefy. The cell apparently is a bone-digesting cell, or __(12)__, and since you are unsure whether or not its enzymes can also liquefy you, you slither backward hurriedly and begin your trek back to your retrieval site.

CHALLENGING YOURSELF

At the Clinic

1. Mrs. Bruso, a woman in her 80s, is brought to the clinic with a fractured hip. X rays reveal compression fractures in her lower vertebral column and extremely low bone density in her vertebrae, hip bones, and femurs. What are the condition, cause, and treatment?

2. A child with a "red-hot" infected finger is brought into the clinic a week after the infection has set in. Why does the doctor order an X ray?

3. An X ray of the arm of an accident victim reveals a faint line curving around and down the shaft. What kind of fracture might this indicate?

4. Johnny, a child with unusually short stature, comes in for diagnostic tests. The tests reveal that he has abnormal collagen production. What name is given to this condition?

5. A woman taking triple the recommended calcium supplement is suffering from lower back pains. X rays reveal that she has kidney stones. What is the name of the condition that causes this problem?

6. Larry, a 6-year-old boy, is already 5 feet tall. For what hormone will the pediatrician order a blood test?

7. What disorder causes the formation of abnormally thick, lumpy bony areas?

8. A young man is donating bone marrow to his sister, whose bone marrow is suppressed. From what bone will the marrow most likely be taken?

Stop and Think

1. Which appears first, the blood vessels within an osteon or the lamellae of bone matrix around the central canal?

2. Is there any form of interstitial growth in bone tissue? (Explain)

3. Contrast the arrangement and articulation of long bones and short bones.

4. Which type of cell, osteoclast or osteoblast, is more active in the medullary cavity of a growing long bone? On the internal surface of cranial bones?

5. Compare the vascularity of bone to that of the fibrous and cartilaginous tissues that precede it.

6. While walking home from a meeting of his adult singles support group, Ike broke a bone and damaged a knee cartilage in a fall. Assuming no special tissue grafts are made, which will probably heal faster, the bone or the cartilage? Why?

7. Eric, a radiologic tech student, was handed an X ray of the right femur of a 10-year-old boy. He noted that the area of the epiphyseal plate was beginning to show damage from osteomyelitis (the boy's diagnosis). Eric expressed his concern about the future growth of the boy's femur to the head radiologist. Need he have worried? Why or why not?

8. In several early cultures, bone deformation was a sign of social rank. For example, in the Mayan culture, head binding was done to give the skull a flattened forehead and a somewhat pointed top. Explain this in terms of Wolff's law.

9. Signs of successful (or at least nonfatal) brain surgery can be found on skulls many thousands of years old. How can an anthropologist tell that the patient didn't die from the surgery?

10. Explain (a) why cartilages are springy and (b) why cartilage can grow so quickly in the developing skeleton.

COVERING ALL YOUR BASES

Multiple Choice

1. Important bone functions include:
   A. support of the pelvic organs
   B. protection of the brain
   C. movement of the limbs
   D. protection of the skin and limb musculature
   E. storage of water

2. Which of the following are correctly matched?
   A. Short bone—wrist
   B. Long bone—leg
   C. Irregular bone—sternum
   D. Flat bone—cranium

3. Terms that can be associated with any type of bone include:
   A. periosteum
   B. diaphysis
   C. diploë
   D. cancellous bone
   E. medullary cavity

4. Which would be common locations of osteoblasts?
   A. Osteogenic layer of periosteum
   B. Lining of red marrow spaces
   C. Covering articular cartilage
   D. Lining central canals
   E. Aligned with Sharpey's fibers

5. Which of the listed bone markings are sites of muscle or ligament attachment?
   A. Trochanter
   B. Meatus
   C. Facet
   D. Spine
   E. Condyle

6. Which of the following are openings or depressions?
   A. Fissure
   B. Tuberosity
   C. Meatus
   D. Fossa
   E. Tubercle

7. A passageway connecting neighboring osteocytes in an osteon is a:
   A. central canal
   B. lamella
   C. lacuna
   D. canaliculus
   E. perforating canal

8. Between complete osteons are remnants of older, remodeled osteons known as:
   A. circumferential lamellae
   B. concentric lamellae
   C. interstitial lamellae
   D. lamellar bone
   E. woven bone

9. Which of these could be found in cancellous bone?
   A. Osteoid
   B. Trabeculae
   C. Canaliculi
   D. Central canals
   E. Osteoclasts

10. Elements prominent in osteoblasts include:
    A. rough ER
    B. secretory vesicles
    C. lysosomes
    D. smooth ER
    E. heterochromatin

11. Which of the following are prominent in osteoclasts?
    A. Golgi apparatus
    B. Lysosomes
    C. Microfilaments
    D. Exocytosis

12. A fetus has ossification centers in the bones of the skull, but not yet in the long bones. What is its approximate gestational age?
    A. 5 weeks   C. 13 weeks
    B. 9 weeks   D. 16 weeks

13. Which precede(s) intramembranous ossification?
    A. Chondroblast activity
    B. Mesenchymal cells
    C. Woven bone
    D. Collagen formation
    E. Osteoid formation

14. Which of the following is (are) part of the process of endochondral ossification and growth?
    A. Vascularization of the fibrous membrane surrounding the cartilage template
    B. Formation of diploë
    C. Destruction of cartilage matrix
    D. Appositional growth
    E. Mitosis of chondroblasts

15. What is the earliest event (of those listed) in endochondral ossification?
    A. Ossification of proximal epiphysis
    B. Appearance of the epiphyseal plate
    C. Invasion of the shaft by the periosteal bud
    D. Cavitation of the cartilage shaft
    E. Formation of secondary ossification centers

16. Which zone of the epiphyseal plate is most influenced by sex hormones?
    A. Zone of resting cartilage
    B. Zone of hypertrophic cartilage
    C. Zone of proliferating cartilage
    D. Zone of calcification

17. The region active in appositional growth is:
    A. osteogenic layer of periosteum
    B. within central canals
    C. endosteum of red marrow spaces
    D. internal callus
    E. epiphyseal plate

18. Deficiency of which of the following hormones will cause dwarfism?
    A. Growth hormone
    B. Sex hormones
    C. Thyroid hormones
    D. Calcitonin
    E. Parathyroid hormone

19. A remodeling unit consists of:
    A. osteoblasts   D. osteoclasts
    B. osteoid   E. chondroblasts
    C. osteocytes

20. The calcification front marks the location of:
    A. newly formed osteoid
    B. newly deposited hydroxyapatite
    C. actively mitotic osteoblasts
    D. active osteoclasts
    E. the activity of alkaline phosphatase

21. A deficiency of calcium in the diet would lead to:
    A. an increase of parathyroid hormone in the blood
    B. an increase in calcitonin secretion
    C. an increase in somatomedin levels in the blood
    D. increased secretion of growth hormone

22. Ionic calcium plays a role in:
    A. the transmission of nerve impulses
    B. blood clotting
    C. muscle contraction
    D. cytokinesis
    E. the activity of sudoriferous glands

23. Which of the following is not associated with Wolff's law?
    A. Compression
    B. Gravity
    C. Growth hormone
    D. Orientation of trabeculae
    E. Bone atrophy following paralysis

24. The initial event following a bone fracture is:
   A. formation of granulation tissue
   B. ossification of internal callus
   C. hemorrhage and hematoma formation
   D. remodeling   **E.** endochondral ossification

25. Women suffering from osteoporosis are frequent victims of _____ fractures of the vertebrae.
   A. compound   **D.** compression
   B. spiral   **E.** depression
   C. comminuted

26. Which of the listed bone disorders is (are) caused by hormonal imbalances?
   A. Osteomalacia   **D.** Achondroplasia
   B. Osteoporosis   **E.** Paget's disease
   C. Gigantism

27. At birth, ossification has progressed to the point where:
   A. only intramembranous ossification has begun
   B. endochondral ossification is complete
   C. some secondary ossification centers have appeared
   D. only major long bones have primary centers of ossification
   E. appositional growth has yet to begin

28. The growth spurt of puberty is triggered by:
   A. high levels of sex hormones
   B. the initial, low levels of sex hormones
   C. growth hormone
   D. parathyroid hormone
   E. calcitonin

Word Dissection

For each of the following word roots, fill in the literal meaning and give an example, using a word found in this chapter.

Word root	Translation	Example
1. call		
2. cancel		
3. clast		
4. fract		
5. lamell		
6. malac		
7. myel		
8. physis		
9. poie		
10. soma		
11. trab		

7 The Skeleton

Student Objectives

When you have completed the exercises in this chapter, you will have accomplished the following objectives:

1. Name the major parts of the axial and appendicular skeletons and describe their most important functions.

PART I: THE AXIAL SKELETON

The Skull

2. Name, describe, and identify the bones of the skull. Identify their important markings.

3. Distinguish between the major functions of the cranium and the facial skeleton.

4. Define the bony boundaries of the orbits, nasal cavity, and paranasal sinuses.

The Vertebral Column

5. Describe the general structure of the vertebral column and list its components.

6. Indicate a common function of the spinal curvatures and the intervertebral discs.

7. Discuss the structure of a typical vertebra and then describe the special characteristics of cervical, thoracic, and lumbar vertebrae.

The Bony Thorax

8. Name and describe the bones of the thorax.

9. Differentiate true from false ribs.

PART II: THE APPENDICULAR SKELETON

The Pectoral (Shoulder) Girdle and the Upper Limb

10. Identify the bones forming the pectoral girdle and relate their structure and arrangement to the function of this girdle.

11. Identify important bone markings on the pectoral girdle.

12. Identify or name the bones of the upper limb and their important markings.

The Pelvic (Hip) Girdle and the Lower Limb

13. Name the bones contributing to the os coxa and relate the strength of the pelvic girdle to its function.

14. Describe differences in the anatomy of the male and female pelves and relate these to functional differences.

15. Identify the bones of the lower limb and their important markings.

16. Name the three supporting arches of the foot and explain their importance.

Developmental Aspects of the Skeleton

17. Describe how skeletal proportions change during childhood and adolescence.

18. Compare and contrast the skeleton of an aged person with that of a young adult. Discuss how age-related skeletal changes may affect health.

Ready, set, go!

The human skeleton is composed of 206 bones, which form a strong, flexible framework that supports the body and protects the vital organs. The bones articulate at joints, forming a complex system of levers stabilized by ligaments and activated by muscles. This system provides an extraordinary range of body movements.

The skeleton is, by evolution and appearance, constructed of two coordinated divisions. The axial skeleton is made up of the bones located along the body's longitudinal axis and center of gravity. The appendicular skeleton consists of the bones of the shoulders, hips, and limbs, appended to the axial division.

Topics for student study and review in Chapter 7 include the identification, location, and function of anatomically important bones, as well as the changes that occur in the bones of the skeleton throughout life.

BUILDING THE FRAMEWORK

1. Figure 7.1 on the opposite page illustrates an articulated skeleton. Identify all bones or groups of bones by writing the correct labels at the leader lines. Then, using two different colors, color the bones of the axial and appendicular skeletons.

   ◯ Axial skeleton   ◯ Appendicular skeleton

2. Take a moment to go over some basic skull "geography" by matching the key choices to the skull features described below.

   _____ **1.** The superolateral parts of the skull (2 choices)

   _____ **2.** The floor of the skull

   _____ **3.** Air-filled cavities that lighten the skull

   _____ **4.** Concavities in the skull floor that support parts of the brain

   _____ **5.** House the eyes

   _____ **6.** Present most special sense organs in the anterior position

   _____ **7.** Encloses the brain

KEY CHOICES

A. Base

B. Calvaria

C. Cranial cavity

D. Facial bones

E. Fossae

F. Orbits

G. Paranasal sinuses

H. Vault

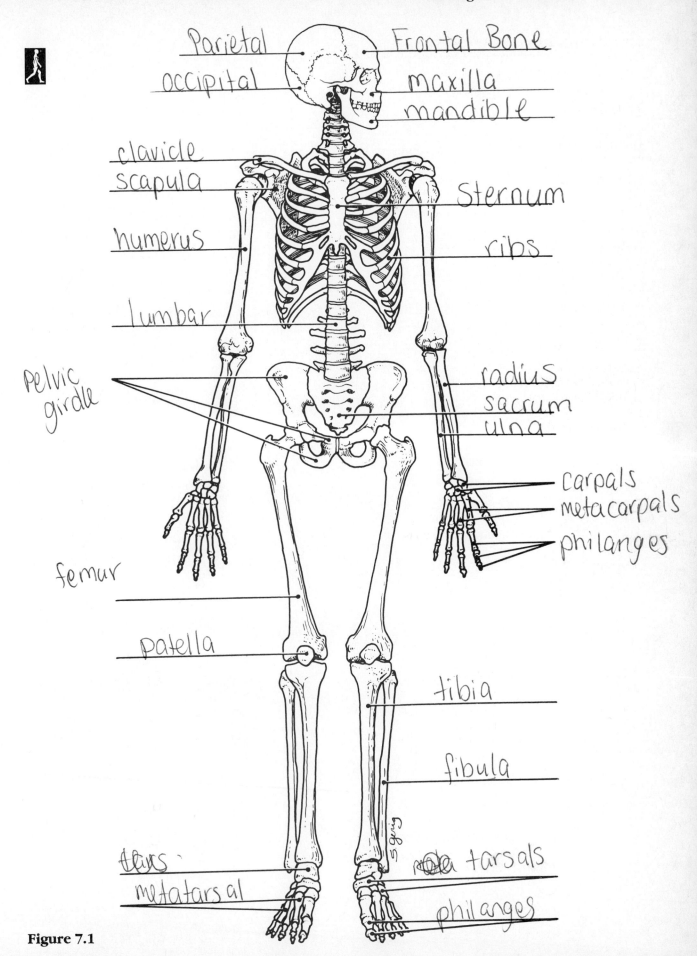

Parietal
Frontal Bone
occipital
maxilla
mandible
clavicle
scapula
Sternum
humerus
ribs
lumbar
Pelvic girdle
radius
sacrum
ulna
carpals
metacarpals
philanges
femur
patella
tibia
fibula
tars
metatarsal
meta tarsals
philanges

Figure 7.1

PART I: THE AXIAL SKELETON

The Skull

1. Circle all of the bone names that represent *cranial* bones.

A. Ethmoid   **F.** Mandible   **K.** Parietal

B. Frontal   **G.** Maxillary   **L.** Sphenoid

C. Hyoid   **H.** Nasal   **M.** Temporal

D. Inferior conchae   **I.** Occipital   **N.** Vomer

E. Lacrimal   **J.** Palatine   **O.** Zygomatic

2. Figure 7.2 shows the lateral (A), inferior (B), and anterior (C) views of the skull. Select different colors for the bones below, and color the coding circles and corresponding bones. Complete views A, B, and C by labeling the bone markings indicated by leader lines.

○ Frontal

○ Parietal

○ Mandible

○ Maxilla

○ Sphenoid

○ Ethmoid

○ Temporal

○ Zygomatic

○ Palatine

○ Occipital

○ Nasal

○ Lacrimal

○ Vomer

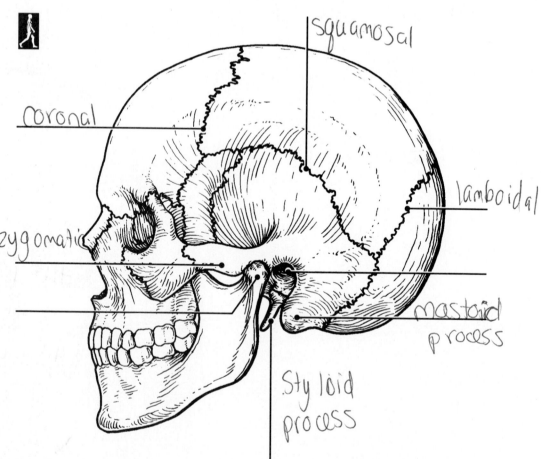

A

Figure 7.2

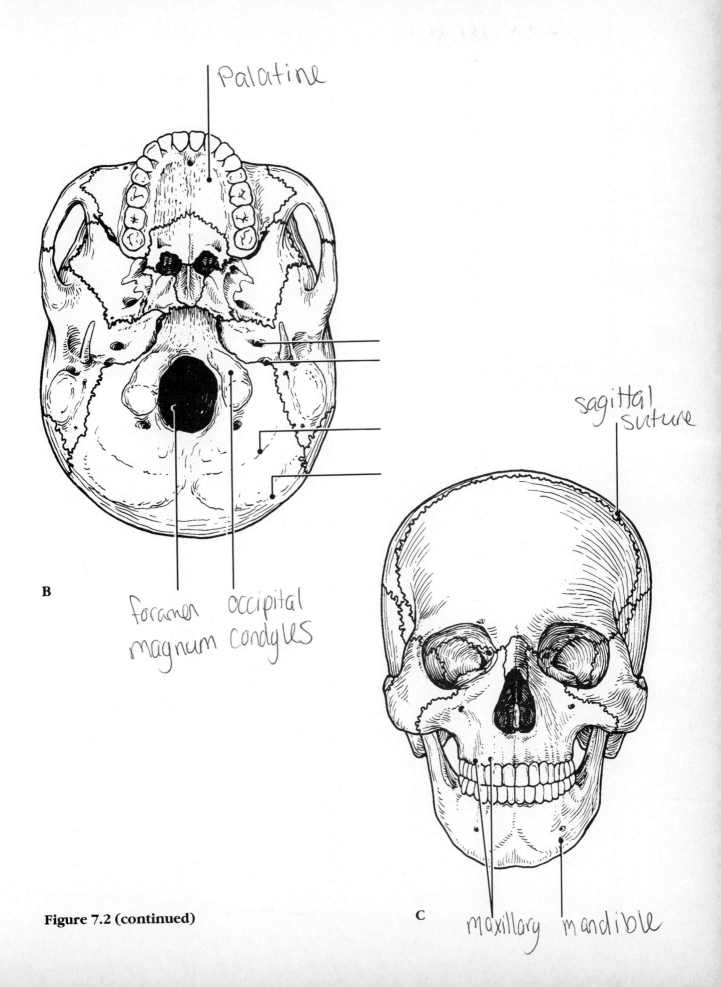

Palatine

B

foramen occipital
magnum condyles

sagittal suture

Figure 7.2 (continued)

C

maxillary mandible

3. Match the bone names listed in the key choices with the following bone descriptions. (Note that some descriptions apply to more than one bone.)

_____ **1.** Connected by the frontal suture

_____ **2.** Connected by the lambdoidal suture

_____ **3.** Connected by the squamosal suture

_____ **4.** Connected by the sagittal suture

_____ **5.** Cheekbone

_____ **6.** Superolateral part of the cranium

_____ **7.** Contains olfactory foramina

_____ **8.** Posterior part of the hard palate

_____ **9.** Posteriormost part of the cranium

_____ **10.** Has two turbinates as part of its structure; also contributes to the nasal septum

_____ **11.** Foramen magnum contained here

_____ **12.** Site of the sella turcica

_____ **13.** Houses hearing and equilibrium receptors

_____ **14.** Forms the bony eyebrow ridges and roofs of orbits

_____ **15.** Forms the chin

_____ **16.** The only bone connected to the skull by a freely movable joint

_____ **17.** Site of the mastoid process

_____ **18.** Contains the mental foramina

_____ **19.** Neither a cranial nor a facial bone

_____ **20.** _____ **21.** Four bones containing paranasal sinuses

_____ **22.** _____ **23.**

_____ **24.** Bears an upward protrusion called the crista galli

_____ **25.** Keystone bone of cranium

_____ **26.** Tiny bones with openings for the tear ducts

_____ **27.** Bony part of nasal septum

KEY CHOICES

A. Ethmoid

B. Frontal

C. Hyoid

D. Inferior conchae

E. Lacrimal

F. Mandible

G. Maxillary

H. Nasal

I. Occipital

J. Palatine

K. Parietal

L. Sphenoid

M. Temporal

N. Vomer

O. Zygomatic

4. Anterior and lateral views of the skull, showing the positions of the sinuses, are shown in Figure 7.3. First, select different colors for each of the sinuses and use them to color the figure. Then, briefly answer the following questions concerning the sinuses.

1. What are sinuses? _____

2. What purpose do they serve in the skull? _____

    _____

3. Why are they so susceptible to infection? _____

    _____

4. What is the function of the mucus-secreting mucosae? _____

    _____

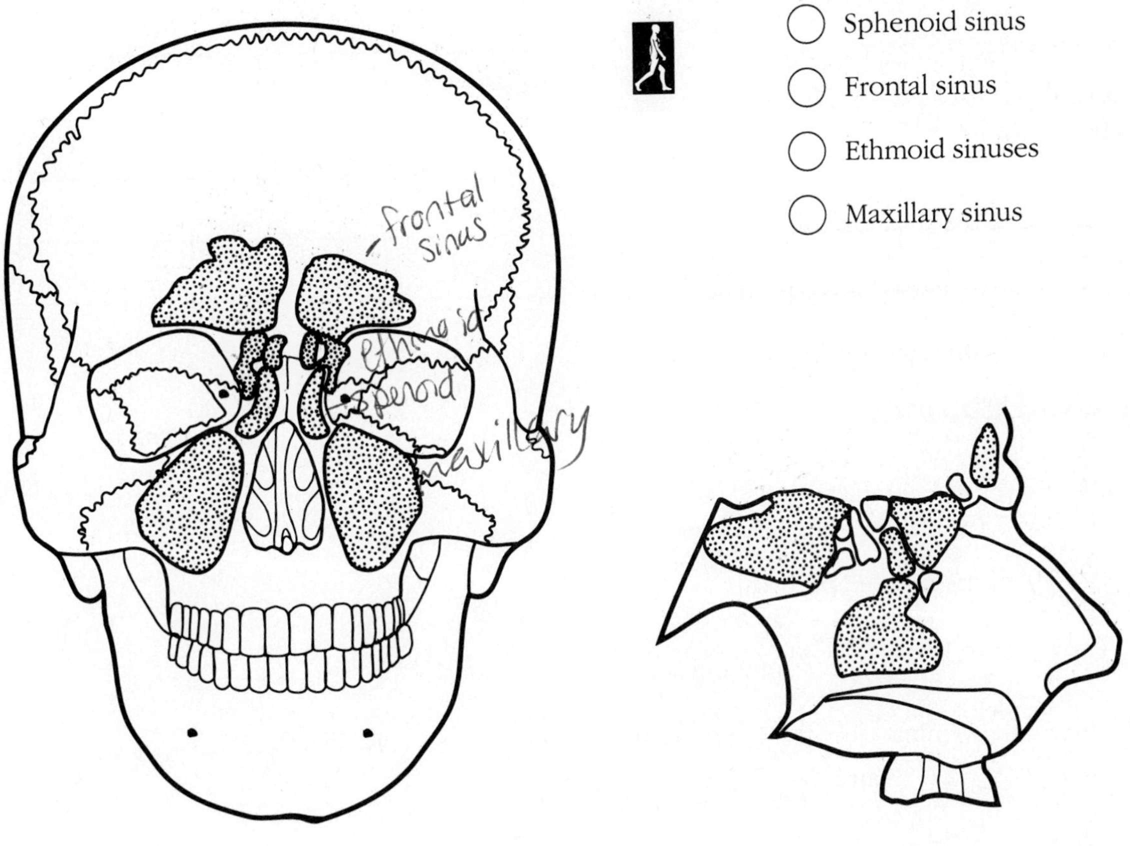

○ Sphenoid sinus

○ Frontal sinus

○ Ethmoid sinuses

○ Maxillary sinus

A

B

Figure 7.3

5. Complete the following table, which pertains to important bone markings on skull bones. Insert your answers in the spaces provided under each heading.

Bone marking(s)	Skull bone	Function of bone marking
	Mandible	Attachment for temporalis muscle
Carotid canal		
		Supports, partly encloses pituitary gland
Jugular foramen		
		Form anterior two-thirds of hard palate
		Attachment site for several neck muscles
Optic foramina		
		Pass olfactory nerve fibers
		Attachment for ligamentum nuchae and neck muscles
Mandibular condyles	Mandibles	
Mandibular fossa		
Stylomastoid foramen		

The Vertebral Column

1. Using the key choices, correctly identify the vertebral parts and areas described below. Enter the appropriate answers in the answer blanks.

KEY CHOICES

A. Body   **C.** Spinous process   **E.** Transverse process

B. Intervertebral foramina   **D.** Superior articular process   **F.** Vertebral arch

_____ **1.** Portion enclosing the nerve cord

_____ **2.** Weight-bearing part; also called the centrum

_____ **3.** Provide levers for the muscles to pull against

_____ **4.** Provide articulation points for the ribs

_____ **5.** Openings allowing spinal nerves to pass

2. Figure 7.4 is a lateral view of the vertebral column. Identify each numbered region of the column by listing in the answer blanks the region name first and then the specific vertebrae involved (for example, sacral region, S# to S#). Also identify the modified vertebrae indicated by numbers 6 and 7 in Figure 7.4. Select different colors for each vertebral region and use them to color the coding circles and the corresponding regions.

_____   1. ◯

_____   2. ◯

_____   3. ◯

_____   4. ◯

_____   5. ◯

_____   6. ◯

_____   7. ◯

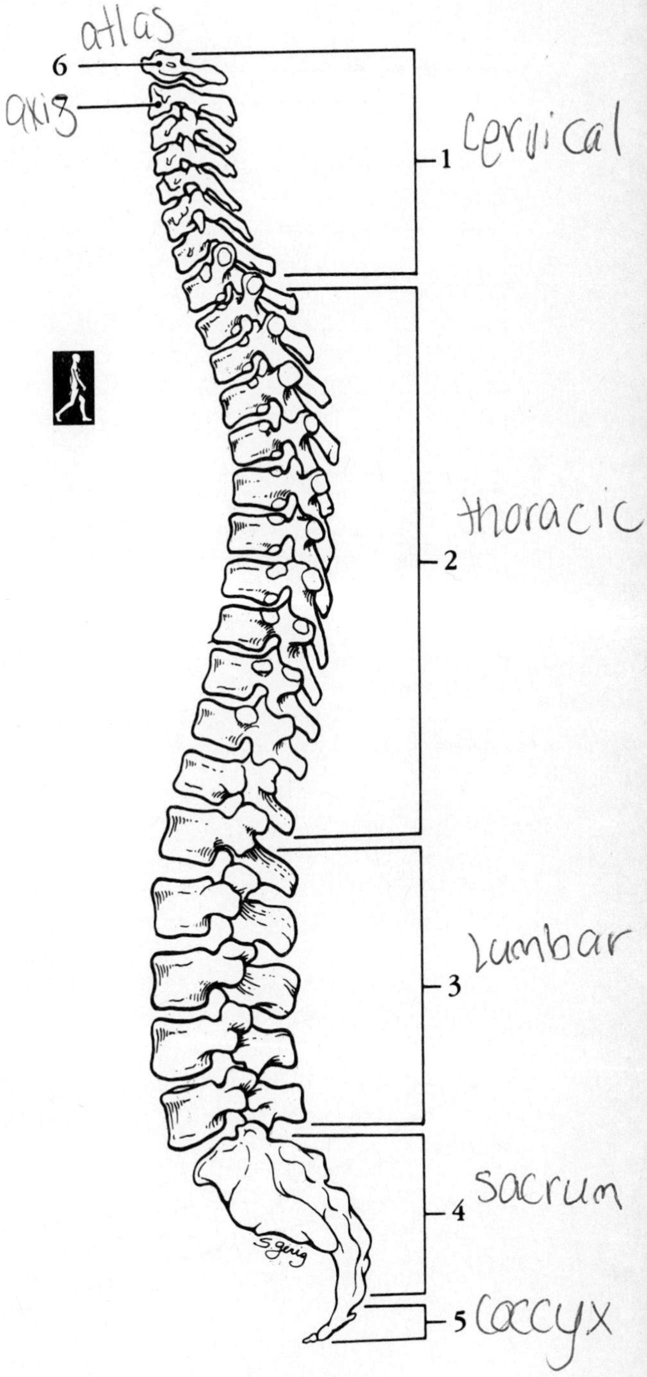

Figure 7.4

3. Match the vertebral column structures in Column B with the appropriate descriptions in Column A.

Column A

Column B

_____ **1.** Contain foramina in the transverse processes

A. Atlas (C₁)

B. Axis (C₂)

_____ **2.** Have articular facets for the ribs on their bodies and transverse processes

C. Cervical vertebrae (C₃–C₇)

_____ **3.** Bifed spinous process

_____ **4.** A circle of bone; articulates superiorly with the occipital condyles

D. Coccyx

E. Lumbar vertebrae

_____ **5.** Shield-shaped composite bone; has alae

F. Sacrum

_____ **6.** Thickest bodies; short blunt spinous processes

G. Thoracic vertebrae

_____ **7.** Bears a peg-shaped dens that acts as a pivot

_____ **8.** Fused rudimentary vertebrae; tailbone

_____ **9.** Are 5 in number; not fused

_____ **10.** Are 12 in number; not fused

_____ **11.** No intervertebral disc between these two bones

4. Figure 7.5 shows superior views of four types of vertebrae.

First, in the answer blank below each figure, indicate in which region of the spinal column it would be found. In addition, specifically identify Figure 7.5A.

Second, select four different colors and use them to color the coding circles and the corresponding structures, if present, for each type of vertebra.

Third, complete the statements relating to articulating surfaces by selecting the correct letters from Figure 7.5 and inserting your responses in the answer blanks.

◯ Body (centrum)   ◯ Transverse process   ◯ Transverse foramen

◯ Spinous process   ◯ Vertebral foramen

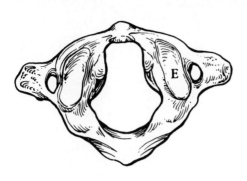

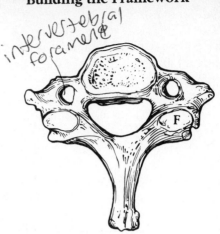

intervertebral foramen

A. _Atlas_

B. _Cervical vertebrae_

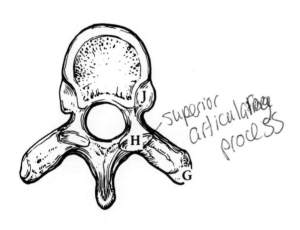

superior articulating process

C. _thoracic_

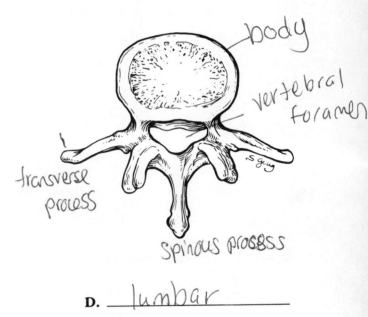

body

vertebral foramen

s guy

transverse process

spinous process

D. _lumbar_

Figure 7.5

_____ 1. Surface __(1)__ articulates with the head of a rib.

_____ 2. Surface __(2)__ articulates with the inferior process of the superior cervical vertebra.

_____ 3. Surface __(3)__ articulates with the tubercle of a rib.

_____ 4. Surface __(4)__ receives the occipital condyle of the skull.

_____ 5. Surface __(5)__ articulates with the inferior process of the superior thoracic vertebra.

5. Three views of spinal curvatures are shown in Figure 7.6. Identify each as being an example of one of the following conditions: lordosis, scoliosis, or kyphosis.

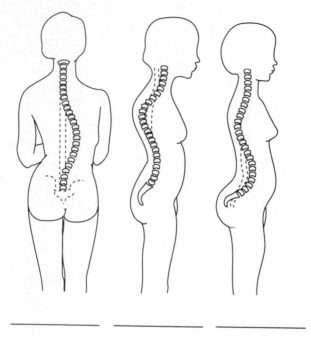

_____   _____   _____

Figure 7.6

The Bony Thorax

1. Complete the following statements referring to the thorax by inserting your responses in the answer blanks.

_____ **1.**

_____ **2.**

_____ **3.**

_____ **4.**

_____ **5.**

_____ **6.**

_____ **7.**

_____ **8.**

The organs protected by the thoracic cage include the **(1)** and the **(2)** . Ribs 1–7 are commonly called **(3)** ribs, whereas ribs 8–12 are called **(4)** ribs. Ribs 11 and 12 are also called **(5)** ribs. All ribs articulate posteriorly with the **(6)** , and most connect anteriorly to the **(7)** , either directly or indirectly.

The general shape of the thoracic cage is **(8)** .

2. Figure 7.7 is an anterior view of the thorax. Select different colors to identify the structures below and color the coding circles and corresponding structures. Then label the subdivisions of the sternum indicated by leader lines.

○ Vertebrosternal ribs   ○ Vertebrochondral ribs

○ Costal cartilages   ○ Sternum

○ Vertebral ribs

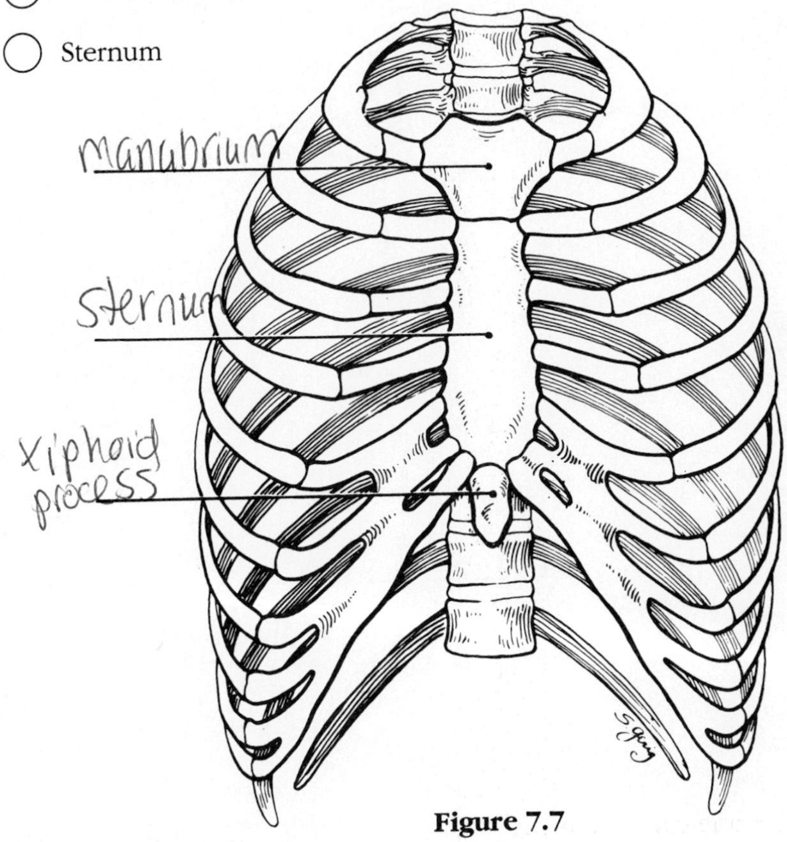

Manubrium

Sternum

Xiphoid process

Figure 7.7

PART II: THE APPENDICULAR SKELETON

The Pectoral (Shoulder) Girdle and the Upper Limb

1. Complete the following statements. Insert your answers in the answer blanks.

_____ 1.

_____ 2.

_____ 3.

_____ 4.

_____ 5.

_____ 6.

The lower limb bones are attached to the axial skeleton by the **(1)** . The upper limb bones are attached to the axial skeleton by the **(2)** . The two major functions of the axial skeleton are **(3)** and **(4)** . The two major functions of the appendicular skeleton are **(5)** and **(6)** .

2. Figure 7.8 is a posterior view of one bone of the pectoral girdle. Identify this bone by inserting its name on the line beside the figure. Select different colors for the structures listed below and use them to color the coding circles and the corresponding structures. Then, label the borders and angles indicated by leader lines.

◯ Spine

◯ Glenoid cavity

◯ Acromion

◯ Coracoid process

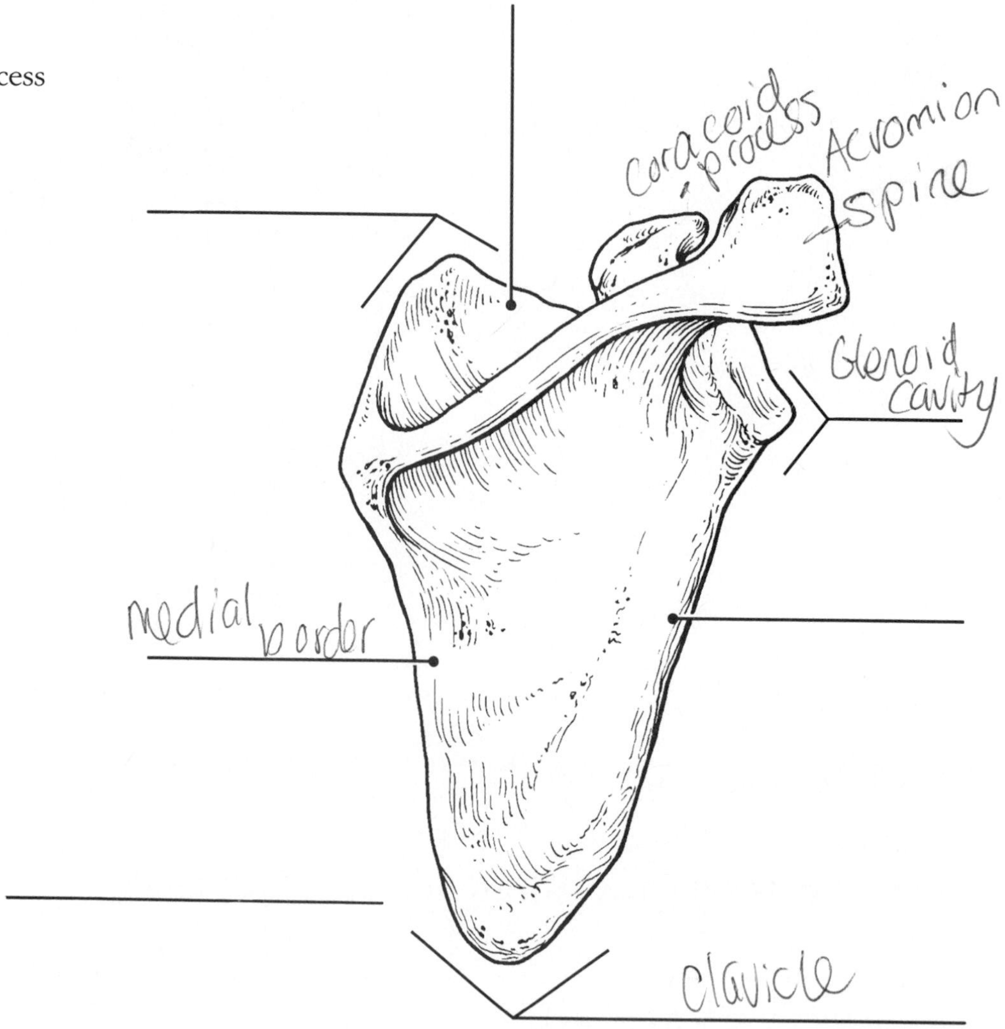

Coracoid process

Acromion

spine

Glenoid cavity

medial border

clavicle

Figure 7.8

3. Figure 7.9 illustrates the anterior views of the three bones (A, B, C) of the left arm and forearm in the anatomical position. Identify each bone (A, B, and C) by writing its name at the leader line. Then label all the bone markings listed below by inserting leader lines and the names of the bone markings on Figure 7.9. Finally, color the coding circles and the bone markings.

○ Trochlear notch   ○ Capitulum   ○ Coronoid process

○ Trochlea   ○ Deltoid tuberosity   ○ Olecranon process

○ Radial tuberosity   ○ Head (three)   ○ Greater tubercle

                      ○ Styloid process   ○ Lesser tubercle

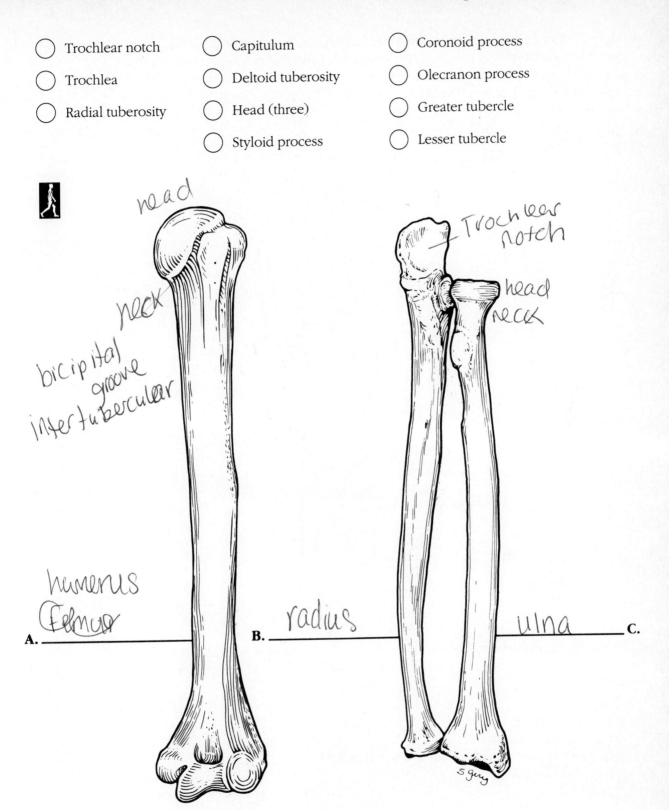

head

neck

bicipital
intertubercular groove

humerus
femur

A. _____   B. _____ radius _____   ulna _____ C.

Trochlear notch

head
neck

S guy

Figure 7.9

4. Figure 7.10 is a diagram of the hand. Select different colors for the following structures and use them to color the coding circles and diagram.

   ◯ Carpals   ◯ Phalanges   ◯ Metacarpals

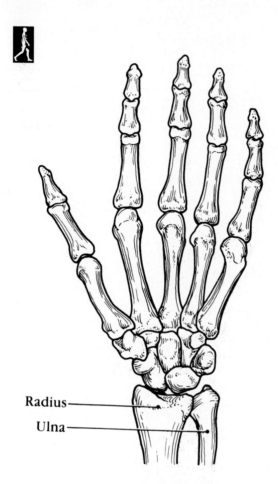

Radius——

Ulna——

Figure 7.10

5. Compare the pectoral and pelvic girdles by choosing appropriate descriptive terms from the key choices. Insert the appropriate key letters in the answer blanks.

   KEY CHOICES

   A. Flexibility   **D.** Shallow socket for limb attachment

   B. Massive   **E.** Deep, secure socket for limb attachment

   C. Lightweight   **F.** Weight-bearing

   1. Pectoral: ____ ____ ____   **2.** Pelvic: ____ ____ ____

6. Using the key choices, identify the bones or bone markings according to the descriptions that follow. Insert the appropriate answers in the answer blanks.

KEY CHOICES

A. Acromion	**F.** Coronoid fossa	**K.** Olecranon fossa	**P.** Scapula
B. Capitulum	**G.** Deltoid tuberosity	**L.** Olecranon process	**Q.** Sternum
C. Carpus	**H.** Glenoid cavity	**M.** Phalanges	**R.** Styloid process
D. Clavicle	**I.** Humerus	**N.** Radial tuberosity	**S.** Trochlea
E. Coracoid process	**J.** Metacarpus	**O.** Radius	**T.** Ulna

_____ **1.** Projection on lateral surface of humerus to which the deltoid muscle attaches

_____ **2.** Arm bone

_____ **3.** _____ **4.** Bones of the shoulder girdle

_____ **5.** _____ **6.** Bones of the forearm

_____ **7.** Point where the scapula and clavicle connect

_____ **8.** Pectoral girdle bone that is freely movable

_____ **9.** Pectoral girdle bone that articulates anteriorly with the sternum

_____ **10.** Socket in the scapula for the arm bone

_____ **11.** Process above the glenoid fossa that permits muscle attachment

_____ **12.** Commonly called the collarbone

_____ **13.** Distal medial process of the humerus; joins the ulna

_____ **14.** Medial bone of the forearm in anatomical position

_____ **15.** Rounded knob on the humerus that articulates with the radius

_____ **16.** Anterior depression, superior to the trochlea, that receives part of the ulna when the forearm is flexed

_____ **17.** Forearm bone most involved in formation of the elbow joint

_____ **18.** _____ **19.** Bones that articulate with the clavicle

_____ **20.** Bones of the wrist

_____ **21.** The fingers have three of these but the thumb has only two

_____ **22.** Form the palm of the hand

The Pelvic (Hip) Girdle and the Lower Limb

1. Figure 7.11 is a diagram of the articulated pelvis.

First, identify the bones and bone markings indicated by leader lines on the figure. Also, label the dashed lines showing the dimensions of the true pelvis and the diameter of the false pelvis.

Second, select different colors for the structures listed at the right and use them to color the coding circles and the corresponding structures in the figure.

Third, complete the illustration by labeling the following bone markings: obturator foramen, anterior superior iliac spine, iliac crest, ischial spine, pubic ramus, and pelvic brim.

Fourth, list three ways the female pelvis differs from the male pelvis in the answer blanks below the illustration.

◯ Os coxae

◯ Sacrum

◯ Pubic symphysis

◯ Acetabulum

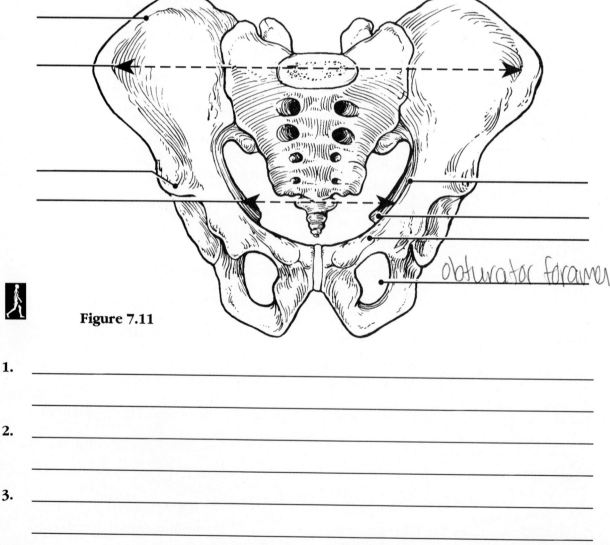

obturator foramen

Figure 7.11

1. _____

_____

2. _____

_____

3. _____

_____

2. The bones of the thigh and leg are shown in Figure 7.12. Identify each and put your answers in the blanks below the diagrams. Select different colors for those structures below with a color-coding circle and use them to color the corresponding structures on the diagrams. Complete the illustration by inserting the following terms at the ends of the appropriate leader lines in the figure.

◯ Femur   ◯ Tibia   ◯ Fibula

  Head of femur   Anterior crest   Head of fibula

  Intercondylar eminence   Lesser trochanter   Medial malleolus

  Tibial tuberosity   Greater trochanter   Lateral malleolus

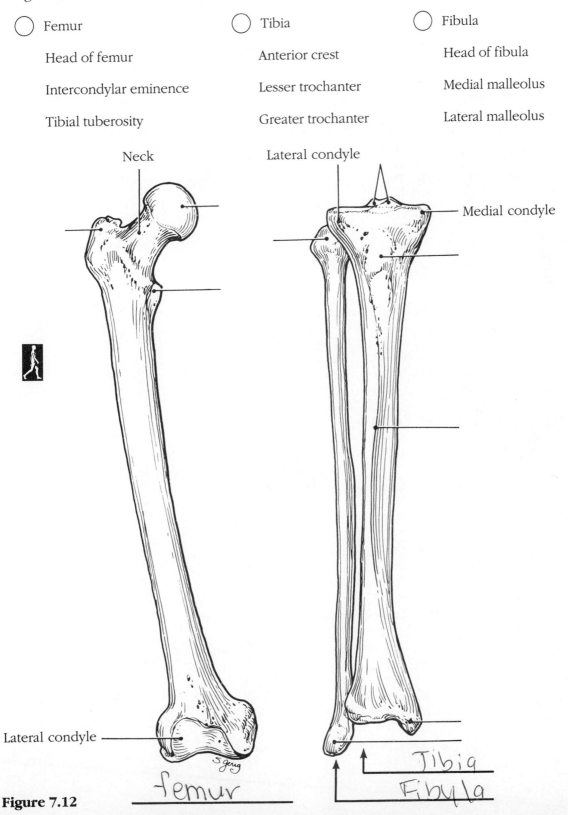

Figure 7.12

3. Match the bone names and markings in Column B to the descriptions in Column A.

Column A

Column B

_____   **1.** Fuse to form the coxal bone (hip bone)

A. Acetabulum

_____   **2.** Receives the weight of the body when sitting

B. Calcaneus

C. Femur

_____   **3.** Point where the coxal bones join anteriorly

D. Fibula

_____   **4.** Upper margin of the iliac bones

E. Gluteal tuberosity

_____   **5.** Deep socket in the coxal bone that receives the head of the thigh bone

F. Greater sciatic notch

G. Greater and lesser trochanters

_____   **6.** Point where the axial skeleton attaches to the pelvic girdle

H. Iliac crest

_____   **7.** Longest bone in the body, articulates with the coxal bone

I. Ilium

_____   **8.** Lateral bone of the leg

J. Ischial tuberosity

_____   **9.** Medial bone of the leg

K. Ischium

_____   **10.** Bones forming the knee joint

L. Lateral malleolus

_____   **11.** Point where the patellar ligament attaches

M. Lesser sciatic notch

N. Medial malleolus

_____   **12.** Kneecap

O. Metatarsus

_____   **13.** Shinbone

P. Obturator foramen

_____   **14.** Distal process on medial tibial surface

Q. Patella

_____   **15.** Process forming the outer ankle

R. Pubic symphysis

_____   **16.** Heel bone

S. Pubis

_____   **17.** Bones of the ankle

T. Sacroiliac joint

_____   **18.** Bones forming the instep of the foot

U. Talus

_____   **19.** Opening in a coxal bone formed by the pubic and ischial rami

V. Tarsals

W. Tibia

_____   **20.** Sites of muscle attachment on the proximal end of the femur

X. Tibial tuberosity

_____   **21.** Tarsal bone that articulates with the tibia

_____ **22.** Its tubercle serves as an attachment point for the inguinal ligament

_____ **23.** Sports the greater and lesser trochanters and the linea aspera

Developmental Aspects of the Skeleton

1. Complete the following statements concerning fetal and infant skeletal development. Insert the missing words in the answer blanks.

_____ **1.**

_____ **2.**

_____ **3.**

_____ **4.**

_____ **5.**

_____ **6.**

_____ **7.**

_____ **8.**

_____ **9.**

"Soft spots," or membranous joints, called __(1)__ in the fetal skull, allow the skull to be __(2)__ slightly during the birth passage. They also allow for continued brain __(3)__ during the later months of fetal development and early infancy. Eventually these soft spots are replaced by immovable joints called __(4)__ .

The two spinal curvatures present at birth are the __(5)__ and __(6)__ curvatures: Because they are present at birth, they are called __(7)__ curvatures. The secondary curvatures develop as the baby matures. The __(8)__ curvature develops as the baby begins to lift his or her head. The __(9)__ curvature develops when the baby begins to walk or assume the upright posture.

2. For the following statements that are true, insert T in the answer blanks. For false statements, correct the underlined words and insert your corrections in the answer blanks.

_____ **1.** Fontanels are replaced by <u>hyaline cartilage</u> after birth.

_____ **2.** The embryonic body of the atlas fuses with the axis and becomes the <u>dens</u>.

_____ **3.** The <u>frontal and mandibular</u> bones are paired in the fetus and fuse after birth.

_____ **4.** At the beginning of the birth process, the fetal skull enters the mother's <u>true pelvis</u> in a sideways position.

_____ **5.** Failure of the palatine processes of the maxillae to fuse medially is an abnormality called <u>cleft lip</u>.

_____ **6.** Bones of the newborn generally lack <u>cavities.</u>

_____ **7.** The spine of the newborn has the shape of a single <u>arch</u>.

_____ **8.** The vertebral column of the newborn is composed of <u>24</u> separate bones.

_____ **9.** Secondary curvatures in the infant result from reshaping or modifications of the <u>vertebrae</u>.

_____ **10.** Flat feet from fallen arches can result from <u>running on hard surfaces</u>.

_____ **11.** The large frontal fontanel persists to about age <u>2 years</u>.

_____ **12.** At birth the UL ratio is <u>1 to 1</u>.

_____ **13.** Enlargement of the jaws in children parallels development of the <u>permanent teeth</u>.

_____ **14.** Facial growth is nearly complete by the age of <u>2 years</u>.

_____ **15.** Fusion of the three pelvic bones occurs at the beginning of <u>adolescence</u>.

_____ **16.** The abnormality called dysplasia of the hip is the result of a <u>nonfusion of the bones forming the</u> acetabulum.

_____ **17.** Adult height is reached when the <u>epiphyseal lines</u> in long bones ossify.

_____ **18.** During middle age, the intervertebral discs are thinner, harder, and less elastic because of the loss of <u>protein</u>.

_____ **19.** In the elderly, a broken hip means a fracture of the <u>crest of the ilium</u>.

_____ **20.** An exaggerated thoracic curvature known as "dowager's hump" is an abnormal condition called <u>scoliosis</u>.

CHALLENGING YOURSELF

At the Clinic

1. Jack, a young man, is treated at the clinic for an accident in which he hit his forehead. When he returns for a checkup, he complains that he can't smell anything. A hurried X ray of his head reveals a fracture. What part of which bone was fractured to cause his loss of the sense of smell?

2. The pediatrician at the clinic explains to parents of a newborn that their child suffers from cleft palate. She tells them that the normal palate fuses in an anterior-to-posterior pattern. The child's palatine processes have not fused. Have his palatine bones fused normally?

3. Janet, a 10-year-old girl, is brought to the clinic after falling out of a tree. An X ray shows she has small fractures of the transverse processes of T3 to T5 on the right side. Janet will be watched for what abnormal spinal curvature over the next several years?

4. Examination of an infant at birth shows that the spinous process on the fifth lumbar vertebra is absent. What condition does this indicate?

5. A young woman is brought to the emergency room after falling on her outstretched arm. Her forearm is bruised both anteriorly and posteriorly along most of its length, but there is no fracture. The doctor explains that she dislocated the head of the radius, but it has relocated spontaneously. What persistent injury is responsible for the bruising?

6. Bart, a truck driver, tells his physician that he has a shooting pain in his right leg. What is a likely diagnosis?

7. Mr. Ogally, a heavy beer drinker with a large potbelly, complained of severe lower back pains. X rays showed displacement of his lumbar vertebrae. What is the condition called and what probably caused it in Mr. Ogally?

8. A man in his late 50s comes to the clinic, worried about a "hard spot" right below his breastbone. An X ray reveals nothing abnormal, although the man insists the spot wasn't there until the last few years. What is the explanation?

9. An infant was born with feet turned medially and toes pointed inferiorly. What name is given to this condition?

10. Antonio is hit in the face with a football during practice. An X ray reveals multiple fractures of the bones around an orbit. Name the bones that form the margins of the orbit.

11. Malcolm was rushed to the emergency room after trying to break a fall with outstretched arms. The physician took one look at his shoulder and saw that he had a broken clavicle (and no other injury). Describe the position of Malcolm's shoulder. Malcolm was worried about injury to the main blood vessels to his arm (the subclavian vessels), but was told such injury was unlikely. Why could the doctor predict this with some assurance?

12. A 75-year-old woman and her 9-year-old granddaughter were in a train crash in which both sustained trauma to the chest. X rays showed that the grandmother had several fractured ribs but her granddaughter had none. Explain these surprisingly (?) different findings.

13. After having a severe cold accompanied by nasal congestion, Helen complained that she had a frontal headache and the right side of her face ached. What bony structures probably became infected by the bacteria or viruses causing the cold?

Stop and Think

1. Name five bones (other than long bones) having cavities in them that make them lighter, and name the cavities.

2. What purpose is served by the unique structure of the turbinates?

3. Name all the parts of the temporal bone that play a part in hearing.

4. Snakes can swallow prey with diameters larger than that of the snake's open mouth. What design modifications of the snake's jaw might allow this?

5. Does nodding the head "yes" involve movement at the same joint as shaking the head "no"? Explain your answer.

6. Look at a skeleton or a picture of the bony thorax and try to figure out which ribs are most easily fractured.

7. The bony thorax doesn't provide nearly as much protection for the heart as the cranium does for the brain. However, its more open structure is advantageous. How so?

8. The acetabulum is much deeper than the glenoid cavity. What is/are the structure-function relationships here?

9. What are fontanels and what advantage do they confer on the fetus? The mother?

COVERING ALL YOUR BASES

Multiple Choice

1. The type of tissue in vertebral ligaments is:
   A. fibrocartilage
   B. elastic cartilage
   C. elastic connective tissue
   D. dense regular connective tissue
   E. dense irregular connective tissue

2. Which of the following are classified as short bones?
   A. Distal phalanx
   B. Cuneiform
   C. Pisiform
   D. Patella
   E. Calcaneus

3. Flat bones include:
   A. ribs
   B. maxilla
   C. nasal
   D. sternum
   E. sphenoid

4. Which of the following bones are part of the axial skeleton?
   A. Vomer
   B. Clavicle
   C. Sternum
   D. Hyoid
   E. Coxal bone

5. Bone pain behind the external auditory meatus probably involves the:
   A. maxilla
   B. ethmoid
   C. sphenoid
   D. temporal
   E. lacrimal

6. Bones that articulate with the sphenoid include:
   A. parietal
   B. vomer
   C. maxilla
   D. zygomatic
   E. ethmoid

7. Which parts of the temporal bone contribute to the calvaria?
   A. Squamous region
   B. Petrous region
   C. Mastoid region
   D. Tympanic region

8. Which of the following are bony landmarks of the anterior skull surface?
   A. Mental foramina
   B. Anterior cranial fossa
   C. Glabella
   D. Supraorbital foramen
   E. Frontal sinuses

9. Points of muscle, tendon, or ligament attachment include:
   A. superior nuchal lines
   B. greater wings of the sphenoid
   C. external occipital crest
   D. mastoid process
   E. coronoid process

10. Which of the following are part of the sphenoid?
    A. Crista galli
    B. Sella turcica
    C. Petrous portion
    D. Pterygoid process
    E. Lesser wings

11. Which of the following are found within the cranial cavity?
    A. Perpendicular plate of ethmoid
    B. Internal acoustic meatus
    C. Crista galli
    D. Cranial fossae
    E. Lacrimal fossa

12. Bones helping to form the oral cavity include:
    A. mandible   D. sphenoid
    B. vomer   E. palatine
    C. maxilla

13. Which of the following bones are at least partly covered by a mucous membrane?
    A. Zygomatic   D. Ethmoid
    B. Vomer   E. Occipital
    C. Temporal

14. Which of these start as separate bones in the fetus, but are fused in an adult skeleton?
    A. Coxal bone   D. Coccyx
    B. Vertebral ribs E. Cuneiforms
    C. Mandible

15. The protective bony ring around the spinal cord is formed by several parts of the vertebra including:
    A. laminae   D. pedicles
    B. centrum   E. facet
    C. transverse process

16. Ribs articulate with the:
    A. transverse processes of lumbar vertebrae
    B. manubrium
    C. spinous processes of thoracic vertebrae
    D. xiphoid process
    E. centrum of cervical vertebrae

17. Structural characteristics of *all* cervical vertebrae are:
    A. small body
    B. bifid spinous process
    C. transverse foramina
    D. small vertebral foramen
    E. costal facets

18. What types of movement are possible between thoracic vertebrae?
    A. Flexion
    B. Rotation
    C. Some lateral flexion
    D. Circumduction
    E. Extension

19. Which humeral process articulates with the radius?
    A. Trochlea   D. Capitulum
    B. Great tubercle E. Olecranon fossa
    C. Lesser tubercle

20. A coronoid process is found on the:
    A. maxilla   D. humerus
    B. mandible   E. ulna
    C. scapula

21. Which of the following bones exhibit a styloid process?
    A. Hyoid   D. Radius
    B. Temporal   E. Ulna
    C. Humerus

22. Which bones articulate with metacarpals?
    A. Hamate   D. Capitate
    B. Pisiform   E. Cuneiform
    C. Proximal phalanges

23. Hip bone markings include:
    A. ala   D. pubic ramus
    B. sacral hiatus E. fovea capitis
    C. gluteal surface

24. Which of the following bones or bone parts articulate with the femur?
    A. Ischial tuberosity D. Fibula
    B. Pubis   E. Tibia
    C. Patella

25. The UL ratio becomes 1 to 1:
    A. at birth
    B. by 10 years of age
    C. at puberty
    D. when the epiphyseal plates fuse
    E. never

26. Which pair of ribs inserts on the sternum at the sternal angle?
    A. First   C. Third
    B. Second   D. Fourth

27. The inferior angle of the scapula is at the same level as the spinous process of vertebra:
    A. C_7 D. T_7
    B. C_5 E. L_4
    C. T_3

28. An important bony landmark that can be recognized by a distinct dimple in the skin is:
    A. posterior superior iliac spine
    B. styloid process of the ulna
    C. shaft of the radius
    D. acromion of the scapula

29. A blow to the cheek is most likely to break what superficial bone or bone part?
    A. Superciliary arches   C. Mandibular ramus
    B. Zygomatic process   D. Styloid process

Word Dissection

For each of the following word roots, fill in the literal meaning and give an example, using a word found in this chapter.

Word root	Translation	Example
1. acetabul		
2. alve		
3. append		
4. calv		
5. clavicul		
6. crib		
7. den		
8. ethm		
9. glab		
10. ham		
11. ment		
12. odon		
13. pal		

Word root	Translation	Example
14. pect	_____	_____
15. pelv	_____	_____
16. pis	_____	_____
17. pter	_____	_____
18. scaph	_____	_____
19. skeleto	_____	_____
20. sphen	_____	_____
21. styl	_____	_____
22. sutur	_____	_____
23. vert	_____	_____
24. xiph	_____	_____

 Joints

Student Objectives

When you have completed the exercises in this chapter, you will have accomplished the following objectives:

1. Define *joint* or *articulation*.

Classification of Joints

2. Classify joints structurally and functionally.

Fibrous Joints

3. Describe the general structure of fibrous joints. Name and give an example of each of the three common types of fibrous joints.

Cartilaginous Joints

4. Describe the general structure of cartilaginous joints. Name and give an example of each of the two common types of cartilaginous joints.

Synovial Joints

5. Describe the structural characteristics shared by all synovial joints.

6. List three natural factors that stabilize synovial joints.

7. Compare the structures and functions of bursae and tendon sheaths.

8. Name and describe (or perform) the common types of body movements.

9. Name six types of synovial joints based on the type of movement allowed. Provide examples of each type.

10. Describe the elbow, knee, hip, and shoulder joints. Consider in each case the articulating bones, anatomical characteristics of the joint, types of movement allowed, and relative joint stability.

Homeostatic Imbalances of Joints

11. Name the most common joint injuries and discuss the symptoms and problems associated with each.

12. Compare and contrast osteoarthritis, rheumatoid arthritis, and gouty arthritis with respect to the population affected, possible causes, structural joint changes, disease outcome, and therapy.

Developmental Aspects of Joints

13. Discuss briefly the factors that promote or disturb joint homeostasis.

Ready, set, go!

Joints are structures that connect adjoining bones. Except for the hyoid bone, and sesamoid bones like the patella, each bone contacts at least one other bone at a joint. A typical joint includes the adjacent surfaces of the bones and the fibrous tissue or ligaments that bind the bones together. In addition to binding the bones together, joints provide the skeleton with the flexibility to permit body movements.

Since joints vary in structure and range of motion, it is convenient to classify them on the basis of their structure (fibrous, cartilaginous, or synovial) and function (the degree of joint movement allowed).

Topics for review in Chapter 8 include the classifications of joints, the structures of selected synovial joints, joint impairment, and changes in the joints throughout life.

BUILDING THE FRAMEWORK

Classification of Joints

1. Write your answers to the following questions in the answer blanks.

    1. What are the two major functions of joints? _____

    _____

    2. List three criteria used to classify joints. _____

    _____

    _____

    3. For each of the structural joint categories—fibrous, cartilaginous, and synovial—give the *most common* functional classification and describe the degree of movement.

    _____

    _____

    _____

Fibrous, Cartilaginous, and Synovial Joints

1. For each joint described below, select an answer from Key A. Then, if the Key A selection *is other than C* (synovial joint), classify the joint further by making a choice from Key B.

KEY A: **A.** Cartilaginous   **KEY B:**   **1.** Gomphosis   **4.** Syndesmosis

          B. Fibrous   **2.** Suture   **5.** Synchondrosis

          C. Synovial   **3.** Symphysis   **6.** Synostosis

_____ 1. Characterized by hyaline cartilage connecting the bony portions

_____ 2. All have a fibrous capsule lined with a synovial membrane surrounding a joint cavity

_____ 3. Bone regions united by fibrous connective tissue

_____ 4. Joints between skull bones

_____ 5. Joint between atlas and axis

_____ 6. Hip, elbow, knee, and intercarpal joints

_____ 7. Intervertebral joints (between vertebral bodies)

_____ 8. Pubic symphysis

_____ 9. All are reinforced by ligaments

_____ 10. Costosternal joints 2–7

_____ 11. Joint providing the most protection to underlying structures

_____ 12. Often contains a fluid-filled cushion

_____ 13. Child's epiphyseal plate made of hyaline cartilage

_____ 14. Most joints of the limbs

_____ 15. Teeth in body alveolar sockets

_____ 16. Joint between first rib and manubrium of sternum

_____ 17. Ossified sutures

_____ 18. Distal tibiofibular joint

2. Which structural joint type is *not* commonly found in the axial skeleton and

why not? _____

_____

_____

3. Figure 8.1 shows the structure of a typical synovial joint. Select different colors to identify and color the following areas. Then label the following: proximal epiphyseal line, distal epiphyseal line, spongy bone, periosteum.

○ Articular cartilage of bone ends   ○ Synovial membrane

○ Fibrous capsule   ○ Joint cavity

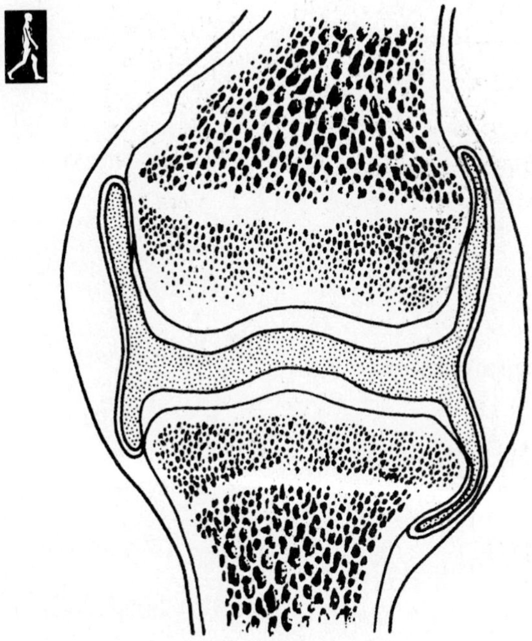

Figure 8.1

4. Match the key choices with the descriptive phrases below, to properly characterize selected aspects of synovial joints.

KEY CHOICES

A. Articular cartilage   **C.** Ligaments and fibrous capsule

B. Synovial fluid   **D.** Muscle tendons

_____ 1. Keeps bone ends from crushing when compressed; resilient

_____ 2. Resists tension placed on joints

_____ 3. Lubricant that minimizes friction and abrasion of joint surfaces

_____ 4. Keeps joints from overheating

_____ 5. Helps prevent dislocation

5. Match the body movement terms in Column B with the appropriate description in Column A. (More than one choice may apply.)

Column A

Column B

_____ **1.** Movement along the sagittal plane that decreases the angle between two bones

A. Abduction

B. Adduction

_____ **2.** Movement along the frontal plane, away from the body midline; raising the arm laterally

C. Circumduction

D. Depression

_____ **3.** Circular movement around the longitudinal bone axis; shaking the head "no"

E. Dorsiflexion

_____ **4.** Slight displacement or slipping of bones, as might occur between the carpals of the wrist

F. Elevation

G. Eversion

_____ **5.** Describing a cone-shaped pathway with the arm

H. Extension

I. Flexion

_____ **6.** Lifting or raising a body part; shrugging the shoulders

J. Gliding

_____ **7.** Moving the hand into a palm-up (or forward) position

K. Inversion

L. Plantarflexion

_____ **8.** Movement of the superior aspect of the foot toward the leg; standing on the heels

M. Pronation

_____ **9.** Turning the sole of the foot medially

N. Protraction

O. Retraction

_____ **10.** Movement of a body part anteriorly; jutting the lower jaw forward

P. Rotation

_____ **11.** Common angular movements

Q. Supination

6. Figure 8.2 illustrates types of movements allowed by synovial joints. Match the letters on the figure with the types of movements listed below. Insert your answers in the answer blanks. Then color the drawing to suit your fancy.

_____ **1.** Flexion

_____ **6.** Protraction

_____ **2.** Plantarflexion

_____ **7.** Circumduction

_____ **3.** Abduction

_____ **4.** Rotation

_____ **8.** Adduction

_____ **5.** Pronation

_____ **9.** Extension

_____ **10.** Dorsiflexion

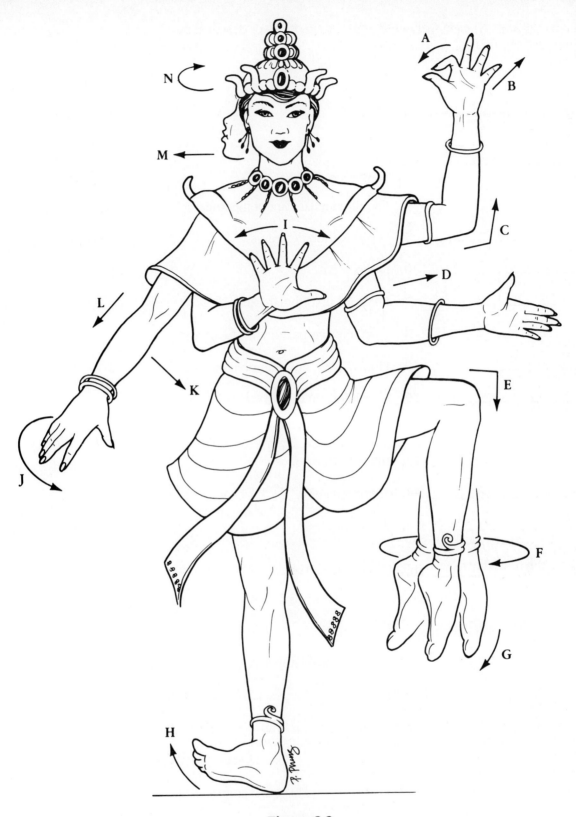

Figure 8.2

7. Match the joint types in Column B with the examples or descriptions of joints listed in Column A.

Column A   Column B

_____ **1.** Knuckles   **A.** Ball and socket

_____ **2.** Sacroiliac joints   **B.** Condyloid

_____ **3.** Intercarpal joints   **C.** Hinge

_____ **4.** Femoropatellar joint of the knee   **D.** Pivot

_____ **5.** Hip and shoulder joints   **E.** Plane

_____ **6.** Radiocarpal   **F.** Saddle

_____ **7.** Proximal tibiofibular joint **G.** Synchondrosis

_____ **8.** Carpometacarpal joint of the thumb

_____ **9.** Elbow, knee, and interphalangeal joints

_____ **10.** Joint between C_1 and C_2

_____ **11.** Uniaxial joints

_____ **12.** Biaxial joints

_____ **13.** Multiaxial joints

_____ **14.** Nonaxial joints

8. Circle the term that does not belong in each of the following groupings.

1. Pivot joint   Uniaxial joint   Multiaxial joint   Atlas/axis joint

2. Ball and socket joint   Multiaxial joint   Hip joint   Saddle joint

3. Tibiofibular joints   Sutures   Synostoses   Gomphoses

4. Articular discs   Multiaxial joint   Largest joint in the body   Knee joint

5. Saddle joint   Carpometacarpal joint of thumb   Elbow joint   Biaxial joint

6. Amphiarthrotic   Intercarpal joints   Plane joints   Nonaxial joints

7. Intervertebral joints   Bursae   Cartilaginous joints   Symphyses

8. Hinge joint   Condyloid joint   Elbow joint   Uniaxial joint

9. Muscle tendon reinforcement   Shoulder joint   Biaxial joint
Most freely moving joint in the body

9. Several characteristics of specific synovial joints are described below. Identify each of the joints described by choosing a response from the key choices.

KEY CHOICES

A. Elbow   **B.** Hip   **C.** Knee   **D.** Shoulder

_____ 1. Rotator cuff muscles are important in stabilizing this joint; capsule reinforced only anteriorly by ligaments; articular surfaces shallow.

_____ 2. Three joints in one; capsule incomplete anteriorly; has menisci and intracapsular cruciate ligaments.

_____ 3. Capsule is loose; reinforced by medial and lateral collateral ligaments; articular surfaces most important in ensuring joint stability.

_____ 4. Articular surfaces deep and secure; capsule heavily reinforced by ligaments and muscle tendons; intracapsular ligamentum teres. Extremely stable joint.

10. Figure 8.3 shows diagrams of two synovial joints. Identify each joint by inserting the name of the joint in the blank below each diagram. Then select different colors and use them to color the coding circles and the structures in the diagrams. Finally, add labels and leader lines on the diagrams to identify the following structures: the ligamentum teres, the anterior cruciate ligament, the posterior cruciate ligament, the suprapatellar bursa, the subcutaneous prepatellar bursa, and the deep infrapatellar bursa.

○ Fibrous capsule (lined with synovial membrane)   ○ Joint cavity

○ Articular cartilage of bone ends   ○ Acetabular labrum

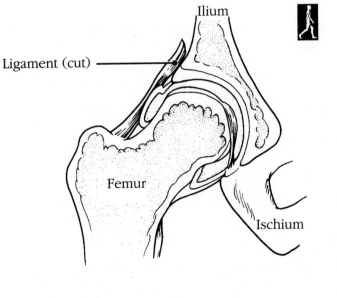
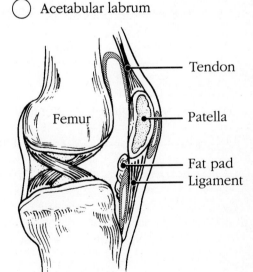

A. _____   B. _____

Figure 8.3

Homeostatic Imbalances of Joints

1. For the following statements that are true, insert T in the answer blanks. For false statements, correct the underlined words and insert your corrections in the answer blanks.

_____ 1. The term <u>arthritis</u> refers to bone ends that are forced out of their normal positions in a joint cavity.

_____ 2. In a <u>sprain</u>, the ligaments reinforcing a joint are excessively stretched or torn.

_____ 3. Erosion of articular cartilages and formation of painful bony spurs are characteristic of <u>gouty arthritis</u>.

_____ 4. "Housemaid's knee" is an example of inflammation of the <u>tendon sheaths</u>.

_____ 5. <u>Chronic</u> arthritis usually results from bacterial invasion.

_____ 6. Healing of a partially torn ligament is slow because its hundreds of fibrous strands are poorly <u>aligned</u>.

_____ 7. <u>Acute arthritis</u> is an autoimmune disease.

_____ 8. Hyperuricemia may lead to <u>rheumatoid arthritis</u>.

_____ 9. Torn menisci of the knee rarely heal because cartilage is <u>avascular</u>.

_____ 10. In rheumatoid arthritis, the initial pathology is inflammation of the <u>bursae</u>.

_____ 11. <u>Gouty arthritis</u> involves several joints that are affected in a bilateral manner.

_____ 12. Most cartilage injuries of the knee involve torn <u>ligaments</u>.

_____ 13. The function of the bursae is to <u>absorb the shock</u> between bony structures in a joint.

_____ 14. The most common form of chronic arthritis is <u>rheumatoid arthritis</u>.

_____ 15. The local swelling seen in bursitis is caused by excessive production of <u>plasma</u> fluid.

2. Explain why sprains and injuries to joint cartilages are particularly troublesome.

_____

_____

Developmental Aspects of Joints

1. Complete each statement by inserting your responses in the answer blanks.

_____ 1.

_____ 2.

_____ 3.

_____ 4.

_____ 5.

Joints develop from the embryonic tissue called __(1)__ . By age __(2)__ weeks, the synovial joints are formed, appearing much like adult joints in structure and function. Usually, uninjured joints withstand normal usage well through __(3)__ . Excessive or abusive use of joints commonly leads to the chronic condition of __(4)__ . Although exercise cannot prevent joint impairment, it helps to maintain joint integrity by increasing the strength of __(5)__ supporting the joints.

The Incredible Journey: A Visualization Exercise for the Knee Joint

This place is like an underground cave, and you flip on your back to swim in the shallow pool.

1. Complete the narrative by inserting the missing words in the answer blanks.

_____ 1.

_____ 2.

_____ 3.

_____ 4.

_____ 5.

_____ 6.

_____ 7.

_____ 8.

_____ 9.

_____ 10.

For this journey, you are miniaturized and injected into the knee joint of a football player who is resting before the game. You find yourself standing on a firm, white, flexible, C-shaped structure, a __(1)__ , which prevents side-to-side rocking of the __(2)__ on the __(3)__ . The space you are in, the __(4)__ , is filled with a clear, viscous liquid, the __(5)__ , which serves to __(6)__ the glassy-smooth articulating surfaces. You note that this place is like an underground cave, and you flip on your back to swim in the shallow pool. As you gaze upward, you see, almost touching your head, two enormous boulderlike structures. At first, you're afraid they might fall on you, but then, of course, you recognize them as the __(7)__ of the femur. After becoming accustomed to this environment, you begin to wonder what this football player is doing with his knee joint. Because you see that his __(8)__ ligament is taut, you surmise that his knee joint must be in extension.

Suddenly, the liquid around you becomes less viscous, and you are swept upward, past a rather small, flat, roundish bone, which you identify as the __(9)__ , into an enormous ocean of that same viscous liquid. You note that it is surrounded by a membrane. After some puzzlement, you realize that this must be the suprapatellar __(10)__ !

Knowing that your football player is no longer at rest, you swim, with great effort, back to the joint cavity to watch his

_____ **11.**

_____ **12.**

_____ **13.**

_____ **14.**

_____ **15.**

knee joint in action. Within a moment, a hard, lateral blow to the knee sends him (and you) spinning. You examine his joint and note that the meniscus has been torn from its attachment at the __(11)__ margins of the __(12)__ and that the extracapsular lateral collateral __(13)__ has also suffered some damage. Since you know that cartilage rarely heals because it has no __(14)__ for nourishment, you predict that your football player will most likely undergo surgery for removal of his __(15)__ . Since you prefer not to wait until then to leave your host, you signal headquarters that you are ready to be picked up by the syringe.

CHALLENGING YOURSELF

At the Clinic

1. Jenny's father loved to hold his 4-year-old daughter by her hand and swing her around in great circles. One day, Jenny's glee was suddenly replaced by tears and she screamed that her left elbow hurt. When examined, the little girl was seen to hold her elbow semiflexed and her forearm pronated. What is your diagnosis?

2. A patient complains of pain starting in the jaw and radiating down the neck. Upon questioning, he states that when he is under stress he grinds his teeth. What joint is causing his pain?

3. In the condition cleidocranial dysostosis, ossification of the skull is delayed and clavicles may be completely absent. What effect would you predict that this condition has on the function and appearance of the shoulders?

4. Mrs. Carlisle, an elderly woman, suffered sharp pain in her shoulder while trying to raise a stuck window. Examination reveals degeneration and rupture of the rotator cuff. What is the function of the rotator cuff, and how does its rupture affect the function of the shoulder joint?

5. A complete fracture of the femoral neck will not result in death of the femoral head, unless the ligamentum teres is ruptured. What does this ligament contribute to maintain the integrity of the femoral head?

6. In a near-fatal accident in which he fell off an embankment, the famous neurologist Oliver Sacks ruptured his patellar ligament. What abnormal movements of his knee allowed him to assess his injury?

7. A surgeon has removed part of the shaft of the radius of a patient suffering from bone cancer. What bone will he likely use to obtain a graft to replace the excised bone?

8. What type of cartilage is damaged in a case of "torn cartilage"?

9. Marjorie, a middle-aged woman, comes to the clinic complaining of stiff, painful joints. A glance at her hands reveals knobby, deformed knuckles. What condition will be tested for?

Stop and Think

1. Name all the diarthrotic joints in the axial skeleton.

2. What joints are involved in pronation and supination? How are these joints classified?

3. To what joint classification would the developing coxal bone be assigned? How is the adult coxal bone classified?

4. Explain the significance of the interdigitating and interlocking surfaces of sutures.

5. What microscopic feature marks the end of the periosteum and the beginning of an intrinsic ligament?

6. Is the lining of a joint cavity an epithelial membrane? If so, what type? If not, what is it?

7. What is the role of menisci in the jaw and sternoclavicular joints?

8. What is the effect of years of poor posture on the posterior vertebral ligaments?

9. Only one joint flexes posteriorly. Which one?

10. Is the head capable of circumduction? Why or why not?

11. How does the knee joint differ from a typical hinge joint?

COVERING ALL YOUR BASES

Multiple Choice

Select the best answer or answers from the choices given.

1. Which of the joints listed would be classified as synarthrotic?
   - **A.** Vertebrocostal
   - **B.** Sternocostal (most)
   - **C.** Acromioclavicular
   - **D.** Distal tibiofibular
   - **E.** Epiphyseal plate

2. Cartilaginous joints include:
   - **A.** syndesmoses
   - **B.** symphyses
   - **C.** synostoses
   - **D.** synchondroses

3. Which of the following may be an amphiarthrotic fibrous joint?
   - **A.** Syndesmosis
   - **B.** Suture
   - **C.** Sternocostal
   - **D.** Intervertebral

4. Joints that eventually become synostoses include:
   - **A.** fontanels
   - **B.** first rib—sternum
   - **C.** ilium-ischium-pubis
   - **D.** epiphyseal plates

5. Which of the following joints have a joint cavity?
   - **A.** Glenohumeral
   - **B.** Acromioclavicular
   - **C.** Intervertebral (at bodies)
   - **D.** Intervertebral (at articular processes)

6. Joints that *contain* fibrocartilage include:
   - **A.** symphysis pubis
   - **B.** knee
   - **C.** acromioclavicular
   - **D.** atlanto-occipital

7. Considered to be part of a synovial joint are:
   - **A.** bursae
   - **B.** articular cartilage
   - **C.** tendon sheath
   - **D.** capsular ligaments

8. Synovial fluid:
   - **A.** is indistinguishable from fluid in cartilage matrix
   - **B.** is formed by synovial membrane
   - **C.** thickens as it warms
   - **D.** nourishes and lubricates

9. Which joint type would be best if the joint receives a moderate to heavy stress load and requires moderate flexibility in the sagittal plane?
   - **A.** Gliding
   - **B.** Saddle
   - **C.** Condyloid
   - **D.** Hinge

10. In comparing two joints of the same type, what characteristic(s) would you use to determine strength and flexibility?
    - **A.** Depth of the depression of the concave bone of the joint
    - **B.** Snugness of fit of the bones
    - **C.** Size of bone projections for muscle attachments
    - **D.** Presence of menisci

11. Which of the following is a multiaxial joint?
    - **A.** Atlanto-occipital
    - **B.** Atlantoaxial
    - **C.** Glenohumeral
    - **D.** Metacarpophalangeal

12. Plane joints allow:
    - **A.** pronation
    - **B.** flexion
    - **C.** rotation
    - **D.** gliding

13. The temporomandibular joint is capable of which movements?
    - **A.** Elevation
    - **B.** Protraction
    - **C.** Hyperextension
    - **D.** Inversion

14. Abduction is:
    A. moving the right arm out to the right
    B. spreading out the fingers
    C. wiggling the toes
    D. moving the sole of the foot laterally

15. Which of the joints listed is classified as a hinge joint functionally but is actually a condyloid joint structurally?
    A. Elbow   C. Knee
    B. Interphalangeal   D. Temporomandibular

16. Which of the following joints has the greatest freedom of movement?
    A. Interphalangeal
    B. Saddle joint of thumb
    C. Distal tibiofibular
    D. Coxal

17. In what condition would you suspect inflammation of the superficial olecranal bursa?
    A. Student's elbow
    B. Housemaid's knee
    C. Target shooter's trigger finger
    D. Teacher's shoulder

18. Contributing substantially to the stability of the hip joint is the:
    A. ligamentum teres
    B. acetabular labrum
    C. patellar ligament
    D. iliofemoral ligament

19. The ligament located on the posterior aspect of the knee joint is the:
    A. fibular collateral   C. oblique popliteal
    B. tibial collateral   D. posterior cruciate

20. Which of the following apply to a sprain?
    A. Dislocation
    B. Slow healing
    C. Surgery for ruptured ligaments
    D. Grafting

21. Erosion of articular cartilage and subsequent formation of bone spurs occur in:
    A. ankylosis   C. torn cartilage
    B. ankylosing spondylitis   D. osteoarthritis

22. An autoimmune disease resulting in inflammation and eventual fusion of diarthrotic joints is:
    A. gout
    B. rheumatoid arthritis
    C. degenerative joint disease
    D. pannus

23. Which of the following is correlated with gouty arthritis?
    A. High levels of antibodies
    B. Hyperuricemia
    C. Hypouricemia
    D. Hammertoe

24. An embryo is examined and found to have completely formed synovial joints. What is its approximate age?
    A. 2 weeks   C. 6 weeks
    B. 4 weeks   D. 8 weeks

25. Synovial joints are richly innervated by nerve fibers that:
    A. monitor how much the capsule is stretched
    B. supply articular cartilages
    C. cause the joint to move
    D. monitor pain if the capsule is injured

26. Which specific joint does the following description identify? "Articular surfaces are deep and secure, multiaxial; capsule heavily reinforced by ligaments; labrum helps prevent dislocation; the first joint to be built artificially; very stable."
    A. Elbow   C. Knee
    B. Hip   D. Shoulder

27. How does the femur move with respect to the tibia when the knee is locking during extension?
    A. Glides anteriorly   D. Rotates medially
    B. Flexes   E. Rotates laterally
    C. Glides posteriorly

Word Dissection

For each of the following word roots, fill in the literal meaning and give an example, using a word found in this chapter.

Word root	Translation	Example
1. ab		
2. ad		
3. amphi		
4. ankyl		
5. arthro		
6. artic		
7. burs		
8. cruci		
9. duct		
10. gompho		
11. labr		
12. luxa		
13. menisc		
14. ovi		
15. pron		
16. rheum		
17. spondyl		
18. supine		

9 Muscles and Muscle Tissue

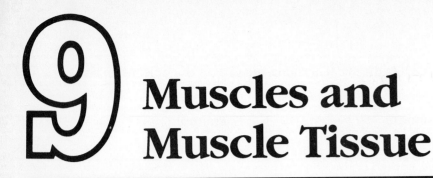

Student Objectives

When you have completed the exercises in this chapter, you will have accomplished the following objectives:

Overview of Muscle Tissues

1. Compare and contrast the basic types of muscle tissue.

2. List four important functions of muscle tissue.

Skeletal Muscle

3. Describe the gross structure of a skeletal muscle with respect to location and names of its connective tissue coverings.

4. Describe the microscopic structure and functional roles of the myofibrils, sarcoplasmic reticulum, and T tubules of skeletal muscle fibers (cells).

5. Explain how muscle fibers are stimulated to contract, and explain the sliding filament mechanism of skeletal muscle contraction.

6. Define muscle twitch and describe the events occurring during its three phases.

7. Explain how smooth, graded contractions of a skeletal muscle are produced.

8. Differentiate between isometric and isotonic contractions.

9. Describe three ways in which ATP is regenerated during skeletal muscle contraction.

10. Define oxygen debt and muscle fatigue. List possible causes of muscle fatigue.

11. List and describe factors that influence the force, velocity, and duration of skeletal muscle contraction.

12. Name and describe three types of skeletal muscle fibers. Explain the relative value of each fiber type in the body.

13. Compare and contrast the effects of aerobic and resistance exercise on skeletal muscles and on other body systems.

Smooth Muscle

14. Compare the gross and microscopic anatomy of smooth muscle fibers to that of skeletal muscle fibers.

15. Compare and contrast the contractile mechanisms and the means of activation of skeletal and smooth muscles in the body.

16. Distinguish between single-unit and multiunit smooth muscle structurally and functionally.

Developmental Aspects of Muscles

17. Describe briefly the embryonic development of muscle tissues and the changes that occur in skeletal muscles with age.

Ready, set, go!

The specialized muscle tissues are responsible for virtually all body movements, both internal and external. Most of the body's muscle is of the voluntary type and is called skeletal muscle because it is attached to the skeleton. Voluntary muscles allow us to manipulate the external environment, move our bodies, and express emotions on our faces. The balance of the body's muscle is smooth muscle and cardiac muscle, which form the bulk of the walls of hollow organs and the heart, respectively. Smooth and cardiac muscles are involved in the transport of materials within the body.

Study activities in Chapter 9 deal with the microscopic and gross structures of skeletal and smooth muscle. (Cardiac muscle is discussed more thoroughly in Chapter 19.) Important concepts of muscle physiology are also examined in this chapter.

BUILDING THE FRAMEWORK

Overview of Muscle Tissues

1. Nine characteristics of muscle tissue are listed below. Identify each muscle type by choosing the correct key choices and writing the letters in the answer blanks.

KEY CHOICES

A. Cardiac   **B.** Smooth   **C.** Skeletal

A+B **1.** Involuntary

A+C **2.** Banded appearance

B **3.** Longitudinally and circularly arranged layers

C **4.** Dense connective tissue packaging

A B **5.** Gap junctions

A **6.** Coordinated activity allows it to act as a pump

C **7.** Moves bones and the facial skin

C **8.** Referred to as the muscular system

C **9.** Voluntary

B **10.** Best at regenerating when injured

2. Identify the type of muscle in each of the illustrations in Figure 9.1. Color the diagrams as you wish.

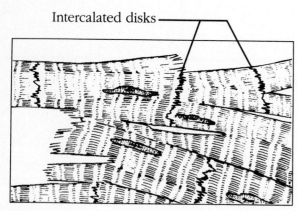

A. smooth

Intercalated disks

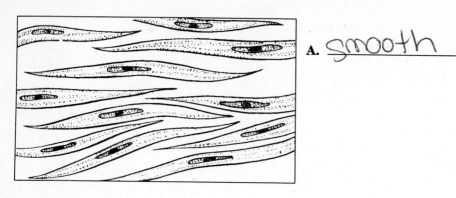

B. cardiac

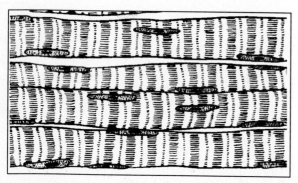

C. skeletal

Figure 9.1

3. Regarding the functions of muscle tissues, circle the term in each of the groupings that does not belong with the other terms.

1. Urine   Foodstuffs   Blood   Bones   Smooth muscle

2. Heart   Cardiac muscle   Blood pump   Promotes labor during birth

3. Excitability   Response to a stimulus   Contractility   Action potential

4. Elasticity   Recoil properties   Contractility   Resumption of resting length

5. Ability to shorten   Contractility   Pulls on bones   Stretchability

6. Posture maintenance   Movement   Promotes growth   Generates heat

Skeletal Muscle

1. Identify the structures described in Column A by matching them with the terms in Column B. Enter the correct letters (and terms if desired) in the answer blanks. Then, select a different color for each of the terms in Column B that has a color-coding circle and color the structures in Figure 9.2.

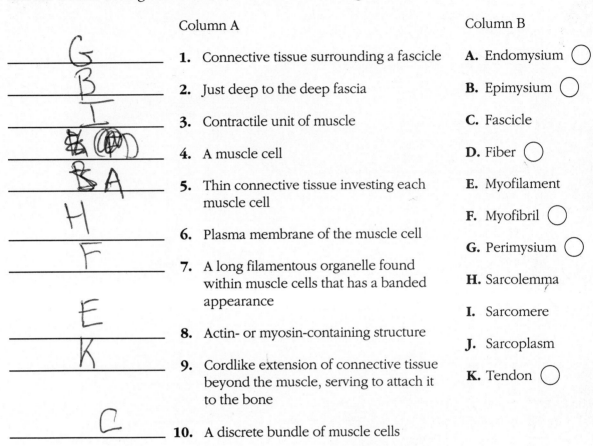

Column A

G   **1.** Connective tissue surrounding a fascicle

B   **2.** Just deep to the deep fascia

I   **3.** Contractile unit of muscle

$ (D)   **4.** A muscle cell

BA   **5.** Thin connective tissue investing each muscle cell

H   **6.** Plasma membrane of the muscle cell

F   **7.** A long filamentous organelle found within muscle cells that has a banded appearance

E   **8.** Actin- or myosin-containing structure

K   **9.** Cordlike extension of connective tissue beyond the muscle, serving to attach it to the bone

C   **10.** A discrete bundle of muscle cells

Column B

A. Endomysium ◯

B. Epimysium ◯

C. Fascicle

D. Fiber ◯

E. Myofilament

F. Myofibril ◯

G. Perimysium ◯

H. Sarcolemma

I. Sarcomere

J. Sarcoplasm

K. Tendon ◯

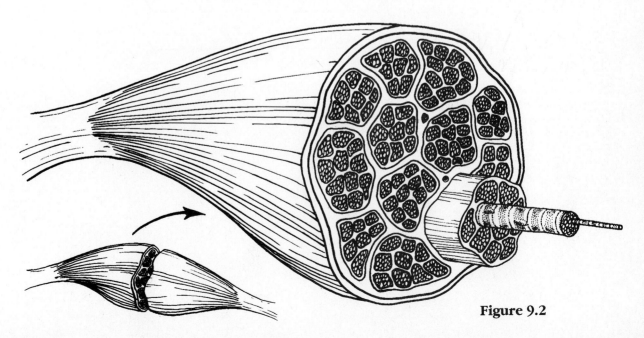

Figure 9.2

2. Figure 9.3 is a diagrammatic representation of a small portion of a relaxed muscle cell (the bracket indicates the portion that has been enlarged). First, select a different color for each of the structures with a coding circle. Color the coding circles and the corresponding structures on Figure 9.3. When you have finished, *bracket* and *label* an A band, an I band, and a sarcomere. Then, match the numbered lines (1, 2, and 3) in part B to the cross sections in part C.

◯ Thin myofilaments   ◯ Thick myofilaments   ◯ Z lines

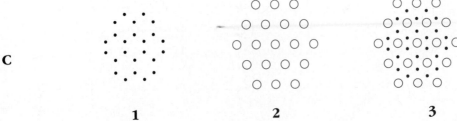

Figure 9.3

3. After referring to Figure 9.3 showing the configuration of a relaxed sarcomere, at the right, draw a contracted sarcomere. Label the thin myofilaments, the thick myofilaments, and the Z lines. Draw an arrow beside one myosin head on each end of the A band to indicate the direction of the power stroke.

4. Fill in the blanks at the right below, giving the location of calcium during the stated phases of muscle contraction. Use the key choices provided at the left.

KEY CHOICES

A. Being transported by calsequestrin

B. Attached to troponin

C. In terminal cisternae

_____ **1.** Unstimulated fiber

_____ **2.** Contracting fiber

_____ **3.** Recovering fiber

5. Figure 9.4 shows the intimate relationship between the sarcolemma and two important muscle cell organelles—the sarcoplasmic reticulum (SR) and the myofibrils—in a small segment of a muscle cell. Identify each structure below by coloring the coding circles and the corresponding structures on the diagram. Then bracket and label the composite structure called the triad.

◯ Mitochondrion   ◯ Myofibrils   ◯ Sarcolemma

◯ SR   ◯ T tubule

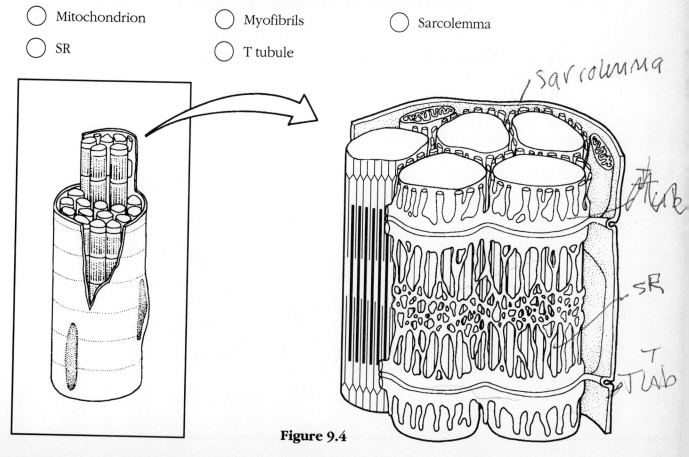

Figure 9.4

6. Correctly relate the story of contraction events in a muscle fiber by *numbering* each event below. The first step is indicated (number 1).

____4____ **1.** Myosin heads bind to active sites on actin molecules.

____7____ **2.** ATP is hydrolyzed.

____11____ **3.** Myosin heads return to their high-energy shape (cocked), ready for the next power stroke.

____1____ **4.** Calcium ions bind to troponin.

____9____ **5.** Cycling continues until calcium ions are sequestered by the SR.

____8____ **6.** Myosin cross bridges detach from actin.

____10____ **7.** Troponin changes shape.

____6____ **8.** ADP and P_i (inorganic phosphate) are released from the thick filament.

____2____ **9.** Myosin heads pull on the thin filaments (power stroke) and slide them toward the center of the sarcomere.

____3____ **10.** ATP binds to the thick filament.

____5____ **11.** Tropomyosin is moved into the groove between the F-actin strands exposing active sites on actin.

7. Relative to general terminology concerning muscle activity, first, label the following structures on Figure 9.5: insertion, origin, tendon, resting muscle, and contracting muscle. Next, identify the two structures named below by choosing different colors for the coding circles and the corresponding structures in the figure.

○ Movable bone

○ Immovable bone

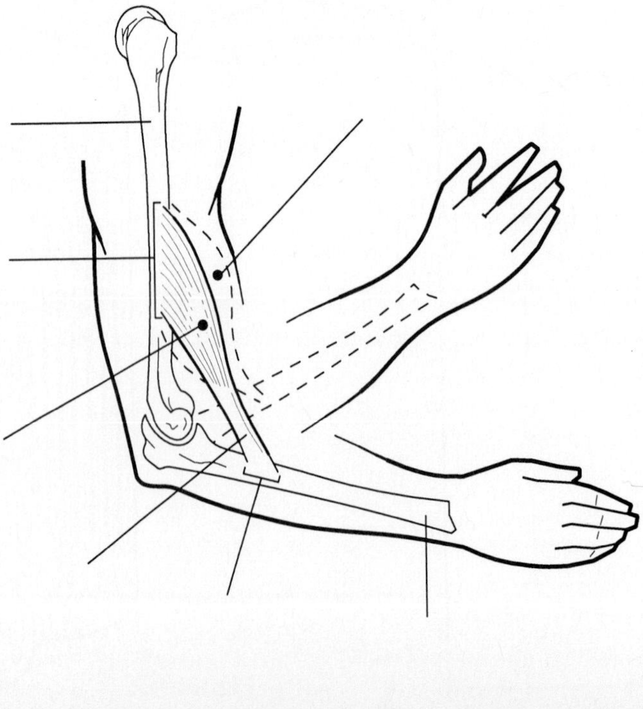

Figure 9.5

8. Figure 9.6 shows the components of a neuromuscular junction. Identify the parts by coloring the coding circles and the corresponding structures in the diagram. Add small circles to indicate the location of the ACh receptors and label appropriately.

○ Mitochondrion   ○ Junctional folds

○ Synaptic vesicles   ○ Synaptic cleft

○ T tubule   ○ Sarcomere

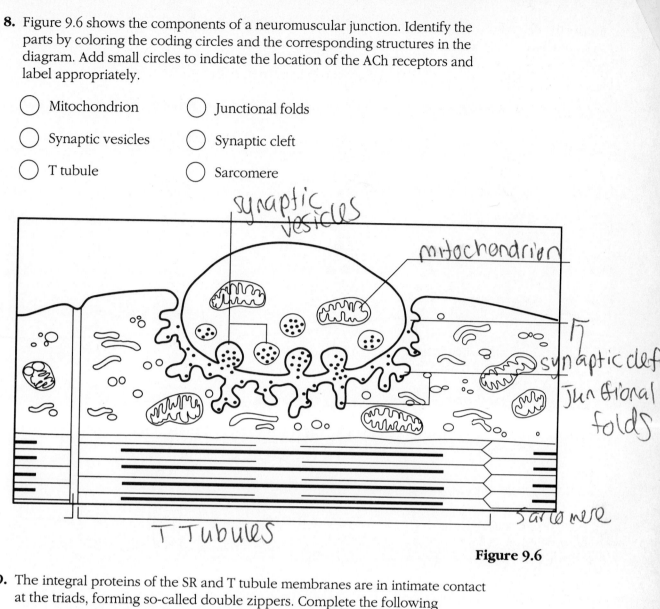

Figure 9.6

9. The integral proteins of the SR and T tubule membranes are in intimate contact at the triads, forming so-called double zippers. Complete the following statements relating to this relationship.

_____ 1.   The name given to the integral proteins of the SR is __(1)__, and these proteins seem to function as __(2)__. The corresponding

_____ 2.   T tubule proteins are believed to function as __(3)__, which change their configuration in response to a(n) __(4)__.

_____ 3.

_____ 4.

10. Figure 9.7 diagrams the elements involved in excitation-contraction coupling.
    Color the coding circles and the corresponding structures.

○ Axonal ending   ○ Tropomyosin   ○ T tubule

○ Actin   ○ Myosin   ○ Sarcolemma

○ Troponin   ○ SR   ○ Mitochondria

○ Synaptic vesicles containing acetylcholine   ○ Calcium ions

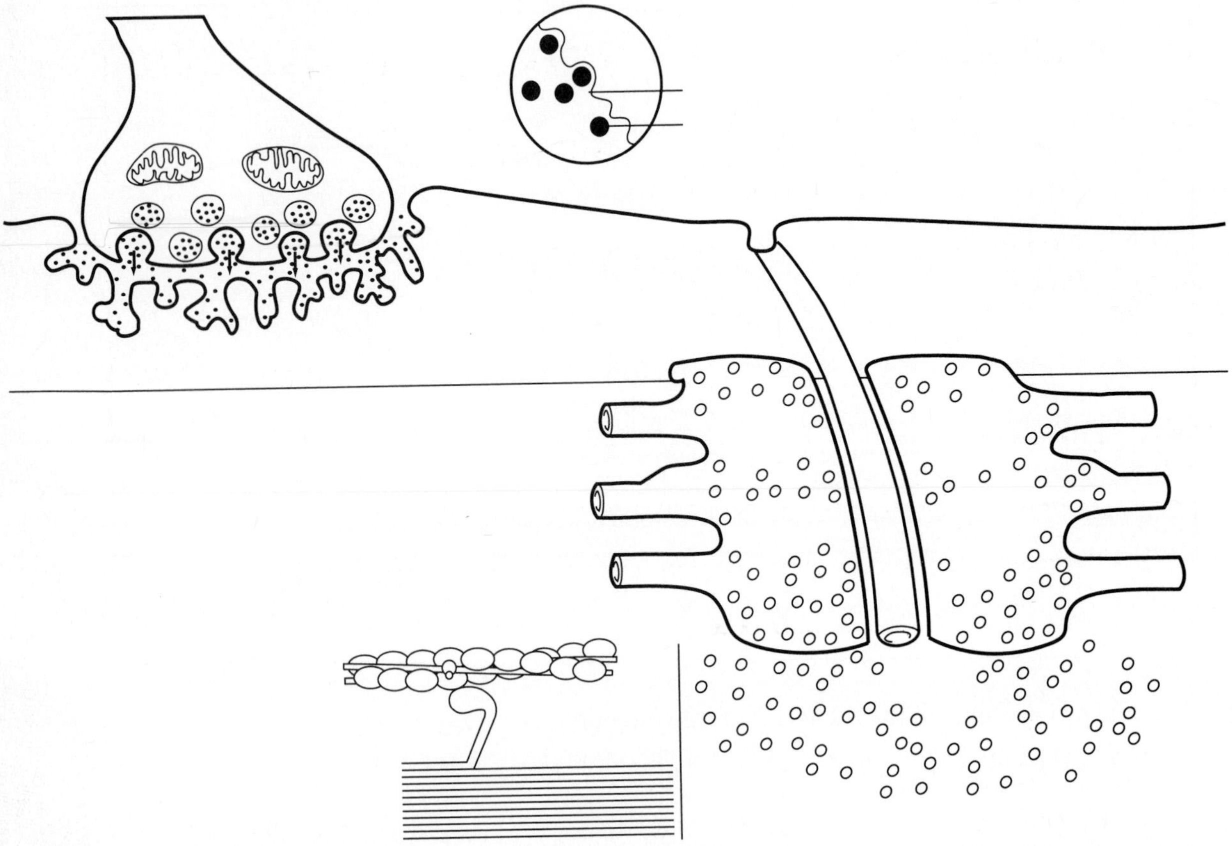

Figure 9.7

11. Number the following statements in their proper sequence to describe
    excitation-contraction coupling in a skeletal muscle cell. The first step has
    already been identified as number 1.

____1____ **1.** Acetylcholine is released by the axonal ending, diffuses to the muscle cell, and attaches
           to ACh receptors on the sarcolemma.

_____ **2.** The action potential, carried deep into the cell via the T tubules, causes the SR to
           release calcium ions.

_____ **3.** AChE breaks down ACh, which separates from its receptors.

_____ **4.** The muscle cell relaxes and lengthens.

_____ **5.** The calcium ion concentration at the myofilaments increases; the myofilaments slide past one another, and the cell shortens.

_____ **6.** Depolarization occurs, and the action potential is generated along the sarcolemma.

_____ **7.** Within 30 ms after the action potential ends, Ca^{2+} concentration at the myofilaments decreases.

12. Define *motor unit*. _a motor neuron + the muscle fibers it stimulates from 2 to 200_____

13. Circle the term in each grouping that does not belong with the others.

1. Treppe Increased enzyme efficiency Multiple motor unit summation

Basis of warmup Increased Ca^{2+}

2. 10–15 seconds High power muscle activity CP + ATP

Relies on aerobic respiration Immediate transfer of PO_4^{2-} to ADP

3. Oxygen Mitochondria $CO_2 + H_2O$ Oxidative phosphorylation

End product is lactic acid Jogging

4. Anaerobic metabolism Lactic acid Glycogen reserves

Fuels extended periods of strenuous activity Energizes 100-m swim

5. Muscle fatigue Increased ADP Increased lactic acid K^+ lost from muscle cells

Psychological factors

6. Series elastic elements stretched Internal tension = external tension

Twitch contraction Prolonged stimulation Increased force exerted

7. Aerobic exercise Increases muscle vascularization Muscle hypertrophy

Cycling More efficient body metabolism Increased cardiovascular efficiency

8. Weight lifting Resistance exercise Increased muscle size

Increased endurance Good muscle definition Nearly immovable loads

9. Disuse atrophy Immobilization Increased muscle strength

Loss of nerve stimulation Increased connective tissue in muscles

14. In the appropriate graph spaces below, draw the indicated myograms. Be sure to include arrows at the bottom to indicate each stimulus. For the twitch myogram, label the latent, contraction, and relaxation periods.

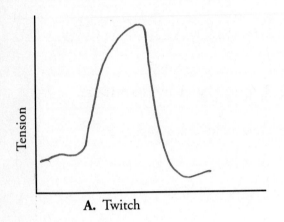

A. Twitch

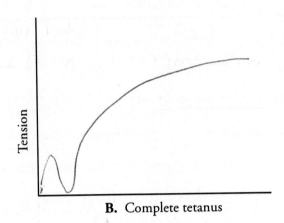

B. Complete tetanus

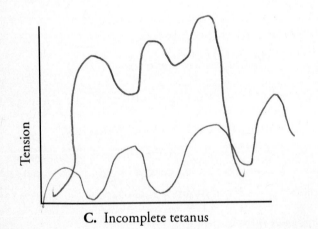

C. Incomplete tetanus

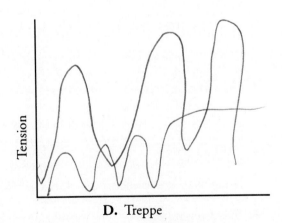

D. Treppe

Figure 9.8

15. Complete the following statements by choosing the correct responses from the key choices and entering the appropriate letters in the answer blanks.

KEY CHOICES

A. Fatigue   **E.** Isometric contraction   **I.** Many motor units

B. Isotonic contraction   **F.** Whole muscle   **J.** Relax

C. Muscle cell   **G.** Tetanus

D. Muscle tone   **H.** Few motor units

_____G_____ 1. __(1)__ is a continuous contraction that shows no evidence of relaxation.

_____B_____ 2. A(n) __(2)__ is a contraction in which the muscle shortens and work is done.

_____I_____ 3. To accomplish a strong contraction, __(3)__ are stimulated at a rapid rate.

_____C_____ 4. The "all-or-none" law applies to skeletal muscle function at the __(4)__ level.

_____H_____ 5. When a weak but smooth muscle contraction is desired, __(5)__ are stimulated at a rapid rate.

_____A_____ 6. When a muscle is being stimulated but is not able to respond, the condition is called __(6)__.

_____E_____ 7. A(n) __(7)__ is a contraction in which the muscle does not shorten, but tension in the muscle keeps increasing.

_____J_____ 8. Wave summation results in stronger contractions at the same stimulus strength because the muscle does not have time to completely __(8)__ between successive stimuli.

_____B_____ 9. __(9)__ includes both concentric and eccentric contractions.

16. Using the key choices, identify the factors requested below.

KEY CHOICES

A. Load   **C.** Series-elastic elements   **E.** Degree of muscle stretch

B. Muscle fiber type   **D.** Number of fibers activated

_____ 1. All factors that influence contractile force

_____ 2. All factors that influence the velocity and duration of muscle contraction

17. Fill in the missing steps in the following flowchart of muscle cell metabolism by writing the appropriate terms in the blanks below it.

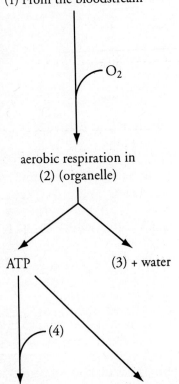

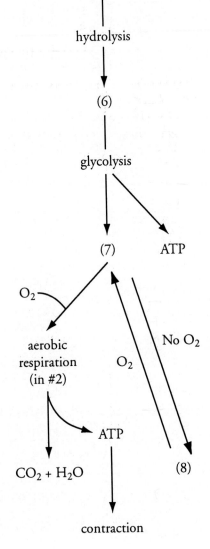

Resting Muscle

(1) From the bloodstream

O_2

aerobic respiration in
(2) (organelle)

ATP

(3) + water

(4)

CP + ADP

other energy
requirements

Contracting Muscle

(5) From muscle reserves

hydrolysis

(6)

glycolysis

(7)   ATP

O_2

aerobic
respiration
(in #2)

O_2

No O_2

$CO_2 + H_2O$

ATP

(8)

contraction

1. _____   5. _____

2. _____   6. _____

3. _____   7. _____

4. _____   8. _____

18. Check the appropriate column in the chart to characterize each type of skeletal muscle fiber.

Characteristics	Slow-twitch fatigue-resistant	Fast-twitch fatigable	Fast-twitch fatigue-resistant
Rapid twitch rate			
Fast myosin ATPases			
Use mostly aerobic metabolism			
Large myoglobin stores			
Large glycogen stores			
Fatigue slowly			
Fibers are white			
Fibers are small			
Fibers contain many capillaries and mitochondria			

19. Explain why muscles contract weakly or not at all when:

1. they are excessively stretched. _____

_____

2. the sarcomeres are strongly contracted. _____

_____

3. excessive levels of lactic acid and ADP have accumulated in the muscle.

_____

_____

20. Which of the following occur within a muscle cell during oxygen debt? Place a check mark (✔) by the correct choices.

✓ **1.** Decreased ATP   _____ **5.** Increased oxygen

_____ **2.** Increased ATP   **6.** Decreased carbon dioxide

✓ **3.** Increased lactic acid   ✓ **7.** Increased carbon dioxide

✓ **4.** Decreased oxygen   _____ **8.** Increased glucose

21. Briefly describe how you can tell you are repaying the oxygen debt. _____

_____

_____

Smooth Muscle

1. Smooth muscle cells differ from skeletal muscle cells in several ways. Indicate which of the following elements seen in skeletal muscle are *absent* in smooth muscle by placing a check mark in the appropriate answer blanks. Circle elements seen in smooth muscle but not in skeletal muscle.

_____ **1.** Dense bodies   _____ **9.** SR

_____ **2.** Diffuse junctions   _____ **10.** Striations

_____ **3.** Gap junctions   _____ **11.** Tropomyosin

_____ **4.** Intermediate filaments   _____ **12.** Troponin

_____ **5.** Elaborate neuromuscular junctions   _____ **13.** T tubules

_____ **6.** Myofibrils   _____ **14.** Varicosities

_____ **7.** Myofilaments   _____ **15.** Z lines

_____ **8.** Sarcomeres

2. Complete the following statements concerning smooth muscle characteristics by inserting the correct terms in the answer blanks.

_____ **1.**

_____ **2.**

_____ **3.**

_____ **4.**

_____ **5.**

_____ **6.**

_____ **7.**

_____ **8.**

_____ **9.**

_____ **10.**

_____ **11.**

_____ **12.**

Whereas skeletal muscle exhibits elaborate connective tissue coverings, smooth muscle has a scant connective tissue __(1)__, which is secreted by the __(2)__ cells. The most common variety of smooth muscle, called __(3)__ smooth muscle, is arranged in opposing layers in the walls of hollow organs. Most often, there are __(4)__ such layers, one running __(5)__ and the other running __(6)__. As these layers alternately contract and relax, substances are squeezed along the internal passages. This phenomenon is called __(7)__. The electrical coupling of the smooth muscle cells by __(8)__ allows smooth muscle to contract as a unit because the __(9)__ spread from cell to cell. Smooth muscle cells, referred to as __(10)__ cells, set the pace for the entire muscle sheet in some cases. Nerve endings called __(11)__ release neurotransmitter into wide synaptic clefts called __(12)__. The function of Z lines in skeletal muscle is assumed by __(13)__ in smooth muscle. Ionic calcium, which triggers contraction, enters the cytosol from both the SR and the __(14)__.

_____ **13.**

_____ **14.**

3. For each of the following descriptions, indicate whether it is more typical of (A) skeletal muscle or (B) smooth muscle (single-unit variety).

_____ **1.** Ionic calcium binds to the thick filaments (perhaps to calmodulin)

_____ **2.** Myosin ATPases are very rapid-acting

_____ **3.** Total length change possible is 150%

_____ **4.** Contraction is slow, seemingly tireless, which saves energy

_____ **5.** Contraction duration is always less

_____ **6.** When stretched, contracts vigorously

_____ **7.** ATP is _routinely_ generated by anaerobic pathways

_____ **8.** Excited by autonomic nerves

_____ **9.** Excited by acetylcholine

_____ **10.** Excited by norepinephrine

_____ **11.** Excited by chemical stimuli, including many hormones

4. Describe the importance of the stress-relaxation response in single-unit

smooth muscle. _____

_____

_____

_____

5. 1. How is multiunit smooth muscle like skeletal muscle? _____

_____

2. How is it different? _____

_____

_____

Developmental Aspects of Muscles

1. Complete the following statements concerning the embryonic development of muscles and their functioning throughout life. Insert your answers in the answer blanks.

_____ 1.

_____ 2.

_____ 3.

_____ 4.

_____ 5.

_____ 6.

_____ 7.

_____ 8.

_____ 9.

_____ 10.

_____ 11.

_____ 12.

_____ 13.

_____ 14.

_____ 15.

_____ 16.

_____ 17.

All muscle tissue derives from cells called **(1)** , which are descendants of embryonic mesoderm. Once skeletal muscle cells have formed, they produce myofilaments and are invaded by fibers of the **(2)** division of the nervous system. Gap junctions form in developing **(3)** muscle but not in **(4)** muscle tissue. Of the three muscle tissue types, only **(5)** maintains its ability to divide throughout life.

A baby's control over muscles progresses in a **(6)** direction as well as in a **(7)** direction. In addition, **(8)** muscular control (that is, waving of the arms) occurs before **(9)** control (pincer grasp) does.

Men are generally stronger than women because of their greater muscle development, due mainly to the influence of **(10)** .

Muscles are amazingly resistant to infection throughout life largely because of their rich **(11)** . However, there are important muscle pathologies. An important congenital muscular disease that results in degeneration of the skeletal muscles by young adulthood is called **(12)** .

Muscles will ordinarily stay healthy if they are **(13)** regularly. However, without normal stimulation they **(14)** , and, with age, our skeletal muscles generally decrease in mass. This leads to a decline in body **(15)** and in muscle **(16)** . Muscle tissue that is lost is replaced by noncontractile **(17)** tissue.

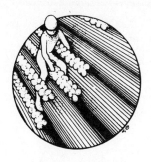

The Incredible Journey:
A Visualization Exercise
for Skeletal Muscle Tissue

As you straddle this structure, you wonder what is happening.

1. Complete the narrative by inserting the missing words in the answer blanks.

_____	1.
_____	2.
_____	3.
_____	4.
_____	5.
_____	6.
_____	7.
_____	8.
_____	9.
_____	10.

On this incredible journey, you will enter a skeletal muscle cell to observe the events that occur during muscle contraction. You prepare yourself by donning a wetsuit and charging your ion detector. Then you climb into a syringe to prepare for injection. Your journey will begin when you see the gleaming connective tissue covering of a single muscle cell, the __(1)__ . Once injected, you monitor your descent through the epidermis and subcutaneous tissue. When you reach the muscle cell surface, you see that it is punctuated with pits at relatively regular intervals. Looking into the distant darkness, you can see a leash of fibers ending close to a number of muscle cells. Since all these fibers must be from the same motor neuron, this functional unit is obviously a __(2)__ . You approach the fiber ending on your muscle cell and scrutinize the __(3)__ junction there. As you examine the junction, minute fluid droplets leave the nerve endings and attach to doughnut-shaped receptors on the muscle cell membrane. This substance released by the nerve ending must be __(4)__ . Then, as a glow falls over the landscape, your ion detector indicates that ions are disappearing from the muscle cell exterior and entering the muscle pits. The needle drops from high to low as the __(5)__ ions enter the pits from the watery fluid outside. You should have expected this, since these ions must enter to depolarize the muscle cells and start the __(6)__ .

Next you begin to explore one of the surface pits. As the muscle jerks into action, you topple deep into the pit. Sparkling electricity lights up the wall on all sides. You grasp for a handhold. Finally successful, you pull yourself laterally into the interior of the muscle cell and walk carefully along what seems to be a log. Then, once again, you notice an eerie glow as your ion detector reports that __(7)__ ions are entering the cytoplasm rapidly. The "log" you are walking on suddenly "comes to life" and begins to slide briskly in one direction. Unable to keep your balance, you fall. As you straddle this structure, you wonder what is happening. On all sides, cylindrical structures—such as the one you are astride—are moving past other similar but larger structures. Suddenly you remember, these are the myofilaments, __(8)__ and __(9)__ , that slide past one another during muscle contraction.

Seconds later, the forward movement ends, and you begin to journey smoothly in the opposite direction. The ion detector now indicates low __(10)__ ion levels. Since you cannot ascend the smooth walls of one of the entry pits, you climb from one myofilament to another to reach the underside of the sarcolemma. Then you travel laterally to enter a pit close to the surface and climb out onto the cell surface. Your journey is completed, and you prepare to leave your host once again.

CHALLENGING YOURSELF

At the Clinic

1. In preparation for tracheal intubation, curare (or a derivative) is given to relax the laryngeal muscles. What specific mechanism does curare inhibit?

2. Gregor, who works at a pesticide factory, comes to the clinic complaining of muscle spasms that interfere with his movement and breathing. A blood test shows that he has become contaminated with organophosphate pesticide. The doctor states that this type of pesticide is an acetylcholinesterase inhibitor. How would you explain to Gregor what this means?

3. Maureen Hamel, a young woman, comes to the clinic with symptoms of muscle weakness. The nurse notes on the record that her eyelids droop and her speech is slurred. What do you suspect is Maureen's problem, and what blood test would you run to support the diagnosis?

4. A young man appears for a routine physical before participating in his high school athletics program. A cursory examination shows striae on his thighs, which are much larger than his calves and arm muscles. What has caused the striae, and what would you suggest as a modification of his workout routine?

5. Fifty-seven-year old Charlie, who admits to a very sedentary lifestyle, has just completed a stress test at the clinic. The test results indicate that he has very little aerobic capacity, and his pulse and blood pressure are at the high end of the normal ranges. Taking into account the effects of aging on bones, joints, and muscles, what exercise program would you recommend for Charlie?

6. A man calls the clinic, worried about painful cramping in his calf he had experienced during the previous night. When asked about his activities that day, he reports having worked in the heat most of the previous day. What would you suggest to help prevent a recurrence?

7. Samantha, a young gymnast, is brought to the clinic with a pulled calf muscle. What treatment will she receive?

8. Mrs. Sanchez says her 6-year-old son seems to be clumsy and tires easily. The doctor notices his calf muscles do not appear atrophied; if anything, they seem enlarged. For what condition must the boy be checked? What is the prognosis?

9. Charlie, our patient from Question 5, has returned, very discouraged about his regimen of a 2-mile daily walk. He reports excruciating pain whenever he tries to walk at more than a "snail's pace." What condition is probably causing his pain? Should he stop his exercise program?

Stop and Think

1. Do all types of muscle produce movement? Maintain posture? Are all three types important in generating heat?

2. Would movement be possible if muscles were not extensible?

3. Chickens carry out only brief bursts of flight and their flying muscles consist of white fibers. The breast muscles of ducks, by contrast, consist of red and intermediate type muscle fibers. What can you deduce about the flying ability of ducks?

4. If you cut across a muscle fiber at the level of the T tubules, what would the cross section look like?

5. Which is a cross bridge attachment more similar to: a precision rowing team or a person pulling a rope and bucket out of a well?

6. How is calcium removed from the sarcomere when relaxation is initiated?

7. When a muscle fiber relaxes, do the thick and thin myofilaments remain in their contracted arrangement or slide back apart? If the latter, what makes them slide? What does this do to the length of the resting sarcomere?

8. All of the membranous structures involved in excitation-contraction coupling have a very large surface area. Explain how this is accomplished for each of these structures: synaptic vesicles, sarcolemma, T tubules, SR. Why is this membrane elaboration important in speeding up the processes involved?

9. What would happen if the refractory period were abnormally prolonged well into the period of relaxation?

10. Asynchronous motor unit summation reduces muscle fatigue. How so? How does this relate to muscle tone?

11. What factors decrease oxygen delivery in very active muscles? How do active muscles offset this effect?

12. Why would paralyzed muscles shorten as they atrophy? Why are physical therapy and braces essential to prevent tightening and contortion at joints?

13. Contrast the roles of calcium in initiating contraction in skeletal muscle and smooth muscle.

14. How does the rudimentary nature of the SR in smooth muscle affect the contraction and relaxation periods of a smooth muscle twitch?

15. Heart transplant operations are becoming ever more common. What controls the rate of cardiac muscle contraction in a heart that has been transplanted?

COVERING ALL YOUR BASES

Multiple Choice

Select the best answer or answers from the choices given.

1. Select the type(s) of muscle tissue that fit the following description: self-excitable, pacemaker cells, gap junctions, extremely extensible, limited SR, calmodulin-activated.
   A. Skeletal muscle   C. Smooth muscle
   B. Cardiac muscle   D. Involuntary muscle

2. Which of the following would be associated with a skeletal muscle fascicle?
   A. Perimysium
   B. Nerve
   C. Artery and at least one vein
   D. Epimysium

3. Skeletal muscle is *not* involved in:
   A. movement of skin
   B. propulsion of a substance through a body tube
   C. heat production
   D. inhibition of body movement

4. Which of the following are part of a thin myofilament?
   A. ATP-binding site   C. Globular actin
   B. TnI   D. Calcium

5. Factors involved in calcium release and uptake during contraction and relaxation include:
   A. calsequestrin   D. terminal cisternae
   B. T tubules   E. sarcolemma
   C. SR

6. In comparing electron micrographs of a relaxed skeletal muscle fiber and a fully contracted muscle fiber, which would be seen only in the *relaxed* fiber?
   A. Z lines   D. A bands
   B. Triads   E. H zones
   C. I bands

7. After ACh attaches to its receptors at the neuromuscular junction, the next step is:
   A. potassium-gated channels open
   B. calcium binds to troponin
   C. the T tubules depolarize
   D. cross bridges attach
   E. ATP is hydrolyzed

8. Detachment of the cross bridges is directly triggered by:
   A. hydrolysis of ATP
   B. repolarization of the T tubules
   C. the power stroke
   D. attachment of ATP to myosin heads

9. Transmission of the stimulus at the neuromuscular junction involves:
   A. synaptic vesicles   C. ACh
   B. TnT   D. motor end plate

10. Which of the following locations are important sites for calcium activity?
    A. SR
    B. T tubules
    C. Axonal ending
    D. Junctional folds of sarcolemma

11. Acetylcholinesterase hydrolyzes ACh during which phase(s) of a muscle twitch?
- **A.** Latent period
- **B.** Period of contraction
- **C.** Period of relaxation
- **D.** Refractory period

12.–13. Use the following graph to answer Questions 12 and 13:

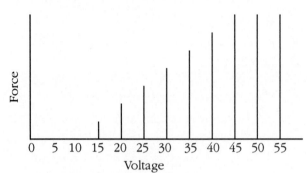

Voltage
(The muscle is stimulated once at each voltage indicated)

12. What is the threshold stimulus for this muscle?
- **A.** 5 volts
- **B.** 10 volts
- **C.** 15 volts
- **D.** 40 volts
- **E.** 45 volts

13. What is the maximal stimulus for this muscle?
- **A.** 15 volts
- **B.** 40 volts
- **C.** 45 volts
- **D.** 55 volts

14. Small motor units are associated with:
- **A.** precision
- **B.** power
- **C.** fine control
- **D.** weak contraction

15. Holding up the corner of a heavy couch to vacuum under it involves which type(s) of contraction?
- **A.** Tetanic
- **B.** Isotonic
- **C.** Isometric
- **D.** Tone

16. Your ability to lift that heavy couch would be increased by which type of exercise?
- **A.** Aerobic
- **B.** Endurance
- **C.** Resistance
- **D.** Swimming

17. Which of the following activities depends most on anaerobic metabolism?
- **A.** Jogging
- **B.** Swimming a race
- **C.** Sprinting
- **D.** Running a marathon

18. The first energy source used to regenerate ATP when muscles are extremely active is:
- **A.** fatty acids
- **B.** glucose
- **C.** creatine phosphate
- **D.** pyruvic acid

19. ATP directly powers:
- **A.** exocytosis of synaptic vesicles
- **B.** sodium influx at the sarcolemma
- **C.** calcium release into the sarcoplasm
- **D.** recocking of the myosin head

20. Which statement is true about contractures?
- **A.** Depletion of ATP is absolute.
- **B.** Contraction is weak.
- **C.** Muscle cramping is due to depletion of electrolytes.
- **D.** The neuromuscular junction has been affected.

21. When paying back the oxygen debt:
- **A.** lactic acid is formed
- **B.** lactic acid is reconverted to pyruvic acid
- **C.** ATP formation requires creatine phosphate
- **D.** muscle cells utilize glycogen reserves

22. Factors affecting force of contraction include:
- **A.** number of motor units stimulated
- **B.** cross-sectional area of the muscle/muscle fibers
- **C.** stretching of series-elastic elements
- **D.** degree of muscle stretch prior to the contraction

23. The ideal length-tension relationship is:
    A. maximal overlap of thin and thick myofilaments
    B. minimal overlap of thin and thick myofilaments
    C. no overlap of thin and thick myofilaments
    D. intermediate degree of overlap of thin and thick myofilaments

24. When the velocity of contraction is zero, the contraction is:
    A. strongest   C. greater than the load
    B. equal to the load D. isometric

25. Which of these bands or lines does not narrow when a skeletal muscle fiber contracts?
    A. H zone   C. I band
    B. A band   D. M line

26. Characteristics of red, but not white, muscle are:
    A. abundant mitochondria
    B. richly vascularized
    C. fast contraction
    D. slow to fatigue

27. Which of the following is (are) unique to smooth muscle cells, compared to the other types of muscle cells?
    A. Production of endomysium
    B. Utilize calmodulin
    C. Can contract even when maximally stretched
    D. Self-excitable

28. Smooth muscle contraction is stimulated by:
    A. hormones
    B. inhibition of calmodulin
    C. neurotransmitters
    D. stretching

29. Which of the following is (are) true of single-unit but not multiunit smooth muscle?
    A. Presence of gap junctions
    B. Each muscle fiber has its own nerve ending
    C. Sheetlike arrangement in hollow organs
    D. Contracts in a corkscrewlike manner

30. The ancestry of muscle cells comprises:
    A. myoblasts
    B. satellite cells
    C. mesenchyme
    D. mesoderm

31. When a muscle is stimulated once (a twitch):
    A. the internal tension exceeds the external tension
    B. the internal and external tensions are equal
    C. the external tension exceeds the internal tension
    D. the series-elastic elements do not have sufficient time to stretch fully before the end of the period of contraction

32. What fiber types would be most useful in the leg muscles of a long-distance runner?
    A. White fast-twitch
    B. White slow-twitch
    C. Intermediate fast-twitch
    D. Red slow-twitch

33. Relative to differences between eccentric and concentric contractions, eccentric contractions:
    A. put the body in a position to contract concentrically
    B. occur as the muscle is actually increasing in length
    C. are more likely to cause muscle soreness the day after a strenuous activity
    D. involve unchanging muscle length and tension

Word Dissection

For each of the following word roots, fill in the literal meaning and give an example, using a word found in this chapter.

Word root	Translation	Example
1. fasci	_____	_____
2. lemma	_____	_____
3. metric	_____	_____
4. myo	_____	_____
5. penna	_____	_____
6. raph	_____	_____
7. sarco	_____	_____
8. stalsis	_____	_____
9. stria	_____	_____
10. synap	_____	_____
11. tetan	_____	_____

10 The Muscular System

Student Objectives

When you have completed the exercises in this chapter, you will have accomplished the following objectives:

Muscle Mechanics: The Importance of Leverage and Fascicle Arrangement

1. Name the three types of lever systems and indicate the arrangement of elements (effort, fulcrum, and load) in each. Also note the advantages of each type of lever system.

2. Name the common patterns of fascicle arrangement and relate these to power generation by muscles.

Interactions of Skeletal Muscles in the Body

3. Explain the function of prime movers, antagonists, synergists, and fixators, and describe how each promotes normal muscular function.

Naming Skeletal Muscles

4. List and define the criteria used in naming muscles. Provide an example to illustrate the use of each criterion.

Major Skeletal Muscles of the Body

5. Name and identify (on an appropriate diagram or torso model) each of the muscles described in Tables 10.1–10.15. State the origin and insertion for each and describe the action of each.

Ready, set, go!

All contractile tissues—smooth, cardiac, and skeletal muscle— are muscles, but the *muscular system* is concerned only with skeletal muscles, the organs composed of skeletal muscle tissue, and connective tissue wrappings. The skeletal muscles clothe the bony skeleton and make up about 40% of body mass. Skeletal muscles, also called voluntary muscles, help shape the body and enable you to smile, run, shake hands, grasp things, and otherwise manipulate your environment.

The topics of Chapter 10 include principles of leverage, classification of muscles on the basis of their major function, identification of voluntary muscles, and a description of their important actions and interactions.

BUILDING THE FRAMEWORK

Muscle Mechanics: The Importance of Leverage and Fascicle Arrangement

1. Complete the following short paragraph, which considers the operation of lever systems in the body. Insert your responses in the answer blanks.

_____ 1.

_____ 2.

_____ 3.

_____ 4.

In the body, lever systems involve the interaction of __(1)__, which provide the force, and __(2)__, which provide the lever arms. The force is exerted where __(3)__ . The site at which movement occurs in each case is the intervening __(4)__, which acts as the fulcrum.

2. Circle all of the elements that characterize a lever that operates at a mechanical disadvantage:

Fast   Moves a large load over a small distance   Requires maximal shortening

Slow   Moves a small load over a large distance   Requires minimal shortening

Muscle force greater than the load   Muscle force less than load moved

3. Figure 10.1 illustrates the elements of three different lever systems on the left and provides examples of each lever system from the body on the right. Select different colors for the lever elements listed above the figure and use them to color the diagrams of each lever system. Complete this exercise by following the steps below for each muscle diagram.

   1. Add a red arrow to indicate the point at which the effort is exerted.

   2. Draw a circle enclosing the fulcrum.

   3. Indicate whether the lever shown is operating at a mechanical advantage or disadvantage by filling in the line below the diagram.

   4. Color the load yellow.

   ◯ Effort   ◯ Fulcrum   ◯ Load (resistance)

A. First-class lever

B. Second-class lever

C. Third-class lever

Figure 10.1

4. The brachialis is a short stout muscle inserting close to the fulcrum of the elbow joint; the brachioradialis is a slender muscle inserting distally, well away from the fulcrum of the elbow joint. Which muscle is associated with mechanical advantage and which with mechanical disadvantage?

5. Four fascicle arrangements are shown in Figure 10.2; these same arrangements are described below. First, correctly identify each fascicle arrangement by (1) writing the appropriate term in the answer blank and (2) writing the letter of the corresponding diagram in the answer blank. Second, on the remaining lines beside the diagrams, name two muscles that have the fascicle arrangement shown.

_____ **1.** Fascicles insert into a midline tendon from both sides.

_____ **2.** Muscle fibers are arranged in concentric array around an opening.

_____ **3.** Fascicles run with the long axis of the muscle.

_____ **4.** Fascicles angle from a broad origin to a narrow insertion.

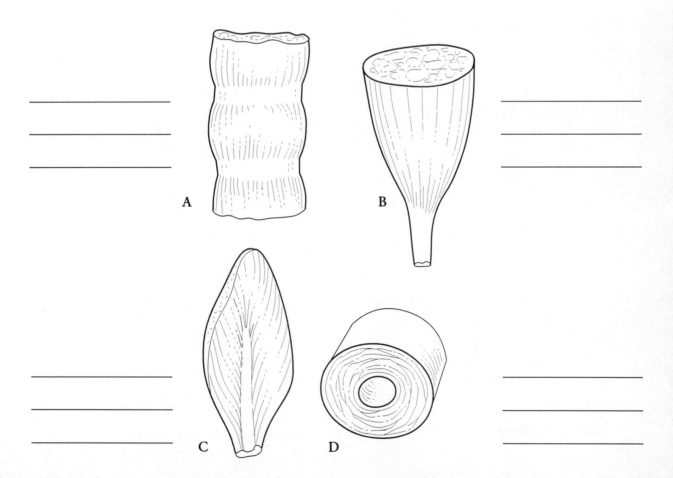

Figure 10.2

Interactions of Skeletal Muscles in the Body

1. The terms in the key choices are often applied to the manner in which muscles interact with other muscles. Select the terms that apply to the following definitions and insert the correct answers in the answer blanks.

KEY CHOICES

A. Antagonist   B. Fixator   C. Prime mover   D. Synergist

_____ 1. Holds parts of the body in proper position for the action of other muscles, primarily postural muscles

_____ 2. Bears the major responsibility for producing a particular movement

_____ 3. Acts to reverse or act against the action of another muscle

_____ 4. Assists an agonist by a like contraction movement or by holding a joint over which an agonist acts immobile

Naming Skeletal Muscles

1. From the key choices, select the criteria that are used in naming the muscles listed below. In some cases, more than one criterion applies.

KEY CHOICES

A. Action   E. Number of origins

B. Direction of fibers   F. Relative muscle size

C. Location of the muscle   G. Shape of the muscle

D. Location of origin and insertion

_____ 1. Adductor brevis   _____ 6. Levator scapulae

_____ 2. Brachioradialis   _____ 7. Orbicularis oculi

_____ 3. Deltoid   _____ 8. Pectoralis major

_____ 4. Extensor digitorum longus   _____ 9. Quadriceps femoris

_____ 5. Internal oblique   _____ 10. Sternohyoid

Major Skeletal Muscles of the Body

1. Match the names of the suprahyoid and infrahyoid muscles listed in Column B with their descriptions in Column A. (Note that not all muscles in Column B are described.) Also, color the coding circles and the corresponding muscles on Figure 10.3 as you like, but color the hyoid bone yellow and the thyroid cartilage blue.

Column A

○ _____   **1.** Most medial muscle of the neck; thin; depresses larynx

○ _____   **2.** Slender muscle that parallels the posterior belly of the digastric muscle; elevates and retracts hyoid bone

○ _____   **3.** Consists of two bellies united by an intermediate tendon; prime mover to open the mouth; also depresses mandible

○ _____   **4.** Lateral and deep to sternohyoid; pulls thyroid cartilage inferiorly

○ _____   **5.** Flat triangular muscle deep to digastric; elevates hyoid, enabling tongue to force food bolus into the pharynx

_____   **6.** Narrow muscle in the midline running from chin to hyoid; widens pharynx for receiving food as it elevates the hyoid bone

○ _____   **7.** Superior continuation of the sternothyroid

Column B

A. Digastric

B. Geniohyoid

C. Mylohyoid

D. Omohyoid

E. Sternohyoid

F. Sternothyroid

G. Stylohyoid

H. Thyrohyoid

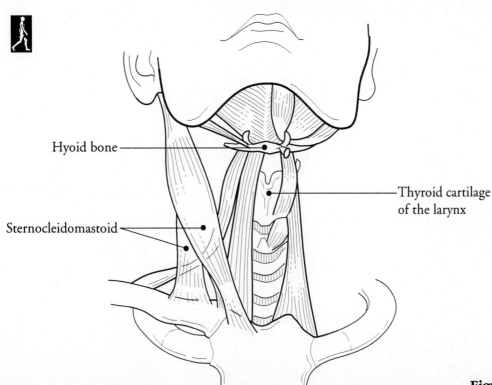

Hyoid bone

Thyroid cartilage of the larynx

Sternocleidomastoid

Figure 10.3

2. Match the muscle names in Column B to the facial muscles described in Column A.

Column A

_____ **1.** Squints the eyes

_____ **2.** Furrows the forehead horizontally

_____ **3.** Smiling muscle

_____ **4.** Puckers the lips

_____ **5.** Draws the corners of the lips laterally and downward

_____ **6.** Pulls the scalp posteriorly

_____ **7.** Tensed during shaving of the chin and neck

Column B

A. Corrugator supercilii

B. Depressor anguli oris

C. Frontalis

D. Occipitalis

E. Orbicularis oculi

F. Orbicularis oris

G. Platysma

H. Zygomaticus

3. Relative to muscles of the head, name the major muscles described here. Select a different color for each muscle and color the coding circles and corresponding muscles on Figure 10.4. Notice that the last muscle to be identified lacks a coding circle.

◯ _____ **1.** Used to show you're happy

◯ _____ **2.** Compresses the cheek; holds food between the teeth

◯ _____ **3.** Used in winking

◯ _____ **4.** Used to raise your eyebrows

◯ _____ **5.** The "kissing" muscle

◯ _____ **6.** Prime mover of jaw closure

◯ _____ **7.** Synergist muscle for jaw closure; elevates and retracts the mandible

◯ _____ **8.** Posterior neck muscle, called the "bandage" muscle

◯ _____ **9.** Prime mover of head flexion, a two-headed neck muscle

_____ **10.** Protrudes the mandible; side-to-side grinding movements

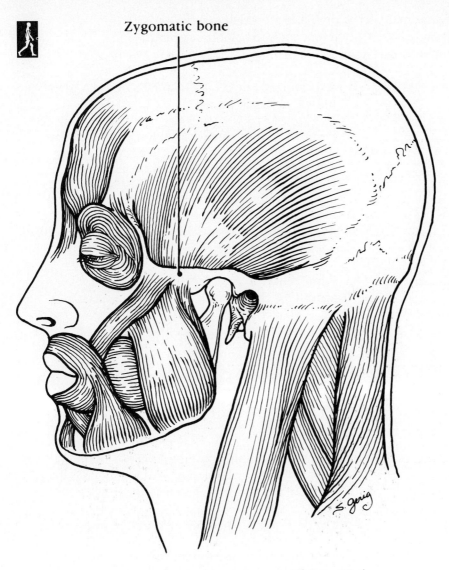

Zygomatic bone

Figure 10.4

4. Describe briefly the function of the pharyngeal constrictor muscles.

_____

_____

5. Name the anterior trunk muscles described below. Then, for each muscle name that has a color-coding circle, select a different color to color the coding circle and the corresponding muscle on Figure 10.5.

_____   **1.** A major spine flexor; the name means "straight muscle of the abdomen"

◯ _____   **2.** Prime mover for shoulder flexion and adduction

◯ _____   **3.** Prime mover for shoulder abduction

_____   **4.** Forms the external lateral walls of the abdomen

◯ _____   **5.** Acting alone, each muscle of this pair turns the head toward the opposite shoulder

◯ _____   **6.** Prime mover to protract and hold the scapula against the thorax wall; the "boxer's muscle"

◯ _____   **7.** Four muscle pairs that together form the so-called abdominal girdle

◯ _____

◯ _____

◯ _____

_____   **8.** A tendinous "seam" running from the sternum to the pubic symphysis that indicates the midline point of fusion of the abdominal muscle sheaths

_____   **9.** The deepest muscle of the abdominal wall

_____   **10.** Deep muscles of the thorax that promote the inspiratory phase of breathing

_____   **11.** An unpaired muscle that acts in concert with the muscles named immediately above to accomplish inspiration

_____   **12.** A flat, thoracic muscle deep to the pectoralis major that acts to draw the scapula inferiorly or to elevate the rib cage

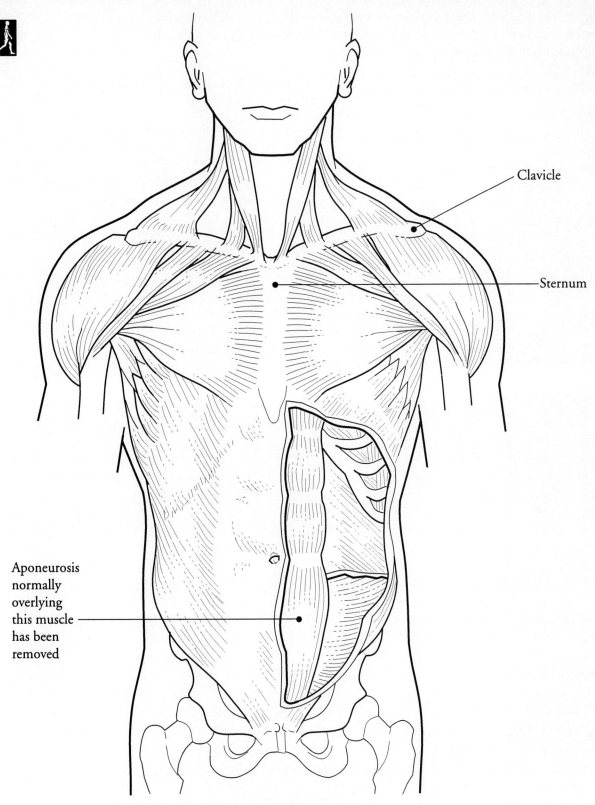

Clavicle

Sternum

Aponeurosis
normally
overlying
this muscle
has been
removed

Figure 10.5

6. Name the posterior trunk muscles described below. Select a different color for each muscle with a color-coding circle and color the coding circles and the corresponding muscles on Figure 10.6.

◯ _____   **1.** A muscle that enables you to shrug your shoulders or extend your head

◯ _____   **2.** A muscle that adducts the shoulder and causes extension of the shoulder joint

◯ _____   **3.** The shoulder muscle that is the antagonist of the muscle described in Question 2 above

_____   **4.** Prime mover of back extension; a composite muscle consisting of three columns

_____   **5.** A fleshy muscle forming part of the posterior abdominal wall that helps to maintain upright posture

◯ _____   **6.** Acting individually, small rectangular muscles that rotate the glenoid cavity of the scapulae inferiorly

_____   **7.** Synergist of the trapezius in scapular elevation; act to flex the head to the same side

◯ _____   **8.** Synergist of latissimus dorsi in extension and adduction of the humerus

◯ _____   **9.** A rotator cuff muscle; prevents downward dislocation of the humerus

◯ _____   **10.** A rotator cuff muscle; rotates the humerus laterally

◯ _____   **11.** A rotator cuff muscle; lies immediately inferior to the infraspinatus

7. Several muscles that act to move and/or stabilize the scapula are listed in the key choices. Match them to the appropriate descriptions below.

KEY CHOICES

A. Levator scapulae   **C.** Serratus anterior

B. Rhomboids   **D.** Trapezius

_____   **1.** Muscle that holds the scapula tightly against the thorax wall

_____   **2.** Kite-shaped muscle pair that elevates, stabilizes, and depresses the scapulae

_____   **3.** Small rectangular muscles that square the shoulders as they act together to retract the scapula

_____   **4.** Small muscle pair that elevates the scapulae

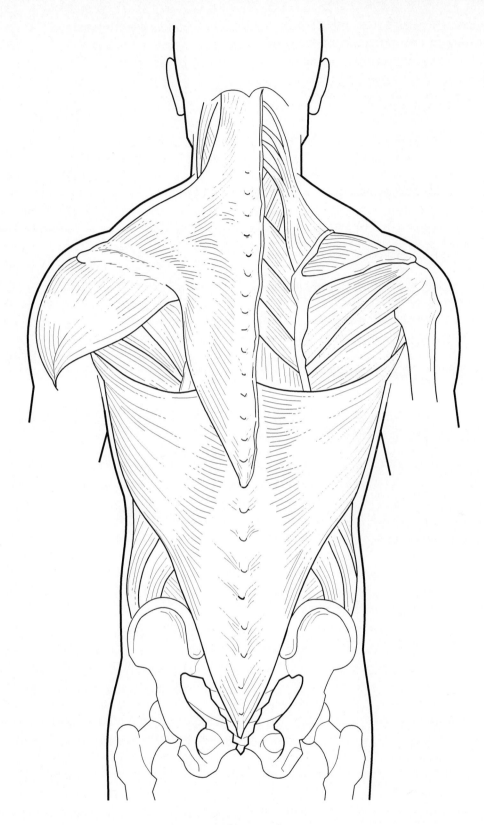

Figure 10.6

8. The muscles described in Column A comprise the pelvic floor and perineum and help support the abdominopelvic organs. Match the muscle names in Column B with the appropriate descriptions in Column A by filling in the answer blanks with the correct letters.

Column A   Column B

_____ _____ **1.** Two paired muscles forming the bulk of the pelvic diaphragm   **A.** Bulbospongiosus

B. Coccygeus

_____ _____ **2.** Muscles that form the urogenital diaphragm

C. Deep transverse perineus

_____ _____ _____ **3.** Muscles forming the superficial space

_____ **4.** A muscle that constricts the urethra   **D.** Ischiocavernosus

_____ **5.** Empties the urethra; assists in penile erection   **E.** Levator ani

_____ **6.** The most important muscle pair in supporting the pelvic viscera; forms sphincters at the anorectal junction   **F.** Sphincter urethrae

G. Superficial transverse perineus
_____ **7.** Retards venous drainage and helps maintain penile erection

9. Identify the neck or vertebral column muscles described below by selecting answers from the key choices. Put each answer in the appropriate answer blank, then color the coding circles and the corresponding muscles on Figure 10.7.

KEY CHOICES

A. Erector spinae   **C.** Scalenes   **E.** Splenius

B. Quadratus lumborum   **D.** Semispinalis   **F.** Sternocleidomastoid

_____ **1.** Elevates the first two ribs

◯ _____ **2.** Prime mover of back extension; consists of three muscle columns (iliocostalis, longissimus, and spinalis)

◯ _____ **3.** One flexes the vertebral column laterally; the pair extends the lumbar spine and fixes the 12th rib

◯ _____ **4.** Extends the vertebral column and head and rotates them to the opposite side

_____ **5.** Prime mover of head flexion; spasms of one causes torticolis

◯ _____ **6.** Acting together, the pair extends the head; one rotates the head and bends it laterally

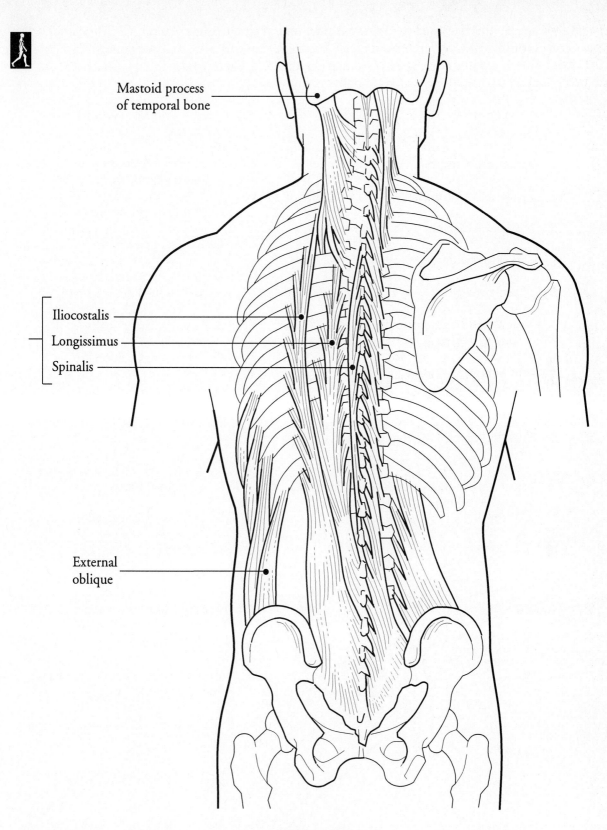

Mastoid process
of temporal bone

Iliocostalis

Longissimus

Spinalis

External
oblique

Figure 10.7

10. Match the forearm muscles listed in Column B with their major action in Column A. Then choose colors and color the muscles provided with coding circles on Figure 10.8. Also identify (by labeling) —two muscles that act on the elbow rather than the wrist and hand—the biceps brachii and the brachioradialis—and the thenar and lumbrical muscles of the hand. All of these muscles have leader lines.

Column A

_____ **1.** Flexes the wrist and middle phalanges

_____ **2.** Pronate the forearm

_____ **3.** Flexes the distal interphalangeal joints

_____ _____ **4.** Powerful wrist flexors

_____ **5.** Thumb flexor

_____ _____ **6.** Extend and abduct the wrist (three muscles)

_____ **7.** Prime mover of finger extension

_____ **8.** Extend the thumb

_____ **9.** Puts the forearm and hand in the anatomical position

_____ **10.** Small muscle that provides a guide to locate the median nerve at the wrist

Column B

◯ **A.** Extensor carpi radialis longus (and brevis)

B. Extensor carpi ulnaris

C. Extensor digitorum

D. Extensor pollicis longus and brevis

◯ **E.** Flexor carpi radialis

◯ **F.** Flexor carpi ulnaris

G. Flexor digitorum profundus

◯ **H.** Flexor digitorum superficialis

◯ **I.** Flexor pollicis longus

◯ **J.** Palmaris longus

◯ **K.** Pronators teres and quadratus

L. Supinator

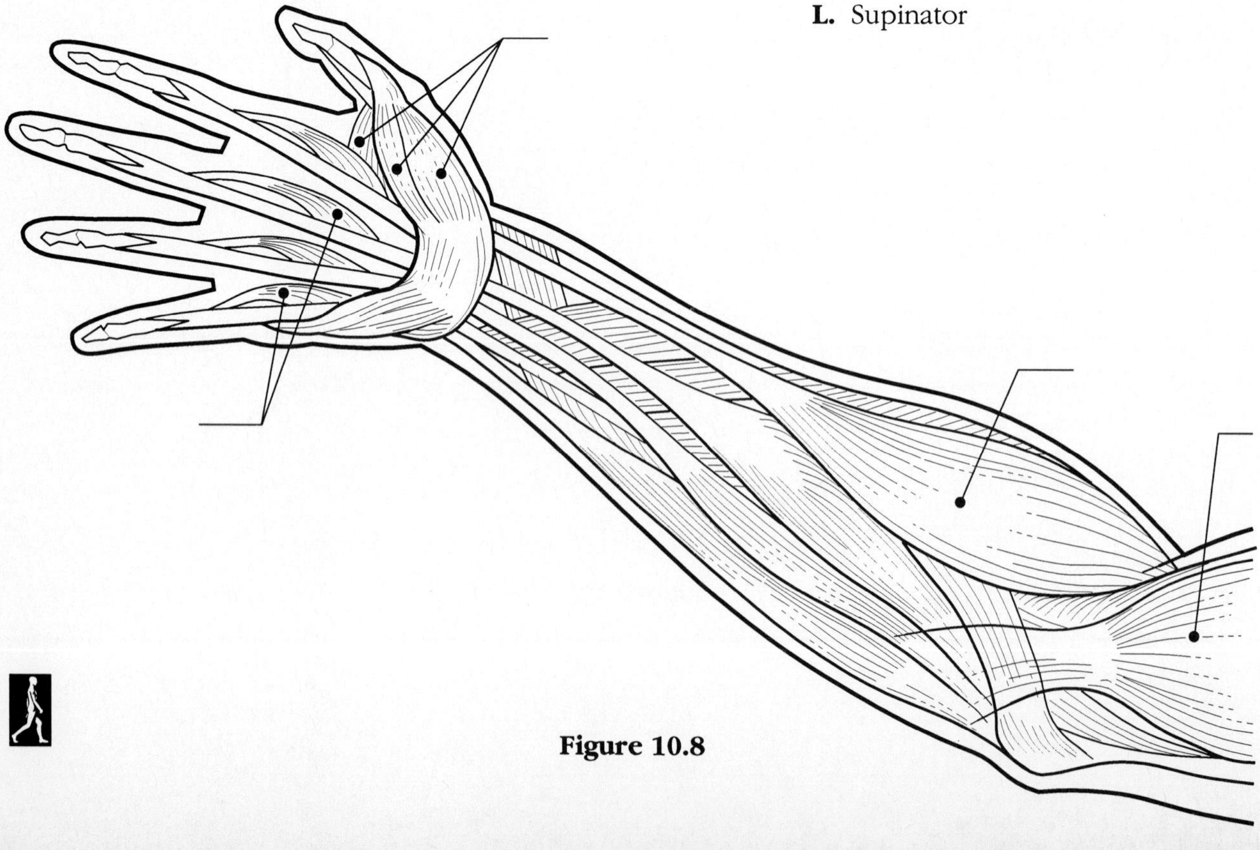

Figure 10.8

11. Name the upper limb muscles described below. Then select a different color for each muscle that has a color-coding circle and color the muscles on Figure 10.9.

1. Wrist flexor that follows the ulna

   ○ _____

2. The muscle that extends the fingers

   ○ _____

3. Two elbow flexor muscles: the first also supinates;

   ○ _____

4. the second is a synergist at best

   ○ _____

5. The muscle that extends the elbow: prime mover

   ○ _____

6. A short muscle; synergist of triceps brachii

   ○ _____

7. A stocky muscle, deep to the biceps brachii; a prime mover of elbow flexion

   _____

8.–9. Two wrist extensors that follow the radius

   ○ _____

   ○ _____

10. The muscle that abducts the thumb

   ○ _____

11. The muscle that extends the thumb

   ○ _____

12. A powerful shoulder abductor; used to raise the arm overhead

   ○ _____

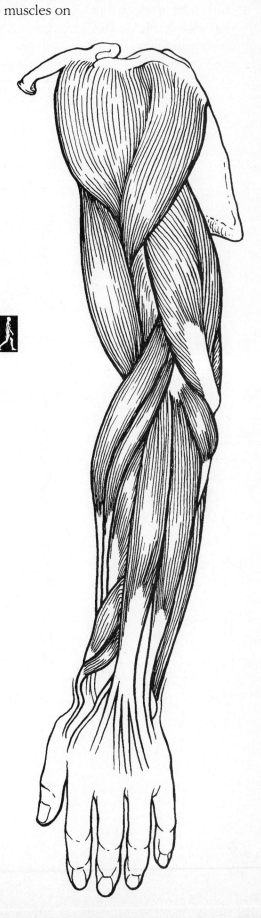

Figure 10.9

12. Name the muscles of the lower limb described below. Select a different color for each muscle that has a color-coding circle and color the circles and the muscles on Figure 10.10. Complete the illustration by labeling those muscles with leader lines.

_____ **1.** Strong hip flexor, deep in pelvis; a composite of two muscles

○ _____ **2.** Power extensor of the hip; forms most of buttock mass

○ _____ **3.** Prime mover of plantar flexion; a composite of two muscles

○ _____ **4.** Inverts and dorsiflexes the foot

○ _____ **5.** The group that enables you to draw your legs to the midline of your body, as when standing at attention

○ _____ **6.** The muscle group that extends the knee

○ _____ **7.** The muscle group that extends the thigh and flexes the knee

○ _____ **8.** The smaller hip muscle commonly used as an injection site

○ _____ **9.** The thin superficial muscle of the medial thigh

○ _____ **10.** A muscle enclosed within fascia that blends into the iliotibial tract; a synergist of the iliopsoas

○ _____ **11.** Dorsiflexes and everts the foot; prime mover of toe extension

○ _____ **12.** A superficial muscle of the lateral leg; plantar flexes and everts foot

_____ **13.** A muscle deep to the soleus; prime mover of foot inversion; stabilizes the medial longitudinal arch of the foot and plantar flexes the ankle

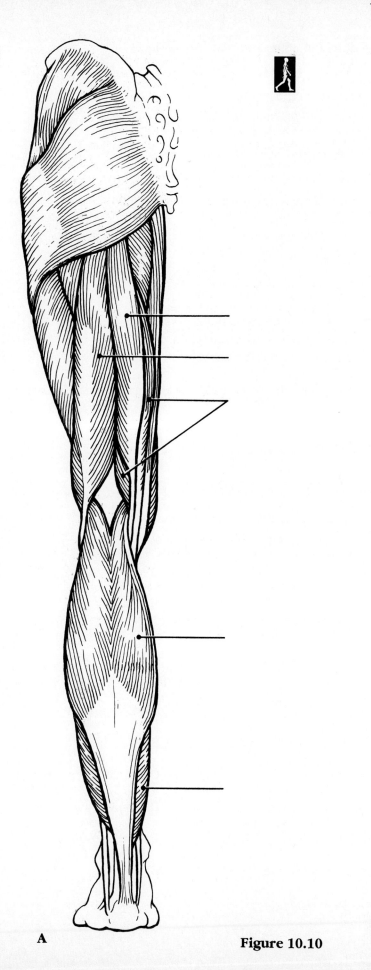

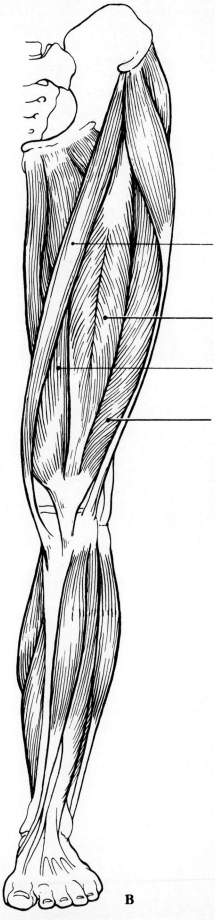

A

Figure 10.10

B

13. Complete the following statements describing muscles. Write the correct answers in the answer blanks.

_____ 1.

_____ 2.

_____ 3.

_____ 4.

_____ 5.

_____ 6.

_____ 7.

_____ 8.

_____ 9.

_____ 10.

_____ 11.

1. Three muscles, __(1)__, __(2)__, and __(3)__, are commonly used for intramuscular injections.

The insertion tendon of the __(4)__ group contains a large sesamoid bone, the patella.

The triceps surae insert in common into the __(5)__ tendon.

The bulk of the tissue of a muscle tends to lie __(6)__ to the part of the body it causes to move.

The extrinsic muscles of the hand originate on the __(7)__.

Most flexor muscles are located on the __(8)__ aspect of the body; most extensors are located __(9)__. An exception to this generalization is the extensor-flexor musculature of the __(10)__.

The pectoralis major and deltoid muscles act synergistically to __(11)__ the arm.

14. When kicking a football, at least three major actions of the lower limb are involved. Name the major muscles (or muscle groups) responsible for the following:

1. Flexing the hip joint: _____

2. Extending the knee: _____

3. Dorsiflexing the foot: _____

_____

15. Now that you have begun your study of the skeletal muscles of the body, use what you have learned to match a specific muscle to its fascicle arrangement:

_____ 1. Deltoid   **A.** Convergent

_____ 2. Rectus femoris   **B.** Fusiform

_____ 3. Gracilis   **C.** Circular

_____ 4. Gluteus maximus   **D.** Bipennate

_____ 5. Vastus medialis   **E.** Multipennate

_____ 6. Sphincter urethrae   **F.** Parallel

_____ 7. Biceps brachii   **G.** Unipennate

16. Circle the term that does not belong in each of the following groupings.

1. Vastus lateralis   Vastus medialis   Knee extension   Biceps femoris

2. Latissimus dorsi   Pectoralis major   Synergists   Adduction of shoulder

   Antagonists

3. Gluteus minimus   Lateral rotation   Gluteus maximus   Piriformis

   Quadratus femoris

4. Adductor magnus   Vastus medialis   Rectus femoris   Origin on os coxa

   Iliacus

5. Lateral rotation   Teres minor   Teres major   Supraspinatus   Infraspinatus

6. Tibialis posterior   Flexor digitorum longus   Peroneus longus   Foot inversion

7. Supraspinatus   Rotator cuff   Teres major   Teres minor   Subscapularis

17. Identify the numbered muscles in Figure 10.11. Match each number with one of the following muscle names. Then select a different color for each muscle that has a color-coding circle and color each muscle group on Figure 10.11.

○ _____ **1.** Orbicularis oris

○ _____ **2.** Pectoralis major

○ _____ **3.** External oblique

○ _____ **4.** Sternocleidomastoid

○ _____ **5.** Biceps brachii

○ _____ **6.** Deltoid

○ _____ **7.** Vastus lateralis

○ _____ **8.** Frontalis

○ _____ **9.** Rectus femoris

○ _____ **10.** Sartorius

○ _____ **11.** Gracilis

○ _____ **12.** Adductor group

○ _____ **13.** Peroneus longus

○ _____ **14.** Temporalis

○ _____ **15.** Orbicularis oculi

○ _____ **16.** Zygomaticus

○ _____ **17.** Masseter

○ _____ **18.** Vastus medialis

○ _____ **19.** Tibialis anterior

○ _____ **20.** Transversus abdominus

○ _____ **21.** Tensor fascia lata

○ _____ **22.** Rectus abdominis

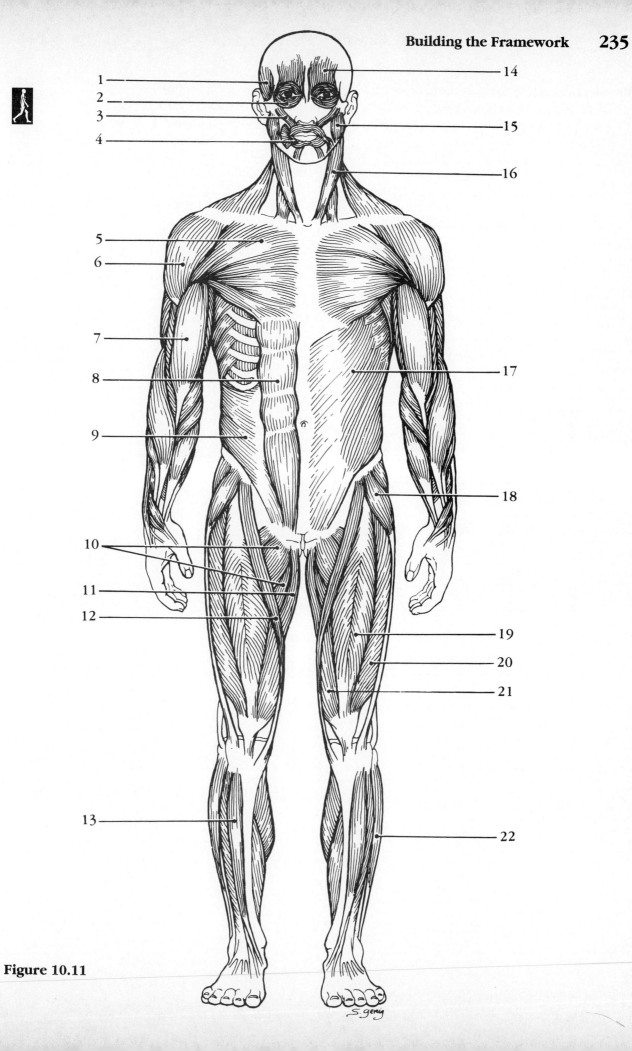

1

2

3

4

5

6

7

8

9

10

11

12

13

14

15

16

17

18

19

20

21

22

S. gerig

Figure 10.11

18. Identify each of the numbered muscles in Figure 10.12. Match each number with one of the following muscle names. Then select different colors for each muscle and color the coding circles and corresponding muscles on Figure 10.12.

○ _____ **1.** Gluteus maximus

○ _____ **2.** Adductor muscles

○ _____ **3.** Gastrocnemius

○ _____ **4.** Latissimus dorsi

○ _____ **5.** Deltoid

○ _____ **6.** Semitendinosus

○ _____ **7.** Trapezius

○ _____ **8.** Biceps femoris

○ _____ **9.** Triceps brachii

○ _____ **10.** External oblique

○ _____ **11.** Gluteus medius

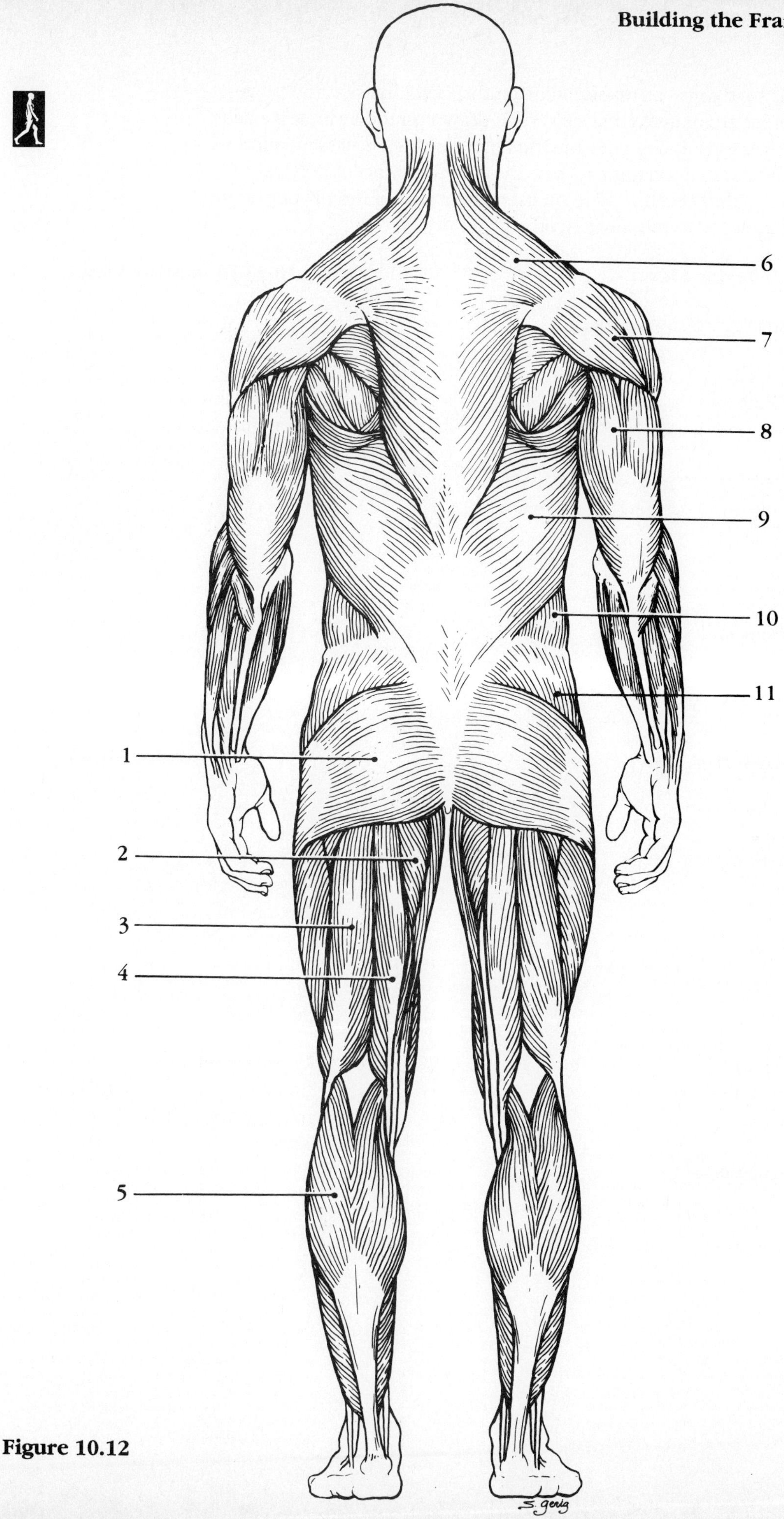

6

7

8

9

10

11

1

2

3

4

5

Figure 10.12

S. gerig

19. On the following two pages are skeleton diagrams, both anterior and posterior views. Listed below are muscles that should be drawn on the indicated sides of the skeleton diagrams. Using your muscle charts and, if necessary, referring back to bone landmarks in Chapter 7, draw the muscles as accurately as possible. Be sure to draw each muscle on the correct side of the proper figure; the muscles are grouped to minimize overlap.

Figure 10.13 (Anterior View)

Right side of body

1. Orbicularis oculi
2. Sternocleidomastoid
3. Deltoid
4. Brachialis
5. Vastus medialis
6. Vastus lateralis

Left side of body

1. Rectus femoris
2. Tibialis anterior
3. Quadratus lumborum
4. Frontalis
5. Platysma

Figure 10.14 (Anterior View)

Right side of body

1. Pectoralis major
2. Psoas major
3. Iliacus
4. Adductor magnus
5. Gracilis

Left side of body

1. Pectoralis minor
2. Rectus abdominis
3. Biceps brachii
4. Sartorius
5. Vastus intermedius

Figure 10.13 (Posterior View)

Right side of body

1. Deltoid
2. Gluteus minimus
3. Semimembranosus
4. Occipitalis

Left side of body

1. Trapezius
2. Biceps femoris

Figure 10.14 (Posterior View)

Right side of body

1. Latissimus dorsi
2. Gluteus maximus
3. Gastrocnemius

Left side of body

1. Triceps brachii
2. Gluteus medius
3. Semitendinosus
4. Soleus

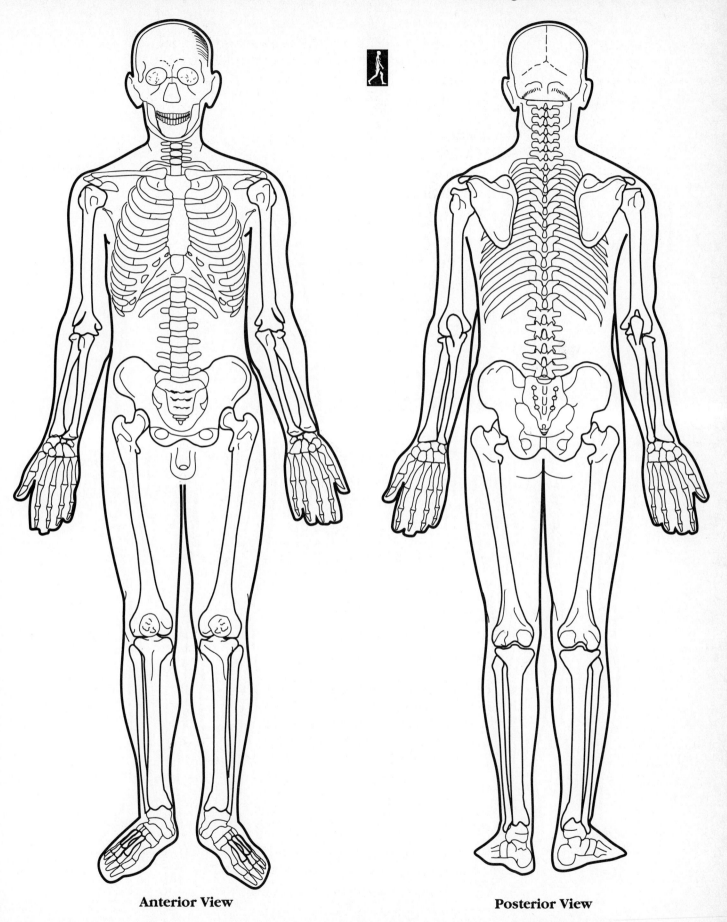

Anterior View   **Posterior View**

Figure 10.13

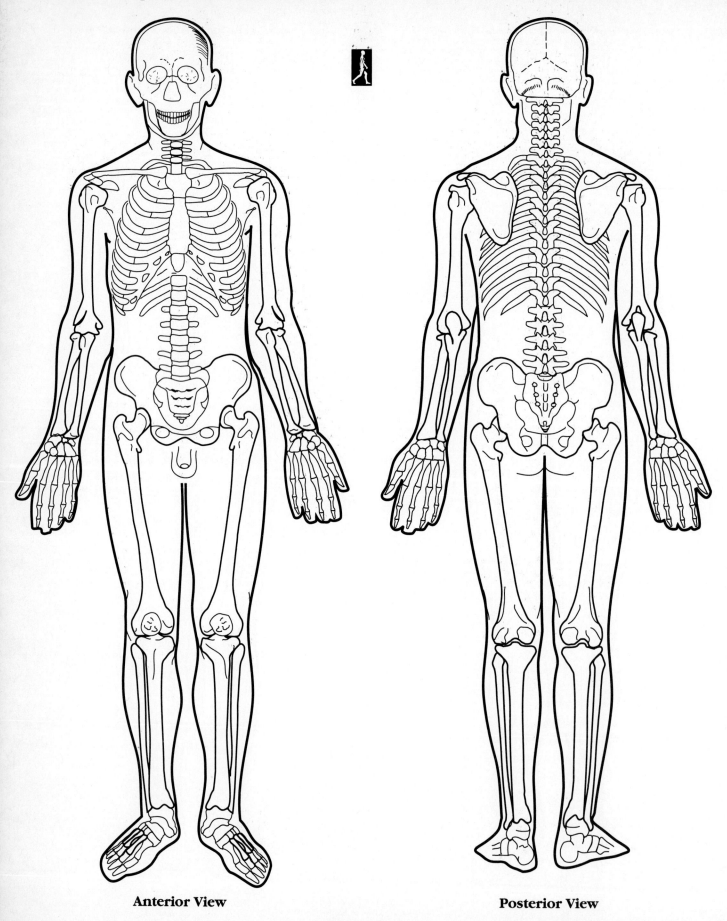

Anterior View   **Posterior View**

Figure 10.14

CHALLENGING YOURSELF

At the Clinic

1. An elderly man is brought to the clinic by his distraught wife. Among other signs, the nurse notices that the muscles on the right side of his face are slack. What nerve is not functioning properly?

2. Pete, who has been moving furniture all day, arrives at the clinic complaining of painful spasms in his back. He reports having picked up a heavy table by stooping over. What muscle group has Pete probably strained, and why are these muscles at risk when one lifts objects improperly?

3. An accident victim (who was not wearing a seat belt) was thrown from a vehicle and pronounced dead at the scene. The autopsy reveals the cause of death to be spinal cord injury resulting in paralysis of the phrenic and intercostal nerves. Why is this injury fatal?

4. Mr. Posibo has had gallbladder surgery. Now he is experiencing weakness of the muscles on his right side only, the side in which the incision was made through the abdominal musculature. Consequently, the abdominal muscles on his left side contract more strongly, throwing his torso into a lateral flexion. Mr. Posibo needs physical therapy. What abnormal spinal curvature will result if he doesn't get it and why?

5. An emergency appendectomy is performed, with the incision made at the lateral edge of the right iliac abdominopelvic region. Was the rectus abdominis cut?

6. In some women who have borne several children, the uterus prolapses (everts through the weakened pelvic diaphragm). The weakening of what muscles allows this to happen?

7. What muscles must be immobilized to prevent movement of a broken clavicle?

8. Out of control during a temper tantrum, Malcolm smashed his fist through a glass door and severed several tendons at the anterior wrist. What movements are likely to be lost if tendon repair is not possible?

9. During an overambitious workout, a high school athlete pulls some muscles by forcing his knee into extension when his hip is already fully flexed. What muscles does he pull?

10. Susan, a massage therapist, was giving Mr. Graves a back rub. What two broad superficial muscles of the back were receiving the "bulk" of her attention?

11. An elderly woman with extensive osteoarthritis of her left hip joint entered the hospital to have total hip joint replacement (prosthesis implantation). After surgery, her left hip was maintained in adduction to prevent dislocation of the prosthesis while healing was occurring. Physical therapy was prescribed to prevent atrophy of the gluteal muscles during the interval of disuse. Name the gluteal muscles and describe which of their actions were being prevented while the hip was adducted.

Stop and Think

1. Do all skeletal muscles attach to a bone? If not, what else do they attach to, and how is the result different from muscle attachment to bone?

2. Are the *stated* origins and insertions always the *functional* origins and insertions? That is, is the stated origin always fixed, and does the stated insertion always move?

3. State, in general terms, how to tell by a muscle's *location* what its *action* will be.

4. How can the latissimus dorsi, which is on the back, aid in medial, instead of lateral, rotation of the humerus?

5. Where is the insertion of the tibialis anterior, which is able to invert the foot? Where is the insertion of the peroneus longus, a foot everter?

6. What is the functional reason why the muscle group on the dorsal leg (calf) is so much larger than the muscle group in the ventral region of the leg?

Closer Connections: Checking the Systems— Covering, Support, and Movement

1. People with chronic back pain occasionally get relief from a "tummy tuck." How does this help?

2. Which of the systems in this unit has the most crucial need for calcium? Why?

3. Where in each system's tissues would you find collagen fibers?

4. Which system in this unit has the most mitotically active cells? Which has the least active?

5. Which systems play a role in homeostasis of body temperature, and how do they do so?

6. Which systems have protection as a basic function? How is this function carried out?

COVERING ALL YOUR BASES

Multiple Choice

Select the best answer or answers from the choices given.

1. Which one of the following muscles has the most motor units relative to its size?
   A. Flexor pollicis longus   C. Soleus
   B. Rectus femoris   D. Latissimus dorsi

2. Which muscle has the largest motor units?
   A. Peroneus brevis   C. Frontalis
   B. Gluteus maximus   D. Platysma

3. Head muscles that insert on a bone include the:
   A. zygomaticus   C. buccinator
   B. masseter   D. temporalis

4. Which muscles change the position or shape of the lips?
   A. Zygomaticus   C. Platysma
   B. Orbicularis oris   D. Mentalis

5. Chewing muscles include the:
   A. buccinator   C. lateral pterygoid
   B. medial pterygoid   D. sternothyroid

6. Which of the following plays a role in swallowing?
   A. Lateral pterygoid   C. Digastric
   B. Styloglossus   D. Geniohyoid

7. The hyoid bone provides an insertion for the:
   A. digastric
   B. pharyngeal constrictor muscles
   C. sternothyroid
   D. thyrohyoid

8. Muscles that function in head rotation include:
   A. scalenes
   B. sternocleidomastoid
   C. splenius
   D. semispinalis capitis

9. Lateral flexion of the torso involves:
   A. erector spinae
   B. rectus abdominis
   C. quadratus lumborum
   D. external oblique

10. Muscles attached to the vertebral column include:
    A. quadratus lumborum
    B. external oblique
    C. diaphragm
    D. latissimus dorsi

11. Which of the following muscles attach to the hip bones?
    A. Transversus abdominis
    B. Rectus femoris
    C. Vastus medialis
    D. Longissimus group of erector spinae

12. Muscles that attach to the rib cage include:
    A. scalenes
    B. internal oblique
    C. trapezius
    D. psoas major

13. Which muscles are part of the pelvic diaphragm?
    A. Superficial transverse perineus
    B. Deep transverse perineus
    C. Levator ani
    D. Sphincter urethrae

14. Which of the following insert on the arm?
    A. Biceps brachii
    B. Triceps brachii
    C. Trapezius
    D. Coracobrachialis

15. Rotator cuff muscles include:
    A. supraspinatus
    B. infraspinatus
    C. teres minor
    D. teres major
    E. subscapularis

16. Muscles that help stabilize the scapula and shoulder joint include:
    A. triceps brachii
    B. biceps brachii
    C. trapezius
    D. rhomboids

17. Which muscles insert on the radius?
    A. Triceps brachii
    B. Biceps brachii
    C. Flexor carpi radialis
    D. Pronator teres

18. Which muscles insert on the femur?
    A. Quadratus lumborum
    B. Iliacus
    C. Gracilis
    D. Tensor fasciae latae

19. Which of these thigh muscles causes movement at the hip joint?
    A. Rectus femoris
    B. Biceps femoris
    C. Vastus lateralis
    D. Semitendinosus

20. Leg muscles that can cause movement at the knee joint include:
    A. tibialis anterior
    B. peroneus longus
    C. gastrocnemius
    D. soleus

21. Muscles that function in lateral hip rotation include:
    A. gluteus medius
    B. gluteus maximus
    C. obturator externus
    D. tensor fasciae latae

22. Hip adductors include:
    A. sartorius
    B. gracilis
    C. vastus medialis
    D. pectineus

23. Which muscles are in the posterior compartment of the leg?
    A. Flexor digitorum longus
    B. Flexor hallucis longus
    C. Extensor digitorum longus
    D. Peroneus tertius

24. Which of the following insert distal to the tarsus?
    A. Soleus
    B. Tibialis anterior
    C. Extensor hallucis longus
    D. Peroneus longus

25. Which muscles are contracted while standing at attention?

    A. Iliocostalis   **C.** Tensor fascia latae

    B. Rhomboids   **D.** Flexor digitorum longus

26. The main muscles used when doing chin-ups are:

    A. triceps brachii and pectoralis major

    B. infraspinatus and biceps brachii

    C. serratus anterior and external oblique

    D. latissimus dorsi and brachialis

27. The major muscles used in doing push-ups are:

    A. biceps brachii and brachialis

    B. supraspinatus and subscapularis

    C. coracobrachialis and latissimus dorsi

    D. triceps brachii and pectoralis major

28. Someone who sticks out a thumb to hitch a ride is ___ the thumb.

    A. extending   **C.** adducting

    B. abducting   **D.** opposing

29. Which are ways in which muscle names have been derived?

    A. Attachments   **C.** Function

    B. Size   **D.** Location

Word Dissection

For each of the following word roots, fill in the literal meaning and give an example, using a word found in this chapter.

Word root	Translation	Example
1. agon		
2. brevis		
3. ceps		
4. cleido		
5. gaster		
6. glossus		
7. pectus		
8. perone		
9. rectus		

11 Fundamentals of the Nervous System and Nervous Tissue

Student Objectives

When you have completed the exercises in this chapter, you will have accomplished the following objectives:

1. List the basic functions of the nervous system.

Organization of the Nervous System

2. Explain the structural and functional divisions of the nervous system.

Histology of Nervous Tissue

3. List the types of supporting cells and cite their functions.

4. Describe the important anatomical structures of a neuron and relate each structure to a physiological role.

5. Explain the importance of the myelin sheath and describe how it is formed in the central and peripheral nervous systems.

6. Classify neurons, both structurally and functionally.

7. Differentiate between a nerve and a tract, and between a nucleus and a ganglion.

Neurophysiology

8. Define *resting membrane potential* and describe its electrochemical basis.

9. Compare and contrast graded and action potentials.

10. Name the two major classes of graded potentials.

11. Explain how action potentials are generated and propagated along neurons.

12. Define *absolute* and *relative refractory periods*.

13. Define *saltatory conduction* and contrast it to conduction along unmyelinated fibers.

14. Define *synapse*. Distinguish between electrical and chemical synapses structurally and in their mechanisms of information transmission.

15. Distinguish between excitatory and inhibitory postsynaptic potentials.

16. Describe how synaptic events are integrated and modified.

17. Define *neurotransmitter* and name several classes of neurotransmitters.

Basic Concepts of Neural Integration

18. Describe common patterns of neuron organization and neuronal processing.

19. Distinguish in a general sense between serial and parallel processing.

Developmental Aspects of Neurons

20. Describe the role of astrocytes and nerve cell adhesion molecules (N-CAMs) in neuronal differentiation.

Ready, set, go!

The nervous system is the master coordinating system of the body. Every thought, action, and sensation reflects its activity. Because of its complexity, the anatomical structures of the nervous system are considered in terms of two principal divisions—the central nervous system (CNS), consisting of the brain and spinal cord, and the peripheral nervous system (PNS). The PNS, consisting of cranial nerves, spinal nerves, and ganglia, provides the communication lines between the CNS and the body's muscles, glands, and sensory receptors. It is most important to recognize that the nervous system acts in an integrated manner both structurally and functionally.

In Chapter 11, we discuss the organization of the nervous system and the histology of nervous tissue, but the primary focus is the structure and function of neurons. Because every body system is controlled, at least in part, by the nervous system, a basic comprehension of how it functions is essential to understanding overall body functioning.

BUILDING THE FRAMEWORK

1. List the three major functions of the nervous system.

1. _____

_____

2. _____

_____

3. _____

_____

Organization of the Nervous System

1. Only one part of the structural classification of the nervous system is depicted in Figure 11.1. Identify that part by color coding it and coloring it on the diagram, and label its two parts. Then, draw in and color code the other structural subdivision of the nervous system, and add labels and leader lines to identify the main parts.

○ Central nervous system (CNS) ○ Peripheral nervous system (PNS)

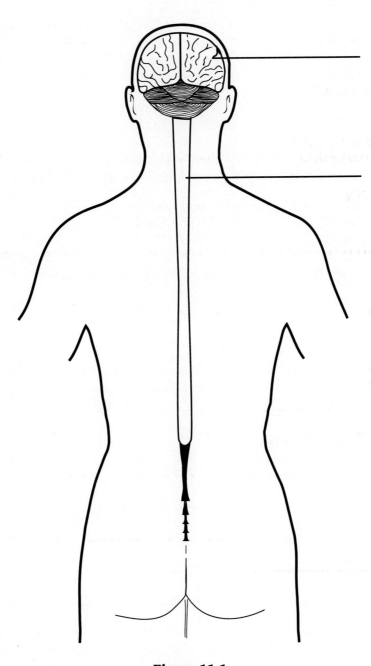

Figure 11.1

2. Choose the key choices that best correspond to the following descriptions. Insert the appropriate answers in the answer blanks.

KEY CHOICES

A. Autonomic nervous system   **C.** Peripheral nervous system

B. Central nervous system   **D.** Somatic nervous system

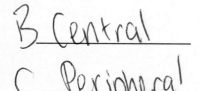

B _Central nervous_ **1.** Nervous system subdivision that is composed of the brain and spinal cord

D _Somatic nervous_ **2.** Subdivision of the PNS that controls voluntary activities such as the activation of skeletal muscles

C _Peripheral_ **3.** Nervous system subdivision that is composed of the cranial and spinal nerves and ganglia

A _Autonomic_ **4.** Subdivision of the PNS that regulates the activity of the heart and smooth muscle and of glands; also called the involuntary nervous system

B _Central_ **5.** A major subdivision of the nervous system that interprets incoming information and issues orders

C _Peripheral_ **6.** A major subdivision of the nervous system that serves as the communication lines, linking all parts of the body to the CNS

Histology of Nervous Tissue

1. This exercise emphasizes the difference between neurons and supporting cells. If a statement is true, write the letter T in the answer blank. If a statement is false, change the underlined word(s) and write the correct word(s) in the answer blank.

_____ **1.** Supporting cells found in the CNS are called <u>neuroglia</u>.

_____ **2.** <u>Neurons</u> are mitotic and therefore are responsible for most brain neoplasms.

_____ **3.** Schwann cells and satellite cells are found only in the <u>CNS</u>.

_____ **4.** <u>Ependymal cells</u> show irritability and conductivity.

_____ **5.** Almost 50% of the volume of neural tissue in the CNS is made up of <u>neurons</u>.

_____ **6.** In the CNS, <u>oligodendrocytes</u> engulf invading microorganisms and dead neural tissue.

_____ **7.** <u>Astrocytes</u> line the central cavities of the brain.

_____ **8.** <u>Schwann cells</u> wrap their cytoplasmic extensions around thick neuron fibers in the CNS.

_____ **9.** The bulbous ends of <u>the axons of neurons</u> cling to capillaries.

2. Relative to neuron anatomy, match the anatomical terms in Column B with the appropriate descriptions of function in Column A. Place the correct answers in the answer blanks.

	Column A		Column B
Axon Terminal	**1.** Releases neurotransmitters		**A.** Axon
Dendrite	**2.** Conducts local currents toward the soma		**B.** Axonal terminal
Myelin Sheath	**3.** Increases the speed of impulse transmission		**C.** Cell body
Cell body	**4.** Location of the nucleus		**D.** Dendrite
Axon	**5.** Conducts impulses away from the cell body		**E.** Myelin sheath
Cell body	**6.** Most are located and protected within the CNS		**F.** Nissl bodies
Dendrite	**7.** Short, tapering, diffusely branched extension from the cell body		
Axon	**8.** The process called a nerve fiber		
Myelin sheath	**9.** Formed by Schwann cells in the PNS		
Nissl bodies	**10.** Clustered ribosomes and rough ER		
Myelin Sheath	**11.** Patchy disappearance in multiple sclerosis		
Cell body	**12.** Site of biosynthetic activities		

3. Circle the term that does not belong in each of the following groupings.

1. Nucleus   Soma   Centrioles   Owl's eye nucleolus

2. Mitochondria   Rough ER   Ribosomes   Nissl bodies

3. Melanin   Glycogen   Lipofuscin   Pigment

4. Dendritic spine   Input   Output   Receptive

5. Axonal terminal   Synaptic knob   Bouton   Axon collateral

4. Figure 11.2 is a diagram of a neuron. First, label the parts with leader lines on the illustration. Then, choose different colors for each of the structures listed below and color the illustration. Finally, draw arrows on the figure to indicate the direction of impulse transmission along the neuron's membrane.

○ Axon   ○ Nerve cell body   ○ Neurilemma

○ Dendrites   ○ Myelin sheath

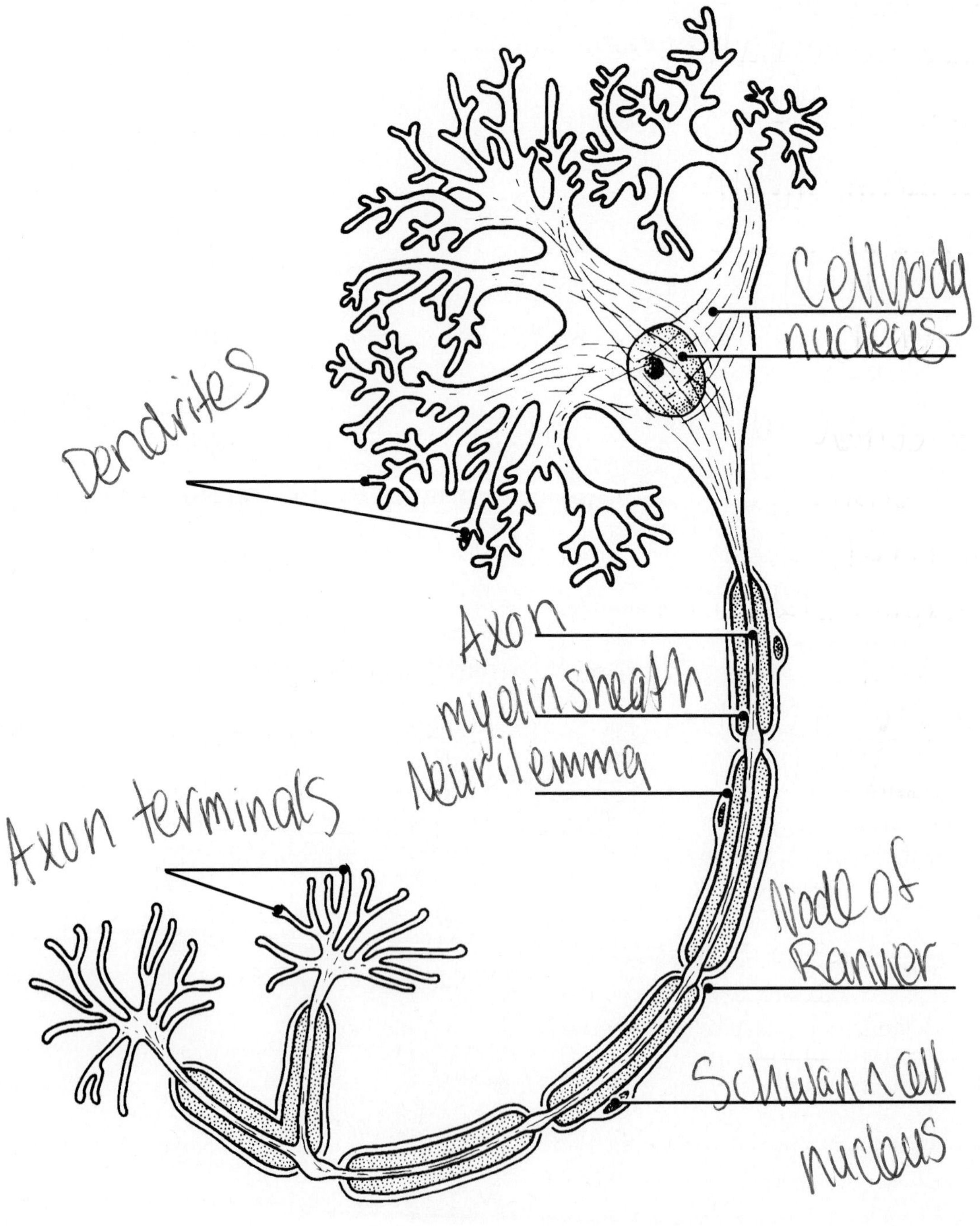

Figure 11.2

5. Match the key choices to their functional designation.

KEY CHOICES

A. Axon   **B.** Axon hillock   **C.** Axonal terminals   **D.** Dendrite

_____ **1.** Receptive region

_____ **2.** Trigger zone

_____ **3.** Conducting component

_____ **4.** Secretory components

6. List the three special characteristics of neurons._____

_____

_____

7. Using the key choices, select the terms that match the following descriptions. Insert the correct answers in the answer blanks.

KEY CHOICES

A. Ganglion   **D.** Nerve   **G.** Synapse

B. Neuroglia   **E.** Nodes of Ranvier   **H.** Stimuli

C. Neurotransmitters   **F.** Nucleus   **I.** Tract

_____Synapse_____ **1.** Junction or point of close contact between neurons

neurotransmitters **2.** Chemicals released by neurons that stimulate other neurons, muscles, or glands

Nodes of Ranvier **3.** Gaps in a myelin sheath

_____ **4.** Bundle of axons in the CNS

_____ **5.** Collection of cell bodies found outside the CNS

_____ **6.** Collection of cell bodies found within the CNS

_____Stimuli_____ **7.** Changes, occurring inside or outside the body, that activate the nervous system

_____ **8.** Bundle of axons in the PNS

8. Circle the term that does not belong in each of the following groupings.

1. Nodes of Ranvier   Myelin sheath   Cell body   Axon

2. Cell body   Centrioles   Nucleolus   Ribosomes

3. Ganglia   Clusters of cell bodies   PNS   Clusters of glial cells

4. Dendrites   Neurotransmitters in vesicles   Telodendria   Synaptic knobs

5. Gray matter   Myelin   Fiber tracts   White matter

9. Three diagrams of neurons are shown in Figure 11.3. On the lines below each diagram, indicate the neuron's structural and (most likely) functional classification. Color the diagram according to your fancy.

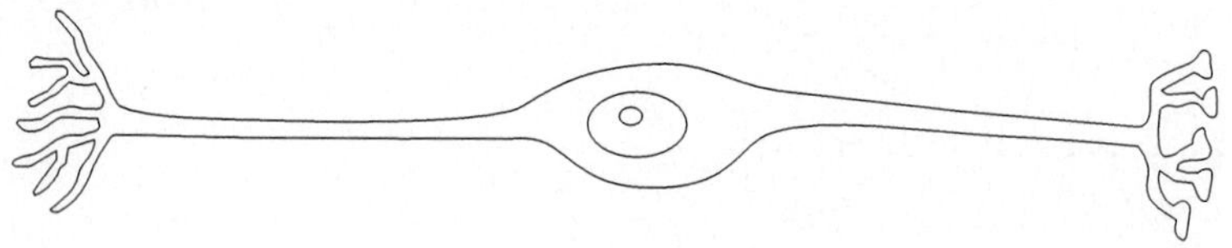

A. Classification: Structural _____

                      Functional _____

Schwann cell

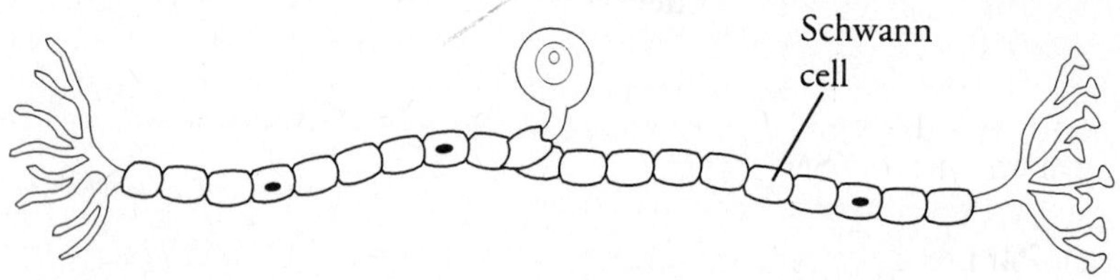

B. Classification: Structural _____

                      Functional _____

Schwann cell

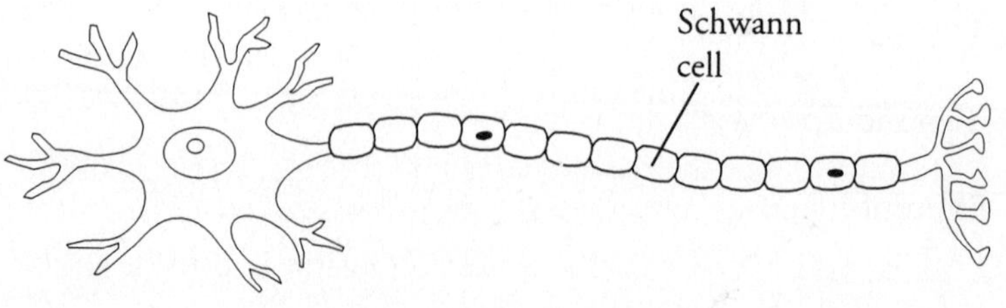

C. Classification: Structural _____

                      Functional _____

Figure 11.3

10. Several descriptions of neurons are given below. Two keys are provided. Key A lists terms that classify neurons structurally; Key B lists terms that classify neurons functionally. Choose responses from one or both keys and insert the terms in the answer blanks as needed to complete the descriptive statements.

KEY A: Structural

Bipolar
Multipolar
Unipolar

KEY B: Functional

Sensory or afferent
Motor or efferent
Association

_____ **1.**

_____ **2.**

A neuron that excites skeletal muscle cells in your biceps muscle is functionally a(n) __(1)__ neuron and structurally a(n) __(2)__ neuron.

_____ **3.**

_____ **4.**

Neurons that reside entirely within the CNS are __(3)__ neurons. Structurally, most of these neurons belong to the __(4)__ classification.

_____ **5.**

_____ **6.**

A structural class of neurons that is very rare in the body is the __(5)__ type.

_____ **7.**

Structurally, the most common neurons are __(6)__ .

_____ **8.**

A neuron that transmits impulses from pain receptors in your skin to your spinal cord is classified as a(n) __(7)__ neuron. Structurally, this type of neuron is __(8)__ .

11. Classify the following inputs and outputs as somatic sensory (SS), visceral sensory (VS), somatic motor (SM), or visceral motor (VM).

SS **1.** Pain from skin

VS **2.** Proprioception

VM **3.** Efferent innervation of a gland

SM **4.** Efferent innervation of your gluteus maximus

VS **5.** A stomachache

SS **6.** A sound you hear

VM **7.** Efferent innervation of the muscle of the urinary bladder wall

12. Two diagrams showing a small section of PNS axon(s) are provided in Figure 11.4. First, identify which axons are *myelinated* and which are *unmyelinated* by writing one of these terms below the appropriate diagram. Then, color code and color the structures listed below. Finally, bracket the neurilemma.

○ Axon   ○ Schwann cell cytoplasm   ○ Schwann cell plasma membrane

○ Axolemma   ○ Schwann cell nucleus

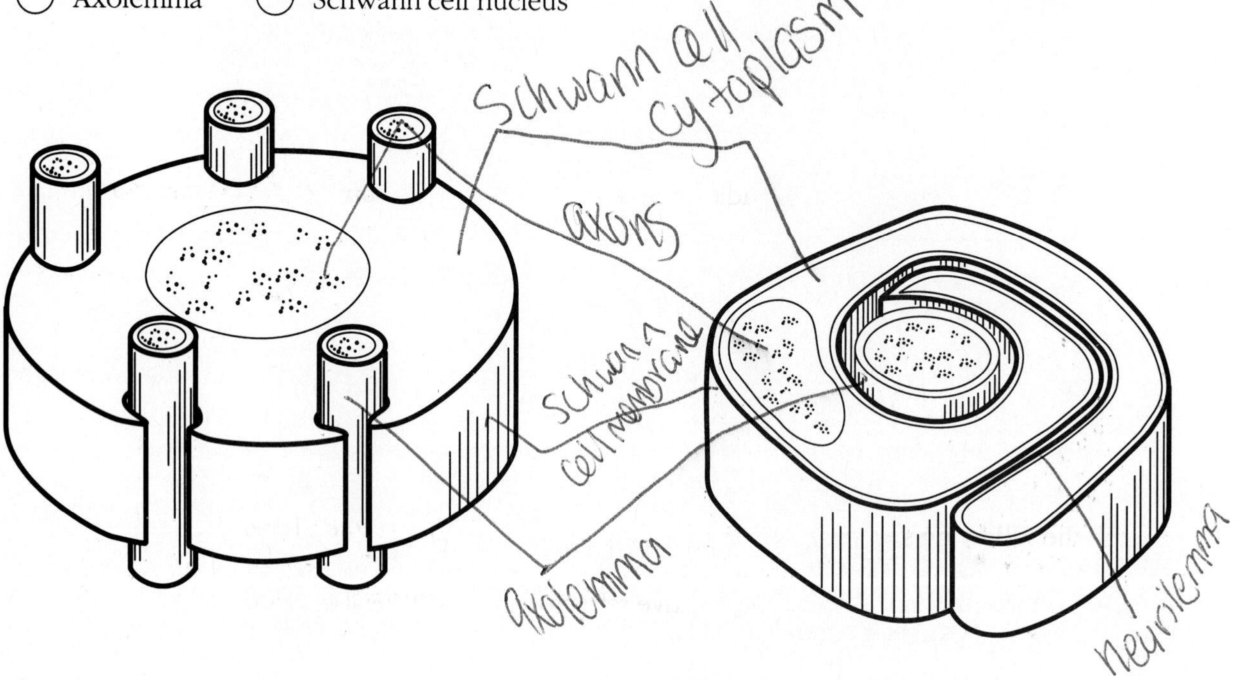

A. _____   **B.** _____

Figure 11.4

Neurophysiology

1. Complete the following statements, which refer to the resting membrane potential. Write the missing terms in the answer blanks.

_____ **1.**

_____ **2.**

_____ **3.**

_____ **4.**

_____ **5.**

_____ **6.**

_____ **7.**

_____ **8.**   _____ **9.**   _____ **10.**

The relationship of voltage (V), current (I), and resistance (R) is known as __(1)__ law and is expressed by the formula __(2)__ . In the body, currents across cellular membranes are carried by __(3)__ . The difference of charge on the two sides of the membrane is called the __(4)__ . The resistance to current flow is provided by the __(5)__ . Ion channels in plasma membranes are constructed of __(6)__ . Channels that remain open are called __(7)__ channels. Channels that open and close in response to various stimuli are called __(8)__ channels. Ions move across the membrane in response to a difference in concentration, which is called the __(9)__ gradient, and also in response to a difference in charge, which is called the __(10)__ gradient.

2. Check (✔) all descriptions that apply to a resting neuron.

　✔　**1.** Its inside is negative relative to its outside.

　＿＿＿　**2.** Its outside is negative relative to its inside.

　＿＿＿　**3.** The cytoplasm contains more sodium and less potassium than does the
extracellular fluid.

　✔　**4.** The cytoplasm contains more potassium and less sodium than does the
extracellular fluid.

　✔　**5.** A charge separation exists at the membrane.

　✔　**6.** The electrochemical gradient for the movement of sodium across the membrane
is greater than that for potassium.

　＿＿＿　**7.** The electrochemical gradient for the movement of potassium across the
membrane is greater than that for sodium.

　＿＿＿　**8.** The membrane is more permeable to sodium than potassium.

　✔　**9.** The membrane is more permeable to potassium than sodium.

3. Graph the following set of data on the axes provided in Figure 11.5, then label
the following: absolute refractory period, action potential (AP), depolarization,
graded potential, hyperpolarization, relative refractory period, repolarization,
resting membrane potential (RMP).

Voltage (mV)	−70	−70	−65	−70	−70	−60	−70	−70	−50	+30	−65	−75	−78	−71	−70	−70
Time (ms)	2.0	0.5	0.7	1.0	1.5	1.7	2.0	2.5	3.0	3.5	4.0	4.2	4.5	5.0	5.5	6.0

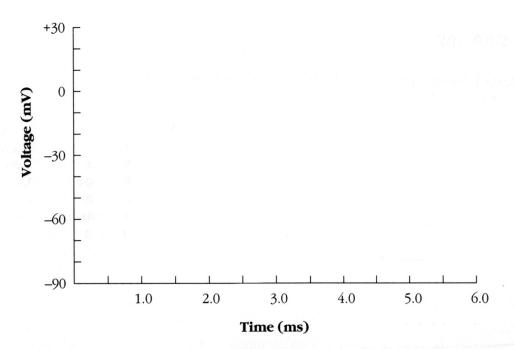

Figure 11.5

4. Using the key choices, select the terms defined in the following statements. Write the correct answers in the answer blanks.

KEY CHOICES

A. Absolute refractory period	**E.** Graded potential	**I.** Repolarization
B. Action potential	**F.** Hyperpolarization	**J.** Sodium-potassium pump
C. Depolarization	**G.** Polarized	**K.** Subthreshold
D. Frequency of impulses	**H.** Relative refractory period	**L.** Threshold

H hyperpolarization **1.** Corresponds to the period of repolarization of the neuron

C Depolarization **2.** Process by which the resting potential is decreased as sodium ions move into the axon

G Polarized **3.** State of an unstimulated neuron's membrane

I Repolarization **4.** Period (event) during which potassium ions move out of the axon

B Action potential **5.** Also called the nerve impulse

A Absolute refractory period **6.** Period when a neuron cannot be restimulated because its sodium gates are open

J sodium-potassium pump **7.** Mechanism by which ATP is used to move sodium ions out of the cell and potassium ions into the cell; completely restores and maintains the resting conditions of the neuron

L Threshold **8.** Point at which an axon "fires"

K Subthreshold **9.** Term for a weak stimulus

B Action potential **10.** Self-propagated depolarization

D Depolarization Frequency of impulses **11.** Codes for intensity of the stimulus

L Threshold **12.** Membrane potential at which the outward current carried by K^+ is exactly equal to the inward current carried by Na^+

F **13.** A voltage change that reduces the ability of a neuron to conduct an impulse; the membrane potential becomes more negative

E **14.** A local change in membrane potential in which current flow is quickly dissipated, that is, decremental

B **15.** An all-or-none electrical event

C **16.** A voltage change that brings a neuron closer to its threshold for firing; the membrane potential becomes less negative and moves toward 0

B **17.** Results from the opening of voltage-regulated ionic gates

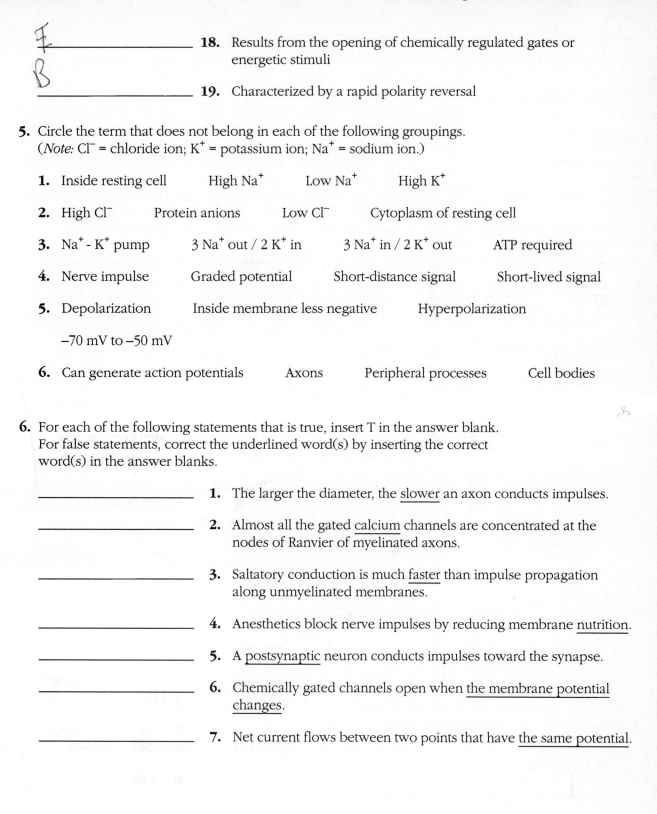

F

B

_____ **18.** Results from the opening of chemically regulated gates or energetic stimuli

_____ **19.** Characterized by a rapid polarity reversal

5. Circle the term that does not belong in each of the following groupings. (*Note:* Cl⁻ = chloride ion; K⁺ = potassium ion; Na⁺ = sodium ion.)

1. Inside resting cell   High Na⁺   Low Na⁺   High K⁺

2. High Cl⁻   Protein anions   Low Cl⁻   Cytoplasm of resting cell

3. Na⁺ - K⁺ pump   3 Na⁺ out / 2 K⁺ in   3 Na⁺ in / 2 K⁺ out   ATP required

4. Nerve impulse   Graded potential   Short-distance signal   Short-lived signal

5. Depolarization   Inside membrane less negative   Hyperpolarization

−70 mV to −50 mV

6. Can generate action potentials   Axons   Peripheral processes   Cell bodies

6. For each of the following statements that is true, insert T in the answer blank. For false statements, correct the underlined word(s) by inserting the correct word(s) in the answer blanks.

_____ **1.** The larger the diameter, the <u>slower</u> an axon conducts impulses.

_____ **2.** Almost all the gated <u>calcium</u> channels are concentrated at the nodes of Ranvier of myelinated axons.

_____ **3.** Saltatory conduction is much <u>faster</u> than impulse propagation along unmyelinated membranes.

_____ **4.** Anesthetics block nerve impulses by reducing membrane <u>nutrition</u>.

_____ **5.** A <u>postsynaptic</u> neuron conducts impulses toward the synapse.

_____ **6.** Chemically gated channels open when <u>the membrane potential changes</u>.

_____ **7.** Net current flows between two points that have <u>the same potential</u>.

7. Complete the following statements referring to chemical synapses by writing the correct terms in the answer blanks.

_____ 1.

_____ 2.

_____ 3.

_____ 4.

_____ 5.

_____ 6.

_____ 7.

_____ 8.

_____ 9.

_____ 10.

_____ 11.

_____ 12.

_____ 13.

_____ 14.

_____ 15.

_____ 16.

_____ 17.

_____ 18.

_____ 19.

When the nerve impulse reaches the axon terminal, __(1)__ ions enter the terminal through voltage-regulated gates. The effect of the entry of these ions is to promote fusion of synaptic vesicles with the __(2)__ . Exocytosis of __(3)__ molecules follows. These diffuse across the __(4)__ and bind to __(5)__ on the postsynaptic membrane, causing ion channels to open. The current flows that result cause local changes in the __(6)__ of the postsynaptic membrane. The neurotransmitter is inactivated enzymatically or otherwise removed from the synaptic cleft.

At __(7)__ synapses, neurotransmitter binding opens a single type of channel, allowing both __(8)__ ions to diffuse through the membrane. Since accumulation of __(9)__ charges inside the cell is prevented, local depolarization events, instead of __(10)__ , occur. The function of these graded potentials is to initiate an __(11)__ distally at the axon __(12)__ of the postsynaptic neuron.

At __(13)__ synapses, potassium and/or chloride permeability of the postsynaptic membrane is increased. Consequently, the charge inside the cell becomes relatively more __(14)__ , and the resting membrane potential changes from −70 mV toward −90 mV, indicating __(15)__ of the postsynaptic membrane.

A single EPSP (excitatory postsynaptic potential) is insufficient to generate an action potential, but threshold depolarization can be achieved by __(16)__ of many EPSPs. Monitoring and integration of all excitatory and/or inhibitory neurotransmitters is accomplished by the postsynaptic membranes, which __(17)__ EPSPs with IPSPs (inhibitory postsynaptic potentials). Sometimes, the release of excitatory neurotransmitters is inhibited by the activity of another neuron via an __(18)__ synapse or by the action of chemicals called __(19)__ .

8. Use the key choices to identify the types of synapses described below. Write the correct answers in the answer blanks.

KEY CHOICES

A. Chemical synapse   **B.** Electrical synapse

_____ 1. Releases neurotransmitters from vesicles

_____ 2. Very rapid transmission

_____ 3. Protein channels connect cytoplasm of adjacent neurons

_____ 4. Postsynaptic neuron has receptor region

_____ **5.** Transmission may be uni- or bidirectional

_____ **6.** Presynaptic neuron has a knoblike axon terminal

_____ **7.** Synchronizes activities of all interconnected neurons

_____ **8.** Allows the flow of ions between neurons

_____ **9.** Fluid fills synaptic space between neurons

9. Using the key choices, first match the neurotransmitters listed in Column A with the numbered descriptions given below. Then, select the appropriate classes of these neurotransmitters from Column B. Finally, if applicable, select the correct effect of each neurotransmitter from Column C. Write the letters of all your choices in the answer blanks.

KEY CHOICES

Column A	Column B	Column C
A. Acetylcholine	**E.** Excitatory	**I.** Affects emotions
B. Amino acids	**F.** Inhibitory	**J.** Painkillers
C. Biogenic amines	**G.** Direct acting	**K.** Reduce anxiety
D. Peptides	**H.** Indirect acting	**L.** Regulate biological clock

_____ **1.** An example is gamma-aminobutyric acid (GABA)

_____ **2.** Examples are beta-endorphins and enkephalins

_____ **3.** Released at neuromuscular junctions of skeletal muscles

_____ **4.** Typically are broadly distributed in the brain

10. Circle the term that does not belong in each of the following groupings.

1. Ion channels open   Change in membrane potential   Slow response
Direct action neurotransmitter

2. Indirect action neurotransmitter   Short-lived effect   G protein linked
Second messenger

3. Receptor potential   Weak stimuli   Threshold strength
More nerve impulses per second

4. Pain receptors   Receptor membrane response declines   Adaptation
Light pressure receptor

5. Nitric oxide   ATP   Novel neurotransmitter   ACh

11. In Figure 11.6, identify by coloring and by labeling if leader lines are provided the following structures, which are typically part of a chemical synapse. Also, bracket the synaptic cleft, and identify the arrows showing (1) the direction of the presynaptic impulse and (2) the direction of net neurotransmitter movements.

- ⬡ Axonal terminal
- ⬡ Postsynaptic membrane
- ⬡ Presynaptic membrane
- ⬡ Mitochondria
- ⬡ Na$^+$ ions
- ⬡ Ca^{2+} ions
- ⬡ K$^+$ ions
- ⬡ Chemically gated channels
- ⬡ Synaptic vesicles
- ⬡ Postsynaptic neurotransmitter receptors
- ⬡ Neurotransmitter molecules

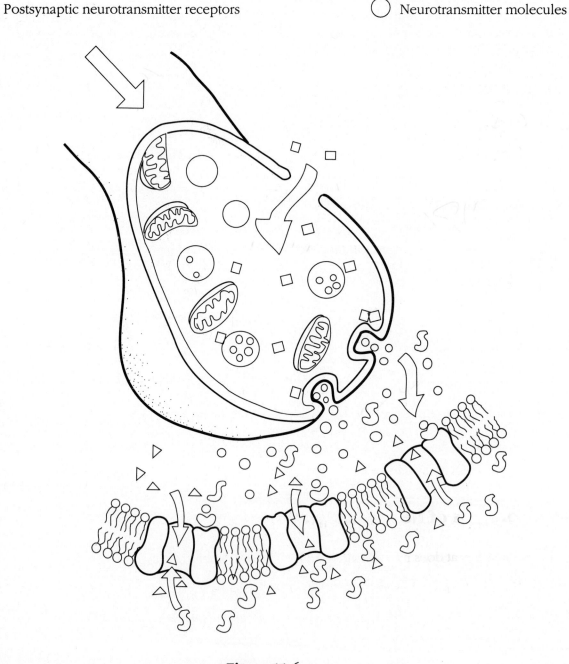

Figure 11.6

12. On Figure 11.7, several types of chemical synapses are illustrated. Identify each type, using the key choices. Color the diagram as you wish.

KEY CHOICES

A. Axoaxonic   **B.** Axodendritic   **C.** Axosomatic   **D.** Dendrodendritic

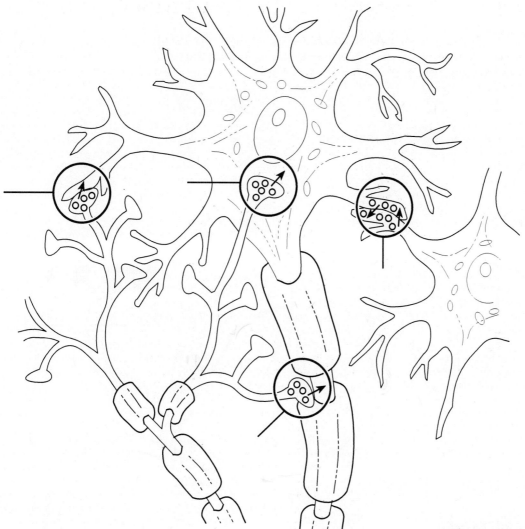

Figure 11.7

Basic Concepts of Neural Integration

1. Circle the term that does not belong in each of the following groupings.

1. PNS   Neural integration   Neuronal pools   Circuits

2. Parallel processing   Several unique responses   CNS integration

    All-or-nothing response

3. Reflex arc   Variety of stimuli   Serial processing   One anticipated response

2. Refer to Figure 11.8, showing a reflex arc, as you complete the following exercise. First, briefly answer the following questions by writing your answers in the answer blanks.

1. What is the stimulus? ___*pin-prick pain*___

2. What tissue is the effector? ___*skeletal muscle*___

3. How many synapses occur in this reflex arc? ___*2 (3rd w/ muscle)*___

Next, select different colors for each of the following structures, and use them to color the diagram. Finally, draw arrows on the figure indicating the direction of impulse transmission through this reflex pathway.

◉ Receptor region   ◎ Association neuron   ◯ Effector

◯ Sensory neuron   ◉ Motor neuron

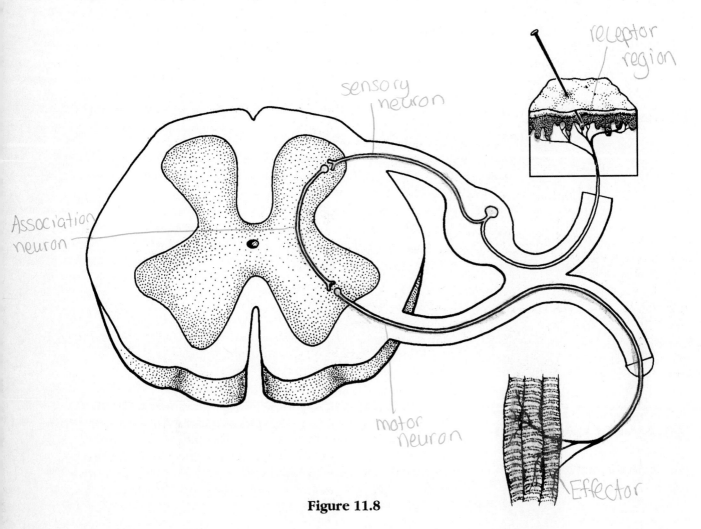

Figure 11.8

3. Using the key choices, match the types of circuits with their descriptions. Write the correct answers in the answer blanks.

KEY CHOICES

A. Converging   **B.** Diverging   **C.** Parallel after-discharge   **D.** Reverberating

_____ **1.** Large number of skeletal muscle fibers stimulated by a few motor neurons

_____ **2.** The sight, sound, and smell of popping popcorn all elicit the same feelings

_____ **3.** One fiber stimulates increasing numbers of neurons

_____ **4.** Impulses repeatedly sent through the same circuit, lasting perhaps a lifetime

_____ **5.** Many presynaptic neurons stimulate a few neurons

_____ **6.** Control of rhythmic activities, such as respiration and the sleep-wake cycle

_____ **7.** Impulses from many neurons reach a common output cell at different times

_____ **8.** Used to solve this conversion for the speed of impulse transmission: 100 m/s = ? mi/hr

Developmental Aspects of Neurons

1. Complete the following statements by writing the missing terms in the answer blanks.

_____ **1.**

_____ **2.**

_____ **3.**

_____ **4.**

_____ **5.**

_____ **6.**

_____ **7.**

_____ **8.**

_____ **9.**

_____ **10.**

The dorsal neural tube, derived from __(1)__, gives rise to all the supporting cells and nerve cells of the nervous system. The one possible exception is the microglia, which may arise from __(2)__. By the fourth week of pregnancy, the fate of neural tube cells in the embryo is fixed, and they become potential neurons called __(3)__. Further organization and development of these cells occur in three phases. The first phase is characterized by cellular __(4)__, after which the neuroblasts become incapable of __(5)__. During the second phase, __(6)__ of cells to various sites of the neural tube occurs. In the third phase, __(7)__ of the neuroblasts begins, a process resulting in __(8)__, __(9)__, and __(10)__.

(continues on next page)

_____ 11.

_____ 12.

_____ 13.

_____ 14.

_____ 15.

_____ 16.

For example, __(11)__ neurons originate as bipolar neurons whose processes converge and fuse. Also, early neuron interactions depend more on __(12)__ synapses than on chemical synapses.

Development of chemical synapses, along with axonal outgrowths, depends in large measure on pathways established by __(13)__ and chemicals, such as __(14)__, which functions to adhere neurons to other cells and structures. Special structures, called __(15)__, appear at the tips of growing axons, enabling them to attach to adjacent structures and direct axonal pathways. Neurons failing to make appropriate synaptic contacts generally __(16)__.

CHALLENGING YOURSELF

At the Clinic

1. A patient taking diuretics comes to the clinic complaining of muscular weakness and fatigue. A deficiency of which ion will be suspected? (*Hint:* you might also want to do a little reading in Chapter 26 for this one.)

2. Being a long-time herpesvirus sufferer, Brad was not surprised to learn (at a herpes information session) that herpesviruses tend to "hide out" in nerve tissue to avoid attack by the body's immune system. However, their "mode of travel" in the body did surprise him. How *do* herpesviruses travel to the neuron cell body?

3. A brain tumor is found in a CT scan of Mr. Childs' head. The physician is assuming that it is not a secondary tumor (i.e., it did not spread from another part of the body) because an exhaustive workup has revealed no signs of cancer elsewhere in Mr. Childs' body. Is the brain tumor more likely to have developed from nerve tissue or from neuroglia? Why?

4. With what specific process does the lack of myelination seen in multiple sclerosis interfere?

5. Sally has been diagnosed as having a certain type of epilepsy associated with a lack of GABA. Why would this deficiency lead to the increased and uncontrolled neuronal activity exhibited by Sally's seizures?

6. Mr. Jacobson, a tax accountant, comes to the clinic complaining of feeling very "stressed out" and anxious. He admits to drinking 10 to 12 cups of coffee daily. His doctor (knowing that caffeine alters the threshold of neurons) suggests he reduce his intake of coffee. What is caffeine's effect on the threshold of neurons, and how might this effect explain Mr. Jacobson's symptoms?

7. Mr. Marple staggered home after a "good night" at the local pub. While attempting to navigate the stairs, he passed out cold and lay (all night) with his right armpit straddling the staircase banister. When he awoke the next morning, he had a severe headache, but what bothered him more was that he had no sensation in his right arm and hand, which also appeared to be paralyzed. Explain.

Stop and Think

1. Histological examination during a brain autopsy revealed a superabundance of microglia in a certain brain area. What abnormal condition might this indicate?

2. What is the benefit of having highly branched cell processes such as those of neurons?

3. Axonal diameter increases noticeably at the nodes of Ranvier. What purpose does this serve?

4. How might a neuron without an axon transmit impulses?

5. Describe ion flow through a neuronal membrane that is simultaneously being excited and inhibited by various presynaptic axons. Assuming an exact balance in excitatory and inhibitory stimuli, how would a graph of membrane potential look?

6. If two parallel pathways were equal in all regards (such as myelination, axon diameter, and total pathway length), except that the axons in one pathway were twice as long as the axons in the neighboring pathway, in which pathway would nerve impulses travel more quickly? To which type of neuronal pool does this type of arrangement apply?

7. Chloride moves fairly freely through the neuronal membrane, and yet the opening of chloride gates can cause inhibition. How can a chloride flux be inhibitory?

8. Neurotransmitter remains in the synaptic cleft only briefly because it is removed by enzymes or other mechanisms. If the mechanism for removal of neurotransmitter is impaired, will the postsynaptic cell remain depolarized and refractory?

9. Each neuron makes and releases one neurotransmitter—true or false? Elaborate on your response.

10. With learning, new synapses can form. How can this affect a neuron's facilitated and discharge zones?

11. Are neurons incorporated into more than one neuronal circuit type?

12. Two anatomists were arguing about the sensory neuron. One said its peripheral process is an axon and gave two good reasons. But the other anatomist called the peripheral process a dendrite and gave one good reason. Cite all three reasons given and state your own opinion.

13. Shortly after birth, no new neurons are formed. This being the case, how can the enhancement of certain pathways, such as those promoting more acute hearing in blind people, develop?

14. Explain why damage to peripheral nerve fibers is often reversible, whereas damage to CNS fibers rarely is.

COVERING ALL YOUR BASES

Multiple Choice

Select the best answer or answers from the choices given.

1. An example of integration by the nervous system is:
   A. the feel of a cold breeze
   B. the shivering and goose bumps that result
   C. the sound of rain
   D. the decision to go back for an umbrella

2. Which of the following functions would utilize a visceral efferent cranial nerve?
   A. Sense of taste
   B. Sense of position of the eye
   C. Movement of the tongue in speech
   D. Secretion of salivary glands

3. The cell type *most immediately affected* by damage to a blood vessel in the brain is a(n):
   A. neuron
   C. astrocyte
   B. microglia
   D. oligondendrocyte

4. Which of the following neuroglial types would be classified as epithelium?
   A. Astrocyte
   C. Oligondendrocyte
   B. Ependymal cells
   D. Microglia

5. Cell types with an abundance of smooth ER include:
   A. neuron
   C. astrocyte
   B. Schwann cell
   D. oligondendrocyte

6. The region of a neuron with voltage-gated sodium channels is the:
   A. axon hillock
   C. dendrite
   B. soma
   D. perikaryon

7. Which of the following would be a direct result of destruction of a neuron's neurofilaments?
   A. Cessation of protein synthesis
   B. Loss of normal cell shape
   C. Inhibition of impulse transmission
   D. Interference with intracellular transport

8. Where might a gray matter nucleus be located?
   A. Alongside the vertebral column
   B. Within the brain
   C. Within the spinal cord
   D. In the sensory receptors

9. Histological examination of a slice of neural tissue reveals a bundle of nerve fibers held together by cells whose multiple processes wrap around several fibers and form a myelin sheath. The specimen is likely to be:
   A. a nucleus
   C. a nerve
   B. a ganglion
   D. a tract

10. Neurotransmitters:
    A. act directly to produce action potentials
    B. produce only graded potentials
    C. have a direct effect on the secretory component of a neuron
    D. can act only at voltage-gated channels

11. What do myelin sheath membranes lack which makes them good insulators?
    A. Lipids
    B. Carbohydrate groups
    C. Channel and carrier proteins
    D. Close contact with adjacent membranes

12. Bipolar neurons:
    A. are found in the head
    B. are always part of an afferent pathway
    C. have two dendrites
    D. have two axons

13. Which of the following would be true of the peripheral process of a unipolar neuron?
    A. Its membrane contains voltage-gated channels.
    B. It is never myelinated.
    C. It connects directly to the central process.
    D. It typically has a wrapping of Schwann cells.

14. Which of the following skin cells would form a junction with a motor neuron?
    A. Keratinocyte
    B. Sudoriferous glandular epithelial cell
    C. Arrector pili muscle cell
    D. Fibroblast

15. The term that refers to a measure of the potential energy of separated electrical charges is:
    A. voltage
    C. resistance
    B. current
    D. conductance

16. Current is inversely related to:
    A. voltage
    C. potential difference
    B. resistance
    D. capacitance

17. An unstimulated plasma membrane is characterized by:
    A. ions flowing along their electrochemical gradients
    B. cytoplasmic side slightly negatively charged
    C. all gated channels open
    D. low concentration of Na^+ in cytoplasm compared to extracellular fluid

18. Which would result from inhibition of a neuron's sodium-potassium pump?
    A. The membrane would lose its polarity.
    B. Potassium would undergo a net movement to the cell's interior.
    C. Cytoplasmic anions would diffuse out of the cell.
    D. Sodium would accumulate outside the cell.

19. Impulse *propagation* is associated with:
    A. graded potentials
    B. chemically gated ion channels
    C. hyperpolarization
    D. voltage-gated sodium channels

20. Hyperpolarization results from:
    A. sodium flux into the cell
    B. chloride flux into the cell
    C. potassium flux into the cell
    D. potassium flux out of the cell

21. Decremental current flow is associated with:
    A. chemically gated channels
    B. postsynaptic potential
    C. saltatory conduction
    D. axolemma

22. Which of these is involved with a positive feedback cycle?
    A. Sodium-potassium pump
    B. Chemically gated channels
    C. Voltage-gated sodium channels
    D. Membrane potential

23. In the graph below, which stimulus (1, 2, 3, or 4) is the strongest? (*Note:* The vertical lines represent action potential tracings.)

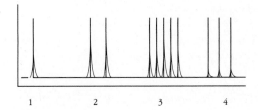

    A. Stimulus 1   C. Stimulus 3
    B. Stimulus 3   D. Stimulus 4

24. If the following receptors were stimulated simultaneously, which signal would reach the brain first?
    A. Pain receptors in the skin of the toe
    B. Sensory receptors in a nearby blood vessel wall
    C. Small touch receptors in the toe's skin
    D. Pressure receptors in the toe's musculature

25. Which is *not* characteristic of a chemical synapse?
    A. Abundant gap junctions
    B. Voltage-gated channels in the postsynaptic membrane
    C. Unidirectional communication
    D. Synaptic delay

26. An IPSP is associated with:
    A. opening of chemically gated channels for K^+
    B. hyperpolarization
    C. decremental conduction
    D. depolarization

27. Which of the following is true concerning tetanic potentiation?
    A. Calcium concentrations increase during stimulation.
    B. Membrane proteins can be altered.
    C. The postsynaptic cell is inhibited.
    D. The presynaptic cell is inhibited.

28. Which of the following are characteristic of the biogenic amine neurotransmitters?
    A. They are direct acting.
    B. Some are catecholamines.
    C. Endorphins are in this class.
    D. At least some are involved in regulation of mood.

29. A neuronal pool that might rely on fatigue to end its impulse propagation is a:
    A. converging circuit
    B. diverging circuit
    C. oscillating circuit
    D. parallel after-discharge circuit

30. A synapse between an axon terminal and a neuron cell body is called:

    A. axodendritic

    B. axoaxonic

    C. axosomatic

    D. axoneuronic

31. Myelin is most closely associated with which of the following cell parts introduced in Chapter 3?

    A. Cell nucleus

    B. Smooth ER

    C. Ribosomes

    D. The plasma membrane

Word Dissection

For each of the following word roots, fill in the literal meaning and give an example, using a word found in this chapter.

Word root	Translation	Example
1. dendr		
2. ependy		
3. gangli		
4. glia		
5. neur		
6. nom		
7. oligo		
8. salta		
9. syn		

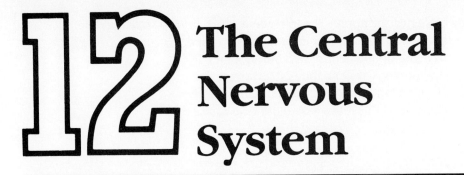 

12 The Central Nervous System

Student Objectives

When you have completed the exercises in this chapter, you will have accomplished the following objectives:

The Brain

1. Describe the process of brain development.
2. Name the major regions of the adult brain.
3. Define the term *ventricle* and indicate the location of the ventricles of the brain.
4. Indicate the major lobes, fissures, and functional areas of the cerebral cortex.
5. Explain lateralization of hemispheric function.
6. Differentiate between commissures, association fibers, and projection fibers.
7. State the general function of the basal nuclei.
8. Describe the location of the diencephalon and name its subdivisions.
9. Identify the three major regions of the brain stem and note the general function of each area.
10. Describe the structure and function of the cerebellum.
11. Localize the limbic system and the reticular formation and explain the role of each functional system.
12. Describe how meninges, cerebrospinal fluid, and the blood-brain barrier protect the central nervous system.
13. Describe the formation of cerebrospinal fluid and follow its circulatory pathway.

14. Distinguish between a concussion and a contusion.
15. Describe the cause (if known) and major signs and symptoms of cerebrovascular accidents and Alzheimer's disease.

The Spinal Cord

16. Describe the embryonic development of the spinal cord.
17. Describe the gross and microscopic structure of the spinal cord.
18. List the major spinal cord tracts and describe each in terms of its origin, termination, and function.
19. Distinguish between flaccid and spastic paralysis and between paralysis and paresthesia.

Diagnostic Procedures for Assessing CNS Dysfunction

20. List and explain several techniques used to diagnose brain disorders.

Developmental Aspects of the Central Nervous System

21. Indicate several maternal factors that can impair development of the nervous systems in an embryo.
22. Explain how true senility and reversible senility differ.

Ready, set, go!

The human brain is a marvelous biological computer. It can receive, store, retrieve, process, and dispense information with almost instantaneous speed. Together, the brain and spinal cord make up the central nervous system (CNS). By way of inputs from the peripheral nervous system, the CNS is advised of changes in both the external environment and internal body conditions. After perception, integration, and coordination of this knowledge, the CNS dispatches instructions to initiate appropriate responses.

Included in Chapter 12 are student exercises on the regions and associated functions of the brain and spinal cord. Traumatic and degenerative disorders and developmental aspects of the central nervous system are also considered.

BUILDING THE FRAMEWORK

The Brain

1. Figure 12.1 is a diagram of the right lateral view of the human brain. First, match the letters on the diagram with the following list of terms and insert the appropriate letters in the answer blanks. Then, select different colors for each of the areas of the brain with a color-coding circle and use them to color the diagram. If an identified area is part of a lobe, use the color you selected for the lobe but use stripes for that area.

○ ___ **1.** Frontal lobe

○ ___ **2.** Parietal lobe

○ ___ **3.** Temporal lobe

○ ___ **4.** Precentral gyrus

___ **5.** Parieto-occipital fissure

○ ___ **6.** Postcentral gyrus

___ **7.** Lateral fissure

___ **8.** Central sulcus

○ ___ **9.** Cerebellum

○ ___ **10.** Medulla

○ ___ **11.** Occipital lobe

○ ___ **12.** Pons

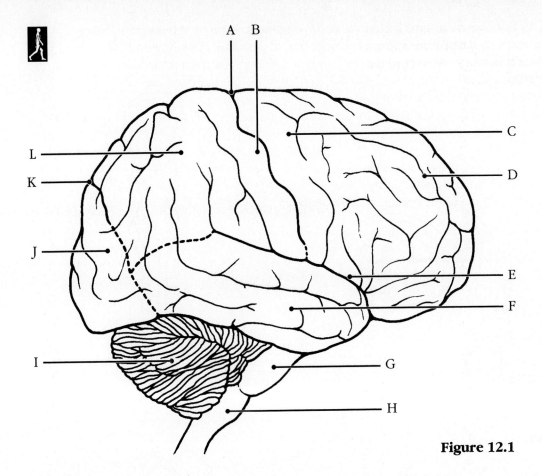

Figure 12.1

2. Fill in the table below by indicating the adult brain structures formed from each of the secondary brain vesicles and the adult neural canal regions. Some of that information has already been entered.

Secondary brain vesicle	Adult brain structures	Neural canal regions
Telencephalon		Lateral ventricles
Diencephalon	Diencephalon:	
Mesencephalon		
Metencephalon	Brain stem: pons; cerebellum	
Myelencephalon		

3. Figure 12.2 shows diagrams of embryonic development. Arrange the diagrams in the correct order by numbering each one. Where possible, insert the time, in days or weeks, of each stage of development. Label all structures that have leader lines.

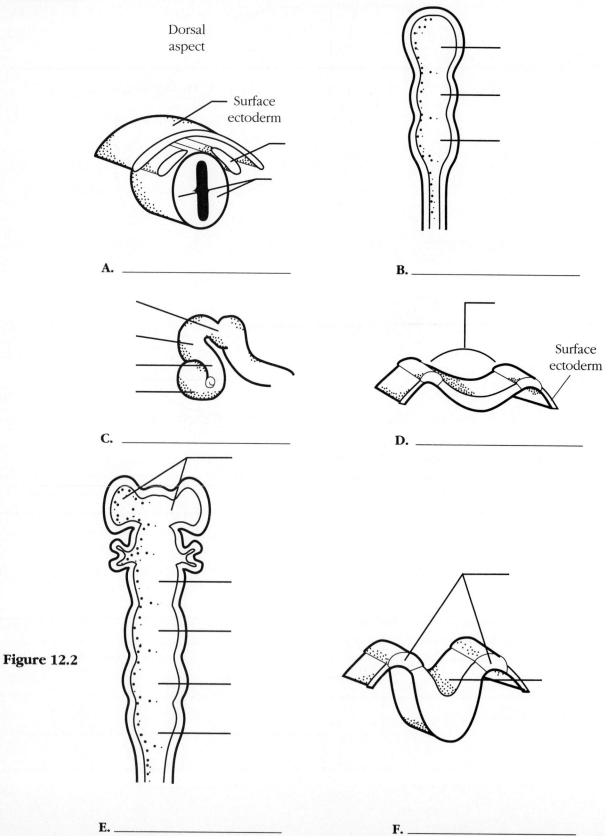

Dorsal aspect

Surface ectoderm

A. _____

B. _____

C. _____

Surface ectoderm

D. _____

Figure 12.2

E. _____

F. _____

4. Figure 12.3 illustrates a "see-through" brain showing the positioning of the ventricles and connecting canals or apertures. Correctly identify all structures having leader lines by using the key choices provided below. One of the lateral ventricles has already been identified. Color the spaces filled with cerebrospinal fluid blue.

KEY CHOICES

A. Anterior horn	**D.** Fourth ventricle	**G.** Lateral aperture
B. Central canal	**E.** Inferior horn	**H.** Third ventricle
C. Cerebral aqueduct	**F.** Interventricular foramen	

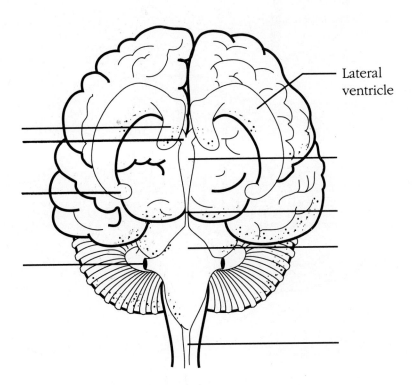

Lateral ventricle

Figure 12.3

5. Figure 12.4 shows a left lateral view of the brain with some of its functional areas indicated by numbers. These areas are listed below. Identify each cortical area by its corresponding number on the diagram. Color the diagram as you wish.

_____ Primary motor cortex   _____ Primary somatosensory cortex

_____ Premotor cortex   _____ Somatosensory association area

_____ Visual cortex   _____ Auditory cortex

_____ Prefrontal cortex   _____ Broca's area

_____ Frontal eye field   _____ Wernicke's area

_____ General interpretation area

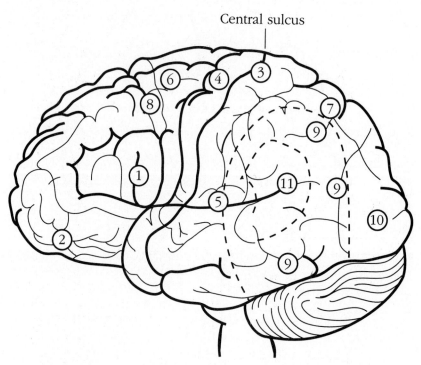

Central sulcus

Figure 12.4

6. Some of the following brain structures consist of gray matter; others are white matter. Write G (for gray) or W (for white) as appropriate.

_____ **1.** Cortex of cerebellum   _____ **5.** Pyramids

_____ **2.** Internal capsule   _____ **6.** Olives

_____ **3.** Anterior commisure   _____ **7.** Thalamic nuclei

_____ **4.** Medial lemniscus   _____ **8.** Cerebellar peduncle

7. If a statement is true, write the letter T in the answer blank. If a statement is false, correct the underlined word(s) and write the correct word(s) in the answer blank.

_____ **1.** The primary somatosensory area of the cerebral hemisphere(s) is found in the <u>precentral</u> gyrus.

_____ **2.** Cortical areas involved in audition are found in the <u>occipital</u> lobe.

_____ **3.** The primary motor area in the <u>temporal</u> lobe is involved in the initiation of voluntary movements.

_____ **4.** The specialized motor speech area is located at the base of the precentral gyrus in an area called <u>Wernicke's</u> area.

_____ **5.** The right cerebral hemisphere receives sensory input from the <u>right</u> side of the body.

_____ **6.** The <u>pyramidal</u> tract is the major descending voluntary motor tract.

_____ **7.** The primary motor cortex is located in the <u>postcentral</u> gyrus.

_____ **8.** Centers for control of repetitious or stereotyped motor skills are found in the <u>primary motor</u> cortex.

_____ **9.** The largest parts of the motor humunculus are the lips, tongue, and <u>toes</u>.

_____ **10.** Sensations such as touch and pain are integrated in the <u>primary sensory cortex</u>.

_____ **11.** The primary visual cortex is in the <u>frontal</u> lobe of each cerebral hemisphere.

_____ **12.** In most humans, the area that controls the comprehension of language is located in the <u>left</u> cerebral hemisphere.

_____ **13.** Elaboration of the <u>visual</u> cortex sets humans apart from other animals.

_____ **14.** Complex sensory memory patterns are stored in an area called the <u>general interpretation</u> area.

_____ **15.** Areas in the cerebral hemisphere opposite the ones containing Broca's and Wernicke's areas are centers for <u>cognitive</u> language.

_____ **16.** Cerebral dominance designates the hemisphere that is dominant for <u>memory</u>.

_____ **17.** The right cerebral hemisphere of <u>left</u>-handed humans is usually involved with intuition, poetry, and creativity.

8. Using the key choices, select the terms identified in the following descriptions by inserting the appropriate letters in the answer blanks.

KEY CHOICES

A. Basal nuclei   **D.** Cerebral hemispheres   **G.** Septum pellucidum

B. Brain stem   **E.** Cortex   **H.** Ventricles

C. Cerebellum   **F.** Diencephalon   **I.** White matter

_____   **1.** The four major subdivisions of the adult brain

_____   _____   _____

_____   **2.** Contain cerebrospinal fluid

_____   **3.** Masses of gray matter embedded deep within the cerebral white matter

_____   **4.** Myelinated fiber tracts

_____   **5.** Consists of the midbrain, pons, and medulla

_____   **6.** Separates the lateral ventricles

_____   **7.** Thin layer of gray matter on outer surface of cerebral hemispheres and cerebellum

_____   **8.** Account for more than 60% of the total brain weight

_____   **9.** Consists of the hypothalamus, thalamus, and epithalamus

9. Figure 12.5 is a diagram of the sagittal view of the human brain. First, match the letters on the diagram with the following list of terms and insert the appropriate letters in the answer blanks. Then, color the brain stem areas blue and the areas where cerebrospinal fluid is found yellow.

_____ **1.** Cerebellum   _____ **10.** Hypothalamus

_____ **2.** Cerebral aqueduct   _____ **11.** Medulla oblongata

_____ **3.** Cerebral hemisphere   _____ **12.** Optic chiasma

_____ **4.** Cerebral peduncle   _____ **13.** Pineal body

_____ **5.** Choroid plexus   _____ **14.** Pituitary gland

_____ **6.** Corpora quadrigemina   _____ **15.** Pons

_____ **7.** Corpus callosum   _____ **16.** Thalamus (intermediate mass)

_____ **8.** Fornix   _____ **17.** Third ventricle

_____ **9.** Fourth ventricle

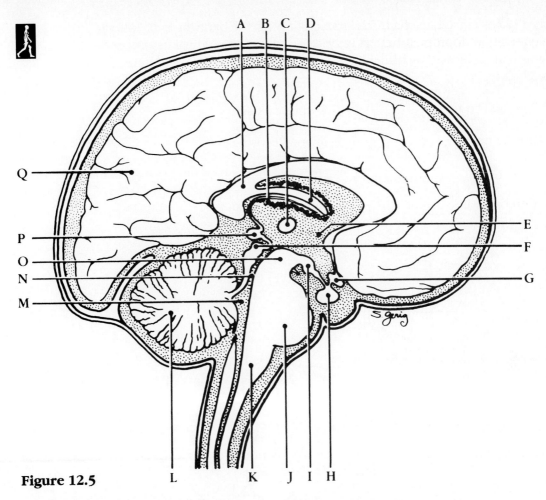

Figure 12.5

10. Referring to the brain areas listed in Exercise 9, match the appropriate brain structures with the following descriptions. Insert the terms selected in the answer blanks.

_____ **1.** Site of regulation of water balance, body temperature, rage, and pain centers; the main visceral (autonomic) center of brain

_____ **2.** Reflex centers involved in regulating respiratory rhythm in conjunction with lower brain stem centers

_____ **3.** Responsible for the regulation of posture and coordination of skeletal muscle movements

_____ **4.** Important relay station for afferent fibers, traveling to the sensory cortex for interpretation

_____ **5.** Contains autonomic centers that regulate blood pressure and respiratory rhythm, as well as coughing and sneezing centers

_____ **6.** Midbrain area consisting of large, descending motor tracts

_____ **7.** Influences body rhythms; interacts with the biological clock

_____ **8.** Location of middle cerebellar peduncles

_____ **9.** Locations of visual and auditory reflex centers

11. Figure 12.6 is a diagram of a frontal section through the brain, as indicated on the orientation diagram. Label the ventricles and the longitudinal fissure, both of which are indicated by leader lines. Color code and color the structures listed below.

- ◯ Cerebral cortex
- ◯ Basal nuclei
- ◯ Thalamus
- ◯ Cerebral white matter
- ◯ Corpus callosum
- ◯ Internal capsule
- ◯ Hypothalamus
- ◯ Septum pellucidum

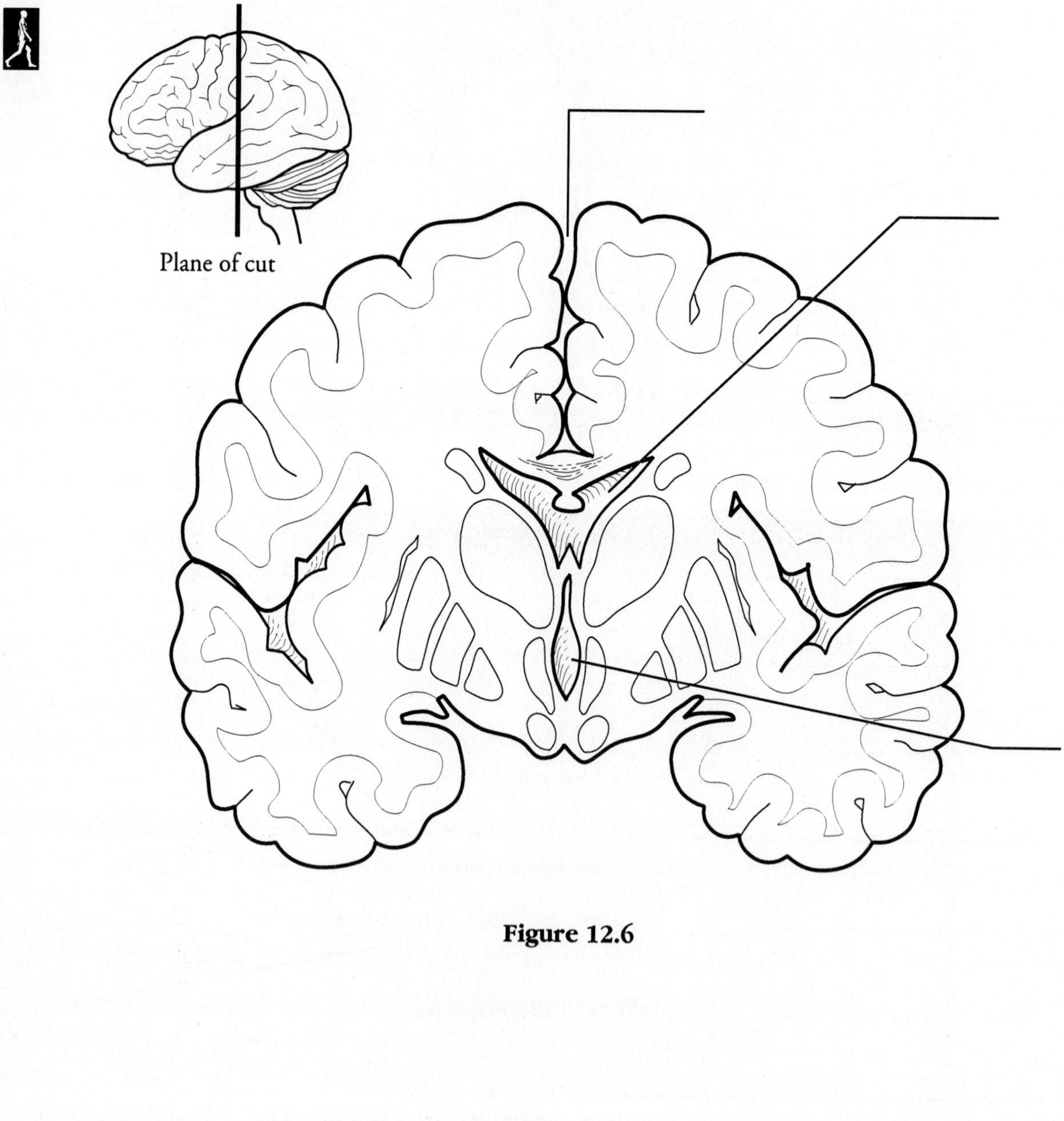

Plane of cut

Figure 12.6

12. In the horizontal section shown in Figure 12.7 (see plane of cut in inset), identify by color coding and coloring, the structures listed below.

◯ Caudate nucleus   ◯ Fornix   ◯ Pineal body

◯ Claustrum   ◯ Inferior horn of lateral ventricle   ◯ Putamen

◯ Corpus callosum   ◯ Insula   ◯ Septum pellucidum

◯ Choroid plexus   ◯ Internal capsule   ◯ Thalamus

                                                                 ◯ Third ventricle

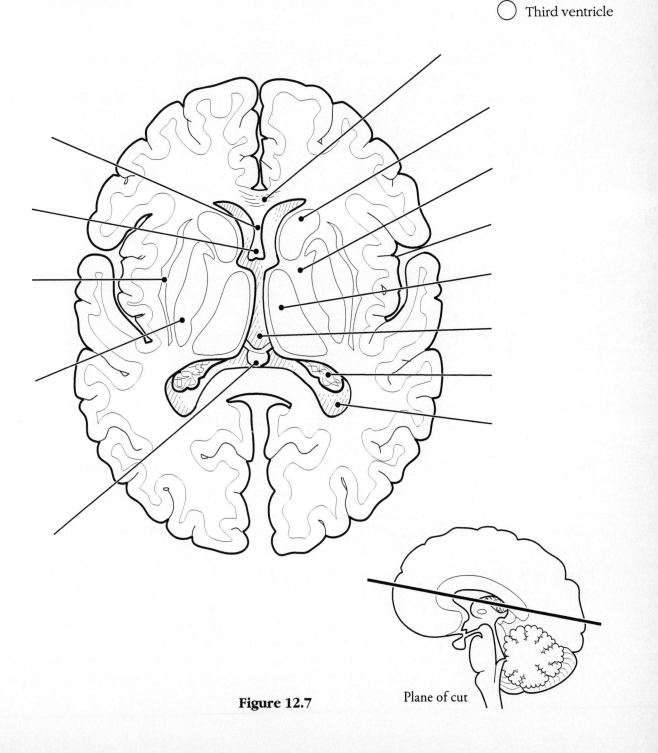

Figure 12.7   Plane of cut

13. Figure 12.8 shows the brain stem and associated diencephalon. Using the key choices, label all structures provided with leader lines. Then, color code and color the following structures or groups of structures.

◯ All cranial nerves   ◯ Medulla oblongata   ◯ Midbrain

◯ Diencephalon   ◯ Pons   ◯ Infundibulum

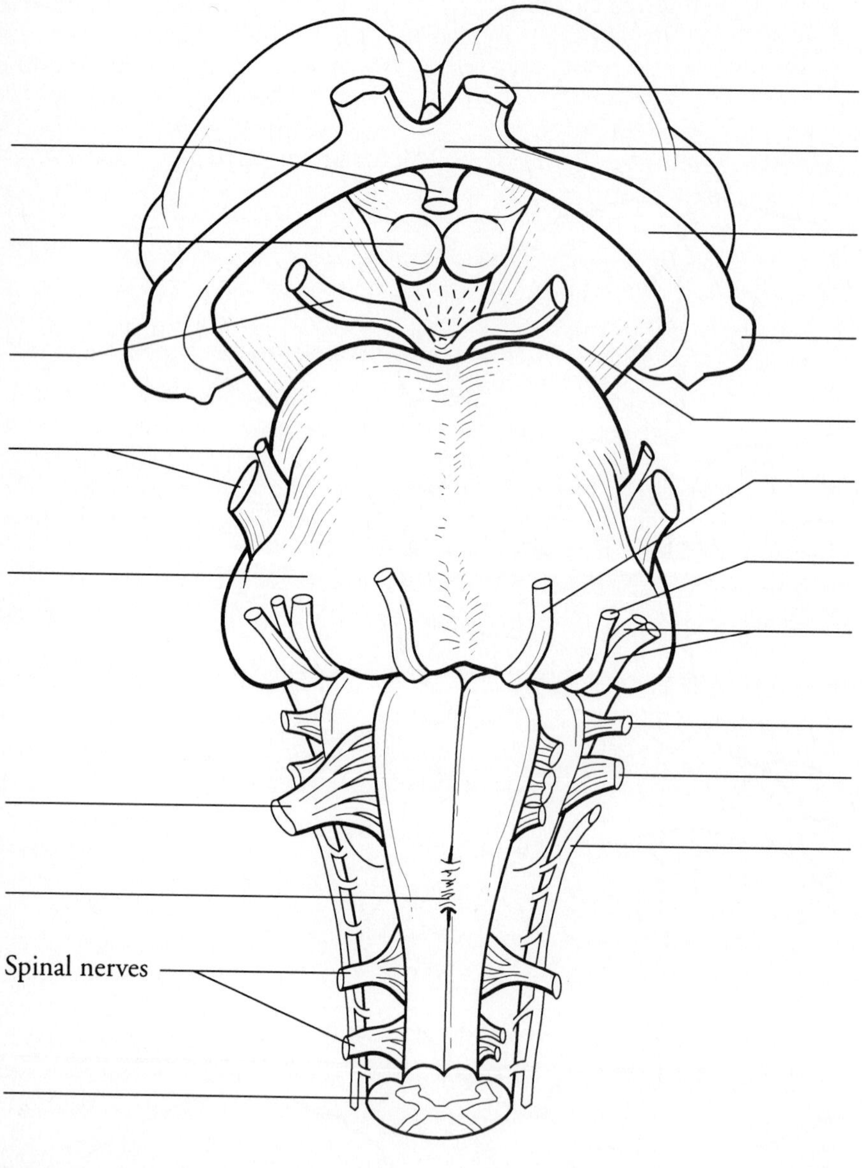

KEY CHOICES

A. Abducens nerve (VI)

B. Accessory nerve (XI)

C. Cerebral peduncle

D. Decussation of the pyramids

E. Facial nerve (VII)

F. Glossopharyngeal nerve (IX)

G. Hypoglossal nerve (XII)

H. Infundibulum

I. Lateral geniculate body

J. Mamillary body

K. Oculomotor nerve (III)

L. Optic chiasma

M. Optic nerve (II)

N. Optic tract

O. Pons

P. Spinal chord

Q. ThalamusN

R. Trigeminal nerve (V)

S. Vagus nerve (X)

T. Vestibulocochlear nerve (VIII)

Spinal nerves

Figure 12.8

14. Circle the term that does not belong in each of the following groupings.

1. Connects cerebral hemispheres   Corpus callosum   Cerebral peduncles

   Commissure

2. Hypophysis cerebri   Pineal gland   Produces hormones   Pituitary gland

3. Medulla oblongata   Pons   Cardiac center   Decussation of pyramids

4. Prefrontal cortex   Distractability   Loss of initiative   Paralysis

5. Aprodosia   Speech disorder   Affective language areas   Loss of vocal expression

6. Lack of cerebral dominance   Reading disorder   Motor impairment   Dyslexia

7. Projection fibers   Internal capsule   Association fibers   Corona radiata

8. Basal nuclei   Internal capsule   Subcortical motor nuclei   Corpora striata

9. Caudate nucleus   Putamen   Globus pallidus   Lentiform nucleus

10. Cerebellum   Vermis   Parietal lobe   Arbor vitae

15. Below is a listing of events involved in voluntary motor activity that also considers the role of the cerebellum. Put the event(s) in their correct temporal (time) sequence by numbering them 1 to 5. (*Note:* There are *two* number 5 events.)

_____ **1.** Cerebellar output to brain stem nuclei initiates subconscious motor output.

_____ **2.** Primary motor cortex sends impulses along pyramidal tracts.

_____ **3.** Feedback from cerebellum to cerebrum.

_____ **4.** Premotor cortex initiates motor activity.

_____ **5.** Pyramidal collaterals signal cerebellum.

_____ **6.** Visual, auditory, equilibrium, and body sensory input to cerebellum is assessed as the motor activity is initiated.

16. Relative to the limbic and reticular systems, identify the correct system for each characteristic described below. Insert L for the limbic system or R for the reticular system in the answer blanks.

_____ **1.** Maintains cortex in a conscious state

_____ **2.** Includes the RAS

_____ **3.** Originates in primitive rhinencephalon

_____ **4.** Its hypothalamus is the gatekeeper for visceral responses

_____ **5.** Site of rage and anger; moderated by cerebral cortex

_____ **6.** Has far-flung axonal connections

_____ **7.** Functioning may be associated with psychosomatic illness

_____ **8.** Helps distinguish and filter out unimportant stimuli

_____ **9.** Depressed by alcohol and some drugs; severe injury may cause coma

_____ **10.** Severe injury may result in personality changes

_____ **11.** Extends through the brain stem

_____ **12.** Includes cingulate gyrus, hippocampus, and some thalamic nuclei

_____ **13.** Contains raphe nuclei, large-celled and small-celled regions

_____ **14.** Located in the medial aspect of both hemispheres

17. Figure 12.9 shows a frontal view of the meninges of the brain at the level of the superior sagittal (dural) sinus. First, label _arachnoid villi_ and _falx cerebri_ on the figure. Then, select different colors for each of the following structures and use them to color the diagram.

◯ Dura mater   ◯ Pia mater

◯ Arachnoid   ◯ Subarachnoid space

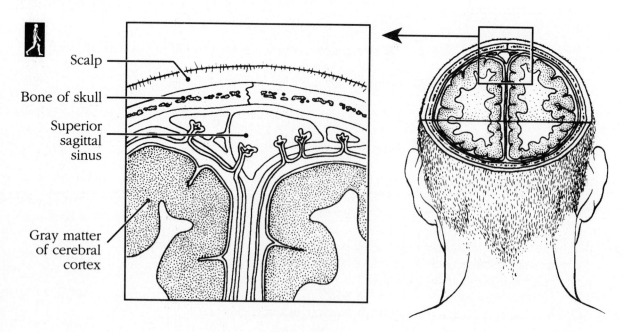

Scalp

Bone of skull

Superior sagittal sinus

Gray matter of cerebral cortex

Figure 12.9

18. Referring to the structures in Figure 12.9, identify the meningeal (or associated) structures described here. Write the correct terms in the answer blanks.

_____ **1.** Innermost covering of the brain; delicate and vascular

_____ **2.** Structures that return cerebrospinal fluid to the venous blood in the dural sinuses

_____ **3.** Its outer layer forms the periosteum of the skull

_____ **4.** Contains cerebrospinal fluid

_____ **5.** Location of major arteries and veins

_____ **6.** Contains venous blood

_____ **7.** Attaches to crista galli of the ethmoid bone

19. Complete the following statements by inserting the missing terms in the answer blanks.

_____ **1.**

_____ **2.**

_____ **3.**

_____ **4.**

_____ **5.**

_____ **6.**

_____ **7.**

_____ **8.**

_____ **9.**

_____ **10.**

_____ **11.**

_____ **12.**

The composition of the cerebrospinal fluid (CSF) is similar to __(1)__, from which it arises. The concentration of __(2)__ in the CSF is important in the control of cerebral blood flow and breathing.

The CSF is formed by capillary knots called __(3)__, which hang from the roof of each __(4)__. Circulation of the CSF is aided by the beating cilia of the __(5)__ cells lining the ventricles.

The CSF ordinarily flows from the lateral ventricles to the third ventricle and then through the __(6)__ to the fourth ventricle. Some of the CSF continues down the __(7)__ of the spinal cord, but most of it circulates into the __(8)__ by passing through three tiny openings in the walls of the __(9)__. The CSF is returned to the blood via the __(10)__, located in the dural sinuses. If drainage of the CSF is obstructed, the fluid accumulates under pressure, causing a condition called __(11)__. In the newborn, the condition results in an enlarged head because the skull can expand, but in adults the increasing pressure may cause __(12)__.

20. Relative to the blood-brain barrier, determine whether the following statements are true or false. For true statements, insert T in the answer blanks. For false statements, correct the underlined word(s) by writing the correct word(s) in the answer blanks.

_____ **1.** <u>Desmosomes</u> are the epithelial structures that make brain capillaries relatively impermeable.

_____ **2.** Water, glucose, and <u>metabolic wastes</u> pass freely from the blood into the brain, but proteins and most drugs are normally prevented from entering the brain.

_____ **3.** <u>Calcium</u> ions are actively pumped from the brain because they modify neural thresholds.

_____ **4.** The blood-brain barrier is absent in the <u>cerebrum</u>.

_____ **5.** Alcohol and other <u>water-soluble</u> molecules easily cross the blood-brain barrier.

_____ **6.** The blood-brain barrier is <u>incompletely developed</u> in newborns.

21. Match the brain disorders listed in Column B with the conditions described in Column A. Place the correct answers in the answer blanks.

Column A

_____ **1.** Slight and transient brain injury

_____ **2.** Traumatic injury that destroys brain tissue

_____ **3.** Total nonresponsiveness to stimulation

_____ **4.** May cause medulla oblongata to be wedged into foramen magnum by pressure of blood

_____ **5.** After head injury, retention of water by brain

_____ **6.** Results when a brain region is deprived of blood or exposed to prolonged ischemia; probably reflects excessive NO release

_____ **7.** Reversible CVA

_____ **8.** Progressive degeneration of the brain with abnormal protein deposits

_____ **9.** Autoimmune disorder with extensive demyelination

Column B

A. Alzheimer's disease

B. Cerebral edema

C. Cerebrovascular accident (CVA)

D. Coma

E. Concussion

F. Contusion

G. Intracranial hemorrhage

H. Multiple sclerosis

I. Transient ischemic attack (TIA)

The Spinal Cord

1. Using the key choices, select the correct terms as defined by the following descriptions. Write the correct letters in the answer blanks.

KEY CHOICES

A. Decussation   **D.** Gray matter   **G.** Pyramidal   **J.** Spinothalamic

B. Extrapyramidal   **E.** Multisynaptic   **H.** Somatotopy

C. Funiculus   **F.** Projection   **I.** Spinocerebellar

_____ 1. Its central location separates white matter into columns

_____ 2. Old terminology indicating any descending tract by which brain stem motor centers communicate with centers in the spinal cord

_____ 3. Crossing of fibers from one side to the other side of the cord

_____ 4. Precise spatial relationship of most spinal cord pathways to an orderly mapping of the body

_____ 5. Major pathways from cerebral motor areas of the brain to association or motor neurons of the cord

_____ 6. A major tract communicating from the spinal cord toward the sensory cortex

_____ 7. The tract conveying information from the spinal cord to the cerebellum

_____ 8. A white column of the cord containing several tracts

_____ 9. The phenomenon by which the brain refers sensation to the usual point of stimulation

2. Complete the following statements by writing the missing terms in the answer blanks.

_____ 1.

_____ 2.

_____ 3.

_____ 4.

_____ 5.

_____ 6.

_____ 7.

_____ 8.   _____ 9.

The spinal cord extends from the __(1)__ of the skull to the __(2)__ region of the vertebral column. The meninges, which cover the spinal cord, extend more inferiorly to form a sac, from which cerebrospinal fluid can be withdrawn without damage to the spinal cord. This procedure is called a __(3)__ . __(4)__ pairs of spinal nerves arise from the cord. Of these, __(5)__ pairs are cervical nerves, __(6)__ pairs are thoracic nerves, __(7)__ pairs are lumbar nerves, and __(8)__ pairs are sacral nerves. The tail-like collection of spinal nerves at the inferior end of the spinal cord is called the __(9)__ .

3. Figure 12.10 is a cross-sectional view of the spinal cord. First identify the areas listed in the key choices by inserting the correct letters next to the appropriate leader lines on parts A and B of the figure. Then, color the bones of the vertebral column in part B gold.

KEY CHOICES

A. Central canal   **E.** Dorsal root   **I.** Ventral (anterior) horn

B. Columns of white matter   **F.** Dorsal root ganglion   **J.** Ventral root

C. Conus medullaris   **G.** Filum terminale

D. Dorsal (posterior) horn   **H.** Spinal nerve

On part A, color the butterfly-shaped gray matter gray, and color the spinal nerves and roots yellow. Finally, select different colors to identify the following structures and use them to color the figure.

◯ Pia mater   ◯ Dura mater   ◯ Arachnoid

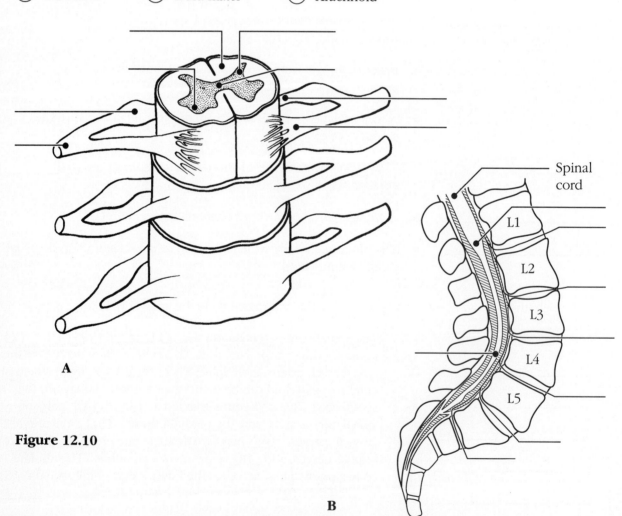

A

Spinal cord

L1

L2

L3

L4

L5

Figure 12.10

B

4. Complete the following statements by writing the missing terms in the answer blanks.

_____ 1.

_____ 2.

_____ 3.

_____ 4.

_____ 5.

_____ 6.

_____ 7.

_____ 8.

_____ 9.   _____ 10.   _____ 11.

The spinal cord sheath is composed of the single __(1)__ layer of the dura mater. Soft, protective padding around the spinal cord fills the __(2)__ space. The space between the arachnoid and pia mater meninges is filled with __(3)__ . The anterior groove extending the length of the cord is called the __(4)__ . The interior central gray matter mass is surrounded by __(5)__ matter. The two sides of the central gray mass are connected by the __(6)__ . Neurons in the __(7)__ horn are somatic motor neurons that serve the skeletal muscles, whereas neurons in parts of the __(8)__ horn are autonomic motor neurons serving the visceral organs. If the dorsal root of a spinal nerve is severed, loss of __(9)__ function follows. If the ventral root is damaged, loss of __(10)__ function occurs. In the embryo, neurons of the dorsal root ganglion arise from the __(11)__ .

5. Match these parts of the adult spinal cord with the embryonic cells that give rise to them. Use the key choices.

KEY CHOICES

A. Alar plate cells   **B.** Basal plate cells   **C.** Neural crest cells

_____ **1.** Sensory neurons   _____ **2.** Motor neurons   _____ **3.** Interneurons

6. Match the terms listed in Column B with the injuries described in Column A.

Column A	Column B
_____ **1.** Loss of sensation	**A.** Flaccid paralysis
_____ **2.** Paralysis without atrophy	**B.** Hemiplegia
_____ **3.** Loss of motor function	**C.** Paralysis
_____ **4.** Traumatic flexion and/or extension of the neck	**D.** Paraplegia
_____ **5.** Result of transection of the cord between T_{11} and L_1	**E.** Paresthesia
	F. Quadriplegia
_____ **6.** Transient period of functional loss induced by trauma to the cord	**G.** Spastic paralysis
_____ **7.** Result of permanent injury to the cervical region of the cord	**H.** Spinal shock
	I. Whiplash
_____ **8.** Paralysis of one side of the body that usually reflects brain injury rather than injury to the spinal cord	

Diagnostic Procedures for Assessing CNS Dysfunction

1. If a statement is true, write the letter T in the answer blank. If a statement is false, change the underlined word(s) and write the correct word(s) in the answer blank.

_____   **1.** An X-ray of gas-filled ventricles is called <u>electroencephalogram</u>.

_____   **2.** A flat electroencephalogram is clinical evidence of brain <u>death</u>.

_____   **3.** Damage to the cerebral arteries of TIA victims is usually assessed by a <u>PET scan</u>.

_____   **4.** The biochemical activity of the brain may be monitored by <u>MRI</u>.

_____   **5.** A lumbar puncture is done below L_1 because the spinal cord ends <u>below</u> that level.

Developmental Aspects of the Central Nervous System

1. Complete the following statements by writing the missing terms in the answer blanks.

_____ **1.**

_____ **2.**

_____ **3.**

_____ **4.**

_____ **5.**

_____ **6.**

During the first month of life, when the prenatal brain and spinal cord are established, the mother's health is critical to her developing infant. __(1)__ can be caused by rubella, and death of developing neurons may result from lack of __(2)__ caused by a mother who smoked during pregnancy. During delivery, lack of oxygen to the newborn's brain may lead to __(3)__ . Normal maturation of the nervous system occurs in a __(4)__ direction, and fine motor control occurs much later than __(5)__ muscle control. In premature infants, body temperature is not well regulated because the __(6)__ is not fully functional.

_____ **7.**

_____ **8.**

_____ **9.**

_____ **10.**

During early embryonic development, several malformations may occur. If the neural folds in the most rostral region fail to fuse, the cerebrum fails to develop, a condition called __(7)__ . If the vertebrae fail to close properly, __(8)__ results. If the resulting cyst contains some spinal cord and spinal nerve roots, it is called a __(9)__ .

_____ **11.**

_____ **12.**

_____ **13.**

_____ **14.**

Growth and maturation of the nervous system accompany progressive __(10)__ of neurons. By adulthood, the brain reaches its maximum weight. Thereafter, because of continual chemical and physical insults, neurons die. Yet learning can continue throughout life because remaining neurons can change their __(11)__ connections. A change in intellect due to a gradual decline in brain-cell function is called __(12)__ . Sudden cessation of blood to a brain area that causes death of brain cells is called __(13)__ . CAT scans of the brains of chronic alcoholics reveal a marked __(14)__ of brain size.

The Incredible Journey:
A Visualization Exercise
for the Nervous System

You climb on the first cranial nerve you see . . .

1. Complete the following narrative by inserting the missing words in the answer blanks.

_____ **1.**

_____ **2.**

_____ **3.**

_____ **4.**

_____ **5.**

_____ **6.**

Nervous tissue is quite densely packed, and it is difficult to envision strolling through its various regions. Imagine instead that each of the various functional regions of the brain has a computerized room in which you might observe what occurs in that particular area. Your assignment is to determine where you are at any given time during your journey through the nervous system.

You begin your journey after being injected into the warm pool of cerebrospinal fluid in your host's fourth ventricle. As you begin your stroll through the nervous tissue, you notice a huge area of branching white matter overhead. As you enter the first computer room, you hear an announcement through the loudspeaker: "The pelvis is tipping too far posteriorly—please correct—we are beginning to fall backward and will soon lose our balance." The computer responds immediately, decreasing impulses to the posterior hip muscles and increasing impulses to the anterior thigh muscles. "How is that, proprioceptor 1?" From this information, you determine that your first stop is the __(1)__ .

At the next stop, you hear "Blood pressure to head is falling; increase sympathetic nervous system stimulation of the blood vessels." Then, as it becomes apparent that your host has not only stood up but is going to run, you hear "Increase rate of impulses to the heart and respiratory muscles—we are going to need more oxygen and a faster blood flow to the skeletal muscles of the legs." You recognize that this second stop must be the __(2)__ .

Computer room 3 presents a problem. There is no loudspeaker here; instead, incoming messages keep flashing across the wall, giving only bits and pieces of information. "Four hours since last meal—stimulate appetite center. Slight decrease in body temperature—initiate skin vasoconstriction. Mouth dry—stimulate thirst center. Um, a stroke on the arm—stimulate pleasure center." Looking at what has been recorded here—appetite, temperature, thirst, and pleasure—you conclude that this has to be the __(3)__ .

Continuing your journey upward toward the higher brain centers, finally you are certain you have reached the cerebral cortex. The first center you visit is quiet, like a library with millions of encyclopedias of facts and recordings of past input. You conclude that this must be the area where __(4)__ are stored and that you are probably in the __(5)__ lobe. The next stop is close by. As you enter the computer center, you once again hear a loudspeaker: "Let's have the motor instructions to say tintinnabulation—we don't want them to think we're tongue-tied." This area is obviously __(6)__ . Your final stop in the cerebral cortex is a very hectic center. Electrical impulses are traveling back and forth between giant neurons, sometimes in different directions and sometimes back and forth between a small number of neurons. Watching intently, you try to make some sense out of these interactions, and

_____ 7.

_____ 8.

_____ 9.

_____ 10.

_____ 11.

_____ 12.

you suddenly realize that this *is* what is happening here. The neurons are trying to make some sense out of something, which helps you decide that this must be the brain area where __(7)__ occurs in the __(8)__ lobe.

You hurry out of this center and retrace your steps back to the cerebrospinal fluid, deciding en route to observe a cranial nerve. You decide to pick one randomly and follow it to the organ it serves. You climb onto the first cranial nerve you see and slide down past the throat. Picking up speed, you quickly pass the heart and lungs and see the stomach and small intestine coming up fast. A moment later you land on the stomach, and now you know that this wandering nerve has to be the __(9)__ . As you look upward, you see that the nerve is traveling almost straight up and decide you'll have to find an alternative route back to the cerebrospinal fluid. You begin to walk posteriorly until you find a spinal nerve, which you follow until you reach the vertebral column. You squeeze between two adjacent vertebrae to follow the nerve to the spinal cord. With your pocket knife, you cut away the tough connective tissue covering the cord. Thinking that the __(10)__ covering deserves its name, you finally manage to cut an opening large enough to get through, and you return to the warm bath of cerebrospinal fluid that it encloses. At this point you are in the __(11)__ , and from here you swim upward until you get to the lower brain stem. Once there, it should be an easy task to find the holes leading into the __(12)__ ventricle, where your journey began.

CHALLENGING YOURSELF

At the Clinic

1. Following a train accident, a woman with an obvious head injury is observed stumbling about the scene. An inability to walk properly and loss of balance are quite obvious. What brain region was injured?

2. A child is brought to the clinic with a high temperature. The doctor states that the child's meninges are inflamed. What name is given to this condition and how is it diagnosed?

3. Jemal, an elderly man with a history of TIAs, complained to his daughter that he had a severe headache. Shortly thereafter, he lapsed into a coma. At the hospital, he was diagnosed as having a brain hemorrhage. Which part of the brain was damaged by the hemorrhage?

4. A young man has just received serious burns resulting from standing with his back too close to a bonfire. He is muttering that he never felt the pain; otherwise, he would have smothered the flames by rolling on the ground. What part of his CNS might be malfunctioning?

5. Elderly Mrs. Barker has just suffered a stroke. She is able to understand verbal and written language, but when she tries to respond, her words are garbled. What cortical region has been damaged by the stroke?

6. An elderly man is brought to the clinic by his wife, who noticed that his speech is slurred, the right side of his face is slack, and he has difficulty swallowing. In which area of the brain would a stroke be suspected?

7. After removal of a brain tumor, diagnostic tests reveal that the patient Hymie cannot distinguish an object in his hand by touch; he must look at the object to name it. However, he can describe the physical characteristics of the object and has no difficulty speaking or writing the correct name of the object. What part of the brain has been damaged?

8. Examining an elderly woman after she suffered a stroke turned out to be frustrating for the clinic staff. Although quite fluent and apparently intelligent, the woman showed absolutely no interest in answering any of their questions or taking any diagnostic tests. In what part of the brain would you suspect damage?

9. Benny, a 7-year-old boy, is referred to the clinic by his teacher, who noticed that he became restless when the class was reading or when watching her write on the board. The clinical staff noted that the boy was left-handed. What condition might you suspect?

10. Mrs. Tonegawa, a new mother, brings her infant to the clinic because it has suffered repeated seizures. Upon questioning, she states that her labor was unusually long and difficult. What condition do you suspect? Will the infant's condition worsen?

11. A semiconscious young woman is brought to the hospital by friends after falling from a roof. She did not lose consciousness immediately, and she was initially lucid. After a while, though, she became confused and then unresponsive. What is a likely explanation of her condition?

12. Why is Alzheimer's disease difficult to diagnose in cases of senility?

13. An elderly man has developed an unstable gait. Blood tests reveal he has syphilis. What name is given to his motor disorder?

14. Beth is brought to the clinic by her frantic husband. He explains that she has gradually lost control of her right hand. No atrophy of the muscles is apparent. Should upper *or* lower motor neuron damage be investigated?

15. Mr. Belsky is complaining of painless muscular weakness in both hands. His speech is not quite clear, and, when asked, he admits to having difficulty swallowing. What chronic condition, which causes hardening of the spinal cord, is suspect in this case?

16. When their second child was born, the Trevors were told the baby had spina bifida and that he would be kept in intensive care for a week, after which time a "shunt would be put in." Also, immediately after the birth, an operation was performed on the infant's lower back. The parents were told that this operation went well but their son would always be a "little weak in the ankles." Explain the statements in quotation marks in more informative and precise language.

Stop and Think

1. Why do you think so much of the cerebral cortex is devoted to sensory and motor connections to the eyes?

2. Attempt to explain, in embryological terms, why the homunculus of the precentral gyrus matches that of the postcentral gyrus so well.

3. Do all tracts from the cerebrum to the spinal cord go through the entire brain stem?

4. Contrast damage to the primary visual cortex with damage to the visual association cortex.

5. Any CSF accumulation will cause hydrocephalus, which in a fetus will lead to expansion of the skull. In a fetus, would blockage of CSF circulation in the cerebral aqueduct have a different effect than blockage of the arachnoid villi?

6. What does mannitol's ability to make capillary cells shrivel tell you about the cells' permeability to mannitol?

7. A victim of a motorcycle accident fractured vertebra T_{12}, and there was concern that the spinal cord was crushed at the level of this vertebra. At what spinal cord segment was the damage expected? Choose and explain: (a) between C_4 and C_8, and the ability to move the arms was tested, (b) at C_1, and the respiratory movements of the diaphragm were tested, (c) spinal-cord level L_3, (d) there could be no deficits because the cord always ends above T_{12}.

8. Why is hemiplegia more likely to be a result of brain injury than of spinal cord injury?

9. Which would more likely result from injury exclusively to the posterior side of the spinal cord—paresthesia or paralysis? Explain your answer.

10. Why does exposure to toxins have more devastating effects early in gestation than late in pregnancy?

11. Every time Spike went to a boxing match, he screamed for a knockout. Then his friend Rudy explained what happens when someone gets knocked out. What is the explanation?

12. Ralph had brain surgery to remove a small intracranial hematoma (blood mass). He was allergic to general anesthesia, so the operation was done under local anesthesia. Ralph remained conscious while the surgeon removed a small part of the skull. The operation went well, and Ralph asked the surgeon to mildly stimulate his (unharmed) postcentral gyrus with an electrode. The surgeon did so. What happened? Choose and explain: (a) Ralph was seized with uncontrollable rage; (b) he saw things that were not there; (c) he asked to see what was touching his hand, but nothing was; (d) he started to kick; (e) he heard his mother's voice from 30 years ago.

COVERING ALL YOUR BASES

Multiple Choice

Select the best answer or answers from the choices given.

1. Which structure specifically gives rise to ganglionic neurons?
   A. Neural plate   C. Neural crest
   B. Rostrum   D. Neural folds

2. The secondary brain vesicle that is least developed in the adult brain is the:
   A. telencephalon   C. metencephalon
   B. mesencephalon   D. rhombencephalon

3. Which is an incorrect association of brain region and ventricle?
   A. Mesencephalon—third ventricle
   B. Cerebral hemispheres—lateral ventricles
   C. Pons—fourth ventricle
   D. Medulla—fourth ventricle

4. The hindbrain does not include the:
   A. cerebellum   C. cerebral aqueduct
   B. pons   D. fourth ventricle

5. Which of the following is not part of the brain stem?
   A. Medulla   C. Pons
   B. Cerebellum   D. Midbrain

6. Connecting a ventricle to the subarachnoid space is the function of the:
   A. interventricular foramen
   B. cerebral aqueduct
   C. lateral aperture
   D. median aperture

7. Which would be associated with the cerebral cortex?
   A. Parieto-occipital sulcus
   B. Corpus callosum
   C. Central sulcus
   D. Cerebral peduncles

8. The discrete correlation of body regions to CNS structures is:
   A. a homunculus   C. lateralization
   B. somatotopy   D. cephalization

9. Which is not associated with the frontal lobe?
   A. Olfaction
   B. Motor control of the eyes
   C. Prefrontal cortex
   D. Skilled motor programs

10. Regions involved with language include:
    A. auditory association area
    B. prefrontal cortex
    C. Wernicke's area
    D. gnostic area

11. When neurons in Wernicke's area send impulses to neurons in Broca's area, the white matter tracts utilized are:
    A. commissural fibers
    B. projection fibers
    C. association fibers
    D. anterior funiculus

12. The basal nuclei include:
    A. hippocampus
    B. caudate nucleus
    C. lentiform nucleus
    D. mammillary bodies

13. Which of the brain areas listed are involved in normal voluntary muscle activity?
    A. Amygdala   C. Medullary pyramids
    B. Putamen   D. Precentral gyrus

14. If impulses to the cerebral cortex from the thalamus were blocked, which of the following sensory inputs would get through to the appropriate sensory cortex?
    A. Visual impulses
    B. Auditory impulses
    C. Olfactory impulses
    D. General sensory impulses

15. Functions controlled by the hypothalamus include:
    A. crude interpretation of pain
    B. regulation of many homeostatic mechanisms
    C. setting of some biological rhythms
    D. secretion of melatonin

16. The pineal gland is located in the:
    A. hypophysis cerebri
    B. mesencephalon
    C. epithalamus
    D. corpus callosum

17. Which of the following are part of the midbrain?
    A. Corpora quadrigemina
    B. Cerebellar peduncles
    C. Substantia nigra
    D. Nucleus of cranial nerve V

18. Cranial nerves with their nuclei in the pons include:
    A. facial   C. trigeminal
    B. vagus   D. trochlear

19. Which of the following brain parts have direct connections to the cerebellum?
    A. Thalamus   C. Pons
    B. Midbrain   D. Medulla

20. Functions that are at least partially overseen by the medulla are:
    A. regulation of the heart
    B. maintaining equilibrium
    C. regulation of respiration
    D. visceral motor function

21. Which of the following are important in cerebellar processing?
    A. Feedback to cerebral motor cortex
    B. Input from body parts
    C. Output for subconscious motor activity
    D. Direct impulses from motor cortex via a separate cerebellar pathway

22. Parts of the limbic system include:
    A. cingulate gyrus   C. fornix
    B. hippocampus   D. septal nuclei

23. Inability to prevent expressing one's emotions most likely indicates damage to the:
    A. affective brain
    B. cognitive brain
    C. affective brain's control of the cognitive brain
    D. cognitive brain's control of the affective brain

24. Which statements concerning the reticular activating system are true?
    A. It consists of neural columns extending through the brain stem.
    B. Its connections reach as far as the cerebral cortex.
    C. It is housed primarily in the thalamus.
    D. It filters sensory input to the cerebrum.

25. Relative to the cranial meninges:
    A. the arachnoid produces CSF
    B. the dura contains several blood sinuses
    C. the pia mater has two layers
    D. three dural folds help support brain tissue

26. Which structures are directly involved with formation, circulation, and drainage of CSF?
    A. Ependymal cilia
    B. Ventricular choroid plexuses
    C. Arachnoid villi
    D. Serous layers of the dura mater

27. A type of brain trauma likely to result in a long-term coma is:
    A. concussion
    B. cortical contusion
    C. brain stem contusion
    D. uncontrolled subdural hemorrhage

28. Which of the following are associated with the conus medullaris?
    A. Filum terminale
    B. The end of the dural sheath
    C. Cauda equina
    D. Location of lumbar puncture

29. The spinal cord feature associated with the leash of nerves supplying the upper limbs is the:
    A. brachial plexus
    B. brachial enlargement
    C. cervical enlargement
    D. lateral gray horns

30. Features associated with the anterior surface of the spinal cord include the:
    A. ventral horns of gray matter
    B. median fissure
    C. median sulcus
    D. root ganglia

31. Poliomyelitis affects the:
    A. posterior white columns
    B. posterior gray horns
    C. anterior white columns
    D. anterior gray horns

32. Which spinal cord tracts carry impulses for conscious sensations?
    A. Fasciculus gracilis
    B. Lateral spinothalamic
    C. Anterior spinocerebellar
    D. Fasciculus cuneatus

33. Damage to which descending tract would be suspected if a person had unilateral poor muscle tone and posture?
    A. Reticulospinal   C. Tectospinal
    B. Rubrospinal   D. Vestibulospinal

34. Damage to the lower motor neurons may result in:
    A. paresthesia   C. spastic paralysis
    B. flaccid paralysis   D. paraplegia

35. Which disorders result from failure of some region of the neural groove to fuse into the neural tube?
    A. Anencephaly   C. Hydrocephalus
    B. Cerebral palsy   D. Spina bifida cystica

36. Which neuron parts occupy the gray matter in the spinal cord?
    A. Tracts of long axons
    B. Motor neuron cell bodies
    C. Sensory neuron cell bodies
    D. Nerves

37. A professor unexpectedly blew a loud horn in his anatomy classroom, and all his students looked up, startled. These reflexive movements of their neck and eye muscles were mediated by:
    A. cerebral cortex   C. raphe nuclei
    B. inferior olives   D. inferior colliculus

Word Dissection

For each of the following word roots, fill in the literal meaning and give an example, using a word found in this chapter.

Word root	Translation	Example
1. campo	_____	_____
2. collicul	_____	_____
3. commis	_____	_____
4. enceph	_____	_____
5. falx	_____	_____
6. forn	_____	_____
7. gyro	_____	_____
8. hippo	_____	_____
9. infundib	_____	_____
10. isch	_____	_____
11. nigr	_____	_____
12. rhin	_____	_____
13. rostr	_____	_____
14. uncul	_____	_____
15. uncus	_____	_____

13

The Peripheral Nervous System and Reflex Activity

Student Objectives

When you have completed the exercises in this chapter, you will have accomplished the following objectives:

Overview of the Peripheral Nervous System

1. Define *peripheral nervous system* and list its components.
2. Classify sensory receptors according to body location, stimulus detected, and structure.
3. Describe receptor potentials and define *adaptation*.
4. Define *nerve* and describe the general structure of a nerve.
5. Distinguish between sensory, motor, and mixed nerves.
6. Define *ganglion* and indicate the general location of ganglia in the body periphery.
7. Describe the process of nerve fiber regeneration.
8. Compare and contrast the motor endings of somatic and autonomic nerve fibers.

Cranial Nerves

9. Name the 12 pairs of cranial nerves and describe the body region and structures innervated by each.

Spinal Nerves

10. Describe the formation of a spinal nerve and distinguish between spinal roots and rami. Describe the general distribution of the ventral and dorsal rami.
11. Define *plexus*. Name the major plexuses, their origin sites, and the major nerves arising from each, and describe the distribution and function of the peripheral nerves.

Reflex Activity

12. Distinguish between autonomic and somatic reflexes.
13. Compare and contrast stretch, flexor, and crossed extensor reflexes.

Developmental Aspects of the Peripheral Nervous System

14. Describe the developmental relationship between the segmented arrangement of peripheral nerves, skeletal muscles, and skin dermatomes.

Ready, set, go!

The part of the nervous system that lies outside the brain and spinal cord is called the peripheral nervous system (PNS). The PNS consists of sensory receptors, all peripheral nerves (12 pairs of cranial nerves, 31 pairs of spinal nerves) and their associated ganglia, and the motor endings. The PNS serves as the two-way communication network between the environment (both inside and outside the body) and the central nervous system (CNS), where the sensory information generated by the PNS is received, interpreted, and used to maintain homeostasis.

The topics for study in Chapter 13 are the sensory receptors and the structure and types of PNS nerves, their pathways, and the body regions they serve. Reflex activities are also studied in this chapter.

BUILDING THE FRAMEWORK

Overview of the Peripheral Nervous System

1. Complete the following statements by writing the missing terms in the answer blanks.

_____ 1.

_____ 2.

_____ 3.

_____ 4.

_____ 5.

_____ 6.

_____ 7.

The two functional divisions of the PNS are the __(1)__ division, in which the nerve fiber type called __(2)__ carries impulses toward the CNS, and the __(3)__ division, in which the nerve fiber type called __(4)__ carries impulses away from the CNS. Nearly all peripheral nerves, however, are __(5)__, carrying both types of fibers. The motor division of the PNS includes two types of fibers, the __(6)__ and __(7)__ motor nerve fibers. Skeletal muscles are innervated by __(8)__ fibers, whereas visceral organs are innervated by __(9)__ fibers.

_____ 8.   _____ 9.

2. By assigning them numbers from 1 to 9, arrange the following elements in order of stimulation. (*Note:* two of the elements will be used twice.)

Spinal nerve   Ventral ramus of spinal nerve   Dorsal root of spinal nerve

Sensory nerve fiber   Motor nerve fiber   Receptor   Effector

3. Classify the receptor types described below by *location* (make a choice from Key A) and then by *stimulus type detected* (make a choice from Key B).

Key A: **A.** Exteroceptor   **Key B:**   **1.** Chemoreceptor   **4.** Photoreceptor

          B. Interoceptor   **2.** Mechanoreceptor   **5.** Thermoreceptor

          C. Proprioceptor   **3.** Nociceptor

_____ , _____   **1.** In skeletal muscles; respond to muscle stretch

_____ , _____   **2.** In the walls of blood vessels; respond to oxygen content of surrounding interstitial fluid

_____ , _____   **3.** In the skin; respond to a hot surface

_____ , _____   **4.** In the eyes; respond to light

_____ , _____   **5.** In the inner ear; respond to sound

_____ , _____   **6.** In the skin; respond to acid splashing on the skin

_____ , _____   **7.** In the wall of blood vessels; respond to blood pressure

_____ , _____   **8.** In the stomach wall; respond to an over-full stomach

_____ , _____   **9.** In the nose; respond to a skunk's spray. (Yuk!)

4. In relation to sensory receptors, circle the term that does not belong in each of the following groupings.

1. Cutaneous receptors   Free dendritic endings   Tendon stretch   Pain and touch

2. Meissner's corpuscle   Encapsulated dendritic endings   Dermal papillae

    Numerous in muscle

3. Largest of corpuscular receptors   Light touch   Pacinian corpuscle

    Multiple layers of Schwann cells

4. Thermoreceptor   Hair movement   Light touch   Root hair plexus

5. Krause's end bulb   Meissner's corpuscle   Ruffini's corpuscle   Touch receptor

6. Proprioceptors   Nociceptors   Sensitive to tendon stretch   Muscle spindle

7. Merkel discs   Sensitive to joint orientation   Articular capsules

    Joint kinesthetic receptors

8. Mechanoreceptor   Thermoreceptor   Interoceptor   Chemoreceptor

9. Free dendritic endings   Root hair plexuses   Merkel discs

    Modified free dendritic endings

5. Identify the highly simplified cutaneous receptors shown in Figure 13.1 and color the figure as it strikes your fancy.

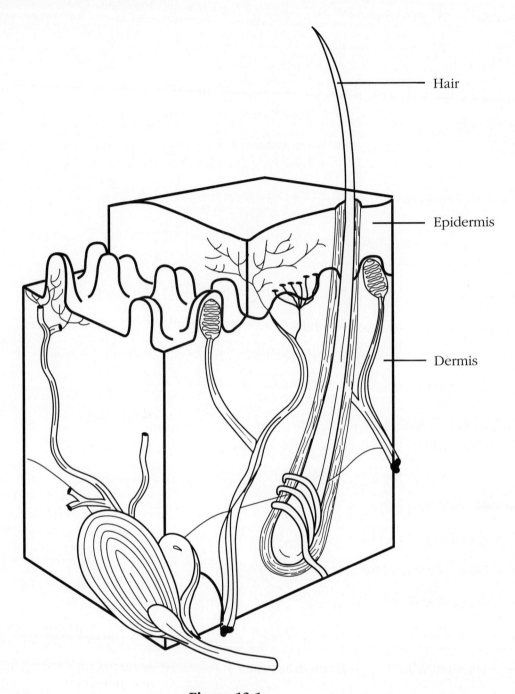

Hair

Epidermis

Dermis

Figure 13.1

6. Certain activities and sensations are listed below. Using the key choices, select the specific receptor type that would be activated by the activity or sensation. Insert the correct letters in the answer blanks. Note that more than one receptor type may be activated in some cases.

KEY CHOICES

A. Free dendritic endings (pain)   **F.** Muscle spindle

B. Golgi tendon organ   **G.** Pacinian corpuscle

C. Krause's end bulb   **H.** Joint kinesthetic receptor

D. Meissner's corpuscle   **I.** Root hair plexuses

E. Merkel discs   **J.** Ruffini's corpuscle

_____ **1.** Standing on hot pavement (identify two)

_____ **2.** Feeling a pinch (identify two)

_____ **3.** Leaning on a shovel (identify four)

_____ **4.** Muscle sensations when rowing a boat (identify three)

_____ **5.** Feeling a caress (identify three)

_____ **6.** Feeling the coldness of an iced-tea glass

7. Complete the following paragraph by inserting the missing words in the appropriate answer blanks.

_____ **1.**   In all cases the stimulus for a sensory receptor is some form of **(1)** . As the receptor responds to the stimulus, changes

_____ **2.**   in the **(2)** of the receptor membrane occur that result in the generation of a(n) **(3)** called a(n) **(4)** . If the stimulus is

_____ **3.**   strong enough, the **(4)** will become a(n) **(5)** and a(n)

_____ **4.**   **(6)** will be transmitted. The entire sequence of events described is referred to as **(7)** .

_____ **5.** _____ **6.** _____ **7.**

8. Several characteristics of motor endings (both autonomic and somatic) are listed below. Differentiaate between the autonomic and somatic motor endings by checking (✔) all those that distinguish the synapses en passant of autonomic fibers.

_____ **1.** Wide synaptic cleft   _____ **5.** Clustered boutons

_____ **2.** Narrow synaptic cleft   _____ **6.** Glycoprotein-rich basal lamina in synaptic cleft

_____ **3.** May contain NE   _____ **7.** Innervate skeletal muscle fibers

_____ **4.** Beadlike varicosities   _____ **8.** Innervate smooth muscle fibers and glands

9. Name the four functional types of nerve fibers.

_____

_____

10. Figure 13.2 is a diagrammatic view of a section of a nerve wrapped in its connective tissue coverings. Select different colors to identify the following structures and use them to color the figure. Then, label these sheaths, indicated by leader lines on the figure.

○ Endoneurium

○ Perineurium

○ Epineurium

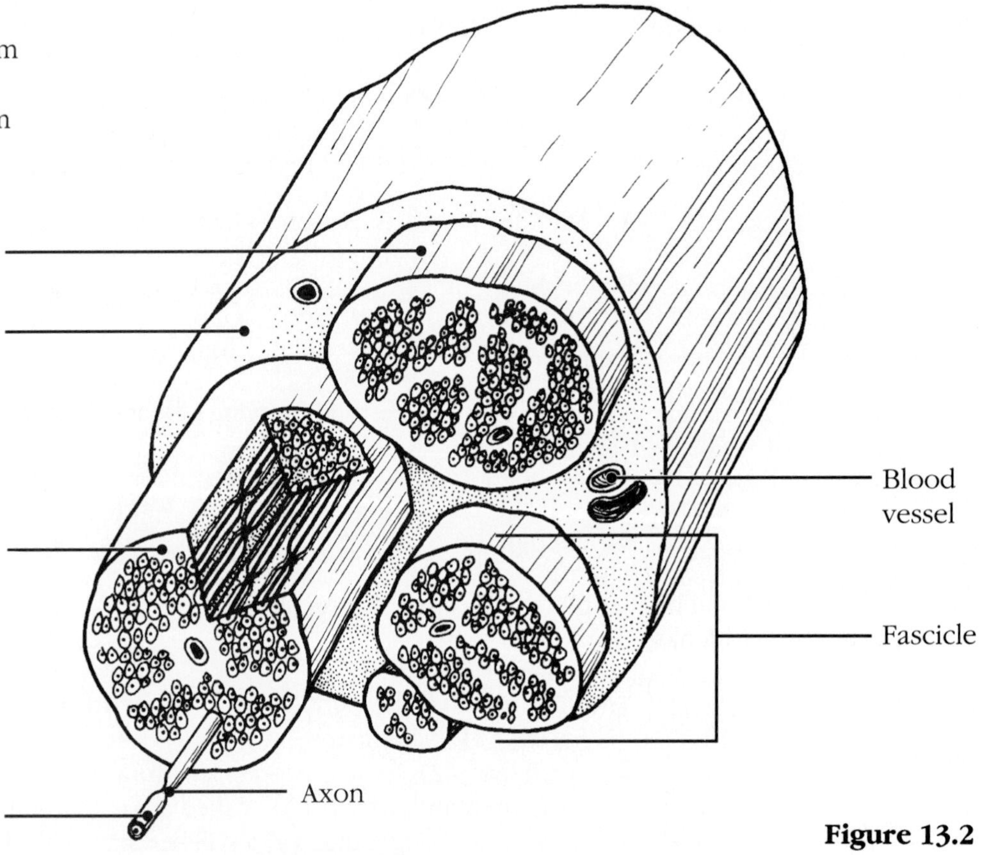

Blood vessel

Fascicle

Axon

Myelin sheath

Figure 13.2

11. Complete the following statements relating to regeneration of nervous tissue. Write the missing terms in the answer blanks.

_____ **1.**

_____ **2.**

_____ **3.**

_____ **4.**

_____ **5.**

_____ **6.**

_____ **7.**   _____ **8.**   _____ **9.**

Lost neurons cannot be replaced because mature neurons cannot __(1)__ . An axon in the __(2)__ nervous system can regenerate successfully if the injury is distal to the __(3)__ , which must remain intact. After the injured axon ends seal off, the axon and its myelin sheath distal to the injury __(4)__ . This debris is phagocytized by __(5)__ and nearby __(6)__ . Channels, formed by __(7)__ that align within the __(8)__ , guide growing axonal processes toward their original contacts. When not properly guided, new axonal growths may form a mass of tissue called a(n) __(9)__ .

_____ 10.

_____ 11.

_____ 12.

_____ 13.

_____ 14.

Nerve fibers located in the __(10)__ nervous system essentially never regenerate. Although the __(11)__ cells clean up the degenerated neural debris, the __(12)__ that form the myelin sheath __(13)__ . Consequently no guiding channels are formed. The myelinating cells (answer 12) are replaced by __(14)__ , which form scar tissue that blocks any new axonal growths.

Cranial Nerves

1. Match the names of the cranial nerves listed in Column B with the appropriate descriptions in Column A by inserting the correct letters in the answer blanks.

Column A

_____ 1. The only nerve that originates from the forebrain

_____ 2. The only cranial nerve that extends beyond the head and neck region

_____ 3. The largest cranial nerve

_____ 4. Fibers arise from the sensory apparatus within the inner ear

_____ 5. Supplies somatic motor fibers to the lateral rectus muscle of the eye

_____ 6. Has five major branches; transmits sensory, motor, and autonomic impulses

_____ 7. Cell bodies are located within their associated sense organs

_____ 8. Transmits sensory impulses from pressure receptors of the carotid artery

_____ 9. Mixed nerve formed from the union of a cranial root and a spinal root

_____ 10. Innervates four of the muscles that move the eye and the iris

_____ 11. Supplies the superior oblique muscle of the eye

_____ 12. Two nerves that supply the tongue muscles

_____ 13. Its _tract_ is frequently misidentified as this nerve

_____ 14. Serves muscles covering the facial skeleton

Column B

A. Accessory

B. Abducens

C. Facial

D. Glossopharyngeal

E. Hypoglossal

F. Oculomotor

G. Olfactory

H. Optic

I. Trigeminal

J. Trochlear

K. Vagus

L. Vestibulocochlear

2. Provide the name and number of the cranial nerves involved in each of the following activities, sensations, or disorders. Write your answers in the answer blanks.

_____ **1.** Hyperextending and flexing the neck

_____ **2.** Smelling freshly baked bread

_____ **3.** Constricting the pupils for reading

_____ **4.** Stimulates the mobility and secretory activity of the digestive tract

_____ **5.** Involved in frowning and puzzled looks

_____ **6.** Crunching an apple and chewing gum

_____ **7.** Tightrope walking and listening to music

_____ **8.** Gagging and swallowing; tasting bitter foods

_____ **9.** Involved in "rolling" the eyes (three nerves; provide numbers only)

_____ **10.** Feeling a toothache

_____ **11.** Watching tennis on TV

_____ **12.** If this nerve is damaged, deafness results

_____ **13.** Stick out your tongue!

_____ **14.** Inflammation of this nerve may cause Bell's palsy

_____ **15.** Damage to this nerve may cause anosmia

_____ **16.** Double vision results if this nerve is paralyzed

3. Match each of the cranial nerves listed in Column A to its terminal connection in the cerebral cortex (an item from Column B).

Column A

_____ **1.** Oculomotor

_____ **2.** Olfactory

_____ **3.** Facial

_____ **4.** Trigeminal

_____ **5.** Optic

_____ **6.** Vagus

Column B

A. Frontal eye field

B. Hypothalamus

C. Parietal lobe

D. Primary motor cortex

E. Occipital lobe

F. Temporal lobe

4. The 12 pairs of cranial nerves are indicated by leader lines on Figure 13.3. First, label each by name and Roman numeral on the figure. Then, color each nerve with a different color.

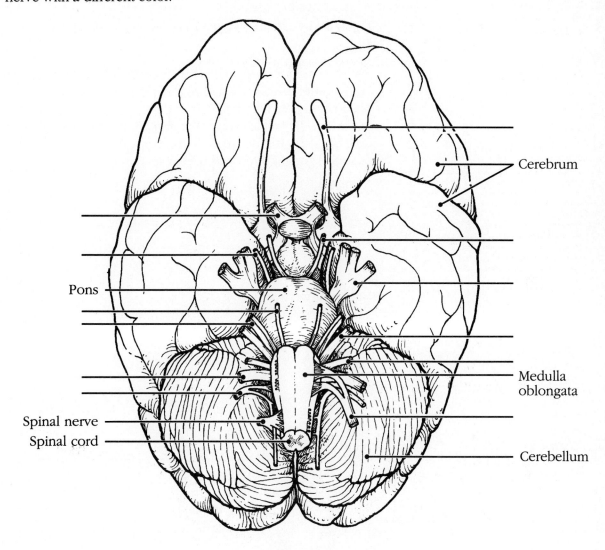

Cerebrum

Pons

Medulla oblongata

Spinal nerve

Spinal cord

Cerebellum

Figure 13.3

5. Name the five major motor branches of the facial nerve.

_____ , _____ , _____ ,

_____ , and _____ .

Spinal Nerves

1. Identify all structures provided with leader lines in Figure 13.4. Color the diagram as you wish and then complete the following statements by inserting your responses in the answer blanks.

_____ 1.

_____ 2.

_____ 3.

_____ 4.

_____ 5.

_____ 6.

_____ 7.

Each spinal nerve is formed from the union of **(1)** and **(2)** . After its formation, the spinal nerve splits into **(3)** . The ventral rami of spinal nerves C_1–T_1 and L_1–S_4 take part in forming **(4)** , which serve the **(5)** of the body. The ventral rami of T_1–T_{12} run between the ribs to serve the **(6)** . The posterior rami of the spinal nerves serve the **(7)** .

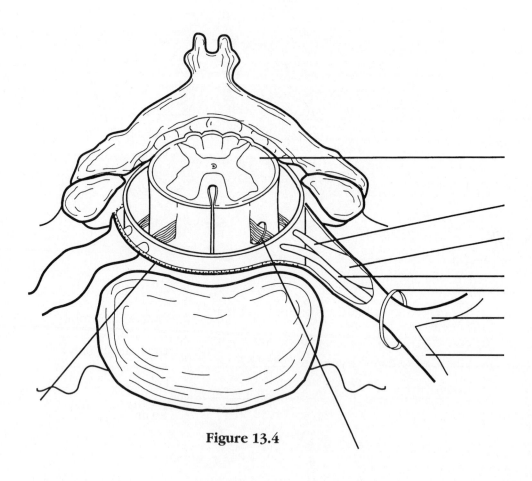

Figure 13.4

2. Referring to the distribution of spinal nerves, circle the term that does not belong in each of the following groupings.

1. Division of spinal nerve   Two rami   Meningeal branch   Two roots

2. Thoracic roots   Lumbar roots   Sacral roots   Cauda equina

3. Activation of skeletal muscles   Ventral root   Motor root   Afferent fibers

4. Anterior ramus   Motor function   Sensory function   Ventral root

5. Autonomic nervous system   Specific skin segment   Dermatome

   All spinal nerves except C_1

6. Ventral ramus   Dorsal ramus   Division of spinal nerve   Two spinal roots

3. Name the major nerves that serve the following body parts. Insert your responses in the answer blanks.

_____ **1.** Head, neck, shoulders (name plexus only)

_____ **2.** Diaphragm

_____ **3.** Hamstrings

_____ **4.** Leg and foot (name two)

_____ **5.** Most anterior forearm muscles

_____ **6.** Flexor muscles of the arm

_____ **7.** Abdominal wall (name plexus only)

_____ **8.** Thigh flexors and knee extensors

_____ **9.** Medial side of the hand

_____ **10.** Plexus serving the upper limb

_____ **11.** Large nerve composed of two nerves in a common sheath

_____ **12.** Buttock, pelvis, lower limb (name plexus only)

_____ **13.** Largest branch of the brachial plexus

_____ **14.** Largest nerve of the lumbar plexus

_____ **15.** Adductor muscles of the thigh

_____ **16.** Gluteus maximus

4. Figure 13.5 is an anterior view of the principal nerves arising
from the cords of the brachial plexus. Select five different
colors and color the coding circles and the nerves
listed below. Also, label each nerve by inserting
its name at the appropriate leader line.

○ Axillary nerve

○ Musculocutaneous nerve

○ Median nerve

○ Radial nerve

○ Ulnar nerve

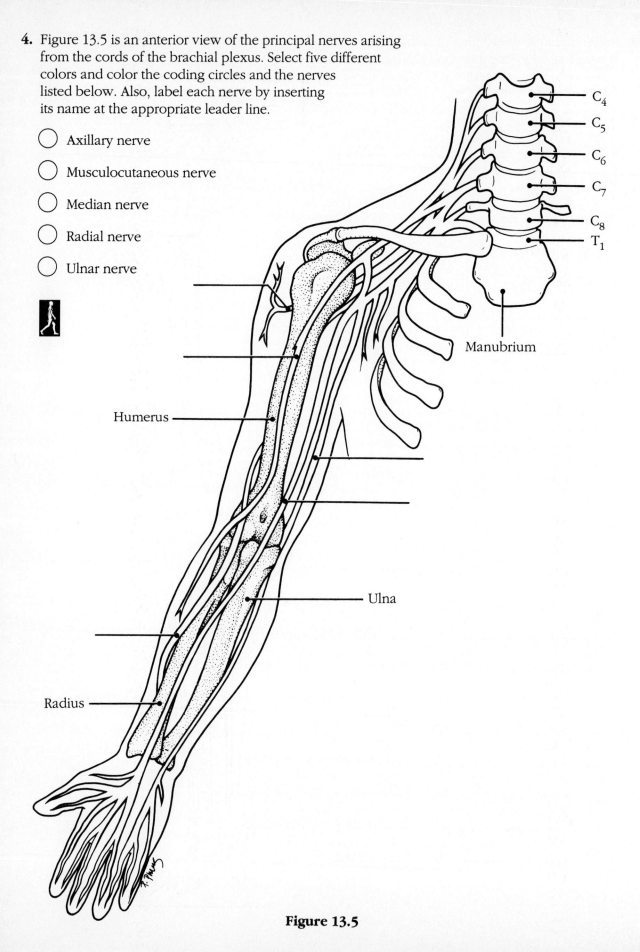

C_4

C_5

C_6

C_7

C_8

T_1

Manubrium

Humerus

Ulna

Radius

Figure 13.5

5. Figure 13.6 is a posterior view of the major
nerves arising from the cords of the sacral plexus.
Choose seven colors and color the coding circles
and the nerves listed below. Then, identify
each nerve by inserting its name at the
appropriate leader line.

○ Common peroneal nerve

○ Inferior gluteal nerve

○ Plantar nerve branches

○ Sciatic nerve

○ Superior gluteal nerve

○ Sural nerve

○ Tibial nerve

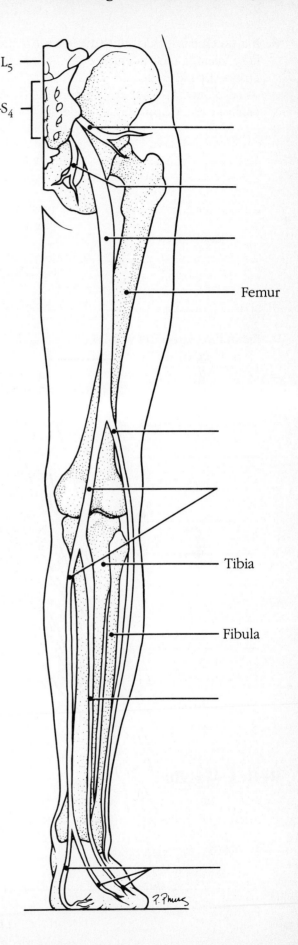

Figure 13.6

6. If a statement is true, write the letter T in the answer blank. If a statement is false, change the underlined word(s) and write the correct word(s) in the answer blank.

_____ **1.** Inability to extend the hand at the wrist (wristdrop) can be caused by injury to the <u>median</u> nerve.

_____ **2.** Injury to the <u>common peroneal</u> nerve causes loss of foot and ankle movements (footdrop).

_____ **3.** Thumb and little finger opposition is lost if the <u>axillary</u> nerve is injured.

_____ **4.** Injury to the <u>radial</u> nerve causes loss of flexion of the fingers at the distal interphalangeal joints (clawhand).

_____ **5.** Irritation of the <u>phrenic nerve</u> may cause hiccups.

7. Fill in the blanks in the following paragraph concerning pain. (Pain is considered in *A Closer Look.*)

_____ **1.**

_____ **2.**

_____ **3.**

_____ **4.**

_____ **5.**

_____ **6.**

_____ **7.**

_____ **8.**

_____ **9.**

Pain receptors are __(1)__ sensory nerve endings that respond to __(2)__ stimuli. The most potent pain-producing chemical known is __(3)__, which also triggers __(4)__. Pain arising from the skin is called __(5)__, while that arising from internal organs is __(6)__. Superficial somatic pain is sharp, __(7)__ pain; deep somatic pain is __(8)__ or __(9)__ pain. The neurotransmitter for pain is __(10)__. The stimulus intensity determines the __(11)__ for pain; pain __(12)__ is determined more by psychological factors. According to the gate control theory, gating occurs in the __(13)__ of the dorsal horn. Stimulation of large __(14)__ fibers inhibits transmission of __(15)__, i.e., it "closes the gate." This gate control theory provides an explanation for the effects of massage.

_____ **10.**   _____ **12.**   _____ **14.**

_____ **11.**   _____ **13.**   _____ **15.**

Reflex Activity

1. Answer the following questions by writing your answers in the answer blanks.

   1. Define the term *reflex.* _____

_____

2. Is all reflex activity inborn? Explain. _____

_____

3. List in order the five essential elements of a reflex arc.

a. _____   c. _____   e. _____

b. _____   d. _____

4. List the two functional classifications of reflexes. _____

_____

5. Name the type of reflex that can occur without the involvement of higher centers.

_____

2. Using the key choices, identify the types of reflexes involved in each of the following situations.

KEY CHOICES

A. Somatic reflex(es)   **B.** Autonomic reflex(es)

_____ **1.** Patellar (knee-jerk) reflex

_____ **2.** Pupillary light reflex

_____ **3.** Effectors are skeletal muscles

_____ **4.** Effectors are smooth muscle and glands

_____ **5.** Flexor reflex

_____ **6.** Regulation of blood pressure

_____ **7.** Salivary reflex

_____ **8.** Clinical testing for spinal cord assessment

3. Complete the following statements by writing the missing terms in the answer blanks:

_____ **1.**

_____ **2.**

_____ **3.**

_____ **4.**

_____ **5.**

_____ **6.**

_____ **7.**

_____ **8.**

_____ **9.**

_____ **10.**

Stretch reflexes are initiated by __(1)__, which monitor changes in muscle length. Each consists of small __(2)__ enclosed in a capsule. There are two types of these cells: the larger __(3)__ have their nuclei in their middle region only, while the smaller __(4)__ have nuclei distributed throughout their length. The nerve endings wrapped around these cells are also of two types, responding to two different aspects of stretch: the __(5)__ endings of the larger __(6)__ are stimulated by amount and rate of stretch. The __(7)__ endings of the small __(8)__ respond only to degree of stretch. Motor innervation to these special muscle cells is via __(9)__, which arise from small motor neurons. The large, contractile __(10)__ are innervated by __(11)__. These nerve fibers' cell bodies, known as __(12)__, are excited in the stretch reflex; antagonists are inhibited via __(13)__.

_____ **11.**   _____ **13.**

_____ **12.**

4. Some of the structural elements described in Exercise 3 are illustrated in Figure 13.7. Identify each of the numbered structures by matching it to the terms listed below the diagram.

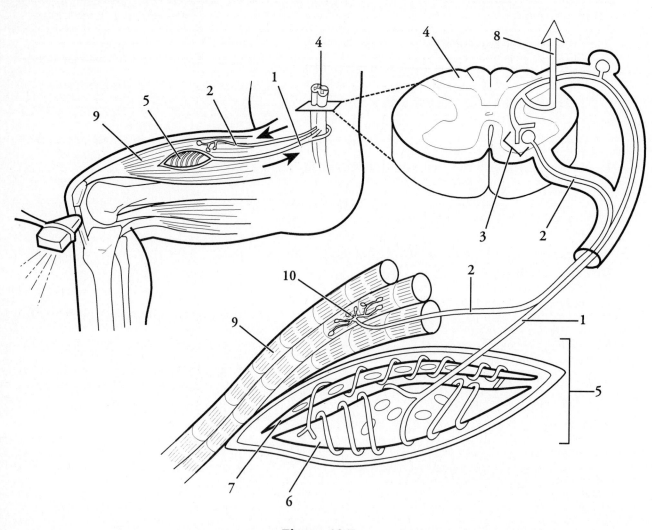

Figure 13.7

_____ Primary sensory afferent

_____ Alpha (α) motor neuron

_____ Spinal cord

_____ Synapse

_____ Muscle spindle

_____ Skeletal muscle (extrafusal fibers)

_____ Motor ending to extrafusal fiber

_____ Nuclear bag fiber

_____ Nuclear chain fiber

_____ Ascending fibers

5. The table below lists several reflex characteristics. Check (✔) each characteristic that relates to a stretch reflex, to a deep-tendon reflex, or to both reflexes.

Characteristic	Stretch	Deep tendon
1. Sensor is the muscle spindle		
2. Sensor is the Golgi tendon organ		
3. Alpha motor neuron innervates agonist muscle		
4. Contraction of agonist muscle is inhibited		
5. Agonist muscle contracts		
6. Contraction of antagonist muscle is inhibited		
7. Antagonist muscle is stimulated		
8. Neuron to antagonist muscle is inhibitory		
9. Type of innervation is reciprocal		
10. Helps ensure smooth onset and termination of muscle contraction		

6. Using the key choices, characterize the types of somatic spinal reflexes according to the descriptions listed below. Insert the letter that best matches each description in the corresponding answer blank.

KEY CHOICES

A. Abdominal   **C.** Deep tendon   **E.** Stretch

B. Crossed extensor   **D.** Flexor   **F.** Superficial

_____ **1.** Maintain normal muscle tone and body posture

_____ **2.** Initiated by painful stimuli

_____ **3.** Particularly important in maintaining balance

_____ **4.** Polysynaptic

_____ **5.** Initiated by proprioceptors

_____ **6.** Monosynaptic and ipsilateral

_____ **7.** Polysynaptic and both ipsilateral and contralateral

_____ **8.** Initiated by cutaneous stimulation

_____ **9.** Polysynaptic and ipsilateral

_____ **10.** Functioning of corticospinal tracts is required

_____ **11.** Protective and overrides the spinal pathways

_____ **12.** The patellar reflex (knee-jerk reflex)

_____ **13.** Plantar reflex

_____ **14.** Stimulation causes the umbilicus to move toward the stimulated site

Developmental Aspects of the Peripheral Nervous System

1. Complete the following statements by writing the missing terms in the answer blanks.

_____ **1.**

_____ **2.**

_____ **3.**

_____ **4.**

_____ **5.**

_____ **6.**

_____ **7.**

_____ **8.**

_____ **9.**

_____ **10.**

Most of the skeletal musculature derives from paired blocks of mesoderm called **(1)**, which align medially. Each spinal nerve, carrying both **(2)** fibers, extends from the spinal cord between two embryonic vertebrae. It becomes associated with a specific muscle mass located adjacent to the neural pathway and helps direct the **(3)** of that muscle mass. **(4)** branches from each spinal nerve (except from **(5)**) supply adjacent dermal segments, or **(6)**. Most of the scalp and facial skin is innervated by sensory fibers of the **(7)** nerves. The uneven growth of the lower body trunk and limbs gives rise to variations in skeletal muscle and skin innervations. Much of the earlier segmental pattern is lost in these areas because of the extensive **(8)** of embryonic muscle cells.

In the elderly, the speed of reflexive responses **(9)**. It has been estimated that 80% of normal reaction time reflects CNS processing, an indication that slower reflexes in the elderly may be the result of a loss of CNS **(10)** rather than age-related changes in the PNS.

CHALLENGING YOURSELF

At the Clinic

1. Brian is brought to the clinic by his parents, who noticed that his right eye does not rotate laterally very well. The doctor explains that the nerve serving the lateral rectus muscle is not functioning correctly. What nerve is involved?

2. Mrs. Hutchinson comes to the clinic because of excruciating pain in the left side of her face whenever she drinks cold fluids. What nerve is likely to be inflamed and what name is given to this condition?

3. A teenage boy has been diagnosed with Bell's palsy. What is the cause of Bell's palsy and what are some of its symptoms?

4. A patient with a CNS tumor has lost his ability to taste sour and bitter substances. What cranial nerve nucleus has been impaired by the tumor?

5. A blow to the neck has caused Alan to lose his voice. What nerve is affected and what would happen if the entire nerve were nonfunctional?

6. In John's checkup, one year after an accident severed his right accessory nerve, severe muscle atrophy was noted. What two prominent muscles have been affected?

7. Three-year-old Samantha is sobbing that her right arm is "gone" and the examination shows her to have little muscle strength in that limb. Questioning the parents reveals that her father had been swinging her by the arms. What part of the PNS has been damaged?

8. A man comes to the clinic for physical therapy after injuring his right elbow. The flexors of the forearm are weak, and he has great difficulty picking up objects with thumb and forefinger. What nerve has been damaged?

9. Kathleen suffered extensive damage to her sciatic nerve, requiring nerve resection (suturing the nerve back together). She now reports that the sensations from the medial and lateral sides of the knee seem to be reversed. Explain how this could have happened.

10. Harry fell off a tall ladder and fractured the anterior cranial fossa of his skull. On arrival at the hospital, a watery, blood-tinged fluid was dripping from his right nostril. Several days later, Harry complained that he could no longer smell. What nerve was damaged in the fall?

11. Frita, a woman in her early 70s, was having problems chewing. She was asked to stick out her tongue. It deviated to the right and its right side was quite wasted. What nerve was injured?

12. Ted is a war veteran who was hit in the back by fragments of an exploding bomb. His skin is numb in the center of his buttocks and along the entire posterior side of a lower limb, but there is no motor problem at all. One of the following choices is the most likely site of his nerve injury. Choose and explain: (a) a few dorsal roots of the cauda equina; (b) spinal cord transection at C_6; (c) spinal cord transection at L_5; (d) femoral nerve transected in the lumbar region.

13. A 67-year-old woman was diagnosed as having advanced nasopharyngeal cancer with cranial infiltration. How would you test her for integrity of cranial nerves 9–11?

14. Can reflex arcs (such as the one triggered when touching something hot) involve (incorporate) pathways for the perception of pain?

Stop and Think

1. When a PNS neuron's axon is cut, several structural changes occur in the cell body (particularly concerning the Nissl bodies [chromatophilic substance] and nucleoli). What are these changes and what is their significance to the repair process?

2. Name an exteroceptor that is not a cutaneous receptor.

3. You wake up in the middle of the night from a sound sleep. Still groggy, you can't feel your body. As you become more awake, you realize this is not a dream. What characteristic of sensation are you experiencing and how could this have happened?

4. Name, in order, the cranial nerves involved in rolling your eyes to follow the second hand on a watch through one complete rotation. Do this separately for each eye.

5. Two of the nerves that control eye movement arise from the midbrain. What other midbrain function involves eye movement?

6. Are the cords of nerve fibers forming the cauda equina truly spinal nerves?

7. What is the benefit of having the nerve supply of the diaphragm, which is located in the thoracic-lumbar area of the spinal cord, arise from cervical nerves?

8. How does accommodation of muscle spindles figure in the importance of stretch routines as a warm-up for exercise?

9. Use Hilton's law to deduce the nerves that innervate the ankle joint.

10. In the sensory receptors called encapsulated dendritic endings, what is the capsule made of?

11. Dr. Omata noticed that her anatomy students always confused the glossopharyngeal (IX) and hypoglossal (XII) nerves because the two names have certain similarities. Thus, every year she made her class explain some basic differences between these two nerves. What would you tell her?

12. Adrian and Abdul, two anatomy students, were arguing about the facial nerve. Adrian said it innervates all the skin of the face, and that is why it is called the facial nerve. Abdul said the facial nerve does not innervate face skin at all. Who was more correct? Explain your choice.

COVERING ALL YOUR BASES

Multiple Choice

Select the best answer or answers from the choices given.

1. In an earthquake, which type of sensory receptor is most likely to sound the *first* alarm?
   A. Exteroceptor   **C.** Mechanoreceptor
   B. Visceroceptor   **D.** Proprioceptor

2. Which of the following can be either somatic or visceral?
   A. Mechanoreceptor   **C.** Chemoreceptor
   B. Photoreceptor   **D.** Nociceptor

3. A cerebral cortical area *not* associated with a special sense is the:
   A. postcentral gyrus   **C.** temporal lobe
   B. occipital lobe   **D.** precentral gyrus

4. Examples of encapsulated nerve endings are:
   A. nociceptors   **C.** Pacinian corpuscles
   B. Merkel's discs   **D.** muscle spindles

5. Structures that can be either exteroceptors or proprioceptors include:
   A. Meissner's corpuscle
   B. Krause's end bulb
   C. Pacinian corpuscle
   D. Ruffini's corpuscle

6. Which of the following are examples of graded potentials?
   A. Generator potential
   B. Action potential
   C. Receptor potential
   D. Synaptic potential

7. The trochlear nerve contains which class of nerve fibers?
   A. Somatic sensory only
   B. Somatic motor and proprioceptor
   C. Somatic sensory, visceral sensory, and visceral motor
   D. Somatic sensory, visceral sensory, somatic motor, and visceral motor

8. Which sheath contains the most collagen?
   - **A.** Endoneurium
   - **B.** Epineurium
   - **C.** Perineurium
   - **D.** Neurilemma

9. Which of the following could be found in a ganglion?
   - **A.** Perikaryon of somatic afferent neuron
   - **B.** Autonomic synapse
   - **C.** Visceral efferent neuron
   - **D.** Somatic interneuron

10. Axonal regeneration involves:
    - **A.** disintegration of the proximal portion of the cut nerve fibers
    - **B.** loss of the oligodendrocyte's myelin sheath
    - **C.** proliferation of the oligodendrocytes
    - **D.** maintenance of the neurilemma

11. Chances of CNS nerve fiber regeneration would (might) be increased by a(n):
    - **A.** oligodendrocyte inhibitor
    - **B.** astrocyte inhibitor
    - **C.** secondary demyelination inhibitor
    - **D.** nerve growth–inhibitor blocking agent

12. Cranial nerves that have some function in vision include the:
    - **A.** trochlear
    - **B.** trigeminal
    - **C.** abducens
    - **D.** facial

13. Eating difficulties would result from damage to the:
    - **A.** mandibular division of trigeminal nerve
    - **B.** facial nerve
    - **C.** glossopharyngeal nerve
    - **D.** vagus nerve

14. If the right trapezius and sternocleido-mastoid muscles were atrophied, you would suspect damage to the:
    - **A.** vagus nerve
    - **B.** motor branches of the cervical plexus
    - **C.** facial nerve
    - **D.** accessory nerve

15. Which nerve has cutaneous branches?
    - **A.** Facial
    - **B.** Trochlear
    - **C.** Trigeminal
    - **D.** Accessory

16. Which nerve is tested by the corneal reflex?
    - **A.** Optic
    - **B.** Oculomotor
    - **C.** Ophthalmic division of the trigeminal
    - **D.** Abducens

17. Dependence on a respirator (artificial breathing machine) would result from spinal cord transection:
    - **A.** between C_1 and C_2
    - **B.** between C_2 and C_3
    - **C.** between C_6 and C_7
    - **D.** between C_7 and T_1

18. Which nerve stimulates muscles that flex the forearm?
    - **A.** Ulnar
    - **B.** Musculocutaneous
    - **C.** Radial
    - **D.** Median

19. Motor functions of arm, forearm, and fingers would be affected by damage to which one of these nerves?
    - **A.** Radial
    - **B.** Axillary
    - **C.** Ulnar
    - **D.** Median

20. An inability to extend the leg would result from a loss of function of the:
    - **A.** lateral femoral cutaneous nerve
    - **B.** ilioinguinal nerve
    - **C.** saphenous branch of femoral nerve
    - **D.** femoral nerve

21. The gastrocnemius muscle is served by the:
    - **A.** branches of the tibial nerve
    - **B.** inferior gluteal
    - **C.** deep branch of the common peroneal nerve
    - **D.** superficial branch of common peroneal nerve

22. Which contains only motor fibers?
    - **A.** Dorsal root
    - **B.** Dorsal ramus
    - **C.** Ventral root
    - **D.** Ventral ramus

23. Which nerve would be blocked by anesthetics administered during childbirth?
   A. Superior gluteal   C. Obturator
   B. Inferior gluteal   D. Pudendal

24. Dermatomes on the posterior body surface are supplied by:
   A. cutaneous branches of dorsal rami of spinal nerves
   B. cutaneous branches of ventral rami of spinal nerves
   C. cutaneous branches of all cranial nerves
   D. cutaneous branch of cranial nerve V

25. After staying up too late the previous night (studying, no doubt), a college student dozed off in his eight o'clock anatomy and physiology lecture. As his head slowly drifted forward to his chest, he snapped it erect again. What type of reflex does this exemplify?
   A. Acquired reflex   C. Deep tendon reflex
   B. Autonomic reflex   D. Stretch reflex

26. Development of the PNS involves:
   A. somites
   B. migration of muscle cells
   C. loss of some of the segmental pattern
   D. dermatomes sharply divided from one another

Word Dissection

For each of the following word roots, fill in the literal meaning and give an example, using a word found in this chapter.

Word root	Translation	Example
1. esthesi	_____	_____
2. glosso	_____	_____
3. kines	_____	_____
4. noci	_____	_____
5. propri	_____	_____
6. puden	_____	_____
7. vagus	_____	_____

14 The Autonomic Nervous System

Student Objectives

When you have completed the exercises in this chapter, you will have accomplished the following objectives:

Overview of the Autonomic Nervous System

1. Compare the somatic and autonomic nervous systems relative to effectors, efferent pathways, and neurotransmitters released.

2. Compare and contrast the general functions of the parasympathetic and sympathetic divisions.

Anatomy of the Autonomic Nervous System

3. Describe the site of CNS origin, locations of ganglia, and general fiber pathways of the parasympathetic and sympathetic divisions.

Physiology of the Autonomic Nervous System

4. Define *cholinergic* and *adrenergic fibers,* and list the different types of cholinergic and adrenergic receptors.

5. Briefly describe the clinical importance of drugs that mimic or inhibit adrenergic or cholinergic effects.

6. State the effects of the parasympathetic and sympathetic divisions on the following organs: heart, blood vessels, gastrointestinal tract, lungs, adrenal medulla, and external genitalia.

7. Describe the levels cf control of autonomic nervous system functioning.

Homeostatic Imbalances of the Autonomic Nervous System

8. Explain the relationship of some types of hypertension, Raynaud's disease, and the mass reflex reaction to disorders of autonomic functioning.

Developmental Aspects of the Autonomic Nervous System

9. Describe the effects of aging on the autonomic nervous system.

Ready, set, go!

The autonomic nervous system (ANS) is the involuntary part of the efferent motor division of the peripheral nervous system (PNS). The ANS is structurally and functionally subdivided into sympathetic and parasympathetic divisions. The two divisions innervate cardiac muscle, smooth muscle, and glands in a coordinated and reciprocal manner to maintain homeostasis of the internal environment.

In Chapter 14, topics for study include structural and functional comparisons between the somatic and autonomic nervous systems and between the sympathetic and parasympathetic divisions. Also included are exercises on the anatomy and physiology of the ANS and impairments and developmental aspects of the ANS.

BUILDING THE FRAMEWORK

Overview of the Autonomic Nervous System

1. Identify, by color coding and coloring, the following structures in Figure 14.1, which depicts the major anatomical differences between the somatic and autonomic motor divisions of the PNS. Also identify by labeling all structures provided with leader lines.

○ Somatic motor neuron

○ ANS preganglionic neuron

○ ANS postganglionic neuron

○ Autonomic ganglion

○ Gray matter of spinal cord (CNS)

○ Effector of the somatic motor neuron

○ Effector of the autonomic motor neuron

○ Intrinsic ganglionic cell

○ Myelin sheath

○ White matter of spinal cord (CNS)

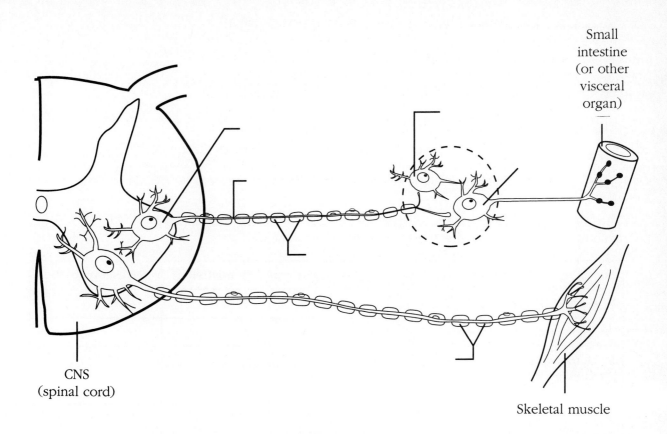

Small
intestine
(or other
visceral
organ)

CNS
(spinal cord)

Skeletal muscle

Figure 14.1

2. Identify the following descriptions as characteristic of the somatic nervous
system (use S), of the autonomic nervous system (use A), or of both systems
(use S, A). Insert your letter responses in the answer blanks.

_____ **1.** Has efferent motor fibers

_____ **2.** Single axonal pathway extends from the CNS to each effector

_____ **3.** Slower conduction of nerve impulses

_____ **4.** Motor neuron cell bodies are located in the CNS

_____ **5.** Typically thick, heavily myelinated motor fibers

_____ **6.** Target organs are skeletal muscles

_____ **7.** Has preganglionic and postganglionic neurons

_____ **8.** Effectors are glands and involuntary muscles

_____ **9.** Neural pathways are found in cranial and spinal nerves

_____ **10.** One ganglion of each motor unit is located outside the CNS

_____ **11.** Typically thin fibers with little or no myelination

_____ **12.** Effect of neurotransmitter on target organ is always excitatory

_____ **13.** Neurotransmitter at effector site is always acetylcholine

_____ **14.** No ganglia are present

_____ **15.** Neurotransmitter effect on target organ may be excitation or inhibition

_____ **16.** Motor activities under control of higher brain centers

3. The following table lists general functions of the ANS. Use a check mark (✔) to show which division of the ANS is involved in each function.

Function	Sympathetic	Parasympathetic
1. Normally in control		
2. "Fight-or-flight" system		
3. More specific local control		
4. Causes a dry mouth, dilates bronchioles		
5. Constricts eye pupils, decreases heart rate		
6. Conserves body energy		
7. Causes increased blood glucose levels		
8. Causes increase in digestive tract mobility		
9. Arector pili muscles contract ("goose bumps")		

Anatomy of the Autonomic Nervous System

1. Figure 14.2 is a highly simplified diagram of the anatomy of the two ANS divisions. Only certain target structures are indicated; for clarity, the spinal cord is depicted twice. Circles represent ganglia, and neural pathways are shown as dotted lines for use as guidelines.

Select eight colors; color the coding circles, structures, and specified boxes listed on the facing page. Use solid lines for preganglionic fibers and dashed lines for postganglionic fibers. Next, label each division in the answer blanks below the figure. Then, insert leader lines and label the four cranial nerves by number.

○ Sites of parasympathetic neurons in the CNS

○ Sites of sympathetic neurons in the CNS

○ Sympathetic fibers

○ Sympathetic trunk of paravertebral ganglia

○ Terminal ganglia

○ Boxes specifying organs that are provided only with sympathetic fibers

○ Parasympathetic fibers

○ Prevertebral ganglia

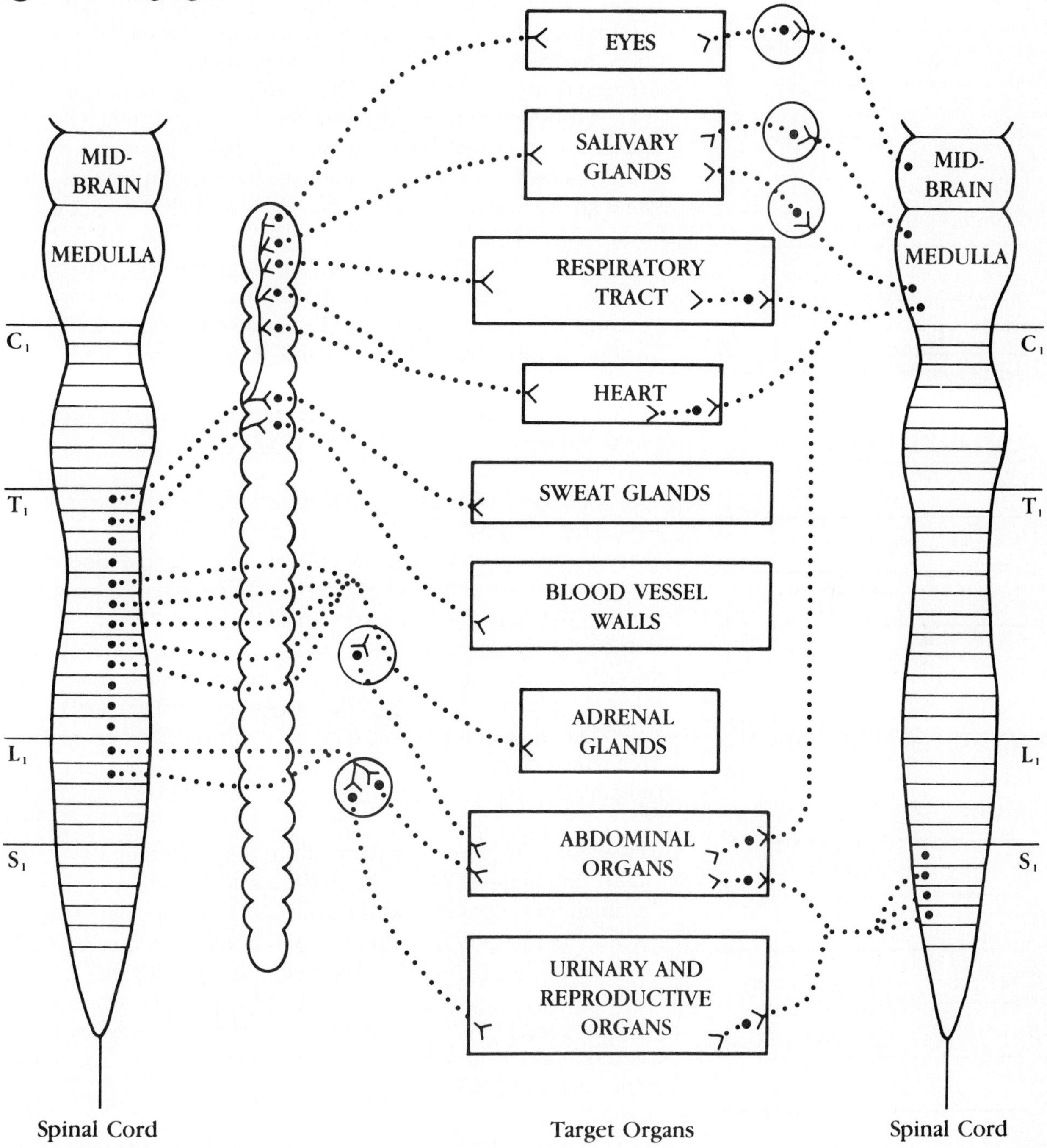

MID-BRAIN

MEDULLA

C_1

T_1

L_1

S_1

Spinal Cord

EYES

SALIVARY GLANDS

RESPIRATORY TRACT

HEART

SWEAT GLANDS

BLOOD VESSEL WALLS

ADRENAL GLANDS

ABDOMINAL ORGANS

URINARY AND REPRODUCTIVE ORGANS

Target Organs

MID-BRAIN

MEDULLA

C_1

T_1

L_1

S_1

Spinal Cord

A. _____ division

B. _____ division

Figure 14.2

2. The following paragraphs trace the sympathetic and parasympathetic
pathways involved in the innervation of selected organs or structures.
Complete the statements by inserting the missing terms in the answer blanks.
Referring to Figure 14.2 should help you complete this exercise.

1. _____

2. _____

3. _____

4. _____

5. _____

6. _____

7. _____

8. _____

9. _____

10. _____

11. _____

12. _____

13. _____

14. _____

15. _____

16. _____

17. _____

18. _____

19. _____

20. _____

21. _____

22. _____

23. _____

24. _____

25. _____

26. _____

Sympathetic preganglionic fibers arise from cell bodies located
in the __(1)__ horns of segments __(2)__ of the __(3)__ . These
fibers leave the ventral root by passing through the __(4)__ rami
communicantes and enter adjoining __(5)__, which are aligned
to form the __(6)__ trunk or chain. Synapses of sympathetic
fibers may or may not occur in the adjoining ganglia. In cases
where synapses are made within the adjoining ganglia, the
__(7)__ axons enter the spinal nerves by way of the gray __(8)__ .
The term *gray* means that the axons are __(9)__ . Examples of
structures with this type of sympathetic innervation are the
sweat glands and the __(10)__ in blood vessel walls.

Some preganglionic axons may travel within the sympathetic
trunk to synapse in other than adjoining ganglia. For example,
synapses within the superior __(11)__ ganglion contribute fibers
innervating the __(12)__ of the eye, the __(13)__ glands, the heart,
and the respiratory tract organs. Though some fibers serve the
heart, most of the postganglionic fibers issuing from the other
two cervical ganglia innervate the __(14)__ .

Some fibers enter and leave the sympathetic chain without
synapsing. Preganglionic fibers T_5–L_2 synapse in __(15)__ ganglia
located anterior to the vertebral column. For example, fibers
T_5–T_{12} contribute to the __(16)__ nerves, which synapse mainly
in the __(17)__ ganglia. From there, postganglionic fibers
distribute to serve most of the __(18)__ organs. A few fibers pass
through the celiac ganglion without synapsing. These fibers
synapse within the medulla of the __(19)__ gland. The __(20)__
(L_1 and L_2) splanchnic nerves synapse in prevertebral ganglia,
from which postganglionic fibers innervate the __(21)__ and
reproductive organs.

Parasympathetic preganglionic neurons are located in the
__(22)__ and in the __(23)__ region of the spinal cord.
Preganglionic axons extend from the CNS to synapse in __(24)__
ganglia close to or within target organs. Some examples of
cranial outflow are the following. In the ciliary ganglion,
preganglionic fibers traveling with the __(25)__ nerve synapse
with postganglionic neurons within the orbits. The salivary
glands are innervated by parasympathetic fibers that travel with
cranial nerves __(26)__ and __(27)__ .

__(28)__ nerve fibers account for about 90% of all preganglionic
parasympathetic fibers in the body. These fibers enter networks
of interlacing nerve fibers called __(29)__, from which arise
several branches that serve the organs located in the __(30)__
and in the __(31)__ . Preganglionic fibers from the __(32)__ of the

_____ **27.**   spinal cord synapse with postganglionic neurons in ganglia
within the walls of the urinary and reproductive organs and the
_____ **28.**   distal half of the large intestine.

_____ **29.**

_____ **30.**

_____ **31.**

_____ **32.**

3. Characterize each of the following anatomical descriptions as it relates to the sympathetic division (use S), the parasympathetic division (use P), or to both divisions (use S, P). Write the correct letter answers in the answer blanks.

_____   **1.**   Short preganglionic and long postganglionic axons

_____   **2.**   Travel within cranial and sacral nerves

_____   **3.**   Called the thoracolumbar division

_____   **4.**   Long preganglionic and short postganglionic axons

_____   **5.**   No fibers in rami communicantes

_____   **6.**   Called the craniosacral division

_____   **7.**   Ganglia are close to the CNS

_____   **8.**   Minimal branching of preganglionic fibers

_____   **9.**   Terminal ganglia are close to or in the visceral organs served

_____   **10.**   Gray and white rami communicantes utilized

_____   **11.**   Extensive branching of preganglionic fibers

_____   **12.**   Nearly all fibers are accompanied by sensory afferent fibers

Physiology of the Autonomic Nervous System

1. Define *cholinergic* fibers, and name the two types of cholinergic receptors below.

1. Cholinergic fibers _____

2. Cholinergic receptors _____

2. Define *adrenergic* fibers, and name the two major classes of adrenergic receptors.

   1. Adrenergic fibers _____

   2. Adrenergic receptors _____

3. Using the key choices, select which CNS centers control the autonomic activities listed below. Write the correct letters in the answer blanks.

KEY CHOICES

A. Brain stem   **B.** Cerebral cortex   **C.** Hypothalamus   **D.** Spinal cord

_____ **1.** The main integration center of the ANS

_____ **2.** Exerts the most direct control over ANS functioning

_____ **3.** Coordinates blood pressure, water balance, and endocrine activity

_____ **4.** Integrates defecation and micturation reflexes

_____ **5.** Controls some autonomic functioning through meditation

_____ **6.** Regulates heart and respiration rates and gastrointestinal reflexes

_____ **7.** Awareness of autonomic functioning through biofeedback training

_____ **8.** Influences autonomic functioning via limbic system connections

4. The following table lists several physiological conditions. Use a check mark (✔) to show which autonomic division is involved for each condition.

Function	Sympathetic	Parasympathetic
1. All neurons secrete acetylcholine		
2. Controls secretions of catecholamines		
3. Control of reflexes that act in thermoregulation		
4. Accelerates metabolism		
5. Short-lived control of effectors		
6. Postganglionic neurons secrete NE		
7. Localized control of effectors; not diffuse		

5. Circle the term that does not belong in each of the following groupings. (*Note:* ACh = acetylcholine; NE = norepinephrine.)

   1. Skeletal muscle contraction   Cholinergic receptors   ACh   NE

   2. NE   Usually stimulatory   Nicotinic receptor   Adrenergic receptor

3. Beta receptor   Cardiac muscle   Heart rate slows   NE

4. Blood pressure control   Sympathetic tone   Parasympathetic tone

    Blood vessels partially constricted

5. Parasympathetic tone   Urinary tract   Digestive tract   Rapid heart rate

6. External genitalia   Sympathetic tone   Blood vessels dilate   Penis erection

Homeostatic Imbalances of the Autonomic Nervous System

1. If a statement is true, write the letter T in the answer blank. If a statement is false, change the underlined word(s) and write the correct word(s) in the answer blank.

_____ 1. Most autonomic disorders reflect the abnormal control of <u>adrenal medulla</u> activity.

_____ 2. The cause of Raynaud's disease is thought to be intense <u>vasoconstriction</u> in response to exposure to cold.

_____ 3. To promote vasodilation in the patient with a severe case of Raynaud's disease, a <u>parasympathectomy</u> is performed.

_____ 4. Hypertension in the overly stressed patient can be controlled with <u>cholinergic</u> blocker medication.

Developmental Aspects of the Autonomic Nervous System

1. Complete the following statements by writing the missing terms in the answer blanks.

_____ 1.

_____ 2.

_____ 3.

_____ 4.

_____ 5.

_____ 6.

_____ 7.

Preganglionic neurons of the ANS arise from the embryonic __(1)__ . Postganglionic neurons, all ANS ganglia, and the medulla of the adrenal gland derive from embryonic __(2)__ tissue. Many structural elements of the ANS, such as the paravertebral and prevertebral ganglia, are the result of the __(3)__ of neural crest cells. Sympathetic neurons within the __(4)__ send axons to these ganglia, from which fibers extend to synapse with their __(5)__ cells in the body's periphery. The chemical called __(6)__ appears to guide these growth processes.

ANS disorders stemming from congenital abnormalities are rare. A notable example is Hirschsprung's disease, in which a plexus of the __(7)__ division, which supplies the distal part of the colon,

_____ 8.

_____ 9.

_____ 10.

_____ 11.

_____ 12.

fails to develop. Consequently, the affected portion of the colon loses __(8)__, and feces accumulate proximal to the affected region.

The ANS serves the body well during youth and middle age. In the elderly, however, the efficiency of the ANS begins to decline. Because of diminished __(9)__ mobility, the elderly are at times troubled with constipation. And, because of reduced __(10)__ formation, they may complain of dry eyes. More troublesome, perhaps, are fainting episodes following sudden standing from a sitting or lying position. These symptoms may indicate that the __(11)__ centers of the __(12)__ division respond more slowly to changes in body position during old age.

CHALLENGING YOURSELF

At the Clinic

1. After surgery, patients are often temporarily unable to urinate, and bowel sounds are absent. What division of the ANS is affected by anesthesia?

2. Stress-induced stomach ulcers are due to excessive sympathetic stimulation. For example, one suspected cause of the ulcers is almost total lack of blood flow to the stomach wall. How is this related to sympathetic function?

3. Barium swallow tests on Mr. Bronson indicate gross distension of the inferior end of his esophagus. What condition does this indicate?

4. A young child has been diagnosed with Hirschsprung's disease. How would you explain the cause of the condition and the surgical treatment to the child's parents?

5. Mrs. Griswold has been receiving treatment at the clinic for Raynaud's disease. Lately her symptoms have been getting more pronounced and severe, and her doctor determines that surgery is her only recourse. What surgical procedure will be performed, and what effect is it designed to produce?

6. Which is the more likely side effect of Mrs. Griswold's surgery—anhidrosis or hyperhidrosis—in the area of concern? (Hidrosis = sweating.)

7. Brian, a young man who was paralyzed from the waist down in an automobile accident, has seen considerable progress in the return of spinal cord reflexes. However, quite unexpectedly, he is brought to the ER because of profuse sweating and involuntary voiding. From what condition is Brian suffering? For what life-threatening complication will the clinical staff be alert?

8. An elderly patient in a nursing home has recurrent episodes of fainting when he stands. An alert nurse notes that this occurs only when his room is fairly warm; on cold mornings, he has no difficulty. What is the cause of the fainting and how does it relate to the autonomic nervous system and to room temperature?

9. Bertha appears at the clinic complaining of intense lower back pain, although she claims she has not done anything to "knock her back out." Questioning reveals painful, frequent urination as well. What is a likely cause of the back pain?

10. John Ryder, a black male in his mid-40s, has hypertension for which no organic cause can be pinpointed. Which class of autonomic nervous system drugs will most likely be prescribed to manage his condition and why?

11. Roweena Gibson, a high-powered marketing executive, develops a stomach ulcer. She complains of a deep abdominal pain that she cannot quite locate, plus a pain in her abdominal wall. Exactly where on the abdominal wall is the superficial pain most likely to be located? (Doing a little research in Chapter 24 may be helpful here.)

12. Imagine that a mad scientist invents a death-ray that destroys a person's ciliary, sphenopalatine, and submandibular ganglia (and nothing else). List all the symptoms the victim would show. Would the victim die, or would the scientist have to go back to the laboratory to invent a more effective death-ray?

Stop and Think

1. Can the autonomic nervous system function properly in the absence of visceral afferent input? Explain your response.

Use the following information for Questions 2–6: Stretch receptors in the bladder send impulses along visceral afferent nerve fibers to the sacral region of the spinal cord. Synapses with motor neurons there initiate impulses along visceral efferent fibers that stimulate contraction of the smooth muscle of the bladder wall and relaxation of the smooth muscle (involuntary) sphincter to allow urination.

2. What division of the autonomic nervous system is involved?

3. What sort of neuronal circuit does this exemplify?

4. What nerves carry the impulses?

5. How is urination inhibited consciously?

6. How would this pathway be affected by spinal cord transection above the sacral region?

7. Migraine headaches are caused by constriction followed by dilation of the vessels supplying the brain. The pain is associated with the dilation phase and is apparently related to the high rate of blood flow through these vessels. Some migraine sufferers seek relief through biofeedback, learning to trigger dilation of the vessels in the hand(s). How does this provide relief?

8. Trace the sympathetic pathway from the spinal cord to the iris of the eye, naming all associated structures, fiber types, and neurotransmitters.

9. Trace the parasympathetic pathway from the brain to the heart, naming all associated structures, fiber types, and neurotransmitters.

COVERING ALL YOUR BASES

Multiple Choice

Select the best answer or answers from the choices given.

1. Which of the following is true of the autonomic, but not the somatic, nervous system?
   A. Neurotransmitter is acetylcholine
   B. Axons are myelinated
   C. Effectors are muscle cells
   D. Has motor neurons located in ganglia

2. Examination of nerve fibers supplying the rectus femoris muscle reveals an assortment of type A fibers and type C fibers. The type A fibers are known to be somatic motor fibers supplying the skeletal muscle. Which of these is likely to be the function of the type C fibers?
   A. Parasympathetic innervation to the muscle spindles
   B. Sympathetic innervation to the muscle spindles
   C. Parasympathetic innervation to the blood vessels
   D. Sympathetic innervation to the blood vessels

3. Which is the best way to describe how the ANS is controlled?
   A. Completely under control of the cerebral cortex
   B. Completely under control of the brain stem
   C. Entirely controls itself
   D. Major control by the hypothalamus and spinal refexes

4. Which of the following disorders is (are) related specifically to sympathetic functions?
   A. Achalasia
   B. Raynaud's disease
   C. Orthostatic hypotension
   D. Hirschsprung's disease

5. Adrenal medulla development involves:
   A. neural crest cells
   B. nerve growth factor
   C. sympathetic preganglionic axons
   D. parasympathetic preganglionic axons

Use the following choices to respond to Questions 6–24:

    A. sympathetic division
    B. parasympathetic division
    C. both sympathetic and parasympathetic
    D. neither sympathetic nor parasympathetic

_____ **6.** Typically has long preganglionic and short postganglionic fibers

_____ **7.** Some fibers utilize gray rami communicantes

_____ **8.** Courses through spinal nerves

_____ **9.** Has nicotinic receptors on its postganglionic neurons

_____ **10.** Has nicotinic receptors on its target cells

_____ **11.** Has splanchnic nerves

_____ **12.** Courses through cranial nerves

_____ **13.** Originates in cranial nerves

_____ **14.** Effects enhanced by direct stimulation of a hormonal mechanism

_____ **15.** Includes otic ganglion

_____ **16.** Includes celiac ganglion

_____ **17.** Contains cholinergic fibers

_____ **18.** Stimulatory impulses from hypothalamus and/or medulla pass through the thoracic spinal cord to connect to preganglionic neurons

_____ **19.** Hyperactivity of this division can lead to ischemia (loss of circulation) to various body parts and hypertension

_____ **20.** Affected by beta blockers

_____ **21.** Hypoactivity of this division would lead to decrease in metabolic rate

_____ **22.** Stimulated by the RAS

_____ **23.** Has widespread, long-lasting effects

_____ **24.** Sets the tone for the heart

Word Dissection

For each of the following word roots, fill in the literal meaning and give an example, using a word found in this chapter.

Word root	Translation	Example
1. adren	_____	_____
2. chales	_____	_____
3. epinephr	_____	_____
4. mural	_____	_____
5. ortho	_____	_____
6. para	_____	_____
7. pathos	_____	_____
8. splanchn	_____	_____

15 Neural Integration

Student Objectives

When you have completed the exercises in this chapter, you will have accomplished the following objectives:

Sensory Integration: From Reception to Perception

1. Describe the role of receptors in sensory processing.

2. Compare and contrast specific and nonspecific ascending somatosensory pathways.

3. Describe the main features of perceptual processing of sensory inputs.

Motor Integration: From Intention to Effect

4. Describe the levels of the motor control hierarchy.

5. Define *central pattern generator* and *command neuron*.

6. Compare the roles of the direct and indirect systems in controlling motor activity.

7. Explain the function of the cerebellum and basal nuclei in somatic sensory and motor integration.

8. Describe symptoms of cerebellar and basal nuclear disease.

Higher Mental Functions

9. Define *EEG* and distinguish between alpha, beta, theta, and delta waves.

10. Explain the importance of slow-wave and REM sleep and indicate how their patterns change through life.

11. Explain what is meant by the term *holistic processing* in relation to human consciousness.

12. Compare and contrast the stages and categories of memory.

13. Indicate the major brain structures believed to be involved in fact and skill memories and describe their relative roles.

14. List some of the brain areas involved in language understanding and articulation, and indicate the role of each as presently understood.

Ready, set, go!

Whereas the previous four chapters dealt separately with components of the nervous system, exercises in this chapter focus on the nervous system as a whole. In life, the nervous system always functions as a complete unit, interconnecting and integrating all of its parts. The nervous system interprets sensations and responds with a myriad of coordinated but diverse activities, ranging from simple physical movements to complex and abstract mental processes.

Topics in Chapter 15 include sensory integration, motor integration, and higher mental functions.

BUILDING THE FRAMEWORK

Sensory Integration: From Reception to Perception

1. Complete the following statements by writing the missing terms in the answer blanks.

_____ 1.

_____ 2.

_____ 3.

_____ 4.

_____ 5.

_____ 6.

_____ 7.

_____ 8.

_____ 9.

_____ 10.

_____ 11.

_____ 12.

   __(1)__ is the awareness of internal and external stimuli. The conscious interpretation of such stimuli is called __(2)__ . The first level of sensory integration (starting with the stimulus) is the __(3)__ level. The second level, the __(4)__ level, is composed of __(5)__ . The highest level, the __(6)__ level, reflects the functioning of the __(7)__ .

Sensory receptors transduce stimulus energy into __(8)__ . As stimulus energy is absorbed by the receptor, the __(9)__ of the receptor membrane changes. This allows ions to flow through the membrane and results in a __(10)__ potential called the receptor potential. If the receptor potential reaches __(11)__ , a(n) __(12)__ potential is generated and transmitted.

The nonspecific ascending pathways are also called the __(13)__ pathways, whereas the specific ascending pathways are also known as the __(14)__ system. Of these, the __(15)__ are more concerned with the emotional aspects of perception. The sensory cortex is arranged in columns of neurons, each representing a __(16)__ of sensory perception. The brain region that acts as a relay station and projects fibers to the sensory cortex is the __(17)__ .

_____ 13.   _____ 15.   _____ 17.

_____ 14.   _____ 16.

2. Figure 15.1 is a simplified diagram of the organization of the pathways of the somatosensory system. For clarity, not all receptors are listed. The ascending pathways are shown as dotted lines to help you color them. Select two colors and color the coding circles and the pathways listed below. Then, at the leader lines on the diagram, label these pathways and also the areas where the somatosensory cortex and the cerebellar cortex are located. Next, bracket the structures that belong in each level and label the brackets to identify the three levels.

◯ Nonspecific ascending pathway ◯ Specific ascending pathway

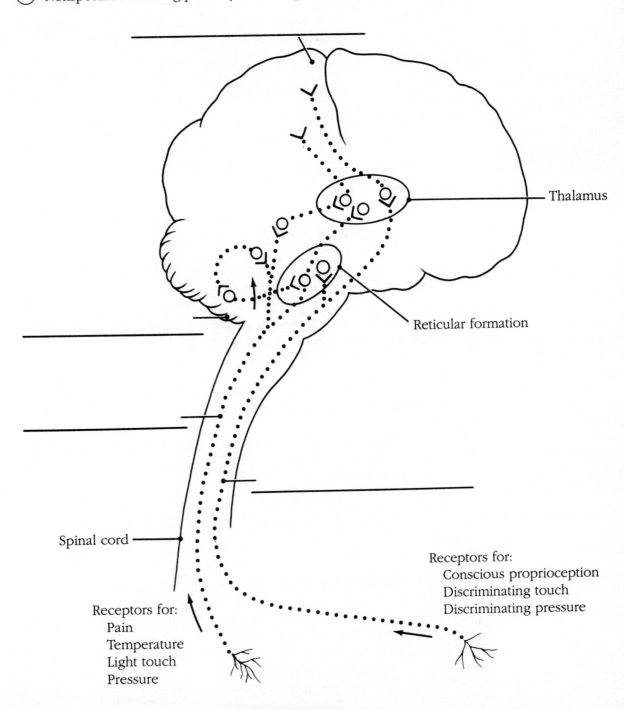

Thalamus

Reticular formation

Spinal cord

Receptors for:
Conscious proprioception
Discriminating touch
Discriminating pressure

Receptors for:
Pain
Temperature
Light touch
Pressure

Figure 15.1

3. If a statement about the three levels of sensory integration is true, write T in the answer blank. If a statement is false, change the underlined word(s) and write the correct word(s) in the answer blank.

_____ 1. Sensation occurs in the <u>sensory association cortex</u>.

_____ 2. The stronger the stimulus, the greater is the <u>velocity</u> of impulse transmission.

_____ 3. <u>Specific</u> ascending pathways are formed by the spinothalamic tracts.

_____ 4. <u>Nonspecific</u> ascending pathways transmit information about pain, touch, pressure, and temperature.

_____ 5. Second-order neurons of both specific and nonspecific ascending circuits terminate in the <u>medulla</u>.

_____ 6. The fasciculus cuneatus, fasciculus gracilis, and median lemniscal tracts form the <u>nonspecific</u> ascending pathways.

_____ 7. The cell bodies of <u>second-order</u> neurons are located in the dorsal root ganglia.

_____ 8. To accommodate precise, discriminatory information, impulses are carried by <u>specific</u> ascending pathways.

_____ 9. The type of neural integration for somatosensory inputs in the cerebral cortex is called <u>reverberating</u>.

4. First, examine the pathway in Figure 15.2 to determine whether it is a *specific* or *nonspecific* pathway. Label accordingly on the left line beneath the pathway. Next, identify the pathway more precisely by selecting one of the choices below, and insert your response on the right line beneath the figure.

Spinocerebellar   Fasciculus cuneatus   Fasciculus gracilis   Spinothalamic

Then, color code the following structures and identify them by coloring them on the diagram.

◯ First-order neuron   ◯ Sensory receptor   ◯ Thalamus

◯ Second-order neuron   ◯ Sensory homunculus   ◯ Postcentral gyrus

◯ Third-order neuron   ◯ Spinal cord

Finally, circle all sites of synapse, and indicate your reasons for identifying this

pathway as you did. _____

_____

_____

Plane of cut

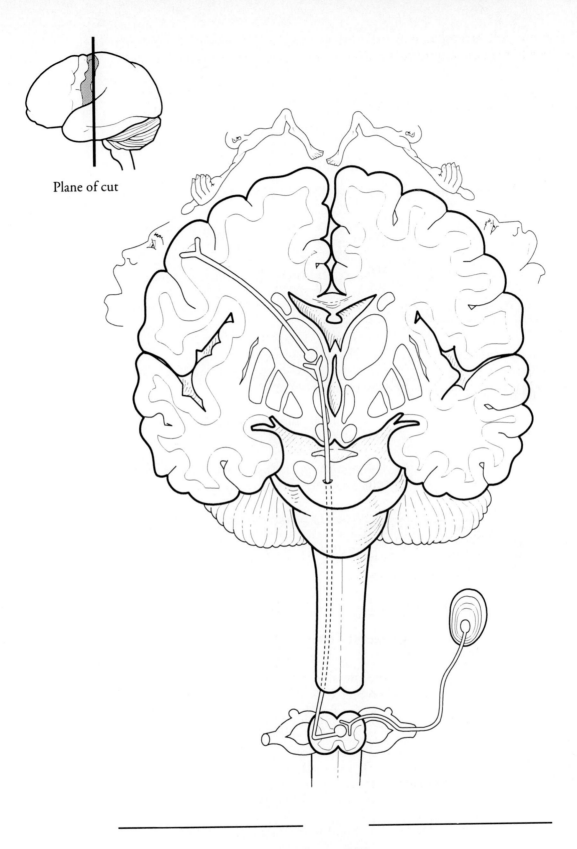

Figure 15.2

5. Using the key choices, identify the correct expression of sensory perception in each of the following descriptions. Write the correct letters in the answer blanks.

KEY CHOICES

A. Feature abstraction   D. Perceptual detection

B. Magnitude estimation   E. Quality discrimination

C. Pattern recognition   F. Spatial discrimination

_____ 1. Depends on the density of touch receptors in various regions of the skin

_____ 2. "That old house by the apple tree reminds me of Grandpa's place"

_____ 3. The result of summation of a minimum number of several receptor impulses

_____ 4. The result of frequency coding

_____ 5. Discerning how much of a stimulus is acting on the body

_____ 6. Discerning texture and shape by touch—for example, that hair is soft, silky, and curly

_____ 7. Distinguishing among different odors when you cook fudge, cabbage, and spaghetti sauce

_____ 8. The result of sensory inputs integrated in parallel circuitry

_____ 9. "In the dark, I feel the rough stubble and the soft dimple on his chin"

Motor Integration: From Intention to Effect

1. Using the key choices, identify the structures, or associated structures, in the following descriptions. Write the correct letters in the answer blanks.

KEY CHOICES

A. Basal nuclei   D. Cerebellum   G. Extrapyramidal (multineuronal) system

B. Brain stem motor areas   E. Command neurons   H. Pyramidal (direct) system

C. Central pattern generators F. Cortical motor area   I. Spinal cord

_____ 1. Activates anterior horn neurons of a spinal cord segment during locomotion

_____ 2. Includes the reticular, red, and vestibular nuclei of the brain stem

_____ **3.** Able to start, stop, or modify the central pattern generators

_____ **4.** The crucial center for sensory-motor integration and control

_____ **5.** Neurons located in the precentral gyri of the frontal lobes

_____ **6.** May involve networks of spinal cord neurons arranged in reverberating circuits

_____ **7.** Functions as a liaison between the cerebral cortex and the cerebellum

_____ **8.** Precommand areas

2. Several neural structures play important roles in the hierarchy of motor control. In the flowchart shown in Figure 15.3, there are several blanks. First, fill in the names for the lowest, middle, and highest levels of the motor hierarchy (put terms inside the appropriate boxes). Then, next to the level terms (lowest, etc.), write the names of the brain regions that contribute to that level of the hierarchy. Also, indentify the boxes representing motor output and sensory input. Finally, color the diagram as you like.

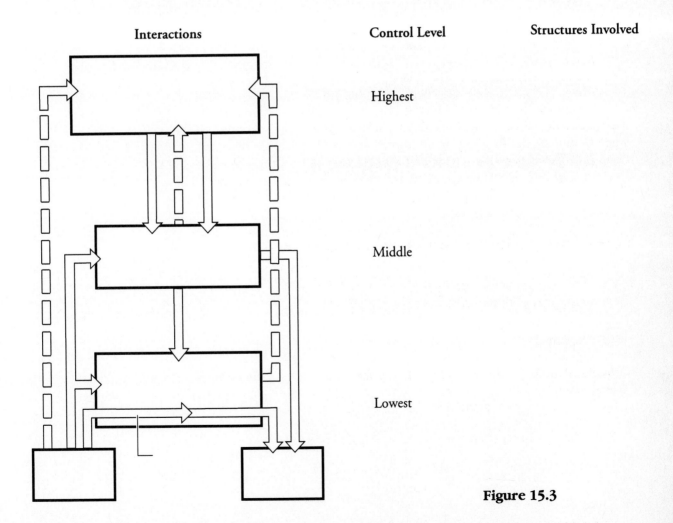

Figure 15.3

3. Circle the term that does not belong in each of the following groupings.

1. Fixed action pattern   Consciously considered actions   Complex motor behavior

   All-or-none sequence of motor actions

2. Cerebral cortex   Pyramidal neurons   Conscious level   Command neurons

3. Cerebellum   Rhythmic bursts of motor impulses   Inherited

   Central pattern generators

4. Command neurons   Breathing rhythms   Interneurons   Precommand areas

5. Reticular nuclei   Vestibular nuclei   Basal nuclei   Red nuclei

6. Anterior horn neurons   Multineuronal tracts   Direct tracts

   Voluntary muscle contractions

7. Cerebellar impairment   Perception disorders   Ataxia   Speech problems

8. Parkinson's disease   Basal nuclei unaffected   "Pill-rolling" tremors

   Dopamine deficit

9. No overshoot   Ataxia   Synergy   Coordination of agonist and antagonist

4. First, examine the diagram in Figure 15.4 to determine whether it is an example of a *direct* or *indirect* motor pathway. Write your response on the line below the figure. Next, determine what specific tracts are depicted by carefully examining the two locations where the fiber tract has been "lassoed" by a leader line, and label the pathways appropriately on the diagram. Then, color code the structures provided with coloring circles and identify them by coloring them on the diagram.

  ◯ Basal nuclei   ◯ Brain stem motor nuclei   ◯ Cerebellum

  ◯ Alpha motor neuron   ◯ Primary motor cortex   ◯ Effector

  ◯ Motor homunculus   ◯ Medullary pyramid   ◯ Internal capsule

  ◯ Thalamus

Finally, indicate on the diagram which of these structures is the point of decussation.

Plane of cut

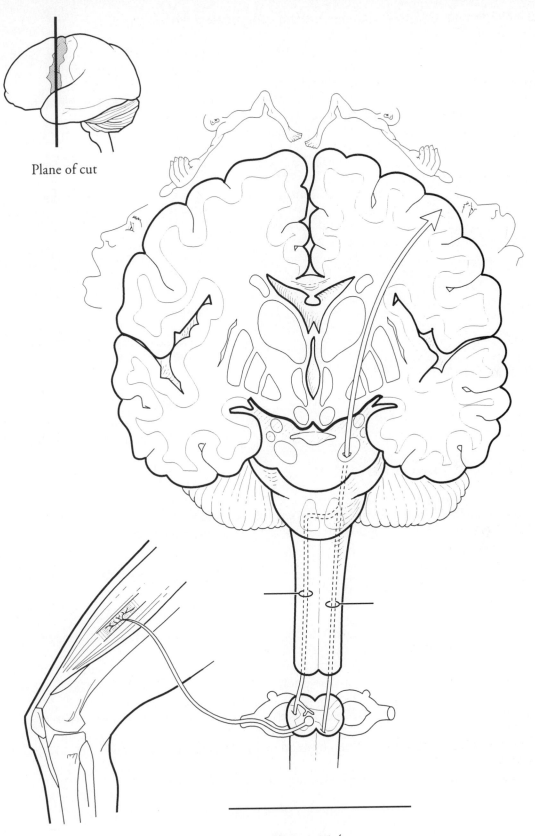

Figure 15.4

5. Match the motor functions in Column A to the appropriate tract and nuclei from Column B.

Column A

_____ **1.** Voluntary control of body

_____ **2.** Voluntary control of head

_____ **3.** Balance, muscle tone

_____ **4.** Equilibrium/gravity support

_____ **5.** Control of flexors

_____ **6.** Head movements/visual reflexes

Column B

A. Precentral gyrus

B. Reticular nuclei

C. Vestibular nuclei

D. Red nuclei

E. Superior colliculi

F. Corticospinal tracts

G. Vestibulospinal tracts

H. Tectospinal tracts

I. Corticobulbar tracts

J. Reticulospinal tracts

K. Rubrospinal tracts

Higher Mental Functions

1. Using the key choices, identify the brain wave patterns described below. Write your answers in the answer blanks. (*Note:* Hz = cycles per second.)

KEY CHOICES

A. Alpha   **B.** Beta   **C.** Delta   **D.** Theta

_____ **1.** Slow, synchronous waves with an average frequency of 10 Hz

_____ **2.** More irregular than beta waves; normal in children but abnormal in awake adults

_____ **3.** Recorded when a person is awake and relaxed, with eyes closed

_____ **4.** High-amplitude waves with a very low frequency (4 Hz or less)

_____ **5.** Recorded during sleep or anesthesia

_____ **6.** Recorded when a person is awake and fully alert

2. Complete the following statements by writing the missing terms in the answer blanks.

_____ 1.
_____ 2.
_____ 3.
_____ 4.
_____ 5.
_____ 6.
_____ 7.
_____ 8.
_____ 9.
_____ 10.

A recording of the electrical activity of the brain is called an __(1)__ . Brain wave patterns are identified by their __(2)__ , which are measured in __(3)__ . When large numbers of neurons fire synchronously, the __(4)__ of the waves increases, a condition typical of the normal adult in a state of __(5)__ . Brain waves that are abnormally slow or fast are typical of a patient in a state of __(6)__ . Very fast brain waves with large spikes are common in a patient who suffers from uncontrollable seizures, a condition called __(7)__ . The mildest form of this disease is known as __(8)__ . The most severe form of this disease is called __(9)__ , during which a sensory "aura" is followed by unconsciousness and intense convulsions. A record showing the complete absence of brain waves is clinical evidence of __(10)__ .

3. For any of the following statements about sleep that are true, write T in the answer blanks. For any false statements, correct the underlined word by writing the correct word in the answer blanks. (*Note:* EEG = electroencephalogram; REM = rapid eye movements; RAS = reticular activating system; NREM = non-rapid eye movements.)

_____ 1. Circadian rhythms recur every 12 hours.

_____ 2. The unconscious state, when arousal is not possible, is called sleep.

_____ 3. During sleep, vital cortical activities continue.

_____ 4. The sleep-wake cycle is most likely timed by the RAS.

_____ 5. During deep sleep, vital signs decline, and EEG waves decrease in amplitude.

_____ 6. During REM sleep, vital signs increase, most muscle movement is inhibited, and the EEG shows alpha waves.

_____ 7. In normal adults, NREM and REM sleep periods alternate, with REM sleep accounting for the greatest proportion of sleeping time.

_____ 8. Most dreaming occurs during REM sleep, when certain brain neurons release more norepinephrine and serotonin.

_____ 9. Deprivation of NREM sleep may lead to emotional instability.

_____ 10. In the elderly, slow wave (stage 4) sleep declines and may disappear.

_____ 11. Two important sleep disorders are insomnia and coma.

4. State the definition of *consciousness* as proposed by cognitive scientists, and list the three underlying suppositions.

_____

_____

_____

_____

5. List the four clinical states of consciousness, starting with the highest state of cortical activity.

_____ , _____ , _____ , _____

6. Categorize the following descriptions as characteristic of either long-term memory (L) or short-term memory (S). Insert your answers in the answer blanks.

_____ **1.** Very brief   _____ **4.** May last a lifetime

_____ **2.** Lasts from seconds to hours   _____ **5.** Limited capacity

_____ **3.** Enormous capacity   _____ **6.** Association process required.

7. List four factors that promote memory consolidation (the transfer of information from short-term to long-term memory).

1. _____

2. _____

3. _____

4. _____

8. Distinguish between fact memory and skill memory by defining each term and by comparing how each type of memory is best remembered.

1. Fact memory _____

_____

_____

2. Skill memory _____

_____

_____

9. Several proposed pathways and regions, including the amygdala, hippo-campus, thalamus, hypothalamus, prefrontal cortex, and basal forebrain, may be involved in the consolidation of fact memory. In terms of these pathways, respond to the questions that follow by writing your answers in the answer blanks.

_____ **1.** According to present assumptions, are the initial connections between old memories and the new perceptions made in the cortex or in the subcortical structures?

_____ **2.** The feedback pathway makes connections between which structure and the cerebral sensory cortex?

_____ **3.** This feedback pathway seems to transform the initial perception into what?

_____ **4.** Which structure appears to be the "gatekeeper" of the memory system?

_____ **5.** What is the result if both the amygdala and the hippocampus are destroyed?

_____ **6.** Can a person with amnesia still learn skills?

_____ **7.** Does widespread cortical damage appear to impair memory?

_____ **8.** Human memory is characterized by great redundancy. If you had to find memory traces or "engrams," where would you look?

_____ **9.** As opposed to fact memory, skill memory is believed to be processed/mediated by which brain regions?

10. Complete the following statements about language by writing the missing terms in the answer blanks.

_____ **1.**
_____ **2.**
_____ **3.**
_____ **4.**
_____ **5.**
_____ **6.**
_____ **7.**
_____ **8.**
_____ **9.**

Human thought, memory, and language are functionally interdependent. These capabilities result directly from the elaboration of the **_(1)_**, in which symbolic information is manipulated. To speak a thought, a sequence of at least three events must occur. These are: selection of the **_(2)_**, which express the thought; finding the **_(3)_** to organize the words into a meaningful sentence; and activating the muscles of **_(4)_**. Several areas in the brain are known to be involved in speech. **_(5)_** is important in sounding out words, while the semantic area (important for assigning meaning to words) is located in **_(6)_**. Lesions in these areas can result in **_(7)_** aphasia, in which the patient can speak but cannot understand written or spoken language. **_(8)_** area and the adjacent premotor area control the muscles used in speaking. Damage to these areas can cause **_(9)_** aphasia, meaning that the patient comprehends written or spoken language but cannot articulate it.

CHALLENGING YOURSELF

At the Clinic

1. Damage to a particular area of the cerebral cortex causes prosopagnosia, an inability to recognize familiar faces. The afflicted individual can still accurately describe the facial features but cannot identify the person depicted, even if the person is standing beside the picture. What aspect of sensory perception has been lost?

2. Tests are run on Mr. Mendoza to determine the site of a suspected brain lesion. His right distal limb muscles are weak and flaccid, and fine motor activity is impaired. There is no abdominal reflex on the right side, and the right foot gives a Babinski reflex response. Where and on which side of the brain might the lesion be? What term is used to indicate weak muscle tone?

3. A second patient with a similar diagnosis (brain tumor) has very different signs. He shows increased muscle tone and spasticity. Where is his lesion likely to be?

4. Marie Nolin exhibits slurred, singsong speech; slow, tentative movements; and a very unstable gait. Examination reveals she cannot touch her finger to her nose with eyes closed. What is the name of this condition and what part of her brain is damaged?

5. A patient exhibits writhing movement of the hands and fingers. What is this abnormal movement called and what class of disorder does it indicate? What part of the brain is probably damaged?

6. A woman brings her elderly father to the clinic. He has been having more and more difficulty caring for himself and now even dressing and eating are a problem. The nurse notes the man's lack of change of facial expression; slow, shuffling gait; and tremor in the arms. What are the likely diagnosis and treatment?

7. Sufferers of Tourette's syndrome have an excess of dopamine in the basal nuclei. They tend to exhibit explosive, uncontrollable episodes of motor activity. The drug haloperidol reduces dopamine levels and returns voluntary control to the patient. What condition would an overdose of haloperidol mimic?

8. Huntington's disease results from a genetic defect that causes degeneration of the basal nuclei and, eventually, the cerebral cortex. Its initial symptoms involve involuntary motor activity, such as arm flapping. Would it be treated with a drug that increases or decreases dopamine levels?

9. Cindy's parents report that she has "spells," lasting 15 to 20 minutes, during which the 4-year-old is unresponsive although apparently awake. What is Cindy's probable diagnosis? Is improvement likely?

10. A young woman comes to the clinic complaining that she is plagued with spells of hand clapping, after which she is extremely disoriented for a while. From what type of epilepsy is she suffering?

11. Gary Suffriti, a man in his early 20s, has begun to have seizures, which an EEG localizes to his right temporal lobe. What is the first course of treatment?

12. Don Johnston, who has been under extreme stress all day, suddenly falls to the floor and is unconscious for about 15 minutes. Friends say he has these "episodes" fairly often. What is the problem and what part of the brain is afflicted?

13. An elderly patient complains that she cannot sleep at night. She says she sleeps no more than 4 hours a night, and she demands sleeping medication. Is her sleep pattern normal? Will a sleeping aid help her?

14. An alcoholic in his 60s is brought to the clinic in a stupor. When he has regained consciousness, a PET scan is ordered. The PET scan shows no activity in parts of the limbic system. The man has no memory of what befell him; in fact, he has no memory of the last 20 years! What type of amnesia does he have? Will it affect his ability to learn new skills?

15. John, a stroke victim, is unable to formulate grammatically correct sentences. His words are garbled, but his apparent frustration indicates that he is aware of his limitations. What part of the brain is a likely site of damage?

16. A psychiatric nurse prepares a bath for a patient and as the tub fills she continuously mixes the water with her hand. As she helps the patient into the tub, he cries out, "What's the matter with you—you're scalding me!" What important aspect of sensory receptor behavior has the nurse forgotten? Explain this phenomenon.

Stop and Think

1. Is all perception at the level of consciousness?

2. Follow the pathway from pricking your finger with a needle to saying "ouch."

3. Follow the neural pathway from the time a mosquito bites your forehead until you smack it.

4. Parkinson's disease involves a deficiency of dopamine, while Tourette's syndrome is caused by an excess of dopamine. What effects do you think these extremes of dopamine have on the mood of victims of the two disorders?

5. Electroconvulsive therapy (ECT) can be used to treat severe clinical depression. After ECT, patients exhibit some memory loss. The electrical current apparently wipes out reverberating circuits, resulting in lifting of the depression as it causes loss of information. Is ECT more likely to affect STM or LTM?

6. Why are the cerebellum and basal nuclei called "*pre*command" areas?

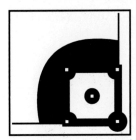

COVERING ALL YOUR BASES

Multiple Choice

Select the best answer or answers from the choices given.

1. Transection of the spinal cord interrupts sensory input at which level?
   - **A.** Receptor level
   - **B.** Perceptual level
   - **C.** Circuit level
   - **D.** Third-order level

2. At the sensory receptor:
   - **A.** transduction occurs
   - **B.** first-order neurons are stimulated
   - **C.** integration occurs
   - **D.** a specific circuit is stimulated

3. Proprioceptors send impulses *primarily* along the:
   - **A.** spinothalamic tracts
   - **B.** spinocerebellar tracts
   - **C.** specific pathways
   - **D.** fasciculus cuneatus

4. Both specific and nonspecific pathways:
   - **A.** transmit impulses to the thalamus
   - **B.** initiate interpretation of emotional aspects of sensation
   - **C.** stimulate the RAS
   - **D.** cross over within the CNS

5. A professional wine or coffee taster relies mostly on:
   A. perceptual detection
   B. feature abstraction
   C. synthetic discrimination
   D. pattern recognition

6. Which of the following is (are) associated with the medial lemniscal pathways?
   A. Fasciculus gracilis
   B. Fasciculus cuneatus
   C. Ventral posterior thalamic nucleus
   D. Trigeminal nerves

7. The ultimate planner of complex motor activities is the:
   A. premotor cortex
   B. primary motor cortex
   C. prefrontal cortex
   D. cerebellum

8. Decerebrate animals can still exhibit fairly normal walking patterns. This is an example of:
   A. simple spinal reflexes
   B. bypass of the spinal cord
   C. fixed-action patterns
   D. learned responses

9. Each segmental circuit can generate rhythmic bursts of impulses even if isolated from sensory and integrative input. These circuits:
   A. are learned
   B. are called central pattern generators if they control locomotion
   C. represent the lowest level of the motor hierarchy
   D. can be controlled by command neurons in the brain stem

10. Command neurons are located in the:
    A. red nuclei   C. cerebellum
    B. precentral gyri   D. basal nuclei

11. Inhibition of the withdrawal reflex stimulated by a pinching crab (when you don't wish to lose it!) occurs:
    A. at the segmental level
    B. at the program level
    C. at the projection level
    D. through the pyramidal system

12. Relaxation of muscles prior to and during sleep is due to:
    A. increased inhibition at the projection level
    B. decreased stimulation by the vestibular nuclei
    C. increased activity of the RAS
    D. inhibition of the reticular nuclei

13. The tectum, which is the roof of the cerebral aqueduct, is associated with the:
    A. red nuclei   C. superior colliculi
    B. vestibular nuclei   D. reticular nuclei

14. The precommand areas are stimulated to plan a series of movements by the:
    A. somatosensory cortex
    B. premotor cortex
    C. command neurons
    D. primary motor area

15. Proper timing and coordination of complex motor activity is carried out by the:
    A. prefrontal cortex   C. cerebellum
    B. premotor cortex   D. basal nuclei

16. Signs of cerebellar damage include:
    A. singsong speech
    B. unsteady gait
    C. overshoot of limb movements
    D. decline in muscle tone

17. Motor abnormalities involving involuntary limb movements include:
    A. Jacksonian seizures   C. ballism
    B. chorea   D. myoclonus

18. Motor disorders for which genetic counseling is recommended include:
- **A.** ataxia
- **B.** Huntington's disease
- **C.** epilepsy
- **D.** hypotonia

19. An adult male is having an EEG. The recording shows irregular waves with a frequency of 4–7 Hz. This is:
- **A.** normal
- **B.** normal in REM sleep
- **C.** normal in deep sleep
- **D.** abnormal in an awake adult

20. Sleep involves:
- **A.** inhibiting the RAS
- **B.** decline in serotonin levels in the brain
- **C.** decreased oxygen use by the brain in slow-wave sleep
- **D.** increase in beta waves

21. During REM sleep:
- **A.** alpha waves appear
- **B.** oxygen utilization of the brain increases
- **C.** arousal threshold is at its lowest
- **D.** active nuclei experience an increase in norepinephrine levels

22. Consciousness involves:
- **A.** discrete, localized stimulation of the cortex
- **B.** activity concurrent with localized activities
- **C.** simultaneous, interconnected activity
- **D.** stimulation of the RAS

23. States of unconsciousness include:
- **A.** sleep
- **B.** petit mal seizure
- **C.** syncope
- **D.** narcoleptic seizure

24. Fact memory involves:
- **A.** the hippocampi
- **B.** rehearsal
- **C.** language areas of the cortex
- **D.** consolidation

25. Language processing probably entails:
- **A.** linear processing
- **B.** activation of Wernicke's area for familiar words
- **C.** semantic assembly
- **D.** activation of Broca's area

26. Which of the following is a possible basis of the long-sought engram?
- **A.** Alterations in the type of RNAs formed
- **B.** Anatomic changes in the presynaptic terminals
- **C.** The role of NMDA channels and NO in LTP
- **D.** Deposit of specific proteins in the synapses

27. Typical of REM sleep is:
- **A.** rapid eye movements
- **B.** average duration of 60 minutes
- **C.** decreased heart and respiratory rates
- **D.** dreaming

28. The sleep neurotransmitter is said to be:
- **A.** serotonin
- **B.** norepinephrine
- **C.** dopamine
- **D.** acetylcholine

Word Dissection

For each of the following word roots, fill in the literal meaning and give an example, using a word found in this chapter.

Word root	Translation	Example
1. alge	_____	_____
2. ceps	_____	_____

Word root	Translation	Example
3. cope	_____	_____
4. duc	_____	_____
5. epilep	_____	_____
6. lemnisc	_____	_____
7. mnem	_____	_____
8. per	_____	_____
9. prax	_____	_____
10. sens	_____	_____
11. somnus	_____	_____
12. trans	_____	_____

16 The Special Senses

Student Objectives

When you have completed the exercises in this chapter, you will have accomplished the following objectives:

The Chemical Senses: Taste and Smell

1. Describe the location, structure, and afferent pathways of taste and smell receptors, and explain how these receptors are activated.

The Eye and Vision

2. Describe the structure and function of accessory eye structures, eye tunics, lens, and humors of the eye.

3. Trace the pathway of light through the eye to the retina and explain how light is focused for distant and close vision.

4. Describe the events involved in the stimulation of photoreceptors by light, and compare and contrast the roles of rods and cones in vision.

5. Note the cause and consequences of astigmatism, cataract, glaucoma, hyperopia, myopia, and color blindness.

6. Compare and contrast light and dark adaptation.

7. Trace the visual pathway to the optic cortex and briefly describe the process of visual processing.

The Ear: Hearing and Balance

8. Describe the structure and general function of the outer, middle, and inner ears.

9. Describe the sound conduction pathway to the fluids of the inner ear and follow the auditory pathway from the organ of Corti to the temporal cortex.

10. Explain how one is able to differentiate pitch and loudness of sounds and to localize the source of sounds.

11. Explain how the balance organs of the semicircular canals and the vestibule help to maintain dynamic and static equilibrium.

12. Note the causes and symptoms of otitis media, deafness, Ménière's syndrome, and motion sickness.

Developmental Aspects of the Special Senses

13. Describe the embryonic development of the special sense organs and list changes that occur in these organs with aging.

Ready, set, go!

The body's sensory receptors react to stimuli or changes occurring within the body and in the external environment. When triggered, these receptors send nerve impulses along afferent pathways to the brain for interpretation. Thus, these specialized parts of the nervous system allow the body to assess and adjust to changing conditions in order to maintain homeostasis.

The minute receptors of general sensation that react to touch—pressure, pain, temperature changes, and muscle tension—are widely distributed in the body. In contrast, receptors of the special senses—sight, hearing, equilibrium, smell, and taste—tend to be localized and in many cases are quite complex.

The structure and function of the special sense organs are the subjects of the student activities in this chapter.

BUILDING THE FRAMEWORK

The Chemical Senses: Taste and Smell

1. Complete the following statements by writing the missing terms in the answer blanks.

_____ 1.

_____ 2.

_____ 3.

_____ 4.

_____ 5.

_____ 6.

The gustatory and olfactory senses rely on __(1)__, meaning that chemicals dissolved in fluid will stimulate these receptors. The receptors for __(2)__ taste, in the back of the tongue, are thought to be protective since many poisons stimulate these receptors. The sensation we call taste is reduced when the nasal passages are swollen; this indicates that "taste" relies heavily on the __(3)__ sense. The loss of smell which accompanies a cold is the result of the less-than-optimal position of the olfactory epithelium, on the __(4)__ of the nasal cavity. The pathway for the sense of smell also runs to the __(5)__ system and thus has emotional ties. It is the only sensory input that does not pass through the __(6)__ in the diencephalon to reach its destination in the cerebral cortex.

2. On Figure 16.1A, label the two types of tongue papillae containing taste buds. Then, using four different colors and appropriate labels, identify the areas of the tongue that are the predominant sites of sweet, salt, sour, and bitter receptors. On Figure 16.1B, color the taste buds green. On Figure 16.1C, color the gustatory cells red, the supporting cells blue, the basal cells purple, and the sensory fibers yellow. Add appropriate labels at the leader lines to identify the *taste pore* and *microvilli* of the gustatory cells.

◯ Sweet   ◯ Salt   ◯ Sour   ◯ Bitter

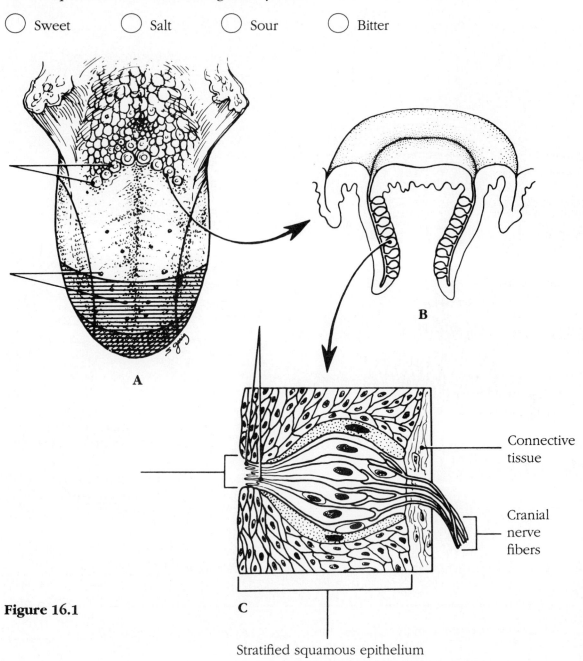

Figure 16.1

A

B

C

Connective tissue

Cranial nerve fibers

Stratified squamous epithelium

3. Figure 16.2 illustrates the site of the olfactory epithelium in the nasal cavity (part A is an enlarged view of the olfactory receptor area). Select different colors to identify the structures listed below and use them to color the illustration. Then add labels and leader lines to identify olfactory nerve fibers, the fibers of the olfactory tract, the mitral cells, and the olfactory "hairs," and add arrows to indicate the direction of impulse transmission. Finally, respond to the questions following the diagrams.

◯ Olfactory neurons (receptor cells)   ◯ Olfactory bulb

◯ Supporting cells   ◯ Cribriform plate of the ethmoid bone

◯ Glomeruli

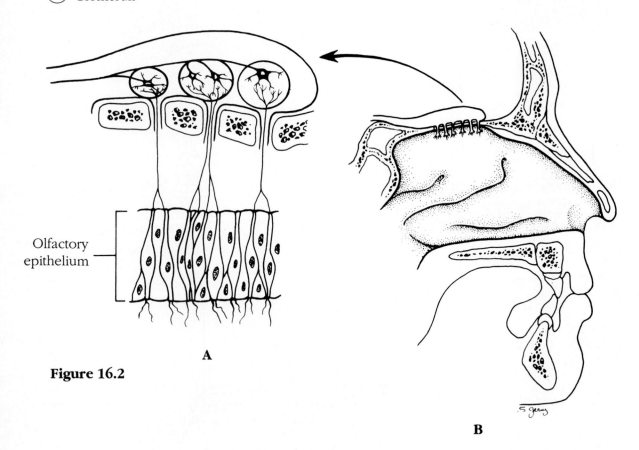

Olfactory epithelium

A

Figure 16.2

B

1. Explain briefly the role of the mitral cells found in the glomeruli. _____

_____

2. What cells, also found in the olfactory bulbs, can act to inhibit the mitral cells? _____

3. What is the result when such inhibition occurs? _____

4. List two brain regions to which olfactory tract fibers project, other than the olfactory cortex in

the temporal lobes. _____

_____

4. Circle the term that does not belong in each of the following groupings.

1. Sweet   Musky   Sour   Bitter   Salty

2. Bipolar neuron   Epithelial cell   Olfactory receptor   Ciliated

3. Gustatory cell   Taste pore   Microvilli   Yellow-tinged epithelium

4. Vagus nerve   Facial nerve   Glossopharyngeal nerve   Olfactory nerve

5. Olfactory receptor   Low specificity   Variety of stimuli   Four receptor types

6. Sugars   Sweet   Saccharine   Metal ions   Amino acids

7. Alkaloids H^+ Nicotine   Quinine   Bitter

8. Olfactory aura   Epileptic seizure   Anosmia   Uncinate fit

The Eye and Vision

1. Complete the following statements by writing the missing terms in the answer blanks.

_____ **1.**

_____ **2.**

_____ **3.**

_____ **4.**

Attached to the eyes are the __(1)__ muscles, which enable us to direct our eyes toward a moving object. The anterior aspect of each eye is protected by the __(2)__, which have eyelashes projecting from their edges. Associated with the eyelids are both typical sebaceous glands and modified sebaceous glands called __(3)__ that help lubricate the eyes. An inflammation of one of these (latter) glands is called a __(4)__ .

2. Three accessory eye structures contribute to the formation of tears and/or aid in lubricating the eyeball. In the table, name each structure and its major secretory product. Indicate the various components of the secretion if applicable.

Accessory eye structures	Secretory product

3. Trace the pathway that the secretion of the lacrimal glands takes from the surface of the eye by assigning a number to each structure.

____ **1.** Lacrimal sac   ____ **3.** Nasolacrimal duct

____ **2.** Nasal cavity   ____ **4.** Lacrimal canals

4. Identify each of the eye muscles indicated by leader lines in Figure 16.3. Color each muscle a different color. Then, in the blanks below, indicate the eye movement caused by each muscle.

1. Superior rectus _____

2. Inferior rectus _____

3. Superior oblique _____

4. Lateral rectus_____

5. Medial rectus_____

6. Inferior oblique _____

_____   _____

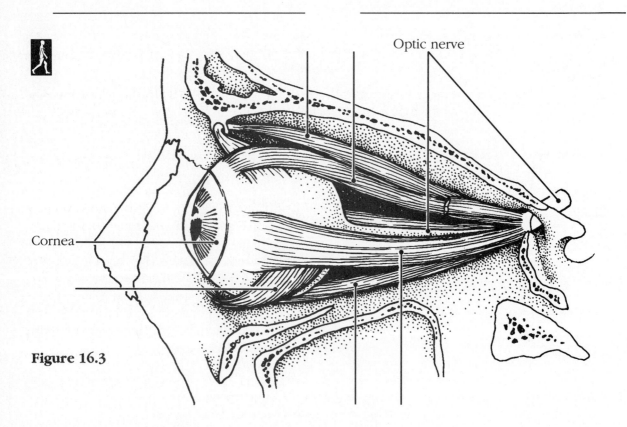

Optic nerve

Cornea

Figure 16.3

5. Choose terms from the key choices to match the following descriptions.

KEY CHOICES

A. Canthus   **C.** Ciliary glands   **E.** Meibomian glands   **G.** Tarsal plate

B. Caruncle   **D.** Conjunctival sac   **F.** Palpebra

_____ **1.** Modified sebaceous (oil) glands embedded in the tarsal plates

_____ **2.** Connective tissue sheet supporting the eyelid internally

_____ **3.** Fleshy elevation at the medial canthus

_____ **4.** Angular point where the eyelids meet

_____ **5.** Recess at the junction of palpebral and ocular conjunctivae

_____ **6.** Another name for the eyelid

6. Circle the term that does not belong in each of the following groupings.

    1. Choroid   Sclera   Vitreous humor   Retina

    2. Inferior oblique   Iris   Superior rectus   Inferior rectus

    3. Pupil constriction   Far vision   Accommodation   Bright light

    4. Mechanoreceptors   Rods   Cones   Photoreceptors

    5. Ciliary body   Iris   Suspensory ligaments   Lens

    6. Uvea   Choroid   Ciliary body   Pigmented retina   Iris

    7. Retina   Neural layer   Pigmented layer   Transparent
    Contains photoreceptors

    8. Optic disc   Blind spot   Lacks photoreceptors   Macula lutea

    9. Cornea   Active sodium ion pump   Transparent   Richly vascular
    Many pain fibers

    10. Inner segment   Outer segment   Photoreceptor pigments   Receptor region

    11. Optic vesicles   Retina   Lens placode   Optic cups

7. Match the functionally corresponding parts of the eye (Column B) to those of a camera (Column A).

	Column A	Column B
_____	**1.** Lenses	**A.** Retina
_____	**2.** Diaphragm	**B.** Sclera
_____	**3.** Light-sensitive layer of film	**C.** Iris
_____	**4.** Focusing by changing the focal distance	**D.** Accommodation
		E. Lens
_____	**5.** Case	**F.** Pupillary constriction
_____	**6.** Reducing light entry by selecting a smaller aperture	

8. In the following table, circle each word that correctly describes events occurring within the eye during close and distant vision.

Vision	Ciliary muscle		Lens convexity		Degree of light refraction	
1. Distant	Relaxed	Contracted	Increased	Decreased	Increased	Decreased
2. Close	Relaxed	Contracted	Increased	Decreased	Increased	Decreased

9. Using the key choices, identify the parts of the eye described in the following statements. Insert the correct letters in the answer blanks.

KEY CHOICES

  ◯ **A.** Aqueous humor   ◯ **F.** Fovea centralis   ◯ **K.** Sclera

     B. Canal of Schlemm   ◯ **G.** Iris   **L.** Suspensory ligaments

  ◯ **C.** Choroid coat   ◯ **H.** Lens   ◯ **M.** Vitreous humor

  ◯ **D.** Ciliary body   ◯ **I.** Optic disk

  ◯ **E.** Cornea   ◯ **J.** Retina

_____ **1.** Secures the lens to the ciliary body

_____ **2.** Fluid that fills the anterior segment of the eye; provides nutrients to the lens and cornea

_____ **3.** Fibrous tunic, white and opaque

_____ **4.** Area of retina that lacks photoreceptors

_____ **5.** Muscular structure that manipulates the lens

_____ **6.** Nutritive (vascular) tunic of the eye

_____ **7.** Drains the aqueous humor of the eye

_____ **8.** Tunic concerned with image formation

_____ **9.** Gel-like substance, filling the posterior segment of the eyeball; helps reinforce the eyeball

_____ **10.** Heavily pigmented tunic that prevents light scattering within the eye

_____ **11.** _____ **12.** Smooth muscle structures (sites of intrinsic eye muscles)

_____ **13.** Area of acute or discriminatory vision

_____ **14.** _____ **15.** _____ **16.** _____ **17.** Refractory media of the eye

_____ **18.** Anteriormost clear part of the fibrous tunic

_____ **19.** Pigmented "diaphragm" of the eye

10. Using the key choices in Exercise 9, identify the structures indicated by leader lines on the diagram of the eye in Figure 16.4. Select different colors for all the structures with a color-coding circle in Exercise 9 and color the structures on the figure.

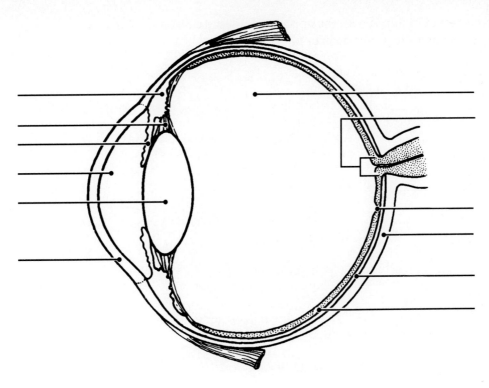

Figure 16.4

11. Explain why it helps to look up and gaze into space after reading for a prolonged period.

_____

_____

_____

12. Match the key choices with the following descriptions concerning optics.

KEY CHOICES

A. Concave lens   **C.** Far point of vision   **E.** Near point of vision

B. Convex lens   **D.** Focal point   **F.** Visible light

_____ **1.** The closest point at which clear focus is possible

_____ **2.** The electromagnetic waves to which the photoreceptors of the eyes respond

_____ **3.** The point at which light rays are converged by a convex lens

_____ **4.** A lens that is thickest at the edges; diverges the light rays

_____ **5.** The point beyond which accommodation is unnecessary

13. Match the terms in Column B with the appropriate descriptions in Column A.
Insert the correct answers in the answer blanks.

Column A

Column B

_____ **1.** Light bending

A. Accommodation

_____ **2.** Ability to focus for close vision (under
20 feet)

B. Accommodation
pupillary reflex

_____ **3.** Normal vision

C. Astigmatism

_____ **4.** Inability to focus well on close objects;
farsightedness

D. Cataract

E. Convergence

_____ **5.** Reflex constriction of pupils when they
are exposed to bright light

F. Emmetropia

_____ **6.** Clouding of lens, resulting in loss of
sight

G. Glaucoma

H. Hyperopia

_____ **7.** Nearsightedness

I. Myopia

_____ **8.** Blurred vision, resulting from unequal
curvatures of the lens or cornea

J. Night blindness

_____ **9.** Condition of increasing pressure inside
the eye, resulting from blocked
drainage of aqueous humor

K. Photopupillary reflex

L. Refraction

_____ **10.** Medial movement of the eyes during
focusing on close objects

_____ **11.** Reflex constriction of the pupils when viewing close objects

_____ **12.** Inability to see well in the dark; often a result of vitamin A
deficiency

14. Complete the following statements by writing the missing terms in the answer blanks.

_____ **1.**

_____ **2.**

_____ **3.**

_____ **4.**

_____ **5.**

_____ **6.**

_____ **7.**

There are __(1)__ varieties of cones. One type responds most
vigorously to __(2)__ light, another to __(3)__ light, and still
another to __(4)__ light. The ability to see intermediate colors
such as purple results from the fact that more than one cone
type is being stimulated __(5)__ . Lack of all color receptors
results in __(6)__ . Because this condition is sex linked, it occurs
most commonly in __(7)__ . Black and white or dim light vision
is a function of the __(8)__ . The density of cones is greatest in
the __(9)__ , whereas rods are densest in the __(10)__ .

_____ **8.** _____ **9.** _____ **10.**

15. Answer the following questions concerning rod photopigment and physiology, or complete the statements as indicated.

1. The bent 11-*cis* form of retinal is combined with a protein called _____

to form the photoreceptor pigment called _____.

2. The light-induced event during which the photoreceptor pigment breaks down to its two

components and retinal assumes its straighter _____ shape is called

_____.

3. Retinal is produced from vitamin _____, which is ordinarily stored in

large amounts by the _____.

4. What ionic and electrical events occur in the photoreceptors (rods) when it is dark?

_____

_____

5. How is this changed in the light? _____

_____

_____

16. Name in sequence the neural elements of the visual pathway, beginning with the retina and ending with the optic cortex.

Retina ➡ _____ ➡ _____ ➡ _____

synapse in thalamus ➡ _____ ➡ optic cortex

17. Check (✔) all of the following conditions that pertain to dark adaptation.

_____ **1.** Retinal sensitivity increases

_____ **2.** Rhodopsin accumulates

_____ **3.** The cones are inactive

_____ **4.** Rods are activated

_____ **5.** Visual acuity increases

_____ **6.** Pupils constrict

_____ **7.** Retinal sensitivity decreases

_____ **8.** Rhodopsin is broken down rapidly

_____ **9.** The cones are active

_____ **10.** Rods are inactivated

_____ **11.** Visual acuity decreases

_____ **12.** Pupils dilate

18. Literally, *binocular vision* means "two-eyed vision." However, in common practice this term means that:

_____

_____

_____

19. Complete the following statements pertaining to visual processing.

_____ **1.**

_____ **2.**

_____ **3.**

_____ **4.**

_____ **5.**

_____ **6.**

_____ **7.**

_____ **8.**

_____ **9.**

_____ **10.**

_____ **11.**

_____ **12.**

_____ **13.**

The only cells of the retina that are capable of generating action potentials are the __(1)__ . These cells receive inputs from the __(2)__ cells, which in turn receive inputs from the __(3)__ . Ganglion cells that depolarize when the center of their receptive field is illuminated are said to have __(4)__ fields, whereas those that hyperpolarize when the center of their receptive field is illuminated are said to have __(5)__ fields. Retinal cells outside the direct pathway to the ganglion cells are the __(6)__ cells and the __(7)__ cells. These cells are involved in retinal processing. As a result of retinal processing, the information flowing from the retina is more concerned with reporting areas of light and dark __(8)__ and bar edges than with transmitting information about individual light points. After being segregated and processed in the __(9)__ , visual inputs flow to the visual cortex. __(10)__ cortical neurons have a receptive field that corresponds to several ganglion cell fields with the same orientation and type of center. __(11)__ cortical neurons, housed in the visual cortex region called the __(12)__ , process inputs concerned with form, color, and __(13)__ to produce dynamic visual images.

20. Label the structures having leader lines in Figure 16.5, which shows the visual pathways to the brain. Then use red to color the left visual field and green to color the right visual field. Use red and green stripes to mark the area of overlap of the two visual fields. Color the entire pathway to the visual cortex in the occipital lobe on both sides: red for the left visual field (*not* for the left eye) and green for the right visual field.

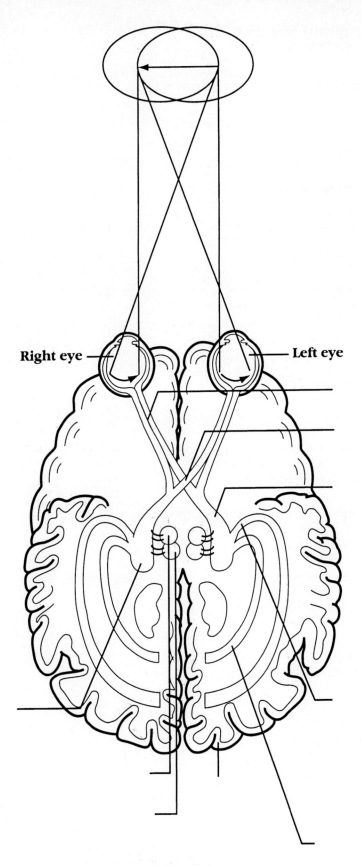

Right eye

Left eye

Figure 16.5

The Ear: Hearing and Balance

1. Using the key choices, select the terms that apply to the following descriptions. Place the correct letters in the answer blanks.

KEY CHOICES

A. Anvil (incus)   E. External auditory canal   I. Pinna   M. Tympanic
                                                                              membrane
B. Auditory tube   F. Hammer (malleus)   J. Round window
                                                                           N. Vestibule
C. Cochlea   G. Oval window   K. Semicircular canals

D. Endolymph   H. Perilymph   L. Stirrup (stapes)

_____ 1. _____ 2.   _____ 3. Structures composing the outer ear

_____ 4. _____ 5.   _____ 6. Elements of the bony or osseous labyrinth

_____ 7. _____ 8.   _____ 9. The ossicles

_____ 10. Ear structures not involved with hearing

_____ 11. Allows middle ear pressure to be equalized with atmospheric pressure

_____ 12. Transmits sound vibrations to the ossicles

_____ 13. Contains the organ of Corti

_____ 14. Passage from the nasopharynx to the middle ear; also called the pharyngotympanic tube

_____ 15. _____ 16. House receptors for the sense of equilibrium

_____ 17. Transfers vibrations from the stirrup to the fluid in the inner ear

_____ 18. Fluid inside the membranous labyrinth

_____ 19. Fluid within the osseous labyrinth, and surrounding the membranous labyrinth

_____ 20. Fits into the oval window

2. Figure 16.6 is a diagram of the ear. Use anatomical terms (as needed) from the key choices in Exercise 1 to correctly identify all structures in the figure with leader lines. Color all external ear structures yellow, color the ossicles red, color the equilibrium areas of the inner ear green, and color the inner ear structures involved with hearing blue.

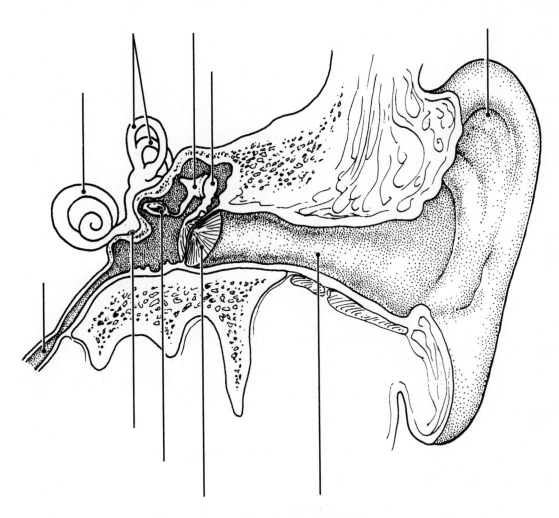

Figure 16.6

3. Sound waves hitting the eardrum set it into vibration. Trace the pathway through which vibrations and fluid currents travel to finally stimulate the hair cells in the organ of Corti. Name the appropriate ear structures in their correct sequence.

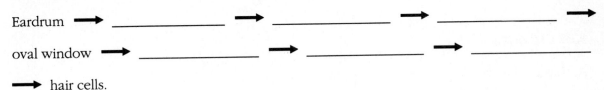

Eardrum ➔ _____ ➔ _____ ➔ _____ ➔

oval window ➔ _____ ➔ _____ ➔ _____ ➔

➔ hair cells.

4. Match the terms in Column B concerning cochlear structures with the appropriate descriptions in Column A.

Column A

Column B

_____ **1.** Bony pillar supporting the coiled cochlea

A. Basilar membrane

B. Helicotrema

_____ **2.** Superior cavity of the cochlea

C. Modiolus

_____ **3.** Inferior cavity of the cochlea

_____ **4.** The cochlear duct

D. Scala media

E. Scala tympani

_____ **5.** The apex of the cochlea; where the perilymph-containing chambers meet

F. Scala vestibuli

_____ **6.** Shelflike extension of the modiolus

G. Spiral lamina

_____ **7.** Forms the endolymph

H. Spiral organ of Corti

_____ **8.** Roof of the cochlear duct

I. Stria vascularis

_____ **9.** Supports the organ of Corti; membranous

J. Vestibular membrane

5. Two tiny skeletal muscles are associated with the bones of the middle ear. Name these two muscles. Note their bony points of attachment and describe their common function.

_____

_____

_____

6. Figure 16.7 is a view of the structures of the membranous labyrinth. Correctly identify and label the following major areas of the labyrinth on the figure: membranous semicircular canals, saccule and utricle, and the cochlear duct. Next, correctly identify and label each of the receptor types shown in enlarged views (organ of Corti, crista ampullaris, and macula). Finally, using terms from the key choices, identify all receptor structures with leader lines. (Some of these terms may need to be used more than once.) Color the diagram as you wish.

KEY CHOICES

A. Basilar membrane

E. Hair cells

B. Cochlear nerve fibers

F. Otoliths

C. Cupula

G. Tectorial membrane

D. Otolithic membrane

H. Vestibular nerve fibers

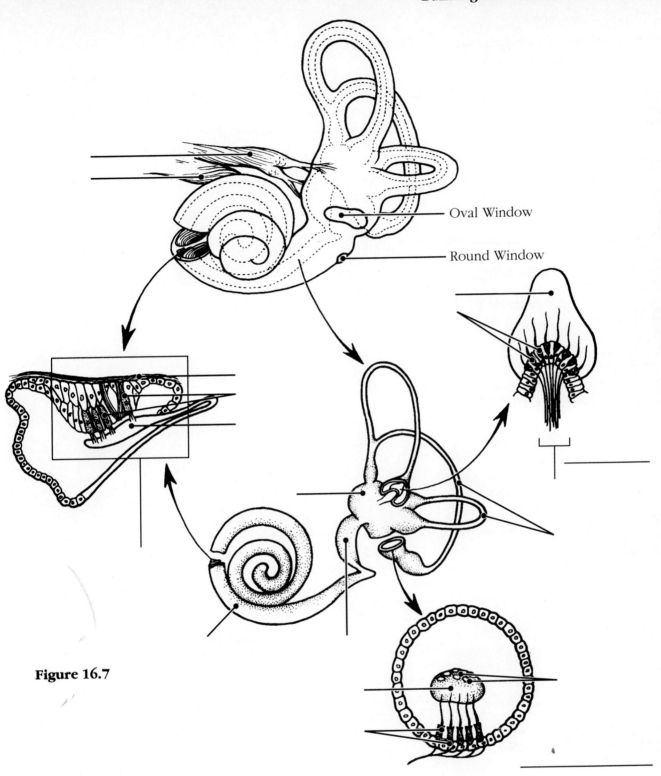

Oval Window

Round Window

Figure 16.7

7. Why is the lever system constructed by the ear ossicles sometimes compared with a hydraulic press?

_____

_____

_____

8. Complete the following statements referring to the physics of sound.

_____ 1.

_____ 2.

_____ 3.

_____ 4.

_____ 5.

_____ 6.

_____ 7.

_____ 8.

_____ 9.

_____ 10.

The source of sound is a __(1)__ . In order for sound to be propagated, there must be a medium, and it must be __(2)__ so that alternating regions of __(3)__ and __(4)__ can be produced. Sound waves are periodic, and the distance between successive peaks of a sound sine wave is referred to as the __(5)__ of sound. As a rule, sounds of high frequency have __(6)__ wavelengths, whereas sounds of low frequency have __(7)__ wavelengths. Different frequencies are perceived as differences in __(8)__ , whereas differences in amplitude or intensity of a sound are perceived as differences in __(9)__ . Sound intensities are measured in units called __(10)__ .

9. The fibers of the basilar membrane vary in length and thickness. Those nearest the oval window are short and stiff, whereas those near the cochlear apex are long and floppy. How is the structure of the basilar membrane related to the excitation of cochlear hair cells?

_____

_____

_____

_____

10. Indicate whether the following conditions relate to conduction deafness (C) or sensorineural (central) deafness (S). Place the correct letters in the answer blanks.

_____ 1. Can result from a bug wedged in the external auditory canal

_____ 2. Can result from a stroke (CVA)

_____ 3. Sound is heard in both ears during bone conduction, but only in one ear during air conduction

_____ 4. Not improved much by a hearing aid

_____ 5. Can result from otitis externa

_____ 6. Can result from otosclerosis, excessive earwax, or a perforated eardrum

_____ 7. Can result from a lesion on the auditory nerve

11. Complete the following description of the functioning of the static and dynamic equilibrium receptors by writing the key choices in the answer blanks.

KEY CHOICES

A. Angular/rotatory   **E.** Gravity   **I.** Semicircular canals

B. Cupula   **F.** Perilymph   **J.** Static

C. Dynamic   **G.** Proprioception   **K.** Utricle

D. Endolymph   **H.** Saccule   **L.** Vision

_____ 1.

_____ 2.

_____ 3.

_____ 4.

_____ 5.

_____ 6.

_____ 7.

_____ 8.

_____ 9.

_____ 10.

_____ 11.

The receptors for __(1)__ equilibrium are found in the crista ampullaris of the __(2)__ . These receptors respond to changes in __(3)__ motion. When motion begins, the __(4)__ fluid lags behind and the __(5)__ is bent, which excites the hair cells. When the motion stops suddenly, the fluid flows in the opposite direction and again stimulates the hair cells. The receptors for __(6)__ equilibrium are found in the maculae of the __(7)__ and __(8)__ . These receptors report on the position of the head in space. Tiny stones found in a gel overlying the hair cells roll in response to the pull of __(9)__ . As they roll, the gel moves and tugs on the hair cells, exciting them. Besides the equilibrium receptors of the inner ear, the senses of __(10)__ and __(11)__ are also important in helping maintain equilibrium.

12. List three problems about which a person with equilibrium imbalance might complain.

1. _____   **2.** _____   **3.** _____

13. Circle the term that does not belong in each of the following groupings.

1. Hammer   Anvil   Pinna   Stirrup

2. Tectorial membrane   Crista ampullaris   Semicircular canals   Cupula

3. Gravity   Angular motion   Sound waves   Rotation

4. Utricle   Saccule   Pharyngotympanic tube   Vestibule

5. Vestibular nerve   Optic nerve   Cochlear nerve   Vestibulocochlear nerve

6. Crista ampullaris   Maculae   Retina   Proprioceptors   Pressure receptors

Developmental Aspects of the Special Senses

1. Complete the following statements by writing the missing terms in the answer blanks.

_____ 1.

_____ 2.

_____ 3.

_____ 4.

_____ 5.

_____ 6.

_____ 7.

The eye beginnings are seen in the two lateral outgrowths, called the **(1)**, that protrude from the **(2)** by the fourth week of development. Soon the tips of these outgrowths indent, forming the **(3)**, which become the **(4)** of the eye. Their proximal parts eventually become the **(5)** . The optic cup induces the surface **(6)** to form the **(7)**, which becomes the lens.

2. Match the key choices with the following descriptions relating to embryonic development of the ear.

KEY CHOICES

A. Mesoderm   **B.** Pharyngeal pouches   **C.** Otic placode

_____ 1. Form the ossicles and the bony cavity of the middle ear

_____ 2. Thickening of the surface ectoderm that forms the membranous structures of the inner ear

_____ 3. Forms the bony labyrinth

3. Complete the following statements by writing the missing terms in the answer blanks.

_____ 1.

_____ 2.

_____ 3.

_____ 4.

_____ 5.

_____ 6.

_____ 7.

_____ 8.

The special sense organs are actually part of the **(1)** and are formed very early in the embryo. Maternal infections, particularly **(2)**, may cause both deafness and **(3)** in the developing child. Of the special senses, the sense of **(4)** requires the most learning and takes longest to mature. All infants are **(5)**, but generally by school age, emmetropic vision has been established. Beginning sometime after the age of 40, the eye lenses start to become less **(6)** and cannot bend properly to refract the light. As a result, a condition of farsightedness, called **(7)**, begins to occur. In old age, a gradual hearing loss, called **(8)**, occurs. A declining efficiency of the chemical senses is also common in the elderly.

The Incredible Journey:
A Visualization Exercise
for the Special Senses

You . . . see a discontinuous sea of glistening, white rock slabs . . .

1. Complete the narrative by inserting the missing words in the answer blanks.

_____ **1.**

_____ **2.**

_____ **3.**

_____ **4.**

_____ **5.**

_____ **6.**

_____ **7.**

_____ **8.**

_____ **9.**

_____ **10.**

_____ **11.**

_____ **12.**

Your present journey will take you through your host's inner ear to observe and record events documenting what you have learned about how hearing and equilibrium receptors work.

This is a very tightly planned excursion. Your host has been instructed to move his head at specific intervals and will be exposed to various sounds so that you can make specific observations. For this journey you are injected into the bony cavity of the inner ear, the __(1)__, and are to make your way through its various chambers in a limited amount of time.

Your first observation is that you are in a warm sea of __(2)__ in the vestibule. To your right are two large sacs, the __(3)__ and __(4)__. You swim over to one of these membranous sacs, cut a small semicircular opening in the wall, and wiggle through. Since you are able to see very little in the dim light, you set out to explore this area more fully. As you try to move, however, you find that your feet are embedded in a thick, gluelike substance. The best you manage is slow-motion movements through this __(5)__.

It is now time for your host's first scheduled head movement. Suddenly your world tips sharply sideways. You hear a roar (rather like an avalanche) and look up to see a discontinuous sea of glistening, white rock slabs sliding toward you. You protect yourself from these __(6)__ by ducking down between the hair cells that are bending vigorously with the motion of the rocks. Now that you have seen and can document the operation of a(n) __(7)__, a sense organ of __(8)__ equilibrium, you quickly back out through the hole you made.

Keeping in mind the schedule and the fact that it is nearly time for your host to be exposed to tuning forks, you swim quickly to the right, where you see what looks like the opening of a cave with tall seaweed waving gently in the current. Abruptly, as you enter the cave, you find that you are no longer in control of your movements, but instead are swept along in a smooth undulating pattern through the winding passageway of the cave, which you now know is the cavity of the __(9)__. As you move up and down with the waves, you see hair cells of the __(10)__, the sense organ for __(11)__, being vigorously disturbed below you. Flattening yourself against the chamber wall to prevent being carried further by the waves, you wait for the stimulus to stop. Meanwhile you are delighted by the electrical activity of the hair cells below you. As they depolarize and send impulses along the __(12)__ nerve, the landscape appears to be alive with fireflies.

_____ 13.

_____ 14.

_____ 15.

_____ 16.

Now that you have witnessed the events for this particular sense receptor, you swim back through the vestibule toward your final observation area at the other end of the bony chambers. You recognize that your host is being stimulated again because of the change in fluid currents, but since you are not close to any of the sensory receptors, you are not sure just what the stimulus is. Then, just before you, three dark openings appear, the **(13)** . You swim into the middle opening and see a strange structure that looks like the brush end of an artist's paint brush; you swim upward and establish yourself on the soft brushy top portion. This must be the **(14)** of the **(15)** , the sensory receptor for **(16)** equilibrium. As you rock back and forth in the gentle currents, a sudden wave of fluid hits you. Clinging to the hairs as the fluid thunders past you, you realize that there will soon be another such wave in the opposite direction. You decide that you have seen enough of the special senses and head back for the vestibule to leave your host once again.

CHALLENGING YOURSELF

At the Clinic

1. After brain surgery, Ray complains that the hospital food tastes unbearably repugnant. His ability to taste salt and sugar are normal, but he firmly maintains that the food is rotten and inedible. From what condition is he suffering, and why?

2. Bernice, a long-time smoker, always seasons her food very highly, claiming it has no taste otherwise. From what condition is she suffering? For what mineral deficiency would she be checked if she has no other contributing factors such as smoking?

3. A woman comes to the clinic with a small red cyst between the eyelashes on her lower eyelid. What gland is probably inflamed and what is this cystlike structure called?

4. Nine children attending the same day-care center developed red, inflamed eyes and eyelids. What is the most likely cause and name of the condition? Would antibiotics be an appropriate treatment?

5. An infant girl is brought to the clinic with strabismus, and tests show that she can control both eyes independently. What noninvasive procedure will be tried before surgery?

6. A man in his early 60s comes to the clinic complaining of fuzzy vision. An eye examination reveals clouding of his lenses. What is his problem and what factors might have contributed to it?

7. Janie is referred to the eye clinic by her teacher, who suspects a need for glasses. Examination demonstrates that Janie is myopic. Will she need concave or convex lenses? Explain.

8. Mr. Daysh is experiencing night blindness and seeks help at the clinic. What is the technical name for this disorder? What dietary supplement will be recommended? If the condition has progressed too far, what retinal structure will be degenerated?

9. A child is brought to the speech therapist because she does not pronounce high-pitched sounds (like "s"). If it is determined that the spiral organ of Corti is the source of the problem, which region of the organ would be defective? Is this conduction or sensorineural deafness?

10. A woman is experiencing disturbances in equilibrium: dizziness, vertigo, and nausea. What part of the ear is probably malfunctioning?

11. When Mrs. House visits her ophthalmologist, she complains of pain in her right eye. The intraocular pressure of that eye is found to be abnormally elevated. What is the name of Mrs. House's probable condition? What causes it? What might be the outcome if the problem is not rectified?

12. Chelsea went to a heavy metal rock concert and sat right in front of a speaker. She enjoyed the concert for about an hour, but then it seemed to get much louder. She knew this was impossible because the amplifier had already been set at maximum. Luckily, Chelsea escaped without permanent damage to her hearing, but she asked her doctor why the music seemed to get louder. Try to deduce what may have happened.

13. Little Biff's uncle tells the physician that 3-year-old Biff gets many headaches. Upon questioning, the uncle reveals that Biff has not had a sore throat for a long time and that he is happily learning to swim. Does Biff have otitis media or otitis externa and does he need ear tubes? Explain your reasoning.

14. Dr. Nakvarati used an instrument to press on Mr. Cruz's eye during his annual physical examination. The eye deformed very little. What is the probable reason (physical) for this situation and what medical term is used to name the disorder?

15. Lionel suffered a ruptured artery in his middle cranial fossa. As a consequence, a pool of blood compressed his left optic tract, destroying its fibers. What part of the visual field was blinded?

Stop and Think

1. Using word roots, explain why disorders of the chemical senses are called "anosmias"?

2. Why are olfactory auras and hallucinations called "uncinate" fits?

3. How is vision affected by albinism?

4. Why do children's eyes appear larger than the eyes of adults?

5. What is the only structure between the conjunctiva and the retina's photoreceptors that contains blood vessels?

6. Why does the eye require more refraction for near vision?

7. Why do stars "disappear" when you look directly at them?

8. What is happening in the photoreceptors from the time you first enter a dark room from the bright outdoors, until your vision returns?

9. What wavelengths of light are absorbed (and utilized for photosynthesis) by plants?

10. What difference in the length of the organ of Corti of an elephant allows it to hear low-pitch sounds that are beyond the range of human hearing? What about a dog's ability to hear high-frequency sounds?

11. All of the special senses have some kind of "hairs" involved in stimulus reception, except vision. What is the true organellar structure of the "hairs" for each of these senses? Do the receptors for vision use one of these "hair" types of organelles for something other than sensory reception? If so, what?

12. Name two special senses whose receptor cells are replaced throughout life and two whose receptor cells are never replaced.

COVERING ALL YOUR BASES

Multiple Choice

Select the best answer or answers from the choices given.

1. Gustatory cells are:
   A. bipolar neurons
   B. multipolar neurons
   C. unipolar neurons
   D. specialized receptor cells

2. Alkaloids excite gustatory hairs at the:
   A. tip of the tongue
   B. back of the tongue
   C. circumvallate papillae
   D. fungiform papillae

3. Cranial nerves that are part of the gustatory pathway include:
   A. trigeminal   C. hypoglossal
   B. facial   D. glossopharyngeal

4. Which of the following parasympathetic responses can be triggered reflexively by the activation of taste receptors?
   A. Coughing   C. Secretion of saliva
   B. Gagging   D. Secretion of gastric juice

5. The receptors for olfaction are:
   A. the ends of dendrites of bipolar neurons
   B. cilia
   C. specialized non-neural receptor cells
   D. olfactory hairs

6. Which cranial nerve controls contraction of the circular smooth muscle of the iris?
   A. Trigeminal   C. Oculomotor
   B. Facial   D. Abducens

7. Second-order neurons for olfaction are located in the:
   A. olfactory bulb
   B. olfactory epithelium
   C. uncus of limbic system
   D. primary olfactory cortex

8. Nociceptors in the nasal mucosa stimulate the:
   A. olfactory nerve
   B. trigeminal nerve
   C. facial nerve
   D. glossopharyngeal nerve

9. Nonfunctional granule cells would result in:
   A. excessive GABA secretion in the olfactory bulb
   B. extremely rapid olfactory adaptation
   C. diminished olfactory adaptation
   D. anosmia

10. Removal of a tumor in a woman has necessitated destruction of the right facial nerve's branch to the lacrimal gland. Which of the following right eye problems will result from the loss of innervation?
    A. Lack of lysozyme
    B. Chalazion
    C. Sty
    D. Reduced lubrication of the conjunctiva

11. The eye movements involved in reading are:
    A. saccades   C. nystagmus
    B. scanning movements   D. strabismus

12. Which of the following would be found in the fovea centralis?
   A. Ganglion neurons   C. Cones
   B. Bipolar neurons   D. Rhodopsin

13. The vitreous humor:
   A. helps support the lens
   B. holds the retina in place
   C. contributes to intraocular pressure
   D. is constantly replenished

14. Blockage of which of the following is suspected in glaucoma?
   A. Ciliary processes
   B. Retinal blood vessels
   C. Choroid vessels
   D. Scleral venous sinus

15. The cornea is nourished by:
   A. corneal blood vessels
   B. aqueous humor
   C. vitreous humor
   D. scleral blood vessels

16. In an emmetropic eye:
   A. the real image is right side up
   B. the focal point is on the fovea centralis
   C. convex lenses are required for correction
   D. light refraction occurs

17. Refraction can be altered for near or far vision by the:
   A. cornea
   B. ciliary muscles
   C. vitreous humor
   D. neural layer of the retina

18. Convergence:
   A. requires contraction of the medial rectus muscles of both eyes
   B. is needed for near vision
   C. involves transmission of impulses along the abducens nerves
   D. can promote eye strain

19. The near point of vision:
   A. occurs when the lens is at its maximum thickness
   B. gets closer with advancing age
   C. changes due to loss of lens elasticity
   D. occurs when the ciliary muscles are totally relaxed

20. In focusing for far vision:
   A. the lens is at its thinnest
   B. the ciliary muscles contract
   C. the light rays are nearly parallel
   D. suspensory fibers are slack

21. An unequal corneal curvature:
   A. can be corrected with concave lenses
   B. is a type of astigmatism
   C. is treated by transplant
   D. results in cataracts

22. Objects in the periphery of the visual field:
   A. stimulate cones
   B. cannot have their color determined
   C. can be seen in low light intensity
   D. appear fuzzy

23. Vitamin A deficiency leads to depletion of:
   A. retinal   C. scotopsin
   B. rhodopsin   D. photopsins

24. Which of the following statements apply to rhodopsin?
   A. Rhodopsin consists of opsin and retinal.
   B. Rhodopsin is identical to the visual pigment in the cones.
   C. The concentration of rhodopsin increases in the rods during dark adaptation.
   D. A solution of rhodopsin looks reddish-purple but becomes colorless after illumination.

25. When moving from darkness to bright light:
   A. rhodopsin breakdown accelerates
   B. adaptation inhibits cones
   C. retinal sensitivity declines
   D. visual acuity increases

26. When light strikes the lateral aspect of the left retina, activity increases in the:
    A. left optic tract
    B. superior colliculus
    C. pretectal nucleus
    D. right primary visual cortex

27. Receptive fields of retinal ganglion cells:
    A. are smaller in the fovea than in the periphery of the retina
    B. are equal in size, over the entire retina
    C. can be subdivided into functionally distinct center and surround regions
    D. are found only in the fovea

28. Excitation of a retinal bipolar cell:
    A. can result from excitation of its photoreceptor cells
    B. causes excitation of its ganglion cell
    C. always results from light striking its photoreceptor cells
    D. is modified by lateral inhibition

29. Depth perception is due to all of the following factors except which one(s)?
    A. The eyes are frontally located.
    B. There is total crossover of the optic nerve fibers at the optic chiasma.
    C. There is partial crossover of the optic nerve fibers at the optic chiasma.
    D. Each visual cortex receives input from both eyes.

30. Which structures are contained within the petrous portion of the temporal bone?
    A. Tympanic cavity
    B. Mastoid air cells
    C. External auditory meatus
    D. Stapedius muscle

31. Auditory processing is:
    A. analytic
    B. synthetic
    C. complex and involves local processing
    D. interpreted in subjective terms such as pitch and loudness

32. Movement of the _____ membrane triggers bending of hairs of the hair cells in the spiral organ of Corti.
    A. tympanic   C. basilar
    B. tectorial   D. vestibular

33. Sounds entering the external auditory canal are eventually converted to nerve impulses via a chain of events including:
    A. vibration of the eardrum
    B. vibratory motion of the ossicles against the round window
    C. stimulation of hair cells in the organ of Corti
    D. resonance of the basilar membrane

34. In the cochlea:
    A. high-frequency sounds resonate close to the helicotrema
    B. low-frequency sounds resonate farther from the oval window than high-frequency sounds
    C. amplitude determines the intensity of movements of the basilar membrane
    D. sound signals are mechanically processed by the basilar membrane before reaching the receptor cells

35. Transmission of impulses from the sound receptors along the cochlear nerve includes which of the following "way stations"?
    A. Spiral ganglion
    B. Superior olivary nucleus
    C. Cochlear nuclei of the medulla
    D. Auditory cortex

36. Which of the following is a space within the membranous labyrinth?
    A. Scala vestibuli   C. Scala media
    B. Scala tympani   D. Helicotrema

37. Sound localization is possible if:
    A. the sound source is in front of the head or slightly to the side; otherwise not
    B. there is a slight difference in the amplitude of the sound entering the two ears
    C. there is a slight difference in the time sound reaches the two ears
    D. the sound is originating at a point exactly equidistant between the ears

38. According to the place theory:
   A. a sound stimulus excites hair cells at a single site on the basilar membrane
   B. the fibers of the cochlear nerve all arise from the same site in the organ of Corti
   C. each frequency component of a sound excites hair cells at particular (and different sites) along the basilar membrane
   D. each sound is interpreted at a specific "place" in the auditory cortex

39. A singer's high C, in the physical sense, is:
   A. noise
   B. a pure tone like the sound of a tuning fork
   C. a sound with a fundamental tone and harmonics
   D. music

40. Sound:
   A. loses energy as it travels
   B. is an example of a longitudinal wave phenomenon
   C. is periodic
   D. requires an elastic medium

41. If the loudness of a (20 dB) sound is doubled, the resulting sound's pressure level is:
   A. 0 dB   **C.** 40 dB
   B. 22 dB   **D.** 10 dB

42. Which of the following structures is involved in static equilibrium?
   A. Maculae   **C.** Crista ampullaris
   B. Saccule   **D.** Otoliths

43. Which lies closest to the posterior pole of the eye?
   A. Cornea   **C.** Macula lutea
   B. Optic disc   **D.** Central artery

44. Which of the following are paired incorrectly?
   A. Cochlear duct—cupula
   B. Saccule—macula
   C. Ampulla—otoliths
   D. Semicircular duct—ampulla

45. Elements associated with static equilibrium include:
   A. sensing the position of the head in space
   B. response of calcium carbonate crystals to the pull of gravity
   C. hair cells, stereocilia, and kinocilia
   D. rotary motion

46. The vestibular apparatus:
   A. responds to changes in linear acceleration
   B. responds to unchanging acceleratory stimuli
   C. does not usually contribute to conscious awareness of its activity
   D. is helpful, but not crucial, to maintaining balance

47. Taste receptor cells are stimulated by:
   A. chemicals binding to the nerve fibers supplying them
   B. chemicals binding to their microvilli
   C. stretching of their microvilli
   D. impulses from the sensory nerves supplying them

48. The equilibrium pathway to the brain ends at which of the following brain centers?
   A. Vestibular nuclei of brain stem
   B. Flocculonodular node of cerebellum
   C. Inferior colliculi of midbrain
   D. Cerebral cortex

Word Dissection

For each of the word roots on the following page, fill in the literal meaning and give an example, using a word found in this chapter.

Word root	Translation	Example
1. ampulla		
2. branchi		
3. caruncl		
4. cer		
5. cochlea		
6. fove		
7. glauc		
8. gust		
9. lut		
10. macula		
11. metr		
12. modiol		
13. olfact		
14. palpebra		
15. papill		
16. pinn		
17. presby		
18. sacc		
19. scler		
20. tars		
21. trema		
22. tympan		

17 The Endocrine System

Student Objectives

When you have completed the exercises in this chapter, you will have accomplished the following objectives:

1. Indicate important differences between endocrine and neural controls of body functioning.

The Endocrine System and Hormone Function: An Overview

2. List the major endocrine organs and describe their locations in the body.

Hormones

3. Describe how hormones are classified chemically.

4. Describe the two major mechanisms by which hormones bring about their effects on their target tissues and explain how hormone release is regulated.

Major Endocrine Organs of the Body

5. Describe the structural and functional relationships between the hypothalamus and the pituitary gland.

6. List and describe the chief effects of adenohypophyseal hormones.

7. Discuss the role of the neurohypophysis and describe the effects of the two hormones it releases.

8. Describe the important effects of the two groups of hormones produced by the thyroid gland. Follow the process of thyroxine formation and release.

9. Note the general functions of parathyroid hormone.

10. List the hormones produced by the cortical and medullary regions of the adrenal gland and cite their physiological effects.

11. Compare and contrast the effects of the two major pancreatic hormones.

12. Describe the functional roles of the hormone products of the testes and ovaries.

13. Briefly describe the importance of thymic hormones in the operation of the immune system.

Other Endocrine Structures

14. Name a hormone produced by the heart and localize enteroendocrine cells.

15. Briefly explain the endocrine functions of the placenta, kidney, and skin.

Developmental Aspects of the Endocrine System

16. Describe the effect of aging on endocrine system functioning.

Ready, set, go!

The endocrine system is vital to homeostasis and plays an important role in regulating the activity of body cells. Acting through blood-borne chemical messengers called hormones, the endocrine system organs orchestrate cellular changes that lead to growth and development, reproductive capability, and the physiological homeostasis of many body systems.

Activities in this chapter concern the localization of the various endocrine organs in the body and explaining the general function of the various hormones and the results of their hypersecretion or hyposecretion.

BUILDING THE FRAMEWORK

The Endocrine System and Hormone Function: An Overview

1. Complete the following statements by choosing answers from the key choices. Record the answers in the answer blanks.

KEY CHOICES

A. Circulatory system   E. Metabolism   I. Nutrient

B. Electrolyte   F. More rapid   J. Reproduction

C. Growth and development   G. Nerve impulses   K. Slower and more prolonged

D. Hormones   H. Nervous system   L. Water

_K_ _slower and more prolonged_   1.

_H_ _Nervous system_   2.

_D_ _hormones_   3.

_G_ _nerve impulses_   4.

_A_ _circulatory system_   5.

_B_ _metabolism / electrolyte_   6.

_I_ _nutrient_   7.

_C_ _growth and development_   7.

_L_ _water_   8. _E_ _metabolism_   9. _J reproduction_ 10.

The endocrine system is a major controlling system in the body. Its means of control, however, is much __(1)__ than that of the __(2)__, the other major body system that acts to maintain homeostasis. Perhaps the reason for this is that the endocrine system uses chemical messengers, called __(3)__, instead of __(4)__. These chemical messengers enter the blood and are carried throughout the body by the activity of the __(5)__.

The endocrine system has several important functions: It helps maintain __(6)__, __(7)__, and __(8)__ balance; regulates energy balance and __(9)__; and prepares the body for childbearing or __(10)__.

2. Figure 17.1 is a diagram of the various endocrine organs of the body. Next to each letter on the diagram, write the name of the endocrine-producing organ (or area). Then select different colors for each and color the illustration. To complete your identification of the hormone-producing organs, name the organs (not illustrated) described in J, K, and L.

J. Small glands that ride "horseback" on the thyroid

_____

K. Endocrine-producing organ present only in pregnant women

_____

L. B and C hang from the floor of this neuro-endocrine organ

_____

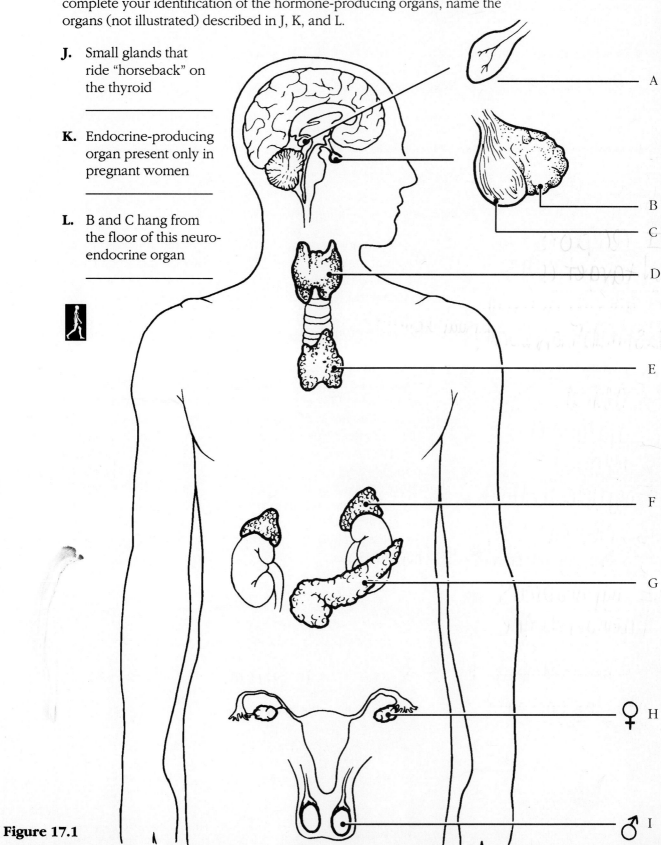

Figure 17.1

Hormones

1. Complete the following statements by choosing answers from the key choices.
   Record the answers in the answer blanks.

KEY CHOICES

A. Altering activity	**F.** Negative feedback	**K.** Steroid or amino acid-based
B. Anterior pituitary	**G.** Neural	**L.** Stimulating new or unusual activities
C. Hormonal	**H.** Neuroendocrine	**M.** Sugar or protein
D. Humoral	**I.** Receptors	**N.** Target cell(s)
E. Hypothalamus	**J.** Releasing and inhibiting factors (hormones)	

I receptors 1.

N target cells 2.

A altering activity 3.

L stimulating new or unusual activities 4.

K steroid or amino acid based 5.

G neural 6.

C hormonal 7.

D Humoral 8.

F negative feedback 9.

B anterior pituitary 10.

J releasing inhibiting factors (hormones) 11.

E hypothalamus 12.

H neuroendocrine 13.

All cells do not respond to endocrine system stimulation. Only those that have the proper __(1)__ on their cell membranes are activated by the chemical messengers. These responsive cells are called the __(2)__ of the various endocrine glands.

Hormones promote homeostasis by __(3)__ of body cells rather than by __(4)__ . Most hormones are __(5)__ molecules. The various endocrine glands are prodded to release their hormones by nerve fibers (a __(6)__ stimulus), by other hormones (a __(7)__ stimulus), or by the presence of increased or decreased levels of various other substances in the blood (a __(8)__ stimulus). The secretion of most hormones is regulated by a __(9)__ system, in which increasing levels of that particular hormone "turn off" its stimulus. The __(10)__ is called the master endocrine gland because it regulates so many other endocrine organs. However, it is in turn controlled by __(11)__ secreted by the __(12)__ . The structure identified in item 12 above is also part of the brain, so it is appropriately called a __(13)__ organ.

2. Indicate the major stimulus for release of each of the hormones listed below. Choose your response from the key choices.

KEY CHOICES

A. Hormonal   **B.** Humoral   **C.** Neural

A Hormonal   **1.** Adrenocorticotropic hormone

B humoral   **2.** Parathyroid hormone

B humoral   **3.** Insulin

A hormonal   **4.** Thyroxine and triiodothyronine

C neural   **5.** Epinephrine

C neural   **6.** Oxytocin and antidiuretic hormone

A hormonal   **7.** Estrogen and progesterone

B humoral   **8.** Calcitonin

3. Complete the following description of a second-messenger system by writing the missing words in the answer blanks.

_____ **1.**

_____ **2.**

_____ **3.**

_____ **4.**

_____ **5.**

_____ **6.**

_____ **7.**

_____ **8.**

_____ **9.**

_____ **10.**

_____ **11.**   _____ **12.**

The cyclic AMP mechanism is a good example of a second-messenger system. In this mechanism, the hormone, acting as the __(1)__ messenger, binds to target cell membrane receptors coupled to a signal transducer called __(2)__. This signal transducer molecule, in turn, acts as an intermediary to activate the enzyme __(3)__, which catalyzes the conversion of intracellular __(4)__ to cyclic AMP. Cyclic AMP then acts as the __(5)__ messenger to initiate a cascade of reactions in the target cell. Most of the subsequent events are mediated by the activation of enzymes called __(6)__, which in turn activate or inactivate other enzymes by adding a __(7)__ group to them. The sequence of events initiated by the second messenger depends on the __(8)__. In addition to cyclic AMP, many other molecules are known to act as second messengers, including __(9)__, __(10)__, and __(11)__. Additionally, the ion __(12)__ can act intracellularly as a "third messenger" in certain cases.

4. Explain why the persistence of a hormone in the blood is so limited.

_____

_____

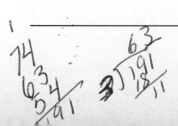

5. Match the terms or phrases in Column B with the descriptions in Column A.

Column A

_____ 1. _____ 2. _____ 3. The extent of target cell activation by hormone-receptor binding depends equally on these three factors

_____ 4. The mechanism by which most steroid-based hormones influence their target cells

_____ 5. _____ 6. _____ 7. _____ 8. Four ways hormones may alter cellular activity (depending on target cell type)

_____ 9. Target cell responds to continued high hormone levels by forming more receptors capable of binding the hormone

_____ 10. Reduced target cell response to continued high hormone levels in the blood

_____ 11. Period of persistence of a hormone in the bloodstream

Column B

A. Affinity of the receptor for the hormone

B. Change in membrane permeability and/or voltage

C. Direct gene activation

D. Down-regulation

E. Enzyme activation/inactivation

F. Half-life

G. Hormone blood levels

H. Initiation of secretory activity

I. Relative number of hormone receptors on the target cells

J. Second-messenger system

K. Synthesis of regulatory molecules such as enzymes

L. Up-regulation

Major Endocrine Organs of the Body

1. For each of the following hormones, indicate specifically its chemical nature by choosing from the key choices.

KEY CHOICES

A. Steroid   B. Catecholamine   C. Peptide   D. Iodinated amino acid derivative

D iodinated amino acid derivative **1.** Thyroxine   _A steroid_ **3.** Aldosterone

B catecholamine **2.** Epinephrine   _C peptide_ **4.** Insulin

2. Figure 17.2 depicts the anatomical relationships between the hypothalamus and the anterior and posterior parts of the pituitary in a highly simplified way. First, identify each of the structures listed below by color coding and coloring them on the diagram. Then, on the appropriate lines write in the names of the hormones that influence each of the target organs shown at the bottom of the diagram. Color the target organ diagrams as you like.

◯ Hypothalamus

◯ Anterior pituitary

◯ Sella turcica of the sphenoid bone

◯ Posterior pituitary

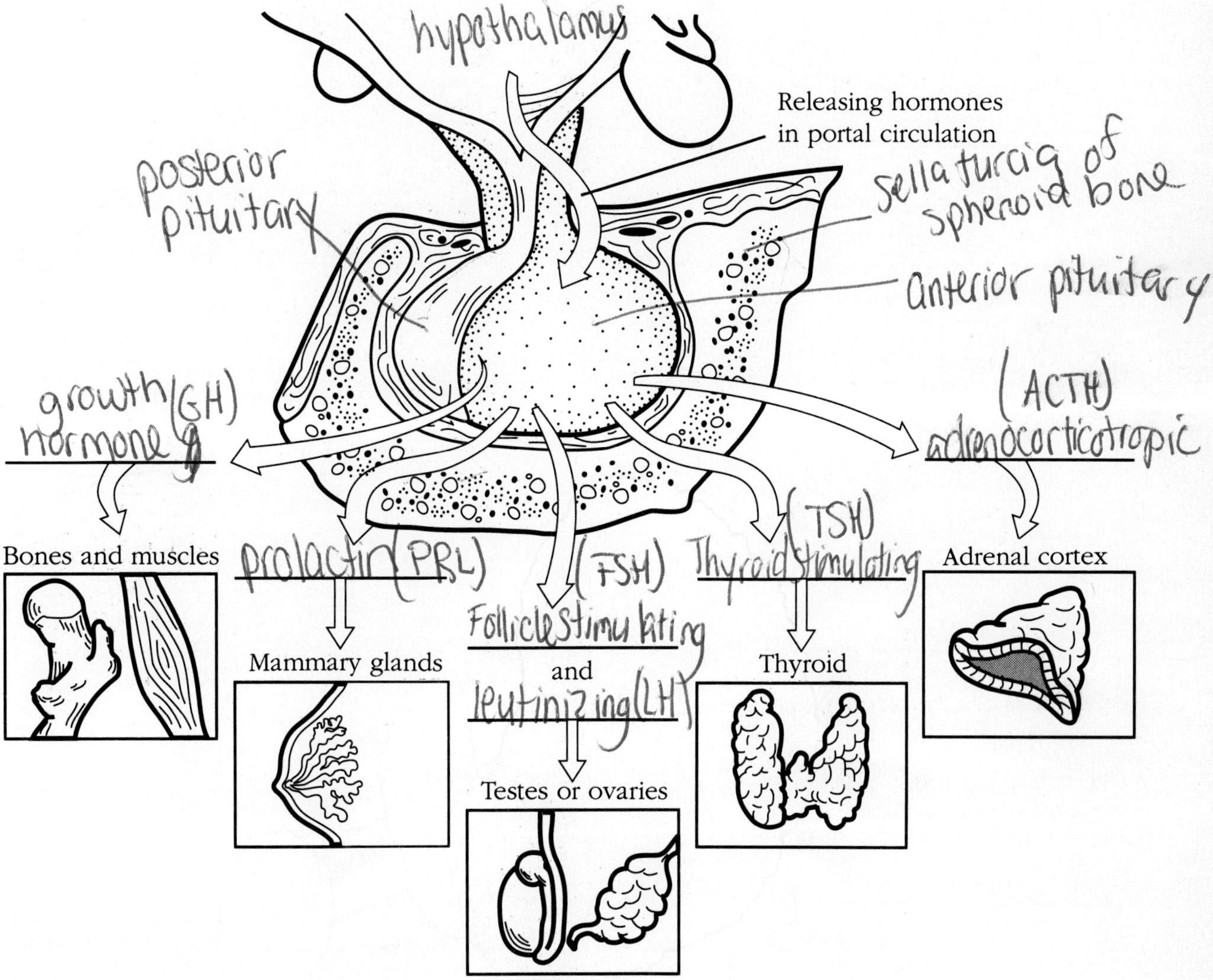

hypothalamus

posterior pituitary

Releasing hormones in portal circulation

sella turcia of sphenoid bone

anterior pituitary

(ACTH) adrenocorticotropic

growth (GH) hormone

Bones and muscles

prolactin (PRL)

Mammary glands

(FSH) Follicle Stimulating and leutinizing (LH)

Testes or ovaries

(TSH) Thyroid Stimulating

Thyroid

Adrenal cortex

Figure 17.2

3. Indicate the organ (or organ part) producing or releasing each of the hormones listed below by inserting the appropriate answers from the key choices in the answer blanks.

KEY CHOICES

A. Adrenal gland (cortex)   **E.** Ovaries   **I.** Placenta

B. Adrenal gland (medulla)   **F.** Pancreas   **J.** Testes

C. Anterior pituitary   **G.** Parathyroids   **K.** Thymus

D. Hypothalamus   **H.** Pineal   **L.** Thyroid

_____ **1.** ACTH   _____ **12.** Melatonin

_____ **2.** ADH   _____ **13.** MSH

_____ **3.** Aldosterone   _____ **14.** Oxytocin

_____ **4.** Cortisone   _____ **15.** Progesterone

_____ **5.** Epinephrine   _____ **16.** Prolactin

_____ **6.** Estrogens   _____ **17.** PTH

_____ **7.** FSH   _____ **18.** Testosterone

_____ **8.** Glucagon   _____ **19.** Thymosin

_____ **9.** Growth   _____ **20.** Thyrocalci-
                   hormone   tonin/calci-
                                                                   tonin

_____ **10.** Insulin   _____ **21.** Thyroxine

_____ **11.** LH   _____ **22.** TSH

4. Name the hormone that best fits each of the following descriptions.

_____ **1.** Basal metabolic hormone

_____ **2.** "Programs" T lymphocytes

_____ **3.** Most important hormone regulating the amount of calcium circulating in the blood; released when blood calcium levels drop

_____ **4.** Helps protect the body during long-term stressful situations, such as extended illness and surgery

_____ **5.** Short-term stress hormone; aids in the fight-or-flight response; increases blood pressure and heart rate, for example

_____ **6.** Necessary if glucose is to be taken up by most body cells

_____ 7. _____ **8.** Regulate the function of another
                                                                 endocrine organ; four tropic
_____ 9. _____ **10.** hormones

_____ **11.** Acts antagonistically to insulin; produced by the same endocrine
                                      organ

_____ **12.** Hypothalamic hormone important in regulating water balance

_____ 13. _____ **14.** Anterior pituitary hormones that
                                                                  regulate the ovarian cycle

_____ 15. _____ **16.** Directly regulate the menstrual or
                                                                  uterine cycle

_____ **17.** Adrenal cortex hormone involved in regulating salt levels of body
                                      fluids

_____ 18. _____ **19.** Necessary for milk production and
                                                                  ejection

5. Circle the term that does not belong in each of the following groupings.

   1. Posterior lobe   Neurohypophysis   Nervous tissue   Anterior lobe

   2. Posterior lobe   Adenohypophysis   Glandular tissue   Anterior lobe

   3. Sympathomimetic amines   Norepinephrine   Epinephrine   Cortisol

   4. Neurohormones   Hypophyseal portal system   Axonal transport

   Hypothalamic-hypophyseal tract

   5. Growth hormone   Prolactin   Oxytocin   ACTH   Pro-opiomelanocortin

   6. Calcitonin   Increases blood Ca^{2+} levels   Thyroid gland   Enhances Ca^{2+} deposit

   7. Thyroxine   Increases BMR   Calorigenic effect   Depresses glucose uptake

   Enhances sympathetic nervous system activity

   8. Glucocorticoids   Steroids   Aldosterone   Sex hormones   Thyroxine

6. Parathyroid hormone has multiorgan effects. Indicate its effects on the organs
   listed below.

   1. Kidneys _____

   2. Intestine _____

   3. Bones _____

7. Explain why antidiuretic hormone is also called vasopressin.

_____

_____

8. Growth hormone has numerous effects, both direct and indirect. Indicate the direct effects of GH by checking (✔) the appropriate choices. Circle the choices that indicate effects mediated indirectly by somatomedins.

____ **1.** Encourages lipolysis

____ **2.** Promotes amino acid uptake from the blood and protein synthesis

____ **3.** Promotes an increase in size and strength of the skeleton

____ **4.** Promotes elevated blood sugar levels

____ **5.** Stimulates cartilage synthesis

____ **6.** Prompts the liver to produce somatomedins

____ **7.** Promotes glucose sparing

____ **8.** Promotes cell division

____ **9.** Causes muscle cell growth

____ **10.** Inhibits glucose metabolism

____ **11.** Increases blood levels of fatty acids

9. Relative to thyroxine synthesis and release, put the following events in their correct time sequence by numbering them from 1 to 9. Event number 1 is already designated.

_____ **1.** Within the Golgi apparatus, sugar molecules are attached to the thyroglobulin protein, and the glycolated molecules are packaged in membranous sacs.

_____ **2.** Iodinated thyroglobulin is taken up into the follicle cells by endocytosis and combined with lysosomes.

_____ **3.** T_3 and T_4 are cleaved out of the colloid by lysosomal enzymes.

_____ **4.** Iodine is attached to tyrosine residues of the colloid, forming DIT and MIT.

_____ **5.** T_4 and T_3 diffuse into the bloodstream.

___1___ **6.** Thyroglobulin protein is synthesized on the ribosomes of the follicle cell.

_____ **7.** Iodides are oxidized and transformed into iodine.

_____ **8.** Thyroglobulin is discharged into the lumen of the follicle.

_____ **9.** Enzymes in the colloid link DITs and MITs together to form T_4 and T_3.

10. List the cardinal symptoms of diabetes mellitus and provide the rationale for the occurrence of each symptom.

1. _____

_____

2. _____

_____

3. _____

_____

11. Compare the magnitude of the conditions in Column A with those immediately opposite in Column B. Circle the phrase indicating the condition that is always or usually greater. If the conditions are essentially always of the same magnitude, put an S on the dotted line between them.

Column A   Column B

1. Plasma levels of gonadotropins during . . . . . . .Plasma levels of gonadotropins during
   childhood   puberty

2. Activity of T$_3$ at target cells . . . . Activity of T$_4$ at target cells

3. Protein wasting and depressed immunity . . . . .Protein wasting and depressed immunity with
   with high levels of glucocorticoids   low levels of glucocorticoids

4. Stimulation of aldosterone release by. . . . . Stimulation of aldosterone release by elevated
   elevated Na$^+$ levels in plasma   K$^+$ levels in plasma

5. Urine volume in a diabetes mellitus patient . . . Urine volume in a diabetes insipidus patient

6. Renin release by the kidneys when blood . . . . Renin release by the kidneys when blood
   pressure is low   pressure is elevated

7. Iodine atoms in a thyroxine molecule . . . . . . .Iodine atoms in a triiodothyronine molecule

8. Mineralocorticoid release by zona . . . . Mineralocorticoid release by zona glomerulosa
   reticularis cells   cells

9. PTH release when blood calcium levels . . . . . PTH release when blood calcium levels are low
   are high (above 11 mg/100 ml)   (below 9 mg/100 ml)

10. Hyperglycemia with high levels of . . . . .Hyperglycemia with high levels of glucagon in
    insulin in plasma   plasma

11. ACTH release with high plasma levels of . . . . . ACTH release with low plasma levels of
    cortisol   cortisol

12. Synthesis of insulin by beta cells of . . . . .Synthesis of insulin by alpha cells of pancreatic
    pancreatic islets   islets

12. For each of the hormones listed below, indicate its effect on blood glucose, blood calcium, and/or blood pressure by using the key choices.

KEY CHOICES

A. Increases blood glucose   **D.** Decreases blood calcium

B. Decreases blood glucose   **E.** Increases blood pressure

C. Increases blood calcium   **F.** Decreases blood pressure

_____   **1.** Cortisol

_____   **2.** Insulin

_____   **3.** Parathormone

_____   **4.** Aldosterone

_____   **5.** Growth hormone

_____   **6.** Antidiuretic hormone

_____   **7.** Glucagon

_____   **8.** Thyroxine

_____   **9.** Epinephrine

_____   **10.** Calcitonin

13. The structure of endocrine cells often allows recognition of the type of hormone product they secrete. Use the key choices to characterize the endocrine cells listed below.

KEY CHOICES   **A.** Has a well developed rough ER and secretory granules

                   B. Has a well developed smooth ER; prominent lipid droplets

_____   **1.** Interstitial cell of the testis

_____   **2.** Chief cell in the parathyroid gland

_____   **3.** Zona fasciculata cell

_____   **4.** Parafollicular cells of the thyroid

_____   **5.** Beta cell of a pancreatic islet

14. Concerning the histology of the pure endocrine glands, match each endocrine gland in Column B with the best approximation of its histology in Column A.

Column A

_____ 1. Spherical clusters of cells

_____ 2. Parallel cords of cells

_____ 3. Branching cords of cells

_____ 4. Follicles

_____ 5. Nervous tissue

Column B

A. Posterior pituitary

B. Zona glomerulosa of the adrenal cortex

C. Thyroid gland

D. Zona fasciculata of adrenal cortex

E. Parathyroid gland

15. Name the hormone that would be produced in *inadequate* amounts in each of the following conditions.

_____ 1. Maturation failure of reproductive organs

_____ 2. Tetany (death due to respiratory paralysis)

_____ 3. Polyuria without high blood glucose levels; causes dehydration and tremendous thirst

_____ 4. Goiter

_____ 5. Cretinism, a type of dwarfism in which the individual retains childlike proportions and is mentally retarded

_____ 6. Excessive thirst, high blood glucose levels, acidosis

_____ 7. Abnormally small stature, normal proportions, a "Tom Thumb"

_____ 8. Spontaneous abortion

_____ 9. Myxedema in the adult

16. Name the hormone that would be produced in *excessive* amounts in each of the following conditions.

_____ 1. Acromegaly in the adult

_____ 2. Bulging eyeballs, nervousness, increased pulse rate, weight loss (Graves' disease)

_____ 3. Demineralization of bones; spontaneous fractures

_____ 4. Cushing's syndrome: moon face, hypertension, edema

_____ 5. Abnormally large stature, relatively normal body proportions

_____ 6. Abnormal hairiness; masculinization

Other Endocrine Structures

1. Besides the major endocrine organs, isolated clusters of cells produce hormones within body organs that are usually not associated with the endocrine system. A number of these hormones are listed in the table below. Complete the missing information on these hormones by filling in the blank spaces in the table.

Hormone	Chemical makeup	Source	Effects
Gastrin	Peptide		
Secretin		Duodenum	
Cholecystokinin	Peptide		
Erythropoietin		Kidney in response to hypoxia	
Cholecalciferol (vitamin D₃)		Skin; activated by kidneys	
Atrial natriuretic factor (ANF)	Peptide		

Developmental Aspects of the Endocrine System

1. Complete the following statements by writing the missing terms in the answer blanks.

_____ 1.

_____ 2.

_____ 3.

_____ 4.

Hormone-producing glands arise from __(1)__ embryonic germ layer(s). Under ordinary conditions, the endocrine organs operate smoothly until old age. However, a __(2)__ in an endocrine organ may lead to hypersecretion of its hormones. A lack of __(3)__ in the diet may result in undersecretion of thyroxine. Later in life, a woman experiences a number of symptoms, such as hot flashes and mood changes, which result from decreasing levels of __(4)__ in her system. This period of a

_____ 5.

_____ 6.

_____ 7.

woman's life is referred to as __(5)__, and it results in a loss of her ability to __(6)__ . Because __(7)__ production tends to decrease in an aging person, adult-onset diabetes mellitus is common.

The Incredible Journey:
A Visualization Exercise
for the Endocrine System

. . . you notice charged particles, shooting pell-mell out of the bone matrix . . .

1. Complete the following narrative by writing the missing words in the answer blanks.

_____ 1.

_____ 2.

_____ 3.

_____ 4.

_____ 5.

_____ 6.

_____ 7.

_____ 8.

_____ 9.

For this journey, you will be miniaturized and injected into a vein of your host. Throughout the journey, you will be traveling in the bloodstream. Your instructions are to record changes in blood composition as you float along and to form some conclusions as to why these changes are occurring (that is, which hormone is being released).

Bobbing gently along in the slowly moving blood, you realize that there is a sugary taste to your environment; however, the sweetness begins to decrease quite rapidly. As the glucose levels of the blood have just decreased, obviously __(1)__ has been released by the __(2)__ so that the cells can take up glucose.

A short while later, you notice that the depth of the blood in the vein you are traveling in has diminished substantially. To remedy this potentially serious situation, the __(3)__ will have to release more __(4)__ , so the kidney tubules will reabsorb more water. Within a few minutes the blood becomes much deeper; you wonder if the body is psychic as well as wise.

As you circulate past the bones, you notice charged particles, shooting pell-mell out of the bone matrix and jumping into the blood. You conclude that the __(5)__ glands have just released PTH because the __(6)__ levels have increased in the blood. As you continue to move in the bloodstream, the blood suddenly becomes sticky sweet, indicating that your host must be nervous about something. Obviously, his __(7)__ has released __(8)__ to cause this sudden increase in blood glucose.

Sometime later, you become conscious of a humming activity around you, and you sense that the cells are very busy. Obviously your host's __(9)__ levels are sufficient since his cells are certainly not sluggish in their metabolic activities. You record this observation and prepare to end this journey.

CHALLENGING YOURSELF

At the Clinic

1. Pete is very short for his chronological age of 8. What physical features will allow you to determine quickly whether to check GH or thyroxine levels?

2. A young girl is brought to the clinic by her father. He complains that his daughter fatigues easily and seems mentally sluggish. You notice a slight swelling in the anterior neck. What condition do you suspect? What are some possible causes and their treatments?

3. Lauralee, a middle-aged woman, comes to the clinic and explains in an agitated way that she is "very troubled" by excessive urine output and consequent thirst. What two hormones might be causing the problem and what urine tests will be done to identify the problem?

4. A 2-year-old boy is brought to the clinic by his anguished parents. He is developing sexually and shows an obsessive craving for salt. Blood tests reveal hyperglycemia. What endocrine gland is hypersecreting?

5. Lester, a 10-year-old, has been complaining of severe lower back pains. The nurse notices that he seems weak and a reflex check shows abnormal response. Kidney stones are soon diagnosed. What abnormality is causing these problems?

6. Bertha Wise, age 40, comes to the clinic, troubled by swelling in her face and unusual fat deposition on her back and abdomen. She reports that she bruises easily. Blood tests show elevated glucose levels. What is your diagnosis and what glands might be causing the problem?

7. A middle-aged man comes to the clinic, complaining of extreme nervousness, insomnia, and weight loss. The nurse notices that his eyes bulge and his thyroid is enlarged. What is the man's hormonal imbalance and what are two likely causes?

8. Mr. Holdt brings his wife to the clinic, concerned about her nervousness, heart palpitations, and excessive sweating. Tests show hyperglycemia and hypertension. What hormones are probably being hypersecreted? What is the cause? What physical factors allow you to rule out thyroid problems?

9. Phyllis, a type-I diabetic, is rushed to the hospital. She had been regulating her diabetes extremely well, with no chronic problems, when her mother found her unconscious. Will blood tests reveal hypoglycemia or hyperglycemia? What probably happened?

10. A woman calls for an appointment at the clinic because she is not menstruating. She also reports that her breasts are producing milk, although she has never been pregnant. What hormone is being hypersecreted and what is the likely cause?

11. An accident victim who had not been wearing a seat belt received severe trauma to his forehead when he was thrown against the windshield. The physicians in the emergency room worried that his brain stem may have been driven inferiorly through the foramen magnum. To help assess this possibility, they quickly took a standard X ray of his head and searched for the position of the pineal gland. How could anyone expect to find a tiny, boneless gland like the pineal in an X ray?

Stop and Think

1. Compare and contrast protein and steroid hormones with regard to the following: (a) organelles involved in their manufacture; (b) ability to store the hormones within the cell; (c) rate of manufacture; (d) method of secretion; (e) means of transport in the bloodstream; (f) location of receptors in/on target cell; (g) use of second messenger; (h) relative time from attachment to receptor until effects appear; and (i) whether effects persist after the hormone is metabolized.

2. (a) Hypothalamic factors act both to stimulate and to inhibit the release of certain hormones. Name these hormones.

    (b) Blood levels of certain humoral factors are regulated both on the "up" and the "down" side by hormones. Name some such humoral factors.

3. What are enteroendocrine cells and why are they sometimes called paraneurons?

4. Would drug tolerance be due to up-regulation or down-regulation?

5. Why do the chemical structures of thyroxine and triiodothyronine require complexing with a large protein to allow long-term storage?

6. The brain is "informed" when we are in a stressful situation, and the hypothalamus responds to stressors by secreting a releasing hormone called corticotropin-releasing hormone. This hormone helps the body deal with the stress through a long sequence of events. Outline this entire sequence, starting with corticotropin-releasing hormone and ending with the release of cortisol. (Be sure to trace the hormones through the hypophyseal portal system and out of the pituitary gland.)

7. Joshua explained to his classmate Jennifer that the thyroid gland contains parathyroid cells in its follicles and the parathyroid cells secrete parathyroid hormone and calcitonin. Jennifer told him he was all mixed up again. Can you correct Josh's mistakes?

8. When the carnival was scheduled to come to a small town, health professionals who felt that the sideshows were cruel and exploitive joined with the local consumer groups to enforce truth-in-advertising laws. They demanded that the fat man, the dwarf, the giant, and the bearded lady be billed as "people with endocrine system problems" (which of course removed all the sensationalism usually associated with these attractions). Identify the endocrine disorder in each case and explain how (or why) the disorder produced the characteristic features of these four showpeople.

Closer Connections: Checking the Systems— Regulation and Integration of the Body

1. How does the central nervous system exert control over the endocrine system?

2. How does the endocrine system exert control over the nervous system?

3. Which of the two regulatory systems is designed to respond to rapid environmental changes?

4. Which system exerts a greater degree of control over the skeletal system? Over metabolism?

5. Explain the differences in control of the muscular system by the nervous system and the endocrine system.

COVERING ALL YOUR BASES

Multiple Choice

Select the best answer or answers from the choices given.

1. Relative to the cyclic AMP second-messenger system, which of the following is not accurate?
   A. The activating hormone interacts with a receptor site on the plasma membrane.
   B. Binding of the galvanizing hormone directly activates adenylate cyclase.
   C. Activated adenylate cyclase catalyzes the transformation of AMP to cyclic AMP.
   D. Cyclic AMP acts within the cell to alter cell function as is characteristic for that specific hormone.

2. Which of the following hormones is (are) secreted by neurons?
   A. Oxytocin   **C.** ADH
   B. Insulin   **D.** Cortisol

3. The paraventricular nucleus of the hypothalamus is named for its proximity to the:
   A. lateral ventricles   **C.** third ventricle
   B. cerebral aqueduct   **D.** fourth ventricle

4. Which of the following might be associated with a second-messenger system?
   A. Phosphatidyl inositol   **C.** Glucagon
   B. Corticosteroids   **D.** Calmodulin

5. ANF, the hormone secreted by the heart, has exactly the opposite function to this hormone secreted by the zona glomerulosa:
   A. epinephrine   **C.** aldosterone
   B. cortisol   **D.** testosterone

6. Hormones that act to elevate blood glucose include:
   A. GH   **C.** insulin
   B. cortisol   **D.** CRH

7. The release of which of the following hormones will be stimulated via the hypothalamic-hypophyseal tract?
   A. ACTH
   B. TSH
   C. ADH
   D. GH

8. Cells sensitive to the osmotic concentration of the blood include:
   A. chromaffin cells
   B. paraventricular neurons
   C. supraoptic neurons
   D. parafollicular cells

9. The gland derived from embryonic throat tissue known as Rathke's pouch is the:
   A. posterior pituitary
   B. anterior pituitary
   C. thyroid
   D. thymus

10. The primary capillary plexus is located in the:
    A. hypothalamus
    B. anterior pituitary
    C. posterior pituitary
    D. infundibulum

11. Pro-opiomelanocortin is the precursor of:
    A. cortisol
    B. corticotropin
    C. melatonin
    D. opium

12. Which of the following are direct or indirect effects of growth hormone?
    A. Stimulates cells to take in amino acids
    B. Increases synthesis of chondroitin sulfate
    C. Increases blood levels of fatty acids
    D. Decreases utilization of glucose by most body cells

13. Which of the following are tropic hormones secreted by the anterior pituitary gland?
    A. LH
    B. ACTH
    C. TSH
    D. FSH

14. Hormones secreted by females include:
    A. estrogens
    B. progesterone
    C. prolactin
    D. testosterone

15. Smooth muscle contractions are stimulated by:
    A. testosterone
    B. FSH
    C. prolactin
    D. oxytocin

16. Nerve input regulates the release of:
    A. oxytocin
    B. epinephrine
    C. melatonin
    D. cortisol

17. Hypertension may result from hypersecretion of:
    A. thyroxine
    B. cortisol
    C. aldosterone
    D. antidiuretic hormone

18. In initiating the secretion of stored thyroxine, which of these events occurs first?
    A. Production of thyroglobulin
    B. Discharge of thyroglobulin into follicle
    C. Attachment of iodine to thyroglobulin
    D. Lysosomal activity to cleave hormone from thyroglobulin

19. Hypothyroidism can cause:
    A. myxedema
    B. Cushing's syndrome
    C. cretinism
    D. exophthalmos

20. Calcitonin targets the:
    A. kidneys
    B. liver
    C. bone
    D. small intestine

21. Imbalances of which hormones will affect nerve function?
    A. Thyroxine
    B. Parathormone
    C. Insulin
    D. Aldosterone

22. Hormones that regulate mineral levels include:
    A. calcitonin
    B. aldosterone
    C. atrial natriuretic factor
    D. glucagon

23. Which of the following is given as a drug to reduce inflammation?
    A. Epinephrine
    B. Cortisol
    C. Aldosterone
    D. ADH

24. After menopause, steroids that maintain anabolism come from the:
    A. ovaries
    B. adrenal cortex
    C. anterior pituitary
    D. thyroid

25. Derivatives of embryonic mesoderm include the
   A. adenohypophysis
   B. neurohypophysis
   C. adrenal cortex
   D. thyroid

26. Which is generally true of hormones?
   A. Exocrine glands produce them.
   B. They travel throughout the body in the blood.
   C. They affect only nonhormone-producing organs.
   D. All steroid hormones produce very similar physiological effects in the body.

27. The major endocrine organs of the body:
   A. tend to be very large organs
   B. are closely connected with each other
   C. all contribute to the same function (digestion)
   D. tend to lie near the midline of the body

28. Which type of cell secretes releasing hormones?
   A. Neuron   **C.** Chromaffin cell
   B. Parafollicular cell   **D.** Adenohypophysis cell

29. Of the following endocrine structures, which develops from the brain?
   A. Neurophyophysis   **C.** Thyroid gland
   B. Adenohypophysis   **D.** Thymus gland

Word Dissection

For each of the following word roots, fill in the literal meaning and give an example, using a word found in this chapter.

Word root	Translation	Example
1. adeno		
2. crine		
3. dips		
4. diuresis		
5. gon		
6. hormon		
7. humor		
8. mell		
9. toci		
10. trop		

18 Blood

Student Objectives

When you have completed the exercises in this chapter, you will have accomplished the following objectives:

Overview: Composition and Functions of Blood

1. Describe the composition and physical characteristics of whole blood. Explain why it is classified as a connective tissue.

2. List six functions of blood.

Blood Plasma

3. Discuss the composition and functions of plasma.

Formed Elements

4. Describe the structural characteristics, function, and production of erythrocytes.

5. List the classes, structural characteristics, and functions of leukocytes. Also describe leukocyte genesis.

6. Describe the structure and function of platelets.

7. Give examples of disorders caused by abnormalities of each of the formed elements. Explain the mechanism of each disorder.

Hemostasis

8. Describe the processes of hemostasis. Indicate the factors that limit clot formation and prevent undesirable clotting.

9. Give examples of hemostatic disorders. Note the cause of each condition.

Transfusion and Blood Replacement

10. Describe the ABO and Rh blood groups. Explain the basis of transfusion reactions.

11. List several examples of blood expanders. Describe their function and the circumstances in which they are usually used.

Diagnostic Blood Tests

12. Explain the importance of blood testing as a diagnostic tool.

Developmental Aspects of Blood

13. Describe changes in the sites of blood production and in the type of hemoglobin produced after birth.

14. Name some blood disorders that become more common with age.

Ready, set, go!

Blood, the indispensable "life fluid" that courses through the body's blood vessels, provides the means by which the body's cells receive essential nutrients and oxygen and dispose of their metabolic wastes. As blood flows past the tissue cells, exchanges continually occur between the blood and the cells so that vital functions are maintained.

This chapter provides an opportunity to review the general characteristics of whole blood and plasma, to identify the various formed elements (blood cells), and to recall their functions. Blood groups, transfusion reactions, clotting, and various types of blood abnormalities are also reviewed.

BUILDING THE FRAMEWORK

Overview: Composition and Functions of Blood

1. Complete the following description of the components of blood by writing the missing words in the answer blanks.

Connective tissue   1.

formed elements   2.

plasma   3.

blood clotting   4.

red blood cells   5.

hematocrit   6.

plasma   7.

wbc's   8.

platelets   9.

one   10.

oxygen   11.

In terms of its tissue classification, blood is classified as a __(1)__ because it has living blood cells, called __(2)__, suspended in a nonliving fluid matrix called __(3)__ . The "fibers" of blood only become visible during __(4)__ .

If a blood sample is centrifuged, the heavier blood cells become packed at the bottom of the tube. Most of this compacted cell mass is composed of __(5)__, and the volume of blood accounted for by these cells is referred to as the __(6)__ . The less dense __(7)__ rises to the top and constitutes about 45% of the blood volume. The so-called "buffy coat" composed of __(8)__ and __(9)__ is found at the junction between the other two blood elements. The buffy coat accounts for less than __(10)__ % of blood volume.

Blood is scarlet red in color when it is loaded with __(11)__ ; otherwise, it tends to be dark red.

2. List four delivery functions of blood, two regulatory functions, and two protection functions.

1. Delivery (distribution) functions _____

_____

_____

2. Regulatory functions _____

_____

_____

3. Protection functions _____

_____

_____

Blood Plasma

1. List three classes of substances normally found dissolved in plasma.

1. ___nutrients___   **2.** ___electrolytes___   **3.** ___plasma proteins___

2. Complete the following table relating to the proteins found in plasma.

Constituent	Description/importance
_____	60% of plasma proteins; important for osmotic balance
Fibrinogen	____% of plasma proteins; important in _____
_____ _____ _____	36% of plasma proteins: transport proteins antibodies
Nonprotein nitrogenous substances	_____ _____ (list 5)
_____	Organic chemicals absorbed from the digestive tract
Respiratory gases	_____ , _____

Formed Elements

1. Number the following cell types to indicate the sequence of erythrocyte maturation in the red bone marrow. Circle the cell type released to the blood; underline the cell type that ejects its nucleus and most organelles.

_____ **1.** Reticulocyte   _____ **5.** Erythrocyte

_____ **2.** Hemocytoblast   _____ **6.** Early erythroblast

_____ **3.** Late erythroblast   _____ **7.** Proerythroblast

_____ **4.** Normoblast

2. Figure 18.1 on the opposite page depicts in incomplete form the erythropoietin mechanism for regulating the rate of erythropoiesis. Several statements are incomplete. Complete the statements that have answer blanks and then choose colors (other than yellow) to identify the structures with color-coding circles. Color all arrows on the diagram yellow. Finally, respond to the following questions to complete this exercise.

◯ The kidney   ◯ Red bone marrow   ◯ RBCs

1. What is the normal life span of erythrocytes? ____120 days____ days

2. What three food nutrients (other than the normally required proteins and carbohydrates) are essential for erythropoiesis?

    ___iron, vitamin B12, + Folic acid___

3. What is the fate of aged or damaged red blood cells? ___They r engulfed by macrophages___ ___of liver spleen, or bone marrow___

4. What is the fate of the released hemoglobin? _____

    _____

    _____

3. Check (✔) all the factors that would serve as stimuli for erythropoiesis.

_____ **1.** Hemorrhage   _____ **3.** Living at a high altitude

_____ **2.** Aerobic exercise   _____ **4.** Breathing pure oxygen

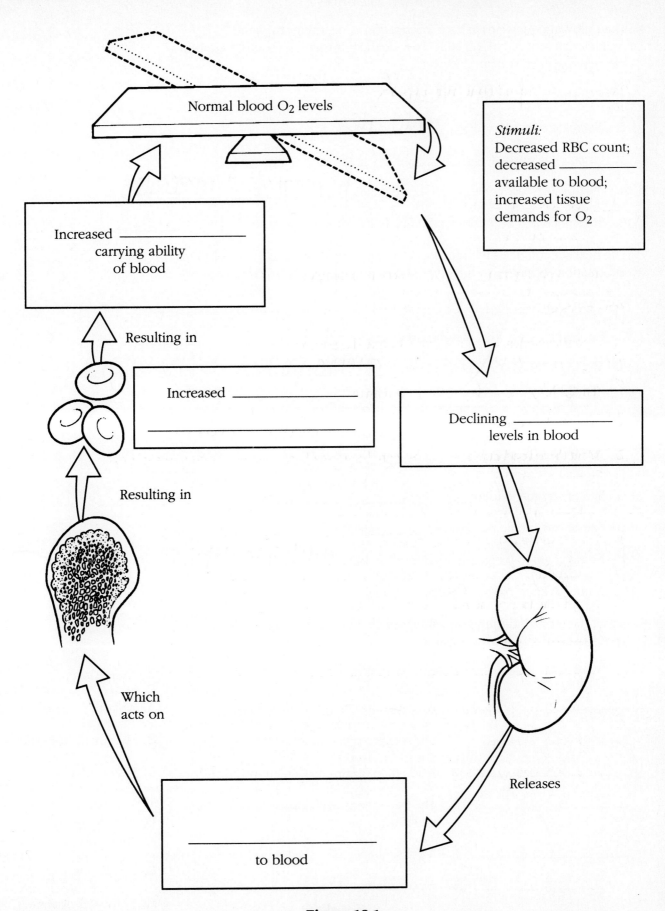

Normal blood O$_2$ levels

Stimuli:
Decreased RBC count; decreased _____ available to blood; increased tissue demands for O$_2$

Increased _____ carrying ability of blood

Resulting in

Increased _____ _____

Declining _____ levels in blood

Resulting in

Which acts on

Releases

_____ to blood

Figure 18.1

4. The following questions ask for information concerning structural characteristics of red blood cells. Provide the information requested to fully characterize each structural aspect.

1. Cell shape __biconcave disk__

2. Nucleate or anucleate? __anucleate__

3. Organelles present? _____

4. Major molecular content of the cytoplasm __hemoglobin molecules__

5. Substance that accounts for the flexibility of the membrane _____

The next section deals with functional characteristics of red blood cells. Provide the information requested to fully describe each functional characteristic.

6. Type of metabolism, aerobic or anaerobic? _____

7. Explain why the RBC metabolism is of this type. _____

8. Molecular makeup of hemoglobin __globin protein + 4 heme groups__
   _____

9. The portion of hemoglobin that binds oxygen __heme (iron)__

10. The portion of hemoglobin that binds carbon dioxide _____

11. Site of oxygen loading _____

12. Site of carbon dioxide loading _____

13. Name given to oxygen-loaded hemoglobin __oxyhemoglobin__

14. Name given to oxygen-depleted hemoglobin _____

5. Match the terms indicating specific types of anemias in Column B with the appropriate descriptions in Column A.

Column A	Column B
_____ **1.** A genetic disorder in which abnormal hemoglobin is produced and becomes spiky under hypoxic conditions	**A.** Aplastic anemia
	B. Hemolytic anemia
_____ **2.** A common occurrence after transfusion error	**C.** Hemorrhagic anemia
_____ **3.** The bone marrow is destroyed or severely inhibited	**D.** Iron-deficiency anemia
_____ **4.** A consequence of acute blood loss	
_____ **5.** A possible consequence of sickle-cell anemia	**E.** Pernicious anemia
	F. Sickle-cell anemia

_____ **6.** Results from inadequate intake of iron-rich foods or conditions involving chronic types of bleeding, such as gastric ulcers

_____ **7.** A common problem of individuals who have a portion of their stomach removed to manage bleeding ulcers

6. If a statement is true, write the letter T in the answer blank. If a statement is false, change the underlined word(s) and write the correct word(s) in the answer blank.

_____ **1.** White blood cells (WBCs) move into and out of blood vessels by the process of <u>positive chemotaxis</u>.

_____ **2.** A count of white blood cells that provides information on the relative number of each WBC type is the <u>total</u> WBC count.

_____ **3.** When blood becomes too acid or too basic, both the respiratory system and the <u>liver</u> may be called into action to restore it to its normal pH range.

_____ **4.** Carbaminohemoglobin is formed when carbon dioxide binds to the <u>heme groups</u> of hemoglobin.

_____ **5.** The cardiovascular system of an average adult contains approximately <u>4</u> liters of blood.

_____ **6.** Blood is circulated through the blood vessels by the pumping action of the <u>heart</u>.

_____ **7.** The only leukocyte type to arise from the lymphoid stem cells is the <u>lymphocyte</u>.

_____ **8.** Normal <u>hemoglobin</u> values are in the range of 42% to 47% of the volume of whole blood.

_____ **9.** An anemia resulting from a decreased RBC number causes the blood to become <u>more</u> viscous.

_____ **10.** The leukocytes particularly important in the immune response are <u>monocytes</u>.

_____ **11.** B-complex vitamins are necessary for <u>DNA</u> synthesis in RBCs.

_____ **12.** In plasma, iron is transported bound to <u>hemosiderin</u>.

7. Using the key choices, identify the cell types or blood elements that fit the following descriptions.

KEY CHOICES

A. Basophil   **D.** Lymphocyte   **G.** Neutrophil

B. Eosinophil   **E.** Megakaryocyte   **H.** Platelets

C. Formed elements   **F.** Monocyte   **I.** Red blood cell

G _neutrophil_   **1.** Granulocyte with the smallest granules

A _Basophil_   **2.** Granular leukocytes (#2– #4)

B _eosinophil_   **3.** G _neutrophil_   **4.**

I _RBC_   **5.** Also called an erythrocyte, anucleate

F _Monocyte_   **6.** G _neutrophil_   **7.** Phagocytic leukocytes that avidly engulf bacteria

D _lymphocyte_   **8.** F _monocyte_   **9.** Agranular leukocytes

E _megakaryocyte_ **10.** Fragments to form platelets

C _formed elements_ **11.** All the others are examples of these

B _Eosinophils_ **12.** Increases in number during allergy attacks

A _Basophil_ **13.** Releases histamine during inflammatory reactions

D _lymphocyte_ **14.** Descendants may be formed in lymphoid tissue

I _RBC_ **15.** Contains hemoglobin; therefore involved in oxygen transport

H,I _platelets, RBC's_ **16.** Does not use diapedesis

F _monocyte_ **17.** Increases in number during chronic infections; the largest WBC

H _platelets_ **18.** I _RBC's_   **19.** The only formed elements that are not spherical

A _Basophils_ **20.** B _Eosinophil_ **21.** D _lymphocyte_ **22.**

F _Monocyte_ **23.** G _Neutrophil_ **24.** Also called white blood cells (#20– #24)

B _Eosinophils_ **25.** Destroys parasitic worms

G _neutrophil_ **26.** Granulocyte with two types of granules

D _lymphocyte_ **27.** A T cell or a B cell

F _monocyte_ **28.** G _Neutrophil_ **29.** Kills by the respiratory burst

B _Eosinophil_ **30.** Secretes major basic protein

8. Four leukocytes are diagrammed in Figure 18.2. First, follow the directions for coloring each leukocyte as it appears when stained with Wright's stain. Then, identify each leukocyte type by writing the correct name in the blank below each illustration.

A. Color the granules pale violet, the cytoplasm pink, and the nucleus dark purple.

B. Color the nucleus deep blue and the cytoplasm pale blue.

C. Color the granules bright red, the cytoplasm pale pink, and the nucleus red/purple.

D. For this smallest white blood cell, color the nucleus deep purple-blue and the sparse cytoplasm pale blue-green.

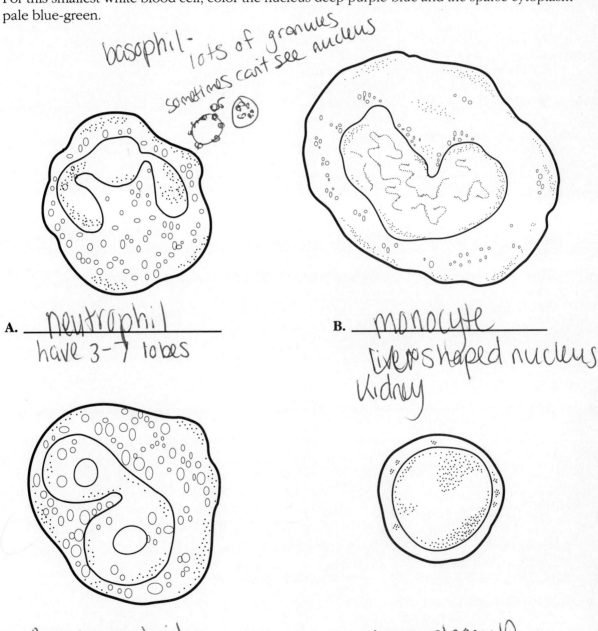

basophil- lots of granules
sometimes can't see nucleus
or

A. ___neutrophil___
have 3-7 lobes

B. ___monocyte___
liver shaped nucleus
kidney

C. ___eosinophil___
2-lobes
1g. red grains

Figure 18.2

D. ___lymphocyte___
1g nucleus
cytoplasm is like halo

9. Circle the term that does not belong in each of the following groupings.

1. Erythrocytes   Lymphocytes   Monocytes   Eosinophils

2. Neutrophils   Monocytes   Basophils   Eosinophils

3. Histamine   Heparin   Basophil   Antibodies

4. Hemoglobin   Lymphocyte   Oxygen transport   Erythrocytes

5. Platelets   Monocytes   Phagocytosis   Neutrophils

6. Thrombus   Aneurysm   Embolus   Clot

7. Increased hemoglobin   Decreased hemoglobin   Red blood cell lysis   Anemia

8. Plasma   Nutrients   Hemoglobin   Wastes   Clotting proteins

9. Myeloid stem cell   Lymphocyte   Monocyte   Basophil   Eosinophil

10. What are *colony-stimulating factors?* _____

_____

Name two cells that are sources of CSFs: _____

_____

11. Give the full names for the CSFs abbreviated below:

1. M-CSF _____

2. G-CSF _____

3. GM-CSF _____

4. IL-3 _____

5. #4 is also known as _____

Hemostasis

1. Using the key choices, correctly complete the following brief description of
the blood-clotting process. Insert the answers in the answer blanks.

KEY CHOICES

A. Break   **D.** Fibrinogen   **G.** Prothrombin   ~~**J.** Thromboxane~~

B. Erythrocytes   **E.** PF$_3$ ~~**H.** Serotonin~~

C. Fibrin   **F.** Platelets   **I.** Thrombin

_____ A _____ F _____ 1. Clotting begins when a __(1)__ occurs in a blood vessel wall.
Almost immediately, __(2)__ cling to the blood vessel wall and
_____ 2. release __(3)__ and __(4)__, which help decrease blood loss by
constricting the vessel. __(5)__ on the platelet surface promotes
_____ 3. the pathway leading to formation of prothrombin activator,
which causes __(6)__ to be converted to __(7)__. Once present,
_____ 4. thrombin acts as an enzyme to attach __(8)__ molecules together
to form long, threadlike strands of __(9)__, which then traps
_____ E _____ 5. __(10)__ flowing by in the blood.

_____ G _____ 6.

_____ I _____ 7.

_____ D _____ 8. _____ C _____ 9. _____ B _____ 10.

2. Several characteristics of the coagulation process are listed below. If the factor
or effect is more typical of the intrinsic pathway, write I in the answer blank; if
it is more typical of the extrinsic pathway, write E in the answer blank. If it
applies equally to both pathways, write B in the answer blank.

_____ 1. Can occur outside the body in a test tube

_____ 2. Involves many more procoagulants

_____ 3. Entails the conversion of prothrombin to thrombin

_____ 4. Requires calcium ions

_____ 5. Occurs much more rapidly

_____ 6. Intermediates of the reaction are activated on the surfaces of platelets

_____ 7. Involves factors III and VII

_____ 8. Involves factors IX, XI, and XII

_____ 9. Tissue factor (TF) is released by injured cells extrinsic to the bloodstream

_____ 10. Clot retraction occurs

_____ 11. Prothrombin activator is a pivotal intermediate

_____ 12. A pivotal factor is PF_3, a phospholipid associated with the external surfaces of
aggregated platelets

3. Match the terms in Column B with the descriptions in Column A.

Column A

_____ **1.** _____ **2.** Two circumstances that prevent unnecessary enlargement of blood clots

_____ **3.** Natural "clot-buster" chemical; causes fibrinolysis

_____ **4.** Hereditary bleeder's disorder resulting from a lack of factor VIII

_____ **5.** _____ **6.** Two factors that prevent undesirable clotting in an unbroken blood vessel

_____ **7.** Free-floating blood clot

_____ **8.** _____ **9.** Plasmin activators (2)

_____ **10.** Clot formed in an unbroken blood vessel

_____ **11.** Bleeder's disease that is a consequence of too few platelets

_____ **12.** Hereditary disease resulting from a deficiency of factor IX

Column B

A. Activated clotting factors inhibited by heparin and antithrombin III.

B. Activated factor XII

C. Coagulation factors washed away by flowing blood

D. Embolus

E. Hemophilia A

F. Hemophilia B

G. Intact endothelium

H. Plasmin

I. Endothelial cell secretions (prostacyclin, PGI$_2$, and others)

J. Thrombin

K. Thrombocytopenia

L. Thrombus

Transfusion and Blood Replacement

1. Complete the following table concerning ABO blood groups.

Blood type	Agglutinogens or antigens	Agglutinins or antibodies in plasma	Can donate blood to type	Can receive blood from type
Type A	A	Anti-B	A, AB	A,O
Type B	B	Anti-A	B, AB	B,O
Type AB	A, B	no anti A or B	AB	A,B,AB,O
Type O	none	none / Anti A, Anti B	A,B,AB,O	O

2. Ms. Pratt is claiming that Mr. X is the father of her child. Ms. Pratt's blood type is O negative. Her baby boy has type A positive blood. Mr. X's blood is typed and found to be type B positive. Could he be the father of her child? If not, what blood type would the father be expected to have?

3. When a person is given a transfusion of mismatched blood, a transfusion reaction occurs. Define the term *transfusion reaction*.

_____

_____

Diagnostic Blood Tests

1. Fill in the table of normal blood values; if appropriate, include units for both males and females. Also rank the leukocytes (listed under the heading "Differential WBC count") in order of their relative abundance (1 to 5, least to most) in the blood of a healthy person.

Characteristic	Normal value or normal range
% of body weight	
Blood volume	
Arterial pH	
Blood temperature	
RBC count	
Hematocrit	
Hemoglobin	
WBC count	
Differential WBC count	
neutrophils	
eosinophils	
basophils	
lymphocytes	
monocytes	
Platelet count	

2. A laboratory examination of blood yields valuable information that can be used to assess a patient's health status. Provide the names of diagnostic tests that fit the categories listed below.

Tests for certain nutrients   Tests for clotting

1. _____   1. _____

2. _____   2. _____

Tests for mineral levels

1. _____

2. _____

3. _____

Tests for anemia

1. _____

2. _____

3. _____

Developmental Aspects of Blood

1. Complete the following statements by writing the missing terms in the answer blanks.

_____ 1.

_____ 2.

_____ 3.

_____ 4.

_____ 5.

_____ 6.

_____ 7.

_____ 8.

Blood cells develop from collections of mesenchymal cells called __(1)__ . A fetus has a special type of hemoglobin, hemoglobin __(2)__ , that has a particularly high affinity for oxygen. After birth, the infant's fetal RBCs are rapidly destroyed and replaced by hemoglobin A-containing RBCs.

When the immature infant liver cannot keep pace with the demands to rid the body of hemoglobin breakdown products, the infant's tissues become yellowed, or __(3)__ .

Diet is important to normal blood formation. Women are particularly prone to __(4)__ -deficiency anemia because of their monthly menses. A decreased efficiency of the gastric mucosa makes elderly individuals particularly susceptible to __(5)__ anemia as a result of a lack of intrinsic factor, which is necessary for vitamin __(6)__ absorption. An important problem in the elderly is their tendency to form undesirable clots, or __(7)__ . Both the young and the elderly are at risk for cancer of the blood, or __(8)__ .

The Incredible Journey: A Visualization Exercise for the Blood

Once inside, you quickly make a slash in the vessel lining . . .

1. Complete the following narrative by inserting the missing words in the answer blanks.

_____ 1.

_____ 2.

_____ 3.

_____ 4.

For this journey, you will be injected into the external iliac artery and will be guided by a fluorescent monitor into the bone marrow of the iliac bone. You will observe and report events of blood cell formation, also called __(1)__ , seen there and then move out of the bone into the circulation to initiate and observe the process of blood clotting, also called __(2)__ . Once in the

_____	**5.**
_____	**6.**
_____	**7.**
_____	**8.**
_____	**9.**
_____	**10.**
_____	**11.**
_____	**12.**
_____	**13.**
_____	**14.**
_____	**15.**
_____	**16.**
_____	**17.**
_____	**18.**
_____	**19.**
_____	**20.**
_____	**21.**
_____	**22.**
_____	**23.**
_____	**24.**
_____	**25.**

bone marrow, you observe several large dark-nucleated stem cells, or __(3)__, as they begin to divide and produce daughter cells. The daughter cells eventually formed have tiny cytoplasmic granules and very peculiarly shaped nuclei that look like small masses of nuclear material connected by thin strands of nucleoplasm. You note that you have just witnessed the formation of a type of white blood cell, called the __(4)__. You describe its appearance and make a mental note to try to observe its activity later. Meanwhile you can tentatively report that this cell type functions as a __(5)__ to protect the body.

At another site, daughter cells arising from the division of a stem cell are difficult to identify initially. As you continue to observe the cells, you see that they, in turn, divide. Eventually some of their daughter cells eject their nuclei and flatten out to assume a disk shape. You conclude that the kidneys must have released the hormone __(6)__ because those cells are __(7)__. That dark material filling their interior must be __(8)__ because those cells function to transport __(9)__ in the blood.

Now you turn your attention to the daughter cells being formed by the division of another stem cell. They are small round cells with relatively large round nuclei. In fact, their cytoplasm is very sparse. You record your observation of the formation of __(10)__. They do not remain in the marrow very long after formation but seem to enter the circulation almost as soon as they are produced. Some of those cells will produce __(11)__, which are released to the blood; others will act in other ways in the immune response. At this point, although you have yet to see the formation of __(12)__, __(13)__, __(14)__, or __(15)__, you decide to proceed into the circulation to observe blood clotting.

You maneuver yourself into a small venule to enter the general circulation. Once inside, you quickly make a slash in the vessel lining, or __(16)__. Almost immediately what appear to be hundreds of jagged cell fragments swoop into the area and plaster themselves over the freshly made incision. You record that __(17)__ have just adhered to the damaged site. As you are writing, your chemical monitor flashes the message, "Vasoconstrictor substance released." You record that __(18)__ has been released, based on your observation that the vessel wall seems to be closing in. Peering out at the damaged site, you see that long ropelike strands are being formed at a rapid rate and are clinging to the site. You report that the __(19)__ mesh is forming and is beginning to trap RBCs to form the basis of the __(20)__. Even though you do not have the equipment to monitor the intermediate steps of this process, you know that the platelets must have also acted as activation sites for the formation of __(21)__, which then catalyzes the conversion of __(22)__ to __(23)__. This second enzyme then joined the soluble __(24)__ molecules together to form the network of strands you can see.

You carefully back away from the newly formed clot. You do *not* want to disturb the area because you realize that if the clot detaches, it might become a life-threatening __(25)__. Your mission here is completed, and you return to the entrance site.

CHALLENGING YOURSELF

At the Clinic

1. A patient on dialysis has a low RBC count. What hormone, secreted by the kidney, can be assumed to be deficient?

2. A bone marrow biopsy of Mr. Bongalonga, a man on a long-term drug therapy, shows an abnormally high percentage of nonhemopoietic connective tissue. What condition does this indicate? If the symptoms are critical, what short-term and long-term treatments may be necessary? Will infusion of whole blood or packed red cells be more likely?

3. Mrs. Francis comes to the clinic complaining of fatigue, shortness of breath, and chills. Blood tests show anemia and a bleeding ulcer is diagnosed. What type of anemia is this?

4. A patient is diagnosed with bone marrow cancer and has a hematocrit of 70%. What is this condition called?

5. The very concerned parents of a young boy named Teddy bring him to the clinic because he has chronic fatigue, bleeding, fever, and weight loss. A blood test reveals abnormally high numbers of lymphoblasts, anemia, and thrombocytopenia. What is his diagnosis?

6. What vitamin must be supplemented in cases of steatorrhea (inability to absorb fat) to avoid clotting problems?

7. After multiple transfusions, a weak transfusion reaction occurs. Is this likely to be caused by the ABO antigens or other antigens? What should be done to avoid this type of reaction?

8. Mrs. Carlyle is pregnant for the first time. Her blood type is Rh negative, her husband is Rh positive, and their first child has been determined to be Rh positive. Ordinarily, the first such pregnancy causes no major problems, but baby Carlyle is born blue and cyanotic.

   1. What is this condition, a result of Rh incompatibility, called?

   2. Why is the baby cyanotic?

   3. Since this is Mrs. Carlyle's first pregnancy, how can you account for the baby's problem?

   4. Assume that baby Carlyle was born pink and healthy. What measures should be taken to prevent the previously described situation from happening in the second pregnancy with an Rh positive baby?

   5. Mrs. Carlyle's sister has had two miscarriages before seeking medical help with her third pregnancy. Blood typing shows that she, like her sister, is Rh negative; her husband is Rh positive. What course of treatment will be followed?

9. A red marrow biopsy is ordered for two patients—one a child and the other an adult. The specimen is taken from the tibia of the child but from the iliac crest of the adult. Explain why different sites are used to obtain marrow samples in adults and children.

10. The Jones family let their dog Rooter lick their faces and kissed it on the mouth, not realizing that it had just explored the neighborhood dump and trash cans. Later the same day the veterinarian diagnosed Rooter as having pinworms. Three weeks later, blood tests ordered for routine physicals for camp indicated that both the family's daughters had blood eosinophil levels of over 3000 per cubic millimeter. Might there be a connection between these events?

11. An elderly man has been receiving weekly injections of vitamin B_{12} ever since nearly all of his stomach was removed 6 months previously (he had stomach cancer). Why is he receiving the vitamin injections? Why can't the vitamin be delivered in tablet form? What would be the result if he refuses the B_{12} injections?

12. List two blood tests that might be ordered if infectious mononucleosis is suspected and three that might be done if a person has bleeding problems.

13. Jenny, a young woman of 38, has had several episodes of excessive bleeding. Her physician orders tests to determine whether her problem is due to vitamin K deficiency or thrombocytopenia. What test results will be important in making this diagnosis?

14. A man of Greek ancestry goes to his doctor with the following symptoms. He is tired all the time and has difficulty catching his breath after even mild exercise. His doctor orders the following tests: complete blood count, hematocrit, differential WBC count. The test show small, pale, immature erythrocytes, fragile erythrocytes, and around 2 million erythrocytes per cubic millimeter. What is the tentative diagnosis?

Stop and Think

1. Why is someone more likely to bleed to death when an artery is cleanly severed than when an artery is crushed and torn?

2. Knowing the pathways of hemopoiesis, explain why leukemia reduces RBC and platelet counts.

3. Dissect these words: polycythemia, diapedesis.

4. What are the functional advantages of the "shortcut" of the extrinsic pathway *and* of the multiple steps of the intrinsic pathway?

5. Sodium EDTA removes calcium ions from solution. What would be the effect of adding sodium EDTA to blood?

6. What is the functional importance of the fact that hemoglobin F has a higher oxygen affinity than hemoglobin A?

7. Explain the phenomenon called "athlete's anemia."

8. Relative to human blood groups, what are "private antigens"?

9. John, a novice to cigarette smoking, is trying to impress his buddies by smoking two packs a day—inhaling every drag. What do you think will happen to his reticulocyte count? Explain your reasoning.

COVERING ALL YOUR BASES

Multiple Choice

Select the best answer or answers from the choices given.

1. Which of the following are true concerning erythrocytes?
   A. They rely strictly on anaerobic respiration.
   B. About one-third of their volume is hemoglobin.
   C. Their precursor is called a megakaryoblast.
   D. Their shape increases membrane surface area.

2. A serious bacterial infection leads to more of these cells in the blood.
   A. Erythrocytes and platelets
   B. Mature and band neutrophils
   C. Erythrocytes and monocytes
   D. All formed elements

3. Each hemoglobin molecule:
   A. contains either alpha or beta polypeptides
   B. can reversibly bind up to four oxygen molecules
   C. has one heme group
   D. has four iron atoms

4. If hemoglobin were carried loose in the plasma, it would:
   A. increase viscosity of blood plasma
   B. decrease blood osmotic pressure
   C. be unable to bind oxygen
   D. leak out of the bloodstream quite easily

5. Which of the following is/are true of hemopoiesis?
   A. All blood formed elements are derived from the hemocytoblast.
   B. Hemopoiesis occurs in myeloid tissue.
   C. The processes of hemopoiesis include production of plasma proteins by the liver.
   D. Irradiation of the distal limbs of an adult probably will not affect hemopoiesis.

6. Which cell stage directly precedes the reticulocyte stage in erythropoiesis?
   A. Basophilic erythroblast
   B. Proerythroblast
   C. Polychromatophilic erythroblast
   D. Normoblast

7. Which would lead to increased erythropoiesis?
   A. Chronic bleeding ulcer
   B. Reduction in respiratory ventilation
   C. Decreased level of physical activity
   D. Reduced blood flow to the kidneys

8. Deficiency of which of the following will have a direct negative impact on RBC production?
   A. Transferrin   C. Folic acid
   B. Iron   D. Calcium

9. One possible test of liver function would measure the level in the bloodstream of:
   A. erythropoietin   C. bilirubin
   B. vitamin K   D. intrinsic factor

10. A possible cause of aplastic anemia is:

   A. deficient production of intrinsic factor

   B. ionizing radiation

   C. vitamin B_{12} deficiency

   D. thalassemia

11. A child is diagnosed with sickle-cell anemia. This means that:

   A. one parent had sickle-cell anemia

   B. one parent carried the sickle-cell gene

   C. both parents had sickle-cell anemia

   D. both parents carried the sickle-cell gene

12. Polycythemia vera will result in:

   A. overproduction of WBCs

   B. exceptionally high blood volume

   C. abnormally high blood viscosity

   D. abnormally low hematocrit

13. Which of the following does not characterize leukocytes?

   A. Ameboid

   B. Phagocytic (some)

   C. Nucleated

   D. Cells found in largest numbers in the bloodstream

14. The colony-stimulating factor that has the greatest effect on the most types of WBCs is:

   A. G-CSF   **C.** GM-CSF

   B. IL-3   **D.** M-CSF

15. The first type of WBC whose development diverges from the common leukopoietic line is the:

   A. monocyte   **C.** granulocyte

   B. lymphocyte   **D.** basophil

16. The WBC type with the longest (possible) life span is the:

   A. neutrophil   **C.** lymphocyte

   B. monocyte   **D.** eosinophil

17. In leukemia:

   A. the cancerous WBCs function normally

   B. the cancerous WBCs fail to specialize

   C. production of RBCs and platelets is decreased

   D. infection and bleeding can be life-threatening

18. Platelet formation:

   A. involves mitosis without cytokinesis

   B. includes specialization from the myeloid stem cell

   C. requires stimulation by thrombopoietin

   D. results in small, nucleated thrombocytes

19. A deficiency of albumin would result in:

   A. increased blood volume

   B. increased blood osmotic pressure

   C. loss of water by osmosis from the bloodstream

   D. a pH imbalance

20. All leukocytes share the following features except:

   A. diapedesis

   B. disease-fighting

   C. distorted, lobed nuclei

   D. more active in connective tissues than in blood

21. Which platelet factors attract more platelets?

   A. Serotonin   **C.** ADP

   B. Thromboxane A_2 **D.** Prostacyclin

22. After activation of factor X, the next step in the coagulation sequence is:

   A. activation of factor XI

   B. activation of factor IX

   C. formation of prothrombin activator

   D. formation of thrombin

23. Fibrinolysis is increased by:
    A. activation of thrombin
    B. activation of plasminogen
    C. release of t-PA
    D. heparin

24. A condition resulting from thrombocytopenia is:
    A. thrombus formation
    B. embolus formation
    C. petechiae
    D. hemophilia

25. Which of the following can cause problems in a transfusion reaction?
    A. Donor antibodies attacking recipient RBCs
    B. Clogging of small vessels by agglutinated clumps of RBCs
    C. Lysis of donated RBCs
    D. Blockage of kidney tubules

26. The erythrocyte count increases when an individual goes from a low to a high altitude because:
    A. the concentration of oxygen and/or total atmospheric pressure is lower at high altitudes
    B. the basal metabolic rate is higher at high altitudes
    C. the concentration of oxygen and/or total atmospheric pressure is higher at high altitudes
    D. the temperature is lower at high altitudes

27. A hematocrit value that would be normal for females but not males is:
    A. 70%
    B. 39%
    C. 45%
    D. 51%

28. Sickling of RBCs can be induced in those with sickle-cell anemia by:
    A. blood loss
    B. vigorous exercise
    C. stress
    D. fever

29. If an Rh– mother becomes pregnant, when can erythroblastosis fetalis *not possibly* occur in the child?
    A. If the child is Rh–
    B. If the child is Rh+
    C. If the father is Rh+
    D. If the father is Rh–

30. What is the difference between a thrombus and an embolus?
    A. One occurs in the bloodstream, whereas the other occurs outside the bloodstream.
    B. One occurs in arteries, the other in veins.
    C. One is a blood clot, while the other is a parasitic worm.
    D. A thrombus must travel to become an embolus.

31. The plasma component that forms the fibrous skeleton of a clot is:
    A. platelets
    B. fibrinogen
    C. thromboplastin
    D. thrombin

32. A substance that may one day be used to boost the oxygen-carrying capacity of the blood in someone who has lost substantial blood volume is:
    A. Fluosol
    B. neohemocytes
    C. plasminate
    D. Hemopure

Word Dissection

For each of the following word roots, fill in the literal meaning and give an example, using a word found in this chapter.

Word root	Translation	Example
1. agglutin	_____	_____
2. album	_____	_____
3. bili	_____	_____
4. embol	_____	_____
5. emia	_____	_____
6. erythro	_____	_____
7. ferr	_____	_____
8. hem	_____	_____
9. karyo	_____	_____
10. leuko	_____	_____
11. lymph	_____	_____
12. phil	_____	_____
13. poiesis	_____	_____
14. rhage	_____	_____
15. thromb	_____	_____

19 The Cardiovascular System: The Heart

Student Objectives

When you have completed the exercises in this chapter, you will have accomplished the following objectives:

Heart Anatomy

1. Describe the size and shape of the heart and indicate its location and orientation in the thorax.

2. Name the coverings of the heart.

3. Describe the structure and function of each of the three layers of the heart wall.

4. Describe the structure and functions of the four heart chambers. Name each chamber and provide the name and general route of its associated great vessel(s).

5. Trace the pathway of blood through the heart.

6. Name the heart valves and describe their location, function, and mechanism of operation.

Blood Supply to the Heart: Coronary Circulation

7. Name the major branches of the coronary arteries and describe their distribution.

Properties of Cardiac Muscle Fibers

8. Describe the structural and functional properties of cardiac muscle, and explain how it differs from skeletal muscle.

9. Briefly describe the events of cardiac muscle cell contraction.

Heart Physiology

10. Name the components of the conduction system of the heart and trace the conduction pathway.

11. Draw a diagram of a normal electrocardiogram tracing; name the individual waves and intervals and indicate what each represents. Name some of the abnormalities that can be detected on an ECG tracing.

12. Describe the timing and events of the cardiac cycle.

13. Describe normal heart sounds and explain how heart murmurs differ from normal sounds.

14. Name and explain the effects of the various factors involved in regulation of stroke volume and heart rate.

15. Explain the role of the autonomic nervous system in regulating cardiac output.

Developmental Aspects of the Heart

16. Describe the formation of the fetal heart and indicate how the fetal heart differs from the adult heart.

17. Provide examples of age-related changes in heart function.

Ready, set, go!

As part of the cardiovascular system, the heart has a single function—to pump the blood into the blood vessels so that it reaches the trillions of tissue cells of the body. Survival of tissue cells depends on constant access to oxygen and nutrients and removal of carbon dioxide and wastes. This critical homeostatic requirement is met by a continuous flow of blood. The heart is designed to maintain this continuous flow by pumping blood simultaneously through two circuits: the pulmonary circulation to the lungs and the systemic circulation to all body regions.

Topics for study in Chapter 19 include the microscopic and gross anatomy of the heart, the related events of the cardiac cycle, the regulation of cardiac output, malfunctions of the heart, and changes in the heart throughout life.

BUILDING THE FRAMEWORK

Heart Anatomy

1. Complete the following statements by writing the missing terms in the answer blanks.

mediastinum **1.**

diaphragm **2.**

second **3.**

midsternal line **4.**

fibrous **5.**

visceral **6.**

epicardium **7.**

friction **8.**

myocardium **9.**

cardiac muscle **10.**

fibrous skeleton **11.**

endocardium **12.**

endothelial **13.**

_4_ **14.**

The heart is a cone-shaped muscular organ located within the __(1)__ of the thorax. Its apex rests on the __(2)__ and its superior margin lies at the level of the __(3)__ rib. Approximately two-thirds of the heart mass is seen to the left of the __(4)__.

The heart is enclosed in a serosa sac called the pericardium. The loosely fitting double outer layer consists of the outermost __(5)__ pericardium, lined by the parietal layer of the serous pericardium. The inner __(6)__ pericardium, also called the __(7)__, is the outermost layer of the heart wall. The function of the fluid that fills the pericardial sac is to decrease __(8)__ during heart activity. The middle layer of the heart wall, called the __(9)__, is composed of __(10)__; it forms the bulk of the heart.

Connective tissue fibers that ramify throughout this layer construct the so-called __(11)__ of the heart. The membrane that lines the heart and also forms the valve flaps is the __(12)__. This layer is continuous with the __(13)__ linings of the blood vessels that enter and leave the heart. The heart has __(14)__ chambers. Relative to the roles of these chambers, the __(15)__ are the receiving chambers, whereas the __(16)__ are the discharging chambers.

atria **15.**   _ventricles_ **16.**

2. Figure 19.1A is a transverse section through the thorax. Label the structures that have leader lines. Figure 19.1B is a highly schematic longitudinal section through the heart wall and pericardium. Select colors for each structure listed below; color the corresponding coding circles and the structure on the figure.

○ Fibrous pericardium

○ Myocardium

○ Pericardial cavity

○ Parietal layer of serous pericardium

○ Visceral layer of serous pericardium (epicardium)

○ Endocardium

○ Diaphragm

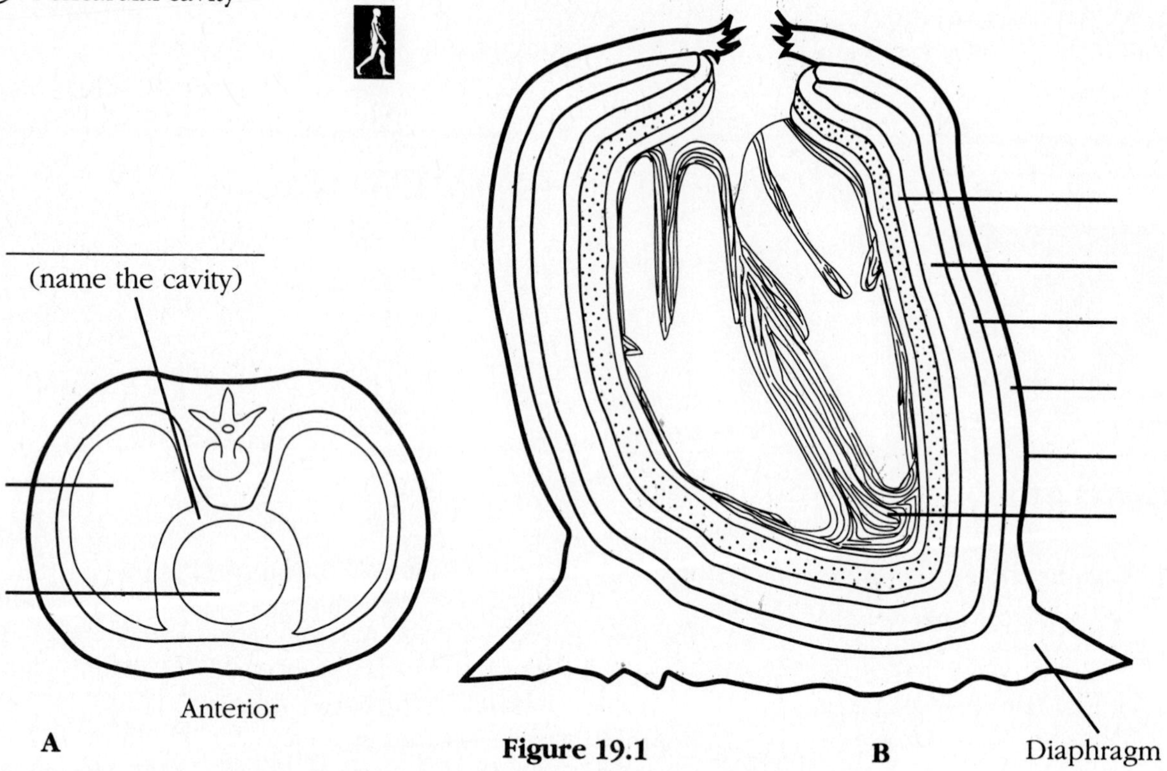

_____
(name the cavity)

Anterior

A

Figure 19.1

B

Diaphragm

3. The heart is called a double pump because it serves two circulations. Trace the flow of blood through both pulmonary and systemic circulations by writing the missing terms in the answer blanks below. Then, identify the various regions of the circulation shown in Figure 19.2 by labeling them using the key choices. Color regions transporting O_2-poor blood blue and regions transporting O_2-rich blood red.

From the atrium through the tricuspid valve to the __(1)__, through the __(2)__ valve to the pulmonary trunk to the right and left __(3)__, to the capillary beds of the __(4)__, to the __(5)__, to the __(6)__ of the heart through the __(7)__ valve, to the __(8)__ through the __(9)__ semilunar valve to the __(10)__, to the systemic arteries, to the __(11)__ of the body tissues, to the systemic veins, to the __(12)__ and __(13)__, which enter the right atrium of the heart.

right ventricle ___ 1.

pulmonary semilunar valve ___ 2.

pulmonary arteries ___ 3.

lungs ___ 4.

r+l pulmonary veins ___ 5.

left atrium ___ 6.

mitral (bicuspid) ___ 7.

left ventricle ___ 8.

aortic ___ 9.

aorta ___ 10.

capillary bed ___ 11.

superior vena cava ___ 12.

inferior vena cava ___ 13.

KEY CHOICES

A. Vessels serving head and upper limbs

B. Vessels serving body trunk and lower limbs

C. Vessels serving the viscera

D. Pulmonary circulation

E. Pulmonary "pump"

F. Systemic "pump"

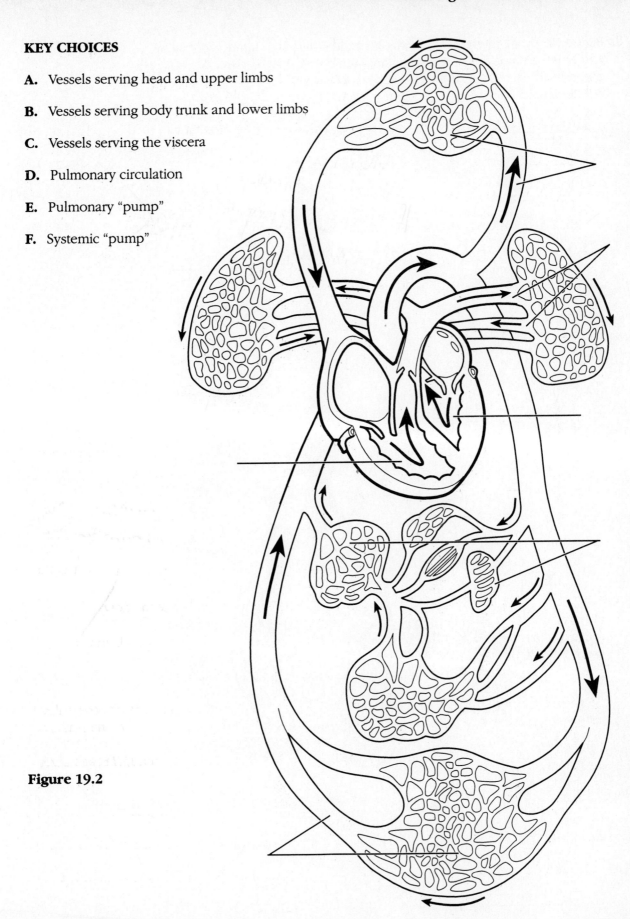

Figure 19.2

4. Figure 19.3 is an anterior view of the heart. Identify each numbered structure and write its name in the corresponding numbered answer blank. Then, select different colors for each structure with a coding circle and color the structures on the figure.

◯ right atrium 1.　◯ inf. vena cava 6.　◯ right pulmonary veins 11.

◯ left atrium 2.　◯ aorta 7.　left pulmonary veins 12.

◯ right ventricle 3.　◯ pulmonary trunk 8.　vessels of coronary circulation 13.

◯ left ventricle 4.　lft pulmonary artery 9.　apex 14.

◯ sup. vena cava 5.　right pulmonary artery 10.　◯ _____ 15.

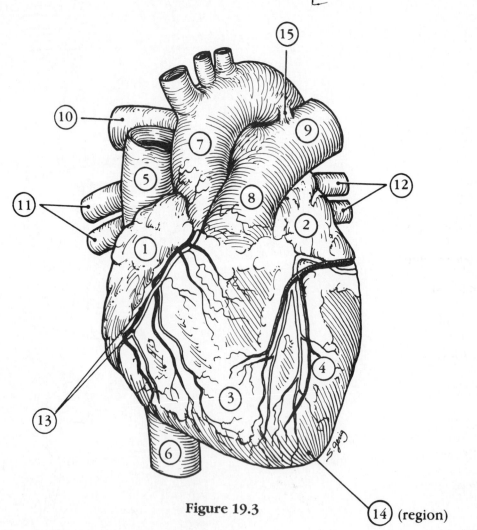

Figure 19.3

Blood Supply to the Heart: Coronary Circulation

1. Figure 19.4 shows the vascular supply of the myocardium. Part A shows the arterial supply; part B shows the venous drainage. In each case, the anterior view is shown and the vessels located posteriorly are depicted as dashed lines. Use a light color to color in the *right atrium* and the *left ventricle* in each diagram. (Leave the left atrium and right ventricle uncolored.) Color the aorta red and the pulmonary trunk and arteries blue. Then color code and color the vessels listed below.

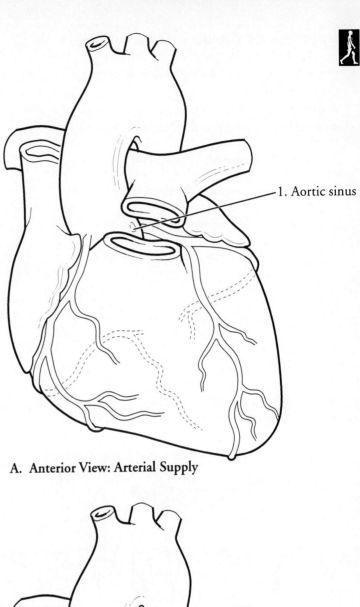

1. Aortic sinus

A. Anterior View: Arterial Supply

Part A

○ Anterior interventricular artery

○ Circumflex artery

○ Left coronary artery

○ Marginal artery

○ Posterior interventricular artery

○ Right coronary artery

Part B

○ Anterior cardiac veins

○ Coronary sinus

○ Great cardiac vein

○ Middle cardiac vein

○ Small cardiac vein

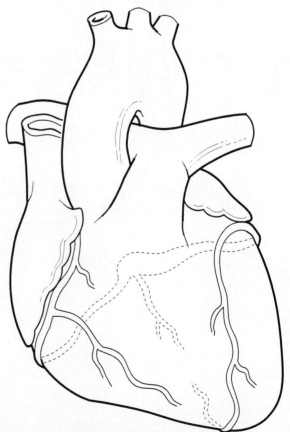

Figure 19.4

B. Anterior View: Venous Drainage

Properties of Cardiac Muscle Fibers

1. A number of characteristics peculiar to skeletal or cardiac muscle are listed
   below. Select which ones refer to each muscle type by writing S in the answer
   blanks preceding descriptions that apply to skeletal muscle and C in the
   answer blanks of those that identify characteristics of cardiac muscle.

_____ **1.** One centrally located nucleus

_____ **2.** More mitochondria per cell

_____ **3.** T tubules line up at the Z lines

_____ **4.** Uses fatty acids more effectively
            for ATP harvest

_____ **5.** Requires stimulation by the nerv-
            ous system to contract

_____ **6.** Has a shorter refractory period

_____ **7.** Contains self-excitable cells

_____ **8.** Has well-developed terminal
            cisternae

_____ **9.** Wide T tubules

_____ **10.** More distinct myofibrils

_____ **11.** All-or-none law applies at the or-
             gan level

_____ **12.** Intercalated discs enhance intercel-
             lular electrical communication

2. Figure 19.5 is a schematic drawing of the microscopic structure of cardiac
   muscle. Using different colors, color the coding circles of the structures listed
   below and the corresponding structures on the figure. Then answer the
   questions that follow the figure. Write your answers in the answer blanks.

◯ Nuclei (with nucleoli)   ◯ Muscle fibers

◯ Intercalated discs   ◯ Striations

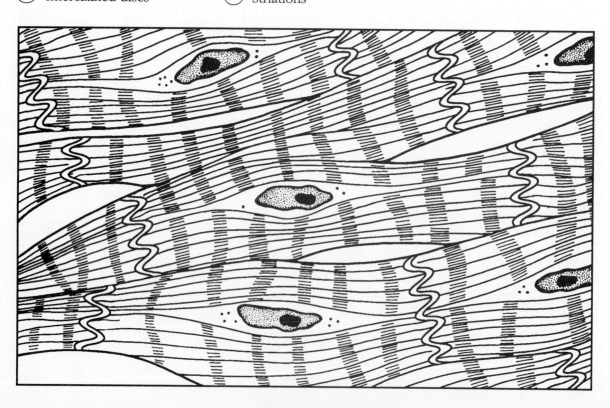

Figure 19.5

_____ **1.** Name the loose connective tissue that fills the intercellular spaces.

_____ **2.** What is the function during contraction of the desmosomes present in the intercalated discs? (*Note:* The desmosomes are not illustrated.)

_____ **3.** What is the function of the gap junctions (not illustrated) also present in the intercalated discs?

_____ **4.** What term describes the interdependent, interconnecting cardiac cells?

_____ **5.** Which structures provide electrical coupling of cardiac cells?

3. On Figure 19.6, indicate (by adding labels) the following changes in membrane permeability and events of an action potential at their approximate point of occurrence. Also indicate by labeling a bracket: (1) the plateau, and (2) the period when Ca^{2+} is pumped out of the cell. Also color code and color the arrows and brackets.

◯ $\uparrow Na^+$ (Na^+ gates open) ◯ $\downarrow Ca^{2+}$ (Ca^{2+} gates close)

◯ $\downarrow Na^+$ (Na^+ gates close) ◯ $\uparrow K^+$ (K^+ gates open)

◯ $\uparrow Ca^{2+}$ (Ca^{2+} gates open) ◯ Plateau

◯ Ca^{2+} pumped from cell

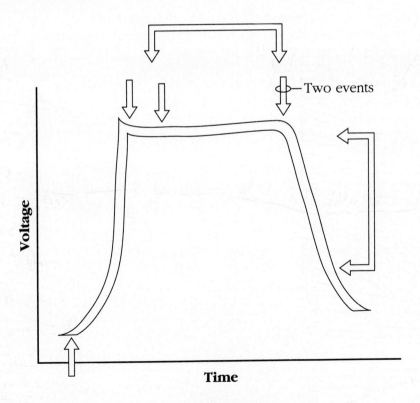

Figure 19.6

Heart Physiology

1. Complete the following statements concerning the cells of the nodal system of the heart. Write the missing terms in the answer blanks.

_____ 1.

_____ 2.

_____ 3.

_____ 4.

_____ 5.

_____ 6.

_____ 7.

_____ 8.

_____ 9.

The nodal cells of the heart, unlike cardiac contractile muscle fibers, have an intrinsic ability to depolarize **(1)** . This reflects their unstable **(2)** , which drifts slowly toward the threshold for firing, that is, **(3)** . These spontaneously changing membrane potentials, called **(4)** , are probably due to reduced membrane permeability to **(5)** . However, **(6)** permeability is unchanged and it continues to diffuse **(7)** the cell at a slow rate. Ultimately, when threshold is reached, gated channels called **(8)** open, allowing extracellular **(9)** to rush into the cells and reverse the membrane potential.

2. Figure 19.7 is a diagram of the frontal section of the heart. Follow the instructions below to complete this exercise, which considers both anatomical and physiological aspects of the heart.

1. Draw arrows to indicate the direction of blood flow through the heart. Draw the pathway of the oxygen-rich blood with red arrows and trace the pathway of oxygen-poor blood with blue arrows.

2. Identify each of the elements of the intrinsic conduction system (numbers 1–5 on the figure) by writing the appropriate terms in the numbered answer blanks. Then, indicate with green arrows the pathway that impulses take through this system.

3. Identify each of the heart valves (numbers 6–9 on the figure) by writing the appropriate terms in the numbered answer blanks. Draw and identify by name the cordlike structures that anchor the flaps of the atrioventricular (AV) valves.

4. Use the numbers from the figure to identify structures (A–H).

6 **A.** 7 **B.** Prevent backflow into the ventricles when the heart is relaxed

8 **C.** 9 **D.** Prevent backflow into the atria when the ventricles are contracting

9 **E.** AV valve with three flaps

8 **F.** AV valve with two flaps

1 **G.** The pacemaker of the Purkinje system

2 **H.** The point in the Purkinje system where the impulse is temporarily delayed

1. SA node
2. AV node
3. AV Bundle (Bundle of His)
4. bundle branches
5. Purkinje fibers
6. pulmonary valve
7. aortic valve
8. mitral (bicuspid) valve
9. tricuspid valve

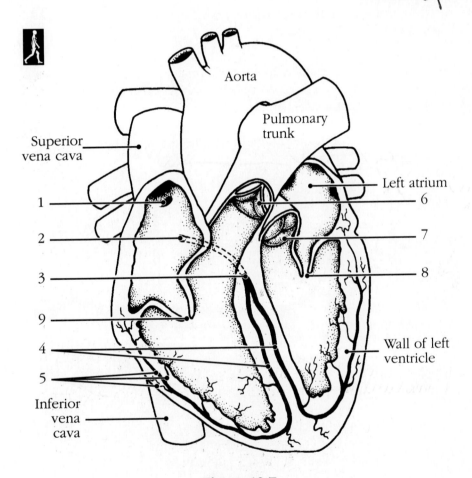

Aorta

Pulmonary trunk

Superior vena cava

Left atrium

1

2

3

9

4

5

6

7

8

Wall of left ventricle

Inferior vena cava

Figure 19.7

3. Respond to the questions below concerning the nodal system.

    1. What name is given to the rate set by the heart's pacemaker? _____

    2. What are the intrinsic contraction rates of the different components of the intrinsic conduction system?

        SA node _____ beats/min   AV node _____ beats/min

        AV bundle _____ beats/min   Purkinje fibers _____ beats/min

    3. The intrinsic conduction system enforces a faster rate of impulse conduction across the heart—at the rate of several meters per second in most parts of the conduction system. What would be the natural speed of impulse transmission across the heart in the absence of such a system? _____ m/s

    4. What is the total time for impulse conduction across the healthy heart, on average? _____ seconds

4. Part of an electrocardiogram is shown in Figure 19.8. On the figure, identify the QRS complex, the P wave, and the T wave. Using a green pencil, bracket the P-Q interval and the Q-T interval. Then, using a red pencil, bracket a portion of the recording equivalent to the length of one cardiac cycle. Using a blue pencil, bracket a portion of the recording in which the *ventricles* would be in diastole.

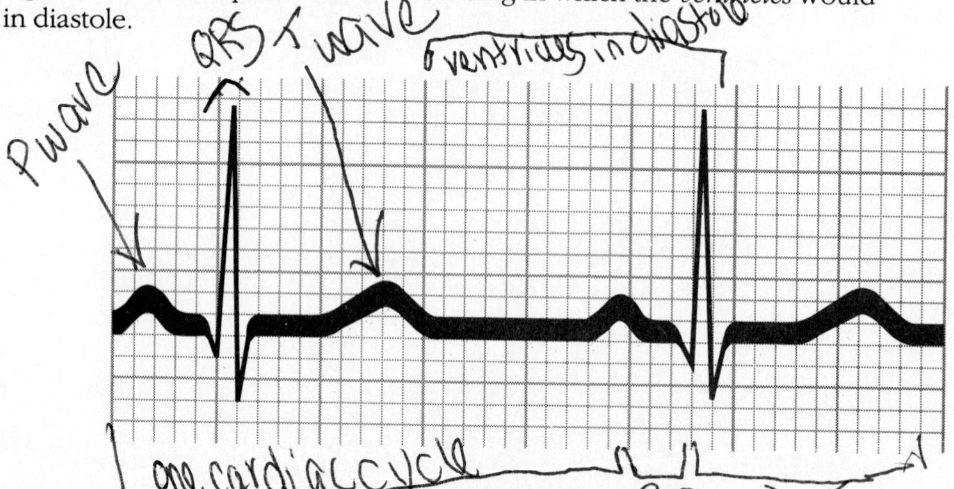

Figure 19.8

5. Examine the abnormal ECG tracings shown in Figure 19.9.

_____ **1.** Which shows extra P waves?

_____ **2.** Which shows tachycardia?

_____ **3.** Which has an abnormal QRS complex?

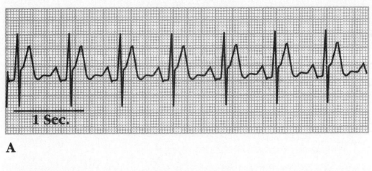

A

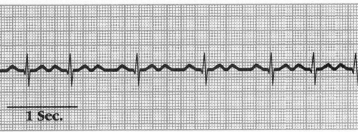

B

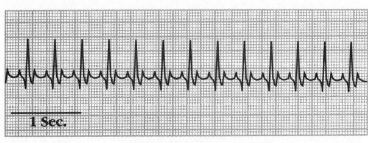

C

Figure 19.9

6. The events of one complete heartbeat are referred to as the cardiac cycle. Complete the following statements that describe these events by writing the missing terms in the answer blanks.

_____ 1.

_____ 2.

_____ 3.

_____ 4.

_____ 5.

_____ 6.

_____ 7.

_____ 8.   _____ 9.   _____ 10.

The contraction of the ventricles is referred to as __(1)__ and the period of ventricular relaxation is called __(2)__ . The monosyllables describing heart sounds during the cardiac cycle are __(3)__ . The first heart sound is a result of closure of the __(4)__ valves; closure of the __(5)__ valves causes the second heart sound. The heart chambers that have just been filled when you hear the first heart sound are the __(6)__ and the chambers that have just emptied are the __(7)__ . Immediately after the second heart sound, the __(8)__ are filling with blood and the __(9)__ are empty. Abnormal heart sounds, or __(10)__ , usually indicate valve problems.

7. The events of one cardiac cycle are graphed in Figure 19.10. First, identify the following by color:

◯ ECG tracing   ◯ Atrial pressure line

◯ Aortic pressure line   ◯ Ventricular pressure line

Then, identify by labeling the following:

- The P, QRS, and T waves of the ECG

- Points of opening and closing of the AV and semilunar values

- Elastic recoil of the aorta

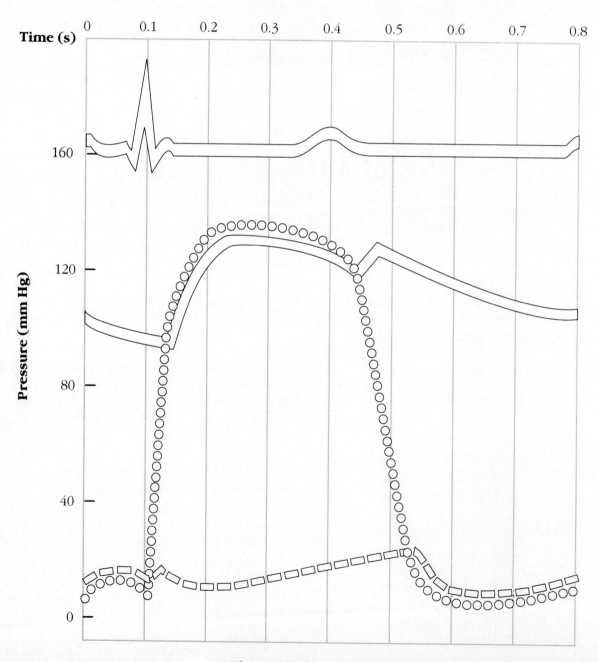

Figure 19.10

8. Circle the term that does not belong in each of the following groupings.

    1. AV valves closed   AV valves open   Ventricular systole   Semilunar valves open

    2. No P wave   SA node is pacemaker   AV node is pacemaker   Junctional rhythm

    3. Ventricles fill   Ventricular systole   AV valves open   Late diastole

    4. Early diastole   Semilunar valves open   Isovolumetric relaxation

       Ventricular pressure drops

    5. Stenotic valve   Restricted blood flow   High-pitched heart sound

       Heart sound after valve closes

    6. Isovolumetric ventricular systole   Blood volume unchanging   AV valve open

       Semilunar valve closed

9. Complete the following statements relating to cardiac output by writing the missing terms in the answer blanks.

_____ 1.	In the relationship CO = HR × SV, CO stands for __(1)__, HR stands for __(2)__, and SV stands for __(3)__. For the normal resting heart, the value of HR is __(4)__ and the value of SV is __(5)__. The normal average adult cardiac output, therefore, is __(6)__. The time for the entire blood supply to pass through the body is once each __(7)__. The normal ventricle holds a volume of blood equal to about __(8)__, of which about __(9)__ of blood remain after each contraction. The heart, however, can exceed its normal cardiac output, a property of the heart called __(10)__. In response to sudden demands, such as __(11)__, cardiac reserve is about four times the normal cardiac output.
_____ 2.	
_____ 3.	
_____ 4.	
_____ 5.	
_____ 6.	
_____ 7.	
_____ 8.	SV represents the difference between the __(12)__, the amount of blood that collects in the ventricle during diastole, and the __(13)__, the volume of blood remaining in the ventricle after contraction. SV is critically related to __(14)__, or the degree of stretch of cardiac muscle before it contracts. Stretching increases the number of __(15)__ interactions between the actin and myosin filaments in the cardiac cells and hence the __(16)__ of heart contraction. Stretching cardiac muscle is accomplished by increasing the amount of __(17)__ returned to the heart, which distends the __(18)__.
_____ 9.	
_____ 10.	
_____ 11.	
_____ 12.	
_____ 13.	
_____ 14.	
_____ 15.	
_____ 16.	
_____ 17. _____ 18.	

10. Stroke volume may be enhanced in two major ways: by increasing venous return and by enhancing contractility of heart muscle.

1. Explain *contractility.* _____

_____

2. Name three factors, one neural and two hormonal, that enhance contractility.

_____

3. Explain how exercise increases venous return. _____

_____

11. Check (✔) all factors that lead to an *increase* in cardiac output by influencing either heart rate or stroke volume.

_____ **1.** Epinephrine

_____ **2.** Thyroxine

_____ **3.** Hemorrhage

_____ **4.** Fear

_____ **5.** Exercise

_____ **6.** Fever

_____ **7.** Depression

_____ **8.** Anxiety

_____ **9.** Activation of the vasomotor center

_____ **10.** Activation of the cardioacceleratory center

_____ **11.** Activation of the cardioinhibitory center

_____ **12.** Activation of the vagus nerves

_____ **13.** Low blood pressure

_____ **14.** High blood pressure

_____ **15.** Increased end diastolic volume

_____ **16.** Prolonged grief

_____ **17.** Acidosis

_____ **18.** Calcium channel blockers

_____ **19.** Atrial reflex

_____ **20.** Hyperkalemia

_____ **21.** Increased afterload

_____ **22.** Glucagon

12. If a statement is true, write the letter T in the answer blank. If a statement is false, change the underlined word(s) and write the correct word(s) in the answer blank.

_____ **1.** Norepinephrine, released by <u>parasympathetic</u> fibers, stimulates the SA and AV nodes and the myocardium itself.

_____ **2.** The resting heart is said to exhibit "vagal tone," meaning that the heart rate slows under the influence of <u>acetylcholine</u>.

_____ **3.** Epinephrine secreted by the adrenal medulla <u>decreases</u> heart rate.

_____ **4.** <u>Low</u> levels of ionic calcium cause prolonged cardiac contractions.

_____ **5.** The resting heart rate is fastest in <u>adult</u> life.

_____ **6.** Because the heart of the highly trained athlete hypertrophies, its <u>stroke volume</u> decreases.

_____ **7.** In congestive heart failure, there is a marked rise in the end <u>diastolic</u> volume.

_____ **8.** If the <u>right</u> side of the heart fails, pulmonary congestion occurs.

_____ **9.** In <u>peripheral</u> congestion, the feet, ankles, and fingers become edematous.

_____ **10.** The pumping action of the healthy heart ordinarily maintains a balance between cardiac output and <u>venous return</u>.

_____ **11.** The <u>cardioacceleratory center</u> in the medulla gives rise to sympathetic nerves supplying the heart.

13. Match the terms in Column B with the statements in Column A. Place the correct letters in the answer blanks.

Column A

Column B

_____ **1.** Results from prolonged coronary blockage

_____ **2.** Abnormal pacemaker

_____ **3.** Allows backflow of blood

_____ **4.** Because of cardiac decompensation, circulation is inadequate to meet tissue needs

_____ **5.** A slow heartbeat, that is, below 60 beats per minute

_____ **6.** A condition in which the heart is uncoordinated and useless as a pump

_____ **7.** A rapid heart rate, that is, over 100 beats per minute

_____ **8.** Damage to the AV node, totally or partially releasing the ventricles from the control of the SA node

_____ **9.** Chest pain, resulting from ischemia of the myocardium

_____ **10.** Result of initial failure of the left side of the heart

A. Angina pectoris

B. Bradycardia

C. Congestive heart failure

D. Ectopic focus

E. Fibrillation

F. Heart block

G. Incompetent valve

H. Myocardial infarction

I. Pulmonary congestion

J. Tachycardia

14. Match the ionic imbalances in Column B to the descriptions of their effects in Column A.

Column A

_____   **1.** Depresses heart

_____   **2.** Spastic contractions

_____   **3.** Inhibits Ca^{2+} transport

_____   **4.** Lowers resting potential

_____   **5.** Feeble contractions, abnormal rhythms

Column B

A. Hypercalcemia

B. Hyperkalemia

C. Hypernatremia

D. Hypocalcemia

E. Hypokalemia

Developmental Aspects of the Heart

1. Complete the following statements by writing the missing terms in the answer blanks.

_____   **1.**

_____   **2.**

_____   **3.**

_____   **4.**

_____   **5.**

_____   **6.**

_____   **7.**

_____   **8.**

_____   **9.**

_____   **10.**

_____   **11.**

_____   **12.**

_____   **13.**

_____   **14.**

The heart forms from a fusion of two pulsating tubes into a single chamber and is actively pumping blood by the __(1)__ week of pregnancy. In just three more weeks it becomes a four-chambered double pump. The fetal heart rate is about __(2)__ . Two fetal structures divert blood away from the collapsed, nonfunctional lungs. The shunt located in the interatrial septum is called the __(3)__ . The other, called the __(4)__ , diverts blood from the __(5)__ to the aorta. Shortly after birth, these bypass structures become __(6)__ . If these shunts remain patent, children exhibit __(7)__ during physical exertion.

The adult heart benefits from regular __(8)__ , which enhances the endurance and strength of the myocardium. Several age-related anatomical changes may occur in the heart. Heart murmurs in the elderly may signal sclerosis and thickening of the __(9)__ , arrhythmias may be a result of fibrosis of the __(10)__ , and atherosclerosis may lead to heart attack if the __(11)__ become occluded. The risk of cardiovascular disease is lowered if three dietary components are decreased. These are __(12)__ , __(13)__ , and __(14)__ .

CHALLENGING YOURSELF

At the Clinic

1. A patient with acute severe pericarditis has a critically low stroke volume. What is the name for the condition causing the low stroke volume and how does it cause it?

2. Mr. Greco, a patient with clotting problems, has been hospitalized with right-sided heart failure. What is his condition? Knowing Mr. Greco's history, what is the probable cause?

3. An elderly man is brought to the clinic because he fatigues extremely easily. An examination reveals a heart murmur associated with the bicuspid valve during ventricular systole. What is the diagnosis? What is a possible treatment?

4. Jimmy is brought to the clinic complaining of a sore throat. His mother says he has had the sore throat for about a week. Culture of a throat swab is positive for strep, and the boy is put on antibiotics. A week later, the boy is admitted to the hospital after fainting several times. He is cyanotic. What is a likely diagnosis?

5. After a bout with bacterial endocarditis, scar tissue often stiffens the edges of the heart valves. What condition will this cause? How would this be picked up in a routine examination?

6. After a fairly severe heart attack, the ECG reveals normal sinus rhythm, but only for the P wave. The QRS and T waves are no longer in synchrony with the P wave, and the ventricular contraction corresponds to junctional rhythm. What is the problem? What part of the heart is damaged?

7. Mary Ghareeb, a woman in her late 50s, has come to the clinic because of chest pains whenever she begins to exert herself. What is her condition called?

8. A man, en route to the hospital ER by ambulance, is in fibrillation. What is his cardiac output likely to be? He arrives at the emergency entrance DOA (dead on arrival). His autopsy reveals a blockage of the posterior interventricular artery. What is the cause of death?

9. Excessive vagal stimulation can be caused by severe depression. How would this be reflected in a routine physical examination?

10. A patient has swollen ankles and signs of degenerating organ functions. What is a likely diagnosis?

11. You are a young nursing student and are called upon to explain how each of the following tests or procedures might be helpful in evaluating a patient with heart disease: blood pressure measurement, determination of blood lipid and cholesterol levels, electrocardiogram, chest X ray. How would you respond?

12. Mr. Langley is telling his friend about his recent visit to his doctor for a checkup. During his story, he mentions that the ECG revealed that he had a defective mitral valve and a heart murmur. Mr. Langley apparently misunderstood some of what the doctor explained to him about his diagnosis process. What has he misunderstood?

13. A 14-year-old girl undergoing a physical examination before being allowed to matriculate at a dancing academy was found to have a loud heart murmur at the second intercostal space to the left side of the sternum. Upon questioning, the girl admitted to frequent "breathlessness," and it was decided to perform further tests. An angiogram showed that the girl had a patent ductus arteriosus. Discuss the location and function of the ductus arteriosus in the fetus and relate the reason for the girl's breathlessness.

Stop and Think

1. The major coronary vessels are on the surface of the heart. What is the advantage of that location?

2. Since the SA node is at the top of the atrial mass, the atria contract from the top down. How does this increase the efficiency of atrial contraction? Do the ventricles have a similar arrangement? If so, how does it work?

3. What parts of the body feel referred pain when there is a major myocardial infarction?

4. What is the purpose of prolonged contraction of the myocardium?

5. How long does it take for a hormone released from the anterior pituitary gland to reach its target organ in the body?

6. What is the functional difference between ventricular hypertrophy due to exercise and hypertrophy due to congestive heart failure?

7. Of the following congenital heart defects—ventricular septal defect, pulmonary stenosis, coarctation of the aorta, and tetralogy of Fallot—which produce mixing of oxygenated and unoxygenated blood, and which increase ventricular work load?

8. What is the effect of hypothyroidism on heart rate?

9. The foramen ovale has a flap that partially covers it, and a groove leads from the opening of the inferior vena cava (which carries freshly oxygenated blood from the umbilical vein) to the foramen ovale. Explain these structural features in terms of directing blood flow.

10. A less-than-respectable news tabloid announced that "Doctors show exercise shortens life. Life expectancy is programmed into a set number of heartbeats; the faster your heart beats, the sooner you die!" Even if this "theory" were true, what is wrong with the conclusion concerning exercise?

COVERING ALL YOUR BASES

Multiple Choice

Select the best answer or answers from the choices given.

1. A knife plunged into the sixth intercostal space will:
   A. pierce the aorta
   B. pierce the atria
   C. pierce the ventricles
   D. miss the heart completely

2. The innermost layer of the pericardial sac is the:
   A. epicardium
   B. fibrous pericardium
   C. parietal layer of the serous pericardium
   D. visceral layer of the serous pericardium

3. Which of the heart layers are vascularized?
   A. Myocardium
   B. Endothelium of endocardial layer
   C. Mesothelium of epicardium
   D. Connective tissue layers of both endocardium and epicardium

4. The fibrous skeleton of the heart:
   A. supports valves
   B. anchors vessels
   C. provides electrical insulation to separate the atrial mass from the ventricular mass
   D. anchors cardiac muscle fibers

5. Which of the following is associated with the right atrium?
   A. Fossa ovalis   C. Chordae tendineae
   B. Tricuspid valve   D. Pectinate muscle

6. A surface feature associated with the circumflex artery is the:
   A. right atrioventricular groove
   B. anterior interventricular sulcus
   C. left atrioventricular groove
   D. posterior interventricular sulcus

7. Atrioventricular valves are held closed by:
   A. papillary muscles
   B. trabeculae carneae
   C. pectinate muscles
   D. chordae tendineae

8. During atrial systole:
   A. the atrial pressure exceeds ventricular pressure
   B. 70% of ventricular filling occurs
   C. the AV valves are open
   D. valves prevent backflow into the great veins

9. Occlusion of which of the following arteries would damage primarily the right ventricle?
   A. Marginal artery
   B. Posterior interventricular artery
   C. Circumflex artery
   D. Anterior interventricular artery

10. A choking sensation in the chest is the result of:
    A. myocardial infarction
    B. death of cardiac cells
    C. temporary lack of oxygen to cardiac cells
    D. possibly spasms of the coronary arteries

11. Which feature(s) is/are greater in number/value/size or more developed in cardiac than in skeletal muscle?
    A. T tubule diameter
    B. Terminal cisternae
    C. Mitochondria
    D. Absolute refractory period

12. Which is/are characteristic of cardiac but not of skeletal muscle?
    A. Gap junctions
    B. All-or-none law
    C. Endomysium
    D. Branching fibers

13. Ca^{2+} slow channels are open in cardiac cells' plasma membranes during:
    A. depolarization
    B. repolarization
    C. plateau
    D. resting membrane potential

14. Which of the following depolarizes next after the AV node?
    A. Atrial myocardium
    B. Ventricular myocardium
    C. Bundle branches
    D. Purkinje fibers

15. Threshold in pacemaker cells is marked by:
    A. opening of Na^+ gates
    B. opening of Ca^{2+} slow channels
    C. opening of Ca^{2+} fast channels
    D. opening of K^+ gates

16. A heart rate of 30 bpm indicates that the _____ is functioning as an ectopic focus.
   A. AV node
   B. bundle branches
   C. Purkinje fibers
   D. AV bundle

17. Stimulation of the cardiac plexus involves stimulation of the:
   A. baroreceptors in the great arteries
   B. cardioacceleratory center
   C. vagus nerve
   D. sympathetic chain

18. Atrial repolarization coincides in time with the:
   A. P wave   C. QRS wave
   B. T wave   D. P-Q interval

19. Soon after the onset of ventricular systole the:
   A. AV valves close
   B. semilunar valves open
   C. first heart sound is heard
   D. aortic pressure increases

20. Which of the following will occur first after the T wave?
   A. Increase in ventricular volume
   B. Decrease in atrial volume
   C. Closure of the semilunar valves
   D. Decrease in ventricular pressure

21. Given an end-diastolic volume of 150 ml, an end-systolic volume of 50 ml, and a heart rate of 60 bpm, the cardiac output is:
   A. 600 ml/min   C. 1200 ml/min
   B. 6 liters/min   D. 3 liters/min

22. The statement "strength of contraction increases intrinsically due to increased stretching of the heart wall" is best attributed to:
   A. contractility
   B. Frank-Starling law of the heart
   C. Bainbridge reflex
   D. aortic sinus reflex

23. An increase in cardiac output is triggered by:
   A. Bainbridge reflex
   B. stimulation of carotid sinus baroreceptors
   C. increase in vagal tone
   D. epinephrine

24. Which of the following ionic imbalances inhibits Ca^{2+} transport into cardiac cells and blocks contraction?
   A. Hyponatremia   D. Hyperkalemia
   B. Hypokalemia   E. Hypercalcemia
   C. Hypernatremia

25. Cardiovascular conditioning results in:
   A. ventricular hypertrophy
   B. bradycardia
   C. increase in SV
   D. increase in CO

26. Conditions known to be associated with congestive heart failure include:
   A. coronary atherosclerosis
   B. diastolic pressure chronically elevated
   C. successive sublethal infarcts
   D. decompensated heart

27. In heart failure, venous return is slowed. Edema results because:
   A. blood volume is increased
   B. venous pressure is increased
   C. osmotic pressure is increased
   D. none of the above

28. Which structures is/are fetal remnant(s) associated with a healthy adult heart?
   A. Foramen ovale
   B. Ligamentum arteriosum
   C. Interventricular septal opening
   D. Patent ductus arteriosus

29. Age-related changes in heart function include:
   A. decrease in cardiac reserve
   B. decrease in resting heart rate
   C. valvular sclerosis
   D. fibrosis of cardiac muscle

30. Which of the following terms refers to an unusually strong heartbeat, so that the person is aware of it?

- **A.** Extrasystole
- **C.** Flutter
- **B.** Tamponade
- **D.** Palpitation

31. The thickest layer of the heart wall is:

- **A.** endocardium
- **C.** epicardium
- **B.** myocardium
- **D.** fibrous pericardium

32. Conditions associated with the tetralogy of Fallot include:

- **A.** interventricular septal opening
- **B.** cyanosis
- **C.** aorta arising from both ventricles
- **D.** treatment with drugs rather than surgery

Word Dissection

For each of the following word roots, fill in the literal meaning and give an example, using a word found in this chapter.

Word root	Translation	Example
1. angina		
2. baro		
3. brady		
4. carneo		
5. cusp		
6. diastol		
7. dicro		
8. ectop		
9. intercal		
10. pectin		
11. sino		
12. stenos		
13. systol		
14. tachy		

20 The Cardiovascular System: Blood Vessels

Student Objectives

When you have completed the exercises in this chapter, you will have accomplished the following objectives:

Overview of Blood Vessel Structure and Function

1. Describe the three layers that typically form the wall of a blood vessel and state the function of each.

2. Define *vasoconstriction* and *vasodilation*.

3. Compare and contrast the structure and function of the three types of arteries.

4. Describe the structure and function of veins and explain how veins differ from arteries.

5. Describe the structure and function of a capillary bed.

Physiology of Circulation

6. Define *blood flow, blood pressure,* and *resistance,* and explain the relationships between these factors.

7. List and explain the factors that influence blood pressure and describe how blood pressure is regulated.

8. Define *hypertension.* Note both its symptoms and its consequences.

9. Explain how blood flow is regulated in the body in general and in its specific organs.

10. Define *circulatory shock.* Note several possible causes.

11. Outline the factors involved in capillary dynamics and explain the significance of each.

Circulatory Pathways: Blood Vessels of the Body

12. Trace the pathway of blood through the pulmonary circuit and state the importance of this special circulation.

13. Describe the general functions of the systemic circuit. Name and give the location of the major arteries and veins in the systemic circulation.

14. Describe the structure and special function of the hepatic portal system.

Developmental Aspects of the Blood Vessels

15. Explain how blood vessels develop in the fetus.

16. Provide examples of changes that often occur in blood vessels as a person ages.

Ready, set, go!

After leaving the heart, the blood is sent through a vast network of vessels that ultimately reach each cell, supplying oxygen and nutrients and removing wastes. Blood flows through an equally extensive vascular network back to the heart. In the adult human, the length of these circulatory pathways is about 60,000 miles. These are not passive tubular structures but dynamic structures that rapidly alter the blood flow in response to changing internal and external conditions.

The focus of Chapter 20 is on the vascular portion of the cardiovascular system. Topics include blood vessel structure and function, the physiology of circulation and blood pressure, the dynamics of the capillary bed, the anatomy of the pulmonary and systemic vasculature, and aspects of the development and diseases of blood vessels.

BUILDING THE FRAMEWORK

Overview of Blood Vessel Structure and Function

1. Assume someone has been injured in an automobile accident and is bleeding profusely. What pressure points could you compress to help stop the bleeding from the following areas?

   _____ **1.** Thigh

   _____ **2.** Forearm

   _____ **3.** Calf

   _____ **4.** Lower jaw

   _____ **5.** Thumb

   _____ **6.** Plantar surface of foot

   _____ **7.** Temple

   _____ **8.** Ankle

2. Select different colors for each of the three blood vessel tunics listed in the key choices and illustrated in Figure 20.1. (*Note:* Elastic laminae are *not* illustrated.) Color each tunic on the three diagrams and correctly identify each vessel type. Identify valves if appropriate. Write your answers in the blanks beneath the illustrations. In the additional blanks, list the structural details that allowed you to make the identifications. Then, using the key choices, identify the blood vessel tunics described in each of the following cases by writing the correct answers in the answer blanks.

KEY CHOICES

○ **A.** Tunica intima   ○ **B.** Tunica media   ○ **C.** Tunica externa

_____A_____   **1.** Single thin layer of endothelium associated with scant connective tissue

_____B_____   **2.** Bulky middle coat, containing smooth muscle and elastin

_____A_____   **3.** Provides a smooth surface to decrease resistance to blood flow

_____A_____   **4.** The only tunic of capillaries

_____C_____   **5.** Called the adventitia in some cases

_____B_____   **6.** The only tunic that plays an active role in blood pressure regulation

_____C_____   **7.** Supporting, protective coat

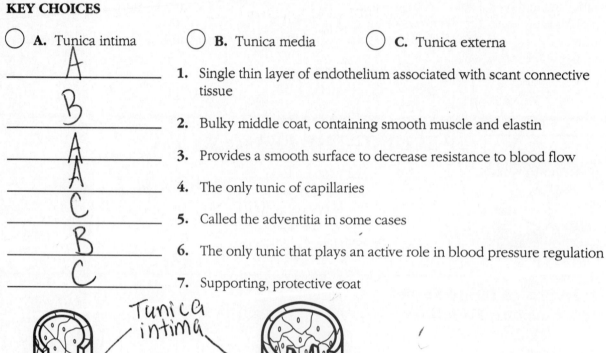

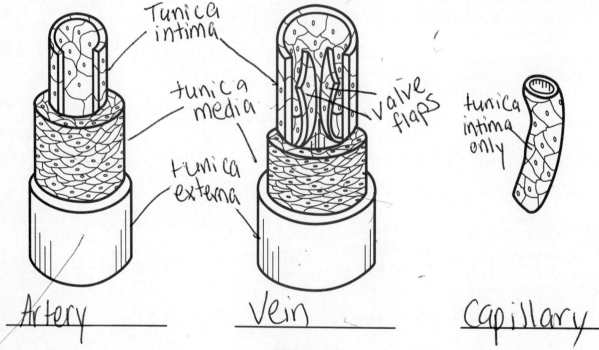

_____Artery_____   _____Vein_____   _____Capillary_____

Figure 20.1

A. artery;
thick media
small round lumen

B. vein
thin media
relatively lger lumen
valves-

C. capillary
single layer of
epithelium

3. Match the terms pertaining to blood vessels in Column B to the appropriate descriptions in Column A.

Column A   Column B

_____ A _____ **1.** Transport blood away from heart   **A.** Arteries

_____ E _____ **2.** Largest arteries, low resistance   **B.** Arterioles

_____ G _____ **3.** Arteries with thickest tunica media; active   **C.** Capillaries
              in vasoconstriction
                                                                 ~~**D.** Continuous~~
_____ B _____ **4.** Control blood flow into individual capillary
              beds   **E.** Elastic

_____ C _____ **5.** Lumen is the size of red blood cells   **F.** Fenestrated

_____ **6.** Capillary type with an uninterrupted lining   **G.** Muscular

_____ **7.** Capillary type with numerous pores and   **H.** Sinusoids
              gap junctions
                                                                 I. Veins
_____ **8.** Capillaries with large intercellular clefts and
              irregular lumen   **J.** Venous sinuses

_____ K _____ **9.** Vessels formed when capillaries unite   **K.** Venules

_____ I _____ **10.** Vessels with thin walls and large lumens
              that often appear collapsed in histologic
              preparations

_____ J _____ **11.** Veins with only a tunica intima; supported
              by surrounding tissues

4. What is the importance of arterial anastomoses?

_____

_____

_____

5. Figure 20.2 shows diagrammatic views of the three types of capillaries. Below each diagram, write in the type of capillary shown. Then, identify all elements and structures that have leader lines. Finally, color code and color the following structures:

⃝ Basal lamina   ⃝ Endothelial cell(s)   ⃝ Macrophage location (if present)

⃝ Capillary lumen   ⃝ Erythrocyte(s)

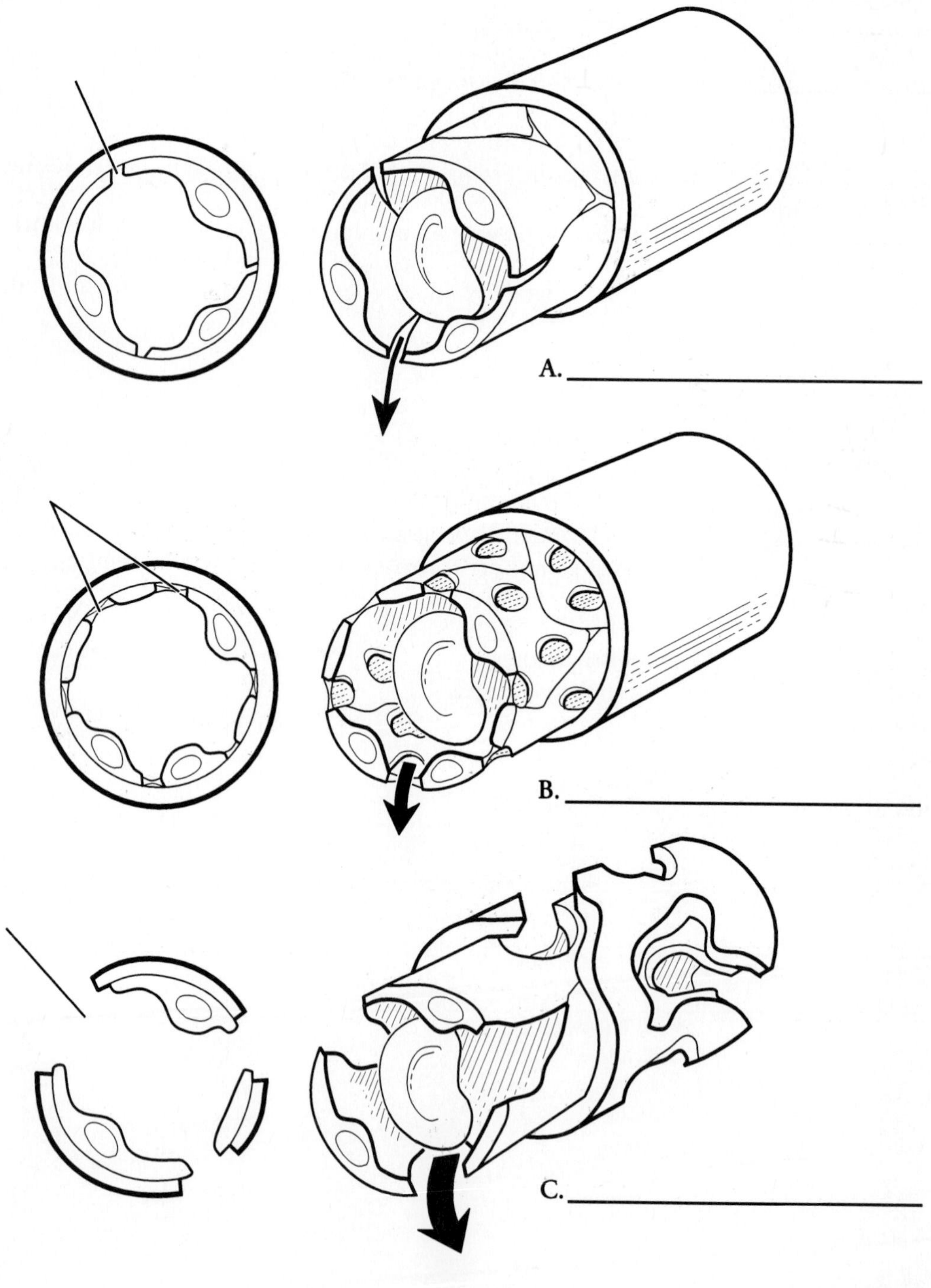

A. _____

B. _____

C. _____

Figure 20.2

6. Circle the term that does not belong in each of the following groupings.

1. Elastic   Conducting   Continuous flow   Muscular   Pulse

2. Distributing   Vasoconstriction   Muscular   Most arteries   Pressure points

3. Cornea   Heart   Cartilage   Epithelium   Tendons

4. Pinocytotic vesicles   Intercellular clefts   Gap junctions   Tight junctions

Continuous capillaries

5. Liver   Bone marrow   Kidney   Lymphoid tissue   Sinusoids

6. High pressure   Veins   Capacitance vessels   Valves

Thick tunica adventitia   Blood reservoirs

7. Anastomoses   End arteries   Collateral channels   Capillary beds

Venous   Arterial

8. Valves   Large lumen   Thick media   Collapsed lumen   Veins

9. Metarteriole   Anastomosis   Thoroughfare channel   Shunt   True capillaries

7. Blood flows from high to low pressure. Hence, it flows from the high-pressure arteries through the capillaries and then through the low-pressure veins. Because blood pressure contributes less to blood propulsion in veins, special measures are required to ensure that venous return equals cardiac output. What role do the venous valves play?

The venous valves prevent backflow of blood

8. Briefly explain why veins are called blood reservoirs and state where in the body venous blood reservoirs are most abundant.

_____

_____

_____

_____

Physiology of Circulation

1. Fill in the blanks in the flowcharts below indicating the relationships between the terms provided. The position of some elements in these schemes has already been indicated.

1. difference in blood pressure   resistance   atherosclerosis

    blood viscosity   vessel length   hematocrit

    vasoconstriction

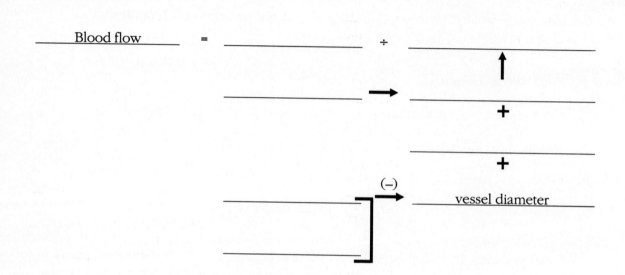

2. cardiac output   peripheral resistance   polycythemia

    blood volume   heart rate   excessive salt intake

    vasoconstriction

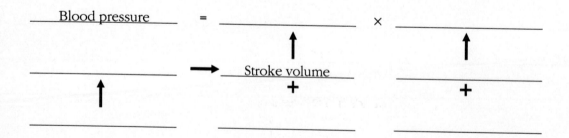

2. Briefly explain why blood flow in large, thick-walled arteries, such as the aorta and its branches, is fairly continuous and does not stop when the heart relaxes.

_____

_____

_____

_____

3. This exercise concerns blood pressure and pulse. Match the items in Column B with the appropriate descriptions in Column A.

Column A

_____ 1. Expansion and recoil of an artery during heart activity

_____ 2. Pressure exerted by the blood against the blood vessel walls

_____ 3. The product of these factors

_____ 4. yields blood pressure (#3–#4)

_____ 5. Event primarily responsible for peripheral resistance

_____ 6. Blood pressure during heart contraction

_____ 7. Blood pressure during heart relaxation

_____ 8. Site where blood pressure determinations are normally made

_____ 9. Points at the body surface where the pulse may be felt

_____ 10. Sounds heard over a blood vessel when the vessel is partially compressed

Column B

A. Over arteries

B. Blood pressure

C. Cardiac output

D. Constriction of arterioles

E. Diastolic blood pressure

F. Peripheral resistance

G. Pressure points

H. Pulse

I. Sounds of Korotkoff

J. Systolic blood pressure

K. Over veins

4. What effects do the following factors have on blood pressure? Use I to indicate an increase in pressure and D to indicate a decrease in pressure.

D **1.** Increased diameter of the arterioles

I **2.** Increased blood viscosity

I **3.** Increased cardiac output

I **4.** Increased pulse rate

I **5.** Anxiety, fear

D **6.** Increased urine output

D **7.** Sudden change in position from reclining to standing

I **8.** Physical exercise

D **9.** Physical training

D **10.** Alcohol

D **11.** Hemorrhage

I **12.** Nicotine

I **13.** Arteriosclerosis

D **14.** Stimulation of arterial baroreceptors

I **15.** Stimulation of carotid body chemoreceptors

I **16.** Release of epinephrine from adrenal medulla

D **17.** Secretion of atrial natriuretic factor

I **18.** Secretion of antidiuretic hormone

I **19.** Release of endothelin

D **20.** Secretion of NO

I **21.** Renin/angiotensin mechanism

I **22.** Secretion of aldosterone

5. Circle the term that does not belong in each of the following groupings.

1. Blood pressure   (mm Hg)   Cardiac output   Force of blood on vessel wall

2. Peripheral resistance   Friction   Arterioles   Blood flow

3. Low viscosity   High viscosity   Blood   Resistance to flow

4. Blood viscosity   Blood pressure   Vessel length   Vessel diameter

5. High blood pressure   Hemorrhage   Weak pulse   Low cardiac output

6. Resistance   Friction   Vasodilation   Vasoconstriction

7. High pressure   Vein   Artery   Spurting blood

8. Elastic arteries   Diastolic pressure   120 mm Hg   Auxiliary pumps

9. Cardiac cycle   Intermittent blood flow   Capillary bed   Precapillary sphincters

10. Muscular pump   Respiratory pump   Inactivity   Venous return

11. Hypotension   Sympathetic activity   Poor nutrition   Training

6. If a statement is true, write the letter T in the answer blank. If a statement is false, change the underlined word(s) and write the correct word(s) in the answer blank.

_____   **1.** Renin, released by the kidneys, causes a <u>decrease</u> in blood pressure.

_____   **2.** The decreasing efficiency of the sympathetic nervous system vasoconstrictor functioning due to aging leads to a type of hypotension called <u>sympathetic</u> hypotension.

_____   **3.** Two body organs in which vasoconstriction rarely occurs are the heart and the <u>kidneys</u>.

_____   **4.** A <u>sphygmomanometer</u> is used to take the apical pulse.

_____   **5.** The pulmonary circuit is a <u>high</u>-pressure circulation.

_____   **6.** Cold has a <u>vasodilating</u> effect.

_____   **7.** <u>Thrombophlebitis</u> is called the silent killer.

_____   **8.** The nervous system controls blood pressure and blood distribution by altering the diameters of the <u>venules</u>.

_____   **9.** The vasomotor center for blood pressure control is located in the <u>left atrium</u>.

_____   **10.** Pressoreceptors in the <u>large arteries</u> of the neck and thorax detect changes in blood pressure.

_____   **11.** ANF, the hormone produced by the atria, causes a <u>rise</u> in blood volume and blood pressure.

_____   **12.** The hormone vasopressin causes intense <u>vasodilation</u>.

_____   **13.** Hypertension in obese people can be promoted by increased <u>viscosity of the blood</u>.

_____   **14.** The velocity of blood flow is slowest in the <u>veins</u>.

_____   **15.** The total cross-sectional area of the vascular bed is least in the <u>capillary bed</u>.

_____   **16.** <u>Hypertension</u> is the local adjustment of blood flow to a given tissue at any particular time.

_____   **17.** In active skeletal muscles, autoregulated vasodilation is promoted by an increase of <u>acetylcholine</u> in the area.

7. Figure 20.3 is a diagram of a capillary bed. Arrows indicate the direction of blood flow. Select five different colors and color the coding circles and their structures on the figure. Then answer the questions that follow by referring to Figure 20.3. Notice that Questions 6 –14 concern fluid flows at capillary beds and the forces (hydrostatic and osmotic pressures) that promote such fluid shifts.

○ Arteriole   ○ Thoroughfare channel   ○ Postcapillary venule

○ Precapillary sphincters ○ True capillaries   ○ Metarteriole

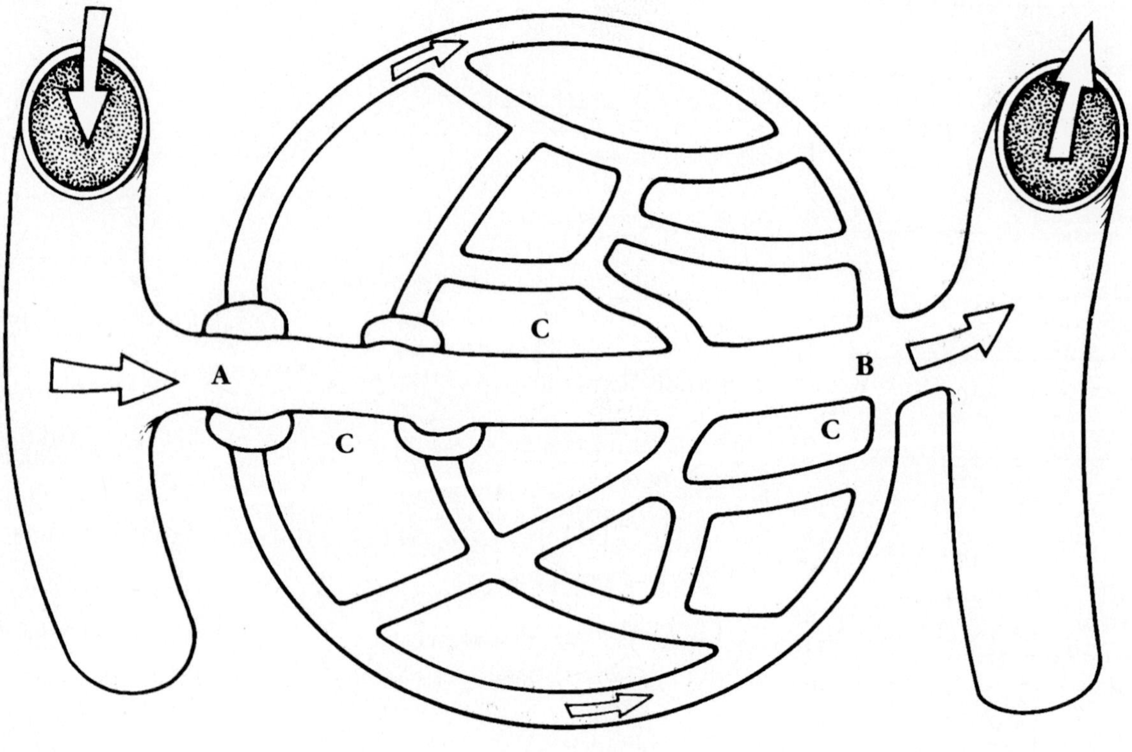

Figure 20.3

1. What is the liquid that surrounds tissue cells called?

___interstitial fluid_____

2. What drives the movement of soluble substances between tissue fluids and the blood?

___concentration gradient_____

3. Which substances pass readily through the endothelial cell plasma membrane?

fat soluble substances like fats + gases

4. Which substances move through fluid-filled capillary clefts?

water + water soluble substances like sugars + amino acids

5. If the precapillary sphincters are contracted, by which route will the blood flow?

   _____

6. Under normal conditions, in which area does hydrostatic pressure predominate,

   A, B, or C? _____

7. Which area has the highest osmotic pressure? _____

8. Which pressure is in excess and causes fluids to move from A to C? (Be specific as to whether the force exists in the capillary or the interstitial space.)

   _____

9. Which pressure causes fluid to move from A to B? _____

10. Which pressure causes fluid to move from C to B? _____

11. Which blood protein is most responsible for osmotic pressure? _____

12. Where does the greater net flow of water out of the capillary occur? _____

13. If excess fluid does not return to the capillary, where does it go?

    _____

14. Which pressure in excess at B, as in congestive heart failure, may cause tissue edema?

    _____

8. Briefly compare the characteristics, causes, and effects of hypovolemic shock and vascular shock.

   _____

   _____

   _____

   _____

   _____

9. Choose the vessel type (arteries, capillaries, veins) with the indicated characteristic.

_____ **1.** Highest total cross-sectional area

_____ **2.** Highest velocity of blood flow

_____ **3.** Lowest velocity of blood flow

_____ **4.** Pulse pressure

_____ **5.** Lowest blood pressure

10. Using the key choices, identify the special circulations described below.

KEY CHOICES

A. Cerebral   **C.** Hepatic   **E.** Skeletal muscle

B. Coronary   **D.** Pulmonary   **F.** Skin

F _____ **1.** Blood flow increases markedly when the body temperature rises

A _____ **2.** The major autoregulatory stimulus is a drop in pH

D _____ **3.** Arteries characteristically have thin walls and large lumens

B _____ **4.** Vessels do not constrict but are compressed during systole

A _____ **5.** Receives constant blood flow whether the body is at rest or strenuously exercising

D _____ **6.** Vasodilation promoted by high oxygen levels

C _____ **7.** Capillary flow markedly sluggish; phagocytes present

E _____ **8.** Prolonged activity places extreme demands on cardiovascular system

B _____ **9.** Additional oxygen can be supplied only by increased blood flow

D _____ **10.** Much lower arterial pressure than that in systemic circulation

C _____ **11.** Large, atypical capillaries with fenestrations

A _____ **12.** Impermeable tight junctions in capillary endothelium

A _____ **13.** One of the most precise autoregulatory systems in the body

A _____ **14.** Venous blood empties into large dural sinuses rather than into veins

F _____ **15.** Abundance of superficial veins

E _____ **16.** During vigorous physical activity, receives up to two-thirds of total blood flow

E _____ **17.** Arterioles have receptors for both acetylcholine and epinephrine

Circulatory Pathways: Blood Vessels of the Body

1. Figure 20.4 shows the pulmonary circuit. Identify all vessels that have leader lines. Color the vessels (and heart chambers) transporting oxygen-rich blood *red*; color those transporting carbon dioxide–rich blood *blue*.

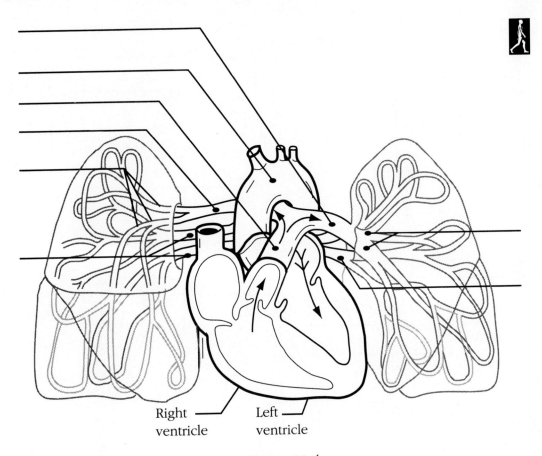

Right
ventricle

Left
ventricle

Figure 20.4

2. Figures 20.5 and 20.6 illustrate the locations of the major systemic arteries and veins of the body. These figures are highly simplified and will serve as a "warm-up" for the more detailed vascular diagrams to come. The arteries are shown in Figure 20.5. Color the arteries red, then identify those indicated by leader lines on the figure. The veins are shown in Figure 20.6. Color the veins blue, then identify each vein that has a leader line on the figure. Or, if you wish, color the individual vessels with different colors to help you to identify their extent.

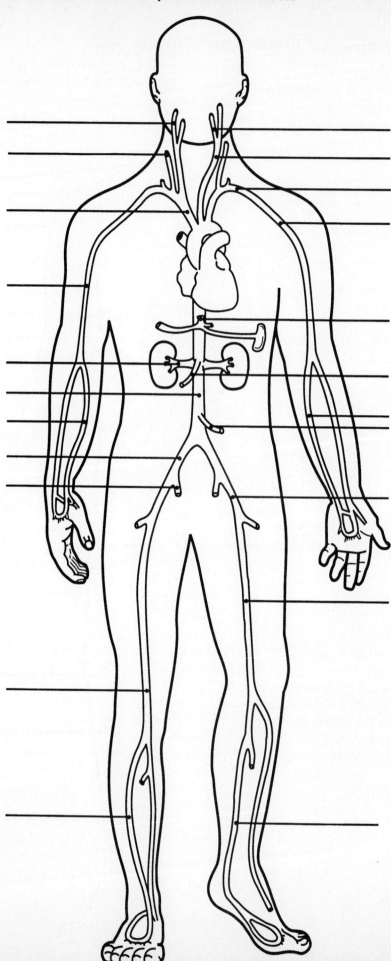

Figure 20.5
Arteries

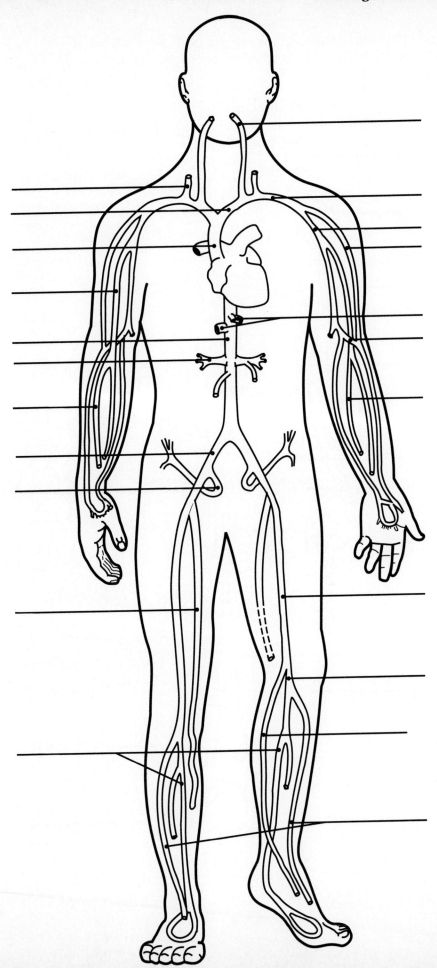

Figure 20.6
Veins

3. Figure 20.7 shows the major arteries of the head and neck. Note that the clavicle is omitted and that dashed lines represent deeper vessels. Color code and color the following vessels.

○ Brachiocephalic   ○ Internal thoracic   ○ Right common carotid

○ Costocervical trunk   ○ Lingual   ○ Right subclavian

○ External carotid   ○ Maxillary   ○ Superior thyroid

○ Facial   ○ Occipital   ○ Thyrocervical trunk

○ Internal carotid   ○ Ophthalmic   ○ Vertebral

Transverse
process of
cervical vertebra

Hyoid bone

Larynx

Thyroid gland

1st rib

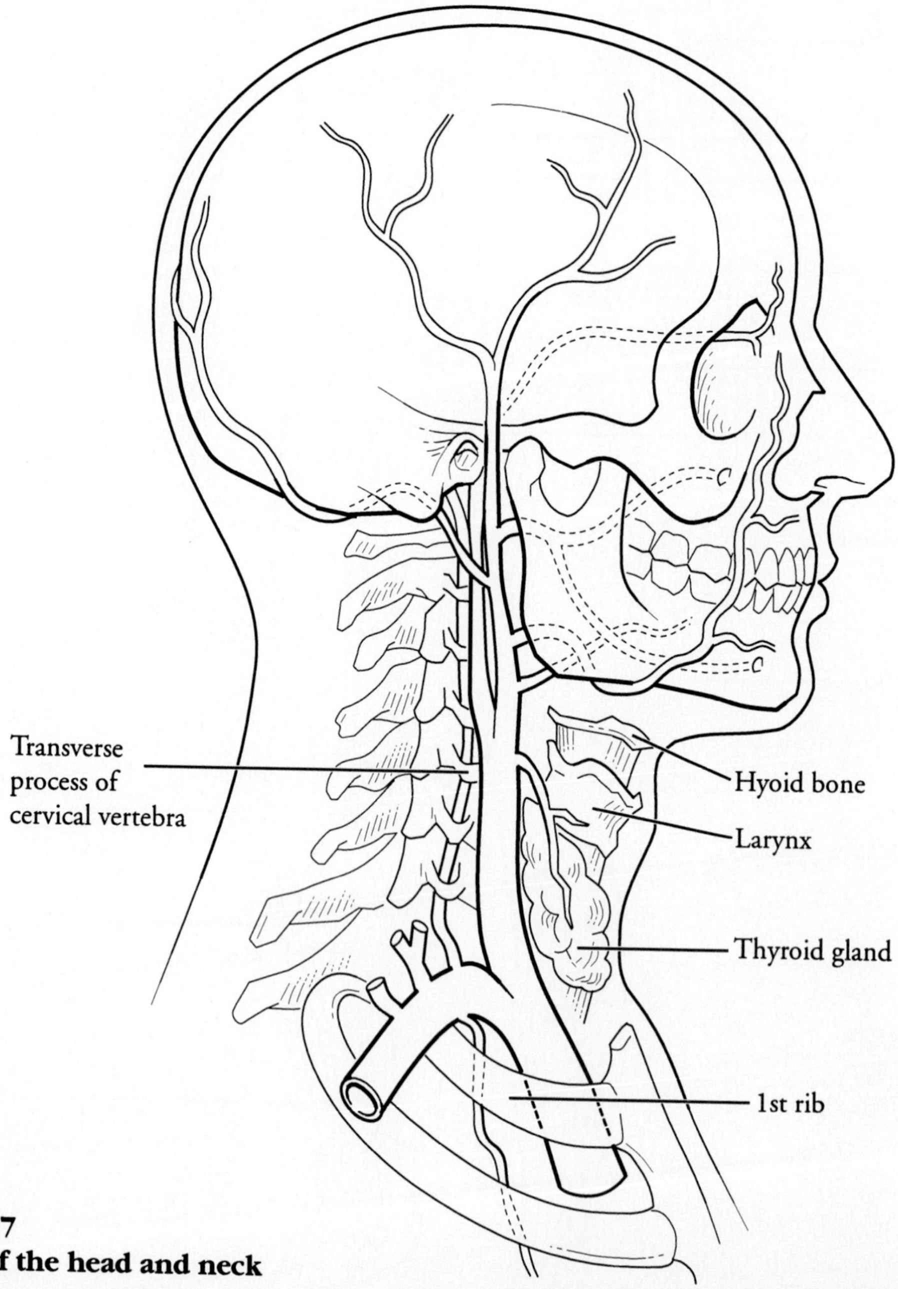

Figure 20.7
Arteries of the head and neck

4. Figure 20.8 illustrates the arterial circulation of the brain. Select different colors for the following structures, and color the diagram. Then respond to the questions following the diagram by choosing responses from the structures to be identified.

    ◯ **A.** Basilar artery   ◯ **D.** Middle cerebral arteries

    ◯ **B.** Communicating arteries   ◯ **E.** Posterior cerebral arteries

    ◯ **C.** Anterior cerebral arteries   **F.** Circle of Willis

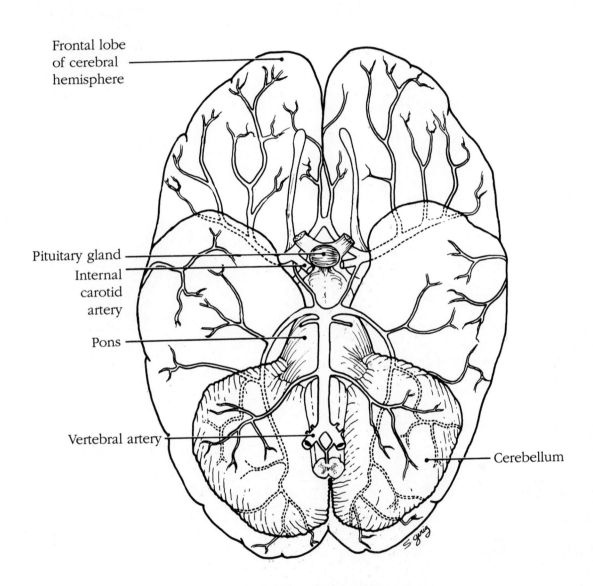

Figure 20.8

____F____   **1.** What is the name of the anastomosis that allows communication between the posterior and anterior blood supplies of the brain?

____C, D____   **2.** What two pairs of arteries arise from the internal carotid artery?

____A____   **3.** The posterior cerebral arteries serving the brain arise from what artery?

5. Using the key choices, identify the *arteries* in the following descriptions.

KEY CHOICES

A. Anterior tibial	**H.** Coronary	**O.** Intercostals	**V.** Renal
B. Aorta	**I.** Deep femoral	**P.** Internal carotid	**W.** Subclavian
C. Brachial	**J.** Dorsalis pedis	**Q.** Internal iliac	**X.** Superior mesenteric
D. Brachiocephalic	**K.** External carotid	**R.** Peroneal	**Y.** Vertebral
E. Celiac trunk	**L.** Femoral	**S.** Phrenic	**Z.** Ulnar
F. Common carotid	**M.** Hepatic	**T.** Posterior tibial	
G. Common iliac	**N.** Inferior mesenteric	**U.** Radial	

_____ **1.** _____ **2.** Two arteries formed by the division of the brachiocephalic artery

_____ **3.** First branches off the ascending aorta; serve the heart

_____ **4.** _____ **5.** Two paired arteries serving the brain

_____ **6.** Largest artery of the body

_____ **7.** Arterial network on the dorsum of the foot

_____ **8.** Serves the posterior thigh

_____ **9.** Supplies the diaphragm

_____ **10.** Splits to form the radial and ulnar arteries

_____ **11.** Auscultated to determine blood pressure in the arm

_____ **12.** Supplies the last half of the large intestine

_____ **13.** Serves the pelvis

_____ **14.** External iliac becomes this artery on entering the thigh

_____ **15.** Major artery serving the arm

_____ **16.** Supplies the small intestine and part of the large intestine

_____ **17.** Terminal branches of the dorsal, or descending, aorta

_____ **18.** Arterial trunk that has three major branches, which serve the liver, spleen, and stomach

_____ **19.** Major artery serving the tissues external to the skull

_____ **20.** _____ **21.** _____ **22.** Three arteries serving the leg

_____ **23.** Artery generally used to feel the pulse at the wrist

6. Figure 20.9 shows the venous drainage of the head. Color code and color each of the drainage veins individually. Label each of the dural venous sinuses that has a leader line, but color all the dural sinuses yellow. Note that the clavicle has been omitted.

Dural venous sinuses

Cavernous sinus
Inferior sagittal sinus
Straight sinus
Superior sagittal sinus
Transverse sinus

Drainage veins

◯ Brachiocephalic   ◯ Ophthalmic

◯ External jugular   ◯ Subclavian

◯ Facial   ◯ Superficial temporal

◯ Internal jugular   ◯ Superior thyroid

◯ Middle thyroid   ◯ Vertebral

Sinuses:

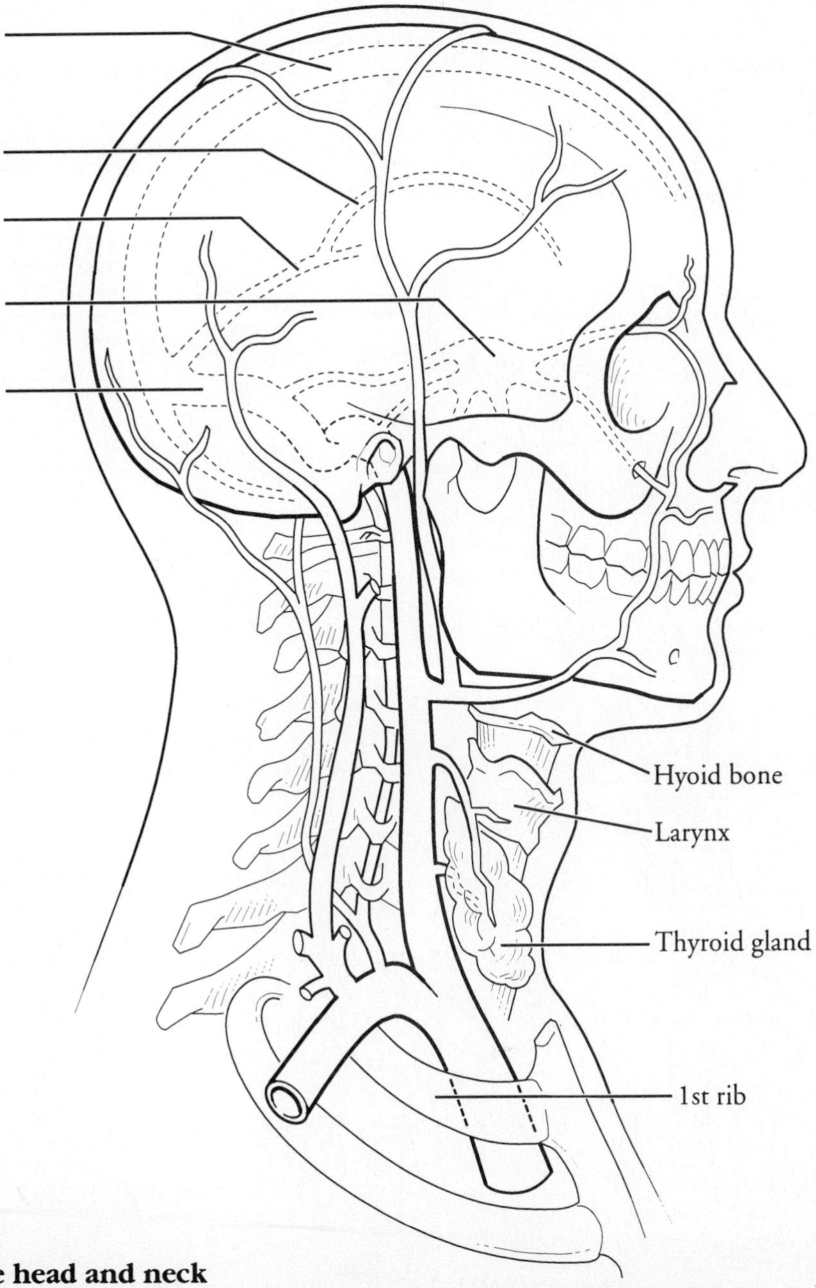

Hyoid bone
Larynx
Thyroid gland
1st rib

Figure 20.9
Veins of the head and neck

7. Figure 20.10 shows the arteries and veins of the upper limb. Using the key choices, identify each vessel provided with a leader line. (*Note:* In many cases, the same term will be used to identify both an artery and a deep vein.)

KEY CHOICES

A. Axillary

B. Basilic

C. Brachial

D. Brachiocephalic

E. Cephalic

F. Circumflex (anterior humeral)

G. Common interosseous

H. Costocervical trunk

I. Deep brachial

J. Deep palmar arch

K. Digital

L. Median cubital

M. Lateral thoracic

N. Radial

O. Subclavian

P. Subscapular

Q. Superficial palmar arch

R. Thoracoacromial trunk

S. Thoracocervical trunk

T. Thoracodorsal

U. Ulnar

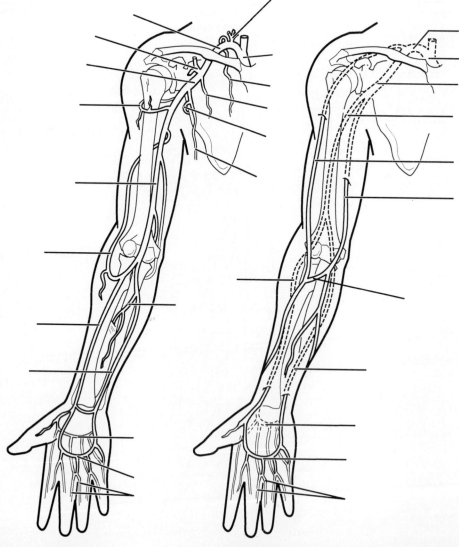

Figure 20.10   A. Arteries   B. Veins

8. The abdominal vasculature is depicted in Figure 20.11. Using the key choices, identify the following vessels by selecting the correct letters. Color the diagram as you wish.

KEY CHOICES

A. Aorta

B. Celiac trunk

C. Common iliac arteries

D. Gonadal arteries

E. Hepatic veins

F. Inferior mesenteric artery

G. Inferior vena cava

H. Lumbar arteries

I. Median sacral artery

J. Superior mesenteric artery

K. Renal arteries

L. Renal veins

M. Left gonadal vein

N. Right gonadal vein

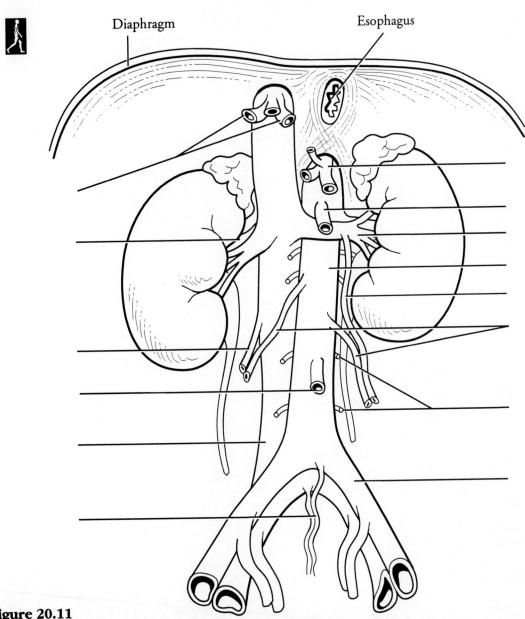

Diaphragm

Esophagus

Figure 20.11

9. On Figure 20.12, identify the thoracic veins by choosing a letter from the key choices.

KEY CHOICES

A. Accessory hemiazygos   **E.** External jugular   **I.** Left subclavian

B. Axillary   **F.** Inferior vena cava   **J.** Posterior intercostals

C. Azygos   **G.** Hemiazygos   **K.** Right subclavian

D. Brachiocephalic   **H.** Internal jugular   **L.** Superior vena cava

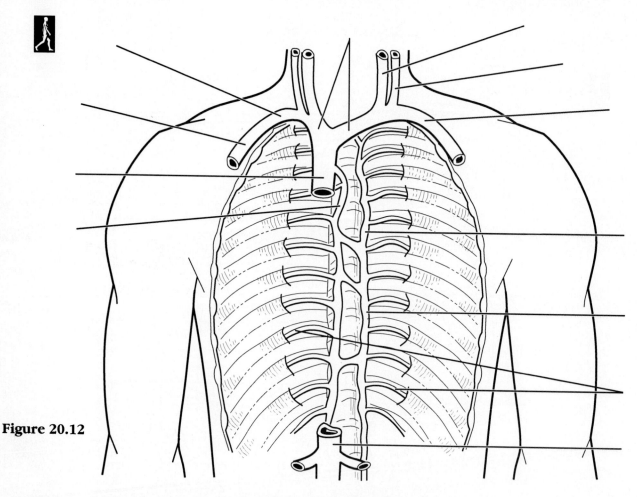

Figure 20.12

10. Figure 20.13, on the opposite page, illustrates the arterial supply of the right lower limb. Correctly identify all arteries that have leader lines by using letters from the key choices. *Notice that the key choices continue onto the opposite page.*

KEY CHOICES

A. Abdominal aorta   **C.** Common iliac   **E.** Digital   **G.** External iliac

B. Anterior tibial   **D.** Deep femoral   **F.** Dorsalis pedis   **H.** Femoral

I. Internal iliac

J. Lateral femoral circumflex

K. Lateral plantar artery

L. Medial plantar artery

M. Peroneal

N. Plantar arch

O. Popliteal

P. Posterior tibial

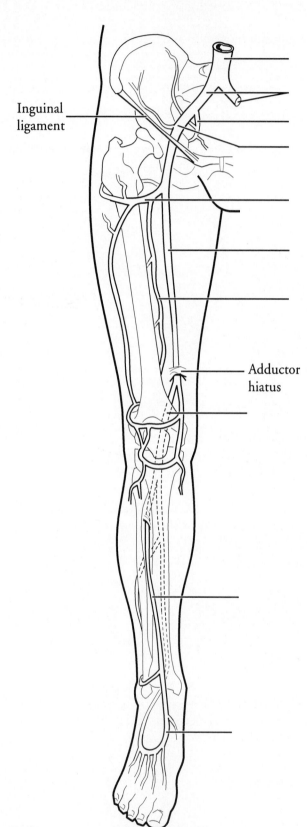

Inguinal
ligament

Adductor
hiatus

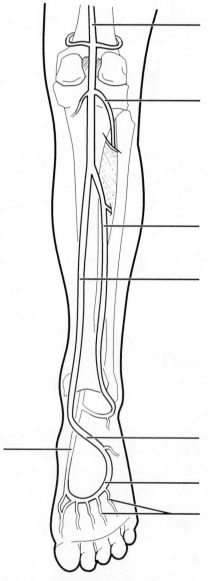

Figure 20.13

**A. Arteries of the pelvis,
thigh, and leg (anterior view)**

**B. Arteries of the leg
(posterior view)**

11. Figure 20.14 is a diagram of the hepatic portal circulation. Select different
colors for the structures listed below and color the structures on the illustration.

○ Inferior mesenteric vein   ○ Superior mesenteric vein

○ Splenic vein   ○ Gastric vein

○ Hepatic portal vein

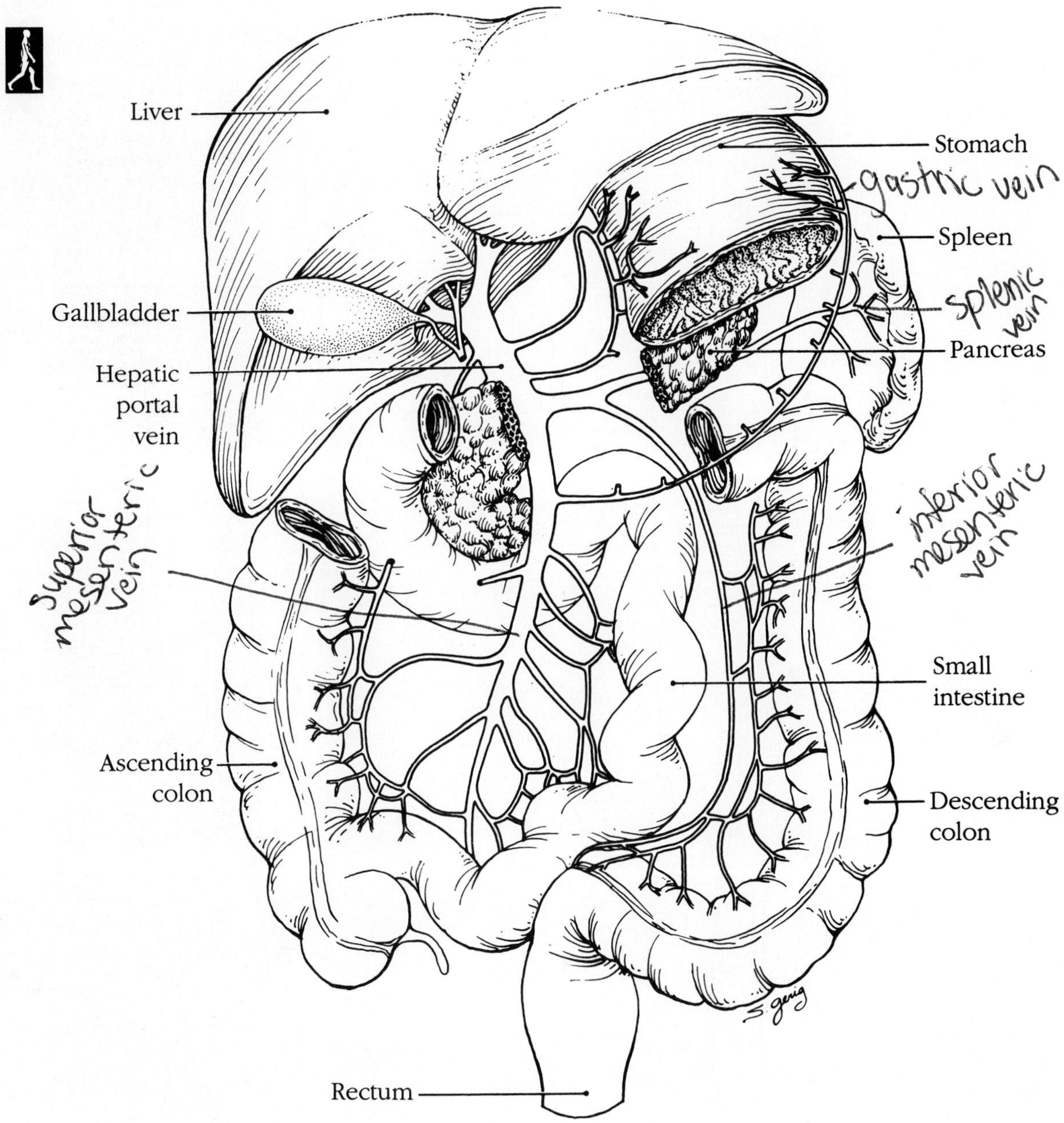

Liver

Stomach

gastric vein

Spleen

Splenic vein

Gallbladder

Pancreas

Hepatic
portal
vein

Superior
mesenteric
vein

inferior
mesenteric
vein

Small
intestine

Ascending
colon

Descending
colon

Rectum

Figure 20.14

12. Using the key choices, identify the veins in the following descriptions.

KEY CHOICES

A. Anterior tibial	**I.** Gastric	**Q.** Internal jugular
B. Azygos	**J.** Gonadal	**R.** Posterior tibial
C. Basilic	**K.** Great saphenous	**S.** Radial
D. Brachiocephalic	**L.** Hepatic	**T.** Renal
E. Cardiac	**M.** Hepatic portal	**U.** Subclavian
F. Cephalic	**N.** Inferior mesenteric	**V.** Superior mesenteric
G. Common iliac	**O.** Inferior vena cava	**W.** Superior vena cava
H. Femoral	**P.** Internal iliac	**X.** Ulnar

_____ **1.** _____ **2.** Deep veins; drain the forearm

_____ **3.** Receives blood from the arm via the axillary vein

_____ **4.** Drains venous blood from the myocardium of the heart into the coronary sinus

_____ **5.** Drains the kidney

_____ **6.** Drains the dural sinuses of the brain

_____ **7.** Join to become the superior vena cava (2)

_____ **8.** _____ **9.** Drain the leg and foot

_____ **10.** Carries nutrient-rich blood from the digestive organs to the liver for processing

_____ **11.** Superficial vein that drains the lateral aspect of the arm

_____ **12.** Drains the ovaries or testes

_____ **13.** Drains the thorax, empties into the superior vena cava

_____ **14.** Largest vein inferior to the thorax

_____ **15.** Drains the liver

_____ **16.** _____ **17.** _____ **18.** Three veins that form/empty into the hepatic portal vein

_____ **19.** Longest superficial vein of the body

_____ **20.** Formed by the union of the external and internal iliac veins

_____ **21.** Deep vein of the thigh

Developmental Aspects of the Blood Vessels

1. Complete the following statements by writing the missing terms in the answer blanks.

_____ 1.

_____ 2.

_____ 3.

_____ 4.

_____ 5.

_____ 6.

_____ 7.

_____ 8.

_____ 9.

_____ 10.

_____ 11.

_____ 12.

_____ 13.

_____ 14.

_____ 15.

_____ 16.

_____ 17.

In the early embryo, primitive cells arising from __(1)__ form the endothelial lining of blood vessels. These cells collect in little masses called __(2)__ throughout the embryo. Reaching toward one another and the heart, they lay down the first circulatory pathways. Meanwhile, the early endothelial tubes are surrounded by __(3)__ cells, which form the muscular and fibrous layers of vessel walls. By the fourth week, the cardiovascular system is functioning. The ductus arteriosus and foramen ovale allow the blood to bypass the nonfunctioning fetal __(4)__ . Another fetal structure, the __(5)__ , allows most of the blood to bypass the liver. The fetus is supplied with oxygen and nutrients via the __(6)__ , which carries blood from the __(7)__ to the __(8)__ . Metabolic wastes and carbon dioxide are removed from the fetus in blood carried by the __(9)__ . In newborns, the blood pressure is __(10)__ than the blood pressure in children and adults.

__(11)__ is a degenerative process that begins in youth but may take its toll in later life by promoting a myocardial infarct or stroke. A vascular problem that affects many in "standing professions" is __(12)__ , in which incompetent valves cause enlargement and twisting of vessels, particularly those in the __(13)__ . The most important precipitating factor in sudden cardiovascular death in middle-aged men is __(14)__ . Three precautions that can impede disease of the cardiovascular system are __(15)__ , __(16)__ , and __(17)__ .

The Incredible Journey:
A Visualization Exercise
for the Circulatory System

All about you are huge white cords, hanging limply from two
flaps of endothelial tissue . . .

1. Complete the following narrative by writing the missing terms in the answer blanks.

_____ 1.

_____ 2.

_____ 3.

_____ 4.

_____ 5.

_____ 6.

_____ 7.

_____ 8.

_____ 9.

_____ 10.

_____ 11.

_____ 12.

Your journey starts in the pulmonary vein and includes a trip to part of the systemic circulation and a special circulation. You ready your equipment and prepare to be injected into your host.

Almost immediately after injection, you find yourself swept into a good-sized chamber the __(1)__ . However, you do not stop in this chamber but continue to plunge downward into a larger chamber below. You land with a large splash and examine your surroundings. All about you are huge white cords, hanging limply from two flaps of endothelial tissue far above you. You report that you are sitting in the __(2)__ chamber of the heart, seeing the flaps of the __(3)__ valve above you. The valve is open, and its anchoring cords, the __(4)__, are lax. Since this valve is open, you conclude that the heart is in the __(5)__ phase of the cardiac cycle.

Suddenly you notice that the chamber walls seem to be closing in. You hear a thundering boom, and the whole chamber vibrates as the valve slams shut above you. The cords, now rigid and strained, form a cage about you, and you feel extreme external pressure. Obviously, the heart is in a full-fledged __(6)__ . Then, high above on the right, the "roof" opens, and you are forced through this __(7)__ valve. A fraction of a second later, you hear another tremendous boom that sends shock waves through the whole area. Out of the corner of your eye, you see that the valve below you is closed, and it looks rather like a pie cut into three wedges.

As you are swept along in a huge artery, the __(8)__, you pass several branch-off points, but continue to career along straight down at a dizzying speed until you approach the __(9)__ artery feeding the small intestine. After entering this artery and passing through successively smaller and smaller subdivisions of it, you finally reach the capillary bed of the small intestine. You watch with fascination as nutrient molecules move into the blood through the single layer of __(10)__ cells forming the capillary wall. As you move to the opposite shore of the capillary bed, you enter a venule and begin to move superiorly once again. The venules draining the small intestine combine to form the __(11)__ vein, which in turn combines with the __(12)__ vein to form the hepatic portal vein that carries you into the

_____ 13.

_____ 14.

_____ 15.

_____ 16.

_____ 17.

_____ 18.

_____ 19.

_____ 20.

_____ 21.

_____ 22.

liver. As you enter the liver, you are amazed at the activity there. Six-sided hepatic cells, responsible for storing glucose and making blood proteins, are literally grabbing __(13)__ out of the blood as it percolates slowly past them. Protective __(14)__ cells are removing bacteria from the slowly moving blood. Leaving the liver through the __(15)__ vein, you almost immediately enter the huge __(16)__, which returns blood from the lower part of the body to the __(17)__ of the heart. From here, you move consecutively through the right chambers of the heart into the __(18)__ artery, which carries you to the __(19)__ .

You report that your surroundings are now relatively peaceful. You catch hold of a red blood cell and scramble aboard. As you sit cross-legged on its concave surface, you muse that it is just like riding in a big, soft rubber boat. Riding smoothly along, you enter vessels that become narrower and narrower. Your red blood cell slows down as it squeezes through the windings of tiny blood vessels. You conclude that you are located in the pulmonary __(20)__ . You report hearing great whooshes of sound as your host breathes, and you can actually see large air-filled spaces! But your little red cell "boat" seems strangely agitated. Its contours change so often that you are almost thrown out. You report that these contortions must be the result of the exchange of __(21)__ molecules because your "boat" is assuming a brilliant red color. This experience has left you breathless, so you poke your head between two thin endothelial cells for some fresh air. You then continue your journey, through increasingly wider vessels. After traveling through the left side of the heart again, you leave your host when you are aspirated out of the __(22)__ artery, which extends from the aorta to the axillary artery of the armpit.

CHALLENGING YOURSELF

At the Clinic

1. Mrs. Gray, age 50, is complaining of dull aching pains in her legs, which she claims have been getting progressively worse since the birth of her last child. (She is the mother of seven children.) During her physical examination, numerous varicosities are seen in both legs. How are varicosities recognized? What veins are most likely involved? What pathologic changes have occurred in these veins and what is the most likely causative factor in this patient's case? What instructions might be helpful to Mrs. Gray?

2. A routine scan on an elderly man reveals partial occlusion of the internal carotid artery, yet blood supply to his cerebrum is unimpaired. What are two possible causes of the occlusion? What compensatory mechanism is maintaining blood supply to the brain?

3. A patient with a bone marrow cancer is polycythemic. Will his blood pressure be high or low? Why?

4. Mrs. Baevich, an elderly, bedridden woman, has several "angry-looking" decubitus ulcers. During the past week, one of these has begun to hemorrhage very slowly. What will occur eventually if this situation is not properly attended to?

5. A blow to the base of the skull has sent an accident victim to the emergency room. Although there are no signs of hemorrhage, his blood pressure is critically low. What is a probable cause of his hypotension?

6. Mr. Grimaldi was previously diagnosed as having a posterior pituitary tumor that causes hypersecretion of ADH. He comes to the clinic regularly to have his blood pressure checked. Would you expect his BP to be chronically elevated or depressed? Why?

7. Renal (kidney) disease in a young man has resulted in blockage of the smaller arteries within the kidneys. Explain why this will lead to secondary hypertension.

8. A man in his 40s was diagnosed as hypertensive. Dietary changes and exercise have helped, but his blood pressure is still too high. Explain to him why his doctor recommended beta blocker and diuretic drugs to treat his condition.

9. An angiogram reveals coronary artery occlusion requiring medical intervention. What measure might be taken to reduce the occlusion in order to avoid bypass surgery?

10. Mrs. Shultz has died suddenly after what seemed to be a low-grade infection. A blood culture reveals that the infection had spread to the blood. What is this condition called and why was it fatal?

11. Progressive heart failure has resulted in insufficient circulation to sustain life. What type of circulatory shock does this exemplify?

12. Mr. Connors has just had a right femoral arterial graft and is resting quietly. Orders are given to palpate his popliteal and pedal (dorsalis pedis) pulses and to assess the color and temperature of his right leg four times daily. Why were these orders given?

13. Len, an elderly man, is bedridden after a hip fracture. He complains of pain in his legs, and phlebitis is diagnosed. What is phlebitis and what life-threatening complication can develop?

14. Examination of Mr. Cummings, a man in his 60s, reveals a blood pressure of 140/120. What is his pulse pressure? Is it normal, high, or low? What does this indicate about the state of his elastic arteries?

15. A worried mother has brought Terry, her 11-year-old son, to the clinic after he complained of feeling dizzy and nauseous. No abnormalities are found during the examination. Upon questioning, however, Terry sheepishly admits that he was smoking with a friend, who had dared him to "take deep drags." Explain how this might have caused his dizziness.

Stop and Think

1. Why is the cardiovascular lining a simple squamous epithelium rather than a thicker layer?

2. Does vasoconstriction increase or decrease the supply of blood to the tissues beyond the vasoconstricted area? Explain your answer.

3. When an entire capillary bed is closed off due to arteriolar vasoconstriction, are the precapillary sphincters open or closed? What is the pattern of blood flow into the capillaries when the arteriole dilates? From this, explain why the face flushes when coming inside on a cold day.

4. Standing up quickly after being in a horizontal position can cause dizziness. Why is this more likely in a warm room than in a cool room?

5. Why shouldn't a pregnant woman sleep on her back late in pregnancy?

6. The dura mater reinforces the walls of the intracranial sinuses. What reinforces the wall of the coronary sinus?

7. Why is necrosis more common in areas supplied by end arteries?

8. What would trigger a change in blood vessel length in the body?

9. Do baroreceptors in the aortic and carotid bodies undergo adaptation, as explained in Chapter 13?

10. Assume there is a clot blocking the right internal carotid artery. Briefly describe another route that might be used to provide blood to the areas served by its branches.

11. Kwashiorkor results from a dietary deficiency in protein. Explain why this causes edema.

12. Describe the positive feedback mechanism between heart and vasomotor center that would be involved in untreated or irreversible circulatory shock.

13. What is the importance of the fetal ductus venosus bypass of the hepatic circulation?

14. Your friend, who knows very little about science, is reading a magazine article about a patient who had an "aneurysm at the base of his brain that suddenly grew much larger." The surgeons' first goal was to "keep it from rupturing" and the second goal was to "relieve the pressure on the brain stem and cranial nerves." The surgeons were able to "replace the aneurysm with a section of plastic tubing," so the patient recovered. Your friend asks you what all this means, and why the condition is life-threatening. What would you tell him?

15. Mr. Brown was distracted while trying to fell a large tree. His power saw whipped around and severed his right arm at the shoulder. Without a limb stump, applying a tourniquet is impossible. Where would you apply pressure to save Mr. Brown from fatal hemorrhage?

16. Freddy was looking for capillaries in his microscope slides of body organs. Although Freddy disagreed, Karl, his lab partner, kept insisting that a venule was a capillary. Finally, the teaching assistant said, "Look, here you see five erythrocytes lined up across the width of the lumen of this vessel so it cannot be a capillary." Explain the logic behind this statement and tell exactly how wide the vessel was (in micrometers).

COVERING ALL YOUR BASES

Multiple Choice

Select the best answer or answers from the choices given.

1. Which of the following is/are part of the tunica intima?
   A. Simple squamous epithelium
   B. Basement membrane
   C. Loose connective tissue
   D. Smooth muscle

2. The tunica adventitia contains:
   A. nerves   C. lymphatics
   B. blood vessels   D. collagen

3. In comparing a parallel artery and vein, you would find that:
   A. the artery wall is thicker
   B. the artery diameter is greater
   C. the artery lumen is smaller
   D. the artery endothelium is thicker

4. Which vessels are conducting arteries?
   A. Brachiocephalic artery
   B. Common iliac artery
   C. Digital artery
   D. Arcuate artery (foot)

5. A pulse would be palpable in the:
   A. anterior cerebral artery
   B. hepatic portal vein
   C. muscular arteries
   D. inferior vena cava

6. Pulse pressure is increased by:
   A. an increase in diastolic pressure
   B. an increase in systolic pressure
   C. an equivalent increase in both systolic and diastolic pressures
   D. vasoconstriction

7. Vessels that do not have all three tunics include the:
   A. smallest arterioles   C. capillaries
   B. smallest venules   D. venous sinuses

8. Structures that are totally avascular are the:
   A. skin
   B. cornea
   C. articular surfaces of long bones
   D. joint capsules

9. Fenestrated capillaries occur in the:
   A. liver   C. cerebrum
   B. kidney   D. intestinal mucosa

10. Kupffer cells are:
    A. true capillary endothelium
    B. found throughout the body
    C. macrophages in the liver
    D. macrophages in the spleen

11. Which of the following is/are part of a capillary bed?
    A. Precapillary sphincter
    B. Metarteriole
    C. Thoroughfare channel
    D. Terminal arteriole

12. Which vessels actively supply blood to associated capillary networks when the body is primarily controlled by the parasympathetic division?
    A. Femoral artery
    B. Superior mesenteric artery
    C. Cerebral arteries
    D. Coronary arteries

13. Which of the following can function as a blood reservoir?
    A. Brachiocephalic artery
    B. Cerebral capillaries
    C. Dural sinuses
    D. Inferior vena cava

14. Collateral circulation occurs in the:
    A. joint capsules   C. kidney
    B. cerebrum   D. myocardium

15. Venous return to the right atrium is increased by:
    A. increasing depth of respiration
    B. vigorous walking
    C. an increase in ventricular contraction strength
    D. activation of angiotensin II

16. The single most important factor in blood pressure regulation is:
    A. feedback by baroreceptors
    B. renal compensatory mechanisms
    C. regulation of cardiac output
    D. short-term changes in blood vessel diameter

17. Blood volume is altered by:
    A. vasodilation
    B. edema
    C. increased salt intake
    D. polycythemia

18. Vasomotor fibers that secrete acetylcholine are (probably) found in:
    A. cerebral arteries   C. skin
    B. skeletal muscle   D. digestive organs

19. The vasomotor center is inhibited by:
    A. impulses from the carotid sinus baroreceptors
    B. impulses from the aortic sinus baroreceptors
    C. stimulation by carotid body chemoreceptors
    D. alcohol

20. Chemical factors that increase blood pressure include:
    A. endothelin   C. ADH
    B. NO   D. ANF

21. Which of the regulatory chemicals listed involve or target the kidneys?
    A. Angiotensin   C. Aldosterone
    B. ADH   D. ANF

22. Mechanisms to decrease systemic arterial blood pressure are invoked by:
    A. standing up
    B. public speaking
    C. losing weight
    D. moving to higher altitude

23. Which of the following interfere with contraction of the smooth muscle of precapillary sphincters?
    A. Adenosine   C. Lactic acid
    B. Oxygen   D. Potassium ions

24. Increase in blood flow to a capillary bed after its supply has been blocked is called:
    A. a myogenic response
    B. reactive hyperemia
    C. active hyperemia
    D. a cardiogenic response

25. An increase in which of the following results in increased filtration from capillaries to the interstitial space?
    A. Capillary hydrostatic pressure
    B. Interstitial fluid hydrostatic pressure
    C. Capillary osmotic pressure
    D. Duration of precapillary sphincter contraction

26. Vessels involved in the circulatory pathway to and from the brain are the:
    A. brachiocephalic artery
    B. subclavian artery
    C. internal jugular vein
    D. internal carotid artery

27. The fight-or-flight response results in the constriction of:
    A. renal arteries
    B. celiac trunk
    C. internal iliac arteries
    D. external iliac arteries

28. Prominent venous anastomoses are formed by the:
    A. azygos vein
    B. median cubital vein
    C. circle of Willis
    D. hepatic portal vein

29. An abnormal thickening of the capillary basement membrane associated with diabetes mellitus is called:
    A. phlebitis
    B. aneurysm
    C. microangiopathic lesion
    D. atheroma

30. Which of the following are associated with aging?
    A. Increasing blood pressure
    B. Weakening of venous valves
    C. Arteriosclerosis
    D. Stenosis of the ductus arteriosus

31. Which layer of the artery wall thickens most in atherosclerosis?
    A. Tunica media   C. Tunica adventitia
    B. Tunica intima   D. Tunica externa

32. Based on the vessels named pulmonary trunk, thyrocervical trunk, and celiac trunk, the term *trunk* must refer to:
    A. a vessel in the heart wall
    B. a vein
    C. a capillary
    D. a large artery from which other arteries branch

33. Which of these vessels is bilaterally symmetrical (i.e., one vessel of the pair occurs on each side of the body)?
    A. Internal carotid artery
    B. Brachiocephalic artery
    C. Azygos vein
    D. Superior mesenteric vein

34. The deep femoral and deep brachial veins drain the:
    A. biceps brachii and hamstring muscles
    B. flexor muscles in the hand and leg
    C. triceps brachii and quadriceps femoris muscles
    D. intercostal muscles of the thorax

35. A stroke that occludes a posterior cerebral artery will most likely affect:
    A. hearing   C. smell
    B. vision   D. higher thought processes

Word Dissection

For each of the following word roots, fill in the literal meaning and give an example, using a word found in this chapter.

Word root	Translation	Example
1. anastomos		
2. angio		

Word root	Translation	Example
3. aort		
4. athera		
5. auscult		
6. azyg		
7. capill		
8. carot		
9. celia		
10. entero		
11. epiplo		
12. fenestr		
13. jugul		
14. ortho		
15. phleb		
16. saphen		
17. septi		
18. tunic		
19. vaso		
20. viscos		

The Lymphatic System

Student Objectives

When you have completed the exercises in this chapter, you will have accomplished the following objectives:

Lymphatic Vessels

1. Describe the structure and distribution of lymphatic vessels and note their important functions.
2. Describe the source of lymph and the mechanism(s) of lymph transport.

Lymphoid Cells, Tissues, and Organs: An Overview

3. Describe the composition of lymphoid tissue (basic structure and cell population) and name the major lymphoid organs.

Lymph Nodes

4. Describe the general location, histological structure, and functions of lymph nodes.

Other Lymphoid Organs

5. Name and describe the other lymphoid organs of the body. Compare and contrast them with lymph nodes, structurally and functionally.

Developmental Aspects of the Lymphatic System

6. Outline the development of the lymphatic system.

Ready, set, go!

The lymphatic system is a rather strange system, with its myriad lymphoid organs and its vessels derived from veins of the cardiovascular system. Although both elements help to maintain homeostasis, these two elements of the lymphatic system have substantially different roles. The lymphatic vessels keep the cardiovascular system functional by maintaining blood volume. The lymphoid organs help defend the body from pathogens by providing operating sites for phagocytes and cells of the immune system. Chapter 21 tests your understanding of the functional roles of the various lymphatic system elements.

BUILDING THE FRAMEWORK

Lymphatic Vessels

1. Complete the following statements by writing the missing terms in the answer blanks.

Pump	**1.**
arteries	**2.**
veins	**3.**
valves	**4.**
lymph	**5.**
3 liters	**6.**

The lymphatic system is a specialized subdivision of the circulatory system. Although the cardiovascular system has a pump (the heart) and arteries, veins, and capillaries, the lymphatic system lacks two of these structures: the **(1)** and **(2)** . Like the **(3)** of the cardiovascular system, the vessels of the lymphatic system are equipped with **(4)** to prevent backflow. The lymphatic vessels act primarily to pick up leaked fluid, now called **(5)** , and return it to the bloodstream. About **(6)** of fluid is returned every 24 hours.

2. Lymphatic capillaries exhibit unique junctions between their endothelial cells, and they are anchored to surrounding tissues by fine filaments. How do these structural modifications aid in their function?

_____

_____

_____

_____

_____

3. Figure 21.1 provides an overview of the lymphatic vessels. In part A, the relationship between lymphatic vessels and the blood vessels of the cardiovascular system is depicted schematically. Part B shows the different types of lymphatic vessels in a simple way. First, color code and color the following structures in Figure 21.1.

○ Heart   ○ Veins   ○ Lymphatic vessels/lymph node

○ Arteries   ○ Blood capillaries   ○ Loose connective tissue around the blood and lymph capillaries

Then, identify by labeling these specific structures in part B:

A. Lymph capillaries   **C.** Lymphatic collecting vessels   **E.** Lymph node

B. Lymph duct   **D.** Lymph trunks   **F.** Valves

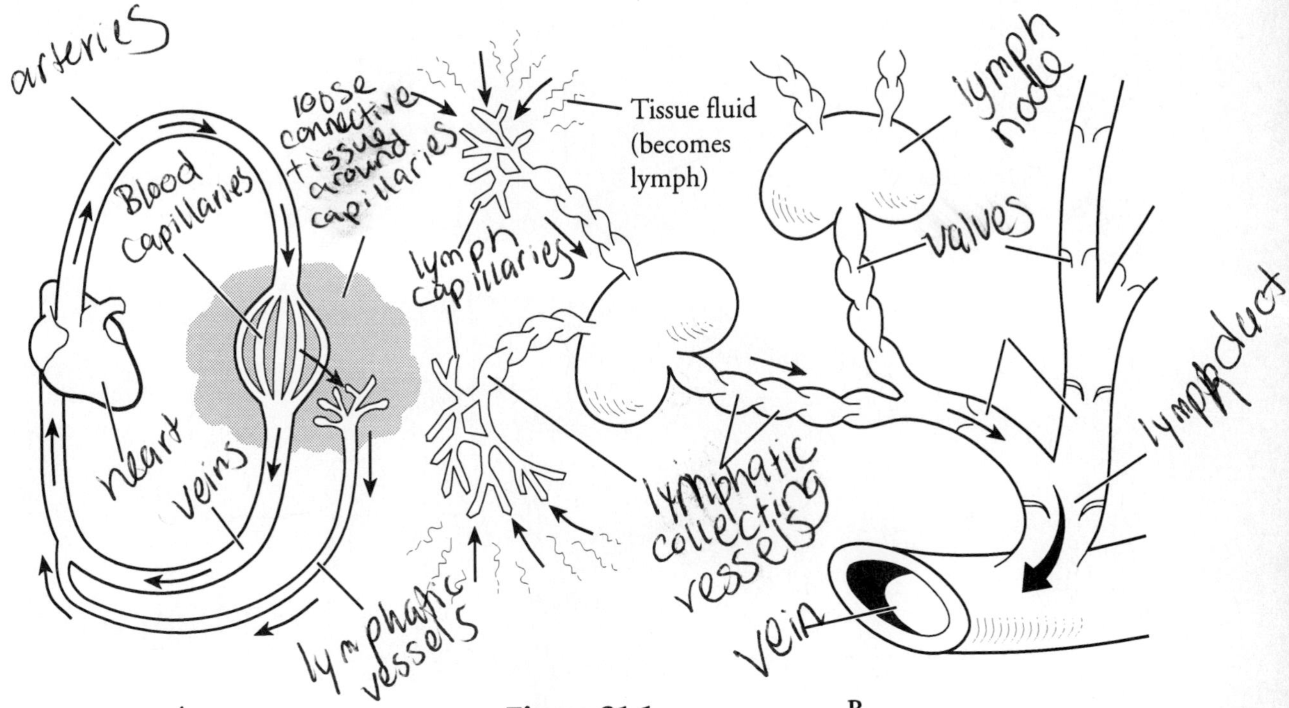

Figure 21.1

4. Circle the term that does not belong in each of the following groupings.

1. Lacteal   Lymphatic capillary   Chyle   Fat-free lymph   Intestinal mucosa

2. Blood capillary   Lymph capillary   Blind-ended   Permeable to proteins

3. Edema   Blockage of lymphatics   Elephantiasis   Inflammation

Abundant supply of lymphatics

4. Skeletal muscle pump   Flow of lymph   Respiratory pump

High-pressure gradient   Action of smooth muscle cells in walls of lymph vessels

5. Minivalves   Endothelial cell overlap   Impermeable   Anchoring filaments

Lymphatic capillaries

5. On the more detailed diagram in Figure 21.2, identify all structures provided with leader lines and color all of the lymphatic vessels (trunks, etc.) green. The terms you will need are listed below. Then color code and color the following structures:

◯ All veins   ◯ Ribs   ◯ Lymph nodes

◯ Trachea   ◯ Muscles of the thorax

Bronchomediastinal trunk(s)   Jugular trunk(s)   Lymphatic collecting vessels

Cisterna chyli   Lumbar trunk(s)   Right lymphatic duct

Intestinal trunk   Subclavian trunk(s)   Thoracic duct

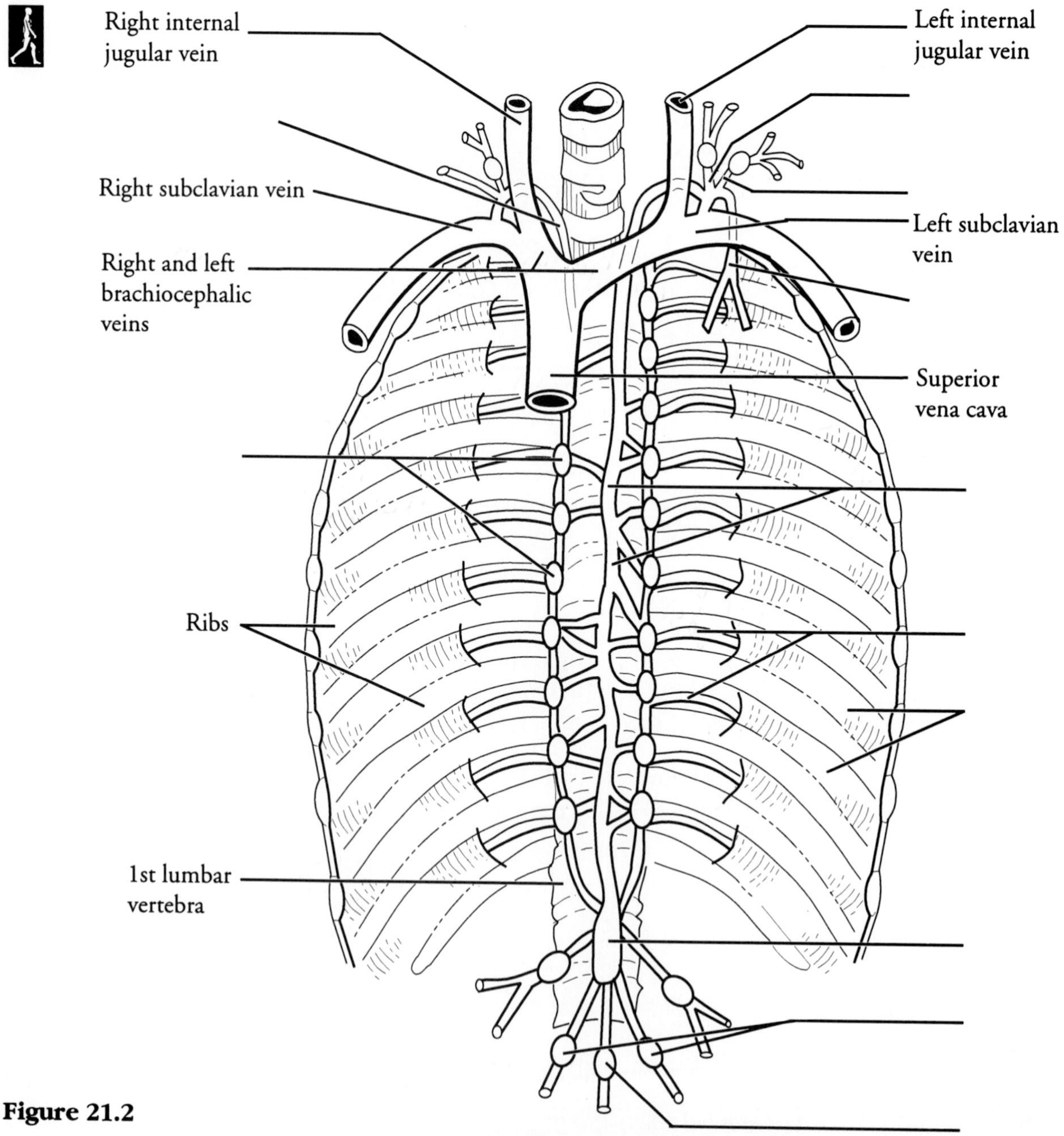

Right internal jugular vein

Left internal jugular vein

Right subclavian vein

Left subclavian vein

Right and left brachiocephalic veins

Superior vena cava

Ribs

1st lumbar vertebra

Figure 21.2

Lymphoid Cells, Tissues, and Organs: An Overview

1. Complete the following statements by writing the missing terms in the answer blanks.

_____ 1.

_____ 2.

_____ 3.

_____ 4.

_____ 5.

_____ 6.

_____ 7.

_____ 8.

_____ 9.

There are three major types of cells in the lymphoid tissues of the body. One category, the __(1)__, contains immuno-competent cells that are able to destroy or immobilize foreign substances in the body; these are called __(2)__. The two varieties in this group are the __(3)__ and the __(4)__. The other class of protective cells are called __(5)__. These defend the body by engulfing, or __(6)__, foreign substances. The third cellular group are the __(7)__, fibroblastlike cells that form the stroma or framework of the lymphoid organs. Lymphoid tissue is specifically classified as a type of connective tissue called __(8)__. When lymphoid tissue is loosely packed, it is called __(9)__; when it is tightly packed into spherical masses, the arrangement is said to be __(10)__.

_____ 10.

2. Figure 21.3 depicts the lymphatic vessels and lymphoid organs. Label all structures indicated by the leader lines. In part A, color the lymphatic vessels dark green. In part B, color the lymphoid organs as you like, and then shade in light green the portion of the body that is drained by the right lymphatic duct.

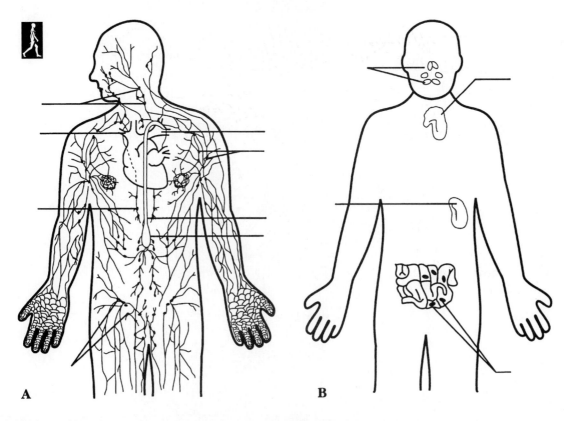

A   B

Figure 21.3

Lymph Nodes

1. Figure 21.4 is a diagram of a lymph node. First, using the terms with color-coding circles, label all structures on the diagram that have leader lines. Color those structures as well. Then, add arrows to the diagram to show the direction of lymph flow through the organ. Circle the region that would approximately correspond to the medulla of the organ. Finally, answer the questions that follow.

○ Germinal centers of follicles

○ Cortex (other than germinal centers)

○ Medullary cords

○ Capsule and trabeculae

○ Hilus

○ Afferent lymphatics

○ Efferent lymphatics

○ Sinuses (subcapsular and medullary)

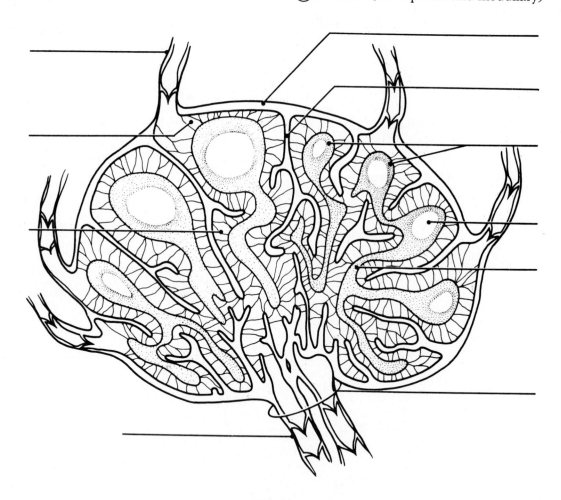

Figure 21.4

1. Which cell type is found in greatest abundance in the germinal centers?

    _____

2. What is the function of their daughter cells, the plasma cells? _____

3. What is the predominant cell type in the cortical regions other than the germinal centers?

    _____

4. What cells cluster around the medullary sinuses? _____

5. Functionally, the cells named in Question 4 act as _____

6. What type of soft connective tissue forms the internal skeletal of the lymph node?

    _____

7. Of what importance is the fact that there are fewer efferent than afferent lymphatics associated with lymph nodes?

    _____

    _____

    _____

8. What structures assure the one-way flow of lymph through the node?

    _____

9. The largest collections of lymph nodes are found in what three body regions?

    _____

10. What is the general function of lymph nodes?

    _____

    _____

Other Lymphoid Organs

1. Match the terms in Column B with the appropriate descriptions in Column A. More than one choice may apply in some cases.

Column A

Column B

_____ 1. The largest lymphatic organ

A. Lymph nodes

_____ 2. Filter lymph

B. Peyer's patches

_____ 3. Particularly large and important during youth; produces hormones that help to program the immune system

C. Spleen

D. Thymus

_____ 4. Collectively called MALT

E. Tonsils

_____ 5. Removes aged and defective red blood cells

_____ 6. Contains red and white pulp

_____ 7. Exhibits Hassall's corpuscles

_____ 8. Includes the adenoids

_____ 9. Acts against bacteria breaching the intestinal wall

_____ 10. Lack a complete capsule and have crypts in which bacteria can become trapped

2. Complete the following "story" by inserting the missing terms in the answer blanks.

_____ 1.

_____ 2.

_____ 3.

_____ 4.

_____ 5.

_____ 6.

_____ 7.

_____ 8.

_____ 9.

The spleen is a versatile organ with many functions. For example, it temporarily stores __(1)__ released during hemoglobin breakdown as well as formed elements called __(2)__ that play a role in blood clotting. It cleanses the blood of both aging __(3)__ and foreign debris. In the fetus, the spleen also carries out __(4)__. The spleen's blood-processing functions are performed by its __(5)__ pulp regions, which contain venous __(6)__ and __(7)__. The spleen's lymphocytes and macrophages, which cluster around the central arteries in areas called __(8)__ pulp, are important in mounting a(n) __(9)__ response against antigens present in the blood percolating through the spleen.

3. Figure 21.5 depicts an enlarged view of a section from the spleen. Identify the following structures or regions by labeling the leader lines with the appropriate terms from the list below.

Hilus   Splenic artery   Splenic vein

Central artery   Splenic capsule   Trabecula

Red pulp   Splenic cords   Venous sinus

Once you have labeled the figure, color it as follows: Color the splenic vein and venous sinuses blue, the splenic artery and central arteries red, the white pulp lavender, and the splenic capsule and trabeculae green.

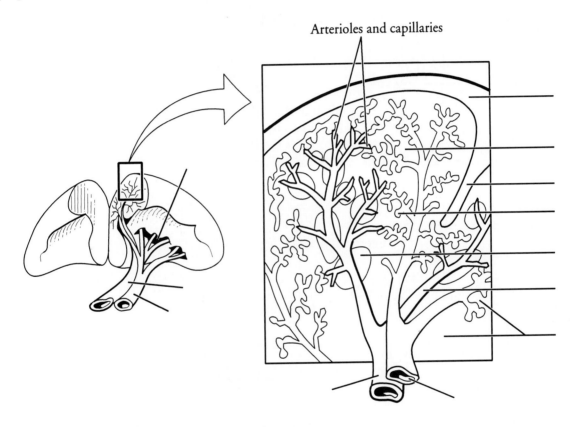

Arterioles and capillaries

Figure 21.5

4. Describe the important function of MALT. _____

_____

_____

5. Name the three major types of tonsils. _____

_____

Developmental Aspects of the Lymphatic System

1. Complete the following paragraph describing the development of the lymphatic system by writing the missing terms in the answer blanks.

_____ 1.	Lymphatic vessels derive from the lymph sacs that "bud" from developing __(1)__. The first of these lymph sacs, the __(2)__
_____ 2.	lymph sacs, arise at the junction of the internal jugular and subclavian veins in the thorax and spread throughout the __(3)__,
_____ 3.	__(4)__, and __(5)__. Abdominal lymphatics bud from the __(6)__ and lymphatics of the lower extremities form from the __(7)__.
_____ 4.	Except for the thymus, which is an endodermal derivative, all lymphoid organs develop from the __(8)__ germ layer. The first
_____ 5.	lymphoid organ to appear is the __(9)__, which forms from an outpocketing of the lining of the __(10)__. The __(11)__ is also
_____ 6.	fairly well formed during embryonic life, but most of the other lymphoid organs are poorly formed before birth. Shortly after
_____ 7.	birth, the lymphoid organs all become heavily populated by __(12)__. The embryonic thymus evidently produces __(13)__,
_____ 8.	which control the development of the lymphoid organs.
_____ 9.	
_____ 10.	
_____ 11.	
_____ 12.	
_____ 13.	

CHALLENGING YOURSELF

At the Clinic

1. Mrs. Burner comes to the clinic with red lines on her left arm, which are painful when touched. She reports being scratched by her cat recently. What is the diagnosis?

2. After surgery to remove lymphatic vessels associated with the removal of a melanoma, what condition can be expected relative to lymph drainage? Is this a permanent problem?

3. A young man appears at the clinic with swollen, tender inguinal lymph nodes. What term is used to describe an infected lymph node? What contagious disease is named for this term?

4. Lynn Oligopoulos, a young woman, has just been told her diagnosis is Hodgkin's disease. Describe, as best you can, the condition of her lymph nodes.

5. Young Joe Chang went sledding; the runner of a sled hit him in the left side and ruptured his spleen. Joe almost died because he did not get to the hospital fast enough. Upon arrival, a splenectomy was performed. What would you deduce is the immediate danger of a spleen rupture? Will Joe require a transplant for spleen replacement?

6. Gary has almost continuous earaches. The pediatrician explains to his parents that his adenoids should be removed to facilitate drainage of the auditory tubes. To which lymphoid organs is she referring?

7. After a trip to the tropics, a young biologist notices a general swelling in his left leg that slowly but progressively increases. What condition might you suspect, given his recent surroundings and signs?

8. Examination of the abdomen of a young girl with unexplained fever identifies a swelling in the upper left quadrant. The doctor suspects splenomegaly. What does that mean? Cite three causes of splenomegaly.

Stop and Think

1. Which lymphoid organs actually filter lymph?

2. Why are metastases common in lymph nodes?

3. Why are macrophages referred to as reticuloendothelial cells?

4. What would result from inability of lymphatic capillaries to pick up proteins from the interstitial fluid?

Closer Connections: Checking the Systems—Circulation and Protection

1. What is the overriding function of the circulatory system?

2. Differentiate between the cardiovascular system and the circulatory system.

3. What kinds of muscle tissue have important circulatory functions? What is the function of each kind?

4. Theoretically we have a closed circulatory system. But, is it *completely* closed? Explain.

5. Cancer spreads in the body primarily via the bloodstream or through the lymphatic system. Why are these *very* effective routes for the dispersal of the cancer cells

6. Other than debris removal and destruction, how do lymphoid tissues help to defend the body?

COVERING ALL YOUR BASES

Multiple Choice

Select the best answer or answers from the choices given.

1. Statements that apply to lymphatic capillaries include:
   A. the endothelial cells are well connected by tight junctions
   B. the arterial end has a higher pressure than the venous end
   C. minivalves prevent the backflow of fluid into the interstitial spaces
   D. the endothelial cells are anchored by filaments to the surrounding structures

2. Chyle flows into the:
   A. lacteals
   B. intestinal lymph nodes
   C. intestinal trunk
   D. cisterna chyli

3. Metastasis of a melanoma to lymph nodes can result in:
   A. lymphedema
   B. lymphadenopathy
   C. Hodgkin's disease
   D. lymphangitis

4. Which region in a lymph node is involved in removing the coal dust inhaled by a miner?
   A. Subcapsular sinus   C. Medullary cords
   B. Trabeculae   D. Follicles

5. Which parts of the lymph node show increased activity when antibody production is high?
   A. Germinal centers   C. Medullary cords
   B. Outer follicle   D. Sinuses

6. Which of the following connect to the lymph node at the hilus?
   A. Afferent lymphatic vessels
   B. Efferent lymphatic vessels
   C. Trabeculae
   D. Anchoring filaments

7. A cancerous lymph node would be:
   A. swollen   C. painful
   B. firm   D. fibrosed

8. The classification lymphoid tissues includes:
   A. the adenoids   C. bone marrow
   B. the spleen   D. the thyroid gland

9. There is no lymphatic supply to the:
    A. brain   **C.** bone
    B. cartilage   **D.** bone marrow

10. The spleen functions to:
    A. remove aged RBCs
    B. house lymphocytes
    C. filter lymph
    D. store some blood components

11. Which characteristics are associated with the thymus?
    A. Providing immunocompetence
    B. Hormone secretion
    C. Hypertrophy in later life
    D. Cauliflowerlike structural organization

12. The tonsils:
    A. have a complete epithelial capsule
    B. have crypts to trap bacteria
    C. filter lymph
    D. contain germinal centers

13. Lymphoid tissue is crowded with both macrophages and lymphocytes in close proximity. This design:
    A. enhances the ability of macrophages to phagocytize worn-out erythrocytes
    B. provides many opportunities for antigen challenge
    C. promotes effective bacterial killing
    D. leads to the production of many memory cells

14. Relative to distinguishing different lymphoid organs from each other in histological sections, how would you tell the thymus from a lymph node?
    A. Only the thymus has a cortex and medulla.
    B. Lymphocytes are far less densely packed in the thymus than in the lymph node.
    C. The thymus contains no blood vessels.
    D. Only the thymus has Hassall's corpuscles.

15. Which of the following are part of MALT?
    A. Tonsils
    B. Thymus
    C. Peyer's patches
    D. Any lymphoid tissue along the digestive tract

16. Developmentally, embryonic lymphatic vessels are most closely associated with the:
    A. veins   **C.** nerves
    B. arteries   **D.** thymus gland

17. Development of the lymphatic system begins with formation of:
    A. jugular sacs   **C.** cisterna chyli
    B. iliac sacs   **D.** tonsils

18. Which of the following lymphoid organs are well populated with lymphocytes at birth?
    A. Spleen   **C.** Tonsils
    B. Thymus   **D.** Lymph nodes

19. Mesoderm derivatives include:
    A. thymus   **C.** lymph nodes
    B. lymphatic vessels   **D.** Peyer's patches

20. Which of the following result from infection?
    A. Lymphangitis   **C.** Tonsillitis
    B. Elephantiasis   **D.** Lymphedema

21. Cells of lymphatic tissue that play a crucial role in helping to activate T cells include:
    A. dendritic cells   **C.** macrophages
    B. reticular cells   **D.** plasma cells

22. Which of the following terms accurately applies to tonsils?
    A. Pharyngeal   **C.** Palatine
    B. Crypts   **D.** Lingual

23. Lymph capillaries:
    A. are open-ended like drinking straws
    B. have continuous tight junctions like the capillaries of the brain
    C. contain endothelial cells separated by flaplike valves that open wide
    D. have special barriers that stop cancer cells from entering

24. Which one of the following lymphoid
organs does not contain lymph follicles or
germinal centers?
- **A.** Lymph nodes
- **B.** Peyer's patches
- **C.** Thymus
- **D.** Spleen

Word Dissection

For each of the following word roots, fill in the literal meaning and give an example, using a word found in this chapter.

Word root	Translation	Example
1. adeno	_____	_____
2. angi	_____	_____
3. chyle	_____	_____
4. lact	_____	_____
5. lymph	_____	_____

22 Nonspecific Body Defenses and Immunity

Student Objectives

When you have completed the exercises in this chapter, you will have accomplished the following objectives:

PART I: NONSPECIFIC BODY DEFENSES

Surface Membrane Barriers

1. Describe the surface membrane barriers and their protective functions.

Nonspecific Cellular and Chemical Defenses

2. Explain the importance of phagocytosis and natural killer cells in nonspecific body defense.

3. Relate the events of the inflammatory process. Identify several inflammatory chemicals and describe their specific roles.

4. Name the body's antimicrobial substances and describe their function.

5. Explain how fever helps protect the body against invading pathogens.

PART II: SPECIFIC BODY DEFENSES: IMMUNITY

Antigens

6. Explain what an antigen is and how it affects the immune system.

7. Define complete antigen, hapten, and antigenic determinant.

Cells of the Immune System: An Overview

8. Compare and contrast the origin, maturation process, and general function of B and T lymphocytes. Describe the role of macrophages in immunity.

9. Define immunocompetence and self-tolerance.

Humoral Immune Response

10. Define humoral immunity.

11. Describe the process of clonal selection of a B cell.

12. Recount the roles of plasma cells and memory cells in humoral immunity.

13. Compare and contrast active and passive humoral immunity.

14. Describe the structure of an antibody monomer and name the five classes of antibodies.

15. Explain the function(s) of antibodies.

16. Describe the development and clinical uses of monoclonal antibodies.

Cell-Mediated Immune Response

17. Define cell-mediated immunity and describe the process of activation and clonal selection of T cells.

18. Describe the functional roles of T cells in the body.

19. Indicate the tests ordered before an organ transplant is done and note methods used to prevent transplant rejection.

Homeostatic Imbalances of Immunity

20. Give examples of immune deficiency diseases and of hypersensitivity states.

21. Note factors involved in autoimmune disease.

Developmental Aspects of the Immune System

22. Describe changes in immunity that occur with aging.

23. Briefly describe the role of the nervous system in regulating the immune response.

Ready, set, go!

The immune system is a unique functional system made up of billions of individual cells, the bulk of which are lymphocytes. The sole function of this highly specialized defensive system is to protect the body against an incredible array of pathogens. In general, these "enemies" fall into three major camps: (1) microorganisms (bacteria, viruses, and fungi) that have gained entry into the body; (2) foreign tissue cells that have been transplanted (or, in the case of red blood cells, infused) into the body; and (3) the body's own cells that have become cancerous. The result of the immune system's activities is immunity, or specific resistance to disease.

The body is also protected by a number of nonspecific defenses provided by intact surface membranes such as skin and mucosae, and a variety of cells and chemicals can quickly mount the attack against foreign substances. The specific and nonspecific defenses enhance each other's effectiveness.

This chapter first focuses on the nonspecific body defenses. It then follows the development of the immune system and describes the mode of functioning of its effector cells—lymphocytes and macrophages. The chemicals that act to mediate or amplify the immune response are also discussed.

BUILDING THE FRAMEWORK

PART I: NONSPECIFIC BODY DEFENSES

1. The three major elements of the body's nonspecific defense system are:

the **(1)** _Surface membrane barriers_, consisting of the skin and _mucosae_;

defensive cells, such as **(2)** _natural killer cells_ and phagocytes; and a whole

deluge of **(3)** _chemicals_.

Surface Membrane Barriers

Lysozyme

1. Indicate the sites of activity of the secretions of the surface membranes by writing the correct terms in the answer blanks.

　1. Lysozyme is found in the body secretions called _tears_ and

　saliva.

2. Fluids with an acid pH are found in the ___Stomach___ and ___vagina___ .

3. Sebum is a product of the ___sebaceous___ glands and acts at the surface of the ___skin___ .

4. Mucus is produced by large mucus-secreting glands and the thousands of ___goblet___ cells found in the respiratory and ___digestive___ system mucosae.

2. Match the terms in Column B with the descriptions of the nonspecific defenses of the body in Column A. More than one choice may apply.

	Column A		Column B
___A, B, E, F___	1.	Have antimicrobial activity	A. Acids
___C, G___	2.	Provide mechanical barriers	B. Lysozyme
___A, B, D, E, F___	3.	Provide chemical barriers	C. Mucosae
___D___	4.	Entraps microorganisms entering the respiratory passages	D. Mucus
			E. Protein-digesting enzymes
___A–G___	5.	Part of the first line of defense	F. Sebum
			G. Skin

3. Describe the protective role of cilia. ___They propel mucus laden w/ trapped debris superiorly away from lungs to the throat, where it can be spit out.___

Nonspecific Cellular and Chemical Defenses

1. Define *phagocytosis* and indicate why efforts at phagocytosis might not succeed.

___

___

___

___

2. Explain why neutrophils are killed in their protective efforts while macro-phages, also phagocytic, are longer-lived.

_____

_____

_____

_____

3. Match the terms in Column B with the descriptions in Column A concerning events of the inflammatory response.

Column A

Column B

local
___F hyperemia__ **1.** Increased blood flow to an area

D __B histamine__ **2.** Inflammatory chemical released by degranulating mast cells

_____ **3.** Promotes release of white blood cells from the bone marrow

__A chemotaxis__ **4.** Cellular migration directed by a chemical gradient

_____ **5.** Fluid leaked from the bloodstream

__G macrophages__ **6.** Phagocytic progeny of monocytes

__B diapedesis__ **7.** Leukocytes pass through the wall of a capillary

__I neutrophil__ **8.** First phagocytes to migrate into the injured area

_____ **9.** White blood cells cling to capillary walls as blood flow slows due to fluid loss from the bloodstream

A. Chemotaxis

B. Diapedesis

C. Exudate

D. Histamine

E. Leukocytosis-inducing factor

F. Local hyperemia

G. Macrophages

H. Margination

I. Neutrophils

4. Circle the term that does not belong in each of the following groupings.

1. Redness   Pain   Swelling   Nausea   Heat

2. Neutrophils   Macrophages   Phagocytes   Natural killer cells

3. Kupffer cells   Neutrophils   Langerhans' cells   Alveolar macrophages

4. Inflammatory chemicals   Histamine   Kinins   Complement   Interferon

5. Interferons   Antiviral   Antibacterial   Host specific   Anticancer

6. Natural killer cells   Null cells   Macrophages   NK cells

7. Defensins   Free radicals   Respiratory burst   NO

5. Figure 22.1 diagrams the events involved in the inflammatory response. Events depicted in A precede those shown in B. Each event is represented by a square with one or more arrows. From the list below, write the correct number in each event square in the figure. Then, color code and color the structures that appear below the numbered list.

1. WBCs are drawn to the injured area by the release of inflammatory chemicals.

2. Tissue repair occurs.

3. Tissue injury occurs and microbes enter the injured tissues.

4. Local blood vessels dilate and the capillaries become engorged with blood.

5. Phagocytosis of microbes occurs.

6. Fluid containing clotting proteins is lost from the bloodstream and enters the injured tissue area.

7. Mast cells degranulate, releasing inflammatory chemicals.

8. Margination and diapedesis occur.

○ Mast cells   ○ Erythrocyte(s)   ○ Endothelium of capillary

○ Mast cell granules   ○ Neutrophil(s)   ○ Microorganisms

○ Monocyte   ○ Macrophage   ○ Fibrous repair tissue

○ Epithelium   ○ Subcutaneous tissue

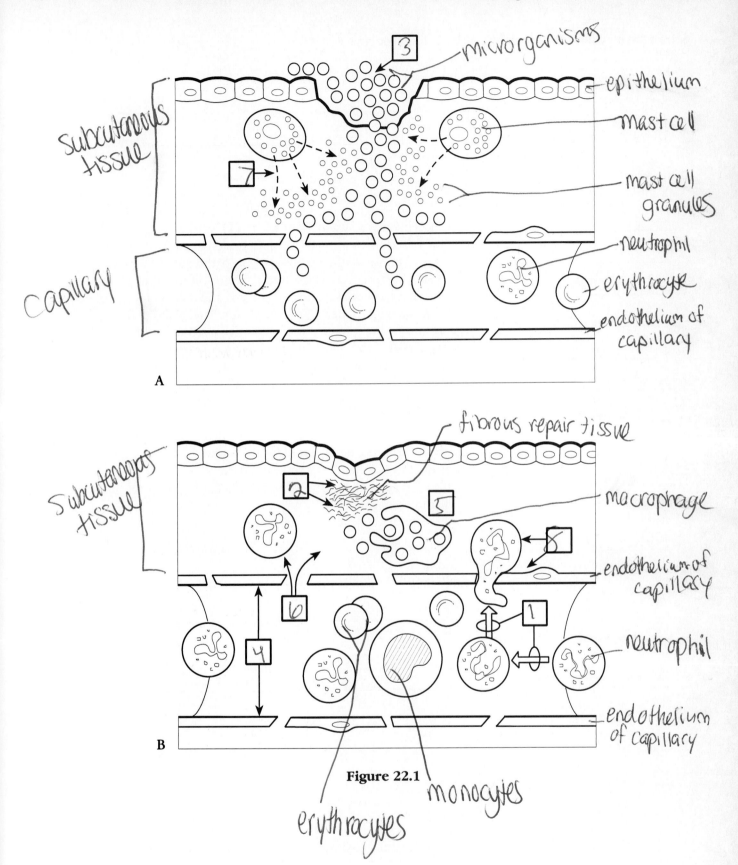

Figure 22.1

6. List three advantages of inflammation-induced edema.

1. _____

2. _____

3. _____

7. Complete the following description of the activation and activity of complement by writing the missing terms in the answer blanks.

_____ **1.**

_____ **2.**

_____ **3.**

_____ **4.**

_____ **5.**

_____ **6.**

_____ **7.**

_____ **8.**

_____ **9.**

_____ **10.**

_____ **11.**

Complement is a system of at least **(1)** proteins that circulate in the blood in an inactive form. Complement may be activated by either of two pathways. The **(2)** pathway involves 11 proteins, designated C1–C9, and depends on the binding of antibodies to invading microorganisms and subsequent binding of complement to the **(3)** complexes. The **(4)** pathway is triggered by the interaction of three plasma proteins, factors, B, D, and **(5)**, with **(6)** molecules present on the surface of certain bacteria and fungi. Both pathways lead to the cleavage of C3 into C3a and C3b. Binding of **(7)** to the target cell's surface results in the incorporation of C5–C9 into the target cell membrane. This incorporated complex is called the **(8)**, and its completed insertion results in target cell **(9)** . The complement fragments bound to the target cell surface also enable it to be phagocytized more readily, a phenomenon called **(10)** . The **(11)** fragment of C3 amplifies the inflammatory response by attracting neutrophils and other inflammatory cells into the area.

8. Describe the event that leads to the synthesis of interferon and indicate the result of its synthesis.

_____

_____

_____

_____

9. Check (✔) all phrases that correctly describe the role of fever in body protection.

_____ **1.** Is a normal response to pyrogens

_____ **2.** Protects by denaturing tissue proteins

_____ **3.** Reduces the availability of iron and zinc required for bacterial proliferation

_____ **4.** Increases metabolic rate

PART II: SPECIFIC BODY DEFENSES: IMMUNITY

Antigens

1. Complete the following statements relating to antigens by writing the missing terms in the answer blanks.

_____ 1.

_____ 2.

_____ 3.

_____ 4.

_____ 5.

_____ 6.

_____ 7.

_____ 8.

_____ 9.

Antigens are substances capable of mobilizing the __(1)__ . Complete antigens can stimulate proliferation of specific lymphocytes and antibodies, a characteristic called __(2)__, and can react with the products of those reactions, a property called __(3)__ . Of all the foreign molecules that act as complete antigens, __(4)__ are the most potent. Small molecules are not usually antigenic, but when they bind to self cell-surface proteins they may act as __(5)__, and then the complex is recognized as foreign. The ability of various molecules to act as antigens depends both on their __(6)__ and on the complexity of their structure. Large, complex molecules may have hundreds of recognizable 3-D structures called __(7)__, to which antibodies or activated lymphocytes can attach. Conversely, large molecules with many regularly repeating units are chemically __(8)__ and thus are poor antigens. Since such substances tend not to be __(9)__ by the body, they are useful for making artificial implants.

2. Complete this section dealing with self-antigens by writing the missing terms in the answer blanks.

_____ 1.

_____ 2.

_____ 3.

Self-antigens, the unique cell-surface glycoproteins called __(1)__, are coded for by genes of the __(2)__ . These self-antigens have a deep groove that allows them to display protein fragments. Protein fragments derived from foreign antigens mobilize the __(3)__.

Cells of the Immune System: An Overview

1. What are three important characteristics of the immune response?

_____

2. Define *immunocompetence.* _____

_____

_____

_____

3. A schematic drawing of the life cycle of the lymphocytes involved in immunity is shown in Figure 22.2. Select different colors for the areas listed below, and color the coding circles and the corresponding regions on the figure. If there is overlap, use stripes of a second color to indicate the second identification. Then respond to the questions following the figure, which relate to the two-phase differentiation process of B cells and T cells.

◯ Area where immature lymphocytes arise

◯ Area seeded by immunocompetent B cells and T cells

◯ Area where T cells become immunocompetent

◯ Area where the antigen challenge and clonal selection are likely to occur

◯ Area where B cells become immunocompetent

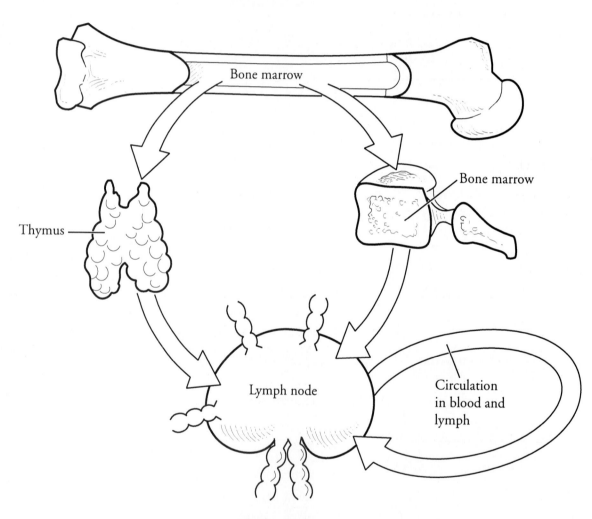

Figure 22.2

1. What signifies that a lymphocyte has become immunocompetent?

_____

2. During what period of life does the development of immunocompetence

   occur? _____

3. What determines which antigen a particular T cell or B cell will be able to

   recognize, its genes or "its" antigen? _____

4. What triggers the process of clonal selection in a T cell or B cell, its genes or

   binding to "its" antigen? _____

5. During the development of immunocompetence (during which B cells or
   T cells develop the ability to recognize a particular antigen), the ability

   to tolerate _____ must also occur if the immune system
   is to function normally.

4. Using the key choices, select the term that correctly completes each statement.
   Insert the appropriate answers in the answer blanks.

KEY CHOICES

A. Antigen(s)	**D.** Cellular immunity	**G.** Lymph nodes
B. B cells	**E.** Humoral immunity	**H.** Macrophages
C. Blood	**F.** Lymph	**I.** T cells

_____ 1.

_____ 2.

_____ 3.

_____ 4.

_____ 5.

_____ 6.

_____ 7.

_____ 8.

_____ 9.

Immunity is resistance to disease resulting from the presence of foreign substances or __(1)__ in the body. When this resistance is provided by antibodies released to body fluids, the immunity is called __(2)__ . When living cells provide the protection, the immunity is called __(3)__ . The major actors in the immune response are two lymphocyte populations, the __(4)__ and the __(5)__ . Phagocytic cells that act as accessory cells in the immune response are the __(6)__ . Because pathogens are likely to use both __(7)__ and __(8)__ as a means of getting around the body, __(9)__ and other lymphoid tissues (which house the immune cells) are in an excellent position to detect their presence.

5. T cells and B cells exhibit certain similarities and differences. Check (✔) the appropriate boxes in the table to indicate the lymphocyte type that exhibits each characteristic.

Characteristic	T cell	B cell
1. Originates in bone marrow from hemocytoblasts		
2. Progeny are plasma cells		
3. Progeny include suppressors, helpers, or killers		
4. Progeny include memory cells		
5. Is responsible for directly attacking foreign cells or virus-infected cells in the body		
6. Produces antibodies that are released to body fluids		
7. Bears a cell-surface receptor capable of recognizing a specific antigen		
8. Forms clones upon stimulation		
9. Accounts for most of the lymphocytes in the circulation		

Humoral Immune Response

1. The basic structure of an antibody molecule is diagrammed in Figure 22.3. Select two different colors and color the heavy chains and light chains. Add labels to the diagram to identify the type of bonds holding the polypeptide chains together, the complement binding site, and the macrophage binding site. Also label the constant (C) and variable (V) regions of the antibody, and add "polka dots" to the variable portions. Then answer the questions following the figure.

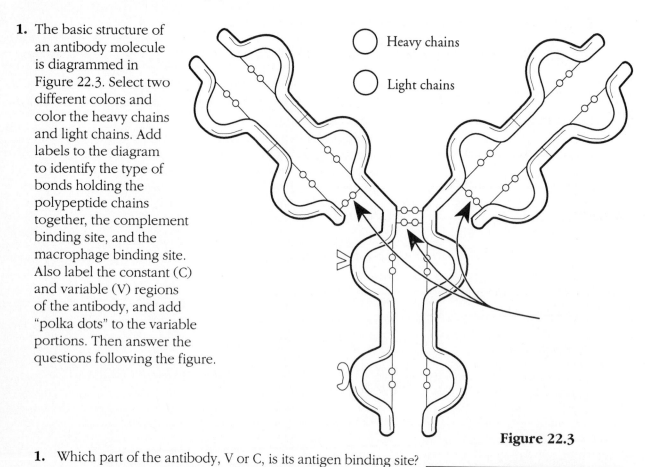

Heavy chains

Light chains

Figure 22.3

1. Which part of the antibody, V or C, is its antigen binding site? _____

2. Which part determines antibody class and specific function? _____

2. Match the antibody classes in Column B with their descriptions in Column A.

Column A   Column B

_____ **1.** Bound to the surface of a B cell; its antigen receptor   **A.** IgA

_____ **2.** Crosses the placenta   **B.** IgD

_____ **3.** The first antibody released during the primary response; indicates a   **C.** IgE
current infection
D. IgG
_____ **4.** Two classes that fix complement
E. IgM
_____ **5.** The most abundant antibody found in blood plasma and the chief
antibody released during secondary responses

_____ **6.** Binds to the surface of mast cells and mediates an allergic response

_____ **7.** Predominant antibody in mucus, saliva, tears, and milk

_____ **8.** Is a pentamer

3. Determine whether each of the following situations provides, or is an example
of, active or passive immunity. If passive, write P in the blank; if active, write A
in the blank.

_____ **1.** An individual receives Sabin polio vaccine

_____ **2.** Antibodies migrate through a pregnant woman's placenta into the vascular system
of her fetus

_____ **3.** A student nurse receives an injection of gamma globulin (containing antibodies to
the hepatitis virus) after she has been exposed to viral hepatitis

_____ **4.** "Borrowed" immunity

_____ **5.** Immunologic memory is provided

_____ **6.** An individual suffers through chicken pox

4. There are several important differences between primary and secondary
immune responses to antigens. If a statement below best describes a primary
response, write P in the blank; if a secondary response, write S in
the blank.

_____ **1.** The initial response to an antigen; "gearing up" stage

_____ **2.** A lag period of several days occurs before antibodies specific to the antigen
appear in the bloodstream

_____ **3.** Antibody levels increase rapidly and remain high for an extended period

_____ **4.** Immunologic memory is established

_____ **5.** The second, third, and subsequent responses to the same antigen

5. Complete the following description of antibody function by writing the missing terms in the answer blanks.

_____ **1.**

_____ **2.**

_____ **3.**

_____ **4.**

_____ **5.**

_____ **6.**

_____ **7.**

Antibodies can inactivate antigens in various ways, depending on the nature of the __(1)__ . __(2)__ is the chief ammunition used against cellular antigens such as bacteria and mismatched red blood cells. The binding of antibodies to sites on bacterial exotoxins or viruses that can cause cell injury is called __(3)__ . The cross-linking of cellular antigens into large lattices by antibodies is called __(4)__ ; Ig __(5)__ , with its 10 antigen binding sites, is particularly efficient in this mechanism. When molecules are cross-linked into lattices by antibodies, the mechanism is more properly called __(6)__ . In virtually all these cases, the protective mechanism mounted by the antibodies serves to disarm and/or immobilize the antigens until they can be disposed of by __(7)__ .

Cell-Mediated Immune Response

1. Several populations of T cells exist. Match the type of cell in Column B with the descriptions in Column A.

Column A

_____ **1.** Binds with and releases chemicals that activate B cells

_____ **2.** Releases chemicals that activate killer T cells, macrophages, and a variety of nonimmune cells, but must be activated itself by recognizing both its antigen and a self protein presented on the surface of a macrophage

_____ **3.** Turns off the immune response when the "enemy" has been routed

_____ **4.** Directly attacks and causes lysis of cellular pathogens

_____ **5.** Plays a major role in chronic inflammations by releasing a deluge of lymphokines

_____ **6.** The major cell type in the CD4 population

_____ **7.** The major cell type(s) in the CD8 population

Column B

A. Cytotoxic T cell

B. Delayed hypersensitivity T cell

C. Helper T cell

D. Suppressor T cell

2. Using the key choices, select the terms that correspond to the following descriptions of substances or events by inserting the appropriate answers in the answer blanks.

KEY CHOICES

A. Antibodies   **E.** Inflammation   **I.** Monokines

B. Antigen   **F.** Interferon   **J.** Perforin

C. Chemotactic factors   **G.** Lymphokine   **K.** Tumor necrosis factor

D. Complement   **H.** Macrophage migration inhibiting factor

_____ **1.** A protein released by macrophages and activated T cells that helps protect other body cells from viral multiplication

_____ **2.** Molecules that attract neutrophils and other protective cells into a region where an immune response is ongoing

_____ **3.** Any substance capable of eliciting an immune response and of binding with the product of that response

_____ **4.** Slow-acting protein that causes cell death; released in large amounts by macrophages

_____ **5.** A consequence of the release of histamine from mast cells and of complement activation by binding to antigen-antibody complexes

_____ **6.** C, F, H, and K are examples of this class of molecules

_____ **7.** A group of plasma proteins that amplifies the immune response by causing lysis of cellular pathogens once it has been "fixed" to their surface

_____ **8.** Chemicals released by activated macrophages; interleukin-1 and interferon, for example

_____ **9.** Proteins released by plasma cells that mark antigens for destruction by phagocytes or complement

_____ **10.** A chemical released by cytotoxic T cells that causes lysis of a cellular antigen

3. Circle the term that does not belong in each of the following groupings.

1. Antibodies   Gamma globulin   Lymphokines   Immunoglobulins

2. Lymph nodes   Liver   Spleen   Thymus   Bone marrow

3. Capping   Clustering of receptors   T cell activation

Endocytosis of antigen-receptor complexes

4. Autograft   Isograft   Allograft   Xenograft   Human

5. Cytotoxic T cells   Cytolytic T cells   Natural killer cells   Killer T cells

6. Perforin Na^+-induced polymerization   Membrane lesions   Lethal hit

   Lysis

7. Macrophage presenters   Helper T cells   Class I MHC proteins

   Class II MHC proteins   Interleukin-1

4. Figure 22.4 is a flowchart of the immune response that tests your understanding of the interrelationships of that process. Several terms have been omitted from this flowchart. First, complete the figure by writing in the boxes the appropriate terms selected from the key choices below. (*Note:* some terms are used more than once. Oval boxes represent cell types and rectangular boxes represent molecules. Also note that solid lines represent stimulatory or enhancing effects, whereas broken lines indicate inhibition.) Then color the coding circles of the cell types listed below and the corresponding oval boxes on the flowchart.

KEY CHOICES

Cell Types	**Molecules**
◯ B cell	Antibodies
◯ Helper T cell	Chemotactic factors
◯ Killer T cell	Complement
◯ Macrophage	Interferon
◯ Memory B cell	Interleukin
◯ Memory T cell	Lymphokines
◯ Neutrophils	Cytotoxins (perforins and others)
◯ Plasma cell	Suppressor factors
◯ Suppressor T cell	

5. Complete the following paragraph about costimulation by inserting the appropriate terms in the answer blanks.

   _____ 1.

   _____ 2.

   _____ 3.

   _____ 4.

   Not emphasized in Figure 22.4 is the importance of costimulatory signals, which may be __(1)__ or __(2)__ . When the required costimulatory signals are absent, T cell activation is incomplete and the T cell __(3)__ the antigen. The two-step signalling sequence (antigen-MHC recognition, then costimulation) is thought to be a safeguard that prevents __(4)__ of our own healthy cells.

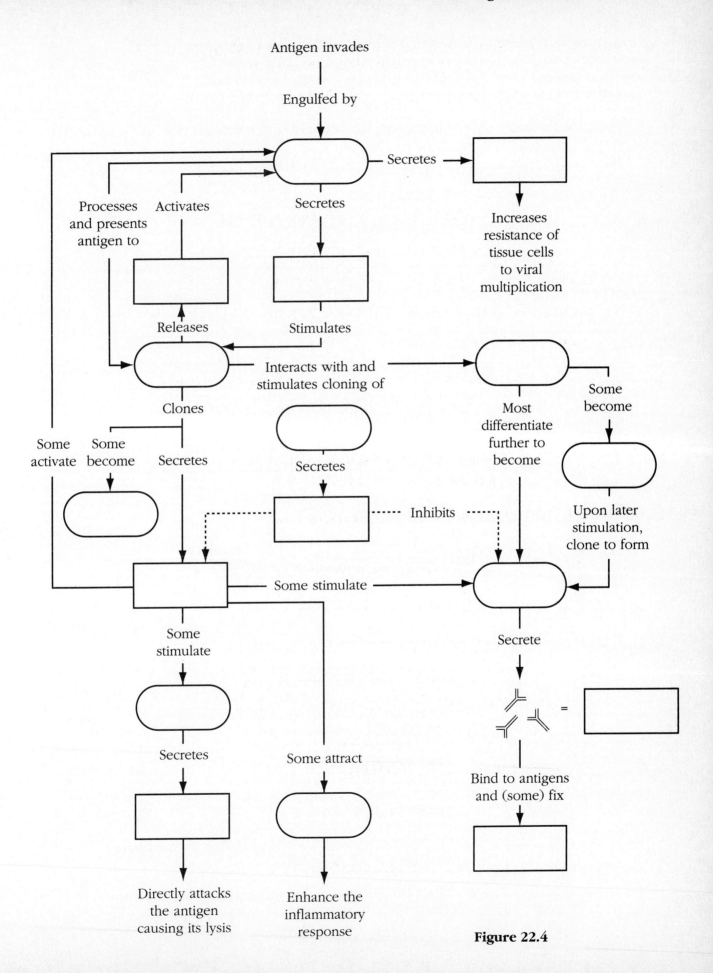

Antigen invades

Engulfed by

Secretes

Secretes

Increases resistance of tissue cells to viral multiplication

Processes and presents antigen to

Activates

Releases

Stimulates

Interacts with and stimulates cloning of

Most differentiate further to become

Some become

Clones

Some activate

Some become

Secretes

Secretes

Inhibits

Upon later stimulation, clone to form

Some stimulate

Secrete

Some stimulate

Some attract

Bind to antigens and (some) fix

Secretes

Directly attacks the antigen causing its lysis

Enhance the inflammatory response

Figure 22.4

6. Organ transplants are often unsuccessful because MHC-encoded proteins vary in different individuals. However, chances of success increase if certain important procedures are followed. The following questions refer to this important area of clinical medicine.

1. Assuming that autografts and isografts are not possible, what is the next most successful graft

   type? _____

2. What is the source of the graft tissue or organ? _____

3. What two cell types are important in rejection phenomena?

   _____

4. Why are immunosuppressive drugs (or therapy) provided after transplant surgery?

   _____

   _____

5. What is the major shortcoming of immunosuppressive therapy?

   _____

   _____

Homeostatic Imbalances of Immunity

1. Using the key choices, identify the types of immunity disorders described below.

KEY CHOICES

A. Allergy   **B.** Autoimmune disease   **C.** Immunodeficiency

_____ **1.** AIDS and congenital thymic aplasia

_____ **2.** The immune system mounts an extraordinarily vigorous response to an otherwise harmless antigen

_____ **3.** A hypersensitivity reaction such as hayfever or contact dermatitis

_____ **4.** Occurs when the production or activity of immune cells or complement is abnormal

_____ **5.** The body's own immune system produces the disorder; a breakdown of self-tolerance

_____ **6.** Affected individuals are unable to combat infections that would present no problem for normally healthy people

_____ 7. Multiple sclerosis, rheumatoid arthritis, and myasthenia gravis

_____ 8. May be treated by bone marrow transplants in some cases

_____ 9. Caused when self-antigens are changed by disease, or when anti-bodies produced against foreign antigens react with self-antigens

2. Match the terms relating to hypersensitivities in Column B with their descriptions in Column A.

Column A

_____ 1. Allergies mediated by antibodies

_____ 2. Occurs when the allergen cross-links to IgE molecules attached to mast cells or basophils, causing them to degranulate; reactions may be local or systemic

_____ 3. Transfusion reaction

_____ 4. Complement activation by circulating or attached immune complexes resulting in intense inflammatory responses

_____ 5. Effective against facultative intracellular pathogens (salmonella and some yeasts)

_____ 6. Can result in death due to circulatory shock in susceptible individuals when the allergen is bee venom

_____ 7. Its major mediator is histamine

_____ 8. Collectively make up the immediate hypersensitivities

_____ 9. Can be passively transferred to others in whole blood transfusions

_____ 10. Serum sickness, farmer's lung, and mushroom grower's lung, for example

_____ 11. Redness, hives, local edema, and wheezing can all be symptoms

Column B

A. Anaphylaxis (type I)

B. Atopy (type I)

C. Cytotoxic (type II) hypersensitivity

D. Delayed hypersensitivity (type IV)

E. Immune complex (type III) hypersensitivity

_____ **12.** Lymphokines released by T cells are the major mediators

_____ **13.** Poison ivy, contact dermatitis, and allergic reactions to detergents are examples

_____ **14.** Together called subacute hypersensitivities

Developmental Aspects of the Immune System

1. Complete the following statements concerning the development and operation of the immune system during the life span by writing the missing terms in the answer blanks.

_____ **1.**

_____ **2.**

_____ **3.**

_____ **4.**

_____ **5.**

_____ **6.**

_____ **7.**

_____ **8.**

_____ **9.**

_____ **10.**

_____ **11.**

_____ **12.**

_____ **13.**

The earliest lymphocyte stem cells that can be identified appear during the first month of development in the embryo's __(1)__. Shortly thereafter, bone marrow becomes the lymphocyte origin site; but, after birth, lymphocyte proliferation occurs in the __(2)__. The __(3)__ is the first lymphoid organ to appear. The development of the other lymphoid organs is believed to be controlled by thymic hormones. The development of immunocompetence has usually been accomplished by __(4)__ and the ability of the immune cells to recognize foreign antigens is __(5)__.

Individuals under severe stress exhibit a __(6)__ immune response, an indication that the __(7)__ system plays a role in immunity. Indeed, investigations have indicated that these two systems share a use of the __(8)__ neuropeptides. Hormones such as __(9)__ and __(10)__ may also serve as chemical links between them. During old age, the effectiveness of the immune system __(11)__ and the elderly are more at risk for __(12)__ and __(13)__.

The Incredible Journey: A Visualization Exercise for the Immune System

Something quite enormous and looking much like an octopus is nearly blocking the narrow tunnel just ahead.

1. Complete the following narrative by writing the missing words in the answer blanks.

_____ 1.	
_____ 2.	
_____ 3.	
_____ 4.	
_____ 5.	
_____ 6.	
_____ 7.	

For this journey, you are equipped with scuba gear before you are miniaturized and injected into one of your host's lymphatic vessels. He has been suffering with a red, raw "strep throat." Your assignment is to travel into a swollen cervical lymph node and observe the activities going on there that reveal that your host's immune system is doing its best to combat the infection.

On injection, you enter the lymph with a "whoosh" and then bob gently in the warm yellow fluid. As you travel along, you see thousands of spherical streptococcus bacteria and a few large globular __(1)__ molecules that, no doubt, have been picked up by the tiny lymphatic capillaries. Shortly thereafter, a large dark mass looms just ahead. This has to be a __(2)__ , you conclude, and you dig in your wetsuit pocket to find the waterproof pen and recording tablet.

As you enter the gloomy mass, the lymphatic stream becomes shallow and begins to flow sluggishly. So that you can explore this little organ fully, you haul yourself to your feet and begin to wade through the slowly moving stream. On each bank you see a huge ball of cells that have large nuclei and such a scant amount of cytoplasm that you can barely make it out. You write, "Sighted the spherical germinal centers composed of __(3)__ ." As you again study one of the cell masses, you spot one cell that looks quite different and reminds you of a nest of angry hornets because it is furiously spewing out what seems to be a horde of tiny Y-shaped "bees." "Ah ha," you think, "another valuable piece of information." You record, "Spotted a __(4)__ making and releasing __(5)__ ."

That done, you turn your attention to scanning the rest of the landscape. Suddenly you let out an involuntary yelp. Something quite enormous and looking much like an octopus is nearly blocking the narrow tunnel just ahead. Your mind whirls as it tries to figure out the nature of this cellular "beast" that appears to be guarding the channel. Then it hits you—this has to be a __(6)__ on the alert for foreign invaders (more properly called __(7)__), which it "eats" when it catches them. The giant cell roars, "Halt, stranger, and be recognized," and you dig frantically in your pocket for your identification pass. As you drift toward the huge cell, you hold the pass in front of you, hands trembling because you know this cell could liquefy you as quick as the blink of an eye. Again the cell bellows at you, "Is this some kind of a security check? I'm on the job, as you can see!" Frantically you shake your head, no, and the cell lifts one long tentacle and allows you to pass. As you squeeze by, the cell says, "Being inside, I've never seen my body's outside. I must say, humans are a rather strange-looking lot!" Still shaking, you decide that you are in no mood for a chat and hurry along to put some distance between yourself and this guard cell.

Immediately ahead are what appear to be hundreds of the same type of cell sitting on every ledge and in every nook and cranny. Some are busily snagging and engulfing unfortunate strep bacteria that float

_____ **8.**

_____ **9.**

_____ **10.**

_____ **11.**

too close. The slurping sound is nearly deafening. Then something grabs your attention—the surface of one of these cells is becoming dotted with some of the same donut-shaped chemicals that you see on the strep bacteria membranes; and a round cell, similar but not identical to those you saw earlier in the germinal centers, is starting to bind to one of these "doorknobs." You smile smugly because you know you have properly identified the octopuslike cells. You then record your observations as follows: "Cells like the giant cell just identified act as __(8)__ . I have just observed one in this role during its interaction with a __(9)__ cell."

You decide to linger a bit to see if the round cell becomes activated. You lean against the tunnel walls and watch quietly, but your wait is brief. Within minutes, the cell that was binding to the octopuslike cell begins to divide, and then its daughter cells divide again and again at a head-spinning pace. You write, "I have just witnessed the formation of a __(10)__ of like cells." Most of the daughter cells enter the lymph stream, but a few of them settle back and seem to go into a light sleep. You decide that the "napping cells" don't have any role to play in helping get rid of your host's present strep infection, but instead will provide for __(11)__ and become active at a later date.

You glance at your watch and wince as you realize that it is already five minutes past the time for your retrieval. You have already concluded that this is a dangerous place, and you are not sure how long your pass is good—so you swim hurriedly from the organ into the lymphatic stream to reach your pickup spot.

CHALLENGING YOURSELF

At the Clinic

1. As an infant receives her first dose of oral polio vaccine, the nurse explains to her parents that the vaccine is a preparation of weakened virus. What type of immunity will the infant develop?

2. After two late-term miscarriages, Mrs. Lyons, who is pregnant again, is having her Rh antibody level checked. What are the Rh types of Mrs. Lyons and her husband and what class of antibodies will be the chief culprit of any problems for the fetus?

3. What xenograft is performed fairly routinely in heart surgery?

4. Nicky, a young SCID sufferer, is exhibiting widespread inflammation after a marrow transplant. Tests show high levels of autoantibodies in Nicky's blood. What name is given to this condition?

5. A young man is rushed to the emergency room after fainting. His blood pressure is alarmingly low and his companion reports the man collapsed shortly after being stung by a wasp. What has caused his hypotension? What treatment will be given immediately?

6. Patty Hourihan is a strict environmentalist and a new mother. Although she is very much against using disposable diapers, she is frustrated by the fact that her infant breaks out in a diaper rash when she uses cloth diapers. Since new cloth diapers do not cause the rash, but washed ones do, what do you think the problem is?

7. A week after starting on aspirin to head off blood clots, Mr. Russell, a scholarly older man, reports to the clinic with large bruises over much of his body. What is his diagnosis, and in what class of immune disorders should it be placed?

8. About six months after an automobile accident in which her neck was severely lacerated, a young woman comes to the clinic for a routine checkup. Visual examination shows a slight swelling beneath the larynx. Upon questioning, the woman reports that she fatigues easily, has been gaining weight, and her hair is falling out. What is your diagnosis?

9. Anna is complaining of cramping, nausea, and diarrhea, which she thinks is a reaction to the dust stirred up by the construction near her home. What is a more likely explanation of her symptoms, assuming an allergy is involved?

10. The mother of a 7-year-old boy with a bad case of mumps is worried that he might become sterile. What will you tell her to allay her fears?

Stop and Think

1. Use of birth control pills decreases the acidity of the vaginal tract. Why might this increase the incidence of vaginitis?

2. Are all pathogens killed by stomach acid?

3. Why does the risk of infection increase when areas such as the auditory tube and sinus openings are swollen?

4. How does chewing sugar-free gum decrease the likelihood of cavities?

5. Dissect the words histamine, diapedesis, and interleukin.

6. The genetic "scrambling" of DNA segments that occurs during the process of gaining immunocompetence provides the best immunological response to the uncountable number of antigenic possibilities. What other part of the body must also respond to large numbers of variable chemicals, and so might also involve genetic "scrambling?" (Think!)

7. Is the allergen in poison ivy sap a water-soluble or lipid-soluble molecule? Explain your reasoning.

8. What is lymphocyte recirculation? What role does it play?

COVERING ALL YOUR BASES

Multiple Choice

Select the best answer or answers from the choices given.

1. Specialized macrophages include:
   A. dendritic cells
   B. Kupffer cells
   C. Langerhans' cells
   D. natural killer cells

2. Which of the following are cardinal signs of inflammation?
   A. Phagocytosis   C. Leukocytosis
   B. Edema   D. Pain

3. Chemical mediators of inflammation include:
   A. pyrogens   C. histamine
   B. prostaglandins D. lymphokines

4. Aspirin inhibits:
   A. kinins   C. chemotactic agents
   B. histamine   D. prostaglandins

5. Which term indicates the WBCs' ability to squeeze their way through the capillary endothelium to get into the tissue spaces?
   A. Diapedesis   C. Margination
   B. Degranulation   D. Positive chemotaxis

6. Neutrophils die in the line of duty because:
   A. they ingest infectious organisms
   B. their membranes become sticky and they are attacked by macrophages
   C. they secrete cellular toxins, which affect them in the same way they affect pathogens
   D. the buildup of tissue fluid pressure causes them to lyse

7. Which of the following is a component of the alternative, but not the classical, pathway of complement activation?
   A. Activation of C1 and C2
   B. Exposure to polysaccharides of certain pathogens
   C. Complement fixation by antibodies
   D. Insertion of MAC in cell membranes of pathogenic organisms

8. Against which of the following will interferon do some good?
   A. Infection of body cells by a virus
   B. Circulating population of free virus
   C. Some types of cancer
   D. Bacterial infection

9. Characteristics of the immune response include:
   A. antigen specificity
   B. memory
   C. systemic protection
   D. species specificity

10. Antigen recognition is due to:
    A. genetic programming
    B. antigen exposure
    C. somatic recombination
    D. immune surveillance

11. Macrophages:
    A. form exudate
    B. present antigens
    C. secrete interleukin 1
    D. activate helper T cells

12. Immunogenicity is a property of:
    A. antibodies   C. antigens
    B. effector T cells   D. macrophages

13. The binding and clustering of B cell receptors by a multivalent antigen is required for:
    A. clonal selection   C. activation
    B. antigen endocytosis   D. degranulation

14. In the primary immune response:
    A. B cell cloning occurs
    B. plasma cell exocytosis is at a high level beginning about 6 days after exposure
    C. IgM antibodies initially outnumber IgG antibodies
    D. plasma cell life span is about 5 days

15. Antibodies secreted in mother's milk:
    A. are IgG antibodies
    B. are IgA antibodies
    C. provide natural active immunity
    D. provide natural passive immunity

16. Conditions for which passive artificial immunity is the treatment of choice include:
    A. measles   C. rabies
    B. botulism   D. venomous snake bite

17. Which parts of an antibody molecule are different for an IgG antibody than for an IgM antibody that attacks the same antigen?
    A. Heavy chain constant region
    B. Heavy chain variable region
    C. Light chain constant region
    D. Light chain variable region

18. Which of the following is important for activation of a B cell during the antigen challenge?
    A. The antigen
    B. A helper T cell
    C. Chemicals that stimulate the lymphocytes to divide
    D. Interleukin 2

19. Which of these antibody classes is usually arranged as a pentamer?
    A. IgG   C. IgA
    B. IgM   D. IgD

20. Which of the following antibody capabilities causes a transfusion reaction with A or B antigens?
    A. Neutralization   C. Complement fixation
    B. Precipitation   D. Agglutination

21. Class II MHC proteins are found on the membranes of:
    A. all body cells
    B. macrophages
    C. mature B lymphocytes
    D. some T cells

22. Effector T cells secrete:
    A. tumor necrosis factor
    B. monokines
    C. perforin
    D. interleukin 2

23. Which of the following terms is applicable to the use of part of the great saphenous vein in coronary bypass surgery?
    A. Isograft   C. Allograft
    B. Xenograft   D. Autograft

24. Which of the following is/are examples of autoimmune disease?
    A. Juvenile diabetes
    B. Multiple sclerosis
    C. Graves disease
    D. Rheumatoid arthritis

25. The predominant cell type in the germinal centers of the lymph nodes are:
    A. proliferating B cells
    B. circulating T cells
    C. macrophages
    D. plasma cells

Word Dissection

For each of the following word roots, fill in the literal meaning and give an example, using a word found in this chapter.

Word root	Translation	Example
1. hapt	_____	_____
2. humor	_____	_____
3. macro	_____	_____
4. opso	_____	_____
5. penta	_____	_____
6. phylax	_____	_____
7. phago	_____	_____
8. pyro	_____	_____
9. vacc	_____	_____

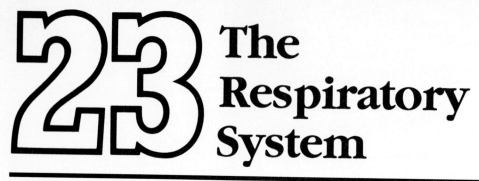

23 The Respiratory System

Student Objectives

When you have completed the exercises in this chapter, you will have accomplished the following objectives:

Functional Anatomy of the Respiratory System

1. Identify the organs forming the respiratory passageway(s) in descending order until the alveoli are reached. Distinguish between conducting and respiratory zone structures.

2. List and describe several protective mechanisms of the respiratory system.

3. Describe the makeup of the respiratory membrane and relate its structure to its function.

4. Describe the gross structure of the lungs and pleural coverings.

Mechanics of Breathing

5. Relate Boyle's law to the events of inspiration and expiration.

6. Explain the relative roles of the respiratory muscles and lung elasticity in effecting volume changes that cause air to flow into and out of the lungs.

7. Explain the functional importance of the partial vacuum that exists in the intrapleural space.

8. Describe several physical factors that influence pulmonary ventilation.

9. Explain and compare the various lung volumes and capacities. Discuss types of information that can be gained from pulmonary function tests.

10. Define dead space.

Gas Exchanges in the Body

11. Describe, in general terms, differences in composition of atmospheric and alveolar air and explain these differences.

12. State Dalton's law of partial pressures and Henry's law. Relate each to events of external and internal respiration.

Transport of Respiratory Gases by Blood

13. Describe how oxygen is transported in the blood, and explain how oxygen loading and unloading is affected by temperature, pH, 2,3-BPG, and P_{CO_2}.

14. Describe carbon dioxide transport in the blood.

Control of Respiration

15. Describe the neural controls of respiration.

16. Compare and contrast the influences of lung reflexes, volition, emotions, arterial pH, and partial pressures of oxygen and carbon dioxide in arterial blood on respiratory rate and depth.

Respiratory Adjustments During Exercise and at High Altitudes

17. Compare and contrast the hyperpnea of exercise with involuntary hyperventilation.

18. Describe the process and effects of acclimatization to high altitude.

Homeostatic Imbalances of the Respiratory System

19. Compare the causes and consequences of chronic bronchitis, emphysema, and lung cancer.

Developmental Aspects of the Respiratory System

20. Follow briefly the development of the respiratory system in an embryo.

21. Note normal changes that occur in respiratory system functioning from infancy to old age.

Ready, set, go!

Body cells require an abundant and continuous supply of oxygen to carry out their activities. As cells use oxygen, they release carbon dioxide, a waste product the body must get rid of. The circulatory and respiratory systems obtain and deliver oxygen to body cells and eliminate carbon dioxide from the body. The respiratory system structures are responsible for gas exchange between the blood and the external environment (that is, external respiration). The respiratory system also plays an important role in maintaining the acid-base balance of the blood.

Questions and activities in Chapter 23 consider both the anatomy and the physiology of the respiratory system structures.

BUILDING THE FRAMEWORK

Functional Anatomy of the Respiratory System

1. What are the four main events of respiration?

   _____

   _____

   _____

   _____

2. The respiratory system is divisible into *conducting zone* and *respiratory zone* structures.

   1. Name the respiratory zone structures. _____

      _____

   2. What is their common function? _____

   3. Name the conducting zone structures. _____

      _____

      _____

3. Use the terms provided to complete the pathway of air to the primary bronchi. Some of the terms have already been filled in. The main pathway—through the upper respiratory organs—is the vertical pathway at the far left. To the right of the name of most of those organs are horizontal lines for listing the sub-divisions of that organ or the structures over which air passes as it moves through that organ.

Adenoids	Internal nares	Nasopharynx	Trachea
Carina	Laryngopharynx	Nose	Vocal folds
Epiglottis	Larynx	Oropharynx	
External nares	Middle meatus	Pharynx	

Nose: external nares ⟶ middle meatus ⟶ internal nares
↓
pharynx: nasopharynx ⟶ adenoids ⟶ oropharynx
↓   laryngopharynx
Larynx: epiglottis ⟶ vocalfolds
↓
Trachea ⟶ carina ⟶ Primary bronchi

4. Figure 23.1 illustrates the skeletal framework of the nose. Identify the various bones and cartilages contributing to that framework by color coding and coloring them on the figure.

○ Dense fibrous connective tissue

○ Greater alar cartilages

○ Lateral cartilage

○ Lesser alar cartilage

○ Maxillary bone (frontal process)

○ Nasal bone

○ Septal cartilage

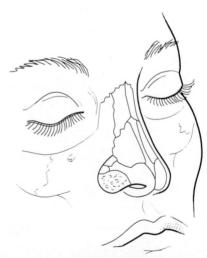

Figure 23.1

5. Figure 23.2 is a sagittal view of the upper respiratory structures. First, correctly identify all structures with leader lines on the figure. Then, select different colors for the structures listed below and color the coding circles and the corresponding structures on the figure.

○ Nasal cavity　　　　○ Larynx　　　　○ Tongue

○ Pharynx　　　　　　○ Paranasal sinuses　　○ Hard palate

○ Trachea

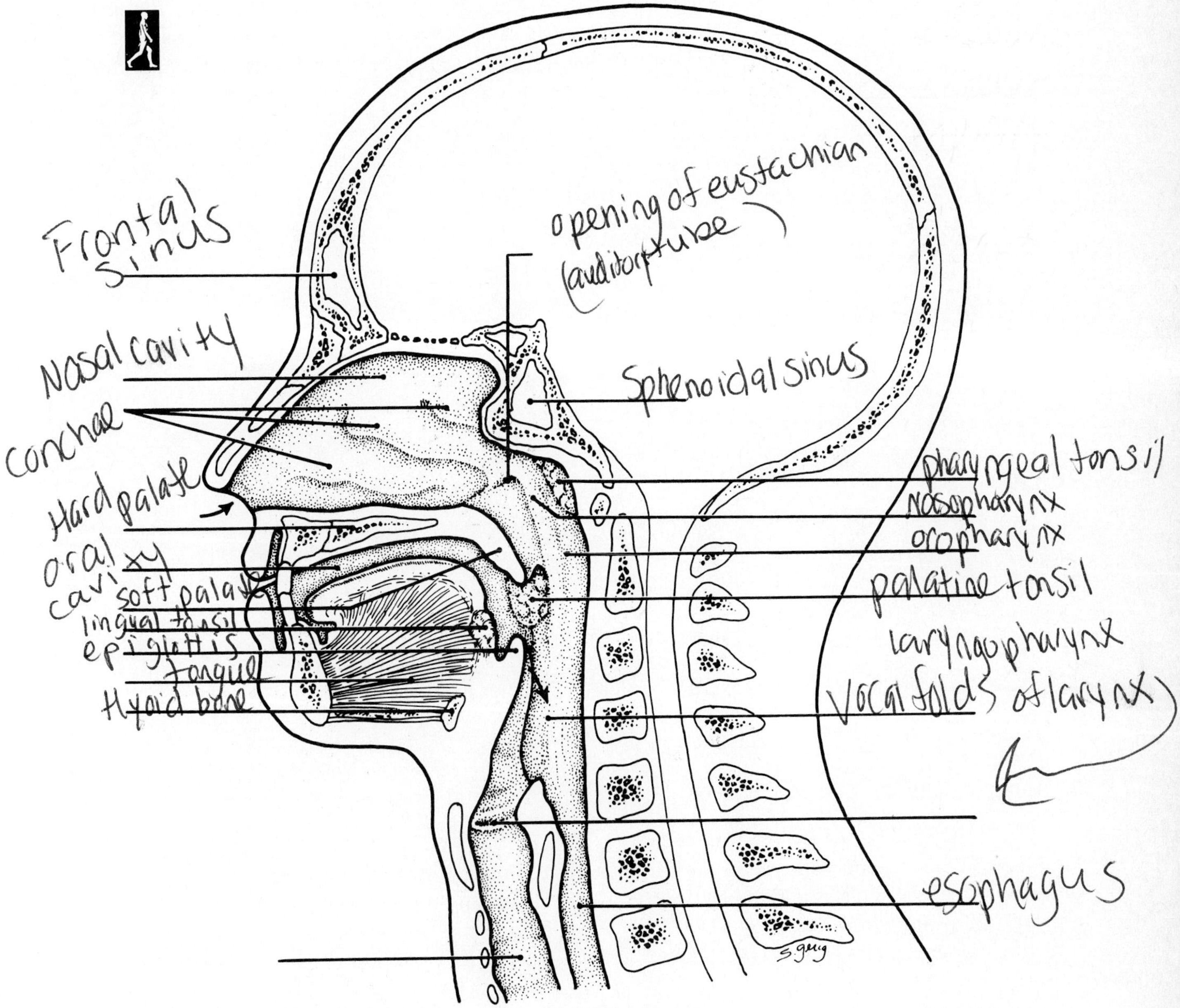

Frontal sinus

Nasal cavity

conchae

Hard palate

oral cavity

soft palate

lingual tonsil

epiglottis

Tongue

Hyoid bone

opening of eustachian (auditory tube)

Sphenoidal sinus

pharyngeal tonsil

nasopharynx

oropharynx

palatine tonsil

laryngopharynx

vocal folds of larynx

esophagus

S geig

Figure 23.2

6. Complete the following statements concerning the nasal cavity and adjacent structures by inserting your answers in the answer blanks.

external nares _____ 1.

nasal septum _____ 2.

warm _____ 3.

moisten _____ 4.

trap debris in _____ 5.

sinuses _____ 6.

speech _____ 7.

pharynx _____ 8.

larynx _____ 9.

tonsils _____ 10.

Air enters the nasal cavities of the respiratory system through the __(1)__ . The nasal cavity is divided by the midline __(2)__ . The nasal cavity mucosa has several functions. Its major functions are to __(3)__ , __(4)__ , and __(5)__ the incoming air. Mucous membrane–lined cavities called __(6)__ are found in several bones surrounding the nasal cavities. They make the skull less heavy and probably act as resonance chambers for __(7)__ . The passageway common to the digestive and respiratory systems, the __(8)__ , is often referred to as the throat; it connects the nasal cavities with the __(9)__ below. Clusters of lymphatic tissue, __(10)__ , are part of the defensive system of the body.

7. Circle the term that does not belong in each of the following groupings.

1. Sphenoidal   Maxillary   Mandibular   Ethmoid   Frontal

2. Nasal cavity   Trachea   Alveolus   Bronchus   Dead air

3. Apex   Base   Hilus   Larynx   Pleura

4. Sinusitis   Peritonitis   Pleurisy   Tonsillitis   Laryngitis

5. External nose   Nasal bones   Lateral cartilages   Nasal septum

6. Conchae   Increase air turbulence   Turbinates   Choanae

7. Cuneiform cartilage   Corniculate cartilage   Tracheal cartilage

Arytenoid cartilage   Epiglottis

8. Laryngopharynx   Oropharynx   Transports air and food   Nasopharynx

8. Figure 23.3 is a diagram of the larynx and associated structures. Select a different color for each structure and color the coding circles and the corresponding structures on the figure. Identify, by adding a label and leader line, the laryngeal prominence. Then, answer the questions below the figure.

◯ Hyoid bone   ◯ Tracheal cartilages   ◯ Thyrohyoid membrane

◯ Cricoid cartilage   ◯ Epiglottis   ◯ Cricothyroid ligament

◯ Thyroid cartilage   ◯ Cricotracheal ligament

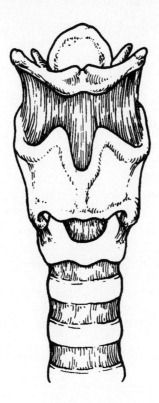

Figure 23.3

1. What are the three functions of the larynx? _____

_____

_____

_____

2. What cartilages, not illustrated, anchor the vocal cords internally?

_____

3. What type of cartilage forms the epiglottis? _____

4. What type of cartilage forms the other eight laryngeal cartilages? _____

5. Explain this difference. _____

_____

_____

6. What is the common name for the laryngeal prominence?

_____

9. Use the key choices to match the following definitions with the correct terms.

KEY CHOICES

A. Alveoli	**E.** Esophagus	**I.** Phrenic nerve	**M.** Visceral pleura
B. Bronchioles	**F.** Glottis	**J.** Primary bronchi	**N.** Uvula
C. Conchae	**G.** Palate	**K.** Trachea	**O.** Vocal folds
D. Epiglottis	**H.** Parietal pleura	**L.** Vagus nerve	

B _____ **1.** Smallest respiratory passageways

G _____ **2.** Separates the oral and nasal cavities

I _____ **3.** Major nerve stimulating the diaphragm

E _____ **4.** Food passageway posterior to the trachea

D _____ **5.** Closes off the larynx during swallowing

K _____ **6.** Windpipe

A _____ **7.** Actual site of gas exchanges

H _____ **8.** Pleural layer covering the thorax walls

L _____ **9.** Autonomic nervous system nerve serving the thorax

F _____ **10.** Lumen of the larynx

C _____ **11.** Fleshy lobes in the nasal cavity that increase its surface area

O _____ **12.** Close the glottis during the Valsalva maneuver

N _____ **13.** Closes the nasopharynx during swallowing

K _____ **14.** The cilia of its mucosa beat upward toward the larynx

10. Complete the following statements by writing the missing terms in the answer blanks.

_____ 1.

_____ 2.

_____ 3.

_____ 4.

_____ 5.

_____ 6.

_____ 7.

_____ 8.

_____ 9.

Within the larynx are the __(1)__, which vibrate with exhaled air and allow an individual to __(2)__ . Speech involves opening and closing the __(3)__ . The __(4)__ muscles control the length of the true vocal cords by moving the __(5)__ cartilages. The tenser the vocal cords, the __(6)__ the pitch. To produce deep tones, the glottis is __(7)__ . The greater the force of air rushing past the vocal cords, the __(8)__ the sound produced. Inflammation of the vocal cords is called __(9)__ .

11. Figure 23.4 shows a cross section through the trachea. First, label the layers indicated by the leader lines. Next, color the following: cilia and epithelium—light pink; lamina propria—dark pink; area containing the submucosal seromucous glands—purple; hyaline cartilage ring—blue; trachealis muscle—orange; and adventitia—yellow. Then, respond to the questions following the figure.

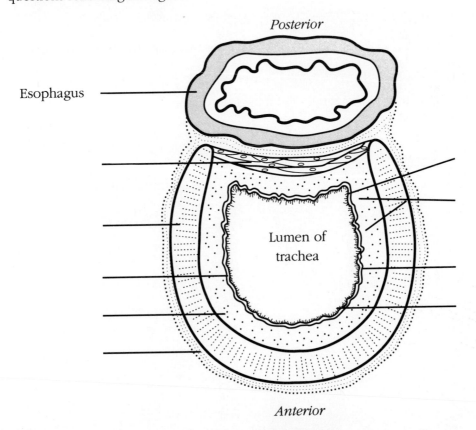

Posterior

Esophagus

Lumen of trachea

Anterior

Figure 23.4

1. What important role is played by the cartilage rings that reinforce the trachea?

   _____

2. Of what importance is the fact that the cartilage rings are incomplete posteriorly?

   _____

3. What occurs when the trachealis muscle contracts and in what activities might this action be

   very helpful? _____

   _____

12. Using the key choices, match the proper type of lining epithelium with each respiratory structure listed below.

KEY CHOICES

A. Stratified squamous

B. Pseudostratified ciliated columnar

C. Simple squamous

D. Stratified columnar

E. Simple cuboidal

_____ **1.** Nasal cavity

_____ **2.** Nasopharynx

_____ **3.** Laryngopharynx

_____ **4.** Trachea and primary bronchi

_____ **5.** Bronchioles

_____ **6.** Walls of alveoli (type I cells)

13. Match the bronchus or bronchiole type (at right) with the lung region supplied by that specific type of air tube (at left).

Lung region

_____ **1.** Bronchopulmonary segment

_____ **2.** Lobule

_____ **3.** Alveolar ducts and sacs

_____ **4.** Entire lung

_____ **5.** Lung lobe

Air tube

A. Primary bronchus

B. Secondary bronchus

C. Tertiary bronchus

D. Large bronchiole

E. Respiratory bronchiole

14. Figure 23.5 illustrates the gross anatomy of lower respiratory structures. Intact structures are shown on the left; isolated respiratory passages are shown on the right. Select a different color for each of the respiratory system structures listed below and color the coding circles and the corresponding structures on the figure. Then complete the figure by labeling the areas and structures with leader lines. Be sure to include the pleural space, mediastinum, apex of right lung, diaphragm, clavicle, and the base of the right lung.

○ Trachea   ○ Primary bronchi   ○ Visceral pleura

○ Larynx   ○ Secondary bronchi   ○ Parietal pleura

○ Intact lung ○ Tertiary bronchi

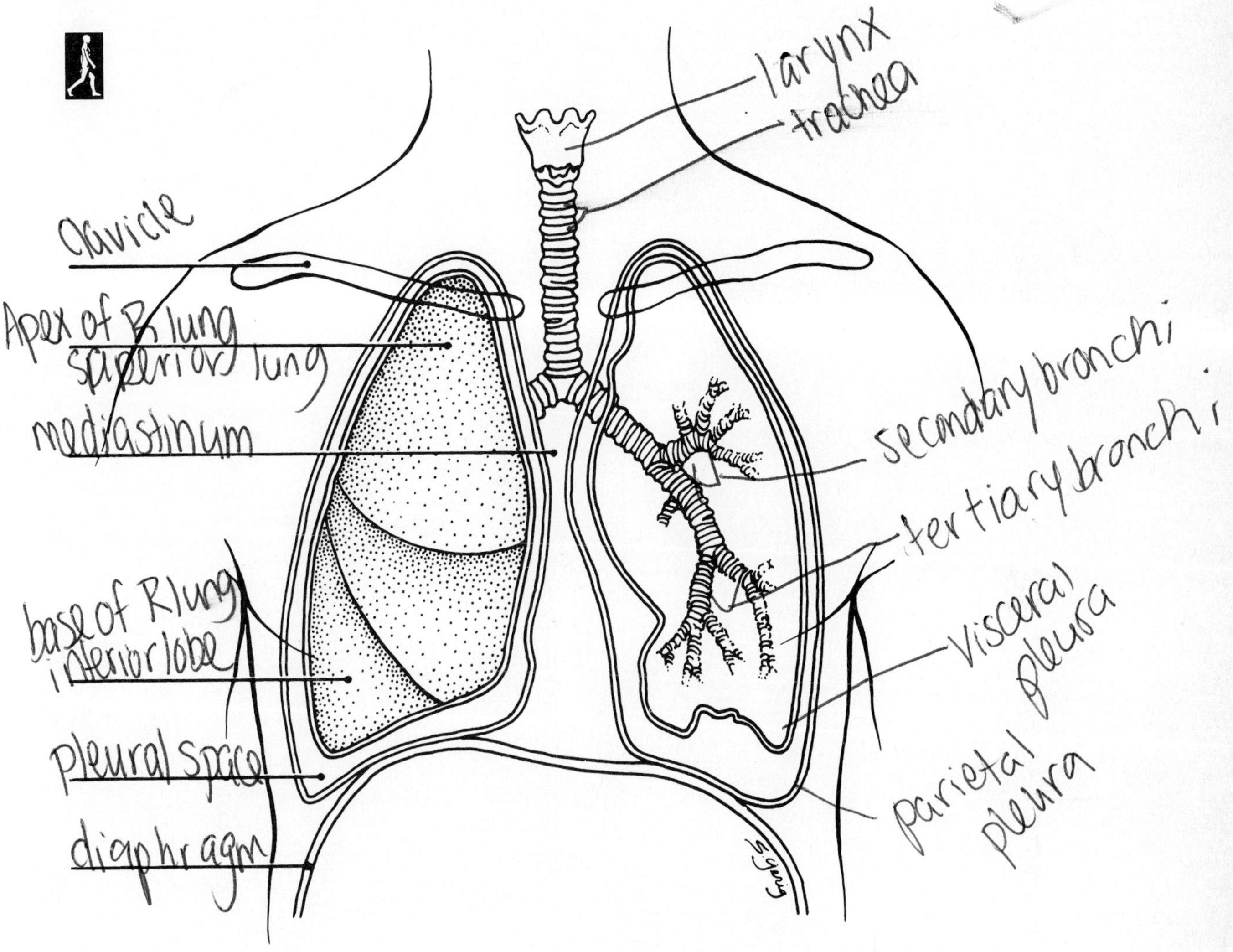

Figure 23.5

15. Figure 23.6 illustrates the microscopic structure of the respiratory unit of lung tissue. The external anatomy is shown in Figure 23.6A. Color the intact alveoli yellow, the pulmonary capillaries red, and the respiratory bronchioles green. Bracket and label an alveolar sac. Also label an alveolar duct.

A cross section through an alveolus is shown in Figure 23.6B. Color the alveolar epithelium yellow, the capillary endothelium pink, and the red blood cells in the capillary red. Also, label the alveolar chamber and color it blue. Finally, add the symbols for oxygen gas (O₂) and carbon dioxide gas (CO₂) in the sites where they would be in higher concentration and add arrows showing their direction of movement through the respiratory membrane.

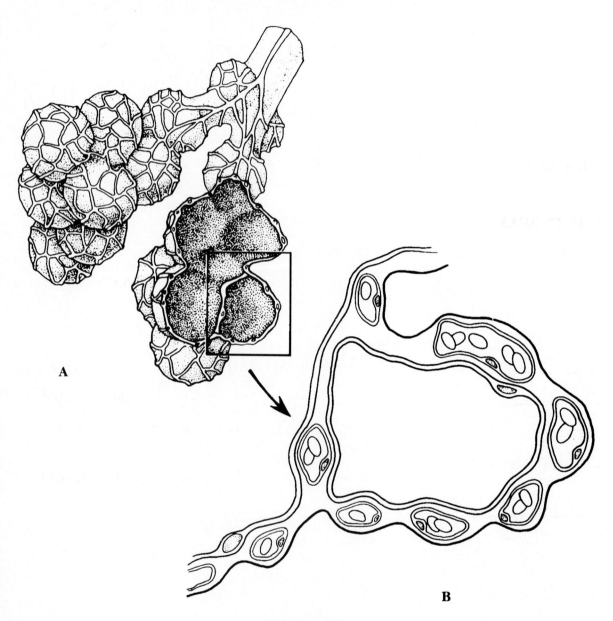

A

B

Figure 23.6

16. Complete the following paragraph on the alveolar cells and their roles by writing the missing terms in the answer blanks.

_____ 1.

_____ 2.

_____ 3.

_____ 4.

_____ 5.

_____ 6.

_____ 7.

With the exception of the stroma of the lungs, which is __(1)__ tissue, the lungs are mostly air spaces, of which the alveoli comprise the greatest part. The bulk of the alveolar walls are made up of squamous epithelial cells called __(2)__ cells. Structurally, these cells are well suited for their __(3)__ function. The cuboidal cells of the alveoli, called __(4)__ cells, are much less numerous. These cells produce a fluid that coats the air-exposed surface of the alveolus and contains a lipid-based molecule called __(5)__ that functions to __(6)__ of the alveolar fluid. Although the pulmonary capillaries spiderweb over the alveolar surfaces, the nutritive blood supply of the lungs is provided by the __(7)__ arteries.

Mechanics of Breathing

1. Using the key choices, match the following facts about pressure with the correct terms.

KEY CHOICES

A. Atmospheric pressure   **B.** Intrapulmonary pressure   **C.** Intrapleural pressure

_____ 1. Baring pneumothorax, this pressure is always lower than atmospheric pressure (that is, is negative pressure)

_____ 2. Pressure of air outside the body

_____ 3. As it decreases, air flows into the passageways of the lungs

_____ 4. As it increases over atmospheric pressure, air flows out of the lungs

_____ 5. If this pressure becomes equal to the atmospheric pressure, the lungs collapse

_____ 6. Rises well over atmospheric pressure during a forceful cough

_____ 7. Also known as the intra-alveolar pressure

2. Many changes occur within the lungs as the diaphragm (and external intercostal muscles) contract and then relax. These changes cause air to flow into and out of the lungs. The activity of the diaphragm is given in the left column of the following table. Several changes in internal thoracic conditions are listed in the column heads to the right. Complete the table by checking (✔) the appropriate column to correctly identify the change that would be occurring in each case relative to the stated diaphragm activity.

Activity of diaphragm (↑ = increased) (↓ = decreased)	Changes in							
	Internal volume of thorax		Internal pressure in thorax		Size of lungs		Direction of air flow	
	↑	↓	↑	↓	⤒	⤓ Å	Into lung	Out of lung
Contracted, moves downward	✓			✓	✓		✓	
Relaxed, moves superiorly		✓	✓			✓		✓

3. Various factors that influence pulmonary ventilation are listed in the key choices. Select the appropriate key choices to match the following descriptions about the lungs.

KEY CHOICES

A. Respiratory passageway resistance   C. Lung elasticity

B. Alveolar surface tension forces   D. Lung compliance

_____ 1. The ease of lung expansion

_____ 2. Gas flow changes inversely with this factor

_____ 3. Essential for normal expiration

_____ 4. Leads to RDS (respiratory distress syndrome) when surfactant is absent

_____ 5. Diminished by age-related lung fibrosis and increasing rigidity of the thoracic cage

_____ 6. Its loss is the major pathology in emphysema

_____ 7. Reflects a state of tension at the surface of a liquid

_____ 8. Dramatically increased by asthma

_____ 9. Greatest in the medium-sized bronchi

4. During forced inspiration, accessory muscles are activated that raise the rib cage more vigorously than occurs during quiet inspiration.

    1. Name three such muscles. _____

    2. Although normal quiet expiration is largely passive due to lung recoil, when expiration must be more forceful (or the lungs are diseased), muscles that increase the abdominal pressure or depress the rib cage are enlisted. Provide two examples of muscles that cause abdominal

    pressure to rise. _____

    3. Provide two examples of muscles that depress the rib cage. _____

5. Figure 23.7 is a diagram showing respiratory volumes. Complete the figure by making the following additions.

    1. Bracket the volume representing the vital capacity and color it yellow; label it VC.

    2. Add green stripes to the area representing the inspiratory reserve volume and label it IRV.

    3. Add red stripes to the area representing the expiratory reserve volume and label it ERV.

    4. Identify and label the respiratory volume, which is *now* yellow. Color the residual volume blue and label it appropriately on the figure.

    5. Bracket and label the inspiratory capacity (IC).

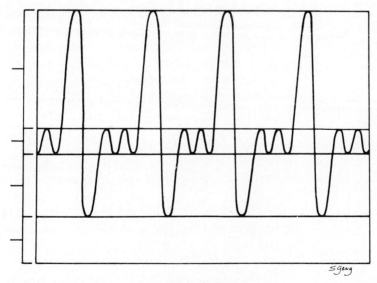

Figure 23.7

6. Check (✔) all factors that *decrease* the AVR.

    ____ **1.** Shallow breathing   ____ **3.** Rapid breathing   ____ **5.** Mucus in the respiratory passageways

    ____ **2.** Deep breathing   ____ **4.** Slow breathing

7. This exercise concerns respiratory volume or capacity measurements. Using the key choices, select the terms that identify the respiratory volumes described by inserting the appropriate answers in the answer blanks.

KEY CHOICES

A. Dead space volume   **D.** Residual volume (RV)   **G.** Vital capacity (VC)

B. Expiratory reserve volume (ERV)   **E.** Tidal volume (TV)

C. Inspiratory reserve volume (IRV) **F.** Total lung capacity (TLC)

_____E_____ **1.** Respiratory volume inhaled or exhaled during normal breathing

_____A_____ **2.** Air in respiratory passages that does not contribute to gas exchange

_____G_____ **3.** Total amount of exchangeable air

_____D_____ **4.** Gas volume that allows gas exchange to go on continuously

_____B_____ **5.** Amount of air that can still be exhaled (forcibly) after a normal exhalation

_____F_____ **6.** Sum of all lung volumes

8. Complete the following statements by writing the missing terms in the answer blanks.

_____ **1.**

_____ **2.**

_____ **3.**

_____ **4.**

_____ **5.**

Pulmonary function tests can be conducted using a __(1)__, a device that measures respired air volumes. This type of test can distinguish between __(2)__, disorders that result from increased airway resistance, and __(3)__, which result from decreased lung capacity caused by changes such as tuberculosis. __(4)__ is decreased by obstructive pulmonary diseases, which reduce the rate of air flow. Restrictive diseases lower __(5)__, because the total inflation capacity is reduced.

9. Identify the three nonrespiratory movements described below.

_____hiccup_____ **1.** Sudden inspiration, resulting from spasms of the diaphragm

_____cough_____ **2.** A deep breath is taken, the glottis is closed, and air is forced out of the lungs against the glottis; clears the lower respiratory passageways

_____sneeze_____ **3.** As just described, but clears the upper respiratory passageways

Gas Exchanges in the Body

1. This exercise describes some ideal gas laws that relate the physical properties of gases and their behavior in liquids. Match the key choices with the descriptions that follow.

   KEY CHOICES

   A. Boyle's law   **B.** Dalton's law   **C.** Henry's law

   _____ 1. Describes the fact that each gas in a mixture exerts pressure according to its percentage in the mixture

   _____ 2. States that $P_1V_1 = P_2V_2$ when temperature is constant

   _____ 3. States that when gases are in contact with liquids, each gas will dissolve in proportion to its partial pressure and solubility in the liquid

2. What therapy is used to treat cases of carbon monoxide poisoning, gas

   gangrene, and tetanus? _____

   This therapy represents a clinical application of which of the ideal gas laws?

   _____

3. What are the mechanism and consequences of oxygen toxicity? _____

   _____

   _____

4. Match the key choices with the following descriptions.

   KEY CHOICES

   A. External respiration   **C.** Inspiration   **E.** Pulmonary ventilation (breathing)

   B. Expiration   **D.** Internal respiration

   ___C___ 1. Period of breathing when air enters the lungs

   ___B___ 2. Period of breathing when air leaves the lungs

   ___E___ 3. Alternate flushing of air into and out of the lungs

   ___A___ 4. Exchange of gases between alveolar air and pulmonary capillary blood

   ___D___ 5. Exchange of gases between blood and tissue cells

5. Rank the following body regions by numbering them in the proper sequence in each column to indicate the relative partial pressure of the gases named in each column heading. Indicate the area of lowest gas partial pressure by the number 1.

P_{O_2}	P_{CO_2}	P_{N_2}	P_{H_2O}
____ **1.** Alveoli	____ **6.** Alveoli	____ **11.** Alveoli	____ **13.** Alveoli
____ **2.** External atmosphere	____ **7.** External atmosphere	____ **12.** External atmosphere	____ **14.** External atmosphere
____ **3.** Arterial blood	____ **8.** Arterial blood		
____ **4.** Venous blood	____ **9.** Venous blood		
____ **5.** Tissue cells	____ **10.** Tissue cells		

6. Use the key choices to complete the following statements about gas exchanges in the body.

KEY CHOICES

A. Active transport

B. Air of alveoli to capillary blood

C. Carbon dioxide-poor and oxygen-rich

D. Capillary blood to alveolar air

E. Capillary blood to tissue cells

F. Diffusion

G. Higher concentration

H. Lower concentration

I. Oxygen-poor and carbon dioxide-rich

J. Tissue cells to capillary blood

_____ **1.**

_____ **2.**

_____ **3.**

_____ **4.**

_____ **5.**

_____ **6.**

_____ **7.**

_____ **8.**

_____ **9.**

All gas exchanges are made by __(1)__ . When substances pass in this manner, they move from areas of their __(2)__ to areas of their __(3)__ . Thus oxygen continually passes from the __(4)__ and then from the __(5)__ . Conversely, carbon dioxide moves from the __(6)__ and from __(7)__ . From there it passes out of the body during expiration. As a result of such exchanges, arterial blood tends to be relatively __(8)__ , while venous blood is relatively __(9)__ .

7. Circle the correct alternative in each of the following statements about alveolar airflow and blood flow coupling.

   1. When alveolar ventilation is inadequate, the P_{O_2} is low/high.

   2. Consequently, the pulmonary vessels dilate/constrict.

   3. Additionally, when alveolar ventilation is inadequate, the P_{CO_2} is low/high.

   4. Consequently, the bronchioles dilate/constrict.

Transport of Respiratory Gases by Blood

1. Complete the following paragraphs concerning gas transport in the blood by writing the missing terms in the answer blanks.

hemoglobin	**1.**
plasma	**2.**
bicarbonate ion	**3.**
carbonic anhydrase	**4.**
chloride ions	**5.**
chloride shift	**6.**
carbamino Hb	**7.**
Haldane effect	**8.**
more	**9.**
hemoglobin	**10.**
Bohr effect	**11.**
oxygen	**12.**

Most oxygen is transported bound to __(1)__ inside the red blood cells. A very small amount is carried simply dissolved in __(2)__ . Most carbon dioxide is carried in the form of __(3)__ in the plasma. The reaction in which CO_2 is converted to this chemical occurs more rapidly inside red blood cells than in plasma because the cells contain the enzyme __(4)__ . After the chemical is generated, it diffuses into the plasma, and __(5)__ diffuse into the red blood cells. This exchange is called the __(6)__ . Smaller amounts of carbon dioxide are transported dissolved in plasma and bound to hemoglobin as __(7)__ .

The amount of CO_2 carried in blood is greatly influenced by the degree of oxygenation of the blood, an effect called the __(8)__ . In general, the lower the amount of oxygen in the blood, the __(9)__ CO_2 that can be transported. Conversely, as CO_2 enters the blood, it prompts oxygen to dissociate from __(10)__ . This effect is called the __(11)__ . Carbon monoxide poisoning is lethal because carbon monoxide competes with __(12)__ for binding sites.

2. List three factors that enhance O_2 loading onto hemoglobin.

high P_{O_2} of bl; low temp; alkalosis; low levels of 2-3 BPG

3. What is the functional importance of the following facts?

   1. Hemoglobin is over 90% saturated with oxygen well before the blood has completed its route through the pulmonary capillaries.

2. Venous blood still has a hemoglobin saturation of over 70%.

_____

_____

_____

4. In the grid provided in Figure 23.8, plot the following points. Then draw a
sigmoid curve to connect the points. Finally, answer the questions following
the grid.

a. P_{O_2} = 100 mm Hg; saturation = 99%   **d.** P_{O_2} = 20 mm Hg; saturation = 35%

b. P_{O_2} = 70 mm Hg; saturation = 98%   **e.** P_{O_2} = 10 mm Hg; saturation = 10%

c. P_{O_2} = 40 mm Hg; saturation = 75%   **f.** P_{O_2} = 0 mm Hg; saturation = 0%

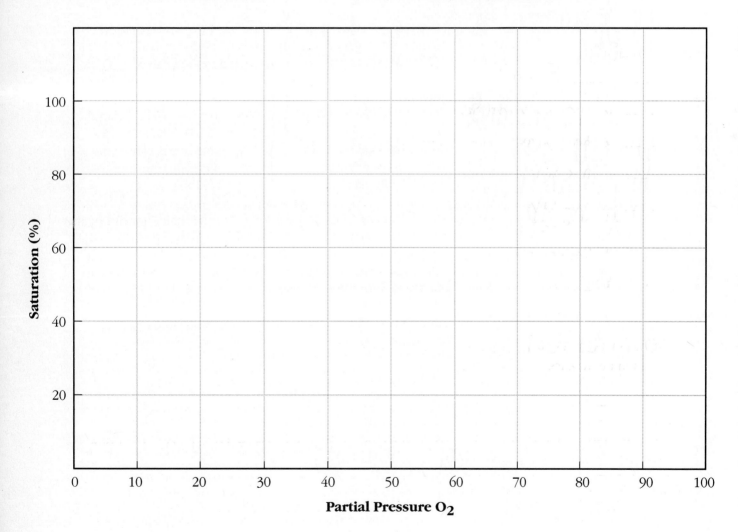

Figure 23.8

1. What is the curve just plotted called? _____

   _____

2. What three factors will decrease the percent saturation of hemoglobin

   in the tissues? _____

   _____

Control of Respiration

1. There are several types of breathing controls. Match the structures in Column B with the descriptions in Column A.

   Column A

   _____ , _____ **1.** Actual or hypothetical respiratory control centers in the pons

   _____ , _____ **2.** Respiratory control centers in the medulla

   _____ **3.** Respond to overinflation and underinflation of the lungs

   _____ **4.** Respond to decreased oxygen levels in the blood

   _____ **5.** Also called the inspiratory center

   _____ **6.** Medullary area that contains approximately even numbers of neurons involved in inspiration and expiration

   _____ **7.** The breathing "rhythm" center

   _____ **8.** Basically acts to inhibit the dorsal respiratory group (DRG)

   Column B

   A. Apneustic center

   B. Chemoreceptors in the aortic arch and carotid body

   C. VRG

   D. DRG

   E. Pneumotaxic center

   F. Stretch receptors in the lungs

2. Why is it inadvisable to administer pure O_2 to a patient who suffers from pulmonary disease and retains CO_2?

   _____

   _____

   _____

   _____

   _____

3. For each of the following, designate whether it primarily affects the central chemoreceptors or the peripheral chemoreceptors:

_____ **1.** Arterial P_{O_2} below 60 mm Hg stimulates hyperventilation.

_____ **2.** Decrease in arterial pH stimulates hyperventilation.

_____ **3.** Increase in P_{CO_2} in the CSF causes increase in respiratory rate and depth.

4. Circle the term that does not belong in each of the following groupings.

1. Acidosis $\uparrow$ Carbonic acid $\downarrow$ pH $\uparrow$ pH

2. Acidosis Hyperventilation Hypoventilation CO_2 buildup

3. Apnea Cyanosis $\uparrow$ Oxygen $\downarrow$ Oxygen

4. $\uparrow$ Respiratory rate $\uparrow$ Exercise Anger $\uparrow$ CO_2 in blood

5. N narcosis The bends N toxicity Rapture of the deep

6. High altitude $\downarrow$ P_{O_2} $\uparrow$ P_{CO_2} $\downarrow$ Atmospheric pressure

Respiratory Adjustments During Exercise and at High Altitudes

1. Answer the following questions concerning respiratory adjustments during exercise:

1. What are two differences between hyperpnea and hyperventilation? _____

_____

_____

2. What are three neural factors that trigger the abrupt increase in

ventilation during exercise? _____

_____

2. List three adaptive responses to high-altitude living.

1. _____

2. _____

3. _____

Homeostatic Imbalances
of the Respiratory System

1. Match the terms in Column B with the pathologic conditions described in Column A.

Column A

_____	**1.** Lack or cessation of breathing
_____	**2.** Normal breathing in terms of rate and depth
_____	**3.** Labored breathing, or "air hunger"
_____	**4.** Chronic oxygen deficiency
_____	**5.** Condition characterized by fibrosis of the lungs and an increase in size of the alveolar chambers
_____	**6.** Condition characterized by increased mucus production, which clogs respiratory passageways and promotes coughing
_____	**7.** Respiratory passageways narrowed by bronchiolar spasms
_____	**8.** Together called COPD
_____	**9.** Incidence strongly associated with cigarette smoking; has increased dramatically in women recently
_____	**10.** Victims become barrel-chested because of air retention
_____	**11.** Infection spread by airborne bacteria; a recent alarming increase in cases among drug users and HIV-infected people
_____	**12.** Temporary cessation of breathing during sleep

Column B

A. Apnea

B. Asthma

C. Chronic bronchitis

D. Dyspnea

E. Emphysema

F. Eupnea

G. Hypoxia

H. Lung cancer

I. Sleep apnea

J. Tuberculosis

Developmental Aspects
of the Respiratory System

1. Complete the following statements by writing the missing terms in the answer blanks.

_____ 1.

_____ 2.

_____ 3.

_____ 4.

_____ 5.

_____ 6.

_____ 7.

_____ 8.

_____ 9.

_____ 10.

The mucosa of the upper respiratory system develops from the ectodermal __(1)__; that of the lower respiratory system develops from the __(2)__ of the foregut.

The respiratory rate of a newborn baby is approximately __(3)__ respirations per minute. In a healthy adult, the respiratory range is __(4)__ respirations per minute.

Most problems that interfere with the operation of the respiratory system fall into one of the following categories: *infections*, such as pneumonia; *obstructive conditions*, such as __(5)__ and __(6)__; and *conditions that destroy lung tissue*, such as __(7)__. With age, the lungs lose their __(8)__ and the __(9)__ of the lungs decreases. Protective mechanisms also become less efficient, causing elderly individuals to be more susceptible to __(10)__.

The Incredible Journey:
A Visualization Exercise
for the Respiratory System

You carefully begin to pick your way down, using cartilages as steps.

1. Complete the following narrative by writing the missing terms in the answer blanks.

_____ 1.

_____ 2.

_____ 3.

_____ 4.

_____ 5.

_____ 6.

_____ 7.

_____ 8.

Your journey through the respiratory system is to be on foot. To begin, you simply will walk into your host's external nares. You are miniaturized, and your host is sedated lightly to prevent sneezing during your initial observations in the nasal cavity and subsequent descent.

You begin your exploration of the nasal cavity in the right nostril. One of the first things you notice is that the chamber is very warm and humid. High above, you see three large, round lobes, the __(1)__, which provide a large mucosal surface area for warming and moistening the entering air. As you walk toward the rear of this chamber, you see a large lumpy mass of lymphatic tissue, the __(2)__ in the __(3)__, or first portion of the pharynx. As you peer down the pharynx, you realize that it will

_____ 9.

_____ 10.

_____ 11.

_____ 12.

_____ 13.

_____ 14.

_____ 15.

_____ 16.

_____ 17.

_____ 18.

_____ 19.

_____ 20.

be next to impossible to maintain your footing during the next part of your journey—it is nearly straight down, and the __(4)__ secretions are like grease. You sit down and dig in your heels to get started. After a quick slide, you land abruptly on one of a pair of flat sheetlike structures that begins to vibrate rapidly, bouncing you up and down helplessly. You are also conscious of a rhythmic hum during this jostling, and you realize that you have landed on a __(5)__ . You pick yourself up and look over the superior edge of the __(6)__ , down into the seemingly endless esophagus behind. You chastise yourself for not remembering that the __(7)__ and respiratory pathways separate at this point. Hanging directly over your head is a leaflike __(8)__ cartilage. Normally, you would not have been able to get this far because it would have closed off this portion of the respiratory tract. With your host sedated, however, this protective reflex did not work.

You carefully begin to pick your way down, using the cartilages as steps. When you reach the next respiratory organ, the __(9)__ , your descent becomes much easier, because the structure's C-shaped cartilages form a ladderlike supporting structure. As you climb down the cartilages, your face is stroked rhythmically by soft cellular extensions, or __(10)__ . You remember that their function is to move mucus laden with bacteria or dust and other debris toward the __(11)__ .

You finally reach a point where the descending passageway splits into the two __(12)__ , and since you want to control your progress (rather than slide downward), you choose the more horizontal __(13)__ branch. If you remain in the superior portion of the lungs, your return trip will be less difficult because the passageways will be more horizontal than steeply vertical. The passageways get smaller and smaller, slowing your progress.

As you are squeezing into one of the smallest of the respiratory passageways, a __(14)__ , you see a bright spherical chamber ahead. You scramble into this __(15)__ , pick yourself up, and survey the area. Scattered here and there are lumps of a substance that looks suspiciously like coal, reminding you that your host is a smoker. As you stand there, a soft rustling wind seems to flow in and out of the chamber. You press your face against the transparent chamber wall and see disclike cells, __(16)__ , passing by in the capillaries on the other side. As you watch, they change from a somewhat bluish color to a bright __(17)__ color as they pick up __(18)__ and unload __(19)__ .

You record your observations and then contact headquarters to let them know you are ready to begin your ascent. You begin your return trek, slipping and sliding as you travel. By the time you reach the inferior edge of the trachea, you are ready for a short break. As you rest on the mucosa, you begin to notice that the air is becoming close and very heavy. You pick yourself up quickly and begin to scramble up the trachea. Suddenly and without warning, you are hit by a huge wad of mucus and catapulted upward and out onto your host's freshly pressed handkerchief! Your host has assisted your exit with a __(20)__ .

CHALLENGING YOURSELF

At the Clinic

1. Barbara is rushed to the emergency room after an auto accident. The 8th through 10th ribs on her left side have been fractured and have punctured the lung. What term is used to indicate lung collapse? Will both lungs collapse? Why or why not?

2. A young boy is diagnosed with cystic fibrosis. What effect will this have on his respiratory system?

3. Len LaBosco, a medical student, has recently moved to Mexico City for his internship. As he examines the lab report of his first patient, he notices that the RBC count is high. Why is this normal for this patient? (*Hint:* Mexico City is located on a high-altitude mesa.) What respiratory adjustments will Len have to make in his own breathing, minute ventilation, arterial P_{CO_2}, and hemoglobin saturation level?

4. Why must patients be forced to cough when recovering from chest surgery?

5. A mother, obviously in a panic, brings her infant who is feverish, hyperventilating, and cyanotic to the clinic. The infant is quickly diagnosed with pneumonia. What aspect of pneumonia has caused the cyanosis?

6. After a long bout of bronchitis, Mrs. Dupee complains of a stabbing pain in her side with each breath. What is her condition?

7. A patient with congestive heart failure has cyanosis coupled with liver and kidney failure. What type of hypoxia is this?

8. The Kozloski family is taking a long auto trip. Michael, who has been riding in the back of a station wagon, complains of a throbbing headache. Then, a little later, he seems confused and his face is flushed. What is your diagnosis of Michael's problem?

9. A new mother checks on her sleeping infant, only to find that it has stopped breathing and is turning blue. The mother quickly picks up the baby and pats its back until it starts to breathe. What tragedy has been averted?

10. Roger Proulx, a rugged four-pack-a-day smoker, comes to the clinic because of painful muscle spasms. Blood tests reveal hypocalcemia and elevated levels of calcitonin. What type of lung cancer is suspected?

11. A young man visiting his father in the hospital hears the clinical staff refer to the patient in the next bed as a "pink puffer." He notes the patient's barrel-shaped chest and wonders what the man's problem is. What is the patient's diagnosis, and why is he called a "pink puffer?"

12. Joanne Willis, a long-time smoker, is complaining that she has developed a persistent cough. What is your first guess as to her condition? What has happened to her bronchial cilia?

13. As a result of a stroke, the swallowing mechanism in an elderly woman is uncoordinated. What effects might this have on her respiratory system?

14. While diapering his 1-year-old boy (who puts almost everything in his mouth), Mr. Gregoire failed to find one of the small safety pins previously used. Two days later, his son developed a cough and became feverish. What probably had happened to the safety pin and where (anatomically) would you expect to find it?

15. Mr. and Ms. Rao took their sick 5-year-old daughter to the doctor. The girl was breathing entirely through her mouth, her voice sounded odd and whiny, and a puslike fluid was dripping from her nose. Which one of the four sets of tonsils was most likely infected in this child?

Stop and Think

1. When dogs pant, they inhale through the nose and exhale through the mouth. How does this provide a cooling mechanism?

2. A deep-sea expedition is being planned. Why is the location of the nearest hyperbaric chamber critically important?

3. Why is it advantageous that fetal hemoglobin has a higher affinity for oxygen than maternal hemoglobin?

4. The genetic defect in cystic fibrosis affects the chloride channels in secretory cells. Explain, in terms of salt and water secretion, why this results in thick mucus in the respiratory tract.

5. Is the partial pressure of oxygen in expired air higher, lower, or the same as that of alveolar air? Explain your answer.

6. Two girls in a high school cafeteria were giggling over lunch, and both accidentally sprayed milk out their nostrils at the same time. Explain in anatomical terms why swallowed fluids can sometimes come out the nose.

A boy in the same cafeteria then stood on his head and showed he could drink milk upside down without any of it entering his nasal cavity or nose. What prevented the milk from flowing downward into his nose?

7. A surgeon had to remove three adjacent bronchopulmonary segments from the left lung of a patient with tuberculosis. Almost half of the lung was removed, yet there was no severe bleeding, and relatively few blood vessels had to be cauterized (closed off). Why was the surgery so easy to perform?

8. The cilia lining the upper respiratory passages (superior to the larynx) beat inferiorly while the cilia lining the lower respiratory passages (larynx and below) beat superiorly. What is the functional "reason" for this difference?

9. What is the function of the abundant elastin fibers that occur in the stroma of the lung and around all respiratory tubes from the trachea through the respiratory tree?

10. A man was choking on a piece of meat and no one was around to offer help. He saw a horizontal railing in his house that ran just above the level of his navel. He hurled himself against the railing in an attempt to perform the Heimlich maneuver on himself. Use logic to deduce whether this was a wise move that might save the man's life.

11. Three terms that are easily confused are *choanae, conchae,* and *carina.* Define each of these, clarifying their differences.

COVERING ALL YOUR BASES

Multiple Choice

Select the best answer or answers from the choices given.

1. Structures that are part of the respiratory zone include:
   A. terminal bronchioles
   B. respiratory bronchioles
   C. tertiary bronchi
   D. alveolar ducts

2. Which of the following have a respiratory function?
   A. Olfactory mucosa   C. Vibrissae
   B. Adenoids   D. Auditory tubes

3. Which structures are associated with the production of speech?
   A. Cricoid cartilage
   B. Glottis
   C. Arytenoid cartilage
   D. Pharynx

4. The skeleton of the external nose consists of:
   A. cartilage and bone
   B. bone only
   C. hyaline cartilage only
   D. elastic cartilage only

5. Which of the following is not part of the conducting zone of the respiratory system?
   A. Pharynx
   B. Alveolar sac
   C. Trachea
   D. Secondary bronchioles
   E. Larynx

6. The function of alveolar type I cells is:
   A. to produce surfactant
   B. to propel mucous sheets
   C. phagocytosis of dust particles
   D. to allow rapid diffusion of respiratory gases

7. An examination of a lobe of the lung reveals many branches off the main passageway. These branches are:
   A. primary bronchi
   B. secondary bronchi
   C. lobar bronchi
   D. segmental bronchi

8. Microscopic examination of lung tissue shows a relatively small passageway with nonciliated simple cuboidal epithelium, abundant elastic fibers, absence of cartilage, and some smooth muscle. What is the classification of this structure?
   A. Secondary bronchus
   B. Tertiary bronchus
   C. Respiratory bronchiole
   D. Terminal bronchiole

9. An alveolar sac:
   A. is an alveolus
   B. relates to an alveolus as a bunch of grapes relates to one grape
   C. is a huge, saclike alveolus in an emphysema patient
   D. is the same as an alveolar duct

10. The respiratory membrane (air-blood barrier) consists of:
    A. alveolar type I cell, basal laminae, endothelial cell
    B. air, connective tissue, lung
    C. type II cell, dust cell, type I cell
    D. pseudostratified epithelium, lamina propria, capillaries

11. Cells responsible for removing foreign particles from inspired air include:
    A. goblet cells
    B. type II cells
    C. dust cells
    D. ciliated cells

12. The root of the lung is at its:
    A. apex
    B. base
    C. hilus
    D. cardiac notch

13. Which of the following are characteristic of a bronchopulmonary segment?
    A. Removal causes collapse of adjacent segments
    B. Fed by a tertiary bronchus
    C. Supplied by its own branches of the pulmonary artery and vein
    D. Separated from other segments by its septum

14. During inspiration, intrapulmonary pressure is:
    A. greater than atmospheric pressure
    B. less than atmospheric pressure
    C. greater than intrapleural pressure
    D. less than intrapleural pressure

15. Lung collapse is prevented by:
    A. high surface tension of alveolar fluid
    B. high surface tension of pleural fluid
    C. high pressure in the pleural cavities
    D. high elasticity of lung tissue

16. Accessory muscles of inspiration include:
    A. external intercostals
    B. scalenes
    C. internal intercostals
    D. pectoralis major

17. The greatest resistance to gas flow occurs at the:
    A. terminal bronchioles
    B. alveolar ducts
    C. medium-sized bronchi
    D. respiratory bronchioles

18. Resistance is increased by:
    A. epinephrine
    B. parasympathetic stimulation
    C. inflammatory chemicals
    D. contraction of the trachealis muscle

19. Which of the following conditions will reduce lung compliance?
    A. Tuberculosis
    B. Aging
    C. IRDS
    D. Osteoporosis

20. Chemically, surfactant is classified as a:
    A. polar molecule
    B. hydrophobic molecule
    C. polysaccharide
    D. lipoprotein

21. Which of the following changes will accompany the loss of elasticity associated with aging?
    A. Increase in tidal volume
    B. Increase in inspiratory reserve volume
    C. Increase in residual volume
    D. Increase in vital capacity

22. What is the approximate alveolar ventilation rate in a 140-pound female with a tidal volume of 400 ml and a respiratory rate of 20 breaths per minute?
    A. 10,800 ml/min
    B. 8000 ml/min
    C. 5200 ml/min
    D. 2800 ml/min

23. FEV is reduced but FVC is not affected in which disorder(s)?
    A. Asthma
    B. Tuberculosis
    C. Polio
    D. Cystic fibrosis

24. Considering an individual in good health, an increase in blood flow rate through the pulmonary capillaries so that blood is in the capillaries for about 0.5 sec instead of 0.75 sec:
    A. decreases the rate of oxygenation
    B. reduces oxygenation of the blood
    C. does not change the level of oxygen saturation of hemoglobin
    D. increases rate of diffusion of oxygen

25. Relatively high alveolar P_{CO_2} coupled with low P_{O_2} triggers:
    A. dilation of the supplying bronchiole
    B. constriction of the supplying bronchiole
    C. dilation of the supplying arteriole
    D. constriction of the supplying arteriole

26. In blood flowing through systemic veins of a person at rest:
    A. hemoglobin is about 75% saturated
    B. the blood is carrying about 20 vol % oxygen
    C. the Bohr effect has reduced hemoglobin's affinity for oxygen
    D. on average, each hemoglobin molecule is carrying two oxygen molecules

27. Which of the following is (are) true concerning CO_2 transport in systemic venous blood in an exercising person?
    A. The level of bicarbonate ion is higher than resting values.
    B. Due to the Haldane effect, the amount of carbaminohemoglobin is increased.
    C. The pH of the blood is higher.
    D. The chloride shift is greater.

28. Stimulation of which of the following stimulates inspiration?
    A. Dorsal respiratory group
    B. Pneumotaxic area
    C. Lung stretch receptors
    D. Medullary chemoreceptors

29. The factor that has the greatest effect on the medullary chemoreceptors is:
    A. acute hypercapnia
    B. hypoxia
    C. chronic hypercapnia
    D. arterial pH

30. Which term refers to increase in depth, but not rate, of ventilation?
    A. Eupnea   C. Hyperpnea
    B. Dyspnea   D. Hyperventilation

31. Adjustment to high altitude involves:
    A. increase in minute respiratory volume
    B. hypersecretion of erythropoietin
    C. dyspnea
    D. increase in hemoglobin saturation

32. Disorders classified as COPDs include:
    A. pneumonia   C. bronchitis
    B. emphysema   D. sleep apnea

33. Arrested development of the respiratory system during the fifth week of development affects formation of the:
    A. olfactory placodes
    B. olfactory pits
    C. laryngotracheal bud
    D. stroma of the lungs

Word Dissection

For each of the following word roots, fill in the literal meaning and give an example, using a word found in this chapter.

Word root	Translation	Example
1. alveol		
2. bronch		

Word root	**Translation**	**Example**
3. capn	_____	_____
4. carin	_____	_____
5. choan	_____	_____
6. crico	_____	_____
7. ectasis	_____	_____
8. emphys	_____	_____
9. flat	_____	_____
10. nari	_____	_____
11. nas	_____	_____
12. pleur	_____	_____
13. pne	_____	_____
14. pneum	_____	_____
15. pulmo	_____	_____
16. respir	_____	_____
17. spire	_____	_____
18. trach	_____	_____
19. ventus	_____	_____
20. vestibul	_____	_____
21. vibr	_____	_____

24 The Digestive System

Student Objectives

When you have completed the exercises in this chapter, you will have accomplished the following objectives:

Overview of the Digestive System

1. Identify the overall function of the digestive system, and differentiate between organs of the alimentary canal and accessory digestive organs.

2. List and define briefly the major processes occurring during digestive system activity.

3. Describe the location and function of the peritoneum and the peritoneal cavity. Define *retroperitoneal* and name the retroperitoneal organs.

4. Describe the tissue composition of the four layers of the alimentary tube wall and note the general function of each layer.

Functional Anatomy of the Digestive System

5. Describe the gross and microscopic anatomy and basic function of each organ and accessory organ of the alimentary canal.

6. Explain the dental formula and differentiate clearly between deciduous and permanent teeth.

7. Describe the composition and functions of saliva and explain how salivation is regulated.

8. Describe the mechanisms of chewing and swallowing.

9. Describe structural modifications of the wall of the stomach and small intestine that enhance the digestive process in these regions.

10. Describe the composition of gastric juice, name the cell types responsible for secreting its various components, and describe the importance of each component in stomach activity.

11. Explain how gastric secretion and motility in the stomach are regulated.

12. Describe the function of local hormones produced by the small intestine.

13. State the roles of bile and pancreatic juice in the digestive process.

14. Describe how entry of pancreatic juice and bile into the small intestine is regulated.

15. List the major functions of the large intestine and describe the regulation of defecation.

Physiology of Chemical Digestion and Absorption

16. List the major enzymes or enzyme groups involved in chemical digestion; name the foodstuffs on which they act; and name the end-products of protein, fat, carbohydrate, and nucleic acid digestion.

17. Describe the process of absorption of digested foodstuffs that occurs in the small intestine.

Developmental Aspects of the Digestive System

18. Describe the embryonic development of the digestive system.

19. Describe important abnormalities of the gastrointestinal tract at different stages of life.

Ready, set, go!

The digestive system processes food so that it can be absorbed and utilized by the body's cells. The digestive organs are responsible for food ingestion, digestion, absorption, and elimination of undigested remains from the body. In one sense, the digestive tract can be viewed as a disassembly line in which food is carried from one stage of its breakdown process to the next by muscular activity, and its nutrients are made available en route to the cells of the body. In addition, the digestive system provides for one of life's greatest pleasures—eating.

Chapter 24 reviews the anatomy of the alimentary canal and accessory digestive organs, the processes of mechanical and enzymatic breakdown, and absorption mechanisms. Cellular metabolism (utilization of foodstuffs by body cells) is considered in Chapter 25.

BUILDING THE FRAMEWORK

Overview of the Digestive System

1. The organs of the alimentary canal form a continuous tube from the mouth to the anus. List the organs of the alimentary canal, including the specific regions of the small and large intestine, from proximal to distal, in the blanks below.

    Mouth → (1)_____ → (2)_____ →

    (3)_____ → small intestine: (4)_____ →

    (5)_____ → (6)_____ → large intestine: cecum,

    appendix, (7)_____ → (8)_____ →

    (9)_____ → (10)_____ →

    (11)_____ → (12)_____ → anus

2. List the accessory structures associated with the mouth and the duodenum of the small intestine; mark each with an I if it is located inside the tract and an O if it is located outside the tract.

    Mouth: _____

    Duodenum: _____

3. Figure 24.1 is a frontal view of the digestive system. First, identify all structures with leader lines. Then, select a different color for each of the following organs and color the coding circles and the corresponding structures on the figure.

○ Esophagus   ○ Colon   ○ Salivary glands   ○ Tongue

○ Liver   ○ Pancreas   ○ Small intestine   ○ Uvula

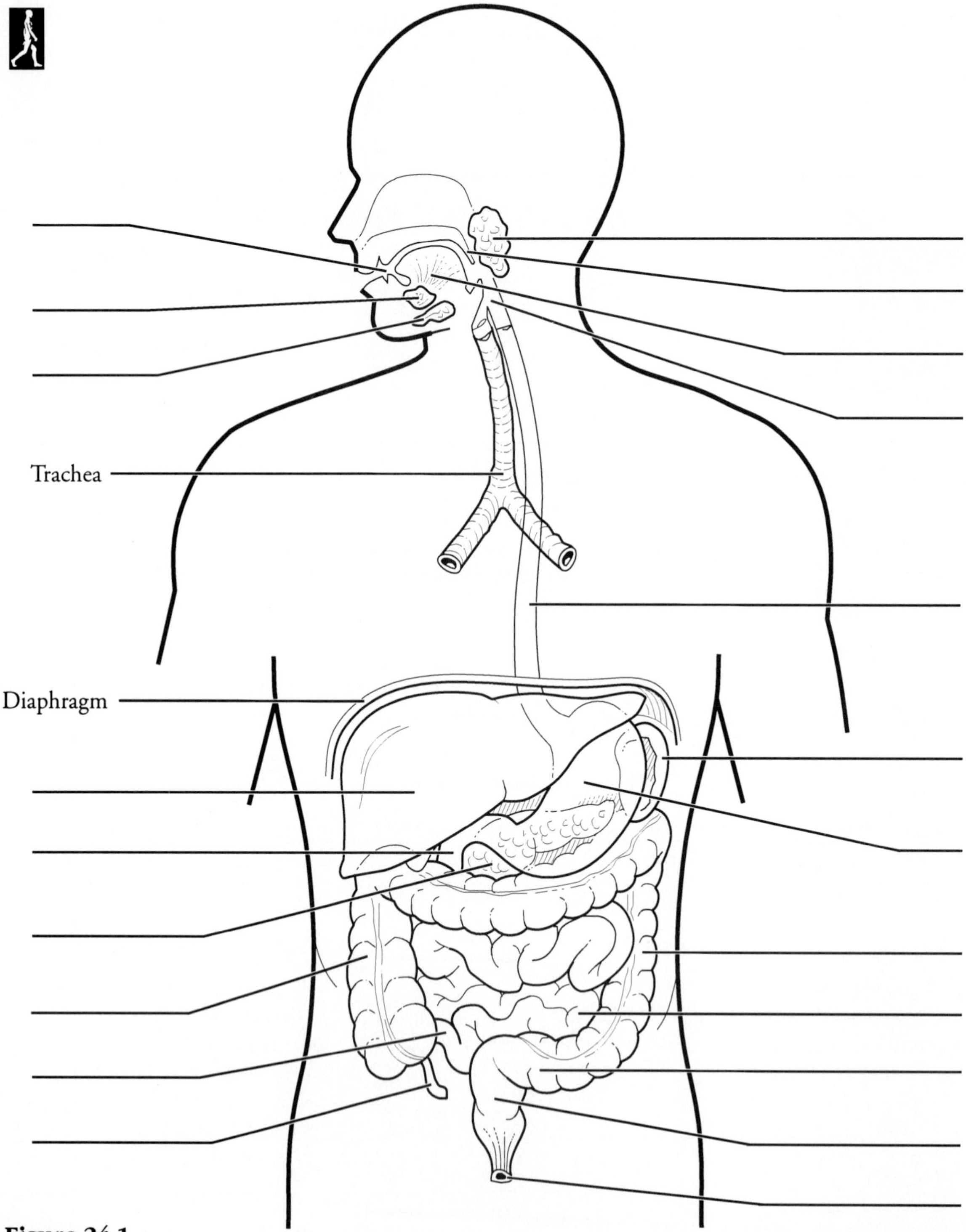

Trachea

Diaphragm

Figure 24.1

4. Match the descriptions in Column B with the appropriate terms referring to digestive processes in Column A.

Column A

_____ **1.** Ingestion

_____ **2.** Propulsion

_____ **3.** Mechanical digestion

_____ **4.** Chemical digestion

_____ **5.** Absorption

_____ **6.** Defecation

Column B

A. Transport of nutrients from lumen to blood

B. Enzymatic breakdown

C. Elimination of feces

D. Eating

E. Chewing

F. Churning

G. Includes swallowing

H. Segmentation and peristalsis

5. Relative to neural controls of digestive activity, answer the following questions in the spaces provided:

1. What are three stimuli that elicit a response from sensors in the wall of

the tract organs? _____

_____

2. What are the two primary nerve plexuses regulating digestive function?

_____

3. What are some differences between long and short reflexes?

_____

_____

_____

6. Fill in the blanks below:

_____ **1.**

_____ **2.**

_____ **3.**

_____ **4.**

The body's most extensive serous membrane is the __(1)__ . Connecting its visceral and parietal layers is a double serous sheet called __(2)__ , which provides a route for blood vessels, lymphatics, and nerves. Organs behind the parietal layer are called __(3)__ ; those within this area are called __(4)__ .

7. Various types of glands secrete substances into the alimentary tube. Match the glands listed in Column B with the functions and locations described in Column A.

Column A   Column B

_____ **1.** Mucus-producing glands located in   **A.** Duodenal glands
                       the submucosa of the small intestine
                                                         B. Gastric glands
_____ **2.** Secretory product contains amylase,
                       a starch-digesting enzyme   **C.** Liver

_____ **3.** Sends a variety of enzymes in   **D.** Pancreas
                       bicarbonate-rich fluid into the small
                       intestine   **E.** Salivary glands

_____ **4.** Produces bile, which is transported to
                       the duodenum via the bile duct

_____ **5.** Produce hydrochloric acid and
                       pepsinogen

8. Using the key choices, match the terms with the descriptions of digestive system organs that follow by inserting the appropriate answers in the answer blanks.

KEY CHOICES

A. Anal canal   **H.** Ileocecal valve   **O.** Pharynx   **V.** Tongue

B. Appendix   **I.** Lesser omentum   **P.** Plicae circulares **W.** Vestibule

C. Colon   **J.** Mesentery   **Q.** Pyloric valve   **X.** Villi

D. Esophagus   **K.** Microvilli   **R.** Rugae   **Y.** Visceral peritoneum

E. Greater omentum   **L.** Oral cavity   **S.** Small intestine

F. Hard palate   **M.** Parietal peritoneum **T.** Soft palate

G. Haustra   **N.** Peyer's patches   **U.** Stomach

_____ **1.** Connects the small intestine to the posterior abdominal wall; looks
                       like lacy curtains

_____ **2.** Projections of the intestinal mucosa that increase the surface area

_____ **3.** Large collections of lymph nodules found in the submucosa of the
                       small intestine

_____ **4.** Folds of the small intestine wall

_____   **5.** Two anatomical regions involved in the mechanical breakdown of food

_____   **6.** Mixes food in the mouth and initiates swallowing

_____   **7.** Common passage for food and air

_____   **8.** Three peritoneal modifications

_____   **9.** A food chute; has no digestive or absorptive role

_____   **10.** Folds of the stomach mucosa and submucosa

_____   **11.** Saclike outpocketing of the large intestine wall

_____   **12.** Projections of the plasma membrane of a cell that increase the cell's surface area

_____   **13.** Prevents food from moving back into the small intestine once it has entered the large intestine

_____   **14.** Responsible for most food and water absorption

_____   **15.** Primarily involved in water absorption and feces formation

_____   **16.** Cul-de-sac between the teeth and lips or cheeks

_____   **17.** Blind-ended tube hanging from the cecum

_____   **18.** Organ in which protein digestion begins

_____   **19.** Membrane that runs from the lesser curvature of the stomach to the liver

_____   **20.** Organ into which the stomach empties

_____   **21.** Sphincter controlling the movement of food from the stomach into the duodenum

_____   **22.** The uvula hangs from its posterior edge

_____   **23.** Receives pancreatic juice and bile

_____   **24.** Serosa of the abdominal cavity wall

_____   **25.** Major site of vitamin (K, B) formation by bacteria

_____   **26.** Region, containing two sphincters, through which feces are expelled from the body

_____   **27.** Anteriosuperior boundary of the oral cavity; supported by bone

_____   **28.** Extends as a double fold from the greater curvature of the stomach; very fat-laden

_____   **29.** Superiorly, its muscle is striated; inferiorly, it is smooth

9. The walls of the alimentary canal have four typical layers, as illustrated in Figure 24.2. Identify each layer by placing its correct name in the answer blank before the appropriate description. Select different colors for each layer and color the coding circles and corresponding structures on the figure. Finally, assume the figure shows a cross-sectional view of the small intestine, and label the three structures with leader lines.

_____ ◯ **1.** The secretory and absorptive layer

_____ ◯ **2.** Layer composed of at least two muscle layers

_____ ◯ **3.** Connective tissue layer, containing blood, lymph vessels, and nerves

_____ ◯ **4.** Outermost layer of the wall; visceral peritoneum

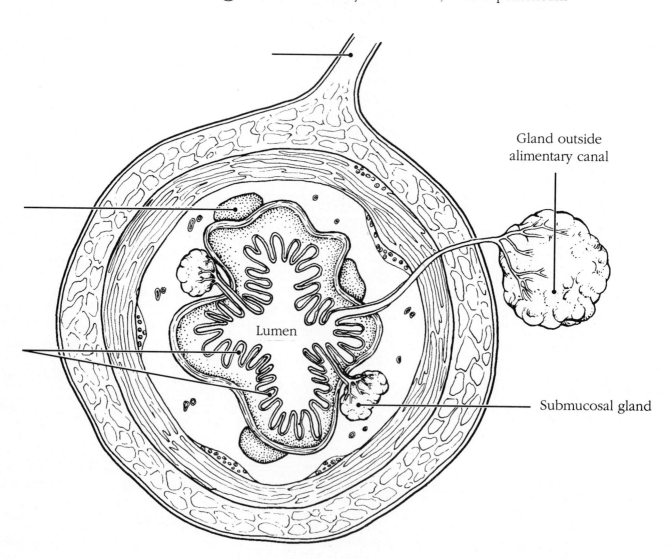

Gland outside alimentary canal

Lumen

Submucosal gland

Figure 24.2

Functional Anatomy
of the Digestive System

1. Figure 24.3 illustrates oral cavity structures. First, identify all structures with leader lines. Then, color the structure that attaches the tongue to the floor of the mouth red, color the portions of the roof of the mouth unsupported by bone blue, color the structures that are essentially masses of lymphatic tissue yellow, and color the structure that contains the bulk of the taste buds pink.

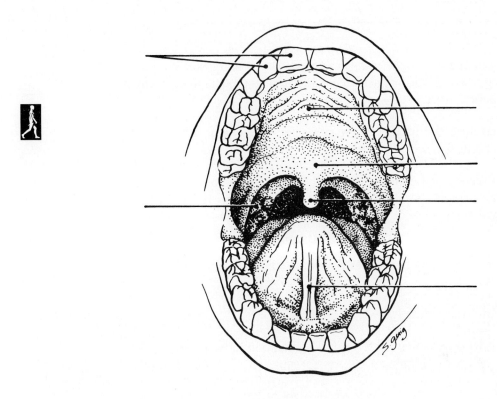

Figure 24.3

2. Answer the following questions concerning the salivary glands and saliva in the spaces provided:

1. What are four functions of saliva? _____

_____

_____

2. What are the three main pairs of salivary glands, and in what part of the

oral cavity do the ducts of each gland open? _____

_____

_____

3. What are the two main cell types found in salivary glands, and what are

their secretions? _____

_____

_____

4. What enzyme found in saliva inhibits bacterial growth?

_____

3. Alternative choices separated by a slash (/) appear in each of the statements below concerning the regulation of salivation. In each case, circle the term that makes the statement correct.

1. Salivation is controlled primarily by the parasympathetic/sympathetic division of the autonomic nervous system.

2. When food is placed in the mouth, it activates chemoreceptors and pressoreceptors to send impulses to the salivatory/gustatory nuclei in the brain stem, which in turn send impulses to the salivary glands via motor fibers of the facial and glossopharyngeal/hypoglossal nerves.

3. Chemoreceptors are strongly activated by bitter/sour substances.

4. Salivation induced by the sight or thought of food is said to be a(n) conditioned/intrinsic reflex.

4. Complete the following statements referring to human dentition by writing the missing terms in the answer blanks.

_____ 1.

_____ 2.

_____ 3.

_____ 4.

_____ 5.

_____ 6.

_____ 7.

_____ 8.

The first set of teeth, called the __(1)__ teeth, begin to appear around the age of __(2)__ and usually have begun to be replaced by the age of __(3)__. The __(4)__ teeth are more numerous; that is, there are __(5)__ teeth in the second set as opposed to a total of __(6)__ teeth in the first set. If an adult has a full set of teeth, you can expect to find two __(7)__, one __(8)__, two __(9)__, and three __(10)__ in one side of each jaw. The most posterior molars in each jaw are commonly called __(11)__ teeth.

_____ 9.

_____ 10.

_____ 11.

5. Using the key choices, identify each tooth area described below and label the tooth in Figure 24.4. Then select different colors to represent the key structures with coding circles and color them on the figure. Finally, add labels to the figure to identify the crown, gingiva, and root of the tooth.

KEY CHOICES

◯ **A.** Cementum   ◯ **C.** Enamel   ◯ **E.** Pulp

◯ **B.** Dentin   ◯ **D.** Periodontal membrane

_____ **1.** Material covering the tooth root

_____ **2.** Hardest substance in the body; covers tooth crown

_____ **3.** Attaches the tooth to bone and surrounding alveolar structures

_____ **4.** Forms the bulk of tooth structure; similar to bone

_____ **5.** A collection of blood vessels, nerves, and lymphatics

_____ **6.** Cells that produce this substance degenerate after tooth eruption

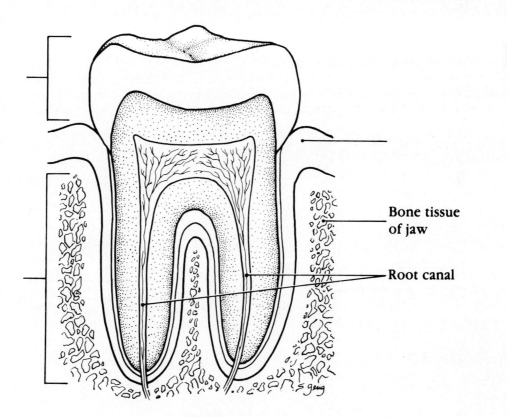

Bone tissue of jaw

Root canal

Figure 24.4

6. Complete the following statements that describe mechanisms of food mixing and movement from the mouth to the stomach. Insert your responses in the answer blanks.

_____ 1.

_____ 2.

_____ 3.

_____ 4.

_____ 5.

_____ 6.

_____ 7.

_____ 8.

_____ 9.

_____ 10.

_____ 11.

_____ 12.

_____ 13.   _____ 14.   _____ 15.

Common pathways for food and air are the __(1)__ , which connects to the mouth, and __(2)__ , which connects to the larynx. Epithelium in this region is __(3)__ , which provides good protection from friction. Three bands of skeletal muscle, the __(4)__ , propel food into the __(5)__ . Entrance to the stomach is guarded by the __(6)__ .

Swallowing, or __(7)__ , occurs in two major phases, the __(8)__ and __(9)__ . During the initial voluntary phase, the __(10)__ is used to push the food into the throat and the __(11)__ rises to close off the nasal passageways. As food is moved involuntarily through the pharynx, the __(12)__ rises to ensure that its passageway is covered by the __(13)__ so that ingested substances do not enter respiratory passages. It is possible to swallow water while standing on your head because the water is carried along the esophagus involuntarily by the process of __(14)__ . The pressure exerted by food on the __(15)__ valve causes it to open so that food can enter the stomach.

7. Figure 24.5A is a longitudinal section of the stomach. Use the following terms to identify the regions with leader lines on the figure.

Body   Pyloric region   Greater curvature   Gastroesophageal valve

Fundus   Pyloric valve   Lesser curvature

Select different colors for each of the following structures or areas and color the coding circles and corresponding structures or areas on the figure.

○ Oblique muscle layer   ○ Longitudinal muscle layer   ○ Serosa

○ Circular muscle layer   ○ Area where rugae are visible

Figure 24.5B shows two types of secretory cells found in gastric glands. Identify the third type, called *chief (or zymogenic) cells,* by choosing a few cells deep in the glands and labeling them appropriately. Then, color the hydrochloric acid–secreting cells red, color the cells that produce mucus yellow, and color those that produce protein-digesting enzymes blue.

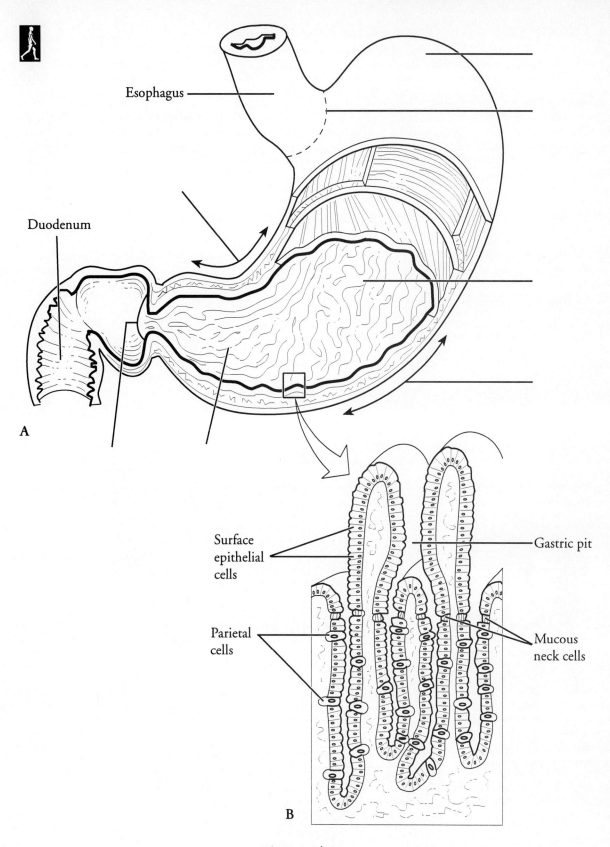

Esophagus

Duodenum

A

Surface
epithelial
cells

Gastric pit

Parietal
cells

Mucous
neck cells

B

Figure 24.5

8. Check (✔) all factors that enhance stomach secretory activity.

_____ **1.** Excessive acidity

_____ **2.** Presence of protein foods in stomach

_____ **3.** Stomach distention

_____ **4.** Aroma of food

_____ **5.** Acetylcholine

_____ **6.** Duodenal distention during gastric phase

_____ **7.** Gastrin

_____ **8.** Emotional upset

9. Figure 24.6 shows three views of the small intestine. First, label the villi in parts B and C and the plicae circulares in parts A and B. Then select different colors to identify the following regions in part C and color them on the figure. Color the tunics in part B as desired.

◯ Simple columnar cells of surface epithelium

◯ Goblet cells of surface epithelium

◯ Lacteal

◯ Capillary network

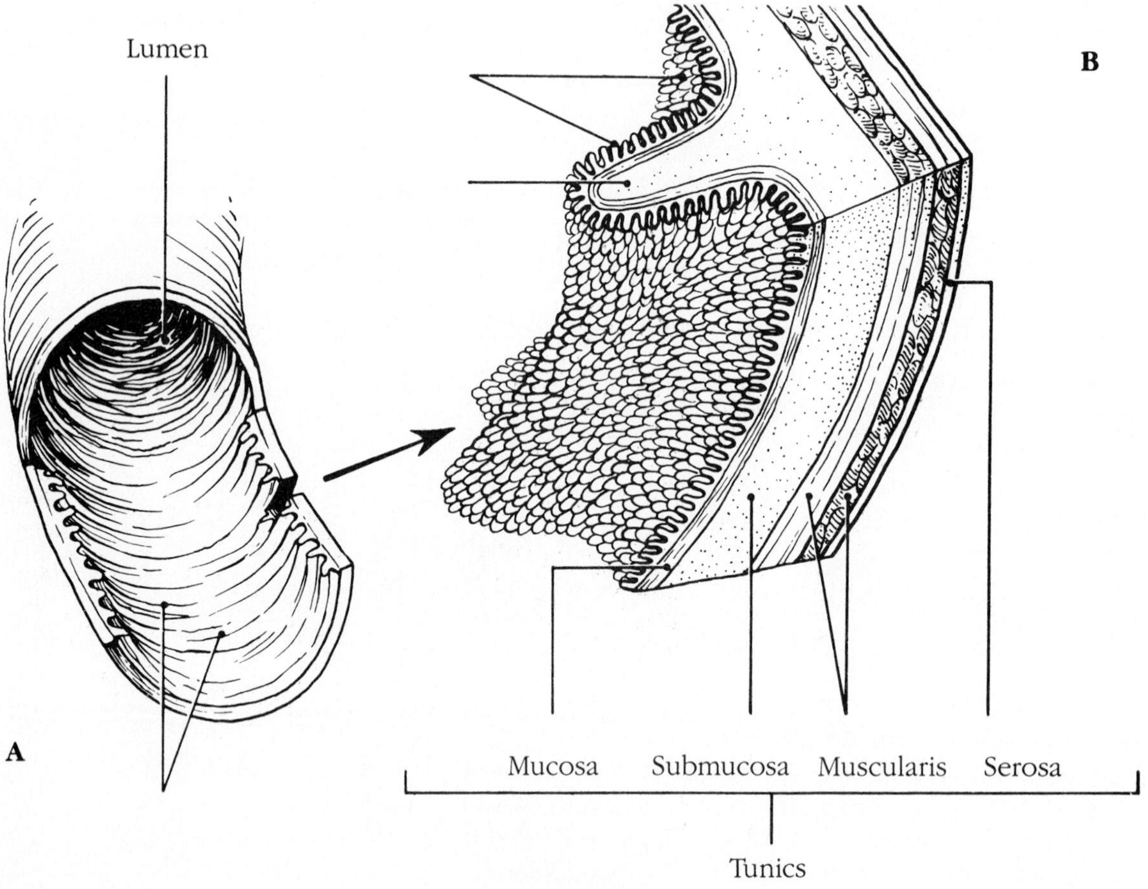

Lumen

A

B

Mucosa   Submucosa   Muscularis   Serosa

Tunics

Figure 24.6

Figure 24.6 (continued)

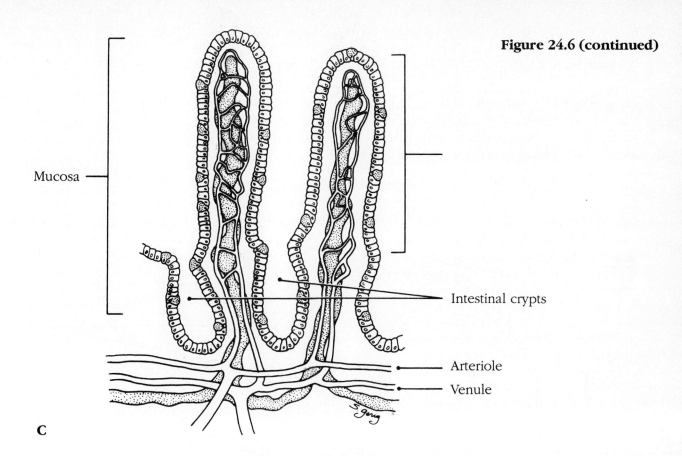

Mucosa

Intestinal crypts

Arteriole

Venule

C

10. Match the mesenteries in Column B with their descriptions in Column A.

Column A

Column B

_____ **1.** Lies closely apposed to serous
lining of anterior abdominal
cavity

A. Falciform ligament

B. Mesentery proper

_____ **2.** Divides liver into lobes

C. Mesocolon

_____ **3.** Fan-shaped intestinal membrane

D. Greater omentum

_____ **4.** Attaches part of large intestine
to body wall

11. Circle the term that does not belong in each of the following groupings.

1. HCl secretion   Acetylcholine   Histamine   Secretin   Gastrin

2. Gastric emptying   Slowed by fats   Enterogastrone   Solid foods

3. Cephalic phase   Reflex phase   Intestinal phase   Psychological stimuli

4. Saliva   IgA   Lysozyme   HCl   Growth factor

5. Local reflexes   Vagovagal reflexes   Parasympathetic   Sympathetic

6.   Pacemaker cells   Small intestine   Esophagus   Stomach

7.   Stomach absorption   Lipid-soluble drugs   Salts   Aspirin   Alcohol

12. Three accessory organs are illustrated in Figure 24.7. Identify each of the three organs, their ducts, and the ligament with leader lines on the figure. Then select different colors for the following structures and color the coding circles and the corresponding structures on the figure.

    ◯   Common hepatic duct   ◯   Bile duct

    ◯   Cystic duct   ◯   Pancreatic duct

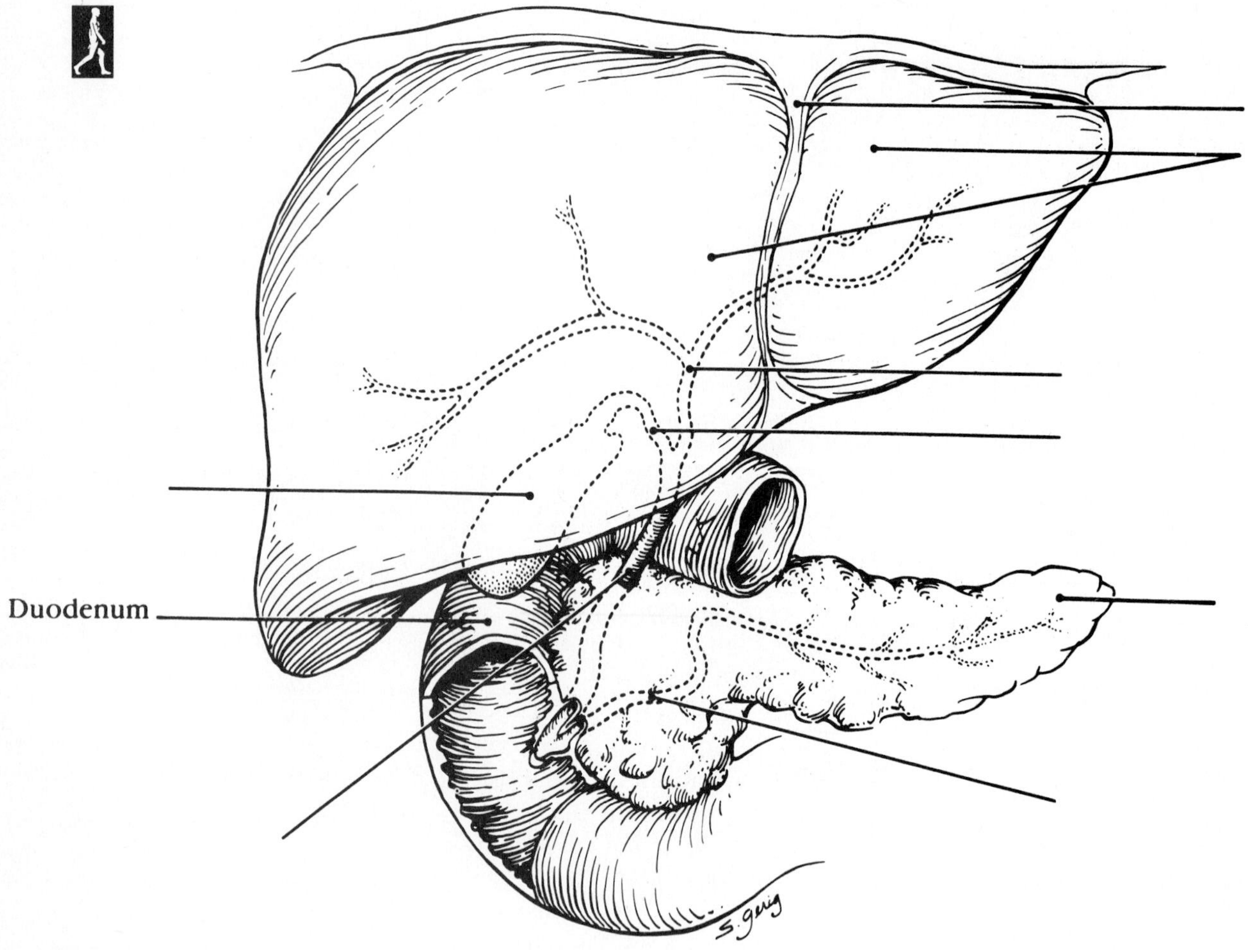

Duodenum

Figure 24.7

13. List the three major vessels of the portal triad: _____ , _____ ,

and _____ . Describe the location of the triads relative to the liver lobules.

_____

_____

14. Hormonal stimuli are important in digestive activities that occur in the stomach and small intestine. Using the key choices, identify the hormones that function as described in the following statements.

KEY CHOICES

A. Cholecystokinin   **B.** Gastrin   **C.** Secretin   **D.** Somatostatin

_____ **1.** These two hormones stimulate the pancreas to release its secretions.

_____ **2.** This hormone stimulates increased production of gastric juice.

_____ **3.** This hormone causes the gallbladder to release stored bile.

_____ **4.** This hormone causes the liver to increase its output of bile.

_____ **5.** These hormones inhibit gastric mobility and secretory activity.

15. This exercise concerns some aspects of food breakdown in the digestive tract. Using the key choices, select the appropriate terms to complete the following statements.

KEY CHOICES

A. Bicarbonate-rich fluid   **D.** HCl (hydrochloric acid)   **G.** Mucus

B. Chewing   **E.** Hormonal stimulus   **H.** Psychological stimulus

C. Churning   **F.** Mechanical stimulus

_____ **1.** The means of mechanical food breakdown in the mouth is __(1)__ .

_____ **2.** The fact that the mere thought of a relished food can make your mouth water is an example of __(2)__ .

_____ **3.** Many people chew gum to increase saliva formation when their mouth is dry. This type of stimulus is a __(3)__ .

_____ **4.** Since living cells of the stomach (and everywhere) are largely protein, it is amazing that they are not digested by the activity of stomach enzymes. The most important means of stomach protection is the __(4)__ it produces.

_____ **5.** The third layer of smooth muscle found in the stomach wall allows mixing and mechanical breakdown by __(5)__ .

_____ **6.** The small intestine is protected from the corrosive action of HCl in chyme by __(6)__, which is ducted in from the pancreas.

16. Figure 24.8 shows a view of the large intestine. First, label a haustrum, the ileocecal valve, the external anal sphincter, the hepatic and splenic flexures, and the transverse mesocolon. Then, select different colors to identify the following parts in the figure:

- ◯ Anal canal
- ◯ Appendix
- ◯ Ascending colon
- ◯ Cecum
- ◯ Rectum
- ◯ Descending colon
- ◯ Sigmoid colon
- ◯ Teniae coli
- ◯ Transverse colon

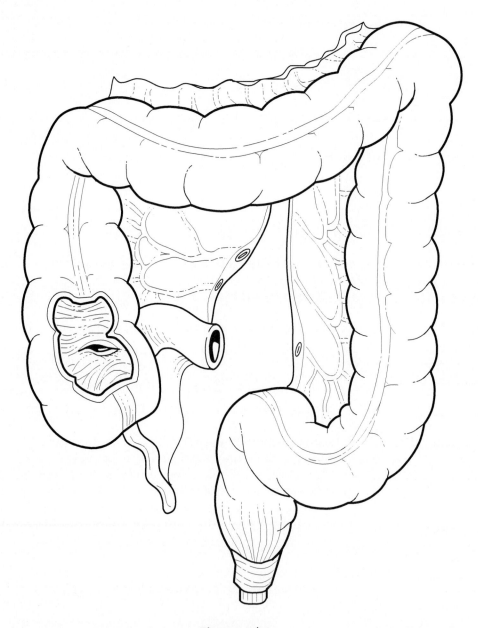

Figure 24.8

17. Complete the paragraph relating to digestive system mobility by writing the missing terms in the answer blanks.

_____ **1.**

_____ **2.**

_____ **3.**

_____ **4.**

_____ **5.**

_____ **6.**

_____ **7.**

_____ **8.**

The two major types of movements that occur in the small intestine are __(1)__ and __(2)__ . One of these movements, __(3)__ , acts to continually mix the food with digestive juices and, strangely, also plays the major role in propelling foods along the tract. Another type of movement seen only in the large intestine, __(4)__ , occurs infrequently and acts to move feces over relatively long distances toward the anus. The presence of feces in the __(5)__ excites stretch receptors so that the __(6)__ reflex is initiated. Irritation of the gastrointestinal tract by drugs or bacteria might stimulate the __(7)__ center in the medulla, causing __(8)__ , which is essentially a reverse peristalsis.

18. Identify the following pathological conditions using the key choices.

KEY CHOICES

A. Ankyloglossia   **C.** Constipation   **E.** Gallstones   **G.** Hiatal hernia   **I.** Peritonitis

B. Appendicitis   **D.** Diarrhea   **F.** Gingivitis   **H.** Jaundice   **J.** Ulcer

_____ **1.** Inflammation of the abdominal serosa

_____ **2.** Condition that may lead to reflux of acidic gastric juice into the esophagus causing heartburn due to esophagitis

_____ **3.** Usually indicates liver problems or blockage of the biliary ducts

_____ **4.** An erosion of the stomach or duodenal mucosa

_____ **5.** Passage of watery stools

_____ **6.** Causes severe epigastric pain; associated with prolonged storage of bile in the gallbladder

_____ **7.** Inability to pass feces; often a result of poor bowel habits

_____ **8.** Inflammation of the gums

_____ **9.** Person is "tongue tied" because of a short lingual frenulum

Physiology of Chemical Digestion and Absorption

1. Various types of foods are ingested in the diet and broken down to their building blocks. Use the key choices to complete the following statements. For some questions, more than one term is correct.

KEY CHOICES

A. Amino acids   E. Fatty acids   H. Glucose   K. Meat/fish

B. Bread/pasta   F. Fructose   I. Lactose   L. Starch

C. Cheese/cream   G. Galactose   J. Maltose   M. Sucrose

D. Cellulose

_____ 1. Examples of carbohydrate foods in the diet

_____ 2. The three common simple sugars in our diet

_____ 3. The most important of the simple sugars; referred to as "blood sugar"

_____ 4. Disaccharides

_____ 5. The only important *digestible* polysaccharide

_____ 6. An indigestible polysaccharide that aids elimination because it adds bulk to the diet

_____ 7. Protein-rich foods include these two food groups

_____ 8. Protein foods must be broken down to these units before they can be absorbed

_____ 9. Fats are broken down to these building blocks and glycerol

2. Dietary substances capable of being absorbed are listed below. If the substance is *most often* absorbed from the digestive tract by active transport processes, write A in the blank. If it is usually absorbed passively (by diffusion or osmosis), write P in the blank. In addition, circle the substance that is *most likely* to be absorbed into a lacteal rather than into the capillary bed of the villus.

____ 1. Water   ____ 3. Simple sugars   ____ 5. Electrolytes

____ 2. Amino acids   ____ 4. Fatty acids

3. This exercise concerns chemical food breakdown and absorption in the digestive tract. Using the key choices, select the appropriate terms to complete the statements.

KEY CHOICES

A. Amino acids	**G.** HCl	**M.** Nucleic acids
B. Bile	**H.** Intrinsic factor	**N.** Pepsin
C. Brush border enzymes	**I.** Lacteal	**O.** Proteases
D. Capillary	**J.** Lipases	**P.** Rennin
E. Fatty acids	**K.** Monosaccharides	**Q.** Salivary amylase
F. Glycerol	**L.** Monoglycerides	**R.** Vitamin D

_____ 1. Starch digestion begins in the mouth when __(1)__ is ducted in by the salivary glands.

_____ 2. Protein foods are largely acted on in the stomach by __(2)__ .

_____ 3. A milk-coagulating enzyme found in children but not usually in adults is __(3)__ .

_____ 4. Intestinal enzymes are called __(4)__ .

_____ 5. For the stomach protein-digesting enzymes to become active, __(5)__ is needed.

_____ 6. A nonenzyme substance that emulsifies fats is __(6)__ .

_____ 7. Trypsin, chymotrypsin, aminopeptidase, and dipeptidase are all __(7)__ .

_____ 8. Chylomicrons are the products that ultimately enter the __(8)__ when fats are being digested and absorbed.

_____ 9. Unlike most other foodstuffs, __(9)__ are broken down even further than their building blocks. The enzyme(s) involved in the final
_____ 10. steps of that breakdown process is/are __(10)__ .

_____ 11. Required for calcium absorption is __(11)__ .

_____ 12.   _____ 13. Transport for __(12–14)__ is
active and often coupled to the Na$^+$ pump.
_____ 14.

4. Circle the term that does not belong in each of the following groupings.

  1. Trypsin   Dextrinase   Maltase   Aminopeptidase   Intestinal enzymes

  2. Micelles   Accelerate lipid absorption   Bile salts   Chylomicrons

  3. Ferritin   Mucosal iron barrier   Lost during hemorrhage   Iron storage

  4. Dextrinase   Glucoamylase   Maltase   Nuclease   Amylase

  5. Hydrolysis   Bond splitting   Removal of water   pH specific

  6. Chymotrypsin   Alkaline pH   Pepsin   Trypsin   Dipeptidase

  7. Vitamin A   Vitamin C   Passively absorbed   Vitamin B$_{12}$ Vitamin D

5. Fill in the table below with the specific information indicated in the headings:

Enzyme	Substrate	Product	Secreted by	Active in
Salivary amylase				
Pepsin				
Trypsin				
Dipeptidase				
Pancreatic amylase				
Lipase				
Maltase				
Carboxypeptidase				
Nuclease				

Developmental Aspects of the Digestive System

1. Using the key choices, match the terms with the following descriptions.

KEY CHOICES

  A. Accessory organs   **F.** Gallbladder problems   **K.** Rooting

  B. Alimentary canal   **G.** Gastroenteritis   **L.** Sucking

  C. Appendicitis   **H.** Hepatitis   **M.** Stomach

  D. Cleft palate/lip   **I.** Periodontal disease   **N.** Tracheoesophageal fistula

  E. Cystic fibrosis   **J.** Peristalsis   **O.** Ulcers

_____ **1.** Internal tubelike cavity of the embryo

_____ **2.** Glands that branch out from the digestive mucosa, which arises from endoderm

_____ **3.** Common congenital defect; aspiration of feeding is common

_____ **4.** Congenital condition(s) characterized by a connection between digestive and respiratory passageways

_____ **5.** Congenital condition in which large amounts of mucus are produced, clogging respiratory passageways and pancreatic ducts

_____ **6.** Common GI problems of middle-aged adults

_____ **7.** Reflex aiding the newborn baby to find the nipple

_____ **8.** Vomiting is common in infants because this structure is small

_____ **9.** Most common adolescent digestive system problem

_____ **10.** Inflammations of the gastrointestinal tract

_____ **11.** Condition of loose teeth and inflamed gums; generally seen in elderly people

_____ **12.** In old age, the production of digestive juices decreases and this action slows, leading to elimination problems

The Incredible Journey: A Visualization Exercise for the Digestive System

. . . the passage beneath you opens, and you fall into a huge chamber with mountainous folds.

1. Complete the following narrative by writing the missing terms in the answer blanks.

_____ **1.**

_____ **2.**

_____ **3.**

In this journey, you are to travel through the digestive tract as far as the appendix and then await further instructions. You are miniaturized as usual and provided with a wetsuit to protect you from being digested during your travels. You have a very easy entry into your host's open mouth. You look around and notice the glistening pink lining, or **(1)**, and the perfectly cared-for teeth. Within a few seconds, the lips part and you find yourself surrounded by bread. You quickly retreat to the safety of the **(2)** between the teeth and the cheek to prevent getting chewed. From there you watch with fascination as a number of openings squirt fluid into the chamber, and the **(3)** heaves and rolls, mixing the bread with the fluid. As the bread begins to disappear, you decide that the fluid contains the enzyme

4. _____

5. _____

6. _____

7. _____

8. _____

9. _____

10. _____

11. _____

12. _____

13. _____

14. _____

15. _____

16. _____

17. _____

(4) . You then walk toward the back of the oral cavity. Suddenly, you find yourself being carried along by a squeezing motion of the walls around you. The name given to this propelling motion is **(5)** . As you are carried helplessly downward, you see two openings, the **(6)** and the **(7)** , below you. Just as you are about to straddle the solid area between them to stop your descent, the structure to your left moves quickly upward, and a trapdoorlike organ, the **(8)** , flaps over its opening. Down you go in the dark, seeing nothing. Then the passage beneath you opens, and you fall into a huge chamber with mountainous folds. Obviously, you have reached the **(9)** . The folds are very slippery, and you conclude that it must be the **(10)** coat that you read about earlier. As you survey your surroundings, juices begin to gurgle into the chamber from pits in the "floor," and your face begins to sting and smart. You cannot seem to escape this caustic fluid, and you conclude that it must be very dangerous to your skin since it contains **(11)** and **(12)** . You reach down and scoop up some of the slippery substance from the folds and smear it on your face, confident that if it can protect this organ it can protect you as well! Relieved, you begin to slide toward the organ's far exit and squeeze through the tight **(13)** valve into the next organ. In the dim light, you see lumps of cellulose lying at your feet and large fat globules dancing lightly about. A few seconds later, your observations are interrupted by a wave of fluid pouring into the chamber from an opening high in the wall above you. The large fat globules begin to fall apart, and you decide that this enzyme flood has to contain **(14)** and the opening must be the duct from the **(15)** . As you move quickly away to escape the deluge, you lose your footing and find yourself on a rollercoaster ride—twisting, coiling, turning, and diving through the lumen of this active organ. As you move, you are stroked by velvety, fingerlike projections of the wall, the **(16)** . Abruptly your ride comes to a halt as you are catapulted through the **(17)** valve and fall into the appendix. Headquarters informs you that you are at the end of your journey. Your exit now depends on your own ingenuity and your sympathy for your host.

CHALLENGING YOURSELF

At the Clinic

1. A young boy is brought to the clinic wincing with pain whenever he opens his mouth. His parents believe he dislocated his jaw, which is obviously swollen on the left side. Luckily, the clinic has seen a number of similar cases in the past week, so the diagnosis is quick. What is the diagnosis?

2. Marv comments to his doctor at the clinic that he gets a "full," very uncomfortable feeling in his chest after every meal, as if the food were lodged there instead of traveling to his stomach. The doctor states that the description fits the condition called achalasia, in which the valve between the esophagus and stomach fails to open. What valve is involved?

3. Mr. Ashe, a man in his mid-60s, comes to the clinic complaining of heartburn. Questioning by the clinic staff reveals that the severity of his attacks increases when he lies down after eating a heavy meal. The man is about 50 pounds overweight. What is your diagnosis? Without treatment, what conditions might develop?

4. A young woman is put through an extensive battery of tests to determine the cause of her "stomach pains." She is diagnosed with gastric ulcers. An antihistamine drug is prescribed and she is sent home. What is the mechanism of her medication? What life-threatening problems can result from a poorly managed ulcer? Why did the clinic doctor warn the woman not to take aspirin?

5. Continuing from the previous question, the woman's ulcer got worse. She started complaining of back pain. The physician discovered that the back pain occurred because the pancreas was now damaged. Use logic to deduce how a perforating gastric ulcer could come to damage the pancreas.

6. An elderly man, found unconscious in the alley behind a local restaurant, is brought to the clinic in a patrol car. His abdomen is distended, his skin and scleras are yellow, and he reeks of alcohol. What do you surmise is the cause of his condition and what terms are used to describe the various aspects of his appearance?

7. A feverish 12-year-old girl named Kelly is brought to the clinic complaining of pain in the lower right abdominal quadrant. According to her parents, for the past week she has been eating poorly and often vomits when she does eat. What condition do you suspect? What treatment will she need? What complication will occur if treatment is given too late?

8. A woman in her 50s complains of bloating, cramping, and diarrhea when she drinks milk. What is the cause of her complaint and what is a solution?

9. Mr. Erickson complains of diarrhea after eating and relates that it is most pronounced when he eats starches. Subsequent dietary variations reveal intolerance to wheat and other grains, but no problem with rice. What condition do you suspect?

10. A new mother is worried about her week-old infant. The baby has begun to turn blue whenever she is fed and chokes during each feeding. What developmental abnormality do you suspect and how will it be corrected?

11. A woman in labor was given a rectal exam and it was determined that the cervix of her uterus was dilated and that her time for giving birth was near. Explain what a rectal exam is.

Stop and Think

1. How do the following terms relate to the function of the smooth muscle of the digestive system? gap junction, stress-relaxation response, nonstriated

2. What happens to salivary amylase in the stomach? Pepsin in the duodenum?

3. How would the release of histamine by gastric enteroendocrine cells improve the *absorption* of nutrients?

4. Which would be more effective: a single "megadose" of calcium supplement once a day or a smaller supplement with each meal?

5. Trace the digestion and absorption of fats from their entrance into the duodenum to their arrival at the liver.

6. Clients are instructed not to eat before having blood tests run. How would a lab technician know if someone "cheated" and ate a fatty meal a few hours before having his blood drawn?

7. Why does liver impairment result in edema?

8. The location of the rapidly dividing, undifferentiated epithelial cells differs in the stomach and intestine. (a) Compare the locations of these cells in the two digestive organs. (b) What is the basic function of these dividing cells?

9. Bianca went on a trip to the Bahamas during spring vacation and did not study enough for her anatomy test that was scheduled early in the following week. On the test, she mixed up the following pairs of structures: (a) serous cells and serous membranes, (b) caries and canaliculi (bile canaliculi), (c) anal canal and anus, (d) diverticulosis and diverticulitis, (e) hepatic vein and hepatic portal vein. Can you help her by defining and differentiating all these sound-alike structures?

10. Name three organelles that are abundant in hepatocytes and explain how each of these organelles contributes to liver functions.

COVERING ALL YOUR BASES

Multiple Choice

Select the best answer or answers from the choices given.

1. Which of the following terms are synonyms?
   A. Gastrointestinal tract
   B. Digestive system
   C. Digestive tract
   D. Alimentary canal

2. A digestive organ that is not part of the alimentary canal is the:
   A. stomach   D. large intestine
   B. liver   E. pharynx
   C. small intestine

3. An organ that is not covered by a visceral peritoneum is the:
   A. stomach   C. thoracic esophagus
   B. jejunum   D. transverse colon

4. The GI tube layer responsible for the actions of segmentation and peristalsis is:
   A. serosa   C. muscularis externa
   B. mucosa   D. submucosa

5. The nerve plexus that stimulates secretion of digestive juices is the:
   A. submucosal plexus
   B. myenteric plexus
   C. subserous plexus
   D. local plexuses

6. Organs that are partly or wholly retroperitoneal include:
   A. stomach   C. duodenum
   B. pancreas   D. colon

7. Which of the following are part of the splanchnic circulation?
   A. Hepatic artery
   B. Hepatic vein
   C. Hepatic portal vein
   D. Splenic artery

8. Which alimentary canal tunic has the greatest abundance of lymph nodules?
   A. Mucosa
   B. Muscularis
   C. Serosa
   D. Submucosa

9. Proteins secreted in saliva include:
   A. mucin
   B. amylase
   C. lysozyme
   D. IgA

10. Which of these tissue types is most cellular?
    A. Enamel
    B. Pulp
    C. Dentin
    D. Cementum

11. The closure of which valve is assisted by the diaphragm?
    A. Ileocecal
    B. Pyloric
    C. Gastroesophageal
    D. Upper esophageal

12. Stratified squamous epithelium lines the:
    A. cardia of stomach
    B. esophagus
    C. anal canal
    D. rectum

13. Smooth muscle is found in the:
    A. tongue
    B. pharynx
    C. esophagus
    D. external anal sphincter

14. Rugae in the stomach encompass which tunics?
    A. Mucosa
    B. Muscularis
    C. Serosa
    D. Submucosa

15. Where in the stomach do the strongest peristaltic waves occur?
    A. Body
    B. Cardia
    C. Fundus
    D. Pylorus

16. The greater omentum:
    A. contains lymph nodes
    B. covers most of the anterior peritoneal cavity
    C. contains fat deposits
    D. attaches to the liver

17. Carbonic anhydrase is contained in:
    A. parietal cells of gastric glands
    B. chief cells of gastric glands
    C. pancreatic acini cells
    D. pancreatic duct cells

18. Sometimes the only treatment for chronic ileitis is surgical removal. What functions will be lost?
    A. Vitamin B_{12} absorption
    B. Absorption of calcium and iron
    C. Normal resistance to bacterial infection
    D. Enterohepatic circulation

19. Which statement is true about the peritoneal cavity?
    A. It is the same thing as the abdomino-pelvic cavity.
    B. This large cavity is filled with air.
    C. Like the pleural and pericardial cavities, it is a potential space containing serous fluid.
    D. It contains the pancreas and all the duodenum.

20. Which of these organs lies in the right hypochondriac region of the abdomen?
    A. Stomach
    B. Spleen
    C. Cecum
    D. Liver

21. Which phases of gastric secretion depend on the vagus nerve?
    A. Cephalic
    B. Gastric
    C. Intestinal (stimulatory)
    D. Intestinal (inhibitory)

22. Which of the following is/are classified as enterogastrone(s)?
    A. Gastrin   C. CCK
    B. GIP   D. Secretin

23. After gastrectomy, it will be necessary to supplement:
    A. pepsin   C. gastrin
    B. vitamin B_{12} D. intrinsic factor

24. Villi:
    A. are composed of both mucosa and submucosa
    B. contain lacteals
    C. can contract to increase circulation of lymph
    D. are largest in the duodenum

25. Brush border enzymes include:
    A. dipeptidase
    B. lactase
    C. enterokinase
    D. lipase

26. The portal triads include:
    A. branch of the hepatic artery
    B. central vein
    C. branch of the hepatic portal vein
    D. bile duct

27. Which of the following are in greater abundance in the hepatic vein than in the hepatic portal vein?
    A. Urea
    B. Aging RBCs
    C. Plasma proteins
    D. Fat-soluble vitamins

28. Jaundice may be a result of:
    A. hepatitis
    B. blockage of enterohepatic circulation
    C. biliary calculi in the common bile duct
    D. removal of the gallbladder

29. Release of CCK leads to:
    A. contraction of smooth muscle in the duodenal papilla
    B. increased activity of hepatocytes
    C. activity of the muscularis of the gallbladder
    D. exocytosis of zymogenic granules in pancreatic acini

30. The pH of chyme entering the duodenum is adjusted by:
    A. bile
    B. intestinal juice
    C. secretions from pancreatic acini
    D. secretions from pancreatic ducts

31. Peristalsis in the small intestine begins:
    A. with the cephalic phase of stomach activity
    B. when chyme enters the duodenum
    C. when CCK is released
    D. when most nutrients have been absorbed

32. Relaxation of the ileocecal sphincter is triggered by:
    A. gastrin
    B. enterogastrones
    C. presence of bile salts in the blood
    D. gastroileal reflex

33. Relaxation of the teniae coli would result in loss of:
    A. haustra
    B. splenic and hepatic flexures
    C. rectal valves
    D. anal columns

34. Diverticulitis is associated with:
    A. too much bulk in the diet
    B. too little bulk in the diet
    C. gluten intolerance
    D. lactose intolerance

35. Which of the following are associated with the complete digestion of starch?
    A. Maltose   C. Amylase
    B. Sucrose   D. Oligosaccharides

36. Which of the following are tied to sodium transport?
    A. Glucose   C. Galactose
    B. Fructose   D. Amino acids

37. Endocytosis of proteins in newborns accounts for:
    A. absorption of casein
    B. absorption of antibodies
    C. some food allergies
    D. absorption of butterfat

38. Cell organelles in the intestinal epithelium that are directly involved in lipid digestion and absorption include:
    A. smooth ER
    B. Golgi apparatus
    C. secretory vesicles
    D. lysosomes

39. Excess iron is stored primarily in the:
    A. liver
    B. bone marrow
    C. duodenal epithelium
    D. blood

40. Which of the following correctly describes the flow of blood through the classical liver lobule and beyond?
    A. Portal vein branch to sinusoids to central vein to hepatic vein to inferior vena cava
    B. Porta hepatis to hepatic vein to portal vein
    C. Portal vein branch to central vein to hepatic vein to sinusoids
    D. Hepatic artery branch to sinusoids to central vein to hepatic vein

41. Mesoderm gives rise to which of the following areas of the stomach wall?
    A. Lining epithelium
    B. Submucosa
    C. Muscularis externa
    D. Serosa

42. The embryonic primitive gut forms from:
    A. endoderm   C. mesoderm
    B. yolk sac   D. ectoderm

43. The opening of the mouth forms when a break occurs in the:
    A. oral membrane   C. cloacal membrane
    B. stomodeum   D. proctodeum

44. Difficult swallowing is called:
    A. ascites   C. ileus
    B. dysphagia   D. stenosis

45. Bulimia has been linked to:
    A. deficiency of CCK
    B. deficiency of gastrin
    C. hypersecretion of enterogastrones
    D. secretion of endorphins

46. The hepatopancreatic ampulla lies in the wall of the:
    A. liver   C. duodendum
    B. pancreas   D. stomach

47. Two structures that produce alkaline secretions that neutralize the acidic stomach chyme as it enters the duodendum are:
    A. intestinal flora and pancreatic acinar cells
    B. gastric glands and gastric pits
    C. lining epithelium of the intestine and Paneth cells
    D. pancreatic ducts and duodenal glands

48. A 3-year-old girl was rewarded with a hug because she was now completely toilet trained. Which muscle had she learned to control?
    A. Levator ani
    B. Internal anal sphincter
    C. Internal and external obliques
    D. External anal sphincter

49. Which cell type fits this description? It occurs in the stomach mucosa, contains abundant mitochondria and many microvilli, and pumps hydrogen ions.
    A. Absorptive cell   C. Goblet cell
    B. Parietal cell   D. Mucous neck cell

50. The only feature in this list that is shared by
both the large and small intestines is:

    A. intestinal crypts   **D.** teniae coli

    B. Peyer's patches   **E.** haustra

    C. plicae circulares   **F.** intestinal villi

Word Dissection

For each of the following word roots, fill in the literal meaning and give an example, using a word
found in this chapter.

Word root	Translation	Example
1. aliment		
2. cec		
3. chole		
4. chyme		
5. decid		
6. duoden		
7. enter		
8. epiplo		
9. eso		
10. falci		
11. fec		
12. fren		
13. gaster		
14. gest		
15. glut		
16. haustr		

Word root	Translation	Example
17. hiat	_____	_____
18. ile	_____	_____
19. jejun	_____	_____
20. micell	_____	_____
21. oligo	_____	_____
22. oment	_____	_____
23. otid	_____	_____
24. pep	_____	_____
25. plic	_____	_____
26. proct	_____	_____
27. pylor	_____	_____
28. ruga	_____	_____
29. sorb	_____	_____
30. splanch	_____	_____
31. stalsis	_____	_____
32. teni	_____	_____

Nutrition, Metabolism, and Body Temperature Regulation

Student Objectives

When you have completed the exercises in this chapter, you will have accomplished the following objectives:

Nutrition

1. Define *nutrient, essential nutrient,* and *calorie.*

2. List the six major nutrient categories. Note important dietary sources and the principal cellular uses of each.

3. Distinguish between nutritionally complete and incomplete proteins.

4. Define *nitrogen balance* and indicate possible causes of positive and negative nitrogen balance.

5. Distinguish between fat- and water-soluble vitamins and list the vitamins belonging to each group.

6. For each vitamin, list its important sources and functions in the body and describe consequences of its deficit or excess.

7. List minerals essential for health; note important dietary sources and describe how each is used in the body.

Metabolism

8. Define *metabolism.* Explain how catabolism and anabolism differ.

9. Define *oxidation* and *reduction* and note the importance of these reactions in metabolism. Explain the role of coenzymes used in cellular oxidation reactions.

10. Explain the difference between substrate-level phosphorylation and oxidative phosphorylation.

11. Follow the oxidation of glucose in body cells. Summarize the important events and products of glycolysis, the Krebs cycle, and the electron transport chain.

12. Define *glycogenesis, glycogenolysis,* and *gluconeogenesis.*

13. Describe the process by which fatty acids are oxidized for energy.

14. Define *ketone bodies* and note the stimulus for their formation.

15. Describe how amino acids are prepared to be oxidized for energy.

16. Describe the need for protein synthesis in body cells.

17. Explain the concept of amino acid or carbohydrate/fat pools and describe pathways by which substances in these pools can be interconverted.

18. List important goals and events of the absorptive and postabsorptive states and explain how these events are regulated.

19. List and describe several metabolic functions of the liver.

20. Differentiate between LDLs and HDLs relative to their structures and major roles in the body.

Body Energy Balance

21. Explain what is meant by body energy balance.

22. Describe some current theories of food intake regulation.

23. Define *basal metabolic rate* and *total metabolic rate.* Name several factors that influence metabolic rate.

24. Explain how body temperature is regulated and describe the common mechanisms regulating heat production/retention and heat loss from the body.

Developmental Aspects of Nutrition and Metabolism

25. Describe the effects of inadequate protein intake on the fetal nervous system.

26. Describe the cause and consequences of the low metabolic rate typical of the elderly.

27. List ways in which medications commonly used by an aged population may influence their nutrition and health.

Ready, set, go!

In this chapter you will complete the studies of digestion and absorption presented in Chapter 24 by examining in detail the nature of nutrients and the roles they play in meeting the body's various needs. The breakdown products resulting from digestion of large nutrient molecules enter the cells. In cells, metabolic pathways may lead to the extraction of energy, the synthesis of new or larger molecules, or the interchange of parts of molecules. Unusable residues of molecules are eliminated as waste.

Subjects for study and review in Chapter 25 include nutrition, metabolism, control of the absorptive and postabsorptive states, the role of the liver, the regulation of body temperature, and important developmental aspects of metabolism.

BUILDING THE FRAMEWORK

Nutrition

1. Respond to the following questions by writing your answers in the answer blanks.

   1. Name the six categories of nutrients. _____

_____

   2. What is meant by the term *essential nutrients*? _____

_____

3. Define the unit used to measure the energy value of foods. _____

_____

4. Which carbohydrate is used by cells as the major ready source of energy? _____

5. Which two types of cells rely almost entirely for their energy needs on the carbohydrate

named in Question 4? _____

6. What is meant by the term *empty calories*? _____

_____

7. Name the major dietary source of cholesterol. _____

8. Which fatty acid(s) cannot be synthesized by the liver and must be ingested as a lipid?

_____

9. Define *complete protein.* _____

_____

_____

10. What is meant by the term *nitrogen balance?* _____

_____

_____

11. Name the two major uses of proteins synthesized in the body. _____

_____

12. What is the function of most vitamins in the body? _____

_____

13. Which organ chemically processes nearly every category of nutrient? _____

14. Which two minerals account for most of the weight of the body's minerals?

_____

2. Several descriptive statements concerning vitamins and the consequences of their excess and deficit appear below. Use the key choices to correctly identify each vitamin considered.

KEY CHOICES

A. Vitamin A   **E.** Vitamin B_{12} **H.** Vitamin E   **K.** Folic acid

B. Vitamin B_1 **F.** Vitamin C   **I.** Vitamin K   **L.** Niacin

C. Vitamin B_2 **G.** Vitamin D   **J.** Biotin   **M.** Pantothenic acid

D. Vitamin B_6

_____ **1.** Most is synthesized in the colon by resident bacteria; essential for the liver's synthesis of clotting proteins.

_____ **2.** The most complex vitamin; contains cobalt; an essential coenzyme for DNA synthesis and production of choline.

_____ **3.** Also called pyridoxine; involved in amino acid metabolism; excess causes neurological defects (loss of sensation, etc.).

_____ **4.** Part of coenzyme cocarboxylase; deficit results in beriberi.

_____ **5.** Chemically related to sex hormones; believed to be an antioxidant that disarms free radicals.

_____ **6.** Best extrinsic source is fortified milk; also made in the skin from cholesterol; deficit results in rickets in children and osteomalacia in adults.

_____ **7.** ___ **8.** ___ **9.** ___ **10.** Fat-soluble vitamins.

_____ **11.** Functions in the body in the form of coenzyme A; essential for oxidation reactions and fat synthesis.

_____ **12.** Part of the coenzyme NAD required for many metabolic reactions; deficit results in pellagra.

_____ **13.** Found in large amounts in citrus fruits, strawberries, and tomatoes; essential for connective tissue matrix synthesis, blood clotting.

_____ **14.** The vitamin of vision; required for photopigment synthesis; deficit results in night blindness, clouding of cornea.

_____ **15.** Part of the coenzyme FAD; named for its similarity to ribose sugar; deficit causes cheilosis.

_____ **16.** Required for erythropoiesis; deficit causes macrocytic anemia.

3. Match the minerals listed in the key choices to the appropriate descriptions. (*Note:* Not all of the trace minerals have been considered.)

KEY CHOICES

A. Calcium   **C.** Iodine   **E.** Magnesium   **G.** Potassium   **I.** Sulfur

B. Chlorine   **D.** Iron   **F.** Phosphorus   **H.** Sodium   **J.** Zinc

_____ **1.** Required for normal growth, wound healing, taste, and smell

_____ **2.** Concentrates in the thyroid gland; essential for T_3 and T_4 synthesis

_____ **3.** Essential component of many proteins (e.g., insulin) as part of -S–S- bonds; found in virtually all of the denser connective tissues as part of the ground substance

_____ **4.** In the anion part of the calcium salts found in bone

_____ **5.** Principal intracellular cation; excesses result in muscle weakness and cardiac abnormalities

_____ **6.** The most abundant cation in extracellular fluid; a major factor in determining fluid shifts in the body; excesses lead to hypertension in some cases

_____ **7.** Its ion is the major extracellular anion

_____ **8.** An essential component of hemoglobin and some cytochromes; excesses cause liver damage

_____ **9.** Part of coenzymes that act in hydrolysis of ATP; excesses lead to diarrhea

4. Circle the term that does not belong in each of the following groupings.

1. Neutral fats   Triglycerides   Trisaccharides   Triacylglycerols

2. Meats   Eggs   Milk   Nuts   Fish

3. Mineral-rich foods   Vegetables   Fats   Milk   Legumes

4. Coenzymes B_2 Niacin   Vitamin C   Biotin

5. K^+ Osmotic pressure of blood Na^+ Cl^-

6. Stored for long-term use   Amino acids   Triacylglycerols   Glycogen

7. Vitamin D   Must be ingested   Ultraviolet radiation   Skin

8. Copper   Iron   Hemoglobin synthesis   Zinc

9. 0.8 g/kg of body weight   About 2 oz   About 16 oz   Minimum daily requirement of protein

10. Catabolic hormones   Anabolic hormones   Sex hormones   Pituitary growth hormones

5. Using the key choices, identify the foodstuffs used by cells in the following descriptions of cellular functions.

KEY CHOICES

A. Carbohydrates **B.** Fats **C.** Proteins

_____ **1.** The most-used substance for producing the energy-rich ATP

_____ **2.** Important in building myelin sheaths and cell membranes

_____ **3.** Tend to be conserved by cells

_____ **4.** The second most important food source for making cellular energy

_____ **5.** Form insulating deposits around body organs and beneath the skin

_____ **6.** Used to make the bulk of structural and functional cell substances such as collagen, enzymes, and hemoglobin

_____ **7.** Building blocks are amino acids

_____ **8.** One of their building blocks, glycerol, is a sugar alcohol

Metabolism

1. Complete the following statements, which provide an overview of the metabolism of energy-containing nutrients, by writing the missing words in the answer blanks.

_____ **1.**

_____ **2.**

_____ **3.**

_____ **4.**

_____ **5.**

_____ **6.**

_____ **7.**

_____ **8.**

_____ **9.**

_____ **10.**

The general term for all chemical reactions necessary to maintain life is __(1)__ . The breakdown of complex molecules to simpler ones is called __(2)__ , and the reverse process, the synthesis of larger molecules from smaller ones, is called __(3)__ . The term that means the extraction of energy from nutrient molecules is __(4)__ . The chemical form of energy that cells generally use to drive their activities is __(5)__ .

There are three states in the metabolism of energy-containing nutrients. Stage 1 occurs in the __(6)__ , where large nutrient molecules are broken down to their absorbable forms (monomers). In Stage 2, the blood transports these monomers to tissue cells, where the monomers undergo catabolic or anabolic reactions in the cytoplasm of the cells.

Stage 3 occurs in the __(7)__ of cells, where generation of ATP requires the gas __(8)__ and where __(9)__ and __(10)__ gas are formed as the final waste products.

_____ 11.

_____ 12.

_____ 13.

_____ 14.

_____ 15.

_____ 16.

Reactions that generate ATP are called oxidation reactions. In most cases, the substrate molecules (the substance oxidized) lose **(11)** and also **(12)** . Reactions in which substrate molecules gain energy by the addition of hydrogen atoms and electrons are called **(13)** reactions. Since the enzymes regulating the oxidation-reduction (OR) reactions cannot accept (bind) hydrogen atoms removed during oxidation, molecules called **(14)** are required. The abbreviations for two of the most important coenzymes in Stage 3 are **(15)** and **(16)** .

2. Figure 25.1 is a simplified, highly schematic diagram of the major catabolic pathways taken by one glucose molecule entering a typical cell. Only the organelles involved with these reactions are included, and only *types* of molecules (*not quantities*) are shown. Dashed arrows indicate two or more steps in a series of reactions, and open arrows indicate the movement of molecules.

Using the key choices, complete the diagram by inserting the correct symbols or words in the empty boxes on the diagram. Also color code and color each box provided with a color-coding circle. Then, referring to Figure 25.1, complete the statements that follow by inserting your answers in the answer blanks.

KEY CHOICES

◯ Acetyl CoA (Acetyl coenzyme A)

◯ ADP (Adenosine diphosphate)

◯ CO_2 (Carbon dioxide)

◯ $FADH_2$ (Reduced FAD)

◯ Glucose-6-phosphate

◯ H_2O (Water)

◯ Krebs cycle

◯ $NADH + H^+$ (Reduced NAD)

◯ O_2 (Oxygen)

◯ P_i (Inorganic phosphate)

◯ Pyruvic acid

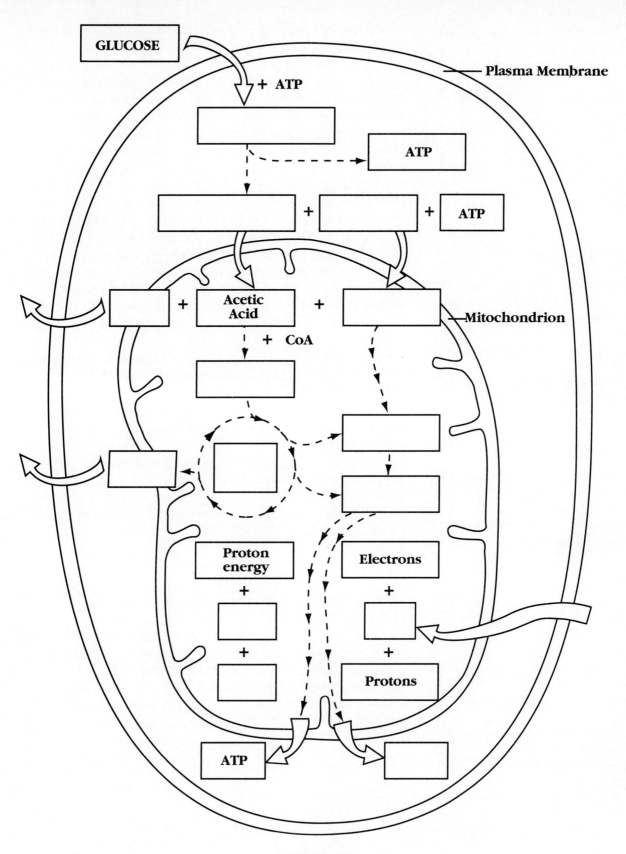

Figure 25.1

_____ 1.

_____ 2.

_____ 3.

_____ 4.

_____ 5.

_____ 6.

_____ 7.

_____ 8.

_____ 9.

_____ 10.

_____ 11.

_____ 12.

_____ 13.

_____ 14.

_____ 15.

_____ 16.

_____ 17.

_____ 18.

_____ 19.

_____ 20.

_____ 21.

_____ 22.

_____ 23.

_____ 24.

_____ 25.

_____ 26.

_____ 27.

_____ 28.

The catabolism of glucose into two pyruvic acid molecules is called __(1)__ . This process occurs in the __(2)__ of the cell and is anaerobic, meaning that oxygen is __(3)__ . Two molecules of reduced __(4)__ and a net gain of __(5)__ ATP molecules also result from this process. The first phase of glycolysis is __(6)__ , in which glucose is converted to fructose-1,6-diphosphate, reactions energized by the hydrolysis of 2 ATP. The second phase, __(7)__ , involves breaking the 6-C sugar into two 3-C compounds. The third phase is __(8)__ , in which 2 pyruvic acid molecules are formed and 4 ATP (2 net ATP) are produced. The fate of the pyruvic acid depends on whether or not oxygen is available. If oxygen is absent, the two hydrogen atoms removed from pyruvic acid and transferred to NAD$^+$ combine again with __(9)__ to form __(10)__ . Some cells, like __(11)__ cells, routinely undergo periods of anaerobic respiration and are not adversely affected, but __(12)__ cells are quickly damaged. Prolonged anaerobic respiration ultimately results in a decrease in blood __(13)__ . __(14)__ cells undergo only glycolysis.

If oxygen is present, carbon is removed from pyruvic acid and leaves the cell as __(15)__ , and pyruvic acid is oxidized by the removal of hydrogen atoms, which are picked up by __(16)__ . The remaining two-carbon fragment (acetic acid) is carried by __(17)__ to combine with a four-carbon acid called __(18)__ , forming the six-carbon acid __(19)__ . This six-carbon acid is the first molecule of the __(20)__ cycle, which occurs in the __(21)__ . The intermediate acids of this cycle are called __(22)__ acids. Each turn of the cycle generates __(23)__ molecules of carbon dioxide, __(24)__ molecules of reduced NAD$^+$, and __(25)__ molecules of reduced FAD. Krebs cycle reactions provide a metabolic pool, where two-carbon fragments from __(26)__ and __(27)__ as well as from glucose may be used to generate ATP.

The final reactions of glucose catabolism are those of the __(28)__ chain, which delivers electrons to molecular oxygen. Oxygen combines with __(29)__ to form __(30)__ . Part of the energy released in the above reactions is used to form ATP by combining __(31)__ with inorganic phosphate (P$_i$). The combined overall reaction is called __(32)__ . In heart muscle and liver cells, for each molecule of glucose that is completely oxidized, __(33)__ ATP molecules are generated.

_____ 29.

_____ 30.

_____ 31.

_____ 32.

_____ 33.

3. Figure 25.2 is a simplified diagram of a cross section of part of a mitochondrion. First, label the intermembrane space and the mitochondrial matrix on the appropriate lines on the diagram. (*Note:* The structures labeled EC [1–4] represent enzyme complexes.) Next, select two colors and color the coding circles and the circles on the diagram that represent the electrically charged particles listed below. Remember that the *types* of particles are indicated, *not* the quantities of each. Then, referring to Figure 25.2, complete the statements that follow by inserting your answers in the answer blanks.

○ Electrons (Color the appropriate circles and insert e¯ in each.)

○ Protons (Color the appropriate circles and insert H⁺ in each.)

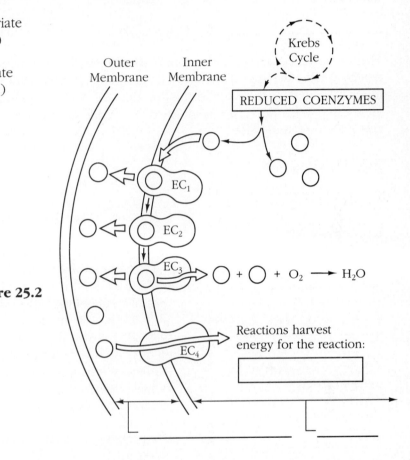

Figure 25.2

1. _____

2. _____

3. _____

4. _____

5. _____

6. _____

7. _____

8. _____

9. _____

10. _____

11. _____

12. _____

13. _____

14. _____

Reduced coenzymes, generated by the Krebs cycle, release __(1)__, which are split into __(2)__ and __(3)__. The electron acceptors are protein–__(4)__ complexes located in the __(5)__ mitochondrial membrane. These respiratory enzyme complexes are labeled __(6)__ in Figure 25.2. The final electron acceptor is __(7)__, which, upon combining with free __(8)__ located in the matrix of the mitochondrion, forms water.

Some of the energy released from the electron transport chain is used by the enzyme complexes to pump protons from the __(9)__ to the __(10)__. As a consequence, the pH of the intermembrane space is __(11)__ than that of the matrix. The electrochemical gradient across the inner membrane is created by __(12)__, which flow through the enzyme complex labeled EC₄ on Figure 25.2. EC₄ is named __(13)__. It uses the energy of proton flow to drive the reaction __(14)__. Also insert your answer to item 14 on the appropriate line in Figure 25.2.

4. Match the metabolic reactions in Column B to the terms in Column A.

	Column A		Column B
_____	**1.** Lipolysis	**A.**	Glycerol + fatty acids → fats
_____	**2.** Ketogenesis	**B.**	Fatty acids → acetyl CoA
_____	**3.** Beta oxidation	**C.**	Fats → glycerol + fatty acids
_____	**4.** Lipogenesis	**D.**	Fatty acids → ketone bodies

5. This exercise considers protein/amino acid metabolism. First, complete the text by writing the appropriate key choices in the answer blanks. Second, use terms from the key to complete the flowcharts shown in Figure 25.3. Third, after filling in the flowchart, respond to Questions 8–12 following the completion section below.

KEY CHOICES

A. Alpha ketoglutaric acid   **D.** Essential   **G.** Urea

B. Ammonia (NH₃)   **E.** Glutamic acid   **H.** Water (H₂O)

C. Oxidative deamination   **F.** Transamination

_____ **1.**

_____ **2.**

_____ **3.**

_____ **4.**

_____ **5.**

_____ **6.**

_____ **7.**

Amino acids are actively accumulated by cells because proteins cannot be made unless all amino acid types are present. The nine amino acids that *must* be taken in the diet are called __(1)__ amino acids. The other amino acids may be synthesized by the process called __(2)__ . When amino acids are oxidized to form cellular energy, their amino groups are removed and liberated as __(3)__ . In the liver, this is combined with carbon dioxide to form __(4)__ , which is removed from the body by the kidneys. The process by which an amine group is removed from an amino acid is called __(5)__ . The keto acid that serves as the acceptor of the amine group is __(6)__ . When this keto acid accepts the amine group, it becomes __(7)__ .

8. Using a key choice, identify the pathway followed by the amino acid labeled I. _____

9. Using a key choice, name the pathway followed by the amino acid labeled II. _____

10. Why must ammonia (from the amine groups) be rapidly removed from the

blood? _____

11. What is the general term for amino acids that lose their amine groups? _____

12. Briefly explain why excess amino acids are oxidized for energy. _____

_____

_____

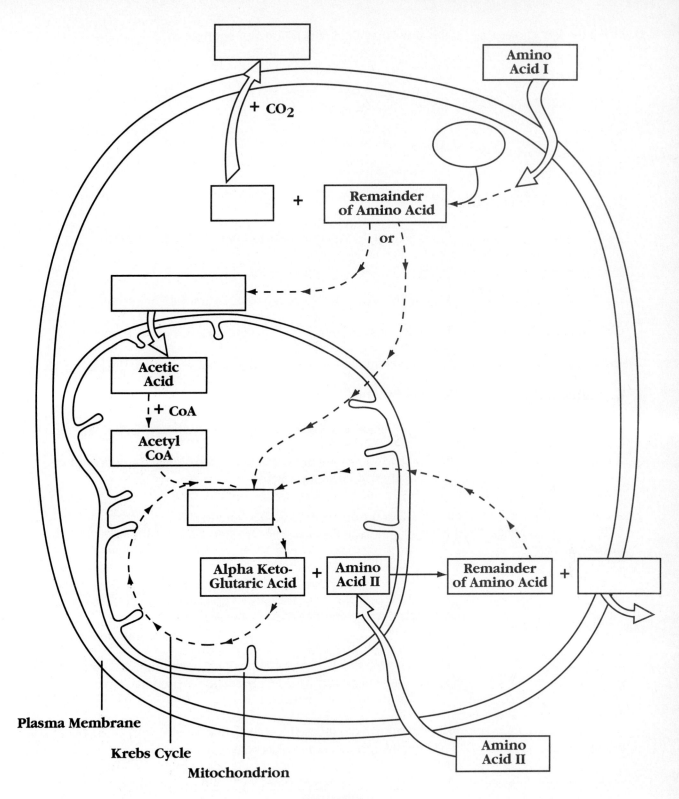

Figure 25.3

6. Using the key choices, select the terms that match the following descriptions.

KEY CHOICES

A. Absorptive state

B. Amino acid pool

C. Blood

D. Carbohydrate-fat pool

E. Glucose sparing

F. Molecular turnover (by degradation and synthesis)

G. Postabsorptive state

_____   **1.** Results in the catabolic-anabolic steady state of the body

_____   **2.** The nutrient-containing transport pool

_____   **3.** Use by body cells of energy sources other than glucose

_____   **4.** Molecules oxidized directly to produce cellular energy

_____   **5.** Period when nutrients move from the GI tract into the bloodstream

_____   **6.** Supply of building blocks available for protein synthesis

_____   **7.** Period when body reserves are used for energy; GI tract is empty

_____   **8.** Excess supply cannot be stored but is oxidized for energy or converted to fat or glycogen

7. If a statement is true, write the letter T in the answer blank. If a statement is false, change the underlined word(s) and write the correct word(s) in the answer blank.

_____   **1.** The efficiency of aerobic cellular respiration is <u>almost 100%</u>.

_____   **2.** Higher concentrations of <u>ADP than ATP</u> are maintained in the cytoplasm.

_____   **3.** If the cytoplasmic concentration of calcium ions (Ca^{2+}) is too high, the gradient energy for ATP production is used instead to pump Ca^{2+} ions <u>out of</u> the mitochondria.

_____   **4.** The storage of glucose molecules in long glycogen chains results from the process called <u>glycolysis</u>.

_____   **5.** The cells of the <u>pancreas and heart muscle</u> are most active in glycogen synthesis.

8. Circle the term that does not belong in each of the following groupings pertaining to triglyceride (fat) metabolism.

    1. 9 kcal/g   Glucose   Fat   Most concentrated form of energy

    2. Chylomicrons   Transported in lymph   Hydrolyzed in plasma

    Hydrolyzed in mitochondria

    3. Cytoplasm   Beta oxidation   Two-carbon fragments   Mitochondria

    4. Triglyceride synthesis   Lipogenesis   ATP deficit   Excess ATP

    5. High dietary carbohydrates   Glucose conversion to fat

    Condensation of acetyl CoA molecules   Low blood sugar level

9. If a statement about the *absorptive* state is true, write the letter T in the answer blank. If a statement is false, change the underlined word(s) and write the correct word(s) in the answer blank.

    _____ **1.** Major metabolic events are <u>catabolic</u>.

    _____ **2.** The major energy fuel is <u>glucose</u>.

    _____ **3.** Excess glucose is stored in muscle cells as <u>glucagon</u> or in adipose cells as fat.

    _____ **4.** The major energy source of skeletal muscle cells is <u>glucose</u>.

    _____ **5.** In a balanced diet, most amino acids pass through the liver to other cells, where they are used for <u>the generation of ATP</u>.

    _____ **6.** The hormone most in control of the events of the absorptive state is <u>insulin</u>.

    _____ **7.** <u>The sympathetic nervous system</u> signals the pancreatic beta cells to secrete insulin.

    _____ **8.** Insulin stimulates the <u>active transport</u> of glucose into tissue cells.

    _____ **9.** Insulin inhibits lipolysis and <u>glycogenesis</u>.

    _____ **10.** Most amino acids in the hepatic portal blood remain in the blood for uptake by <u>hepatocytes</u>.

    _____ **11.** Amino acids move into tissue cells by <u>active transport</u>.

10. Respond to the following questions about the *postabsorptive* state by writing your answers in the answer blanks.

   1. What is the homeostatic range of blood glucose levels? Give units.

   _____

   2. Why must blood glucose be maintained at least at the lower levels?

   _____

   3. Name four sources of glucose in order, from the most to the least available when blood glucose levels are falling. Include the tissues of origin and the name of the metabolic process involved.

   _____

   _____

   _____

   _____

   4. Briefly explain the indirect metabolic pathway by which skeletal muscle cells help maintain blood glucose levels.

   _____

   _____

   _____

   5. Trace the metabolic pathway by which lipolysis produces blood glucose.

   _____

   _____

   _____

   _____

   6. How are tissue proteins used as a source of blood glucose?

   _____

   _____

   7. During the period of glucose sparing, which energy source is available to organs other than the brain? _____

8. After four or five days of fasting, accompanied by low blood glucose levels, the brain begins to use a different energy source. What is this energy source?

_____

9. Name the two major hormones controlling events of the postabsorptive state.

_____

10. Name the origin and targets of glucagon. _____

_____

11. Name two humoral factors that stimulate the release of glucagon.

_____

_____

12. Why is the release of glucagon important when blood levels of amino acids rise (as after a high-protein, low-carbohydrate meal)?

_____

_____

13. Briefly describe the role of the sympathetic nervous system. _____

_____

_____

11. Complete the table below, which indicates the effects of several hormones on the characteristics indicated at the top of each column. Use an up arrow (↑) to indicate an increase in a particular value or a down arrow (↓) to indicate a decrease. No response is required in the boxes with XXX.

Hormone	Blood glucose	Blood amino acids	Glycogenolysis	Lipogenesis	Protein synthesis
Insulin					
Glucagon		XXX			XXX
Epinephrine		XXX			XXX
Growth hormone					
Thyroxine		XXX			

12. The liver has many functions in addition to its digestive function. Complete the following statements that elaborate on the liver's function by inserting the correct terms in the answer blanks.

_____ 1.

_____ 2.

_____ 3.

_____ 4.

_____ 5.

_____ 6.

_____ 7.

_____ 8.

_____ 9.

_____ 10.

_____ 11.

_____ 12.

_____ 13.

_____ 14.

_____ 15.

_____ 16.

_____ 17.

_____ 18.

_____ 19.

_____ 20.

_____ 21.

_____ 22.

_____ 23.

_____ 24.

_____ 25.

_____ 26.

_____ 27.

The liver is the most important metabolic organ in the body. In its metabolic role, the liver uses amino acids from the nutrient-rich hepatic portal blood to make many blood proteins such as __(1)__, which helps to hold water in the bloodstream, and __(2)__, which prevent blood loss when blood vessels are damaged. The liver also makes a steroid substance that is released to the blood. This steroid, __(3)__, has been implicated in high blood pressure and heart disease. Additionally, the liver acts to maintain homeostatic blood glucose levels. It removes glucose from the blood when blood levels are high, a condition called __(4)__, and stores it as __(5)__. Then, when blood glucose levels are low, a condition called __(6)__, liver cells break down the stored carbohydrate and release glucose to the blood once again. This latter process is termed __(7)__. When the liver makes glucose from noncarbohydrate substances such as fats or proteins, the process is termed __(8)__. In addition to its processing of amino acids and sugars, the liver plays an important role in the processing of fats. Other functions of the liver include the __(9)__ of drugs and alcohol, and its __(10)__ cells protect the body by ingesting bacteria and other debris.

The liver forms small complexes called __(11)__, which are needed to transport fatty acids, fats, and cholesterol in the blood because lipids are __(12)__ in a watery medium. The higher the lipid content of the lipoprotein, the __(13)__ is its density. Thus, low-density lipoproteins (LDLs) have a __(14)__ lipid content. __(15)__ are lipoproteins that transport dietary lipids from the GI tract. The liver produces VLDLs (very low density lipoproteins) to transport __(16)__ to adipose tissue.

After releasing their cargo, VLDL residues convert to LDLs, which are rich in __(17)__. The function of LDLs is transport of cholesterol to peripheral tissues, where cells use it to construct their plasma __(18)__ or to synthesize __(19)__. The function of HDLs (high-density lipoproteins) is transport of cholesterol to the __(20)__, where it is degraded and secreted as __(21)__, which are eventually excreted.

High levels of cholesterol in the plasma are of concern because of the risk of __(22)__. The type of dietary fatty acid affects cholesterol levels. __(23)__ fatty acids promote excretion of cholesterol. When cholesterol levels in plasma are measured, high levels of __(24)__ DLs indicate that the fate of the transported cholesterol is catabolism to bile salts.

Two other important functions of the liver are the storage of vitamins (such as vitamins __(25)__) and of the metal __(26)__ (as ferritin) and the processing of bilirubin, which results from the breakdown of __(27)__ cells.

Body Energy Balance

1. Circle the term that does not belong in each of the following groupings.

   1. BMR   TMR   Rest   Postabsorptive state

   2. Thyroxine   Iodine   ↓ Metabolic rate   ↑ Metabolic rate

   3. Obese person   ↓ Metabolic rate   Women   Child

   4. 4 kcal/gram   Fats   Carbohydrates   Proteins

   5. ↑ TMR   Fasting   Muscle activity   Ingestion of protein

   6. Body's core   Lowest temperature   Heat loss surface   Body's shell

   7. Radiation   Evaporation   Vasoconstriction   Conduction

2. Using the key choices, select the factor regulating food intake that best fits the following conditions. Also state whether hunger is depressed (use ↓), or stimulated (use ↑), or both (use ↑↓) by each condition described. Insert your letter answers and the arrows in the answer blanks.

   KEY CHOICES

   A. Body temperature   **C.** Nutrient signals related to total energy stores

   B. Hormones   **D.** Psychological factors

   _____   1. Rise in plasma glucose levels

   _____   2. Ingestion, digestion, and liver activities increase

   _____   3. Secretions of insulin or cholecystokinin

   _____   4. Low plasma levels of amino acids

   _____   5. The sight, taste, smell, or thought of food

   _____   6. Secretions of glucagon or epinephrine

   _____   7. Fat reserves release large amounts of fatty acids and glycerol to blood

3. Respond to the following by writing your answers in the answer blanks.

   1. Briefly explain cell metabolism in terms of the first law of thermodynamics (energy can neither be created nor destroyed).

   _____

   _____

   _____

2. Nearly all the energy derived from food is eventually converted to _____

3. How does the balance between energy input and energy output affect body weight?

_____

_____

4. Using the key choices, select the terms that match the following descriptions pertaining to body temperature regulation. Insert the appropriate answers in the answer blanks.

KEY CHOICES

A. Blood	**D.** Heat	**G.** Hypothermia	**J.** Pyrogens
B. Vasoconstriction in skin	**E.** Hyperthermia	**H.** Perspiration	**K.** Shivering
C. Frostbite	**F.** Hypothalamus	**I.** Radiation	

_____ 1. By-product of cell metabolism

_____ _____ 2. Means of conserving or increasing body heat

_____ 3. Medium that distributes heat to all body tissues

_____ 4. Site of the body's thermostat

_____ 5. Chemicals released by injured tissue cells and macrophages that cause resetting of the thermostat

_____ 6. Death of cells deprived of oxygen and nutrients, resulting from withdrawal of blood from the skin circulation

_____ _____ 7. Means of liberating excess body heat

_____ 8. Extremely low body temperature

_____ 9. Fever

_____ 10. Cause release of hypothalamic prostaglandins

5. If a statement about the regulation of body temperature is true, write the letter T in the answer blank. If a statement is false, change the underlined word(s) and write the correct word(s) in the answer blank.

_____ 1. One of the most important ways the body regulates temperature is by changing <u>visceral organ</u> activity.

_____ 2. The homeostatic range of body temperature is from 35.6°C (96°F) to <u>43°C (110°F)</u>.

_____ **3.** With each 1℃ rise in body temperature, the rate of enzymatic catalysis increases by 1%.

_____ **4.** When the body temperature exceeds the homeostatic range, carbohydrates begin to break down.

_____ **5.** Convulsions occur when the body temperature reaches 41℃ (106℉).

_____ **6.** The cessation of all thermoregulatory mechanisms may lead to heat exhaustion.

Developmental Aspects of Nutrition and Metabolism

1. Complete the following statements by inserting the missing terms in the answer blanks.

_____ **1.**

_____ **2.**

_____ **3.**

_____ **4.**

_____ **5.**

_____ **6.**

_____ **7.**

_____ **8.**

_____ **9.**

_____ **10.**

_____ **11.**

_____ **12.**

_____ **13.**

_____ **14.**

_____ **15.**

_____ **16.**

_____ **17.**

_____ **18.** _____ **19.**

Most serious in fetal nutrition is the lack of __(1)__ required for tissue growth, especially the growth of the __(2)__ . Without adequate nutrition for the first __(3)__ years after birth, mental impairment will occur. The two most common genetic disorders of metabolism are __(4)__ , typified by nonsecretion of pancreatic enzymes, and __(5)__ , characterized by the deficiency of one enzyme in the production of tyrosine. The substrate of this enzyme is an amino acid called __(6)__ , which is present in all protein foods; its accumulation leads to __(7)__ damage and retardation within a few months. Afflicted children are given a diet low in __(8)__ until about age 6. These children are very blond and light-skinned because __(9)__ , a precursor of melanin, is not produced.

Two carbohydrate-related genetic defects are worthy of note. A failed conversion of galactose to glucose in the liver results in __(10)__ . In __(11)__ , the enzyme controlling glycogenolysis is absent, and, as a result, there is enlargement of the glycogen storage organs, the __(12)__ and __(13)__ . The most significant metabolic disorder in children and in the elderly is __(14)__ .

As the body ages, the rate of metabolism __(15)__ . The elderly use many drugs for age-related medical problems, which may negatively affect their nutrition. __(16)__ can be caused by the loss of potassium from the body resulting from the use of __(17)__ . Some antibiotics may cause diarrhea and thus impair the __(18)__ of ingested nutrients. When alcohol is substituted for a proper diet, dietary deficiencies and damage to the pancreas and __(19)__ may occur.

CHALLENGING YOURSELF

At the Clinic

1. Mary Maroon comes to the clinic to get information on a vegetarian diet. What problems may arise when people make uninformed decisions on what to eat for a vegetarian diet? What combination of vegetable foods will provide Mary with all the essential amino acids?

2. A group of new parents is meeting to learn about good nutrition for their children. The topic is "empty calories." What does the term mean and what are some foods with empty calories?

3. Mrs. Piercecci, a woman in her mid-40s, complains of extreme fatigue and faintness about four hours after each meal. What hormone might be deficient and what condition is indicated?

4. Zena, a teenager, has gone to the sports clinic for the past 2 years to have her fat content checked. This year, her percent body fat is up and tissue protein has not increased. Questioning reveals that Zena has been on crash diets four times since the last checkup, only to regain the weight (and more) each time. She also admits sheepishly that she "detests" exercise. How does cyclic dieting, accompanied by lack of exercise, cause an increase in fat and a decrease in protein?

5. Benny, a 6-year-old child who is allergic to milk, has extremely bowed legs. What condition do you suspect and what is the connection to not drinking milk?

6. An anorexic girl shows high levels of acetone in her blood. What is this condition called and what has caused it?

7. There has been a record heat wave lately, and many elderly people are coming to the clinic complaining that they "feel poorly." In most cases, their skin is cool and clammy, and their blood pressure is low. What is their problem? What can be done to alleviate it?

8. During the same period, Bert Winchester, a construction worker, is rushed in unconscious. His skin is hot and dry, and his co-workers say that he suddenly keeled over on the job. What is Bert's condition and how should it be handled?

9. The mother of a 1-year-old confides in the pediatrician (somewhat proudly) that she has started her child on skim milk to head off any future problems with atherosclerosis. Why is this not a good practice?

10. A blood test shows cholesterol levels at about 280 mg/100 ml. What is this condition called? What cholesterol level is safe? What are some complications of high cholesterol?

11. Mrs. Rodriguez has a bleeding ulcer and has lost her appetite. She appears pale and lethargic when she comes in for a physical. She proves to be anemic and her RBCs are large and pale. What mineral supplements should be ordered?

12. Mr. Hodges, pretty much a recluse, unexpectedly appears at the clinic seeking help with a change he has experienced. He has been losing weight, his skin is full of bruises, and his gums are bleeding. However, what really "drove" him to the clinic is a sore that won't heal. What is his problem called? What vitamin deficiency causes it? What dietary changes would you suggest?

Stop and Think

1. Amino acids must be metabolized; they cannot be stored. What would be the osmotic and acid-base effects of accumulating amino acids?

2. What would be the diffusion and osmotic effects of failing to turn glucose into glycogen and adipose tissue for storage?

3. What is the *specific* mechanism by which oxygen starvation causes death?

4. Does urea production increase or decrease when essential amino acids are lacking in the diet? Explain your answer.

5. What is the difference in the effect on blood glucose of a meal of simple sugars versus a meal of an equal amount of complex carbohydrates?

6. Does loss of mental function accompany the deterioration associated with diabetes mellitus? Why or why not?

7. What would happen to blood clotting if excess omega-3 fatty acids were ingested?

8. Why is the factor for calculating BMR higher for males than it is for females?

9. Why might PKU sufferers have poor vision?

10. What supplements would you suggest for someone taking diuretics?

11. What is the metabolic link between diabetes mellitus and arteriosclerosis?

12. Does hypothyroidism result in hyperthermia or hypothermia? Explain.

13. Why do we feel hot when we exercise vigorously?

COVERING ALL YOUR BASES

Multiple Choice

Select the best answer or answers from the choices given.

1. Which of the following are "essential" nutrients?
   - **A.** Glucose
   - **B.** Linoleic acid
   - **C.** Cholesterol
   - **D.** Leucine

2. Which factors will work to promote (and allow) protein synthesis?
   - **A.** Presence of all 20 amino acids
   - **B.** Sufficiency of caloric intake
   - **C.** Negative nitrogen balance
   - **D.** Secretion of anabolic hormones

3. Which of the listed vitamins is not found in plant foods?
   - **A.** Folic acid
   - **B.** Biotin
   - **C.** Cyanocobalamin (B$_{12}$)
   - **D.** Pantothenic acid

4. Deficiency of which of these vitamins results in anemia?
   - **A.** Thiamin
   - **B.** Riboflavin
   - **C.** Biotin
   - **D.** Folic acid

5. Vitamins that act as coenzymes in the Krebs cycle include:
   - **A.** riboflavin
   - **B.** niacin
   - **C.** biotin
   - **D.** pantothenic acid

6. Absorption of which of the following minerals parallels absorption of calcium?
   - **A.** Zinc
   - **B.** Magnesium
   - **C.** Iron
   - **D.** Copper

7. Components of the electron transport system include:
   - **A.** cobalt
   - **B.** copper
   - **C.** iron
   - **D.** zinc

8. Substrate-level phosphorylation occurs during:
   - **A.** glycolysis
   - **B.** beta-oxidation
   - **C.** Krebs cycle
   - **D.** electron transport

9. Which processes occur in the mitochondria?
   - **A.** Glycogenesis
   - **B.** Gluconeogenesis
   - **C.** Beta-oxidation
   - **D.** Formation of acetyl CoA

10. The chemiosmotic process involves:
    - **A.** buildup of hydrogen ion concentration
    - **B.** electron transport
    - **C.** oxidation and reduction
    - **D.** ATP synthase

11. Which of the following molecules has the highest energy content?
    - **A.** Glucose
    - **B.** Acetyl CoA
    - **C.** Dihydroxyacetone phosphate
    - **D.** ATP

12. Krebs cycle intermediates include:
    - **A.** Glutamic acid
    - **B.** Pantothenic acid
    - **C.** Succinic acid
    - **D.** Fumaric acid

13. Chemicals that can be used for gluconeogenesis include:
    A. pyruvic acid
    B. glycerol
    C. fatty acids
    D. alpha ketoglutaric acid

14. Ingestion of saturated fatty acids (as opposed to equal quantities of poly-unsaturated fatty acids) is followed by:
    A. increased cholesterol formation
    B. increased cholesterol excretion
    C. increased production of chylomicrons
    D. increased formation of bile salts

15. Which events occur during the absorptive state?
    A. Use of amino acids as a major source of energy
    B. Lipogenesis
    C. Beta oxidation
    D. Increased uptake of glucose by skeletal muscles

16. Hormones that act to decrease blood glucose level include:
    A. insulin
    B. glucagon
    C. epinephrine
    D. growth hormone

17. During the postabsorptive state:
    A. glycogenesis occurs in the liver
    B. fatty acids are used for fuel
    C. amino acids are converted to glucose
    D. lipolysis occurs in adipose tissue

18. Which processes contribute to gluconeogenesis?
    A. Conversion of pyruvic acid to glucose by skeletal muscle
    B. Release of lactic acid by skeletal muscle
    C. Formation of glucose from triglycerides in adipose tissue
    D. Release of glycerol from adipose tissue

19. Only the liver functions to:
    A. store iron
    B. form urea
    C. produce plasma proteins
    D. form ketone bodies

20. Which transport particles carry cholesterol destined for excretion from the body?
    A. HDL
    B. Chylomicron
    C. LDL
    D. VLDL

21. Transport of triglycerides from the liver to adipose tissue is the function of:
    A. LDL
    B. HDL
    C. VLDL
    D. all lipoproteins

22. Glucose (or its metabolites) can be converted to:
    A. glycogen
    B. triglycerides
    C. nonessential amino acids
    D. starch

23. Basal metabolic rate:
    A. is the lowest metabolic rate of the body
    B. is the metabolic rate during sleep
    C. is measured as kcal per square meter of skin per hour
    D. increases with age

24. Body temperature is higher:
    A. in children than in adults
    B. in cases of hyperthyroidism
    C. after eating
    D. when sleeping

25. Which of the following types of heat transfer involves heat loss in the form of infrared waves?
    A. Conduction
    B. Convection
    C. Evaporation
    D. Radiation

26. Body temperature increases:
    A. during chemical thermogenesis
    B. when pyrogens are released
    C. when sweating and flushing occur
    D. with heat exhaustion

27. PKU is the result of inability to metabolize:

   A. tyrosine   **C.** ketone bodies

   B. melanin   **D.** phenylalanine

28. A wasting disease due to deficiency of both protein and carbohydrates is:

   A. marasmus   **C.** pica

   B. kwashiorkor   **D.** cystic fibrosis

Word Dissection

For each of the following word roots, fill in the literal meaning and give an example, using a word found in this chapter.

Word root	Translation	Example
1. acet		
2. calor		
3. flav		
4. gluco		
5. kilo		
6. lecith		
7. linol		
8. nutri		
9. pyro		

26 The Urinary System

Student Objectives

When you have completed the exercises in this chapter, you will have accomplished the following objectives:

Kidney Anatomy

1. Describe the gross anatomy of the kidney and its coverings.
2. Trace the blood supply through the kidney.
3. Describe the anatomy of a nephron.

Kidney Physiology: Mechanisms of Urine Formation

4. List several kidney functions that help maintain body homeostasis.
5. Identify the parts of the nephron responsible for filtration, reabsorption, and secretion, and describe the mechanisms underlying each of these functional processes.
6. Explain the role of aldosterone and of atrial natriuretic factor in sodium and water balance.
7. Describe the mechanism that maintains the medullary osmotic gradient.
8. Explain the formation of dilute versus concentrated urine.

9. Describe the normal physical and chemical properties of urine.
10. List several abnormal urine components and name the condition when each is present in detectable amounts.

Ureters, Urinary Bladder, and Urethra

11. Describe the general structure and function of the ureters.
12. Describe the general structure and function of the urinary bladder.
13. Describe the general structure and function of the urethra.
14. Compare the course, length, and functions of the male urethra with those of the female.

Micturition

15. Define *micturition* and describe the micturition reflex.

Developmental Aspects of the Urinary System

16. Describe the embryonic development of the organs of the urinary system.
17. List several changes in urinary system anatomy and physiology that occur with age.

Ready, set, go!

Metabolism of nutrients by body cells produces various wastes, such as carbon dioxide and nitrogenous wastes (creatine, urea, and ammonia), as well as imbalances of water and essential ions. The metabolic wastes and excesses must be eliminated from the body. Essential substances are retained to ensure internal homeostasis and proper body functioning.

Although several organ systems are involved in excretory processes, the urinary system is primarily responsible for removal of nitrogenous wastes from the blood. In addition to this purely excretory function, the kidneys maintain the electrolyte, acid-base, and fluid balances of the blood. Kidneys are thus major homeostatic organs of the body. Malfunction of the urinary system, particularly the kidneys, leads to a failure of homeostasis, resulting (unless corrected) in death.

Student activities in Chapter 26 are concerned with identification of urinary system structures, the composition of urine, and the physiological processes involved in urine formation.

 **BUILDING THE FRAMEWORK**

1. Name three metabolic activities of the kidneys unrelated to urine production. _____

   _____

   _____

   _____

2. Define and give the cause of:

   *ptosis:*_____

   _____

   hydronephrosis: _____

   _____

3. Figure 26.1 is a frontal view of the entire urinary system. Excluding red and blue, select different colors to color the coding circles and the corresponding organs on the figure. Color the aorta and its branches red; color the inferior vena cava and its tributaries blue.

◯ Kidney   ◯ Bladder   ◯ Ureters   ◯ Urethra

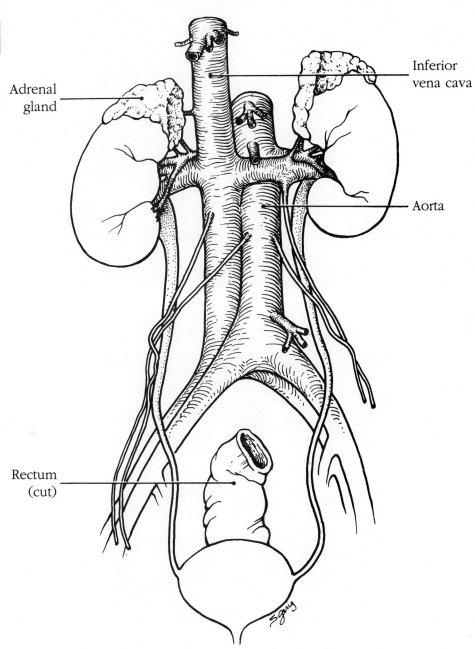

Adrenal gland

Inferior vena cava

Aorta

Rectum (cut)

Figure 26.1

Kidney Anatomy

1. Figure 26.2 is a longitudinal section of a kidney. First, using the correct anatomical terminology, identify and label the following regions and structures indicated by leader lines on the figure.

   1. Fibrous membrane immediately surrounding the kidney

   2. Basinlike area of the kidney that is continuous with the ureter

   3. Cuplike extension of the pelvis that drains the apex of a pyramid

   4. Area of cortexlike tissue running through the medulla

   Then, excluding the color red, select different colors to identify the following areas and structures. Color the coding circles and the corresponding areas and structures on the figure; label these regions using the correct anatomical terms.

   ○ Area of the kidney that contains the greatest proportion of nephron structures

   ○ Striped-appearing structures formed primarily of collecting ducts

   Finally, beginning with the renal artery, draw part of the vascular supply to the cortex on the figure. Include and label the segmental artery, lobar artery, interlobar artery, arcuate artery, and interlobular artery. Color the vessels bright red.

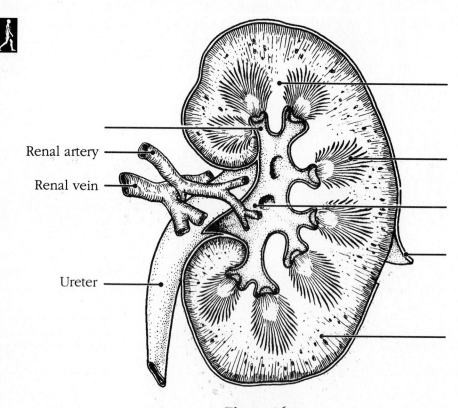

Renal artery

Renal vein

Ureter

Figure 26.2

2. Figure 26.3A shows two views of a renal corpuscle (glomerulus plus glomerular capsule). Label each structure that is indicated by a leader line. Color blood vessels red, the parietal layer of the renal corpuscle and parts of the renal tubule yellow, and the podocytes orange. When you have completed part A of the figure, identify specifically the locations of the cross sections of the renal tubules shown in part B, using one of the following choices. Then, on the lines below, explain the rationale or reasoning behind your choices.

Proximal convoluted tubules (PCT)   Collecting tubule

Distal convoluted tubules (DCT)   Ascending loop of Henle (thin segment)

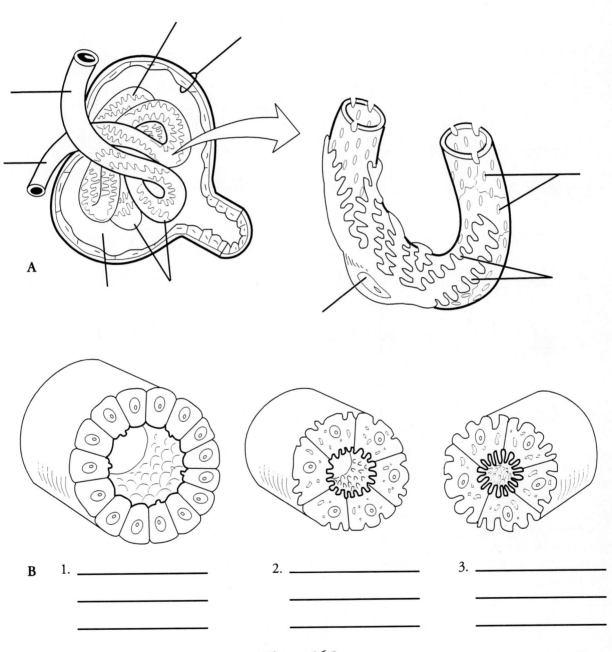

B   1. _____   2. _____   3. _____

_____   _____   _____

_____   _____   _____

Figure 26.3

3. Figure 26.4 is a diagram of the nephron and associated blood supply. First, match each of the numbered structures on the figure to one of the terms following the figure. Write the terms in the numbered answer blanks. Then, color the structure on the figure that contains podocytes green, the filtering apparatus red, the capillary bed that directly receives the reabsorbed substances from the tubule cells blue, the structure into which the nephron empties its urine product yellow, and the tubule area that is the primary site of tubular reabsorption orange.

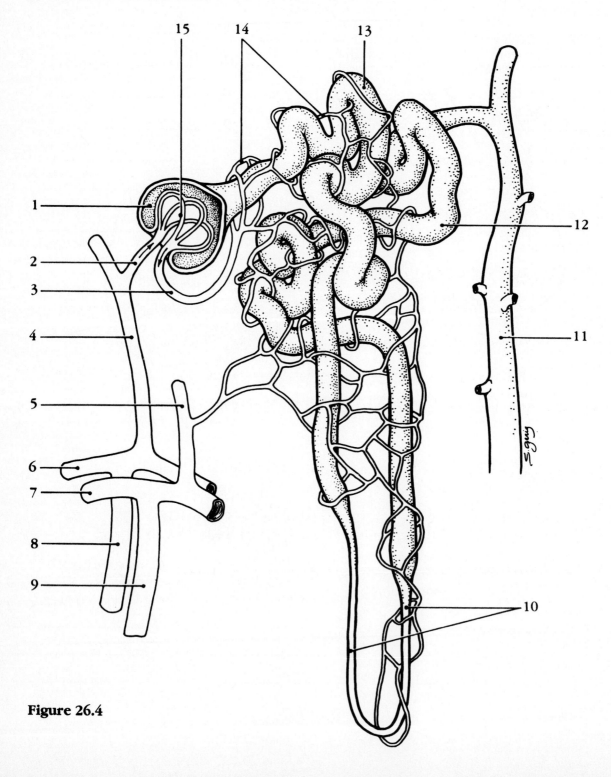

Figure 26.4

Afferent arteriole

Arcuate artery

Arcuate vein

Collecting duct

Distal convoluted tubule (DCT)

Efferent arteriole

Glomerular capsule

Glomerulus

Interlobar artery

Interlobar vein

Interlobular artery

Interlobular vein

Loop of Henle

Proximal convoluted tubule (PCT)

Peritubular capillaries

_____ 1. _____ 6. _____ 11.

_____ 2. _____ 7. _____ 12.

_____ 3. _____ 8. _____ 13.

_____ 4. _____ 9. _____ 14.

_____ 5. _____ 10. _____ 15.

4. Circle the term that does not belong in each of the following groupings.

1. Renal pelvis   Renal sinus   Renal pyramid   Minor calyx   Major calyx

2. Adipose capsule   Renal capsule   Renal corpuscle   Renal fascia

3. Juxtaglomerular apparatus   JG cells   Glomerulus   Macula densa cells

4. Glomerulus   Peritubular capillaries   Vasa recta   Collecting duct

5. Cortical nephrons   Vasa recta   Juxtamedullary nephrons   Long loops of Henle

6. Nephron   Proximal convoluted tubule   Distal convoluted tubule

   Collecting duct

7. Medullary pyramids   Glomeruli   Renal pyramids   Collecting ducts

5. Assign each of these urine-carrying ducts a number from 1 to 5, according to the path taken by the urine (the urine first passes through #1, then #2, etc.).

_____ **1.** renal pelvis   _____ **3.** papillary duct   _____ **5.** minor calyx

_____ **2.** ureter   _____ **4.** major calyx

Kidney Physiology: Mechanisms of Urine Formation

1. Complete the following statements by inserting your answers in the answer blanks.

_____ 1.

_____ 2.

_____ 3.

_____ 4.

_____ 5.

_____ 6.

_____ 7.

_____ 8.

_____ 9.

_____ 10.

_____ 11.

_____ 12.

_____ 13.

_____ 14.

_____ 15.

The glomerulus is a unique high-pressure capillary bed because the __(1)__ arteriole feeding it is larger in diameter than the __(2)__ arteriole draining the bed. Glomerular filtrate is very similar to __(3)__, but it has fewer proteins. Mechanisms of tubular reabsorption include __(4)__ and __(5)__. As an aid for the reabsorption process, the cells of the proximal convoluted tubule have dense __(6)__ on their luminal surface, which increase the surface area dramatically. Other than reabsorption, an important tubule function is __(7)__, which is important for ridding the body of substances not already in the filtrate. Blood composition depends on __(8)__, __(9)__, and __(10)__. In a day's time, 180 liters of blood plasma are filtered into the kidney tubules, but only about __(11)__ liters of urine are actually produced. __(12)__ is responsible for the normal yellow color of urine. The three major nitrogenous wastes found in the blood, which must be disposed of, are __(13)__, __(14)__, and __(15)__. The kidneys determine how much water is to be lost from the body. When water loss via vaporization from the __(16)__ or __(17)__ of perspiration from the skin is excessive, urine output __(18)__. If the kidneys become nonfunctional, __(19)__ is used to cleanse the blood of impurities.

_____ 16.

_____ 17.

_____ 18.

_____ 19.

2. Use the three components of the filtration membrane in the key choices to match the descriptions that follow.

KEY CHOICES

A. Capillary endothelium   **B.** Basement membrane   **C.** Podocyte membrane of glomerular capsule

_____ 1. Repels anions and prevents their filtration

_____ 2. Its pores prevent filtration of blood cells

_____ 3. Its filtration slits appear to play little or no role in restricting passage of substances into the tubule

_____ 4. Prevents passage of all but the smallest proteins

3. Figure 26.5 is a diagram of a nephron. Add colored arrows on the figure to show the location and direction of the following processes.

   1. Black arrows at the site of filtrate formation.

   2. Red arrows at the major site of amino acid and glucose reabsorption.

   3. Green arrows at the sites most responsive to action of ADH (show direction of water movement).

   4. Yellow arrows at the sites most responsive to the action of aldosterone (show direction of Na^+ movement).

   5. Blue arrows at the major sites of tubular secretion.

   Then, label the proximal convoluted tubule (PCT), distal convoluted tubule (DCT), loop of Henle, glomerular capsule, and glomerulus on the figure. Also label the collecting duct (not part of the nephron).

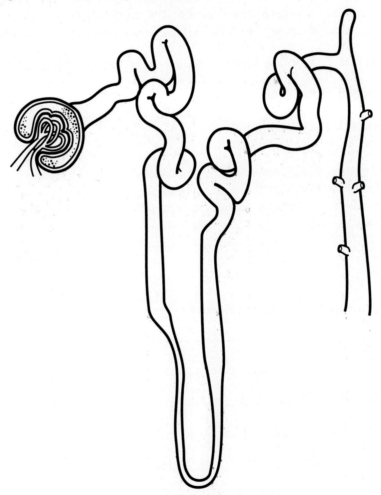

Figure 26.5

4. Complete the following equation referring to net filtration pressure (NFP).

1. NFP = Glomerular hydrostatic pressure – $\Bigl($ _____ $\Bigr)$.

2. NFP = _____ mm Hg

3. Which of these pressures is equivalent to blood pressure? _____

4. Which of these pressures is a consequence of proteins in blood? _____

5. List below two of the suspected intrinsic mechanisms of renal autoregulation, along with the structure(s) responsible for each:

_____

_____

6. Renin release by the JG apparatus of the nephrons plays an important role in increasing systemic blood pressure. Check (✔) all factors that promote renin release.

_____ **1.** Drop in systemic blood pressure

_____ **2.** Slowly flowing renal filtrate

_____ **3.** Increased stretch of the afferent arterioles

_____ **4.** Sympathetic nervous system stimulation of the JG cells

_____ **5.** Low filtrate osmolarity

7. For each of the substances listed, write whether it is actively reabsorbed, passively reabsorbed in response to an electrical gradient, passively reabsorbed in response to an osmotic gradient, or reabsorbed by solvent drag.

1. Amino acids _____

2. Bicarbonate ions _____

3. Chloride ions _____

4. Fatty acids _____

5. Glucose _____

6. Lactate _____

7. Potassium ions _____

8. Sodium ions _____

9. Urea _____

10. Vitamins _____

11. Water _____

8. A glucose tolerance test is given to a woman suspected of having diabetes mellitus. Urine and blood samples are taken periodically (typically before testing and at 30, 60, 120, and 180 minutes) after the patient drinks a liquid containing a known amount of glucose. Graph the data given in the table below, using red for blood glucose and yellow for urine glucose. Then answer the questions that follow.

Sample number	Blood glucose, mg/100ml	Urine glucose, mg/min
1	100	0
2	225	0
3	290	0
4	375	0
5	485	90

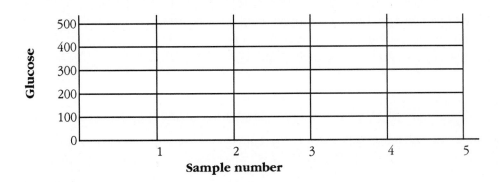

1. According to the graphed information, what is the transport maximum for glucose? _____

2. Is the woman diabetic? _____

9. Name two cations and two nitrogenous wastes that are secreted by the tubule

cells into the urine. _____

_____

10. Of what importance is the medullary osmotic gradient? _____

_____

_____

_____

_____

11. Complete the statements by filling in the blanks below:

_____ **1.**

_____ **2.**

_____ **3.**

_____ **4.**

_____ **5.**

_____ **6.**

_____ **7.**

_____ **8.**

The __(1)__ limb of the loop of Henle is relatively impermeable to solutes but freely permeable to water. The __(2)__ limb of the loop is permeable to sodium and chloride but not to water; its __(3)__ segment, in particular, has an active pump that moves salt (NaCl) out of the filtrate into the medullary interstitial fluid. The medullary regions of the collecting ducts are permeable to __(4)__, a small solute that enhances the osmolarity of the medullary interstitial fluid somewhat. The __(5)__ acts as a countercurrent exchanger to maintain the osmotic gradient while delivering a nutrient blood supply. The result of this mechanism is to maintain a high solute content in the __(6)__ of the medullary pyramids of about __(7)__ mosm/L. The permeability of the collecting ducts to water is controlled by the hormone __(8)__.

12. Figure 26.6 depicts some of the regulatory events that control processes occurring in the collecting ducts, particularly concerning water balance in the body. Use the terms provided in the list just below to fill in the empty boxes in the diagram. Also, identify by coloring and by adding appropriate leader lines and labels: the hypothalamus, the collecting duct, and the posterior pituitary. (*Note:* ⊕ indicates a stimulatory effect and ⊖ indicates an inhibitory effect.)

Rising blood osmolarity   Water reabsorption   Rising blood volume   ADH

Falling blood osmolarity   Increases porosity   Falling blood volume

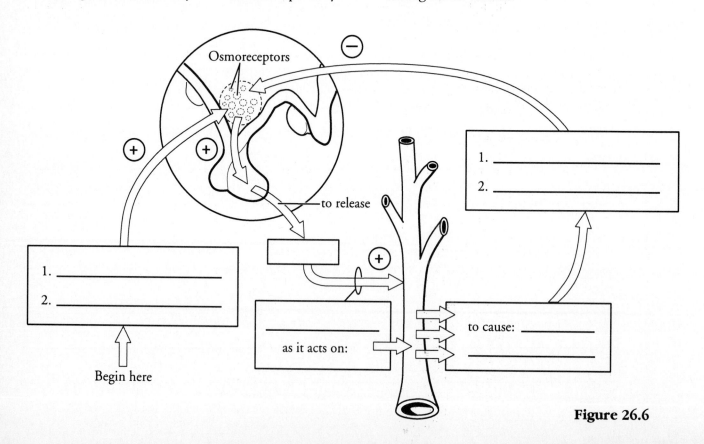

Figure 26.6

13. Circle the term that does not belong in each of the following groupings. (*Note:* BP = blood pressure.)

1. Hypothalamus   ADH   Aldosterone   Osmoreceptors

2. Glomerulus   Secretion   Filtration   Podocytes

3. Aldosterone   ↑ Na^+ reabsorption   ↑ K^+ reabsorption   ↑ BP

4. ADH   ↓ BP   ↑ Blood volume   ↑ Water reabsorption

5. ↓ Aldosterone   Edema   ↓ Blood volume   ↓ K^+ retention

14. Respond to the following by inserting your answer into the blank following each question:

1. What is the glomerular filtration rate of an individual who was injected with inulin at a plasma concentration of 1 mg/ml, a flow rate of urine formation of 2 ml/min, and a urine concentration of inulin of 90 mg/ml? _____

2. Is this a normal, low, or high GFR? _____

Assuming a normal GFR, state whether each of the following substances would be partially reabsorbed or secreted into the urine:

3. creatinine, RC = 145 ml/min _____

4. uric acid, RC = 15 ml/min _____

5. urea, RC = 75 ml/min _____

6. penicillin, RC = 230 ml/min _____

15. Decide whether the following conditions would result in an increase or a decrease in the specific gravity of urine. Write I in the answer blank to indicate an increase and D to indicate a decrease.

_____ **1.** Drinking excessive fluids   _____ **4.** Using diuretics

_____ **2.** Chronic renal failure   _____ **5.** Limited fluid intake

_____ **3.** Pyelonephritis   _____ **6.** Fever

16. Decide whether the following conditions would cause urine to become more acidic or more basic. If more acidic, write A in the blank; if more basic, write B in the blank.

_____ **1.** Protein-rich diet   _____ **4.** Diabetes mellitus

_____ **2.** Bacterial infection   _____ **5.** Vegetarian diet

_____ **3.** Starvation

17. Assuming normal conditions, note whether each of the following substances would be in greater concentration in the urine than in the glomerular filtrate (G), in lesser concentration in the urine than in the glomerular filtrate (L), or absent in both urine and glomerular filtrate (A). Place the correct letter in each answer blank.

_____ 1. Water

_____ 2. Urea

_____ 3. Uric acid

_____ 4. Pus (white blood cells)

_____ 5. Glucose

_____ 6. Albumin

_____ 7. Creatinine

_____ 8. Hydrogen ions

_____ 9. Potassium ions

_____ 10. Red blood cells

_____ 11. Sodium ions

_____ 12. Amino acids

18. Several specific terms are used to indicate the presence of abnormal urine constituents. Identify each of the following abnormalities with the term that names the condition. Then, for each condition, provide one possible cause of the condition.

1. Presence of red blood cells _____ Cause _____

2. Presence of ketones _____ Cause _____

3. Presence of albumin _____ Cause _____

4. Presence of pus _____ Cause _____

5. Presence of bile pigments _____ Cause _____

6. Presence of "sand" _____ Cause _____

7. Presence of glucose _____ Cause _____

19. Glucose and albumin are both normally absent from urine, but the reason for their absence differs.

1. Explain the reason for the absence of glucose in urine. _____

    _____

2. Explain the reason for the absence of albumin in urine. _____

    _____

Ureters, Urinary Bladder, and Urethra

1. Complete the following statements by inserting your answers in the answer blanks.

_____ 1.

_____ 2.

_____ 3.

_____ 4.

_____ 5.

_____ 6.

_____ 7.

_____ 8.

Urine continuously formed by the __(1)__ is routed down the __(2)__ by the mechanism of __(3)__ to a storage organ called the __(4)__ . Eventually the urine is conducted to the body exterior by the __(5)__ . In the male, this tubelike structure is approximately __(6)__ inches long and transports both urine and __(7)__ . In the female, this conduit is approximately __(8)__ inches long and transports only urine.

2. Circle the term that does not belong in each of the following groupings.

1. Bladder   Kidney   Transitional epithelium   Detrusor muscle   Urine storage

2. Trigone   Ureter openings   Urethral opening   Bladder   Forms urine

3. Surrounded by prostate gland   Drains bladder   Continuous with renal pelvis

    Urethra   Contains internal and external sphincters

4. Renal calculi   Urinary retention   Lithotripsy   Calcium salts   Blocks urethra

5. Prostatic   Male   Female   Membranous   Spongy

Micturition

1. Complete the following statements by writing the missing terms in the answer blanks.

_____ 1.

_____ 2.

_____ 3.

_____ 4.

_____ 5.

_____ 6.

_____ 7.

__(1)__ is another term that means voiding. As urine accumulates in the bladder, __(2)__ are activated. This results in a reflex that causes the muscular wall of the bladder to __(3)__ , and urine is forced past the __(4)__ sphincter. The __(5)__ sphincter is controlled __(6)__ ; thus an individual can temporarily postpone emptying the bladder until it has accumulated __(7)__ ml of urine.

(continues on next page)

_____ 8.

_____ 9.

_____ 10.

_____ 11.

_____ 12.

_____ 13.

__(8)__ is a condition in which voiding cannot be voluntarily controlled. It is normal in __(9)__ because nervous control of the voluntary sphincter has not yet been achieved. Other causative conditions include __(10)__ and __(11)__ . __(12)__, a condition in which one is unable to void, is a common problem in elderly men due to __(13)__ enlargement.

Developmental Aspects of the Urinary System

1. Complete the following statements by writing the missing terms in the answer blanks.

_____ 1.

_____ 2.

_____ 3.

_____ 4.

_____ 5.

_____ 6.

_____ 7.

_____ 8.

_____ 9.

_____ 10.

_____ 11.

_____ 12.

_____ 13.

_____ 14.

Although three separate sets of renal tubules develop in the embryo, embryonic nitrogenous wastes are actually disposed of by the __(1)__ . All three sets are derived from the __(2)__ germ layer. Only the last set, called the __(3)__, persists to form the kidneys. The others, called the __(4)__ and the __(5)__, deteriorate.

A congenital condition typified by blisterlike sacs in the kidneys is __(6)__ . __(7)__ is a congenital condition seen in __(8)__, when the urethral opening is located ventrally on the penis.

A newborn baby voids frequently, which reflects the small size of its __(9)__ . Daytime control of the voluntary urethral sphincter is usually achieved by approximately __(10)__ months. In old age, progressive __(11)__ of the renal blood vessels results in the death of __(12)__ cells. The loss of bladder tone leads to __(13)__ . __(14)__, the need to void during the night, is particularly troublesome to elderly people.

2. Match the terms in Column B with the descriptions in Column A. Insert the correct letters in the answer blanks.

Column A

_____ **1.** Inflammatory condition particularly common in women who practice poor toileting habits

_____ **2.** Bedwetting

_____ **3.** Toxic condition due to renal failure

_____ **4.** Inflammation of a kidney

_____ **5.** A condition in which excessive amounts of urine are produced because of a deficiency of antidiuretic hormone (ADH)

_____ **6.** Region of dead or necrotic renal tissue

_____ **7.** Neoplasm of transitional epithelium

_____ **8.** Herniation of the urinary bladder into the vagina

_____ **9.** Degenerative disease that exhibits symptoms in the 60s; enlarged kidneys with sacs of blood, mucus, or urine

_____ **10.** Autoimmune-triggered inflammation of the glomeruli

Column B

A. Acute glomerulonephritis

B. Adult polycystic disease

C. Bladder cancer

D. Cystocele

E. Cystitis

F. Diabetes inspidus

G. Renal infarct

H. Nocturnal enuresis

I. Pyelonephritis

J. Uremia

The Incredible Journey:
A Visualization Exercise
for the Urinary System

You see the kidney looming brownish red through the artery wall.

1. Complete the following narrative by writing the missing terms in the answer blanks.

_____ 1.

_____ 2.

_____ 3.

_____ 4.

_____ 5.

_____ 6.

_____ 7.

_____ 8.

_____ 9.

_____ 10.

_____ 11.

_____ 12.

_____ 13.

_____ 14.

_____ 15.

_____ 16.

_____ 17.

_____ 18.

_____ 19.

For your journey through the urinary system, you must be made small enough to filter through the filtration membrane from the bloodstream into a renal __(1)__ . You will be injected into the subclavian vein and must pass through the heart before entering the arterial circulation. As you travel through the systematic circulation, you have at least 2 minutes to relax before reaching the __(2)__ artery supplying a kidney. You see the kidney looming brownish red through the artery wall. Once you have entered the kidney, the blood vessel conduits become increasingly smaller until you finally reach the __(3)__ arteriole feeding into the filtering device, or __(4)__ .

Once in the filter, you maneuver yourself so that you are directly in front of a pore. Within a fraction of a second, you are swept across the filtration membrane into the __(5)__ part of the renal tubule. Drifting along, you lower the specimen cup to gather your first filtrate sample for testing. You study the readout from the sample and note that it is very similar in composition to __(6)__ , with one exception—there are essentially no __(7)__ . Your next sample doesn't have to be collected until you reach the "hairpin," or, using the proper terminology, the __(8)__ part of the tubule. As you continue your journey, you notice that the tubule cells have dense fingerlike projections extending from their surfaces into the lumen of the tubule. These are __(9)__ , which increase the surface area of tubule cells, because this portion of the tubule is very active in the process of __(10)__ . Soon you collect your second sample, and later, in the distal convoluted tubule, your third sample. When you read the computer's summary of the third sample, you make the following notes in your register.

1. Virtually no nutrients such as __(11)__ and __(12)__ are left in the filtrate.

2. The pH is acid, 6.0. This is quite a change from the pH of __(13)__ recorded for the newly formed filtrate.

_____ 20.

_____ 21.

_____ 22.

_____ 23.

_____ 24.

3. There is a much higher concentration of __(14)__ wastes here.

4. There are fewer __(15)__ ions but more __(16)__ ions than there were earlier.

5. The color of the filtrate is yellow, indicating a high relative concentration of the pigment __(17)__ .

Gradually you become aware that you are moving along much more quickly. You see that the water level has dropped dramatically and that the stream is turbulent and rushing. As you notice this, you realize that the hormone __(18)__ must have been released recently to cause this water level drop. You take an abrupt right turn and then drop straight downward. You realize that you must be in a __(19)__ . Within a few seconds, you land with a splash in what appears to be a large tranquil sea with an ebbing tide toward a dark area at the far shore. You drift toward the dark area, confident that you are in the kidney __(20)__ . As you reach and enter the dark tubelike structure, your progress becomes rhythmic—something like being squeezed through a sausage skin. Then you realize that your progress is being regulated by the process of __(21)__ . Suddenly, you free-fall and land in the previously stored __(22)__ in the bladder, where the air is very close. Soon the walls of the bladder begin to gyrate, and you realize you are witnessing a __(23)__ reflex. In a moment, you are propelled out of the bladder and through the __(24)__ to exit your host.

CHALLENGING YOURSELF

At the Clinic

1. Mimi is brought to the clinic by her parents. Though she is 23 years old and 5′5″ tall, she weighs only 90 pounds. She insists there is nothing wrong, except for some lower back pains. The clinic staff quickly realizes that the young woman is anorexic. What kidney problems might result from extremely low body weight?

2. A man was admitted to the hospital after being trampled by his horse. He received crushing blows to his lower back, on both sides, and is in considerable pain. His chart shows a urine output of 70 ml in the last 24 hours. What is this specific symptom called? What will be required if the effects of his trauma persist?

3. A young woman has come to the clinic with dysuria and frequent urination. What is the most likely diagnosis?

4. The parents of 3-year-old Mitchell bring him to the clinic because his "wee-wee" (urine) is tinged with blood. His face and hands are swollen, and a check of his medical record shows he was diagnosed with strep throat two days before. What is the probable cause of Mitchell's current kidney problem?

5. Mr. Laplante, a heavy smoker who works at the local pesticide plant, has come to the clinic complaining of bloody urine. He has no other complaints. His urine composition is normal (other than the blood). Cystoscopy reveals an abnormal mucosal surface at the trigone. What is the diagnosis?

6. What single test will reveal the cause of polyuria to be insulin-related? What other hormone-related disorder results in polyuria? What will be the specific gravity and nature of the urine in this second disorder?

7. Mrs. David, a new mother, comes to the clinic complaining of pain associated with a full bladder and slow micturition. What condition might you suspect?

8. Four-year-old Eddie is a chronic bedwetter. The clinical name for his disorder, noted on his chart, is _____. What might explain his problem?

9. An elderly diabetic woman is admitted to the hospital with a multitude of vascular problems; uremia indicates involvement of the kidneys. What is a likely problem in this case and what is a common cause?

10. Felicia, a medical student, arrived late to surgery where she was to observe removal of an extensive abscess from the external surface of a patient's kidney. Felicia was startled to find the surgical team working to reinflate the patient's lung and remove air from the pleural cavity. How could renal surgery lead to such an event?

Stop and Think

1. A typical capillary has a high-pressure (arterial) and a low-pressure (venous) end. How does a glomerular capillary compare to this? What provides the low-pressure end for the glomerulus and how does this arrangement increase the ability of the kidney to filter the blood?

2. What will happen to GFR and urine output if plasma protein production decreases (due to hepatitis, cirrhosis, or dietary protein deficit)?

3. Assume you could peer into cells of the proximal convoluted tubule to examine their organelles. What clues could you use to help determine whether movement of solutes through the apical and basolateral cell membranes is passive or active?

4. What happens to the medullary osmotic gradient when ADH is secreted? How is the gradient reestablished?

5. Why does excess alcohol intake result in "cotton mouth" the next morning?

6. Does caffeine cause production of a more dilute urine, as in diabetes insipidus, or loss of excess solute?

7. What happens to the rate of RBC production in a patient on dialysis with total renal failure? What could be given to the patient to counteract such a problem?

8. What is the effect on the micturition reflex of spinal cord transection above the pelvic splanchnic nerves?

9. What is the effect of damage to the sacral region of the spinal cord on the micturition reflex?

COVERING ALL YOUR BASES

Multiple Choice

Select the best answer or answers from the choices given.

1. A radiologist is examining an X ray of the lumbar region of a patient. Which of the following is/are indicative of normal positioning of the right kidney?
   A. Slightly higher than the left kidney
   B. More medial than the left kidney
   C. Closer to the inferior vena cava than the left kidney
   D. Anterior to the 12th rib

2. Which of the following encloses both kidney and adrenal gland?
   A. Renal fascia
   B. Adipose capsule
   C. Renal capsule
   D. Visceral peritoneum

3. Which structures are inflamed in pyelitis?
   A. Renal pelvis
   B. Minor calyces
   C. Renal capsule
   D. Ureter

4. Microscopic examination of a section of the kidney shows a thick-walled vessel with renal corpuscles scattered in the tissue on one side of the vessel but not on the other side. What vessel is this?
   A. Interlobar artery
   B. Interlobular artery
   C. Interlobular vein
   D. Arcuate artery

5. Structures that are at least partly composed of simple squamous epithelium include:
   A. collecting tubules
   B. glomerulus
   C. glomerular capsule
   D. loop of Henle

6. Which structures are freely permeable to water?
   A. Distal convoluted tubule
   B. Thick segment of ascending limb of the loop of Henle
   C. Descending limb of the loop of Henle
   D. Proximal convoluted tubule

7. Cells with large numbers of mitochondria are found in the:
   A. thick segment of the loop of Henle
   B. proximal convoluted tubule
   C. distal convoluted tubule
   D. collecting tubule

8. Low-pressure capillary beds include:
   A. glomerulus
   B. vasa recta
   C. peritubular capillaries
   D. renal plexus

9. Which statement(s) is/are true concerning the JG apparatus?
   A. The macula densa secretes renin.
   B. The JG cells are part of the renal corpuscle.
   C. The macula densa monitors filtrate concentration.
   D. The JG cells are mechanoreceptors.

10. The barriers in the filtration membrane consist of:
    A. fenestrations, which exclude blood cells
    B. anionic basement membrane, which repels plasma proteins
    C. filtration slits, which are smaller than plasma proteins
    D. a "molecular sieve," which excludes nutrients

11. The GFR would be increased by:
    A. glomerular hydrostatic pressure of 70 mm Hg
    B. glomerular osmotic pressure of 35 mm Hg
    C. sympathetic stimulation of the afferent arteriole
    D. secretion of aldosterone

12. Which stimulates the contractile myogenic mechanism of renal autoregulation?
    A. Increased systemic blood pressure
    B. Low filtrate osmolarity
    C. Secretion of renin
    D. Sympathetic stimulation of afferent arterioles

13. Consequences of hypovolemic shock include:
    A. dysuria   C. anuria
    B. polyuria   D. pyelitis

14. Sodium deficiency hampers reabsorption of:
    A. glucose   C. creatinine
    B. albumin   D. water

15. Which substance has a high transport maximum?
    A. Urea   C. Glucose
    B. Inulin   D. Lactate

16. Facultative water reabsorption mainly occurs in the:
    A. glomerular capsule
    B. distal convoluted tubule
    C. collecting duct
    D. proximal convoluted tubule

17. Effects of aldosterone include:
    A. increase in sodium ion excretion
    B. increase in water retention
    C. increase in potassium ion concentration in the urine
    D. higher blood pressure

18. Which of the following is dependent on tubular secretion?
    A. Clearing penicillin from the blood
    B. Removal of nitrogenous wastes that have been reabsorbed
    C. Removal of excess potassium ions from the blood
    D. Control of blood pH

19. Which is/are involved in establishing the medullary osmotic gradient?
    A. Movement of water out of the descending limb of the loop of Henle
    B. Movement of water into the ascending limb of the loop of Henle
    C. Active transport of urea into the interstitial space
    D. Impermeability of the collecting tubule to water

20. To calculate renal clearance, one must know:
    A. the concentration of the substance in plasma
    B. the concentration of the substance in urine
    C. the glomerular filtration rate
    D. the rate of urine formation

21. Normal urine constituents include:
    A. uric acid   C. acetone
    B. urochrome   D. potassium ions

22. Which of the following are normal values?
    A. Urine output of 1.5 L/day
    B. Specific gravity of 1.5
    C. pH of 6
    D. GFR of 125 ml/hour

23. The ureter:
    A. is continuous with the renal pelvis
    B. is lined by the renal capsule
    C. exhibits peristalsis
    D. is much longer in the male than in the female

24. The urinary bladder:
    A. is lined with transitional epithelium
    B. has a thick, muscular wall
    C. receives the ureteral orifices at its superior aspect
    D. is innervated by the renal plexus

25. Which of the following are controlled voluntarily?
    A. Detrusor muscle
    B. Internal urethral sphincter
    C. External urethral sphincter
    D. Levator ani muscle

26. Embryonic structures that persist and develop into adult structures include the:
    A. metanephros   C. ureteric buds
    B. mesonephros   D. pronephric ducts

27. A condition that can either be congenital or develop later in life is:
    A. hypospadias
    B. polycystic condition
    C. exstrophy
    D. cystocele

28. The main function of transitional epithelium in the ureter is:
    A. protection against kidney stones
    B. secretion of mucus
    C. reabsorption
    D. stretching

29. Jim was standing at a urinal in a crowded public rest room and a long line was forming behind him. He became anxious (sympathetic response) and found he could not micturate no matter how hard he tried. Use logic to deduce Jim's problem.
    A. His internal urethral sphincter was constricted and would not relax.
    B. His external urethral sphincter was constricted and would not relax.
    C. His detrusor muscle was contracting too hard.
    D. He almost certainly had a burst bladder.

30. A major function of the collecting ducts is:
    A. secretion
    B. filtration
    C. concentrating urine
    D. lubrication with mucus

31. What is the glomerulus?
    A. The same as the renal corpuscle
    B. The same as the uriniferous tubule
    C. The same as the nephron
    D. Capillaries

32. Urine passes through the ureters by which mechanism?
    A. Ciliary action
    B. Peristalsis
    C. Gravity alone
    D. Suction

Word Dissection

For each of the following word roots, fill in the literal meaning and give an example, using a word found in this chapter.

Word root	Translation	Example
1. azot	_____	_____
2. calyx	_____	_____
3. diure	_____	_____
4. glom	_____	_____
5. gon	_____	_____
6. mictur	_____	_____
7. nephr	_____	_____
8. pyel	_____	_____
9. ren	_____	_____
10. spado	_____	_____
11. stroph	_____	_____
12. trus	_____	_____

27 Fluid, Electrolyte, and Acid-Base Balance

Student Objectives

When you have completed the exercises in this chapter, you will have accomplished the following objectives:

Body Fluids

1. List the factors that determine body water content and describe the effect of each factor.

2. Indicate the relative fluid volume and solute composition of the fluid compartments of the body.

3. Contrast the overall osmotic effects of electrolytes and nonelectrolytes.

4. Describe factors that determine fluid shifts in the body.

Water Balance

5. List the routes by which water enters and leaves the body.

6. Describe the feedback mechanisms that regulate water intake and hormonal controls of water output in urine.

7. Explain the importance of obligatory water losses.

8. Describe possible causes and consequences of dehydration, hypotonic hydration, and edema.

Electrolyte Balance

9. Describe the routes of electrolyte entry and loss from the body.

10. Describe the importance of ionic sodium in fluid and electrolyte balance of the body and note its relationship to normal cardiovascular system functioning.

11. Briefly describe mechanisms involved in regulating sodium and water balance.

12. Explain how potassium, calcium, magnesium, and anion balance of plasma is regulated.

Acid-Base Balance

13. List important sources of acids in the body.

14. Name the three major chemical buffer systems of the body and describe how they operate to resist pH changes.

15. Describe the influence of the respiratory system on acid-base balance.

16. Describe how the kidneys regulate hydrogen and bicarbonate ion concentrations in the blood.

17. Distinguish between acidosis and alkalosis resulting from respiratory and metabolic factors. Describe the importance of respiratory and renal compensations to acid-base balance.

Developmental Aspects of Fluid, Electrolyte, and Acid-Base Balance

18. Explain why infants and the aged are at greater risk for fluid and electrolyte imbalances than are young adults.

Ready, set, go!

Scientists believe that life began in a salty sea and that the migration of living organisms from the sea to the land was accomplished by taking the chemistry of the sea into the body. Our trillions of cells are continually bathed in this "sea of life." (How salty is this sea of life? Taste your tears!) The functioning of each cell depends on the chemical composition of its immediate environment, which must remain roughly constant, for only then can each cell maintain its own composition and fluid volume. When the body's chemistry is in balance, the whole body exists in a steady, stable state, which we call health. This secret of life was discovered by the great physiologist Claude Bernard, who applied the term *homeostasis* to describe the constant state of the body's internal environment.

This chapter focuses on the composition of the internal environment and the organs and mechanisms involved in maintaining its homeostasis. Subjects include the composition of the major fluid compartments of the body, the water/electrolyte/acid-base balance of these fluid compartments and their interrelationships, and the related developmental changes occurring throughout the life span.

BUILDING THE FRAMEWORK

Body Fluids

1. Circle the term that does not belong in each of the following groupings. (*Note:* ECF = extracellular fluid compartment; ICF = intracellular fluid compartment.)

   1. Female adult   Male adult   About 50% water   Less skeletal muscle

   2. Obese adult   Lean adult   Less body water   More adipose tissue

   3. Claude Bernard's "internal environment"   ICF   About two-thirds of total body water

      All cells

   4. ECF   Interstitial fluid   ICF   Plasma   CSF

   5. Electric charge   Nonelectrolyte   Covalent bonding   No electric charge

2. Determine if the following descriptions refer to electrolytes (E) or to nonelectrolytes (N).

    1. Lipids, monosaccharides, and neutral fats

    2. Have greater osmotic power at equal concentrations

    3. The most numerous solutes in the body's fluid compartments

    4. Salts, acids, and bases

    5. Most of the *mass* of dissolved solutes in the body's fluid compartments

    6. Each molecule dissociates into two or more ions

3. Using the data in the table below, calculate how many milliequivalents per liter (mEq/L) of each ion are normally present in plasma. Insert your answers in the spaces in the table. Then, referring to the table, answer the questions that follow.

	Bicarbonate HCO_3^-	Calcium Ca^{2+}	Chloride Cl^-	Magnesium Mg^{2+}	Hydrogen phosphate HPO_4^{2-}	Potassium K^+	Sodium Na^+
Average concentration (mg/L) in plasma	1586	100	3675	36	96	156	3266
Atomic weight (daltons)	61	40	35	24	96	39	23
mEq/L in plasma							

1. Electrolyte concentrations in the two extracellular compartments are similar, except for the high concentration of which *organic* molecules

    in plasma? _____

2. Name the chief cation in the extracellular compartments. _____

3. Name the two major anions in the extracellular compartments.

_____

4. Indicate which electrolytes are in higher concentrations (use ↑) and which are in lower concentrations (use ↓) in intracellular fluids as compared with extracellular fluids.

    ____ HCO_3^- ____ Cl^- ____ HPO_4^{2-} ____ Na^+ ____ Ca^{2+} ____ Mg^{2+} ____ K^+

4. Figure 27.1 illustrates the three major fluid compartments of the body. Arrows indicate direction of fluid flow. First, select three different colors and color the coding circles and the fluid compartments on the figure. Then, referring to Figure 27.1, respond to the statements that follow. If a statement is true, write T in the answer blank. If a statement is false, change the underlined word(s) and write the correct word(s) in the answer blank.

◯ Interstitial fluid   ◯ Intracellular fluid   ◯ Plasma

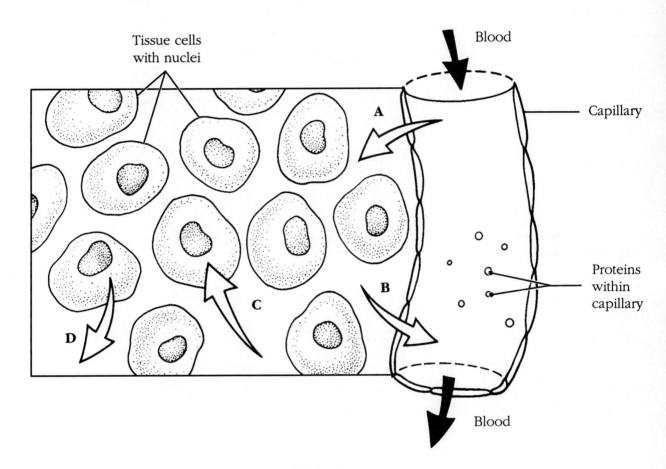

Figure 27.1

_____ **1.** Exchanges between plasma and interstitial fluid compartments take place across the <u>capillary</u> membranes.

_____ **2.** The fluid flow indicated by arrow A is driven by <u>active transport</u>.

_____ **3.** The fluid flow indicated by arrow B is driven by <u>osmotic pressure</u>.

_____ **4.** The excess of fluid flow at arrow A over that at arrow B normally enters the <u>tissue cells</u>.

_____ **5.** Exchanges between the interstitial and intracellular fluid compartments occur across <u>capillary</u> membranes.

_____ **6.** Interstitial fluid serves as the link between the body's external and internal environments.

_____ **7.** The osmolarities of all body fluids are nearly always unequal.

_____ **8.** If the osmolarity of the extracellular fluid is increased, the fluid flow at arrow C will occur.

Water Balance

1. Respond to the following questions about body water by writing your answers in the answer blanks.

    1. Name three sources of body water and specify which source accounts for the bulk of body water.

    _____

    _____

    2. Name four routes by which water is lost from the body and specify which route accounts for the greatest water loss.

    _____

    _____

    _____

    3. Explain the term _insensible water loss_ and give two examples. _____

    _____

    _____

2. Complete the following statements about the regulation of water intake and output by writing the missing terms in the answer blanks.

_____ **1.**

_____ **2.**

_____ **3.**

_____ **4.**

A dry mouth excites the thirst center, located in the **(1)** . **(2)** in the volume and/or **(3)** in the osmolarity of plasma stimulate the **(4)** of the thirst center. After water intake, moistening of the mouth and throat mucosae and activation of

_____ 5.

_____ 6.

_____ 7.

_____ 8.

_____ 9.

_____ 10.

_____ 11.

_____ 12.

_____ 13.

 (5) receptors in the **(6)** send signals that inhibit the thirst center (negative feedback). This mechanism rapidly satisfies thirst and prevents excessive **(7)** of body fluids.

After thirst is quenched, a **(8)** in plasma osmolarity causes the kidneys to begin eliminating excess water. Inhibition of the release of the hormone **(9)** leads to excretion of large volumes of **(10)** urine. Regulation of water balance by kidney activity is closely related to **(11)** ion concentration.

Uncontrollable water losses in urine are called **(12)** water losses. To maintain blood homeostasis, the kidneys must excrete a minimum of **(13)** ml of water per day to get rid of about 1000 mosm of water-soluble solutes.

3. Using the key choices, select the disorder of water balance that applies to each of the following descriptions and write the letter of your answer in the answer blank.

KEY CHOICES

A. Dehydration   **B.** Edema   **C.** Hypotonic hydration

_____ **1.** Abnormal increase of interstitial fluid

_____ **2.** Abnormal increase of intracellular fluid

_____ **3.** May lead to decrease of blood volume and falling blood pressure

_____ **4.** Signs are thirst, sticky oral mucosa, dry skin, and decreased urine volume

_____ **5.** If not remedied, leads to cerebral edema, coma, and death

_____ **6.** Usually follows severe burns or prolonged vomiting

_____ **7.** Extracellular fluid compartment loses water

_____ **8.** Markedly decreased electrolyte concentrations in the ECF, especially Na^+

_____ **9.** May be caused by renal insufficiency

_____ **10.** Causes may be increased capillary hydrostatic pressure and hypoproteinemia

Electrolyte Balance

1. Complete the following statements concerning electrolyte balance by writing the missing terms in the answer blanks.

_____ **1.**

_____ **2.**

_____ **3.**

_____ **4.**

_____ **5.**

_____ **6.**

_____ **7.**

_____ **8.**

Electrolyte balance usually refers to the balance of __(1)__, which enter the body in __(2)__ and leave the body in three ways: __(3)__, __(4)__, and __(5)__. Regulating the electrolyte balance of the body is one of the most important functions of the __(6)__.

The most important and most abundant cation in the ECF is __(7)__. It is the only cation in the ECF that exerts significant __(8)__ pressure. As a consequence, it controls ECF __(9)__ and the distribution of __(10)__ in the body.

_____ **9.**

_____ **10.**

2. Figure 27.2 illustrates one mechanism that controls the release of aldosterone. Complete the flowchart by responding to the questions in the boxes or filling in the missing terms. Write your answers in the answer blanks in each box. Then complete the statements that follow.

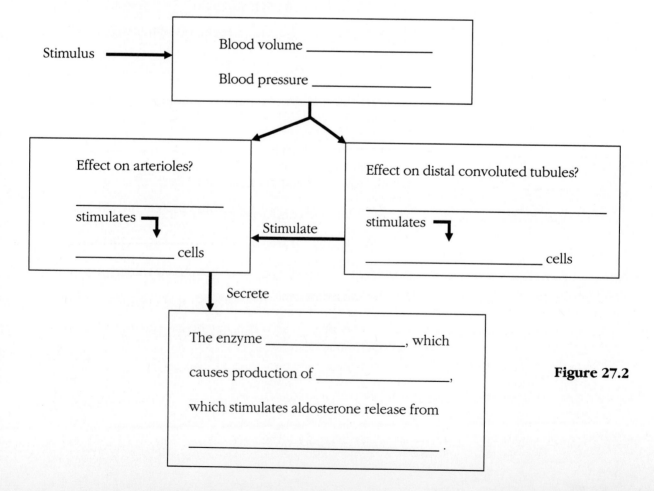

Figure 27.2

_____   **1.**   The juxtaglomerular apparatus is made up of the __(1)__ and
                            __(2)__ cells. Aldosterone promotes more __(3)__ of sodium and
_____   **2.**   __(4)__ . Consequently, blood volume and blood pressure __(5)__ .
                            However, aldosterone acts slowly, and the length of time for
_____   **3.**   the effects of aldosterone to occur is measured in __(6)__ .

_____   **4.**
                            A faster reversal of low ECF volume and low blood pressure is
                            effected by the __(7)__ system. __(8)__ , located in the heart,
_____   **5.**   carotid arteries, and aorta, sense the low blood pressure and
                            alert the __(9)__ , which causes sympathetic impulses to be sent
_____   **6.**   to the __(10)__ of the body. These vessels __(11)__ , increasing the
                            peripheral resistance and, thus, the blood pressure.
_____   **7.**   Sympathetic fibers also stimulate the __(12)__ of the kidneys to
                            release __(13)__ , which promotes an increase in blood pressure
_____   **8.**   by promoting a __(14)__ in blood volume.

_____   **9.**

_____   **10.**

_____   **11.**

_____   **12.**

_____   **13.**

_____   **14.**

3. From the following electrolytes, Ca^{2+}, Cl^-, K^+, Mg^{2+}, HCO_3^-, and Na^+, select the
ions that apply to each of the following descriptions. Insert the symbols of the
ions in the answer blanks.

_____   **1.** Excretion of two ions that is accelerated by aldosterone

_____   **2.** Balance controlled primarily by parathyroid hormone

_____   **3.** During acidosis, is excreted in place of bicarbonate ion

_____   **4.** Renal effect on ion is chiefly excretory; little retention ability

_____   **5.** Activates coenzymes needed for carbohydrate and protein metabolism

_____   **6.** Ion that accompanies Na^+ reabsorption under normal circumstances

_____   **7.** For cation balance, this ion moves across cell membrane in opposition to H^+

_____   **8.** Intestinal absorption of ion enhanced by active form of vitamin D

_____   **9.** Relative ECF/ICF concentrations of ion directly affect resting membrane potential

_____   **10.** Renal tubules conserve this ion while excreting phosphate ion

4. Circle the term that does not belong in each of the following groupings.

1. ↓ H⁺ secretion   ↓ Blood pH   ↑ H⁺ secretion   ↓ K⁺ secretion

2. ↑ Water output   ↓ Plasma volume   ↑ ADH   ↓ ADH

3. Atrial natriuretic factor   Vasodilation   ↑ Na⁺ excretion   ↑ Na⁺ retention

4. ↑ Na⁺ reabsorption   Progesterone   Estrogens   Glucocorticoids

5. Aldosterone   ↑ Na⁺ reabsorption   ↑ K⁺ reabsorption   ↑ BP

6. ADH   ↓ BP   ↑ Blood volume   ↑ Water reabsorption

Acid-Base Balance

1. Match the pH values in Column B with the conditions described in Column A.

Column A		Column B
_____	1. Normal pH of arterial blood	A. pH < 7.00
_____	2. Normal pH of intracellular fluid	B. pH = 7.00
_____	3. Normal pH of venous blood and interstitial fluid	C. pH < 7.35
_____	4. Physiological alkalosis (arterial blood)	D. pH = 7.35
_____	5. Physiological acidosis (arterial blood)	E. pH = 7.40
_____	6. Chemical neutrality; neither acidic nor basic	F. pH > 7.45
_____	7. Chemical acidity

2. Indicate the chief sources of the following acids found in the body.

1. Lactic acid _____

2. Carbonic acid _____

3. Hydrochloric acid _____

4. Fatty acids and ketones _____

5. Phosphoric acid _____

_____

6. Citric acid and acetic acid _____

3. By what three methods is H^+ concentration in body fluids regulated? Give the approximate time for each method to respond to pH changes.

1. _____

2. _____

3. _____

4. Using the key choices, characterize the activity of the following chemical species when they are dissolved in water. Write the appropriate answers in the answer blanks.

KEY CHOICES

A. Neutral   **C.** Strong base   **E.** Weak base

B. Strong acid   **D.** Weak acid

_____ **1.** H_2CO_3 _____ **6.** NaCl

_____ **2.** NH_3 _____ **7.** HCl

_____ **3.** H_2SO_4 _____ **8.** HAc (Acetic)

_____ **4.** Na_2HPO_4 _____ **9.** NaOH

_____ **5.** NaH_2PO_4 _____ **10.** $NaHCO_3$

5. Describe the following substances with respect to their activity related to H^+ ions.

1. Weak acid _____

2. Strong acid _____

3. Weak base _____

4. Strong base _____

5. Buffer system _____

_____

6. Figure 27.3 is a cutaway drawing of a kidney tubule and peritubular capillary showing renal regulation of H^+ secretion, Na^+ and HCO_3^- conservation, and control of ECF pH. Select three colors and color the coding circles. Then, conplete the reactions in the drawing by writing in the answer blanks the appropriate chemical symbols in color. Color the rest of the figure as you wish. Next, referring to Figure 27.3, complete the statements that follow by writing your responses in the answer blanks.

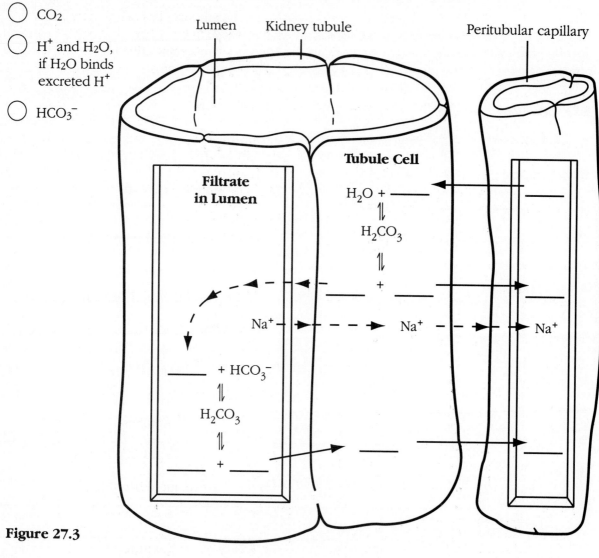

Figure 27.3

_____ 1.
_____ 2.
_____ 3.
_____ 4.
_____ 5.

When the pH of the ECF falls, the concentration of CO_2 in the peritubular capillary __(1)__ . CO_2 diffuses into the tubule cell, shifting the equilibrium to form __(2)__ , which quickly dissociates, liberating H^+ and HCO_3^-. Tubule cells cannot reabsorb HCO_3^- from the __(3)__ , but they can move HCO_3^- formed __(4)__ into the __(5)__ , thus conserving HCO_3^-.

_____ 6.

_____ 7.

_____ 8.

_____ 9.

_____ 10.

_____ 11.

_____ 12.

_____ 13.

_____ 14.

At the same time, the H^+ liberated in the tubule cell is secreted into the __(6)__, where it combines with __(7)__ in the filtrate to form __(8)__, generating both __(9)__. This excess H^+ is therefore tied up in the form of __(10)__. For every H^+ secreted into the lumen of the kidney tubule by the kidney cell, __(11)__ Na^+ is reabsorbed and moved back into the __(12)__. The tubule cells secrete H^+ and reabsorb Na^+ by __(13)__ transport.

When all of the filtered bicarbonate has been reclaimed, additional secreted H^+ begin to be __(14)__, but first they must be buffered. The two most important urine buffers are the __(15)__ and the __(16)__ buffer systems.

_____ 15.

_____ 16.

7. Circle the term that does not belong in each of the following groupings. (*Note:* NH_3 = ammonia; NH_4^+ = ammonium ion.)

1. Large amounts of H^+ excreted H^+ in H_2O Phosphate buffer system

   NH_3-NH_4^+ buffer system

2. Source of NH_3 Amino acid metabolism   Tubule cells   Plasma

3. Na_2HPO_4 H^+ excretion Na_2CO_3 NaH_2PO_4

4. NH_4HCO_3 NH_4Cl $NH_3 + H^+ \rightarrow NH_4^+$ Urine

5. $\uparrow H^+$ in filtrate $\downarrow H^+$ in filtrate $\uparrow NaHCO_3$ reabsorbed $\uparrow NH_4$ secretion

6. $\downarrow$ Urine pH $\uparrow H^+$ in urine $\uparrow HCO_3^-$ in urine $\uparrow$ Ketones

7. Carbonic anhydrase   Plasma   Red blood cells   Rapid CO_2 loading

8. $\downarrow$ Plasma H^+ Respiratory centers in medulla   Deep, rapid breathing

   $\uparrow CO_2$ retention

9. $\downarrow$ Respiratory rate $\downarrow$ in blood CO_2 $\downarrow$ Oxygen   Acidosis

10. Acidosis $\uparrow$ Carbonic acid $\downarrow$ pH $\uparrow$ pH

8. Respond to the following statements concerning buffer systems of the body. If a statement is true, write T in the answer blank. If a statement is false, change the underlined word(s) and write the correct word(s) in the answer blank.

_____ 1. Weak acids are important in chemical buffer systems because they dissociate <u>completely</u>.

_____ 2. The bicarbonate, phosphate, and protein buffer systems act <u>independently</u>.

_____ 3. If NaOH is added to the bicarbonate buffer system, the equilibrium will shift to the <u>right</u>.

_____ 4. When NaOH is added to the bicarbonate buffer system, the chemical species H$_2$O and <u>Na$_2$CO$_3$</u> will be formed.

_____ 5. Addition of HCl to the phosphate buffer system causes an equilibrium shift so that the weak base <u>HPO$_4^{2-}$</u> is produced.

_____ 6. The most abundant and powerful buffers in the ICF and plasma are the <u>phosphates</u>.

_____ 7. The group of atoms on a protein molecule that acts as a base is <u>R-COOH</u>.

_____ 8. <u>Amphoteric</u> molecules can act reversibly to buffer either acids or bases.

_____ 9. The single most important blood buffer is the <u>phosphate</u> buffer system.

_____ 10. The ultimate acid-base regulatory organs are the <u>lungs</u>.

9. Match the key choices with the acid-base balance abnormalities described below. (*Note:* P$_{CO_2}$ = partial pressure of CO$_2$.)

KEY CHOICES

A. Metabolic acidosis   **C.** Normal range   **E.** Respiratory alkalosis

B. Metabolic alkalosis   **D.** Respiratory acidosis

_____ 1. Indicated by plasma HCO$_3^-$ levels above the normal range

_____ 2. P$_{CO_2}$ = 35–45 mm Hg; pH = 7.35–7.45

_____ 3. A common result of hyperventilation

_____ 4. Indicated by plasma P$_{CO_2}$ levels above the normal range

_____ 5. A common cause is excess HCO$_3^-$ loss resulting from prolonged diarrhea

_____ 6. Imbalance where HCO$_3^-$ levels are high; compensated by slow, shallow breathing (therefore high P$_{CO_2}$)

Developmental Aspects of Fluid, Electrolyte, and Acid-Base Balance

1. Complete the following statements by writing the missing terms in the answer blanks.

_____ **1.**

_____ **2.**

_____ **3.**

_____ **4.**

_____ **5.**

_____ **6.**

_____ **7.**

_____ **8.**

_____ **9.**

_____ **10.**

_____ **11.**

_____ **12.**

_____ **13.**

_____ **14.**

_____ **15.**

_____ **16.**

_____ **17.**

_____ **18.**

_____ **19.**

The average body water content for adults is 58%; for infants, it is __(1)__ . This value reflects a relatively higher water content in the __(2)__ compartment of infants. Infants also show different concentrations of __(3)__ in the ICF as compared with adults. At age 2 years, the adult pattern of body water distribution is evident. At adolescence, females have a __(4)__ body water content than males because the volume of the __(5)__ compartment changes as a result of __(6)__ accumulation.

Infants have greater difficulty maintaining fluid homeostasis than adults because of their __(7)__ rates of fluid ingestion and output, large, insensible water losses through the __(8)__ , and large water gains from a high __(9)__ . Infants are also at greater risk for acid-base imbalance because of the very low __(10)__ of the lungs and the immature functioning of the __(11)__ .

In old age, body water content usually __(12)__ , reflecting changes in the __(13)__ compartment. Blood pH and osmotic pressure show __(14)__ changes, but restoration of homeostasis after imbalances occur __(15)__ with age. The elderly may become dehydrated if they fail to respond to __(16)__ stimuli. They are also particularly subject to such diseases as __(17)__ and __(18)__ , which lead to severe fluid, electrolyte, and acid-base imbalances. Because these imbalances become evident when the __(19)__ is highest or lowest, both infants and the elderly are at the highest risk.

CHALLENGING YOURSELF

At the Clinic

1. Jack Tyler exhibits excessive thirst and polyuria, with hypernatremia. Will you check his pituitary or adrenal glands? Do you suspect hyposecretion or hypersecretion, and of what hormone? What is the name of his condition?

2. A patient is diagnosed with bronchogenic carcinoma complicated by ectopic secretion of ADH. Is he more likely to suffer from dehydration or hypotonic hydration? Will he be hyponatremic or hypernatremic?

3. Jessie Fuller, an elderly woman in a nursing home, is reported to have a depraved appetite; she is caught scooping dirt out of potted plants and eating it. What is the name of this condition? What is a likely cause?

4. A woman with Cushing's disease has regular checkups at the clinic. Because of the associated hyperaldosteronism, will she more likely suffer from dehydration, hypotonic hydration, or edema?

5. If a tumor of the glucocorticoid-secreting cells of the adrenal cortex crowds out the cells that produce aldosterone, what is the likely effect on urine composition and volume?

6. Ben Ironfield, an elderly patient, suffers from chronic renal hypertension. What are its effects on his urine output?

7. Ms. Schmid comes to the clinic suffering from extreme premenstrual syndrome; her estrogen level is high. Will a diuretic relieve some of her symptoms?

8. A hypertensive patient named Georg Handel is given potassium replacement to counteract the K^+ depletion caused by his diuretic. He now exhibits hyperventilation. His blood pH is 7.33, his P_{CO_2} is 33 mm Hg, and his blood bicarbonate level is 20 mEq/L. Clinically, what name is given to his state of acid-base balance? What system is compensating? What is the connection with potassium and how does it relate to his blood pH?

9. A young boy has been diagnosed with a hypersecretory tumor of a parathyroid gland. What condition will result?

10. A particularly virulent intestinal virus has triggered repeated vomiting and severe diarrhea in Christine. After 12 hours of feeling helpless, her parents bring her to the clinic. Along with rehydration, what electrolytes must be replaced? What pH adjustment will be necessary if Christine begins to hyperventilate? Shortly after Christine has been diagnosed and treated, her infant cousin starts to exhibit the same symptoms. Three hours later, the baby's anterior fontanel and eyes appear sunken. Should treatment be sought, or is it safe to wait and see if symptoms subside without treatment?

Stop and Think

1. Mannitol is used to reverse hypotonic hydration. Are body cell membranes permeable to mannitol? Explain your response.

2. What are the cardiac effects of dehydration?

3. High levels of stress stimulate secretion of glucocorticoids. What effect will this have on blood volume?

4. When protein synthesis is increased (such as during postsurgical healing), is hypokalemia or hyperkalemia more common?

5. Red blood cells break down in stored blood. What ion imbalance will be seen following rapid transfusion of stored blood?

6. Is calcitonin secretion more important in children or in adults? Why?

7. Much of the calcium carried in the blood is bound to albumin, and calcium and hydrogen ions exchange places when albumin buffers pH changes. That is, as H^+ is bound, Ca^{2+} is liberated. Will alkalosis be correlated with hypocalcemia or hypercalcemia? Will hypoalbuminemia be correlated with hypo- or hypercalcemia?

8. Calcium phosphate is not freely ionizable at high concentrations. How does this fact relate to the increase in phosphate ion excretion when calcium ion is absorbed at the kidneys?

COVERING ALL YOUR BASES

Multiple Choice

Select the best answer or answers from the choices given.

1. Which is a normal value for percent of body weight which is water for a middle-aged man?
    A. 73%   **C.** 45%
    B. 50%   **D.** 60%

2. The smallest fluid compartment is the:
    A. ICF   **C.** plasma
    B. ECF   **D.** IF

3. Which of the following are electrolytes?
    A. Glucose   **C.** Urea
    B. Lactic acid   **D.** Bicarbonate

4. One mEq/L of calcium chloride equals:
    A. 1 mEq/L of sodium chloride
    B. 1 mEq/L of magnesium chloride
    C. 1 mosm
    D. 0.5 mosm

5. The chloride ion concentration in plasma:
    A. equals that in ICF
    B. equals that in IF
    C. equals the sodium ion concentration in plasma
    D. is less than that in IF

6. In movement between IF and ICF:
    A. water flow is bidirectional
    B. nutrient flow is unidirectional
    C. ion flow is selectively permitted
    D. ion fluxes are not permitted

7. Insensible water loss includes:
    A. metabolic water
    B. vaporization from lungs
    C. water of oxidation
    D. evaporation from skin

8. The thirst mechanism involves:
    A. crenation of osmoreceptors
    B. hypoactivity of salivary glands
    C. damping of the mechanism when osmotic balance is reestablished
    D. stretch receptors in the stomach

9. Which of the following will result in hyponatremia?
    A. SIADH
    B. Hyperaldosteronism
    C. Renal compensation for alkalosis
    D. Excessive water ingestion

10. Disorders that result in edema include:
    A. kwashiorkor
    B. cirrhosis
    C. renal insufficiency
    D. acute glomerulonephritis

11. Sodium retention is "uncoupled" from water retention by the action of which hormone?
    A. Aldosterone
    B. Artrial natriuretic factor
    C. Antidiuretic hormone
    D. Estrogen

12. People with Cushing's disease:
   A. retain sodium
   B. hypersecrete aldosterone
   C. suffer from potassium deficiency
   D. are only slightly hypertensive

13. Inhibition of ADH:
   A. is caused by alcohol consumption
   B. results in edema
   C. disturbs acid-base balance
   D. is triggered by hyponatremia

14. Hyperkalemia:
   A. triggers secretion of aldosterone
   B. may result from severe alcoholism
   C. is caused by diarrhea
   D. results from widespread tissue injury

15. Renal tubular secretion of potassium is:
   A. obligatory
   B. increased by aldosterone
   C. balanced by tubular reabsorption
   D. increased in alkalosis

16. Resorption of calcium from the skeleton:
   A. is increased by calcitonin
   B. is paralleled by resorption of phosphate
   C. results in transiently increased buffering capacity of the blood
   D. is enhanced by vitamin D

17. Chloride ion reabsorption:
   A. exactly parallels sodium ion reabsorption
   B. fluctuates according to blood pH
   C. increases during acidosis
   D. is controlled directly by aldosterone

18. Of the named fluid compartments, which has the lowest pH?
   A. Arterial blood
   B. Venous blood
   C. Intracellular fluid
   D. Interstitial fluid

19. Which buffer system(s) is/are not important urine buffers?
   A. Phosphate
   B. Ammonia-ammonium
   C. Protein
   D. Bicarbonate

20. Which of the following is a volatile acid?
   A. Carbonic   C. Lactic
   B. Hydrochloric   D. Phosphoric

21. Respiratory acidosis occurs in:
   A. asthma
   B. emphysema
   C. barbiturate overdose
   D. cystic fibrosis

22. Only the kidneys can remove excesses of:
   A. phosphoric acid   C. keto acids
   B. carbonic acid   D. uric acid

23. Regulation of acid-base balance by the kidneys involves:
   A. activity of carbonic anhydrase
   B. buffering by deamination
   C. secretion of hydrogen ions into the urine
   D. pH-dependent selective reabsorption of chloride or bicarbonate from the urine

24. Given the blood values pH = 7.31, P_{CO_2} = 63 mm Hg, and HCO_3 = 30 mEq/L, it would be accurate to conclude that:
   A. the patient exhibits respiratory acidosis
   B. the patient exhibits metabolic acidosis
   C. renal compensation is occurring
   D. respiratory compensation is occurring

25. Contributing factors to imbalances in infants include:
   A. high residual volume of infant lungs
   B. high fluid output
   C. high rate of production of metabolic wastes
   D. inefficiency of kidneys in removing acids from the blood

26. Bicarbonate ion:
   A. is usually conserved
   B. is usually secreted
   C. is unable to penetrate the luminal border of the tubule cells
   D. is reabsorbed into the blood as H^+ is secreted

27. In the carbonic acid-bicarbonate buffer system, strong acids are buffered by:
   A. carbonic acid
   B. water
   C. bicarbonate ion
   D. the salt of the strong acid

Word Dissection

For each of the following word roots, fill in the literal meaning and give an example, using a word found in this chapter.

Word root	Translation	Example
1. kal	_____	_____
2. natri	_____	_____
3. osmol	_____	_____

28 The Reproductive System

Student Objectives

When you have completed the exercises in this chapter, you will have accomplished the following objectives:

1. Describe the common function of the male and female reproductive systems.

Anatomy of the Male Reproductive System

2. Describe the structure and function of the testes, and explain the importance of their location in the scrotum.

3. Describe the location, structure, and function of the accessory organs of the male reproductive system.

4. Describe the structure of the penis and note its role in the reproductive process.

5. Discuss the sources and functions of semen.

Physiology of the Male Reproductive System

6. Define *meiosis*. Compare and contrast it to *mitosis*.

7. Outline the events of spermatogenesis.

8. Describe the phases of the male sexual response.

9. Discuss hormonal regulation of testicular function and the physiological effects of testosterone on male reproductive anatomy.

Anatomy of the Female Reproductive System

10. Describe the location, structure, and function of the ovaries.

11. Describe the location, structure, and function of each of the organs of the female reproductive duct system.

12. Describe the anatomy of the female external genitalia.

13. Discuss the structure and function of the mammary glands.

Physiology of the Female Reproductive System

14. Describe the process of oogenesis and compare it to spermatogenesis.

15. Describe the phases of the ovarian cycle and relate them to events of oogenesis.

16. Describe the regulation of the ovarian and menstrual cycles.

17. Discuss the physiological effects of estrogens and progesterone.

18. Describe the phases of female sexual response.

Sexually Transmitted Diseases

19. Describe the infectious agents and modes of transmission of gonorrhea, syphilis, chlamydia, and genital herpes.

Developmental Aspects of the Reproductive System: Chronology of Sexual Development

20. Discuss the determination of genetic sex and prenatal development of male and female structures.

21. Note the significant events of puberty and menopause.

Ready, set, go!

The biological function of the reproductive system is to provide the means for producing offspring. The essential organs are those producing the germ cells (testes in males and ovaries in females). The male manufactures sperm and delivers them to the female's reproductive tract; the female, in turn, produces eggs. If the time is suitable, the fusion of egg and sperm produces a fertilized egg, which is the first cell of the new individual. Once fertilization has occurred, the female uterus protects and nurtures the developing embryo.

In this chapter, student activities test understanding of the structures of the male and female reproductive systems, germ cell formation, and the menstrual cycle.

BUILDING THE FRAMEWORK

Anatomy of the Male Reproductive System

1. The location of the testes in the scrotum is essential to provide the ideal temperature for sperm formation, which is approximately 3°C lower than that in the abdominal cavity.

   1. How do the scrotal muscles (dartos and cremaster) help maintain temperature homeostasis of the testes?

   _____

   _____

   _____

   2. How does the pampiniform plexus surrounding the testicular artery help maintain temperature homeostasis of the testes?

   _____

   _____

2. Figure 28.1 is a sagittal view of the male reproductive structures. First, identify the following organs in the figure by writing each term at the appropriate leader line.

Bulbourethral gland	Glans penis	Prepuce	Spongy urethra
Corpus cavernosum	Ejaculatory duct	Prostate gland	Testis
Corpus spongiosum	Epididymis	Scrotum	
Ductus deferens	Penis bulb	Seminal vesicle	

Next, select different colors for the structures that correspond to the following descriptions and color the coding circles and the structures on the figure.

◯ Spongy tissue that is engorged with blood during erection

◯ Portion of the duct system that also serves the urinary system

◯ Structure that provides the ideal temperature conditions for sperm formation

◯ Structure cut or cauterized during a vasectomy

◯ Structure removed in circumcision

◯ Gland whose secretion contains sugar to nourish sperm

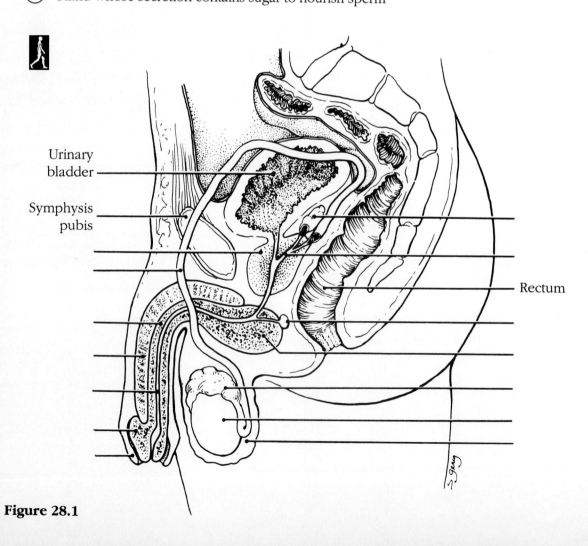

Urinary bladder

Symphysis pubis

Rectum

Figure 28.1

3. Using the key choices, match each term with the appropriate anatomical descriptions. Insert the correct answers in the answer blanks.

KEY CHOICES

A. Bulbourethral glands   **E.** Penis   **I.** Scrotum

B. Epididymis   **F.** Prepuce   **J.** Spermatic cord

C. Ductus deferens   **G.** Prostate gland   **K.** Testes

D. Glans penis   **H.** Seminal vesicles   **L.** Urethra

_____ **1.** Organ that delivers semen to the female reproductive tract

_____ **2.** Site of sperm and testosterone production

_____ **3.** Passageway for conveying sperm from the epididymis to the ejaculatory duct

_____ **4.** Conveys both sperm and urine down the length of the penis

_____   _____ **5.** Organs that contribute to the formation of semen

_____   _____

_____ **6.** External skin sac that houses the testes

_____ **7.** Tubular storage site for sperm; hugs the lateral aspect of the testes

_____ **8.** Cuff of skin encircling the glans penis

_____ **9.** Surrounds the urethra at the base of the bladder; produces a milky alkaline fluid

_____ **10.** Produces over half of the seminal fluid

_____ **11.** Empties a lubricating mucus into the urethra

_____ **12.** Connective tissue sheath enclosing the ductus deferens, blood vessels, and nerves

4. Using the following terms in their proper order, trace the pathway of sperm from the testis to the urethra: rete testis, epididymis, seminiferous tubule, ductus deferens.

_____ → _____ → _____ → _____

5. Figure 28.2 is a longitudinal section of a testis. First, select different colors for the structures that correspond to the following descriptions. Then, color the coding circles and color and label the structures on the figure that have leader lines. Complete the labeling of the figure by adding the following terms: lobule, rete testis, and septum.

○ Site(s) of spermatogenesis

○ Tubular structure in which sperm mature and become motile

○ Fibrous coat protecting the testis

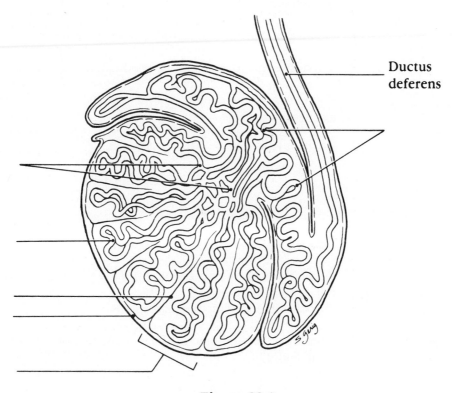

Ductus deferens

Figure 28.2

Physiology of the Male Reproductive System

1. The following statements refer to events that occur during cellular division. Using the key choices, indicate in which type of cellular division each event occurs by writing the correct answers in the answer blanks.

KEY CHOICES

A. Mitosis   **B.** Meiosis   **C.** Both mitosis and meiosis

_____   **1.** Final product is two daughter cells, each with 46 chromosomes

_____   **2.** Final product is four daughter cells, each with 23 chromosomes

_____   **3.** Involves the phases prophase, metaphase, anaphase, and telophase

_____ **4.** Occurs in all body tissues

_____ **5.** Occurs only in the gonads

_____ **6.** Increases the cell number for growth and repair

_____ **7.** Daughter cells have the same number and types of chromosomes as the mother cell

_____ **8.** Daughter cells are different from the mother cell in their chromosomal makeup

_____ **9.** Chromosomes are replicated before the division process begins

_____ **10.** Provides cells for the reproduction of offspring

_____ **11.** Consists of two consecutive divisions of the nucleus; chromosomes are not replicated before the second division

_____ **12.** Involves synapsis and crossing over of the homologous chromosomes

2. Figure 28.3 illustrates a single sperm. On the figure, bracket and label the head and the midpiece, and circle and label the tail. Then, select different colors for the structures described below. Color the coding circles and color and label the structures, using correct terminology, on the figure.

◯ The DNA-containing area

◯ The enzyme-containing sac that aids sperm penetration of the egg

◯ Metabolically active organelles that provide ATP to energize sperm movement

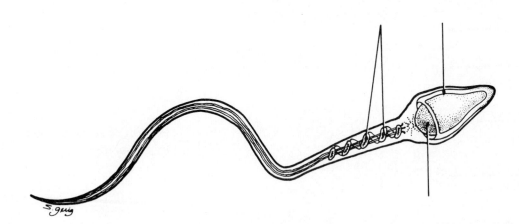

Figure 28.3

3. This exercise considers the process of sperm production in the testis. Figure 28.4 represents a cross-sectional view of a seminiferous tubule in which spermatogenesis is occurring. First, using the key choices, select the terms that match the following descriptions.

KEY CHOICES

    A. Follicle-stimulating hormone   ◯ **D.** Spermatogonium   **G.** Testosterone

◯ **B.** Primary spermatocyte   ◯ **E.** Sperm

◯ **C.** Secondary spermatocyte   ◯ **F.** Spermatid

_____ **1.** Primitive stem cell   _____ **5.** Functional motile gamete

_____ **2.** Contain 23 chromosomes   _____ **6.** Hormones necessary for sperm production

_____ **3.** Product of meiosis I

_____ **4.** Product of meiosis II

Then, select different colors for the cell types with coding circles listed in the key choices and color them on the figure. Label all the structures with leader lines on the figure. In addition, label and color the cells that produce testosterone.

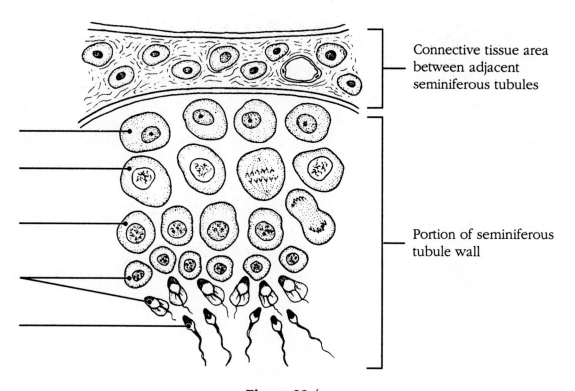

Connective tissue area between adjacent seminiferous tubules

Portion of seminiferous tubule wall

Figure 28.4

4. Match the key terms with the following statements relating to the hormonal regulation of male reproductive function.

KEY CHOICES

A. Androgen-binding protein (ABP)   **D.** Inhibin

B. Follicle-stimulating hormone (FSH)   **E.** Luteinizing hormone (LH)

C. Gonadotropin-releasing hormone (GnRH)   **F.** Testosterone

_____ **1.** Substance released by sustentacular cells that enhances spermatogenesis by binding to testosterone

_____ **2.** Hormone that prompts the interstitial cells to produce testosterone

_____ **3.** Hormone released by the hypothalamus that stimulates the anterior pituitary to release FSH and LH

_____ **4.** Hormones that feed back to inhibit either hypothalamic or anterior pituitary secretory activity

5. The process by which spermatids are converted to mature sperm is called spermiogenesis. Answer the following questions about this process.

1. What exactly does spermiogenesis involve? _____

_____

_____

2. What cells aid in this process by disposing of excess cytoplasm and providing nutrition to the maturing sperm?

_____

3. What is the blood-testis barrier and why is it important?

_____

_____

_____

_____

_____

6. Name four male secondary sex characteristics.

1. _____   **3.** _____

2. _____   **4.** _____

7. Circle the term that does not belong in each of the following groupings.

    1. Semen   Acidic   Relaxin   Fibrinolysin

    2. Spermatogonium   Type B cell   Type A cell   Germ cell

    3. Sertoli cells   Sustentacular cells   Blood-testis barrier   Spermatogenic

    4. Parasympathetic nervous system   Ejaculation   Climax

        Rhythmic contractions of bulbospongiosus muscle

    5. Testosterone   Inhibin   Cholesterol nucleus   Dihydrotestosterone

    6. Nitric oxide   Vasoconstriction   Parasympathetic reflex   Erection

Anatomy of the Female Reproductive System

1. Figure 28.5 is an inferior view of the the female external genitalia. Label the anus, clitoris, labia minora, urethral orifice, hymen, mons pubis, and vaginal orifice on the figure. These structures are indicated with leader lines. Then color the homologue of the male penis blue, color the membrane that partially obstructs the vagina yellow, and color the vestibule red. Mark the extent of the perineum with a dashed line.

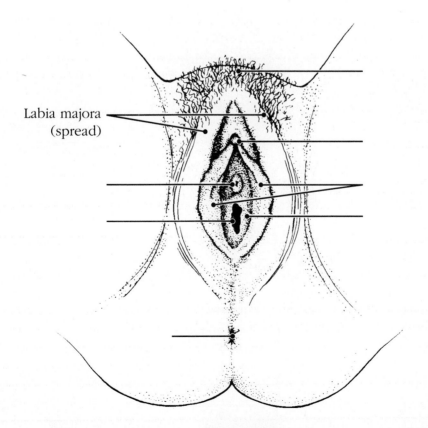

Labia majora
(spread)

Figure 28.5

2. Figure 28.6 is a sagittal view of the female reproductive organs. First, label all structures on the figure with leader lines. Then, select different colors for the following structures and color the coding circles and corresponding structures on the figure.

◯ Lining of the uterus, endometrium

◯ Muscular layer of the uterus, myometrium

◯ Pathway along which an egg travels from the time of its release to its implantation

◯ Ligaments suspending the reproductive organs

◯ Structure forming female hormones and gametes

◯ Homologue of the male scrotum

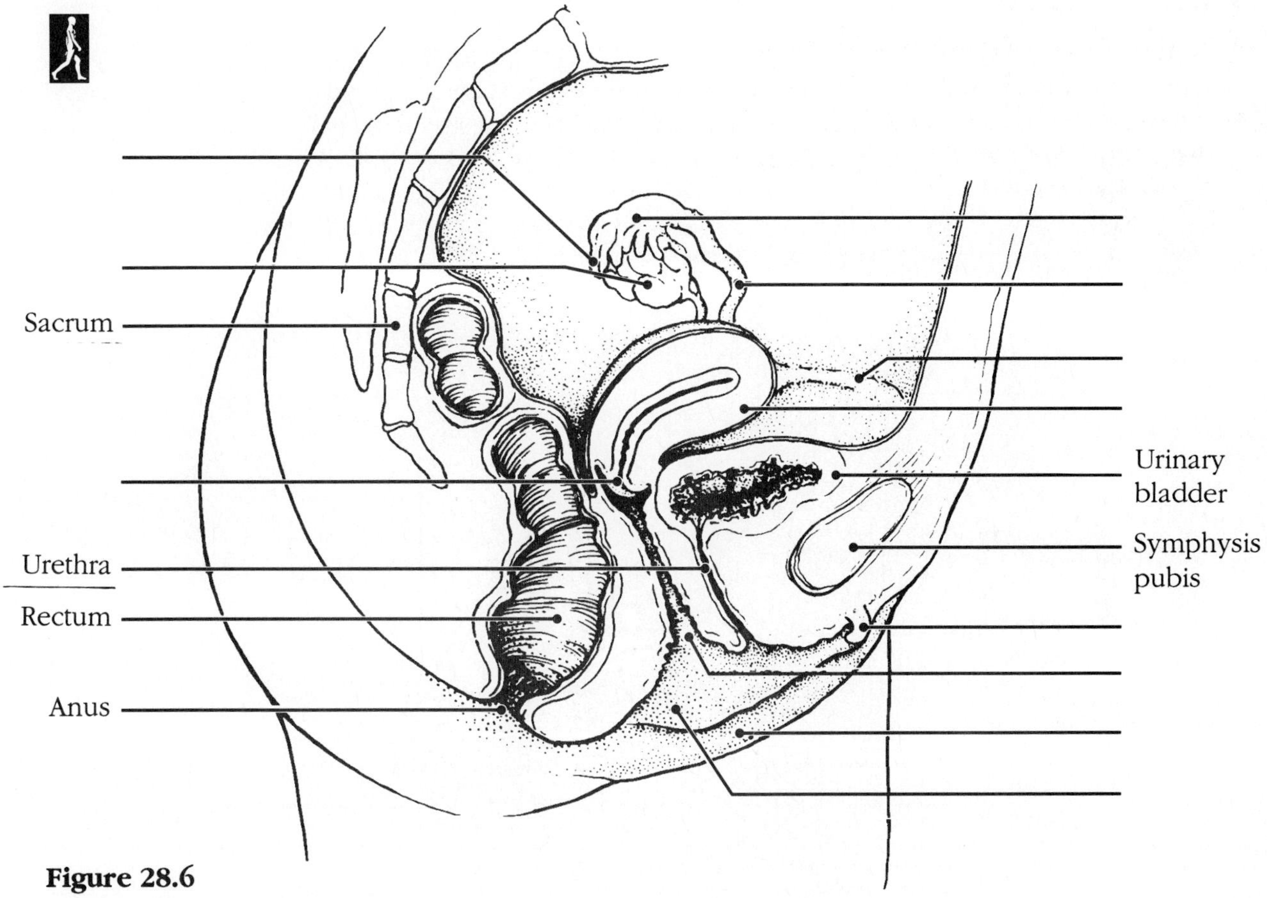

Sacrum

Urethra

Rectum

Anus

Urinary bladder

Symphysis pubis

Figure 28.6

3. What is the significance of the fact that the uterine tubes are not structurally continuous with the ovaries? Address this question from both reproductive and health aspects.

_____

_____

_____

4. Identify the following female anatomical structures.

_____ **1.** Chamber that houses the developing fetus

_____ **2.** Canal that receives the penis during sexual intercourse

_____ **3.** Usual site of fertilization

_____ **4.** Becomes erect during sexual stimulation

_____ **5.** Duct through which the ovum travels to reach the uterus

_____ **6.** Membrane that partially closes the vaginal canal

_____ **7.** Primary female reproductive organ

_____ **8.** Move to create fluid currents to draw the ovulated "egg" into the uterine tube

5. Figure 28.7 is a sagittal section of a breast. First, use the following terms to correctly label all structures with leader lines on the figure: alveolar glands, areola, lactiferous duct and sinus, and nipple. Then color the structures that produce milk blue and the fatty tissue of the breast yellow.

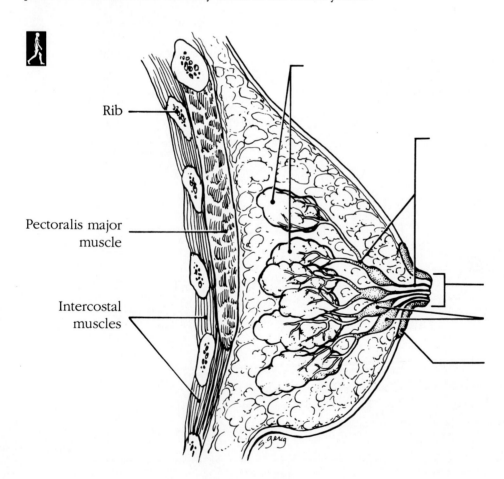

Rib

Pectoralis major muscle

Intercostal muscles

Figure 28.7

Physiology of the
Female Reproductive System

1. The following statements deal with anterior pituitary and ovarian hormonal interrelationships. Name the hormone(s) described in each statement. Place your answers in the answer blanks.

_____	**1.**	Promotes growth of ovarian follicles and production of estrogen
_____	**2.**	Triggers ovulation
_____ and _____	**3.**	Inhibit follicle-stimulating hormone (FSH) release by the anterior pituitary
_____	**4.**	Stimulates luteinizing hormone (LH) release by the anterior pituitary
_____	**5.**	Converts the ruptured follicle into a corpus luteum and causes it to produce progesterone and estrogen
_____	**6.**	Maintains the hormonal production of the corpus luteum

2. Figure 28.8 is a sectional view of the ovary. First, identify all structures with leader lines on the figure. Second, select different colors for the following structures and color the coding circles and corresponding structures on the figure. Then, answer the questions that follow on the next page.

◯ Cells that produce estrogen   ◯ All oocytes   ◯ Structure that produces progesterone

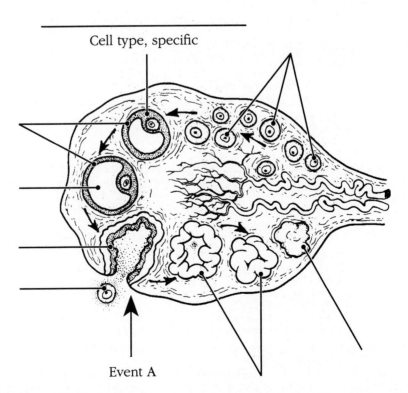

Cell type, specific

Event A

Figure 28.8

1. Name the event depicted as Event A on the figure. _____

2. Into what area is the ovulated cell released? _____

3. When is a mature ovum (egg) produced in humans? _____

4. What structure in the ovary becomes a corpus luteum? _____

5. Name the final cell types and indicate the number of each cell type produced

    by oogenesis in the female. _____

6. What happens to the tiny cells nearly devoid of cytoplasm ultimately produced during
    oogenesis and why does it happen?

    _____

3. Using the key choices, identify the cell type you would expect to find in the
    following structures.

KEY CHOICES

A. Oogonium   **B.** Ovum   **C.** Primary oocyte   **D.** Secondary oocyte

_____   1. Forming part of the primary follicle in the ovary

_____   2. In the uterine tube before fertilization

_____   3. In the mature, or graafian, follicle of the ovary

_____   4. In the uterine tube shortly after sperm penetration

4. The following exercise refers to Figure 28.9.

    In Figure 28.9A, the blood levels of two gonadotropic hormones (FSH and LH)
    of the anterior pituitary are indicated. Identify each hormone by labeling the
    blood level lines on the figure. Then select different colors for each of the
    blood level lines and color them on the figure.

    In Figure 28.9B, identify the blood level lines for the ovarian hormones
    estrogen and progesterone. Then select different colors for each blood level
    line, and color them on the figure.

    In Figure 28.9C, color the following structures, using different colors for each.

    ◯ Primordial follicle   ◯ Secondary follicle   ◯ Ovulating follicle

    ◯ Vesicular follicle   ◯ Corpus luteum

    In Figure 28.9D, identify the endometrial changes occurring during the
    menstrual cycle by coloring the areas depicting the following phases. Use
    different colors to represent each phase.

    ◯ Secretory phase   ◯ Menses   ◯ Proliferative phase

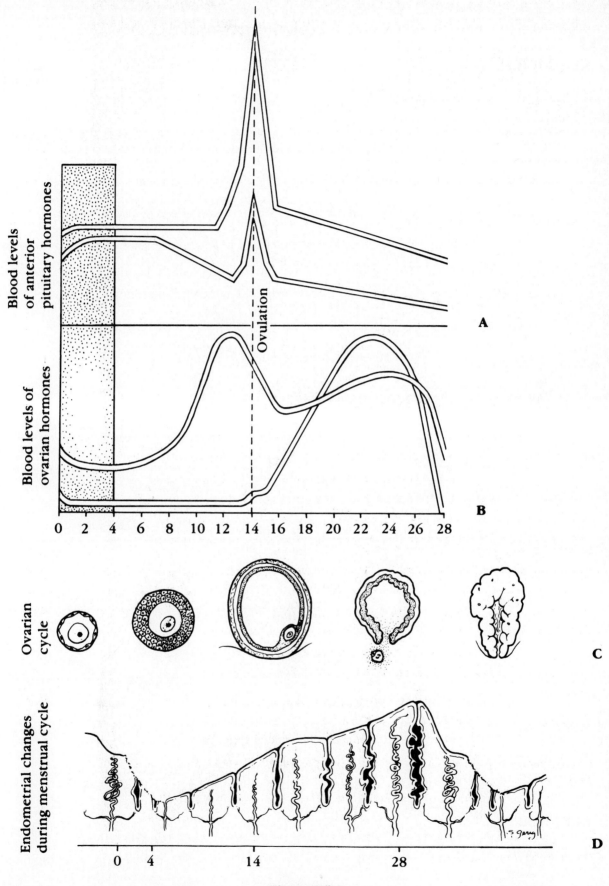

Blood levels of anterior pituitary hormones

Ovulation

A

Blood levels of ovarian hormones

B

Ovarian cycle

C

Endometrial changes during menstrual cycle

D

0 2 4 6 8 10 12 14 16 18 20 22 24 26 28

0 4 14 28

Figure 28.9

5. Use the key choices to identify the ovarian hormone(s) responsible for the following events. Insert the correct answer(s) in the answer blanks.

KEY CHOICES

A. Estrogen   **B.** Progesterone

_____ **1.** Absence causes the spiral arteries of the endometrium to go into spasms and kink and the endometrium to slough off (menses)

_____ **2.** Causes the endometrial glands to begin secreting nutrients

_____ **3.** The endometrium is repaired and grows thick and velvety

_____ **4.** Maintains the myometrium in an inactive state if implantation of an embryo has occurred

_____ **5.** Stimulates gland formation in the endometrium

_____ **6.** Responsible for the secondary sex characteristics of females

_____ **7.** Causes the cervical mucus to become viscous

6. Name four secondary sex characteristics of females.

1. _____   **3.** _____

2. _____   **4.** _____

7. Compare and contrast the two phases of sexual response in females and males.

_____

_____

_____

_____

_____

_____

Sexually Transmitted Diseases

1. Sexually transmitted diseases (STDs) are infectious diseases spread through sexual contact. Match the names of these diseases in the key choices with the appropriate descriptions.

KEY CHOICES

A. Chlamydia   **B.** Genital herpes   **C.** Gonorrhea   **D.** Syphilis   **E.** Vaginitis

_____ **1.** Marked by exacerbations and remissions.

_____ **2.** Its destructive tertiary lesions are called gummas.

_____ **3.** Most common symptom in males is urethritis; may be asymptomatic in females.

_____ **4.** Not recognized as a health problem until the 1970s; accounts for 25%–50% of all pelvic inflammatory disease.

_____ **5.** Congenital forms can cause severe fetal malformations.

_____ **6.** Newborns of infected mothers may have conjunctivitis and respiratory tract infections.

_____ **7.** Like AIDS, a viral rather than a bacterial disease.

_____ **8.** Only diagnosed by cell culture techniques; treated with tetracycline.

_____ **9.** Two diseases routinely treated with penicillin.

_____ **10.** Typically caused by a protozoan; accompanied by severe itching and a copious discharge.

Developmental Aspects of the Reproductive System: Chronology of Sexual Development

1. Complete the following statements by writing the missing terms in the answer blanks.

_____ **1.**
_____ **2.**
_____ **3.**
_____ **4.**
_____ **5.**
_____ **6.**
_____ **7.**
_____ **8.**
_____ **9.**
_____ **10.**
_____ **11.**
_____ **12.**
_____ **13.**
_____ **14.**
_____ **15.**

The gonads of both sexes arise from a mass of mesoderm called the __(1)__ . During early development, the so-called __(2)__ stage, the reproductive structures of both sexes are identical, but by the seventh week, either the male or female gonads begin to be differentiated. Maleness and formation of testes is determined by the fact that males have __(3)__ sex chromosomes, whereas females have __(4)__ . Shortly thereafter, the male duct system begins to form from the __(5)__ duct if the hormone __(6)__ is present. __(7)__ form in the hormone's absence. As a general rule, this hormone is formed by male embryos. The external genitalia derive from a small projection called the __(8)__ . The hormone named above must also be present for development of the male external genitalia. The external genitalia of females develop in the hormone's absence. The gonads descend (the testes into the scrotum) before birth; this process is guided by a structure called the __(9)__ . Within the scrotum, the testes are suspended by the __(10)__ . If the testes fail to descend, the resulting condition is called __(11)__ .

The period of life when the reproductive organs grow to their adult size and become functional is called __(12)__ . In males, the first sign of these events is enlargement of the __(13)__ and the __(14)__ . In females, the first sign is __(15)__ . Most women experience a decline in ovarian function and

_____ 16.

_____ 17.

_____ 18.

_____ 19.

_____ 20.

finally its cessation in their __(16)__ decade of life. This is called __(17)__ . Intense vasodilation of blood vessels in the skin lead to uncomfortable __(18)__ . Additionally, bone mass __(19)__ and blood levels of cholesterol __(20)__ when levels of the hormone __(21)__ wane. In contrast, healthy men are able to father children well into their __(22)__ decade of life.

_____ 21.   _____ 22.

The Incredible Journey:
A Visualization Exercise
for the Reproductive System

. . . you hear a piercing sound coming from the almond-shaped organ as its wall ruptures.

1. Complete the following narrative by writing the missing terms in the answer blanks.

_____ 1.

_____ 2.

_____ 3.

_____ 4.

_____ 5.

_____ 6.

_____ 7.

_____ 8.

_____ 9.

_____ 10.

_____ 11.

_____ 12.

_____ 13.

_____ 14.

_____ 15.

This is your final journey. You are introduced to a hostess who has agreed to have her cycles speeded up by megahormone therapy so that all of your observations can be completed in less than a day. Your instructions are to observe and document as many events of the two female cycles as possible.

You are miniaturized to enter your hostess through a tiny incision in her abdominal wall (this procedure is called a laparotomy, or, more commonly, "belly button surgery"), and you end up in her peritoneal cavity. You land on a large, pear-shaped organ in the abdominal cavity midline, the __(1)__ . You survey the surroundings and begin to make organ identifications and notes of your observations. Laterally and way above you and on each side is an almond-shaped __(2)__ . Each of these almond-shaped organs is suspended by a ligament and almost touched by featherdusterlike projections of a tube snaking across the abdominal cavity toward it. The projections appear to be almost still, which is puzzling because you thought that they were the __(3)__ , or fingerlike projections, of the uterine tubes, which are supposed to be in motion. You walk toward the end of one of the uterine tubes to take a better look. As you study the end of the tube more closely, you discover that the featherlike projections are now moving more rapidly, as if they are trying to coax something into the uterine tube. Then you spot a reddened area on the almond-shaped organ, which seems to be enlarging even as you watch. As you continue to observe the area, you waft gently up and down in the

_____ **16.**   peritoneal fluid. Suddenly you feel a gentle but insistent
sucking current, drawing you slowly toward the uterine tube.

_____ **17.**   You look upward and see that the reddened area now looks
like an angry boil, and the fingerlike projections are gyrating

_____ **18.**   and waving frantically. You realize that you are about to
witness __(4)__ . You try to get still closer to the opening of the
uterine tube, when you hear a piercing sound coming from the
almond-shaped organ as its wall ruptures. Then you see a ball-like structure, with a "halo" of tiny cells
enclosing it, being drawn into the uterine tube. You have just seen the __(5)__ surrounded by its
capsule of __(6)__ cells entering the uterine tube. You hurry into the uterine tube behind it and hold
onto one of the tiny cells to follow it to the uterus. The cell mass that you have attached to has no way
of propelling itself, yet you are being squeezed along toward the uterus by a process called __(7)__ .
You also notice that there are __(8)__ , or tiny hairlike projections of the tubule cells, that are all waving
in the same direction as you are moving.

For a while you are carried tranquilly along, but then you are startled by a deafening noise. Suddenly
there are thousands of tadpolelike __(9)__ swarming all around you and the sphere of cells. Their heads
seem to explode as their __(10)__ break and liberate digestive enzymes. The cell mass now has
hundreds of openings in it, and some of the small cells are beginning to fall away. As you peer
through the rather transparent cell "halo," you see that one of the tadpolelike structures has penetrated
the large central cell. The penetrated cell then begins to divide. You have just witnessed the second
__(11)__ division. The products of this division are one large cell, the __(12)__ , and one very tiny cell, the
__(13)__ , which is now being ejected. This cell will soon be __(14)__ because it has essentially no
cytoplasm or food reserves. As you continue to watch, the sperm nucleus and that of the large central
cell fuse, an event called __(15)__ . You note that the new cell just formed by this fusion is called a
__(16)__ , the first cell of the embryonic body.

As you continue to move along the uterine tube, the central cell divides so fast that virtually no cell
growth occurs between the divisions. Thus the number of cells forming the embryonic body increases,
but the cells become smaller and smaller. Finally, the uterine chamber looms before you. As you drift
into its cavity, you scrutinize its lining, the __(17)__ . You notice that it is thick and velvety in
appearance and that the fluids you are drifting in are slightly sweet. The embryo makes its first contact
with the lining, detaches, and then makes a second contact at a slightly more inferior location. This
time it sticks, and as you watch, the lining of the organ begins to erode away. The embryo is obviously
beginning to burrow into the rich cushiony lining.

You now leave the embryo and propel yourself well away from it. As you float in the cavity fluids,
you watch the embryo disappear from sight beneath the uterine lining. Then you continue to travel
downward through your hostess' reproductive tract, exiting her body at the external opening of the
__(18)__ .

CHALLENGING YOURSELF

At the Clinic

1. A young man with a painless, hard lump on his left testicle has been admitted to the hospital and scheduled for surgery. What do you suspect?

2. Johnny is brought to the clinic by his concerned parents. His scrotum is shriveled on the right side, and only one testis can be palpated. What is Johnny's condition and how is it corrected?

3. Unable to afford a professional mover, Gary enlists the help of his friends to move into a new apartment. Since the apartment was on the third floor, the move was a difficult laborious one, but it was complete by the end of the day. Exhausted, Gary fell into bed but awoke the next morning with pain in his groin. What do you think happened?

4. An elderly man has the problem of urine retention, and a digital rectal examination reveals an enlargement inferior to his urinary bladder. What is this condition and what are some likely causes?

5. Mrs. Ginko's Pap smear result shows some abnormal cells. What possibility should be investigated?

6. Janet Roe is having a fertility assessment done at the GYN clinic. She is frustrated because she has been unable to become pregnant. Her history indicates that her menstrual flows have been very heavy and accompanied by considerable pain for the past several years. What condition might be suspected?

7. Marylou, a sexually active young woman, comes to the clinic after several weeks of more-or-less continuous pelvic pain and heavy bleeding. What is the doctor's diagnosis? What does he need to know before prescribing medication?

8. A 53-year-old woman has come to the clinic complaining of abdominal discomfort, increased flatulence, and bloating. Palpation during a pelvic exam reveals a massive left ovary. What do you think is wrong?

9. After bearing six children in eight years, a woman complains of perineal discomfort whenever she increases the abdominal pressure (sneezing, bowel movement). She says it feels like "her insides are coming out." What is her problem?

10. Julio is diagnosed with gonorrhea and chlamydia. What clinical name is given to the general class of reproductive system infections and why is it crucial to inform his partners of his infection?

Stop and Think

1. In what ways does the reproductive system contribute to the body's homeostasis?

2. After 12 years of a childless marriage, the Dulaps decided to consult a physician concerning their chances of having a family. Examination of Mrs. Dulap revealed abnormal internal genitalia, and the uterus was totally absent. Blood samples showed *testosterone* levels typical of a male. Explain the woman's situation.

3. Why might testicular mumps in an adult, but not in a prepubertal, male result in sterility?

4. What would have to happen to trigger circulating antibodies to sperm antigens?

5. What is the connection between use of the term estrogen for female hormones and its literal meaning ("frenzy")?

COVERING ALL YOUR BASES

Multiple Choice

Select the best answer or answers from the choices given.

1. Which of the following are accessory sex structures in the male?

  A. Gonads   **C.** Broad shoulders

  B. Gametes   **D.** Seminal vesicles

2. Within the spermatic cord are the:

  A. ductus deferens   **C.** testicular artery

  B. dartos muscle   **D.** gubernaculum

3. Which is/are associated with varicocele?

  A. Low sperm count

  B. Malfunctional cremaster muscle

  C. Low testosterone levels

  D. Pampiniform plexus

4. Stereocilia:

  A. are found in the ducts of the epididymis

  B. are elongated cilia

  C. supply nutrients to stored sperm

  D. supply activating enzymes during ejaculation

5. Which of the following structures have a region called the ampulla?

  A. Ductus deferens

  B. Uterine tube

  C. Ejaculatory duct

  D. Lactiferous duct

6. Seminal vesicle secretions have:

  A. a low pH

  B. fructose

  C. prostaglandins

  D. sperm-activating enzymes

7. The structure(s) superior to the urogenital diaphragm is (are):

  A. bulbourethral glands

  B. bulb of the penis

  C. prostate gland

  D. membranous urethra

8. Seminalplasmin:

  A. activates sperm

  B. digests the intercellular glue in the corona radiata

  C. has bacteriostatic properties

  D. neutralizes vaginal acidity

9. A chromosome count in a spermatogenic cell within the adluminal compartment of the seminiferous tubule shows it to be diploid. The cell could be a:
   A. spermatogonium
   B. spermatid
   C. sustentacular cell
   D. primary spermatocyte

10. As a result of crossover:
    A. maternal genes can end up on a paternal chromosome
    B. synapsis occurs
    C. a tetrad is formed
    D. no two spermatids have exactly the same genetic makeup

11. Prominent structures in a cell undergoing spermiogenesis include:
    A. chromosomes
    B. Golgi apparatus
    C. rough ER
    D. mitochondria

12. Treatment for which disorders will reduce fertility?
    A. Chlamydia
    B. Testicular cancer
    C. Cryptorchidism
    D. Gonorrhea

13. Components of the brain-testicular axis include:
    A. parasympathetic impulses of the erection reflex triggered by pleasurable thoughts
    B. stimulation of gonadotropin release
    C. negative feedback from the sustentacular cells
    D. maintenance of normal sperm count

14. Which of the following attach to the ovary?
    A. Fimbriae
    B. Mesosalpinx
    C. Suspensory ligaments
    D. Broad ligament

15. Which structures are involved in estrogen secretion?
    A. Primordial follicle
    B. Theca interna
    C. Granulosa cells
    D. Corpus luteum

16. Each month, only one:
    A. primordial follicle is stimulated
    B. follicle secretes estrogen
    C. vesicular follicle undergoes ovulation
    D. ovary is stimulated

17. On day 17 of a woman's monthly cycle:
    A. FSH levels are rising
    B. progesterone is being secreted
    C. the ovary is in the ovulatory phase
    D. the uterus is in the proliferative phase

18. Cells that are functional on day 8 include:
    A. granulosa cells
    B. theca interna cells
    C. luteal cells
    D. secondary oocyte

19. A sudden decline in estrogen and progesterone levels:
    A. causes spasms of the spiral arteries
    B. triggers ovulation
    C. ends inhibition of FSH release
    D. causes fluid retention

20. Dimensions of which of the following is most important when a diaphragm or cervical cap is being fitted?
    A. The hymen
    B. The vestibule
    C. The fornix
    D. The vaginal diameter

21. Which of the female structures listed below have a homologue in the adult male?
    A. Labia majora
    B. Clitoris
    C. Uterine tubes
    D. Labia minora

22. An STD that is more easily detected in males than females, is treatable with penicillin, and can cause lesions in the nervous and cardiovascular systems is:
    A. gonorrhea
    B. chlamydia
    C. syphilis
    D. herpes

23. Which are present in the sexually indifferent stage of the embryo?
    A. Paramesonephric ducts
    B. Genital tubercle
    C. Primordial germ cells
    D. Labioscrotal swellings

24. If the uterine tube is a trumpet ("salpinx"), what part of it represents the wide, open end of the trumpet?
    A. Isthmus
    C. Infundibulum
    B. Ampulla
    D. Flagellum

25. The myometrium is the muscular layer of the uterus, and the endometrium is the _____layer.
    A. serosa
    C. submucosa
    B. adventitia
    D. mucosa

26. All of the following are true of the gonadotropins except that they are:
    A. secreted by the pituitary gland
    B. LH and FSH
    C. hormones with important functions in both males and females
    D. the sex hormones secreted by the gonads

27. The approximate area between the anus and clitoris in the female is:
    A. peritoneum
    C. vulva
    B. perineum
    D. labia

28. A test to detect cancerous changes in cells of the uterus and cervix is:
    A. pyelogram
    B. Pap smear
    C. D & C
    D. laparoscopy

29. In humans, separation of the cells at the two-cell stage following fertilization may lead to the production of twins, which in this case, would be:
    A. of different sexes
    C. fraternal
    B. identical
    D. dizygotic

30. Human ova and sperm are similar in that:
    A. about the same number of each is produced per month
    B. they have the same degree of motility
    C. they are about the same size
    D. they have the same number of chromosomes

31. A person who has had this type of mastectomy will have great difficulty adducting the arms, as when doing push-ups.
    A. Radical mastectomy
    C. Lumpectomy
    B. Simple mastectomy

32. The embyronic paramesonephric duct gives rise to the:
    A. uterus and uterine tubes
    C. ovary
    B. epididymis and vas deferens
    D. urethra

33. Which of these is mismatched?
    A. Vagina—penis
    B. Testis—ovary
    C. Labia majora—scrotum
    D. Oviduct—ductus deferens

34. After ovulation, the ruptured follicle:
    A. degenerates
    B. becomes a corpus luteum
    C. sloughs off as waste material
    D. mends and produces another oocyte

Word Dissection

For each of the following word roots, fill in the literal meaning and give an example, using a word found in this chapter.

Word root	Translation	Example
1. acro	_____	_____
2. alb	_____	_____
3. apsi	_____	_____
4. arche	_____	_____
5. cremaster	_____	_____
6. didym	_____	_____
7. estro	_____	_____
8. forn	_____	_____
9. gamet	_____	_____
10. gen	_____	_____
11. gest	_____	_____
12. gono	_____	_____
13. gubern	_____	_____
14. gyneco	_____	_____
15. herp	_____	_____
16. hymen	_____	_____
17. hyster	_____	_____
18. inguin	_____	_____
19. lut	_____	_____
20. mamma	_____	_____

Word root	Translation	Example
21. mei		
22. men		
23. menstru		
24. metr		
25. oo		
26. orchi		
27. ov		
28. pampin		
29. penis		
30. pub		
31. pudend		
32. salpin		
33. scrot		
34. semen		
35. uter		
36. vagin		
37. vener		
38. vulv		

 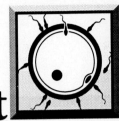

29 Pregnancy and Human Development

Student Objectives

When you have completed the exercises in this chapter, you will have accomplished the following objectives:

From Egg to Embryo

1. Describe the importance of capacitation to the ability of sperm to penetrate an oocyte.

2. Explain the mechanism of the fast and slow blocks to polyspermy.

3. Define *fertilization*.

4. Explain the process and product of cleavage.

5. Describe the processes of implantation and placenta formation and list placental functions.

Events of Embryonic Development

6. Describe the process of gastrulation and its consequence.

7. Name and describe the formation, location, and function of the embryonic membranes.

8. Define *organogenesis* and note the important roles of the three primary germ layers in this process.

9. Describe the unique features of the fetal circulation.

Events of Fetal Development

10. Indicate the duration of the fetal period, and note the major events of fetal development.

Effects of Pregnancy on the Mother

11. Describe changes in maternal reproductive organs and in cardiovascular, respiratory, and urinary system functioning during pregnancy.

12. Note the effects of pregnancy on maternal metabolism and posture.

Parturition (Birth)

13. Explain how labor is initiated and describe the three stages of labor.

Adjustments of the Infant to Extrauterine Life

14. Outline the events leading to the first breath of a newborn.

15. Describe the changes that occur in the fetal circulation after birth.

Lactation

16. Explain how the breasts are prepared for lactation.

Ready, set, go!

We all begin life as a single cell, the fertilized egg, which divides in a seemingly endless manner to produce the trillions of cells that form the human body. In Chapter 29 we review the important events of gestation and birth from the point of view of the developing infant and the mother.

 **BUILDING THE FRAMEWORK**

From Egg to Embryo

1. What is the importance of the fact that sperm are capacitated in the female reproductive tract and not in the male reproductive tract?

   _____

   _____

   _____

2. What event occurs *between* sperm penetration of the secondary oocyte and the formation of the zygote?

   _____

3. Relative to events of sperm penetration:

   1. What portion of the sperm actually enters the oocyte? _____

   2. What is the functional importance of the acrosomal reaction?

      _____

      _____

   3. In terms of binding events, the part of the sperm's binding apparatus called the

      _____ must first bind to an oocyte membrane receptor before the

      _____ of the sperm's binding apparatus can insert through the oocyte membrane.

4. Match the terms in Column B with the developmental events or results described in Column A.

Column A

_____ 1. A fertilized egg

_____ 2. Swollen ovum and sperm nuclei

_____ 3. Process during which a sperm becomes capable of undergoing the acrosomal reaction

_____ 4. Release of digestive enzymes by sperm in the immediate vicinity of an oocyte

_____ 5. Name given to the developing infant until the third week after fertilization

_____ 6. Fusion of the ovum and sperm nuclei

_____ 7. Cells resulting from cleavage

_____ 8. Initiated by sperm contact and fusion with the oocyte membrane

_____ 9. Period of rapid mitotic cell division that results in cells with a high surface-to-volume ratio

_____ 10. A consequence of syncytiotrophoblast activity, during which endometrial cells are digested

_____ 11. Two events that constitute blocks to polyspermy

_____ 12. Term applied to the developing infant after the eighth week

_____ 13. Initiated by rising levels of ionic calcium within the oocyte cytoplasm

_____ 14. Term applied to the developing infant from the third week to the end of the eighth week after fertilization

Column B

A. Acrosomal reaction

B. Blastomeres

C. Capacitation

D. Cleavage

E. Conceptus

F. Cortical reaction

G. Depolarization of the oocyte membrane

H. Embryo

I. Fertilization

J. Fetus

K. Implantation

L. Preembryo

M. Pronuclei

N. Zygote

5. Figure 29.1 depicts early embryonic events. Identify the following events. Then, referring to the figure, complete the descriptive flowchart that follows the diagram.

Part I:

1. Event A _____

2. Cell A _____

3. Process B _____

4. Embryonic structure B₁ _____

5. Completed process C _____

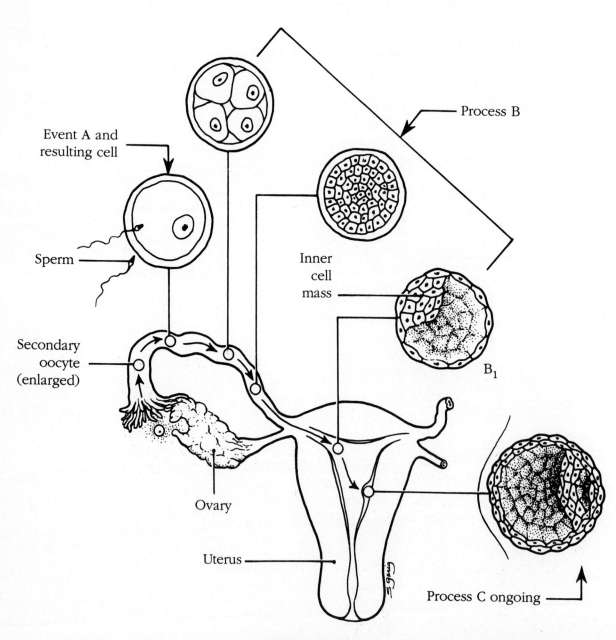

Figure 29.1

Part II:

Fill in the missing terms in the flow diagram below. Blanks beside arrows should be filled by naming a process; other blanks represent stages or structures of development.

(1) _____

    ↓ ← sperm penetration

        meiosis II

(2) _____ + (3) _____

    ↓ fusion of male and female (4) _____

(5) _____

    ↓ (6) _____ (division without growth)

(7) _____

    ↓ differentiation
      (hollows out forming)

(8) _____, with an outer (9) _____

and an inner cell mass   ↓ (extraembryonic mesoderm appears)

    ↓ (10) _____

(11) _____   (12) _____

(now 3 germ layers)   (which forms fetal portion of placenta)

6. Describe the mechanism of implantation: _____

_____

_____

_____

7. Figure 29.2 shows an implanting blastocyst. Choose five colors to identify and color the following structures on the diagram. Notice the wall of the uterus has already been identified.

◯ Cavity of blastocyst   ◯ Inner cell mass   ◯ Wall of uterus

◯ Cytotrophoblast   ◯ Syncytiotrophoblast

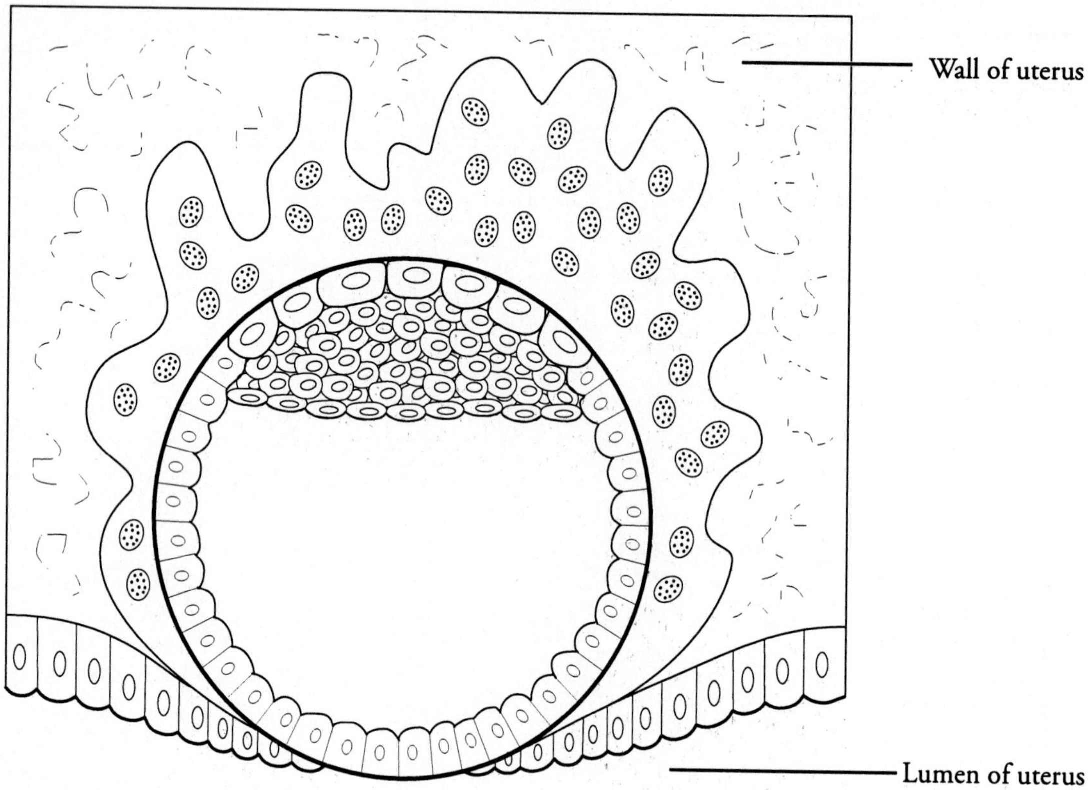

Wall of uterus

Lumen of uterus

Figure 29.2

8. Using the key choices, select the terms that match the following descriptions, most concerned with placentation. Insert the correct answers in the answer blanks.

KEY CHOICES

A. Allantois   **C.** Endometrium   **E.** Placenta   **G.** Yolk sac

B. Chorionic villi   **D.** Inner cell mass   **F.** Umbilical cord

_____   **1.** Part of the blastocyst that forms the embryonic body

_____   **2.** Formed by the delamination of endodermal cells; a ventral sac

_____   **3.** Site of respiratory exchanges

_____   **4.** Attaches the embryo to the placenta

_____   **5.** Fingerlike projections of the trophoblast that are invaded by extraembryonic mesoderm

_____   **6.** After three months, the source of estrogen and progesterone during pregnancy

_____   **7.** The organ that delivers nutrients to and disposes of wastes for the fetus

_____   **8.** Projection of the yolk sac that serves as a depository for wastes in animals that form large-yolked eggs

9. Explain why the corpus luteum does not stop producing its hormones (estrogen and progesterone) when fertilization has occurred.

_____

_____

10. Relative to the process of placentation, complete the statements below by writing the missing terms in the corresponding answer blanks.

_____   **1.**   The placenta is formed from both __(1)__ and __(2)__ tissues. As the extraembryonic mesoderm associates with the proliferating

_____   **2.**   __(3)__, the combined structure becomes the chorion. The chorion then develops elaborate __(4)__, which come to lie in

_____   **3.**   large blood-filled lacunae, or __(5)__ in the stratum __(6)__ layer of the endometrium. Together the __(7)__ and the __(8)__ form

_____   **4.**   the placenta.

_____   **5.**

_____   **6.**

_____   **7.**

_____   **8.**

Events of Embryonic Development

1. Match the descriptions in Column B to the embryonic membranes listed in Column A.

Column A   Column B

_____ **1.** Allantois   **A.** Site of earliest blood cell formation

_____ **2.** Amnion   **B.** Sac of fluid; shock absorber

_____ **3.** Yolk sac   **C.** Source of primordial germ cells

                                                  D. Structural basis of umbilical cord

2. Figure 29.3 shows the relationship between the primitive streak on or about day 15 and the infant's body that forms later. Color the future ectoderm blue and the future endoderm yellow. Identify the following structures by inserting labels at the ends of the appropriate leader lines.

Primitive streak   Embryonic disc   Amniotic sac

Primitive node   Yolk sac

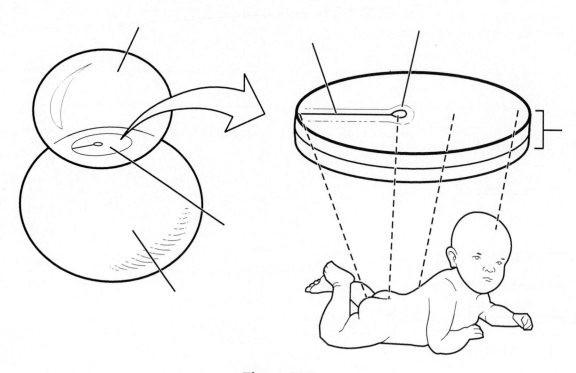

Figure 29.3

3. Figure 29.4 depicts the process of gastrulation. First, color the ectodermal cells blue, endodermal cells yellow, and presumptive mesodermal cells red. Second, correctly identify, by using choices from the terms below, all structures provided with leader lines by writing in the appropriate labels. Third, add arrows to show the direction of migration of cells destined to become mesodermal cells. Finally, use the terms provided to complete the short descriptive passage below the diagram.

◯ Ectoderm   Primitive streak/groove   Cut edge of yolk sac

◯ Endoderm   Primitive node   Embryonic disc

◯ Mesoderm   Cut edge of amnion

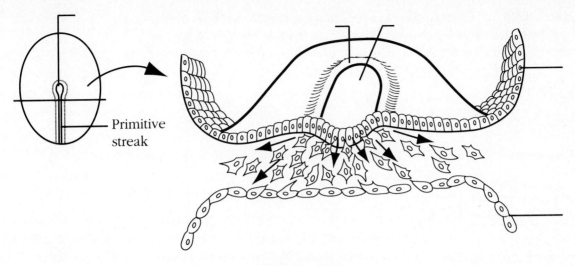

Primitive streak

Figure 29.4

1. _____

2. _____

3. _____

4. _____

5. _____

6. _____

7. _____

8. _____

The process of gastrulation involves vast cell migrations that result in the formation of the three __(1)__ . Essentially, after the __(2)__ has lifted away from the inner cell mass, the balance is called the __(3)__ . The surface cells form the __(4)__ ; the cells that migrate from the surface to the middle layer through the __(5)__ are destined to become the __(6)__ , and the cells at the inferior aspect (forming the roof of the __(7)__) become the __(8)__ .

4. For each of the somite subdivisions listed below, name the major structures that it produces.

1. Dermatome _____

2. Myotome _____

3. Sclerotome _____

5. Complete the following statements concerning gastrulation and the events that occur shortly thereafter by choosing terms from the key choices. Write the answers in the answer blanks.

KEY CHOICES

A. Ectoderm   **D.** Lateral mesoderm   **G.** Organogenesis

B. Endoderm   **E.** Neurulation   **H.** Primitive streak

C. Intermediate mesoderm   **F.** Notochord   **I.** Somites

_____ 1.

_____ 2.

_____ 3.

_____ 4.

_____ 5.

_____ 6.

_____ 7.

_____ 8.

_____ 9.

_____ 10.

_____ 11.

Once gastrulation has been completed, __(1)__, the formation of body systems, begins. Its first event is the formation of a dorsal, rodlike thickening of mesoderm called the __(2)__, which appears immediately deep to the former __(3)__ and establishes the longitudinal axis of the embryo. This is quickly followed by the process called __(4)__, which leads to the formation of the nervous system. This process begins as the __(5)__ lying superior to the notochord is induced to thicken and then to fold and detach as the neural tube. This is quickly followed by the specialization of the __(6)__, which forms the lining of digestive and respiratory tracts and all glands that develop from those mucosae.

Early differentiation of the mesoderm is forecast by its division into three regions, the __(7)__, __(8)__, and __(9)__. The limb buds and parietal serosa are formed by the somatic mesoderm part of the __(10)__, whereas the cardiovascular system forms from the splanchnic layer. The gonads and kidneys arise from the __(11)__.

Events of Fetal Development

1. Briefly explain why the lungs are bypassed to a great extent by the circulating blood in the fetus.

_____

_____

_____

2. Figure 29.5 illustrates the special cardiovascular fetal structures listed below. Select different colors for each structure and color the coding circles and the structures on the diagram.

◯ Foramen ovale   ◯ Ductus venosus   ◯ Umbilical cord

◯ Ductus arteriosus   ◯ Umbilical arteries   ◯ Umbilical vein

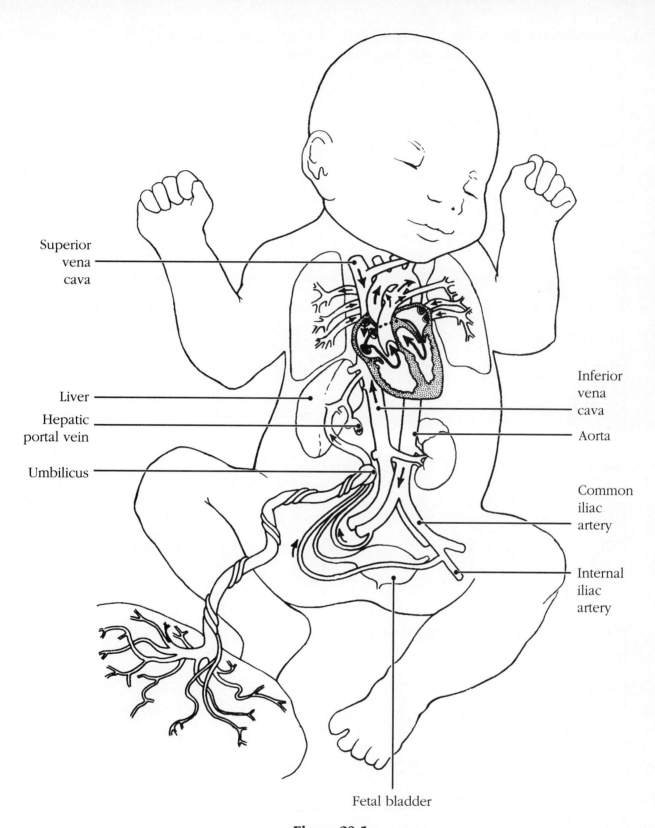

Superior
vena
cava

Liver

Hepatic
portal vein

Umbilicus

Inferior
vena
cava

Aorta

Common
iliac
artery

Internal
iliac
artery

Fetal bladder

Figure 29.5

3. The table below lists the primary germ layers and the systems of the body. Fill in the parts of each fetal system that derive from each germ layer and put an X in the space for any germ layer that does not contribute tissue to the system.

System	Ectoderm	Mesoderm	Endoderm
Nervous			
Circulatory			
Digestive			
Respiratory			
Endocrine			
Muscular			
Skeletal			
Integumentary			
Urogenital			

4. Essentially all organs are present by the time the developing infant is called a fetus, and only two types of activities occur from this time on. Name them.

1. _____   **2.** _____

Effects of Pregnancy on the Mother

1. A pregnant woman undergoes numerous changes during her pregnancy—anatomical, metabolic, and physiological. Several such possibilities are listed below. Check (✔) all that are commonly experienced during pregnancy.

_____ **1.** Chadwick's sign

_____ **2.** Glucose sparing

_____ **3.** Reduction in breast size

_____ **4.** Relaxation of pelvic ligaments by relaxin

_____ **5.** Decrease in vital capacity

_____ **6.** Lordosis

_____ **7.** Reduced blood pressure and pulse rate

_____ **8.** Decline in metabolic rate

_____ **9.** Increased mobility of GI tract

_____ **10.** Increased blood volume and cardiac output

_____ **11.** Nausea, heartburn, constipation

_____ **12.** Possible dyspnea

_____ **13.** Urgency and stress incontinence

_____ **14.** Dry mouth

_____ **15.** Impaired diaphragm descent

2. For each of the hormones listed below, indicate its major actions during pregnancy.

Hormone   *Action*

1. Human chorionic gonadotropin   _____

_____

2. Estrogen   _____

_____

3. Progesterone   _____

_____

4. Relaxin   _____

_____

5. Human placental lactogen   _____

_____

6. Human chorionic
thyrotropin _____

_____

7. Oxytocin _____

_____

8. Prostaglandins _____

_____

9. Prolactin _____

_____

Parturition (Birth)

1. The very simple flowchart in Figure 29.6 illustrates the sequence of events that occur during labor. Complete the flowchart by filling in the missing terms in the boxes. Use color as desired.

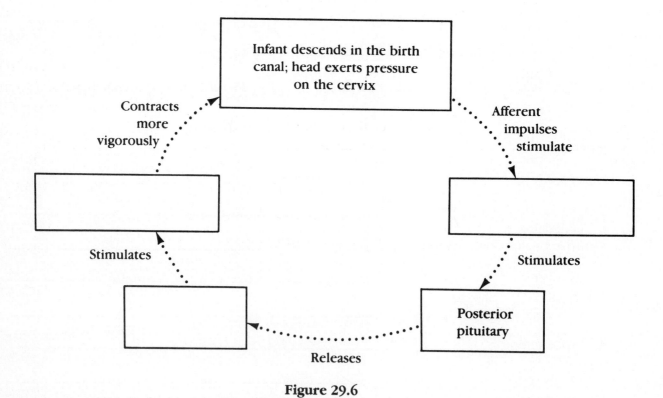

Figure 29.6

2. How long will the cycle illustrated in Figure 29.6 continue to occur?

_____

_____

3. Name the three phases of parturition and briefly describe each phase.

    1. _____

       _____

    2. _____

       _____

    3. _____

       _____

4. What are Braxton Hicks contractions and why do they occur?

_____

_____

_____

_____

Adjustments of the Infant to Extrauterine Life

1. Identify 3 shunts or fetal circulatory modifications that are closed after birth and explain the reasons for their closure.

_____

_____

_____

2. What event stimulates the newborn baby to take the first breath? _____

Lactation

1. Both labor and the milk let-down reflex are positive feedback mechanisms. What does that mean?

_____

_____

2. Compare and contrast colostrum and true milk. _____

_____

CHALLENGING YOURSELF

At the Clinic

1. Mrs. Chang, who has been trying to conceive a child, has been keeping close track of her menstrual cycle. She has come to the clinic to ask for a pregnancy test three days after ovulation. Will a pregnancy test show positive results if she is pregnant? What is the basis of pregnancy tests and what is the earliest date that a meaningful test can be performed?

2. A newborn exhibited microcephaly, and the parents were encouraged to bring him in for further testing as he developed. Successive visits showed the child to be grossly mentally retarded, and his growth is abnormal. The mother mentioned that recently she joined Alcoholics Anonymous. From what condition might you suspect the child suffers?

3. Mrs. Duclos has had two previous miscarriages at about 4 months. She is currently beginning the second month of her third pregnancy, and (sadly) her doctor suspects slow placental development. For what hormones will her blood be tested and what is the correlation with the fourth month?

4. A pregnant woman complains of frequent urination and sweet-smelling urine. What hormone of pregnancy might cause this?

5. Cynthia Powers, in the second month of her first pregnancy, has lost a little weight due to a particularly miserable bout with morning sickness. Will she be given treatment? Why or why not?

6. Why are hemorrhoids a common problem after giving birth?

7. Mrs. Monica is experiencing Braxton Hicks contractions in the sixth month of her pregnancy. Why did her doctor prescribe ibuprofen (with what mechanism does this drug interfere)?

8. After assisting a woman during several hours of hard labor, the nurse notices meconium is being passed. What might this indicate concerning the presentation of the fetus?

9. A woman suffering from dystocia is taken to have her pelvis X rayed. The X ray shows that her ischial spines leave a space too narrow for passage of the fetal head. What birth procedure will be employed?

10. Billy, a full-term newborn, is cyanotic and hyperventilating. His blood gas measurement shows low oxygenation. Persistence of what fetal circulatory features might be suspected?

11. A nursing mother complains that painful menstrual-type cramps occur when her baby is suckling. Explain the positive aspect of this (uncomfortable) event.

12. Mrs. DeAngelo, a new mother, makes an appointment to bring her infant to the clinic because his skin is yellow. What is this condition and does it necessarily require medical intervention?

13. A pregnant woman has just arrived for a routine checkup in her seventh month. The doctor notices that her ankles are swollen and her diastolic blood pressure has risen 15 mm Hg. What condition might these signs suggest?

Stop and Think

1. What size will blastomeres be when cleavage stops?

2. Correlate the escape from the zona pellucida with the size of the preembryo.

3. How are the chances of a successful pregnancy affected by scarring of the uterine tube which slows down, but does not completely block, the passage of the developing morula?

4. Would a teratogen have more effect, less effect, or the same effect if exposure were during the second week of gestation compared to the eighth week? Explain your answer.

5. What causes the motility of the GI tract to decrease during pregnancy?

6. Many women find that their shoe size increases late in their pregnancy. What factors (including hormonal) might contribute to this phenomenon?

7. How can amniotic fluid volume indicate disorders of the fetal urinary system?

8. Does each somite develop into a vertebra? Why or why not?

9. Think about the terminology used for different types of epithelium. Which type would you suspect was originally thought to come from endoderm? From which germ layer does this type actually derive?

10. What fetal vessel(s) carries/carry the most oxygen-rich blood?

11. What *maternal* factor limits prenatal brain development?

12. What is the "blue" in the term *blue baby?*

13. If CPR were required for a near-term pregnant woman, would it be advisable to position her on her back? Why or why not?

14. What cell organelles will be particularly prominent in the cells of the membranes covering the oral and anal openings just prior to their perforation?

15. By what procedure was Julius Caesar born?

16. What circulatory changes associated with birth provide the pressure changes necessary to close the foramen ovale?

17. The citizens of Nukeville brought a class-action lawsuit against the local power company, which operated a nuclear power plant on the outskirts of town. The suit claimed that radiation leaks were causing birth defects. As evidence, they showed that of the 998 infants born in town since the plant opened, 20 had congenital abnormalities. Of these, 6 had heart defects, 3 were mentally retarded, 2 were born without a brain, 2 had spina bifida, 2 had Down syndrome, and 3 had cleft palate. Did the citizens have a case? (You may need to consult an outside reference for this one.) Explain.

COVERING ALL YOUR BASES

Multiple Choice

Select the best answer or answers from the choices given.

1. Which are known to be necessary for sperm activity?
   A. Activation by vaginal acid
   B. Loss of cholesterol from sperm cell membrane
   C. Increased fragility of sperm cell membrane
   D. Time (about 6–8 hours)

2. The acrosomal reaction:
   A. allows degradation of the corona radiata
   B. involves release of hyaluronidase
   C. occurs in the male urogenital tract
   D. involves only one sperm, which penetrates the oocyte membrane

3. Polyspermy is prevented by:
   A. depolarization of the oocyte membrane
   B. acrosomal reaction
   C. cortical reaction
   D. swelling of material between oocyte membrane and zona pellucida

4. The first mitotic division in the zygote occurs as soon as:
   A. male and female pronuclei fuse
   B. male and female chromosomes are replicated
   C. meiosis II in the oocyte nucleus is completed
   D. the second polar body is ejected

5. Which of the following result from cleavage?
   A. Daughter cells having a higher surface-to-volume ratio
   B. Blastomeres
   C. Morula
   D. Blastocyst

6. Which contain cells that ultimately become part of the embryo?
   A. Blastocyst   C. Cytotrophoblast
   B. Trophoblast   D. Inner cell mass

7. The blastocyst:
   A. is the earliest stage at which differentiation is clearly evident
   B. is the stage at which implantation occurs
   C. has a three-layered inner cell mass
   D. can detect "readiness" of uterine endometrium

8. Human chorionic gonadotropin is secreted by the:
   A. trophoblast   C. chorion
   B. 5-month placenta   D. corpus luteum

9. Which of the following is/are part of the placenta?
   A. Chorionic villi   C. Decidua
   B. Lacunae   D. Gastrula

10. A teratogen shown to affect limb bud differentiation is:
    A. alcohol   C. nicotine
    B. thalidomide   D. German measles

11. Which of the following are true of the gastrula?
    - **A.** It develops from the blastocyst.
    - **B.** Human chorionic gonadotropin is first produced.
    - **C.** The syncytiotrophoblast first appears.
    - **D.** Three primary germ layers become evident.

12. Amniotic fluid:
    - **A.** prevents fusion of embryonic parts
    - **B.** contains cells and chemicals derived from the embryo
    - **C.** is derived from embryonic endoderm
    - **D.** helps maintain a constant temperature for the developing fetus

13. The yolk sac:
    - **A.** is derived from the mesoderm of the embryonic disc
    - **B.** is pinched off as the embryonic gut is enclosed
    - **C.** is the site of origin of hemocytoblasts
    - **D.** is the structural framework of the umbilical cord

14. The primitive streak:
    - **A.** forms the notochord
    - **B.** becomes the neural groove
    - **C.** establishes the axis of the embryo
    - **D.** signals the onset of gastrulation

15. The notochord:
    - **A.** develops from the primitive streak
    - **B.** develops from mesoderm beneath the primitive streak
    - **C.** becomes the vertebral column
    - **D.** persists as the nucleus pulposis in the intervertebral discs

16. Which of the following give rise to epithelia?
    - **A.** Mesenchyme
    - **B.** Endoderm
    - **C.** Mesoderm
    - **D.** Ectoderm

17. At the end of the embryonic period:
    - **A.** the crown-rump measurement is about 50 cm (2 inches)
    - **B.** the embryo is about 8 weeks old
    - **C.** all adult organ systems are established
    - **D.** the placenta is secreting estrogens and progesterones

18. The first major event in organogenesis is:
    - **A.** gastrulation
    - **B.** appearance of the notochord
    - **C.** neurulation
    - **D.** development of blood vessels in the umbilical cord

19. Which of the following appears first in the development of the nervous system?
    - **A.** Neural crest cells
    - **B.** Neural folds
    - **C.** Neural plate
    - **D.** Neural tube

20. Which of these digestive structures develops from ectoderm?
    - **A.** Midgut
    - **B.** Liver
    - **C.** Lining of the mouth and anus
    - **D.** Lining of esophagus and pharynx

21. Mesodermal derivatives include:
    - **A.** somites
    - **B.** mesenchyme
    - **C.** most of the intestinal wall
    - **D.** sweat glands

22. Vertebrae arise from:
    - **A.** somites
    - **B.** sclerotome
    - **C.** myotome
    - **D.** intermediate mesoderm

23. Which contribute to endocrine glands?
    - **A.** Endoderm
    - **B.** Ectoderm
    - **C.** Intermediate mesoderm
    - **D.** Lateral mesoderm

24. Somatic mesoderm gives rise to:
   A. splanchnic mesoderm
   B. limb buds
   C. dermis of ventral body region
   D. dermatomes

25. Derivatives of splanchnic mesoderm include:
   A. cardiovascular system
   B. submucosa, muscularis, and serosa of the digestive and respiratory tracts
   C. most of the body's connective tissue
   D. long bones

26. Which of the following is a shunt to bypass the fetal liver?
   A. Ductus arteriosus
   B. Ductus venosus
   C. Ligamentum teres
   D. Umbilical vein

27. Maternal metabolism is stepped up by:
   A. relaxin
   B. somatomammotropin
   C. thyrotropin
   D. lactogen

28. The stage is set for the onset of labor by a(n):
   A. increase in oxytocin receptors in the myometrium
   B. decrease in progesterone production by the placenta
   C. release of prostaglandins from the placenta
   D. secretion of fetal oxytocin

29. The usual and most desirable presentation for birth is:
   A. vertex   **C.** nonvertex
   B. breech   **D.** head first

30. Which of the following are hormones associated with lactation?
   A. Placental lactogen   **C.** Prolactin
   B. Colostrum   **D.** Oxytocin

31. The outer layer of the blastocyst, which attaches to the uterine wall, is the:
   A. yolk sac   **C.** amnion
   B. inner cell mass   **D.** trophoblast

32. The neural crest:
   A. arises from ectoderm
   B. is inside the neural tube
   C. contributes to wandering mesenchyme
   D. produces some neural structures

33. Which of the following embryonic structures is/are segmented?
   A. Somites   **C.** Lateral plate
   B. Endoderm   **D.** Myotomes

34. When it is 1½ months old, an average embryo is the size of:
   A. Lincoln's image on a penny
   B. an adult's fist
   C. its mother's nose
   D. the head of a straight pin

Word Dissection

For each of the following word roots, fill in the literal meaning and give an example, using a word found in this chapter.

Word root	Translation	Example
1. bryo	_____	_____
2. chori	_____	_____

Word root	Translation	Example
3. coel	_____	_____
4. decid	_____	_____
5. episio	_____	_____
6. fetus	_____	_____
7. gest	_____	_____
8. nata	_____	_____
9. noto	_____	_____
10. partur	_____	_____
11. placent	_____	_____
12. terato	_____	_____
13. toc	_____	_____
14. troph	_____	_____
15. ula	_____	_____
16. zygot	_____	_____

30 Heredity

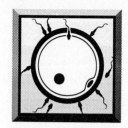

Student Objectives

When you have completed the exercises in this chapter, you will have accomplished the following objectives:

The Vocabulary of Genetics

1. Define *allele*.
2. Differentiate clearly between genotype and phenotype.

Sexual Sources of Genetic Variation

3. Describe events that lead to genetic variability of gametes.

Types of Inheritance

4. Compare and contrast dominant-recessive inheritance with incomplete dominance and co-dominance.
5. Describe the mechanism of sex-linked inheritance.

6. Explain how polygene inheritance differs from that resulting from the action of a single pair of alleles.

The Influence of Environmental Factors on Gene Expression

7. Provide examples illustrating how gene expression may be modified by environmental factors.

Nontraditional Inheritance

8. Describe how genetic imprinting and extrachromosomal (mitochondrial) inheritance differ from classical Mendelian inheritance.

Genetic Screening, Counseling, and Therapy

9. List and explain several techniques used to determine or predict genetic diseases.
10. Describe briefly some approaches of gene therapy.

Ready, set, go!

At conception, the fertilized human egg is so small that it is just visible to the human eye. Yet it contains enough information to guide the development of the zygote from a single cell to an adult body composed of over 50 trillion cells, and on through middle age, old age, and death. The developmental information is stored in the genes, the chemical units of DNA that code and guide the production of polypeptide chains or proteins, which in turn establish the individual's basic appearance and behavior.

Mitotic cell division and the replication of DNA that precedes it occur with fantastic accuracy to ensure that the information present in the zygote is unchanged as it passes from cell to cell throughout life. The transfer of genetic information from generation to generation, however, fosters variation. Production of gametes by a parent results in cells containing 23 chromosomes each, but the genetic content of these cells differs. Therefore, the resulting zygote inherits a unique combination of genetic instructions from both of its parents that guides its development and growth and sets its individual inherited characteristics. These cellular transactions are the basis of heredity.

Topics in Chapter 30 include the terminology and principles of inheritance, the causes of variation among familial offspring, nongenetic factors influencing heredity, and selected applications of the science of genetics.

BUILDING THE FRAMEWORK

The Vocabulary of Genetics

1. Circle the term that does not belong in each of the following groupings.

   1. Genotype   *FFXX*   *FfXX*   Freckled female

   2. Identical chromosomes   Chiasma   Meiosis I   Crossing over

   3. *AA*   *aa*   Heterozygous   Homozygous

   4. *Ff*   Phenotype   Freckles   Flat feet

   5. *GG*   *Hh*   Recessive is expressed   Dominant is expressed

   6. 2^n Number of different gamete types   Haploid   Diploid

   7. Karyotype   Micrograph   Haploid   Arranged in homologous pairs

   8. Same locus   Homologous chromosomes   Alleles   Code for different traits

2. If a statement is true, write the letter T in the answer blank. If a statement is false, change the underlined word(s) and write the correct word(s) in the answer blank.

_____   **1.** Sex chromosomes are not homologous in a male.

_____   **2.** Genomes code for alternate forms of the same trait.

_____   **3.** _HbFFXX_ is a phenotype.

_____   **4.** The probability of the recessive phenotype being expressed in the offspring of heterozygous parents is one-fourth.

_____   **5.** Sex-linked traits are located on the same chromosome and are separated only by crossover.

_____   **6.** A phenotype in between the dominant and recessive phenotypes indicates codominance.

_____   **7.** A phenocopy results from environmental factors.

Sexual Sources of Genetic Variation

1. Figure 30.1A is a diagram of one of the possible metaphase I alignments of the cell under consideration in which the diploid chromosomal complement is 6. The dashed line indicates the metaphase plate. Shaded chromatids or chromosomes indicate paternal inheritance. Unshaded chromatids or chromosomes indicate maternal inheritance. Centromeres are shown as black dots. The smaller circles to the right (B–F) represent gametes that _may or may not_ result from the possible metaphase I alignments of this cell.

Select two colors and color the coding circles and the structures as they appear in all the diagrams, A–I. Then complete the statements that follow by writing your answers in the answer blanks.

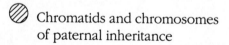

Chromatids and chromosomes   ◯ Chromatids and chromosomes
  of paternal inheritance   of maternal inheritance

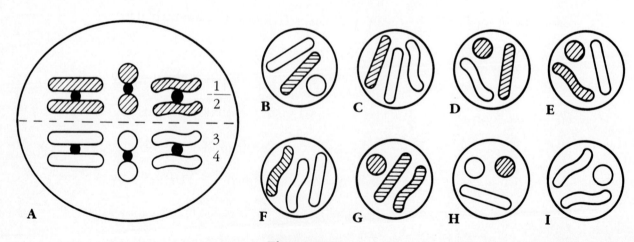

Figure 30.1

_____ 1. Of gametes B–I, only __(1)__ could result from the metaphase I alignment illustrated in part A.

_____ 2. Diagrams __(2)__ correctly represent possible different types of gametes produced from any normal metaphase I alignment of this cell.

_____ 3. In diagram A, crossing over could occur between the chromatids numbered __(3)__ .

_____ 4. In each correctly drawn gamete, the number of alleles for each trait is __(4)__ .

2. Name three normal cellular events that significantly contribute to genetic variation in offspring of the same parents.

1. _____

2. _____

3. _____

Types of Inheritance

1. Using the key choices, select the type of inheritance that matches each description or example given below.

KEY CHOICES

A. Codominance   C. Incomplete dominance   E. Polygenic

B. Dominant-recessive   D. Multiple alleles   F. Sex-linked

_____ 1. Alleles are on the sex chromosomes

_____ 2. Inheritance of sex

_____ 3. Phenotype varies between two extremes of a trait

_____ 4. Blood types A, B, AB, and O

_____ 5. Expression of a recessive allele in the presence of a dominant gene

_____ 6. Albinism, flat feet, and tight thumb ligaments

_____ 7. Hemophilia, red-green color blindness, and hairy ear pinnas

_____ 8. Recessive allele not expressed if a dominant gene is present

_____ 9. Eye color and height

2. Figure 30.2 is a drawing of Peter. Arrows indicate five of his inherited traits. At each arrow, letters designate Peter's alleles for that trait. Select two different colors and color the coding circles and Peter's dominant and recessive features. Then answer the questions that follow.

◯ Peter's dominant traits   ◯ Peter's recessive traits

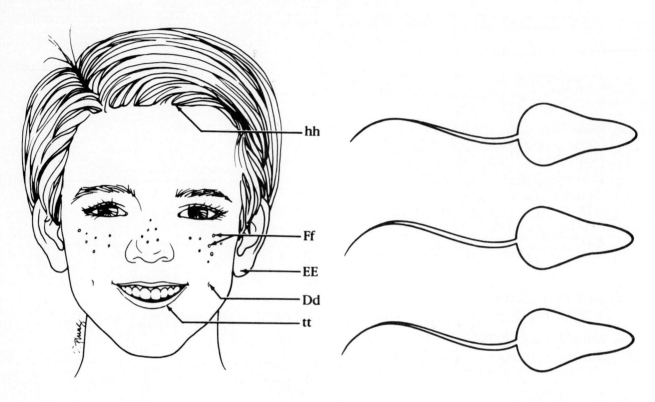

hh

Ff

EE

Dd

tt

Figure 30.2

1. What is Peter's phenotype relative to these five traits? _____

_____

2. What is Peter's genotype for these traits? _____

3. Could Peter have a child with (1) attached earlobes? _____   (2) no freckles? _____

4. The three "tailed" ovals in Figure 30.2 represent sperm. Using the ovals as answer spaces, write the genetic content of three possible types of Peter's sperm that could result in a daughter.

3. In the Punnett square shown in Figure 30.3, circles represent eggs and ovals with tails represent sperm. Consider Andrew and Ella, both of whom have normal foot arches. Their son Bart was born with flat feet, a recessive Mendelian trait. Using F and f to designate the alleles for this trait, complete the Punnett square. Then answer the questions that follow.

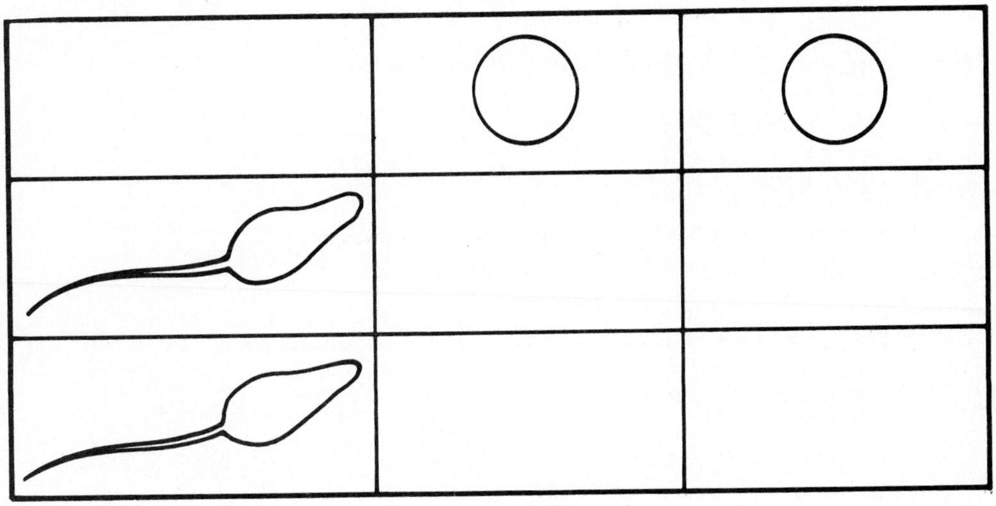

Figure 30.3

_____ **1.** What is the probability that this couple's next child will have flat feet?

_____ **2.** If Bart wants all his children to have normal foot arches, what genotype should he prefer in a mate?

4. Red-green color blindness is a recessive, sex-linked trait, and the alleles are on the X chromosomes. Marjorie and her parents have normal vision, but her brother Leo is color-blind. What are the genotypes of Marjorie, her mother, her father, and Leo? Use the letters B and b to represent the alleles. Indicate alleles _and_ chromosomes. (_Hint:_ Don't forget the X and Y chromosomes!)

1. Marjorie _____   **3.** Father _____

2. Mother _____   **4.** Leo _____

5. Trisha, who has blood type A ($I^A i$), has a baby out of wedlock. The baby has blood type O. Trisha is suing Eddie for the support of this baby. On the basis of his blood type, what proof could Eddie show that he is _not_ the father? Give all possible genotypes that would free Eddie from paternity liability.

_____

_____

6. Sickle-cell anemia is a recessive trait that exhibits incomplete dominance. Using *A* and *a* to designate the alleles, give the genotypes for the following phenotypes.

_____ **1.** Full-blown sickle-cell anemia

_____ **2.** Healthy but subject to crisis if oxygen is lacking

_____ **3.** Normal with no trace of sickling cells

_____ **4.** A carrier with the sickle-cell trait

7. The range of skin color in humans results from polygenic inheritance. The three genes responsible for skin color have alleles designated *A,a; B,b;* and *C,c.* Suppose two brown-skinned parents have a white-skinned child.

1. What is the genotype of the child? _____

2. What is a possible genotype of the mother? _____

3. What is a possible genotype of the father? _____

4. Is the child an albino? Why or why not? _____

_____

_____

5. Could these parents have a child with skin color darker than theirs? Explain.

_____

_____

The Influence of Environmental Factors on Gene Expression

1. If a statement is true, write the letter T in the answer blank. If a statement is false, change the underlined word(s) and write the correct word(s) in the answer blank.

_____ **1.** The direct effect of radiation on heredity can be the alteration of <u>gene expression</u>.

_____ **2.** Crossover of homologous chromosomes is <u>an abnormal</u> event that influences gene distribution.

_____ **3.** Environmental factors such as the maternal ingestion of alcohol or thalidomide can alter <u>normal gene expression</u> of the embryo.

_____ **4.** The mother's <u>intelligence</u> directly affects fetal brain growth and development.

_____ **5.** A deficiency of <u>exercise</u> during adolescence can lead to abnormally stunted growth, a phenotype not dictated by genes coding for height.

Nontraditional Inheritance

1. Decide whether each of the following phrases applies to *A,* genetic imprinting, or to *B,* extrachromosomal inheritance.

_____ **1.** A consequence of reversible methylation of the DNA in gametes

_____ **2.** May cause a form of mental retardation called the fragile X syndrome

_____ **3.** Characteristics transmissible by mitochondrial genes

_____ **4.** Expression of the transmitted genes varies depending on whether the genes are paternal or maternal in origin

_____ **5.** Essentially all genes inherited by this pathway are maternal genes

_____ **6.** May account for some cases of Parkinson's disease

_____ **7.** Responsible for Prader-Willi and Angelman syndromes

Genetic Screening, Counseling, and Therapy

1. Complete the following statements by writing the missing terms in the answer blanks.

_____ **1.** The diagram of an individual's diploid chromosomal complement is called a __(1)__ .

_____ **2.** Down syndrome can be detected in the fetal karyotype because an extra __(2)__ is present.

_____ **3.** Babies at risk for Down syndrome are usually born to mothers who are over __(3)__ years old.

_____ **4.** An individual with one recessive gene, such as that causing cystic fibrosis, is called a __(4)__ .

_____ **5.** If both parents are heterozygous for a recessive genetic abnormality, the child's chance of being homozygous recessive is __(5)__ %.

_____ **6.** __(6)__ is the method used to trace a particular genetic trait through several family generations.

_____ **7.** Screening for detection of the sickling gene in heterozygotes is accomplished by __(7)__ .

_____ **8.** The most common type of fetal testing for genetic abnormalities is called **(8)** .

_____ **9.** The type of testing identified in Question 8 is not usually done unless there is risk of a known **(9)** .

_____ **10.** Two disadvantages of amniocentesis are: the procedure is not usually done before the **(10)** week of pregnancy, and fetal

_____ **11.** cells must be **(11)** for several weeks before they can be karyotyped.

_____ **12.** Another fetal testing procedure entails removal of rapidly dividing cells from the **(12)** of the placenta. This sampling

_____ **13.** procedure can be done as early as **(13)** weeks into the pregnancy.

2. Figure 30.4 shows a pedigree of a family some of whose members are afflicted with a rare genetic enzyme defect that results in death by age 12. After using the key provided to examine the pedigree, answer the questions below in the answer blanks provided.

☐ Normal male   ▲ Afflicted male

◯ Normal female   ⬓ Afflicted female

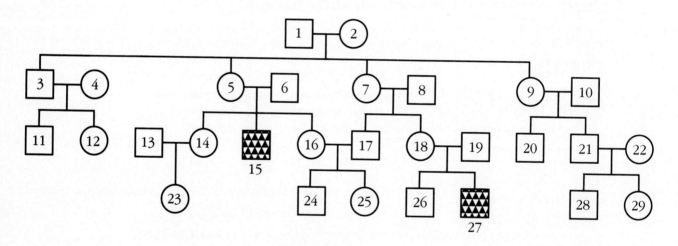

Figure 30.4

_____ **1.** Is the trait sex-linked?

_____ **2.** Is the trait dominant or recessive?

_____ **3.** Identify, by number, any known carriers.

_____ **4.** Identify, by number, any other possible carriers.

_____ **5.** Are there any afflicted females?

_____   **6.** What term is used for the marriage between #16 and #17?

_____   **7.** Do the offspring of #16 and #17 have an increased risk of inheriting the affliction?

Give the probability, as a percentage, that each of the following individuals will carry the recessive allele:

_____ **8.** #2   _____ **9.** #3   _____ **10.** #5   _____ **11.** #14   _____ **12.** #23

CHALLENGING YOURSELF

At the Clinic

1. A young couple, both of whom have achondroplasia, are at the clinic for counseling. The wife has already had two miscarriages. What is the most likely explanation for this? What are their chances of having a normal child?

2. Andre, a young man whose father and maternal grandmother were both stricken with Huntington's disease, has come in for genetic counseling. Does inheritance from both sides of his family increase his risk for inheriting this disorder? Why or why not? Without knowing whether or not he carries the gene, predict the probability that his children will inherit the disorder.

3. Sherry Long, an albino, is marrying a man with no history of albinism in his family. She makes an appointment for counseling because she wants to know the risk of passing the condition on to her children. What is the probability that a child from this marriage will be albino?

4. The Agassis have one son with cystic fibrosis. Mrs. Agassi is pregnant, and she wants to know what the chances are of the next child also being stricken. What would you tell her?

5. The parents of 8-month-old Seymour have brought him to the clinic because he has been having seizures for the past several months. Tests show failing vision and impaired development of his mental and physical abilities. The father relates that Seymour had a brother who died at age 3. What information will you seek from the couple about their ethnic heritage and why?

6. A young boy is stricken with sickle-cell anemia, but his younger brother appears healthy. What precautions and possible limitations will you suggest concerning the younger boy's activity level?

7. Brian is a diagnosed hemophiliac, and his parents are concerned about the possibility of their next child also suffering from this condition. There is a history of hemophilia in the mother's family, but none in the paternal line. The father is normal. Brian's mother is pregnant again, and a sonogram shows that the fetus is a girl. What are the chances of this child's having hemophilia?

8. Blood tests eight days after birth show an infant to have PKU. Is this a genetic disorder? What would you tell the parents about the child's expectations of living a normal life?

9. The Caserios, a middle-aged couple who just got married, visit the clinic to discuss their chances of having a child with Down syndrome. The woman is 40 and the man is 45. Can you predict the probability of this affliction in the same way as for Tay-Sachs disease or Huntington's disease? What test can be done during pregnancy to determine whether the fetus has Down syndrome? What would the karyotype show?

10. Ms. Thurston, a 40-year-old oil company executive, has learned she is pregnant, after accepting a one-year assignment to Saudi Arabia. She is concerned about the risk of Down syndrome, but she will be leaving before the fetus has developed enough so that amniocentesis will be possible. What procedure will you recommend and how long will the woman have to wait before it can be carried out?

Stop and Think

1. Technology now enables discrimination between maternal and paternal inheritance in cases such as trisomy 21. Studies show that the extra chromosome 21 does not always come from the oocyte. What other explanation is possible for the increased incidence of Down syndrome with increasing maternal age?

2. Since the contribution of the father is the sperm nucleus only, what is the source of all mitochondria (and their DNA)? How can this help trace lineages?

3. By what mechanism can an XX genetic female have genetic characteristics of a male? Likewise, how can an XY genetic male have no male characteristics?

4. What effect does the degree of adjacency of gene locus have on the probability that a crossover will occur between them?

5. A blood test is available to determine whether a child has inherited Huntington's disease. Why is the blood test (rather than symptomatic diagnosis) preferred as a means of identification of heterozygotes for this condition? What would a genetic counselor recommend for a child who has inherited this gene? Is it likely that the test would find a homozygous dominant individual? Why or why not?

6. Victoria is a woman in her 40s who is proud to be called a "woman in transition" on her college campus. She is surprised to learn that she has one variety of red-green color blindness when that topic is covered in her anatomy and physiology lab. She thought only males could be color-blind. Her father, her brother, and her mother's uncle are all color-blind. Explain how her genetic heritage occurred.

COVERING ALL YOUR BASES

Multiple Choice

Select the best answer or answers from the choices given.

1. The number of sex chromosomes in a diploid cell is:

   A. 44   **C.** 2

   B. 22   **D.** 1

2. Which of the following is true concerning homologous chromosomes?

   A. They share matched genes at corresponding loci.

   B. Their alleles are identical.

   C. They are homozygous.

   D. They are found together in haploid cells.

3. Dominant traits:

   A. are more common than recessive traits

   B. are found only in the absence of polygenic inheritance

   C. are not possible in cases of multiple alleles

   D. are expressed in heterozygous individuals

4. Which of the following terms is/are descriptive of a genotype?

   A. Dominant   **C.** Homozygous

   B. Polygenic   **D.** Carrier

5. Which of these phenotypes are seen only with the heterozygous genotype?

   A. Achondroplasia

   B. Blood type AB

   C. Schizophrenia

   D. Sickle-cell trait

6. Segregation is associated with:

   A. recombinant chromosomes

   B. independent assortment

   C. crossover

   D. random fertilization

7. Independent assortment in humans results in how many different kinds of gametes?

   A. 2^{46} **C.** 2×46

   B. 2^{23} **D.** 2×23

8. Linked genes are separated by:

   A. crossover

   B. independent assortment

   C. meiosis II

   D. segregation

9. Crossover:

   A. occurs during meiosis I

   B. is associated with synapsis

   C. results in the formation of novel genes

   D. involves only two members of a tetrad

10. In which of the following crosses is there a 50% probability of heterozygous genotype in the offspring?

   A. $Aa \times aa$ **C.** $Aa \times aa$

   B. $AA \times Aa$ **D.** $Aa \times Aa$

11. Phenotypes associated with dominant alleles are:

   A. blood type A

   B. syndactyly

   C. cystic fibrosis

   D. sickle-cell trait

12. The most common blood type is:
    A. type A   C. type AB
    B. type B   D. type O

13. Genetic conditions that impair brain function include:
    A. PKU
    B. achondroplasia
    C. cretinism
    D. Tay-Sachs disease

14. Consanguineous marriages:
    A. increase the risk of passing on dominant alleles
    B. increase the chance of genetic mutation
    C. increase the chance of inheriting two recessive alleles
    D. always result in genetic defects

15. Which of the following is/are examples of multiple-allele inheritance?
    A. Skin pigmentation
    B. Eye color
    C. Blood type
    D. Sickle-cell anemia

16. X-linked traits include:
    A. hemophilia
    B. hairy ears
    C. red-green color blindness
    D. astigmatism

17. Polygenic inheritance:
    A. involves codominant alleles
    B. involves more than one locus
    C. results in continuous variations of the phenotype
    D. can result in children with a more pronounced phenotype than either parent

18. Examples of phenocopies affecting height include:
    A. pituitary giantism
    B. cretinism
    C. fetal alcohol syndrome
    D. achondroplasia

19. A pedigree analysis of a hemophiliac shows that he shares this condition with his maternal uncle, great uncle, and cousin. From this information, which of these statements is true?
    A. His cousin is the child of his mother's brother.
    B. His cousin is the child of his father's sister.
    C. His cousin is male.
    D. His mother's sister is a carrier.

20. Which of the following conditions can be detected by amniocentesis?
    A. Nondisjunction
    B. Deletion
    C. Monosomy
    D. Some enzyme defects

Word Dissection

For each of the following word roots, fill in the literal meaning and give an example, using a word found in this chapter.

Word root	Translation	Example
1. allel	_____	_____
2. centesis	_____	_____
3. chiasm	_____	_____

Word root	Translation	Example
4. gen	_____	_____
5. karyo	_____	_____
6. pheno	_____	_____
7. sanguin	_____	_____

Epilogue: A Day in the Life

Surprise—no preprogrammed exercises for you to do for this chapter! But that doesn't mean you have nothing to do. Some students become frustrated about the wealth of information in an A&P course and have problems seeing relationships between the body's parts and its activities. Perhaps you sometimes feel the same way, too. This is not surprising, considering that we instructors teach in manageable bits and pieces. This approach is necessary because it is impossible to swallow any subject, including A&P, whole. However, the realization that sometimes "the pieces" didn't get put back together into a working understanding of the human body encouraged me to write the Epilogue. I hope that when my students complete the course they can get a real sense of success about what they actually know and how much they can apply from their A&P studies. This is important. You've been learning about your own body, not some inanimate object, and the more you know about it, the better—for both your own health and that of your patients, should you enter the health services field.

After completing the Epilogue, I realized it could be used as another learning tool. For example:

1. It provides a guide for your studying by spelling out what physiological events are occurring during everyday activities and situations, like riding a bicycle. It is certainly clear that riding a bike involves the muscular system (skeletal muscles), but the Epilogue also points out how the nervous system, cardiovascular system, respiratory system, and others are indispensible to muscle activity and maintaining homeostasis.

2. You can expand your understanding of the events described in the Epilogue by figuring out what would happen if one or more of the variables in certain events were changed. For example, what would happen to John's equilibrium if he had more than the one beer? What would happen to his immune system and skin cells if he stayed out all day on the beach instead of just part of the day?

3. Use the type of approach in the Epilogue to examine the impact of what you are doing to your own physiology. For instance, work out the pathways of physiological events and the various system connections involved when you are performing with your town's theater group, swimming, or just relaxing. Or ask yourself questions. "What if I did 75 crunches instead of 20? What would change, and how? What if I ate 5 jelly doughnuts instead of cereal and fruit for breakfast? What effect would that have on my endocrine system, metabolism, and weight?"

Learning is most meaningful when you can apply it to yourself. That's just what the Epilogue will help you to do if you let it—it is meant to be a fun and personal approach to learning that helps you see and understand the body as a functioning whole.

Appendix A
Answers

Chapter 1 The Human Body: An Orientation

BUILDING THE FRAMEWORK

An Overview of Anatomy and Physiology

1. 1. Regional anatomy: All structures (bones, muscles, blood vessels, etc.) in a particular body *region* studied at the same time. Systemic anatomy: All organs of an organ system studied simultaneously; for example, if one is studying the skeletal system, all the bones of the body are studied.   2. Gross anatomy: Studies easily visible structures (bones, muscles, etc.). Microscopic anatomy: Studies structures that cannot be viewed without a microscope—for example, cells and tissues.   3. Developmental anatomy: Studies the changes in body structure that occur throughout life. Embryology: Considers only those changes that occur from conception to birth.   4. Histology: Study of tissues, collections of cells with similar structure and function. Cytology: Study of parts of a cell and the functions of those parts.

2. Physiological study: C, D, E, F, G, H, J   Anatomical study: A, B, I, K, L, M

3. complementarity

The Hierarchy of Structural Organization

1. cells, tissues, organs, organ systems

2. 1. atom   2. epithelium   3. heart   4. digestive system

3.

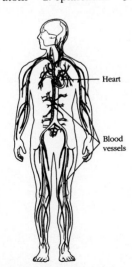

Figure 1.1
Cardiovascular system

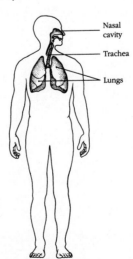

Figure 1.2
Respiratory system

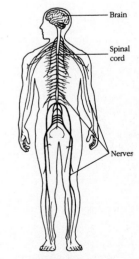

Figure 1.3
Nervous system

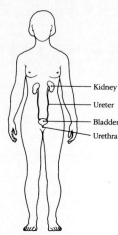

Figure 1.4
Urinary system

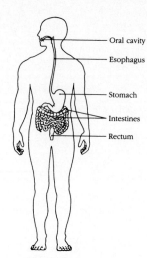

Figure 1.5
Digestive system

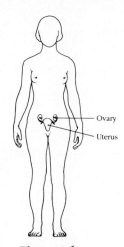

Figure 1.6
Reproductive system

4. 1. K   2. C   3. J   4. A   5. D   6. E   7. B   8. I   9. A   10. F   11. K   12. C, H   13. C   14. D
5. 1. A   2. C   3. K   4. H   5. B   6. J   7. G

Maintaining Life

1. 1. D   2. H   3. C   4. A   5. B   6. G   7. F   8. E   9. D
2. 1. C   2. B   3. E   4. D   5. E   6. A

Homeostasis

1. A dynamic state of equilibrium in which internal conditions vary within narrow limits.
2.

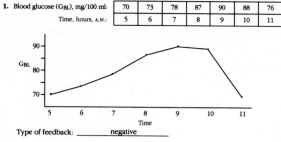

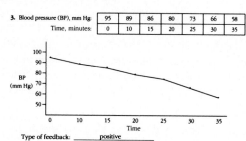

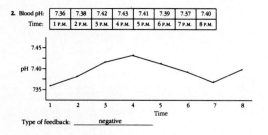

The Language of Anatomy

1. 1. A   2. G   3. D   4. D   5. F   6. A   7. F   8. H   9. B   10. A   11. D   12. I   13. K   14. C   15. K   16. J
17. A
2. 1. elbow   2. shoulder   3. forehead   4. muscles   5. knee   6. small intestine
3. 1. distal   2. antecubital   3. brachial   4. left upper quadrant   5. ventral cavity
4. 1. B   2. H   3. I   4. F   5. P   6. G   7. L   8. J   9. K   10. D

5. The hands are incorrectly depicted; the palms should face anteriorly.   **Section A:** midsagittal   **Section B:** transverse

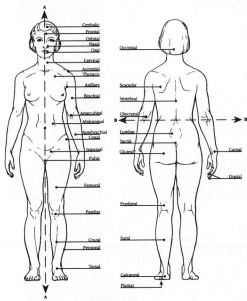

Figure 1.7

6. 1. ventral   2. dorsal   3. dorsal
7. 1. pleura   2. pericardium   3. peritoneum
8. 1. A   2. A   3. A   4. A   5. A   6. C   7. A   8. D   9. D   10. B   11. A   12. A
9. 1. 2, 3, 7, 11, 12   2. 2, 3   3. 2   4. 1, 2, 3, 5   5. 2, 3
10.

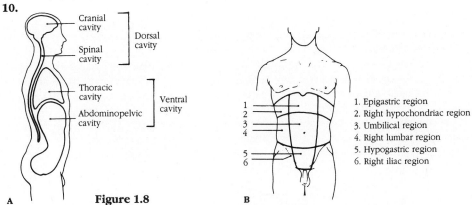

Figure 1.8

1. Epigastric region
2. Right hypochondriac region
3. Umbilical region
4. Right lumbar region
5. Hypogastric region
6. Right iliac region

11.

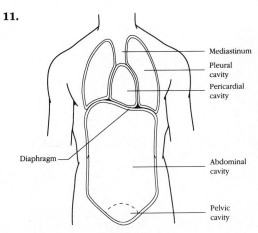

Figure 1.9

12. 1. D   2. E   3. B   4. A   5. A

CHALLENGING YOURSELF

At the Clinic

1. skeletal, muscular, circulatory, integumentary, nervous
2. the need for nutrients and water
3. The high blood pressure increases the work load on the heart. Circulation of blood decreases, and the heart itself begins to receive an inadequate blood supply. As the heart weakens further, the backup in the veins worsens and the blood pressure rises even higher. Without intervention, circulation becomes so sluggish that organ failure sets in. A heart-strengthening medication will increase the force of the heartbeat so that more blood is pumped out with each beat. More blood can then flow into the heart, reducing backflow and blood pressure. The heart can then pump more blood, further reducing the backup and increasing circulation. The blood supply to the heart musculature improves and the heart becomes stronger.
4. Fever, the elevation of body temperature, helps the immune system ward off infection, apparently by inhibiting the reproduction of infectious organisms. The stimulus (presence of infectious agents) triggers a response (fever) that reduces the stimulus (presence of foreign invaders). Thus, the temporary increase in body temperature returns the body to homeostasis.
5. right side, below the rib cage
6. 1 (d), 2 (b), 3 (a), 4 (c)
7. The physician will probably use ultrasound because it will provide the information desired (approximation of head size) with the least danger to the fetus.
8. Perhaps the best choice—CT scans, which can localize structures of the brain very precisely—is not one of the options. Of the options given, MRI is the best choice because it visualizes gray and white matter of the brain very clearly.
9. The anterior and lateral aspects of the abdomen have no bony (skeletal) protection.

Stop and Think

1. 1. protection   2. absorb light   3. absorption   4. thick, hollow, muscular organ   5. thin, elastic tissue
2. integumentary—boils, acne, athlete's foot; digestive—gingivitis, colitis, gastritis; respiratory—cold and flu; urinary—cystitis and kidney infections; reproductive—sexually transmitted diseases
3. maintenance of boundaries, movement, responsiveness, digestion, metabolism, excretion, and growth
4. nutrients, oxygen, and water; nutrients can be stored
5. Negative means opposite: a parameter that moves away from its optimum value will trigger changes that move it in the opposite direction, back toward its optimum.
6. A serous membrane forms as a hollow sphere that wraps around an organ and then continues on to cover the walls of the cavity that houses the organ. Since the organ does not puncture through its covering serous membrane to gain entry to the space between the two serous membrane layers, the result is a double membrane with fluid between the layers. The layers slide over each other with very little friction.
7. Anatomists describe positions of body parts with extreme precision; the smaller, more numerous abdominopelvic regions localize the parts very accurately. Medical personnel are often rushing to locate an injury and often rely on the patient's description of the site of pain. Quadrants give sufficient information and take less time to describe.
8.

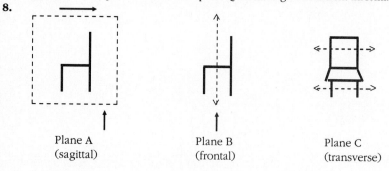

Plane A
(sagittal)

Plane B
(frontal)

Plane C
(transverse)

frontal; sagittal; transverse
9. CT and DSA utilize X rays. MRI employs radio waves and magnetic fields. PET uses radioisotopes. CT, MRI, and PET scans can display body regions in sections.

COVERING ALL YOUR BASES

Multiple Choice

1. D   2. E   3. B   4. C   5. D   6. A, B, C, D, E   7. C   8. C   9. A, B, C, D   10. A, D   11. B   12. A   13. B
14. B   15. B   16. C   17. C, E   18. A   19. D   20. B   21. C   22. C   23. A   24. D   25. A, C, E
26. B, C, D, E   27. A, B, D   28. A, B, D   29. D   30. D

Word Dissection

	Word root	Translation	Example		Word root	Translation	Example
1.	ana	apart	anatomy	**12.**	ology	study of	physiology
2.	chondro	cartilage	hypochondriac	**13.**	org	living	organism
3.	corona	crown	coronal	**14.**	para	near	parasagittal
4.	cyto	cell	cytology	**15.**	parie	wall	parietal
5.	epi	upon	epigastric	**16.**	pathy	disease	pathology
6.	gastr	stomach, belly	hypogastric	**17.**	peri	around	pericardium
7.	histo	tissue	histology	**18.**	stasis	standing still	homeostasis
8.	homeo	same	homeostasis	**19.**	tomy	to cut	anatomy
9.	hypo	below, under	hypogastric	**20.**	venter	belly	ventral
10.	lumbus	loin	lumbar	**21.**	viscus	organ	visceral
11.	meta	change	metabolism				

Chapter 2   Chemistry Comes Alive

BUILDING THE FRAMEWORK

PART I: BASIC CHEMISTRY

Definition of Concepts: Matter and Energy

1.   1. B, D   2. A, B, C, D   3. A, B
2.   1. C   2. B, D   3. C   4. A   5. D

Composition of Matter: Atoms and Elements

1.

	Particle	Location	Electrical charge	Mass
	Proton	Nucleus	+1	1 amu
	Neutron	Nucleus	0	1 amu
	Electron	Orbitals	−1	0 amu

2.   1. O   2. C   3. K   4. I   5. H   6. N   7. Ca   8. Na   9. P   10. Mg   11. Cl   12. Fe
3.   1. E   2. F   3. C   4. B   5. B   6. D   7. A   8. G   9. I   10. J   11. H   12. I
4.   1. Ca, P   2. C, O, H, N   3. Fe   4. Ca, K, Na   5. I   6. P   7. Cl

How Matter Is Combined: Molecules and Mixtures

1.   1. Molarity   2. Mixture   3. Solution   4. Suspension   5. Colloid   6. Colloid   7. Suspension
2.   1. H_2O_2 is a molecule of a compound. $2OH^-$ are two negative ions.   2. $2O^{2-}$ are two negative ions. O_2 is a molecule of an element.   3. $2H^+$ are two positive ions. H_2 is a molecule of an element.
3.   1. water = 18   2. urea = 32   3. carbonic acid = 62
4.   1. Perspiration   2. $Ca_3(PO_4)_2$ 3. Scatters light   4. Water   5. One-molar NaCl   6. Salt water

Chemical Bonds

1.   1. 6   2. 12 amu   3. carbon   4. 4   5. ±4   6. active   7. isotope   8. radioisotope   9. true

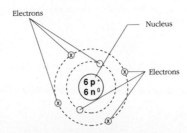

Figure 2.1

2.   A. Bond = electrovalent (ionic); name of compound = LiF (lithium fluoride)
           B. Bond = covalent; name of compound = HF (hydrogen fluoride)

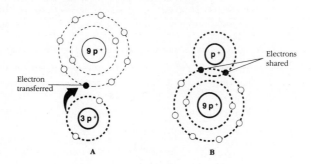

Figure 2.2

3.

Chemical symbol	Atomic number	Atomic mass	Electron distribution
H	1	1	1, 0, 0
C	6	12	2, 4, 0
N	7	14	2, 5, 0
O	8	16	2, 6, 0
Na	11	23	2, 8, 1
Mg	12	24	2, 8, 2
P	15	31	2, 8, 5
S	16	32	2, 8, 6
Cl	17	35	2, 8, 7
K	19	39	2, 8, 8, 1
Ca	20	40	2, 8, 8, 2

4.   (Table completion: Atoms forming ionic bonds)

Chemical symbol	Loss/gain of electrons	Electrical charge
H	Loses 1	+1
Na	Loses 1	+1
Mg	Loses 2	+2
Cl	Gains 1	−1
K	Loses 1	+1
Ca	Loses 2	+2

5.   (Table completion: atoms that form covalent bonds)

Chemical Symbol	Number of electrons shared	Number of bonds made
H	1	1
C	4	4
N	3	3
O	2	2

6.

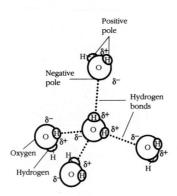

Figure 2.3

7.   Circle B, C, E

Chemical Reactions

1.   1. A   2. C   3. B   4.C and D

2.   1. low reactant concentration   2.large particles   3. reactant   4. electron acceptor   5. simple → complex

PART II: BIOCHEMISTRY: THE COMPOSITION AND REACTIONS OF LIVING MATTER

1.   Organic: 3, 5, 6, 8

Inorganic Compounds

1.   1. heat capacity   2. water (aqueous solution)   3. 70% (60% to 80%)   4. hydrogen bonds   5. heat of vaporization
6. hydrolysis (digestion)   7. synthesis (dehydration)

2.   1. Milk of magnesia   2. Urine (pH 6.2)   3. $NaHCO_3$ 4. Organic   5. pH 4   6. pH 2

3.   1. A   2. B   3. D   4. B   5. A   6. D   7. D   8. A   9. C   10. B

4.   a measure of acidity or H+ ion concentration

5.   Weak acid: B, C, E; strong acid: A, D, E, F, G

Organic Compounds

1.   1. T   2. neutral lipids   3. T   4. polar   5. T   6. ATP   7. T   8. peptide   9. glucose   10. T   11. phosphate

2.   1. G   2. D, E   3. A   4. F   5. H   6. B   7. C   8. G   9. C   10. A   11. F   12. F   13. C   14. H   15. B   16. C
17. H   18. C

3.   1. T   2. substrates   3. -ase   4. T   5. decrease   6. T   7. secondary and possibly primary

4.   1. starch or glycogen   2. A, E   3. tertiary   4. C, D   5. E
    Figure 2.4: A. monosaccharide   B. globular protein   C. polysaccharide   D. fat   E. nucleotide

5.   1. C   2. C   3. B   4. A

Figure 2.5

6.   X all but #3.

7.

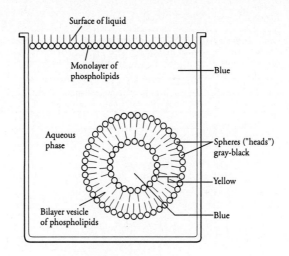

Figure 2.6

8.   Unnamed nitrogen bases: thymine (T) and guanine (G)   1. hydrogen bonds   2. double helix   3. 12
4. complementarity

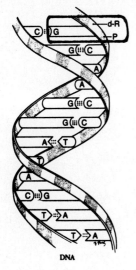

Figure 2.7

9.   1. Glucose   2. Ribose   3. Glycogen   4. Glycerol   5. Glucose   6. Hydrolysis   7. Carbon   8. Denaturation
10.   A. cyclic AMP   B. ATP

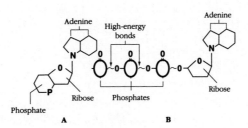

Figure 2.8

11.

Name of protein?	Secondary or tertiary structure?	Molecular shape: fibrous or globular?	Water-soluble or insoluble?	Function of protein?	Location in body?
Collagen	Secondary	Fibrous	Insoluble	Structure	Cartilage and bone
Proteinase	Tertiary	Globular	Soluble	Enzyme-hydrolysis	Stomach
Hemoglobin	Quaternary	Globular	Soluble	Carries O_2	Red blood cells
Keratin	Secondary	Fibrous	Insoluble	Structure; also prevents water loss	Skin and hair

The Incredible Journey

1.   1. negative   2. positive   3. hydrogen bonds   4. red blood cells   5. protein   6. amino acids   7. peptide   8. H^+ and OH^- 9. hydrolysis or digestion   10. enzyme   11. glucose   12. glycogen   13. dehydration or synthesis   14. H_2O 15. collagen   16. hydration (water)   17. increase   18. evaporate   19. thermal   20. vaporization

CHALLENGING YOURSELF

At the Clinic

1.   Acidosis means blood pH is below the normal range. The patient should be treated with something to raise the pH.

2.   Lack of sufficient ventilation would cause CO_2 to accumulate in the blood, thus increasing the amount of carbonic acid formed. This addition of acid would decrease the blood pH, causing acidosis.

3.   As hydrogen ion concentration builds, causing acidosis, Hugo will hyperventilate. This will lower the amount of CO_2 in the blood and reduce the amount of carbonic acid formed. The decrease in carbonic acid will help offset the increase in acid retention by his diseased kidneys.

4.   An inability to absorb fat will reduce vitamin D absorption from the diet. Since cholesterol absorption also will be reduced, the amount of vitamin D formed in the body may also decrease. Without sufficient levels of vitamin D (which aids calcium uptake by intestinal cells), calcium absorption will also decrease. Consequently, the bones will lose calcium, resulting in soft bones.

5.   Each of the 20 amino acids has a different chemical group called the R group. The R group on each amino acid determines how it will fit in the folded, three-dimensional, tertiary structure of the protein and the bonds it may form. If the wrong amino acid is inserted, its R group might not fit into the tertiary structure properly, or required bonds might not be made; hence the entire structure might be altered. Since function depends on structure, this means the protein will not function properly.

6.   Some enzymes require a mineral or a vitamin as a cofactor. In such cases, if the particular vitamin or mineral is not available, the enzymes will not function properly.

7.   Heat increases the kinetic energy of molecules. Vital biological molecules, like proteins and nucleic acids, are denatured (rendered nonfunctional) by excessive heat because intramolecular bonds essential to their functional structure are broken. Since all enzymes are proteins, their destruction is lethal.

Stop and Think

1.   Breaking the ATP down to ADP and P_i releases the amount of energy stored in the bonds. Only part of that potential energy is actually used for cellular use; the rest is lost as heat. Nonetheless, the total amount of energy released (plus activation energy) must be absorbed to remake the bonds of ATP.

2.   Not all chemical bonds are easily explained by the planetary models. The bonds in ozone and carbon monoxide do not fit into the explanation provided by the octet rule, since there is no way to satisfy the proper bond number for each atom in those molecules.

3.   After helium (atomic number 2) would come atomic number 10 (2 + 8), then atomic number 18 (2 + 8 + 8).

4.   The negative surface charges on the proteins make them repel each other. If the proteins settled out, they would come in closer contact than their electrical charges would allow. Thus, the repulsion prevents settling. Additionally, water molecules orient to the charged proteins, keeping them separated by hydration layers. Hence, cellular fluid is a colloid.

5.   Sodium chloride dissociates into two ions when it is dissolved in water; glucose does not dissociate. Since a one-molar solution of any substance has the same number of molecules, the original salt added was comparable to the amount of sugar. Once the salt dissociates, however, there are 2 moles of particles in solution: 1 mole of sodium ions, and 1 mole of chloride ions. Since calcium chloride consists of three ions (one calcium ion and two chloride ions), a one-molar solution will have 3 moles of ions, giving it three times as many particles in solution as a one-molar glucose solution.

6. Calcium carbonate, $CaCO_3$.

7.

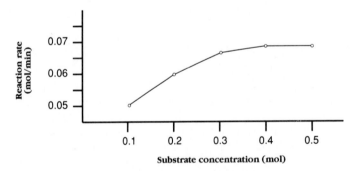

8. If water molecules were closer together in ice than in liquid water, ice would be denser than water and would sink instead of floating. As a lake surface froze, the ice would sink. This would expose additional lake water to freezing conditions, and the additional ice formed would also sink. The process would continue until the lake froze from the bottom up. Complete thawing would be extremely slow because the warm air and sunlight would not reach the lake bottom easily. Since ice floats, a layer of ice insulates the water beneath it, slowing the freezing process and usually preserving some liquid water for aquatic life to survive the winter.

9. Since ammonia is a polar molecule, the large nitrogen atom is slightly electronegative and can attract hydrogen ions.

10. Raising the temperature increases the kinetic energy of molecules, so water dissociates more at 38°C than at room temperature. The increased dissociation results in a greater number of free hydrogen ions, which, since pH is a measure of free hydrogen ions, will be reflected in a lower pH. Thus, although water is neutral by definition (hydrogen ions always equal hydroxyl ions), the numerical standard of pH 7 is not always precisely neutral.

11. They are the most energetic electrons of an atom and the only ones to participate in bonding behavior.

12. The fact that fats float indicates that neutral fats (nonpolar compounds) and water (polar molecules) do not mix; hence fats will not go into aqueous solution (dissolve in water). Thus, they can be stored in an "inactive" state that does not interfere with solute concentrations needed for other chemical reactions.

13. "a" = without; "tom" = cut; atom means "indivisible"

14. Chlorine can make covalent bonds. If an atom of chlorine combines with an atom with similar electronegativity (such as another chlorine atom), one covalent bond will be formed, since it needs only one electron to fill its valence shell.

15. Since the relative concentrations of reactants and products govern the direction of a reversible reaction, a cell can control direction by controlling concentration. Cells can manipulate concentrations of either reactants or products.

16. Water is an acid because it releases a hydrogen ion when it dissociates. Water is a base because the negative pole at the oxygen atom can attract a hydrogen ion, in the same way as ammonia.

17. The pH is 6; this solution is 10 times as acidic as a solution of pH 7.

18.

1. The plateau indicates saturation of enzyme.
2. Interference with active site would decrease reaction rate.

19. Energy does not occupy space and has no mass. Energy is described in terms of its effect on matter, i.e., its ability to do work or put matter into motion.

20. Unsaturated fats are oils; they tend to have shorter fatty acid chains containing one or more carbons that have double or triple bonds. Saturated fats are solid (like butter or lard); their fatty acids are longer and their contained carbon atoms are fully saturated with hydrogen (all single bonds).

21. The reactants are X and YC; the products are XC and Y. All are 1 mole quantities.

22. Radioisotopes are used to destroy cancerous tissue, in scans for diagnostic purposes, for tracing out metabolic pathways, and for carbon dating.

23. Evelyn has less "insulation" (subcutaneous fat) and a larger relative surface area, so she loses heat to the environment much more readily than Barbara.

COVERING ALL YOUR BASES

Multiple Choice

1. B, E   **2.** C   **3.** B, C   **4.** A, C, D   **5.** D   **6.** B   **7.** C, D, E   **8.** A, D   **9.** A, B, E   **10.** A   **11.** E   **12.** D
13. E   **14.** D   **15.** A   **16.** B   **17.** C, E   **18.** C, D   **19.** C   **20.** D   **21.** A, E   **22.** A   **23.** A, B, C, D, E
24. A, D   **25.** B, C   **26.** C

Word Dissection

Word root	Translation	Example	Word root	Translation	Example
1. ana	up	anion, anabolic	**8.** iso	equal	isotope
2. cata	down	cation, catabolic	**9.** kin	to move	kinetic
3. di	two	dipole, disaccharide	**10.** lysis	break	hydrolysis
4. en	in	endergonic	**11.** mono	one	monosaccharide
5. ex	out	exergonic	**12.** poly	many	polysaccharide
6. glyco	sweet	glycogen	**13.** syn	together	synthesis
7. hydr	water	dehydration	**14.** tri	three	triglyceride

Chapter 3   Cells: The Living Units

BUILDING THE FRAMEWORK

Overview of the Cellular Basis of Life

1.   1. A cell is the basic structural and functional unit; an organism's activity is dependent on cellular activity; biochemical activity determines and is determined by subcellular structure; continuity of life is based on cell reproduction.
2. cubelike, tilelike, disk-shaped, spherical, branching, cylindrical   3. plasma membrane, cytoplasm, nucleus
4. A model that describes a cell in terms of common features/functions that all cells share.

The Plasma Membrane: Structure and Functions

1.

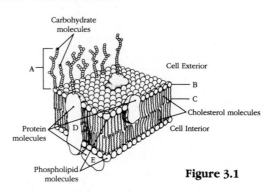

Figure 3.1

1. fluid mosaic model   2. to stabilize the plasma membrane by wedging between the phospholipid "tails"   3. glycocalyx
4. C   5. hydrophobic   6. D: integral; E: peripheral

2.

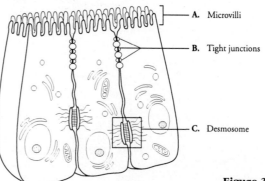

A. Microvilli

B. Tight junctions

C. Desmosome

Figure 3.2

1. Microvilli increase the surface area of the plasma membrane.   2. Microvilli are found on cells involved in secretion and/or absorption.   3. actin   4. the glycocalyx and tongue-in-groove folding of adjacent plasma membranes
5. tight junction   6. desmosome   7. desmosome   8. gap junction   9. connexons

3.

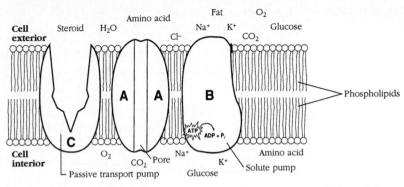

Figure 3.3

Arrows for the amino acid and Na^+ should be red and shown leaving the cell; those for glucose, Cl^-, O_2, fat, and steroid should be blue and entering the cell. CO_2 (blue arrow) should be leaving the cell and moving into the extracellular fluid. Amino acids and K^+ (red arrows) should be entering the cell.

1. fat, steroid, O_2, CO_2 2. glucose   3. H_2O, (probably) Cl^- 4. Na^+, K^+, amino acid.

4.   1. E, F   2. A, B   3. C   4. A, B   5. D

5.   1. A. hypertonic   B. isotonic   C. hypotonic   2. A. crenated   B. normal (discoid)   C. spherical, some hemolyzed   3. same solute concentration inside and outside the cell   4. Water is moving by osmosis into the cell from the site of higher water concentration (cell exterior) to the site of lower water concentration (cell interior).   5. Tonicity deals with the effect of nonpenetrating solutes on the movement of water into or out of cells; osmolarity is a measure of the total solute concentration, both penetrating and nonpenetrating solutes.

6.

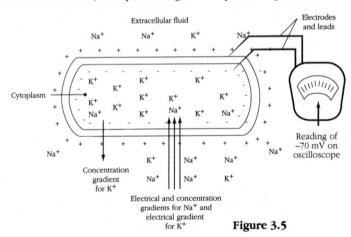

Figure 3.5

7.   1. Communication between adjacent cells   2. Impermeable junction   3. Impermeable intercellular space   4. High extracellular K^+ concentration   5. Protein anions move out of cell   6. More K^+ pumped out than Na^+ carried in   7. Carbohydrate chains on cytoplasmic side of membrane   8. Nonselective   9. Exocytosis   10. CAMs   11. NO

The Cytoplasm

1.   The unstructured gel-like part of the cytoplasm that contains biological molecules in solution.

2.   Organelles: the highly structured functional structures within the cell. Inclusions: stored nutrients, crystals of various types, secretory granules, and waste products in the cell.

3.

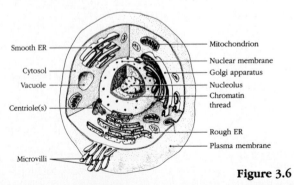

Figure 3.6

4.

Cell structure	Location	Function
Plasma membrane	External boundary of the cell	Confines cell contents; regulates entry and exit of materials
Lysosome	Scattered in cytoplasm	Digests ingested materials and worn-out organelles
Mitochondria	Scattered throughout the cell	Controls release of energy from foods; forms ATP
Microvilli	Projections of the plasma membrane	Increase the membrane surface area
Golgi apparatus	Near the nucleus (in the cytoplasm)	Packages proteins to be exported from the cell; packages lysosomal enzymes
Centrioles	Two rod-shaped bodies near the nucleus	"Spin" the mitotic spindle
Smooth ER	In the cytoplasm	Site of steroid synthesis
Rough ER	In the cytoplasm	Transports proteins (made on its ribosomes) to other sites in the cell; site of membrane lipid synthesis
Ribosomes	Attached to membranes or scattered in the cytoplasm	Synthesize proteins
Cilia	Extensions of cell to exterior	Act collectively to move substances across cell surface in one direction
Microtubules	Internal structure of centrioles; part of the cytoskeleton	Important in cell shape; suspend organelles
Peroxisomes	Throughout cytoplasm; possibly budding from rough ER	Detoxify alcohol and free radicals accumulating from normal metabolism
Microfilaments	Throughout cytoplasm; part of cytoskeleton	Contractile protein (actin); moves cell or cell parts; core of microvilli
Intermediate filaments	Part of cytoskeleton	Act as internal "guy wires"; help form desmosomes
Inclusions	Dispersed in the cytoplasm	Provide nutrients; represent cell waste products, etc.

5.   1. Centrioles   2. Cilia   3. Smooth ER   4. Vitamin A storage   5. Mitochondria
6.   1. Microtubules   2. Intermediate filaments   3. Microtubules   4. Microfilaments   5. Intermediate filaments
6. Microtubules
7.   1. B   2. F   3. D   4. E   5. C, H   6. G   7. A
8.   The nuclear and plasma membranes and essentially all organelles except mitochondria and cytoskeletal elements make up the endomembrane system. The components of this system act together to synthesize, store, and export cell products and to degrade or detoxify ingested or harmful substances.

The Nucleus

1.

Nuclear structure	General location/appearance	Function
Nucleus	Usually in center of the cell; oval or spherical	Storehouse of genetic information; directs cellular activities
Nucleolus	Dark spherical body in the nucleus	Storehouse and assembly site for ribosomes
Chromatin	Dispersed in the nucleus; threadlike	Contains genetic material (DNA); coils during mitosis
Nuclear membrane	Encloses nuclear contents; double membrane penetrated by pores	Regulates entry/exit of substances to and from the nucleus

2.   1. Histone proteins   2. The components are believed to play a role in regulating DNA function by exposing or not exposing certain DNA fragments.

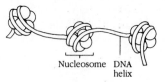

Nucleosome DNA helix

Figure 3.7

Cell Growth and Reproduction

1.   The cell life cycle consists of interphase and the mitotic phase, during which the cell divides.

Phase of interphase	Important events
G_1	Cell grows rapidly and is active in its normal metabolic activities. Centrioles begin replicating.
S	Cell growth continues. DNA is replicated, new histone proteins are made, and chromatin is assembled.
G_2	Brief phase when remaining enzymes (or other proteins) needed for cell division are synthesized; centriole replication completed.

2.

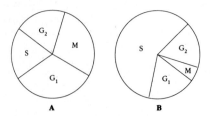

A   B

Figure 3.8

3.   1. C   2. A   3. D   4. D   5. B   6. C   7. C   8. E   9. B, C   10. C   11. E   12. A, B   13. E   14. A

4. Figure 3.9: A. Prophase   B. Anaphase   C. Telophase   D. Metaphase

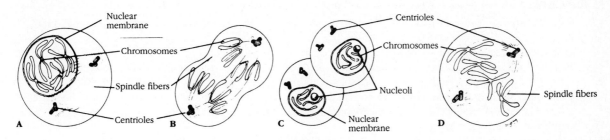

Figure 3.9

Figure 3.10: A. Prophase   B. Metaphase   C. Anaphase (early)   D. Prophase   E. Telophase   F. Anaphase (late)

5.   1. P   2. K   3. O   4. T   5. C   6. B   7. E   8. F   9. V   10. S   11. Q   12. M   13. U   14. L   15. N   16. H
17. I   18. R

6.   1. transcription   2. translation   3. DNA   4. anticodon; triplet   5. One or more segments of a DNA strand that programs one polypeptide chain (or one tRNA or rRNA)

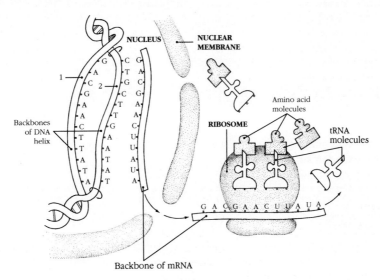

Figure 3.11

7.   1. nucleus   2. cytoplasm   3. coiled   4. centromeres   5. binucleate cell   6. spindle   7. interphase

Extracellular Materials

1. Body fluids: e.g., blood plasma, interstitial fluid, cerebrospinal fluid, eye humors. Important in transport in body and as solvents.

2. Cellular secretions: e.g., saliva, gastric juice, mucus, tears. Important as lubricants and some aid the digestive process.

3. Extracellular matrix: e.g., glycoprotein intercellular "glue," and the matrix (ground substance and fibers) secreted by connective tissue cells. Important in binding cells together and providing strong structures (bones, cartilages, etc.) that can support or protect other body organs.

Developmental Aspects of Cells

1. 1. zygote   2. genes   3. chemical   4. proteins   5. cell replacement   6. tissue repair   7. hyperplasia   8. alcohol
9. carbon monoxide   10. X rays

2. cumulative effects of chemical damage; progressive decrease in immune function; loss of ability for cell division

The Incredible Journey

1. 1. cytosol   2. nucleus   3. mitochondrion   4. ATP   5. ribosomes   6. rough endoplasmic reticulum   7. nuclear pores   8. chromatin   9. DNA   10. nucleoli   11. Golgi apparatus   12. lysosome (or peroxisome)

CHALLENGING YOURSELF

At the Clinic

1. Normally, lactose (a disaccharide) is digested to monosaccharides by the enzyme lactase in the intestine. Lactase deficiency prevents this digestion. Consequently, lactose remains in the intestinal lumen and acts as an osmotic agent to attract water and prevent its absorption. Hence, diarrhea (or watery feces) occurs. Adding lactase to milk hydrolyzes lactose to glucose and galactose, which can move into intestinal cells by facilitated diffusion. Water follows.

2. Increased capillary permeability causes more blood plasma to filter into the tissue spaces from the bloodstream. On a small scale, this causes the localized swelling (edema) associated with inflammation. In anaphylaxis, the systemic reaction causes so *much* fluid to leave the bloodstream that there is insufficient blood volume (fluid) to maintain circulation to the vital organs.

3. LDL is removed from the circulation by receptor-mediated endocytosis.

4. Glucose is reclaimed from the forming urine by carrier molecules in the plasma membranes of cells forming the tubules of the kidneys. If too much glucose is present, the carrier molecules are *saturated*. Hence, some glucose will not be reabsorbed and returned to the blood, but instead will be lost to the body in urine.

5. In blood stasis, the filtration pressure at the capillaries increases. Albumin (being a small protein) can be forced into the interstitial fluid under such conditions, increasing its osmotic pressure. If the osmotic pressure of the interstitial fluid exceeds that of the blood plasma, more of the water filtering out of the capillary will *remain* in the tissue spaces. This increase in interstitial fluid volume leads to edema.

6. Lysosomal destruction releases acid hydrolases into the cytoplasm, killing the cell. When the cell lyses, inflammation is triggered. Hydrocortisone is an antiinflammatory steroid that stabilizes lysosomal membranes. Since hydrocortisone is a steroid, it is soluble in an oil base rather than a water base. Steroids can diffuse through the lipid bilayer of cell membranes, so they can be administered topically.

7. Streptomycin inhibits bacterial protein synthesis. If the cells are unable to synthesize new proteins (many of which would be essential enzymes), they will die.

8. Tissue overgrowth (hyperplasia) occurred in response to the continual irritation. However, there was no evidence of neoplasia (tumor formation).

Stop and Think

1. Because the phospholipids orient themselves so that polar-to-polar and nonpolar-to-nonpolar regions are aligned, any gaps in the membrane are quickly sealed.

2. The molecular weights are ammonium hydroxide, 35; hydrochloric acid, 18. Since ammonium hydroxide is about twice the weight of hydrochloric acid, it diffuses at about half the rate. The precipitate forms about one-third of a meter from the ammonium hydroxide end.

3. Skin cells lose their desmosomes as they age. By the time they reach the free surface, they are no longer tightly bound to each other.

4. Conjugation (attachment) of the lipid product to a non–lipid-soluble substance, such as protein, will trap the product.

5. The plasma membrane folds (microvilli); the internal membranes are folded (ER and Golgi apparatus); the inner membranes of mitochondria are folded (cristae).

6. Since mitochondria contain DNA and RNA, mitochondrial ribosomes could be expected to, and do indeed, exist. Mitochondrial nucleic acids and ribosomes are extremely similar to those of bacteria.

7. A: This cell secretes a protein product, as evidenced by large amounts of rough ER, Golgi apparatus, and secretory vesicles near the cell apex.   B: The cell secretes a lipid or steroid product, as evidenced by the lipid droplets surrounded by the abundant smooth ER.

8. Avascular tissues rely on the diffusion of nutrients from surrounding fluids. Since this is not as efficient or effective as a vascular supply, these tissues do not get very thick. An exception is the epidermis of the skin, which is quite thick, but cells in the skin layers farthest from the underlying blood vessels are dead.

9. 1. Fructose will diffuse into the cell.   2. Glucose will diffuse out of the cell.   3. Water enters the cell by diffusing along its concentration gradient.   4. The cell swells due to water entry.

10.

		After deletion:	
DNA:	TAC GCA TCA CTT TTG ATC	DNA:	TAG ATC
mRNA:	AUG CGU AGU GAA AAC UAG	mRNA:	AUC UAG
amino acids:	Met Arg Ser Glu Asn Stop	amino acids:	none; a nonsense message

COVERING ALL YOUR BASES

Multiple Choice

1. E   2. C   3. B   4. A   5. C, D   6. C   7. B   8. B, D   9. B   10. C   11. C   12. A, D   13. A, B, C, D   14. C
15. E   16. D, E   17. C   18. A   19. E   20. B   21. D   22. C   23. D   24. A   25. A   26. A, B, C   27. A

Word Dissection

Word root	Translation	Example	Word root	Translation	Example
1. chondri	lump	mitochondria	**13.** osmo	pushing	osmosis
2. chrom	color	chromosome	**14.** permea	pass through	permeable
3. crist	crest	cristae	**15.** phag	eat	phagocyte
4. cyto	cell	cytology	**16.** philo	love	hydrophilic
5. desm	bond	desmosome	**17.** phobo	fear	hydrophobic
6. dia	through	dialysis	**18.** pin	drink	pinocytosis
7. dys	bad	dysplasia	**19.** plasm	to be molded	cytoplasm
8. flagell	whip	flagellum	**20.** telo	end	telophase
9. meta	between	metaphase	**21.** tono	tension	isotonic
10. mito	thread	mitosis, mitochondria	**22.** troph	nourish	atrophy
11. nucle	little nut	nucleus	**23.** villus	hair	microvilli
12. onco	a mass	oncogene			

Chapter 4 Tissues: The Living Fabric

BUILDING THE FRAMEWORK

Overview of Body Tissues

1.   1. Areolar   2. Cell   3. Elastic fibers   4. Bones   5. Nervous   6. Blood   7. Vascular   8. Keratin   9. Vascular

2.   A. simple squamous epithelium   B. simple cuboidal epithelium   C. cardiac muscle   D. dense regular connective tissue   E. bone   F. skeletal muscle   G. nervous tissue   H. hyaline cartilage   I. smooth muscle tissue   J. adipose (fat) tissue   K. stratified squamous epithelium   L. areolar connective tissue
The noncellular areas of D, E, H, J, and L are matrix.

3.   1. B   2. C   3. D   4. A   5. B   6. D   7. C   8. B   9. A   10. A   11. C   12. A   13. D

Epithelial Tissue

1.   protection, absorption, filtration, excretion, secretion, and sensory reception

2.   cellularity, specialized contacts (junctions), polarity, avascularity, basement membrane, regeneration

3.   1. B   2. F   3. A   4. D   5. G   6. K   7. H   8. J   9. I   10. L

4.   A=1; C=2; E=3; D=4; B=5

5.   B=1; D=2; C=3; A=4

6.

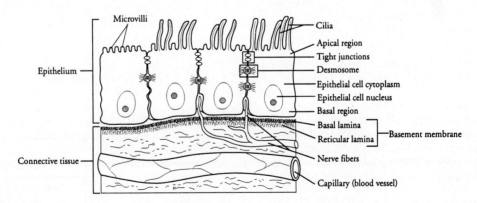

Figure 4.2

7.   1. simple alveolar gland   2. simple tubular gland   3. tubuloalveolar gland   4. compound gland

8.   1. T   2. glandular cells secrete their product   3. merocrine   4. merocrine   5. T   6. repair damage   7. merocrine   8. T

9.   1. B   2. A   3. A   4. B   5. C   6. B

10.   1. epithelial   2. goblet cells   3. mucin   4.6. duct (from epithelium), secretory unit, and supportive connective tissue   7. branching   8. tubular

Connective Tissue

1. 1. C 2. A 3. D 4. J 5. B 6. H 7. A 8. H 9. J 10. G 11. K 12. I 13. E 14. L 15. F
2.

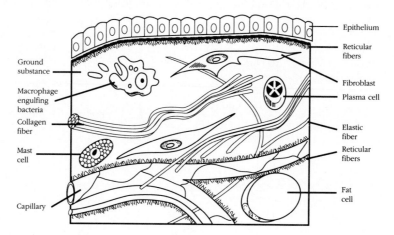

Figure 4.3

3. B=1; C=2; A=3
4. 1. H 2. E 3. C 4. F 5. L 6. G 7. E 8. B 9. D 10. E 11. A 12. I 13. K

Epithelial Membranes

1. In each case, the visceral layer of the serosa covers the external surface of the organ, and the parietal layer lines the body cavity walls.

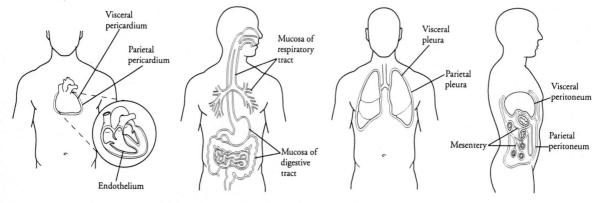

Figure 4.4

2.

Membrane	Tissue type (epithelial/connective)	Common locations	Functions
Mucous	Epithelial sheet with underlying connective tissue lamina propria	Lines respiratory, digestive, and reproductive tracts	Protection, lubrication, secretion, absorption
Serous	Epithelial sheet (mesothelium), scant areolar connective tissue	Lines internal ventral cavities and covers their organs	Lubrication for decreased friction during organ movements
Cutaneous	Epithelial (stratified squamous, keratinized) underlain by dense fibrous connective tissue dermis	Covers the body exterior; is the skin	Protection from external insults; protection from water loss

Muscle Tissue

1.   Skeletal: 1, 3, 5, 6, 7, 11, 13   Cardiac: 2, 3, 4 (typically), 10, 12, 14, 15   Smooth: 2, 4, 8, 9, 14

Nervous Tissue

1.   The neuron has long cytoplasmic extensions that promote its ability to transmit impulses long distances within the body.

2.   Conduct

Tissue Repair

1.   1. inflammation   2. clotting proteins   3. granulation   4. T   5. regeneration   6. T   7. collagen   8. T   9. T
   10. bacterial-inhibiting substances   11. fibrosis

Developmental Aspects of Tissues

1.   1. tissues   2. growth   3.–5. ectoderm, endoderm, and mesoderm   6. nervous   7. muscle

2.   1. nervous tissue   2. connective tissue   3. muscle tissue   4. epithelial tissue

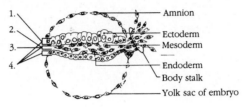

Figure 4.5

3.

Germ layer	Tissue type
Endoderm	Mucosae (epithelial)
Mesoderm	Muscle
Ectoderm	Nerve tissue
Mesoderm	Bone
Ectoderm	Epidermis
Mesoderm	Mesothelium (epithelial)

4.   1. thinning   2. decline in amount   3. atrophy and fibrosis

CHALLENGING YOURSELF

At the Clinic

1.   Individual adipocytes easily lose and gain cell volume by losing or accumulating more fat.

2.   Granulation tissue secretes bacterial-inhibiting substances.

3.   Microglia are the macrophages of the brain. Their abundance is indicative of phagocytic activity to remove dead cells and any infectious organisms present.

4.   Cartilage is avascular, and tendons are poorly vascularized. The more ample the blood supply, the quicker tissue heals.

5.   Hypersecretion of serous fluid and its accumulation in the pleural space displaces lung volume. Because the lungs are not able to expand fully, respiration is more difficult. Additionally, when there is more fluid in the tissue spaces, diffusion of nutrients to the tissue cells (from the bloodstream) takes much longer and can impair tissue health and viability.

6.   The peritoneum will be inflamed and infected. Since the peritoneum encloses so many richly vascularized organs, a spreading peritoneal infection can be life-threatening.

Stop and Think

1.   Simple epithelium is thinner than the diameter of the smallest vessel and must overlie all other tissues. Stratified epithelium forms the protective covering of deeper tissues. If blood vessels were within the epithelial layer, protection of the vessels would be reduced. Cartilage undergoes severe compression, which could damage vessels. Tendons are designed to withstand a strong pull; vessels would weaken the structure of tendons.

2.   By virtue of their elongated shape, columnar cells share more adjacent membrane with neighboring cells. This means more surface area for desmosomes and tongue-in-groove folds, both of which help bind cells together and reduce the chance of their separation.

3. Microvilli are found on virtually all simple cuboidal and simple columnar epithelia. The increase in surface area contributed by the microvilli offsets the decrease in transport rate by the thicker layers. Ciliated tissue can include all three types of simple epithelium but is found predominantly as simple columnar and pseudostratified.

4. Endocrine glands, like exocrine glands, are derived from the surface epithelium. In embryonic stages, these glands grow down from the surface, but eventually they lose their surface connection.

5. A basal lamina is only part of the basement membrane, which consists of epithelial cell secretions (the basal lamina) plus connective tissue fibers (the reticular lamina). A mucous membrane is an epithelial membrane composed of an epithelial sheet and underlying connective tissue. A sheet of mucus is a layer of secreted mucus (a cell product).

6. Large arteries must be able to expand in more than one direction; hence, they have irregularly arranged elastic connective tissue.

7. A bone shaft, like a skeletal cartilage, is surrounded by a sheet of dense irregular connective tissue.

8. Skeletal muscle cells are quite large compared to other muscle types. Typically, the multinucleate condition indicates that the demands for protein synthesis could not be handled by a single nucleus.

9. The nervous system develops by invagination from the ectoderm, much as glands develop internally from superficial epithelium.

COVERING ALL YOUR BASES

Multiple Choice

1. B   **2.** D   **3.** A   **4.** B   **5.** C   **6.** B   **7.** D   **8.** C   **9.** A   **10.** D   **11.** B   **12.** A, C, D, E   **13.** C   **14.** D
15. B   **16.** D   **17.** A   **18.** B   **19.** C   **20.** B   **21.** E   **22.** B, C, D   **23.** C   **24.** B, E   **25.** C   **26.** A, B, D, E
27. A, B, C, D, E   **28.** B, C   **29.** A   **30.** A, C, D

Word Dissection

Word root	Translation	Example	Word root	Translation	Example
1. ap	tip	apical, apocrine	**11.** hormon	excite	hormone
2. areola	space	areolar connective	**12.** hyal	glass	hyaline, hyaluronic
3. basal	foundation	basal lamina	**13.** lamina	thin plate	reticular lamina
4. blast	forming	fibroblast	**14.** mero	part	merocrine
5. chyme	juice	mesenchyme	**15.** meso	middle	mesoderm, mesothelium
6. crine	separate	endocrine	**16.** retic	network	reticular tissue
7. endo	within	endoderm, endothelium	**17.** sero	watery fluid	serous, serosa
8. epi	upon, over	epithelium	**18.** squam	a scale	squamous
9. glia	glue	neuroglia, microglia	**19.** strat	layer	stratified
10. holo	whole	holocrine			

Chapter 5   The Integumentary System

BUILDING THE FRAMEWORK

The Skin

1. 1. stratified squamous epithelium   2. dense irregular connective tissue

2. 1. They are increasingly farther from the blood supply in the dermis.   2. They are increasingly laden with (water-insoluble) keratin and surrounded by glycolipids, which hinder their ability to receive nutrients by diffusion.

3. 1. lines of cleavage   2. striae   3. flexure lines   4. wrinkles

4.

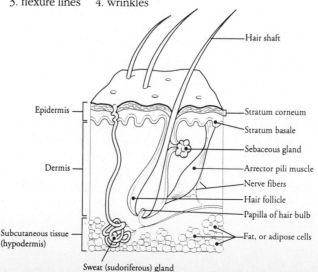

Epidermis
Dermis
Subcutaneous tissue (hypodermis)
Sweat (sudoriferous) gland

Hair shaft
Stratum corneum
Stratum basale
Sebaceous gland
Arrector pili muscle
Nerve fibers
Hair follicle
Papilla of hair bulb
Fat, or adipose cells

Figure 5.1

5.   1. D   2. B, D   3. F   4. I   5. A   6. B   7. I   8. H   9. I   10. J   11. A   12. A   13. C   14. B   15. I   16. B

6.   1. Keratin   2. Wart   3. Stratum basale   4. Keratinocytes   5. Arrector pili   6. Elastin   7. Melanocytes
        8. Laminated granules   9. Fibroblast

7.   1. C   2. A   3. C   4. B   5. C   6. A   7. B

8.   1. A   2. E   3. D   4. C   5. B

Appendages of the Skin

1.   The papilla contains blood vessels that nourish the growth zone of the hair. The sebaceous gland secretes sebum into the follicle. The arrector pili muscle pulls the follicle into an upright position during cold or fright. The bulb is the actively growing region of the hair.

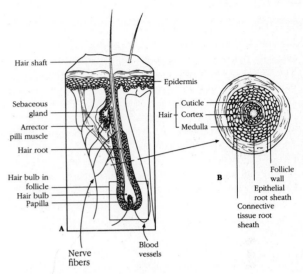

Figure 5.2

2.   1. Poor nutrition   2. Keratin   3. Stratum corneum   4. Eccrine glands   5. Vellus hair   6. Desquamation
        7. Growth phase

3.   Alopecia

4.   1. Genetic factors; emotional trauma; some drugs (chemotherapy agents); ringworm infection

5.   1. cuticle   2. The stratum germinativum is thicker here, preventing the rosy cast of blood from flushing through.

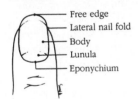

6.   1. A   2. B, C   3. B   4. C (B)   5. A   6. A   7. A, B   8. A   9. B   10. C

Functions of the Integumentary System

1.   1. B   2. M   3. C   4. M (C)   5. C   6. C

2.   All *but* water-soluble substances

3.   It depresses immune response by inhibiting the activity of macrophages that aid in the immune response.

4.   Sweat glands are regulated by the sympathetic division of the nervous system. When it is hot, blood vessels in the skin become flushed with warm blood, allowing heat to radiate from the skin surface, and perspiration output increases. Evaporation of sweat from the body surface removes large amounts of body heat.

5.   1. nervous   2.–5. cold, heat, deep pressure, and light pressure   6. cholesterol   7. epidermis   8. ultraviolet
        9. calcium

Homeostatic Imbalances of Skin

1.   Loss of body fluids containing proteins and essential electrolytes, which can lead to dehydration, shock, and renal shutdown

2.   1. C   2. B   3. A   4. B   5. C   6. C

3.   It allows the estimation of the extent of burns so that fluid volume replacement can be correctly calculated.

4.   1. squamous cell carcinoma   2. basal cell carcinoma   3. malignant melanoma

5.   Pigmented areas that are <u>A</u>ssymetric, have irregular <u>B</u>orders, exhibit several <u>C</u>olors, and have a <u>D</u>iameter greater than 6 mm are likely to be cancerous.

Developmental Aspects of the Integumentary System

1.   1. C   2. D   3. F   4. B   5. A   6. G   7. E

The Incredible Journey

1.   1. collagen   2. elastin   3. dermis   4. phagocyte (macrophage)   5. connective tissue root   6. epidermis   7. stratum basale   8. melanin   9. keratin   10. squamous (stratum corneum) cells

CHALLENGING YOURSELF

At the Clinic

1. The sun's radiation can cause damage to DNA, in spite of the protection afforded by melanin. Without repair mechanisms, the DNA damage is permanent, and the effects are cumulative.

2. The baby has seborrhea, or cradle cap, a condition of overactive sebaceous glands. It is not serious; the oily deposit is easily removed with attentive washing and soon stops forming.

3. A Mongolian spot is a pigmented area that is the result of escaped melanin being scavenged by dermal macrophages. It is not a sign of child abuse.

4. Systemic hives would indeed be worrisome because it might indicate a possibly life-threatening plasma loss and correlated drop in blood volume.

5. Probably not. Organic solvents are lipid-soluble and thus can pass through the skin by dissolving in the lipids of the cells' plasma membranes. Hence, a shower (which would wash away substances soluble in water) will not remove the danger of poisoning. Kidney failure and brain damage can result from poisoning by organic solvents.

6. Fluid/electrolyte replacement and the prevention of massive infection.

7. The most likely diagnosis is basal cell carcinoma, because it causes ulcers and is slow to metastasize. Years of exposure to sunlight probably caused his cancer.

8. Because there are fewer connections to underlying tissues, wounds to the face tend to gape more.

9. Carotene can accumulate in the stratum corneum and the hypodermis, giving the skin an orange cast. Awareness of diet is particularly important in very young children who tend to go on "food jags," refusing all but one food, to determine whether "liked" foods are high in carotene.

10. The drugs used to treat cancer exert their major cell-killing effects on rapidly dividing cells. Like cancer cells, the hair follicles are targets of these drugs, which accounts for the children's baldness.

Stop and Think

1. The term "membrane" refers to the sheetlike *structure* of the skin; all epithelial membranes are at least simple organs. The term "organ" means there are at least *two types of tissue* in a structure's composition, and at least simple functions can be performed.

2. The undulating folds (formed by the dermal papillae) at the surface between the epidermis and the dermis increase the surface area at the site of their union. Increased surface area provides more space for capillary networks (increasing the area for diffusion of nutrients to and wastes from the avascular epidermis) and is more secure, making it harder for the epidermis to tear away from the dermis.

3. There is very little hypodermal tissue underlying the skin of the shins. Consequently, the skin is quite securely and tightly fastened to the underlying structure (bone) there.

4. The stratum corneum is so far removed from the underlying blood supply in the dermis that diffusion of nutrients is insufficient to keep the cells alive. Also, the accumulation of glycolipids in the intercellular spaces retards diffusion of water-soluble nutrients. The accumulated keratin within the cells inhibits life functions. In terms of benefit, bacterial or other infections cannot become established easily in the dead, water-insoluble layer. Dead cells provide a much more effective mechanical barrier than living cells.

5. The nerve endings located closest to the surface are the most sensitive to light touch. Merkel discs, in the epidermis, are very responsive to light touch; root hair plexuses, although not as close, are stimulated by the movement of hairs on the skin, which project from the surface.

6. Both basal keratinocytes and melanocytes are attached to the basement membrane. When basal keratinocytes divide, the axis of division is oriented so that one daughter cell is superficial to the other. The deep daughter cell remains attached to the basement membrane and remains to divide again. The superficial daughter cell becomes part of the stratum spinosum, differentiates, and moves upward to eventually slough off as a cornified cell.

7. When skin peels as a result of sunburn damage, the deeper layers of the stratum corneum, which still maintain their desmosomal connections, detach. The desmosomes hold the cells together within the layers.

8.
- A whitehead is formed by blockage of the duct of a sebaceous gland with sebum. When this sebum oxidizes, it produces a blackhead. When a blocked sebaceous gland becomes infected, it produces a pimple.
- When the arrector pili muscles contract to raise the hair, they also dimple the skin producing "goose bumps."
- Greasy hair and shiny nose both result from the secretion of sebum onto the skin.
- Stretch marks represent small tears in the dermis, as the skin is stretched by obesity or pregnancy.
- One can turn blue from holding one's breath because the blood in the capillaries of the skin changes from scarlet red to purple as it gives up its oxygen.

- Fingerprints are films of sweat, derived from sweat glands that open along the epidermal ridges of the hand.
- Pores on the face are the surface openings of hair follicles and the associated sebaceous glands.
- The sparseness of hair on the body surface of humans may be an adaptation that allows more efficient sweating in hot climates, for abundant body hair would inhibit the evaporation of sweat.
- A blister, caused by severe, short-term friction to the skin, is a separation of the epidermis and dermis by a fluid-filled pocket.
- In a bruise, blood has escaped from the circulation and clotted deep to the skin.

9. Unless there is a protein deficiency, additional dietary protein will not increase protein utilization by hair- and nail-forming cells.

10. The apocrine sweat glands, associated with axillary and pubic hair follicles, appear to secrete chemicals that act as sexually important signals.

11. Because of the waterproofing substance in the epidermis (both in and between the cells), the skin is relatively impermeable to water entry.

COVERING ALL YOUR BASES

Multiple Choice

1. B, D **2.** D **3.** B **4.** A **5.** D **6.** D **7.** A **8.** D **9.** C **10.** D, E **11.** E **12.** A, B, D **13.** C **14.** A, B **15.** A, B **16.** A, C, D **17.** D **18.** C **19.** A, B, D **20.** B, C, E **21.** B **22.** A, D, E **23.** B, C, D **24.** C **25.** C **26.** D **27.** A, C, D **28.** B, C, D, E

Word Dissection

	Word root	Translation	Example		Word root	Translation	Example
1.	arrect	upright	arrector pili	**15.**	lanu	wool, down	lanugo
2.	carot	carrot	carotene	**16.**	lunul	crescent	lunula
3.	case	cheese	vernix caseosa	**17.**	medull	marrow	medulla
4.	cere	wax	cerumen	**18.**	melan	black	melanin
5.	corn	horn, horny	stratum corneum	**19.**	pall	pale	pallor
6.	cort	bark, shell	cortex	**20.**	papilla	nipple	dermal papillae
7.	cutic	skin	cutaneous, cuticle	**21.**	pili	hair	arrector pili
8.	cyan	blue	cyanosis	**22.**	plex	network	plexus
9.	derm	skin	dermis, epidermis	**23.**	rhea	flow	seborrhea
10.	folli	a bag	follicle	**24.**	seb	grease	sebaceous gland
11.	hemato	blood	hematoma	**25.**	spin	spine, thorn	stratum spinosum
12.	hirsut	hairy	hirsutism	**26.**	sudor	sweat	sudoriferous
13.	jaune	yellow	jaundice	**27.**	tegm	cover	integument
14.	kera	horn	keratin	**28.**	vell	fleece, wool	vellus

Chapter 6 Bones and Bone Tissue

BUILDING THE FRAMEWORK

Skeletal Cartilages

1. 1. C 2. A 3. C 4. B 5. A 6. C 7. C
2. 1. T 2. F 3. T 4. F 5. F 6. T 7. F

Functions of the Bones

1. 1. They support the body by providing a rigid skeletal framework. 2. They protect the brain, spinal cord, lungs, and other internal organs. 3. They act as levers and provide attachments for muscles in movement. 4. They serve as a reservoir for fat and minerals. 5. They are the sites of formation of blood cells.

Classification of Bones

1. 1. S 2. F 3. L 4. L 5. F 6. L 7. L 8. F 9. I

Bone Structure

1.

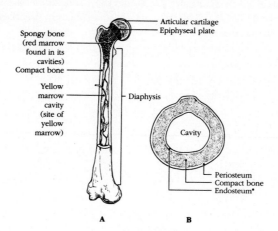

Figure 6.1

2. 1. C   2. A   3. C, D   4. A   5. E   6. B   7. B
3. 1. internal layer of spongy bone   2. endosteum

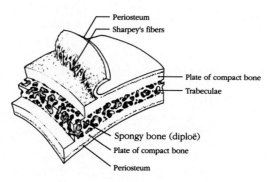

Figure 6.2

4. 1. B   2. C   3. A   4. E   5. D   Matrix is the nonliving part of bone shown as unlabeled white space on the figure.

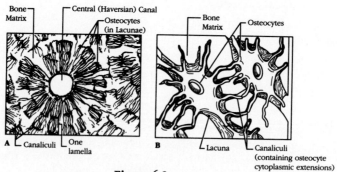

Figure 6.3

5. 1. P   2. P   3. O   4. O   5. P   6. O   7. P   8. P   9. P   10. D   11. P   12. O
6. 1. Collagen   2. Yellow marrow   3. Osteoclasts   4. Marrow cavity   5. Periosteum   6. Perichondrium   7. Lamellar

Bone Development (Osteogenesis)

1. If arranged in the correct sequence, the listed elements would be numbered: 3, 2, 4, 1, 5, 6 (note: events 2 and 3 may occur simultaneously).

2. 1. intramembranous   2. cartilage   3. T   4. calcium salts   5. T   6. mesenchymal cells or osteoblasts   7. T
   8. secondary   9. hyaline cartilage   10. endosteal   11. T

3. 1. resting cartilage cells (chondrocytes)   2. chondroblasts   3. C   4. D   5. osteoblasts   6. bone

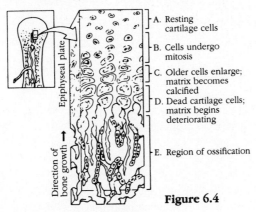

A. Resting cartilage cells

B. Cells undergo mitosis

C. Older cells enlarge; matrix becomes calcified

D. Dead cartilage cells; matrix begins deteriorating

E. Region of ossification

Figure 6.4

Bone Homeostasis: Remodeling and Repair

1. 1. G   2. F   3. A   4. H   5. D   6. B   7. E   8. C

2. 1. Bone resorption   2. Calcium salts   3. Hypocalcemia   4. Blood calcium levels   5. Growth in length

3. Because the points of maximal compression and tension are on opposing sides of the shaft, they cancel each other out, so there is essentially no stress in the middle (center) of the shaft. Consequently, the shaft can contain spongy (rather than compact) bone in its center.

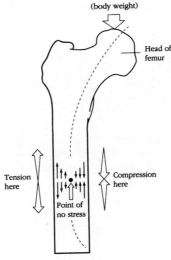

Load (body weight)

Head of femur

Tension here

Compression here

Point of no stress

Figure 6.5

4. 1. I   2. A   3. J   4. F   5. C   6. D   7. G   8. H   9. E   10. I   11. J   12. B

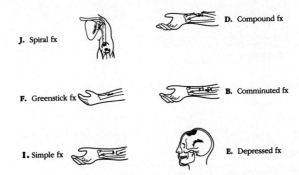

J. Spiral fx

F. Greenstick fx

I. Simple fx

D. Compound fx

B. Comminuted fx

E. Depressed fx

Figure 6.6

5. 1. T   2. T   3. phagocytes (macrophages)   4. fibroblasts   5. periosteum   6. T   7. spongy

Homeostatic Imbalances of Bone

1. 1. Osteoporosis   2. Osteomalacia   3. Paget's disease   4. Porous bones   5. Osteomalacia   6. Calcified epiphyseal discs

Developmental Aspects of Bones: Timing of Events

1. 1. D   2. F   3. A   4. C   5. A, E   6. B
2. 1. B   2. C   3. D   4. A   5. D   6. C
3. 1. B, A, C   2. C, A, B   3. E, B, D, C, A   4. D, B, A, C   5. B, D, A, C

The Incredible Journey

1. 1. femur   2. spongy   3. stress   4. red blood cells   5. red marrow   6. nerve   7. central (Haversian)   8. compact   9. canaliculi   10. lacunae (osteocytes)   11. matrix   12. osteoclast

CHALLENGING YOURSELF

At the Clinic

1. The woman's vertebral compression fractures and low bone density indicate osteoporosis. This condition in postmenopausal women is at least partially due to lack of estrogen production. It can be treated with supplements of calcium and vitamin D, proper nutrition, and possibly an estrogen-progestin protocol or fluoride and etidronate.
2. Infection can spread to the bone. The doctor will examine the X ray for evidence of osteomyelitis, which is notoriously difficult to manage.
3. The curving of the line indicates a spiral fracture.
4. Abnormal collagen production coupled with short stature indicates achondroplasia.
5. The woman is suffering from hypercalcemia, which is causing metastatic calcification (i.e., calcium salt deposits in soft tissues of the body).
6. The doctor will check for hypersecretion of growth hormone.
7. Paget's disease can cause abnormal thickening of bone.
8. The hip bone is the usual site for obtaining marrow for transplants.

Stop and Think

1. The blood vessels determine the pattern for the development of the lamellae, so blood vessels in the osteon region appear first.
2. Bone tissue is too rigid for interstitial growth; there is only appositional growth after ossification. Before ossification, growth can occur in the membrane or in cartilage templates.
3. Long bones tend to be arranged and to articulate end to end. Even if two long bones are side by side, they still articulate most prominently at their ends. Short bones are arranged and articulate side to side as well as end to end.
4. The medullary cavity's diameter must increase to keep pace with the appositional growth of the surrounding bone collar. Osteoclasts will remove the bone on the interior of the collar to widen the cavity. Likewise, osteoclasts are active on the internal surfaces of cranial bones to broaden their curvatures to accommodate the expanding brain.
5. Bone tissue is highly vascular. Fibrous tissue is poorly vascularized, and cartilage is avascular. In fact, the vascularization of the template is part of the process that triggers ossification.
6. The bone will heal faster because bone is vascularized and has an excellent nutrient delivery system (canaliculi). Cartilage lacks both of these characteristics.
7. Yes, he has cause to worry. The boy is about to enter his puberty-adolescence growth spurt, and interference with the growth plate (epiphyseal plate) is probable.
8. According to Wolff's law, growth of a bone is a response to the amount of compression and tension placed on it, and a bone will grow in the direction necessary to best respond to the stresses applied.
9. Signs of healing of the bone at the surgical incision (usually a hole in the skull) indicate that the patient lived at least for a while.
10. The high water content helps explain both points. (a) Cartilage is largely water plus resilient ground substance and fibers (no calcium salts). (b) Water "saves room" for bones to develop in.

COVERING ALL YOUR BASES

Multiple Choice

1. A, B, C   **2.** A, B, D   **3.** A, D   **4.** A, B, D   **5.** A, D   **6.** A, C, D   **7.** D   **8.** C   **9.** A, B, C, E   **10.** A, B
11. A, B, C, D   **12.** B   **13.** B, D   **14.** A, C, D, E   **15.** D   **16.** C   **17.** A   **18.** A, C   **19.** A, D   **20.** B, E
21. A   **22.** A, B, C, D, E   **23.** C   **24.** C   **25.** D   **26.** B, C   **27.** C   **28.** B

Word Dissection

Word root	Translation	Example	Word root	Translation	Example
1. call	hardened	callus	3. clast	break	osteoclast
2. cancel	latticework	cancellous bone	4. fract	break	fracture

Word root	Translation	Example	Word root	Translation	Example
5. lamell	small plate	lamellar bone	**9.** poie	make	hematopoiesis
6. malac	soft	osteomalacia	**10.** soma	body	somatomedin
7. myel	marrow	osteomyelitis	**11.** trab	beam	trabeculae
8. physis	growth	diaphysis			

Chapter 7 The Skeleton

BUILDING THE FRAMEWORK

1. Bones of the skull, vertebral column, and bony thorax are parts of the axial skeleton. All others belong to the appendicular skeleton.

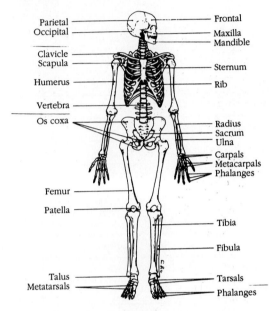

Figure 7.1

2. 1. B, H   2. A   3. G   4. E   5. F   6. D   7. C

Part I: The Axial Skeleton

The Skull

1. A, B, I, K, L, and M should be circled.

2.

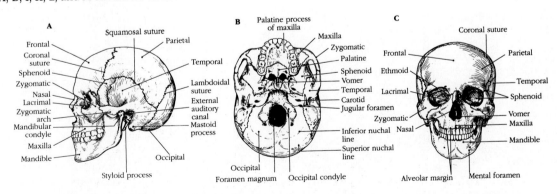

Figure 7.2

3. 1. B, K   2. I, K   3. K, M   4. K   5. O   6. K   7. A   8. J   9. I   10. A   11. I   12. L   13. M   14. B   15. F
    16. F   17. M   18. F   19. C   20.–23. A, B, G, L   24. A   25. L   26. E   27. N

4.   1. Mucosa-lined air-filled cavities in bone.   2. They lighten the skull and serve as resonance chambers for speech.   3. Their mucosa is continuous with that of the nasal passages, into which they drain.   4. To moisten and warm incoming air and trap particles and bacteria in sticky mucus.

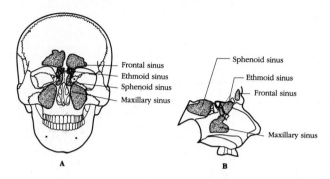

Figure 7.3

5.

Bone marking(s)	Skull bone	Function of bone marking
Coronoid process	Mandible	Attachment for temporalis muscle
Carotid canal	Temporal	Passes internal carotid artery
Sella turcica	Sphenoid	Supports, partly encloses pituitary gland
Jugular foramen	Temporal	Passes jugular vein and cranial nerves IX, X, and XI
Palatine processes	Maxillae	Form anterior two-thirds of hard palate
Styloid process	Temporals	Attachment site for several neck muscles
Optic foramina	Sphenoid	Pass optic nerves and ophthalmic arteries
Cribriform plates	Ethmoid	Pass olfactory nerve fibers
External occipital crest	Occipital	Attachment for ligamentum nuchae and neck muscles
Mandibular condyles	Mandibles	Form freely movable joints with temporal bones
Mandibular fossa	Temporal	Receives condyle of mandible
Stylomastoid foramen	Temporal	Where facial nerve leaves skull

The Vertebral Column

1.   1. F   2. A   3. C, E   4. A, E   5. B
2.   1. Cervical, C_1 to C_7 2. Thoracic, T_1 to T_{12} 3. Lumbar, L_1 to L_5 4. Sacrum, fused   5. Coccyx, fused   6. Atlas, C_1 7. Axis, C_2
3.   1. A, B, C   2. G   3. C   4. A   5. F   6. E   7. B   8. D   9. E   10. G   11. A, B
4.   1. J   2. F   3. G   4. E   5. H   A. Cervical, atlas   B. Cervical   C. Thoracic   D. Lumbar

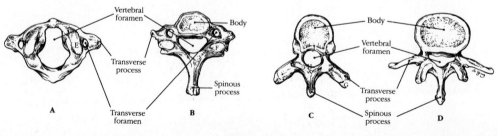

Figure 7.5

5.

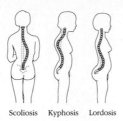

Scoliosis   Kyphosis   Lordosis

Figure 7.6

The Bony Thorax

1.   1. lungs   2. heart   3. true   4. false   5. floating   6. thoracic vertebrae   7. sternum   8. an inverted cone
2.   Ribs 1–7 on each side are vertebrosternal (true) ribs; ribs 8–10 on each side are vertebrochondral ribs; and ribs 11–12 on each side are vertebral ribs. (Ribs 8–12 collectively are also called false ribs.)

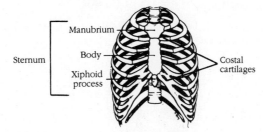

Figure 7.7

Part II: The Appendicular Skeleton

The Pectoral (Shoulder) Girdle and the Upper Limb

1.   1. pelvic girdle   2. pectoral girdle   3. support   4. to protect organs   5. to allow manipulation movements
         6. locomotion
2.

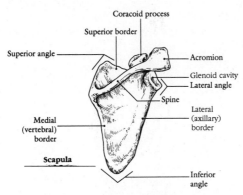

Figure 7.8

3.

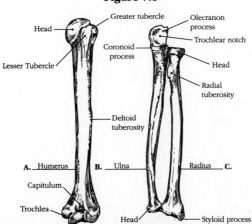

Figure 7.9

4.

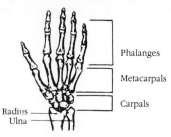

Figure 7.10

5.   1. A, C, D   2. B, E, F
6.   1. G   2. I   3. D   4. P   5. O   6. T   7. A   8. P   9. D   10. H   11. E   12. D   13. S   14. T   15. B   16. F
17. T   18. P   19. Q   20. C   21. M   22. J

The Pelvic (Hip) Girdle and the Lower Limb

1.   1. Female inlet is larger and more circular.   2. Female sacrum is less curved; pubic arch is rounder.   3. Female ischial spines are shorter; pelvis is lighter and shallower.

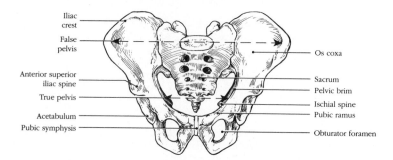

Figure 7.11

2.

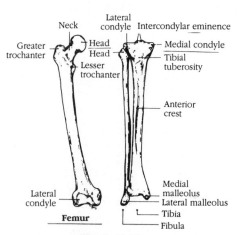

Figure 7.12

3.   1. I, K, S   2. J   3. R   4. H   5. A   6. T   7. C   8. D   9. W   10. C, Q, W   11. X   12. Q   13. W   14. N   15. L
16. B   17. V   18. O   19. P   20. E, G   21. U   22. S   23. C

Developmental Aspects of the Skeleton

1.   1. fontanels   2. compressed   3. growth   4. sutures   5. thoracic   6. sacral   7. primary   8. cervical   9. lumbar
2.   1. bone   2. T   3. T   4. pelvic inlet   5. cleft palate   6. bone markings   7. T   8. 33   9. intervertebral discs
10. T   11. T   12. 1.6 to 1   13. T   14. 12 years   15. adulthood   16. shallow   17. epiphyseal plates   18. water
19. neck of the femur   20. kyphosis

CHALLENGING YOURSELF

At the Clinic

1. The cribriform plates of the ethmoid bone, which surround the olfactory nerves. These plates are quite fragile and are often crushed by a blow to the front of the skull. This severs the nerves, which cannot grow back.
2. No; the palatine bones are posterior to the palatine processes of the maxillae. If the palatine processes do not fuse, then the palatine bones remain unfused as well.
3. The weakening of thoracic vertebrae and corresponding muscles on *one side* of the body favors the development of scoliosis.
4. Spina bifida is indicated; spinous processes will not develop unless the vertebral laminae develop properly.
5. The interosseous membrane, which connects the radial and ulnar shafts, was torn.
6. Truck drivers frequently suffer from herniated discs, since they spend so many hours in a seated position.
7. The condition is called lordosis and it is an accentuated lumbar curvature. The added weight of Mr. Ogally's pot belly changed his center of gravity, forcing the discs of the weight-bearing lumbar vertebrae to adapt to that change.
8. The man is of an age at which the xiphoid process finally ossifies. In its cartilaginous state, the xiphoid process is not as easily felt.
9. The condition is clubfoot.
10. Seven bones contribute to the orbit: frontal, sphenoid, zygomatic, maxilla, palatine, lacrimal, and ethmoid bones.
11. Malcolm's shoulder has "fallen" anteromedially. Because of its curvatures, the clavicle typically fractures anteriorly, thus sparing the subclavian artery, which lies posteriorly.
12. The grandmother's bones have lost bone mass and are more porous and brittle than those of the child, placing her weakened ribs at greater risk for fracture when stressed by physical trauma.
13. Most likely the paranasal sinuses on the right side of the face.

Stop and Think

1. Frontal, sphenoid, ethmoid, maxillae (all have paranasal sinuses) and temporal (mastoid air cells).
2. The turbinates increase the surface area of the mucous membrane of the nasal cavity, which lies over them. This increases the contact between air and membrane for filtering, warming, and humidifying the air.
3. External auditory meatus, petrous region, middle ear cavity, inner ear cavity, internal acoustic meatus.
4. Some snakes have jaws that are "designed" to dislocate; others have double articulations at the jaw.
5. No. Nodding "yes" involves the joint between the occipital condyles and the superior articular surfaces of the atlas. Shaking "no" involves the joint between the atlas and the dens of the axis.
6. The largest ribs are easiest to fracture. They protrude more, and don't have the protection of the clavicle that the first pairs have. The floating ribs do not fracture easily; lacking an anterior bony attachment, they can move (or be displaced) by forceful trauma.
7. The thoracic cage must maintain flexibility for breathing; for that, it must sacrifice the strong but rigid protection afforded by solid bone casing such as that of the cranium.
8. The acetabulum must provide greater strength and resist stress that might dislocate its limb bone. The glenoid cavity must provide more flexibility.
9. Fontanels are membranous areas that connect the developing bones in a fetal skull. The advantage for the mother is that the fontanels allow the fetal skull to be compressed (as adjoining skull bones slip over or under each other slightly). Since the skull is the largest part of the fetus, this eases birth passage somewhat. The advantage for the fetus/infant is that it permits the skull to enlarge as the brain continues to increase in size. Premature closure of fontanels always leads to brain damage.

COVERING ALL YOUR BASES

Multiple Choice

1. C 2. B, C, E 3. A, B, C, D 4. A, C, D 5. D 6. A, B, C, D, E 7. A 8. A, C, D 9. A, C, D, E 10. B, D, E
11. B, C, D 12. A, C, D, E 13. B, C, D 14. A, C, D 15. A, B, D 16. B 17. C 18. B, C 19. D
20. B, E 21. B, D, E 22. A, C, D 23. A, C, D 24. B, C, E 25. B 26. B 27. A 28. A 29. B

Word Dissection

Word root	Translation	Example	Word root	Translation	Example
1. acetabul	wine cup	acetabulum	9. glab	smooth	glabella
2. alve	socket	alveoli, alveolar	10. ham	hooked	hamate
3. append	hang to	appendicular	11. ment	chin	mental foramen
4. calv	bald	calvaria	12. odon	tooth	odontoid process
5. clavicul	a key	clavicle	13. pal	shovel	palate
6. crib	a sieve	cribriform plate	14. pect	the breast	pectoral girdle
7. den	tooth	dens	15. pelv	a basin	pelvic girdle
8. ethm	a sieve	ethmoid	16. pis	pea	pisiform

Word root	Translation	Example	Word root	Translation	Example
17. pter	wing	pterygoid process	**21.** styl	pointed instrument	styloid process
18. scaph	a boat	scaphoid	**22.** sutur	a seam	suture, sutural bone
19. skeleto	dried body	skeleton	**23.** vert	turn, joint	vertebra
20. sphen	a wedge	sphenoid	**24.** xiph	a sword	xiphoid process

Chapter 8   Joints

BUILDING THE FRAMEWORK

Classification of Joints

1.   1. To bind bones together and permit movement   2. The type of material binding the bones; whether a joint cavity is present; degree of movement permitted   3. Fibrous: synarthroses, immovable; cartilaginous: amphiarthroses, slightly movable; synovial: diarthroses, freely movable

Fibrous, Cartilaginous, and Synovial Joints

1.   1. A; 5   2. C   3. B; 1, 2, 4, and 6   4. B; 2 and 6   5. C   6. C   7. A; 3   8. A; 3   9. C   10. C   11. B; 6   12. C
13. A; 5   14. C   15. B; 1   16. A; 5   17. B; 6   18. B; 4

2.   Synovial. Synovial joints are ideal where mobility and flexibility are the goals. Most joints of the axial skeleton provide immobility (or at most slight mobility) because the bones of the axial skeleton typically *protect or support* internal (often fragile) organs.

3.

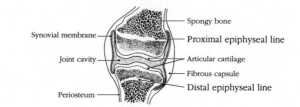

Figure 8.1

Spongy bone — Proximal epiphyseal line — Articular cartilage — Fibrous capsule — Distal epiphyseal line — Periosteum — Joint cavity — Synovial membrane

4.   1. A   2. C, D   3. B   4. B   5. C, D
5.   1. I   2. A   3. P   4. J   5. C   6. F   7. Q   8. E   9. K   10. N   11. A, B, C, H, I
6.   1. A, C, E   2. G   3. D, I   4. N   5. J   6. M   7. F   8. K   9. B, L   10. H
7.   1. B   2. E   3. E   4. E   5. A   6. B   7. E   8. F   9. C   10. D   11. C, D   12. B, F   13. A   14. E
8.   1. Multiaxial joint   2. Saddle joint   3. Tibiofibular joints   4. Multiaxial joint   5. Elbow joint   6. Amphiarthrotic
7. Bursae   8. Condyloid joint   9. Biaxial joint
9.   1. D   2. C   3. A   4. B
10.   A. hip joint   B. knee joint

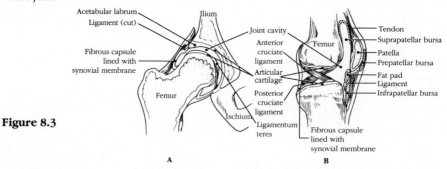

Acetabular labrum — Ligament (cut) — Ilium — Joint cavity — Tendon — Anterior cruciate ligament — Femur — Suprapatellar bursa — Patella — Prepatellar bursa — Fibrous capsule lined with synovial membrane — Articular cartilage — Fat pad — Ligament — Infrapatellar bursa — Femur — Posterior cruciate ligament — Ischium — Ligamentum teres — Fibrous capsule lined with synovial membrane

Figure 8.3

A   B

Homeostatic Imbalances of Joints

1.   1. dislocation   2. T   3. osteoarthritis   4. bursa   5. acute   6. vascularized   7. rheumatoid arthritis
8. gouty arthritis   9. T   10. synovial membrane   11. rheumatoid arthritis   12. menisci   13. reduce friction
14. osteoarthritis   15. synovial
2.   Ligaments and tendons (injured during sprains) are poorly vascularized and cartilages are avascular. Hence, these structures heal slowly and often poorly.

Developmental Aspects of Joints

1.   1. mesenchyme   2. 8   3. middle age   4. osteoarthritis   5. muscles

The Incredible Journey

1.   1. meniscus   2. femur   3. tibia   4. joint cavity   5. synovial fluid   6. lubricate   7. condyles   8. anterior cruciate   9. patella   10. bursa   11. lateral (outer)   12. tibia   13. ligament   14. blood supply   15. meniscus

CHALLENGING YOURSELF

At the Clinic

1.   Jenny's elbow has been dislocated.
2.   The temporomandibular joint is the source of his problem.
3.   Because the clavicles brace the scapulae, without clavicles the scapulae are extremely movable and the shoulders appear droopy. Although the condition is not serious, it reduces the shoulders' resistance to stress.
4.   The rotator cuff consists of ligaments that are the main reinforcers of the shallow shoulder joint. Its rupture reduces outward movements, such as abduction.
5.   The ligamentum teres contains a blood vessel that supplies the head of the femur. The femoral head ordinarily also receives nourishment from vessels reaching it from the diaphysis, but a fracture would cut off that blood supply.
6.   Flexion of his knee was exaggerated and rotation was possible. The patellar ligament is a major factor in joint stabilization.
7.   The fibula is usually used for bone grafts of this sort.
8.   Typically a "torn cartilage" is an articular disc, which is fibrocartilage.
9.   Rheumatoid arthritis, fairly common in middle-aged women, causes this type of deformity.

Stop and Think

1.   Diarthrotic joints in the axial skeleton: temporomandibular joint, atlanto-occipital joint, atlantoaxial joint, intervertebral joints (between articular processes), vertebrocostal joints, sternoclavicular joints (which include one axial bone, the sternum, and one appendicular bone, the clavicle), and sacroiliac joint (which also includes an appendicular bone).
2.   Pronation and supination involve not only rotation at the proximal radioulnar joint (a diarthrotic joint) but also crossing of the shafts of the radius and the ulna (an amphiarthrotic syndesmosis). The wrist bones are not involved.
3.   The coxal bone develops from the ossification centers in the ilium, ischium, and pubis within a fused hyaline cartilage template. Thus the joint is classified as a synchondrosis. After ossification of the entire coxal bone, the joint classification is synostosis.
4.   Interdigitation and interlocking increase the articulating surface area between adjoining bones. Such an increased area of attachment means more ligament between the bones, increasing the resistance to movement at the joints. Further, the interlocking of bony regions prevents separation of the bones.
5.   The periosteum is a dense irregular connective tissue, and ligaments are dense regular connective tissue. Microscopic examination would reveal a transition from collagenous fibers, which are in different orientations to those arranged in parallel bundles.
6.   The synovial membrane, which lines the joint cavity, is composed of loose connective tissue only. Hence, it is not an epithelial membrane.
7.   The purpose of menisci is to more securely seat the bones within the joint and to absorb shock. Although such activities are uncommon, the jaw (temporomandibular) joint can withstand great pressure when you bite the cap off a glass bottle or hang by your teeth in a circus act. The sternoclavicular joint absorbs shock from your entire arm when you use your hand to break a fall.
8.   Chronic slumping forward stretches the posterior intervertebral ligaments; once stretched, ligaments do not resume their original length. Eventually, the overextended ligaments are unable to hold the vertebral column in an erect position.
9.   The knee joint flexes posteriorly.
10.   The head is able to move in a manner similar to circumduction, but the term more accurately applies to movement at a single joint. Circular movement of the head involves the atlanto-occipital and atlantoaxial joints, as well as amphiarthrotic movements of the cervical vertebrae.
11.   A typical hinge joint has a single articulation with a convex condyle fitting into a concave fossa. The knee joint has a double condylar articulation, with tibial (concave) condyles replacing the fossa. The doubling of the condyloid joints prevents movement in the frontal plane (abduction and adduction).

COVERING ALL YOUR BASES

Multiple Choice

1. E and possibly D   **2.** B, D   **3.** A   **4.** A, B, C, D   **5.** A, B, D   **6.** A, B   **7.** B, D   **8.** A, B, D   **9.** D   **10.** A, B, C, D
11. C   **12.** D   **13.** A, B   **14.** A, B   **15.** C   **16.** D   **17.** A   **18.** B, D   **19.** C   **20.** B, C, D   **21.** D   **22.** B
23. B   **24.** D   **25.** A, D   **26.** B   **27.** D

Word Dissection

Word root	Translation	Example		Word root	Translation	Example
1. ab	away	abduction		**10.** gompho	nail	gomphosis
2. ad	toward	adduction		**11.** labr	lip	glenoid labrum
3. amphi	between	amphiarthrosis		**12.** luxa	dislocate	luxation
4. ankyl	crooked	ankylosis		**13.** menisc	crescent	meniscus
5. arthro	joint	arthritis		**14.** ovi	egg	synovial
6. artic	joint	articulation		**15.** pron	bent forward	pronation
7. burs	purse	bursa		**16.** rheum	flux	rheumatism
8. cruci	cross	cruciate ligament		**17.** spondyl	vertebra	spondylitis
9. duct	lead, draw	abduction		**18.** supine	lying on the back	supination

Chapter 9 Muscles and Muscle Tissue

BUILDING THE FRAMEWORK

Overview of Muscle Tissues

1. 1. A, B 2. A, C 3. B 4. C 5. A, B 6. A 7. C. 8. C 9. C 10. B

2. A. smooth muscle B. cardiac muscle C. skeletal muscle

3. 1. Bones 2. Promotes labor during birth 3. Contractility 4. Contractility 5. Stretchability 6. Promotes growth

Skeletal Muscle

1. 1. G 2. B 3. I 4. D 5. A 6. H 7. F 8. E 9. K 10. C

The endomysium is the fine connective tissue surrounding each muscle cell (fiber) in the figure.

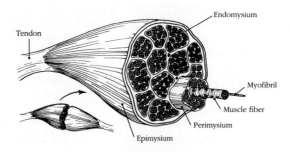

Figure 9.2

2. Matching: B1 = C2 B2 = C3 B3 = C1

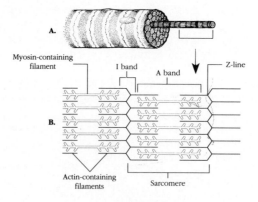

Figure 9.3

3.

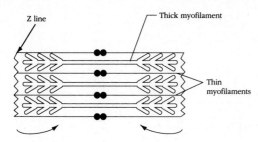

4. 1. C 2. B 3. A

5.

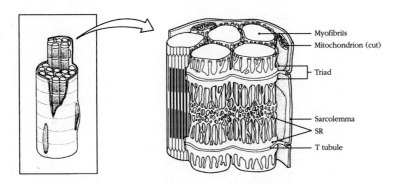

Figure 9.4

6. 1. 4 2. 9 3. 10 4. 1 5. 11 6. 8 7. 2 8. 6 9. 5 10. 7 11. 3

7.

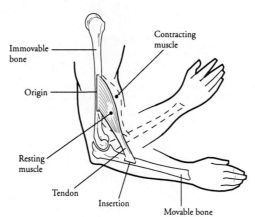

Figure 9.5

8.

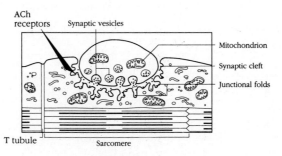

Figure 9.6

9.   1. foot proteins   2. calcium channels   3. voltage sensors   4. action potential

10.

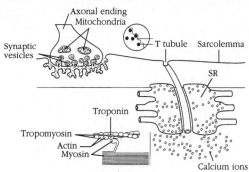

Figure 9.7

11.   1. 1   2. 3   3. 4   4. 7   5. 5   6. 2   7. 6

12.   One motor neuron and all the muscle cells it innervates.

13.   1. Multiple motor unit summation   2. Relies on aerobic respiration   3. End product is lactic acid   4. Fuels extended periods of strenuous activity   5. Psychological factors   6. Twitch contraction   7. Muscle hypertrophy   8. Increased endurance   9. Increased muscle strength

14.

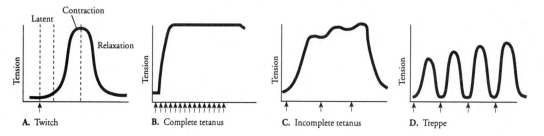

Figure 9.8

15.   1. G   2. B   3. I   4. C   5. H   6. A   7. E   8. J   9. B

16.   1. C, D, E   2. A, B

17.   1. fatty acids   2. mitochondria   3. carbon dioxide   4. creatine   5. glycogen   6. glucose   7. pyruvic acid   8. lactic acid

18.

Characteristics	Slow-twitch fatigue-resistant	Fast-twitch fatigable	Fast-twitch fatigue-resistant
Rapid twitch rate		✔	✔
Fast myosin ATPases		✔	✔
Use mostly aerobic metabolism	✔		✔
Large myoglobin stores	✔		
Large glycogen stores		✔	Intermediate
Fatigue slowly	✔		Intermediate
Fibers are white		✔	
Fibers are small	✔		
Fibers contain many capillaries and mitochondria	✔		✔

19.   1. Myofilaments do not overlap; cross bridges cannot attach to generate sliding.   2. Myofilaments touch and interfere; sliding is not possible.   3. The muscle is fatigued.

20.   Correct choices: 1, 3, 4, 7.

21. You breathe more deeply and possibly more rapidly until the debt has been repaid.

Smooth Muscle

1. Answers checked: 5, 8, 10, 12, 13, 15.   Answers circled: 1, 2, 3, 4, 14.
2. 1. endomysium   2. smooth muscle   3. single-unit (visceral)   4. two   5. longitudinally   6. circularly   7. peristalsis   8. gap junctions   9. action potentials   10. pacemaker   11. varicosities   12. diffuse junctions   13. dense bodies   14. extracellular space
3. 1. B   2. A   3. B   4. B   5. A   6. A   7. B   8. B   9. A, B   10. B   11. B
4. Smooth muscle is in the walls of hollow organs, many of which store substances temporarily (for example, urine, a baby). When smooth muscle is stretched, it responds (by contraction) only briefly and then relaxes to accommodate filling.
5. 1. Both are innervated independently by their "own" axonal endings and exhibit summation.   2. Multiunit muscle is regulated by the autonomic division (rather than the somatic division) of the nervous system and also responds to hormonal stimuli.

Developmental Aspects of Muscles

1. 1. myoblasts   2. somatic   3. smooth and cardiac   4. skeletal   5. smooth muscle   6. cephalocaudal   7. proximal-distal   8. gross   9. fine   10. testosterone   11. blood supply   12. muscular dystrophy   13. exercised   14. atrophy   15. weight   16. mass   17. connective

The Incredible Journey

1. 1. endomysium   2. motor unit   3. neuromuscular (myoneural)   4. acetylcholine   5. sodium   6. action potential   7. calcium   8. actin   9. myosin   10. calcium

CHALLENGING YOURSELF

At the Clinic

1. Curare binds to ACh receptors in the neuromuscular junctions and prevents ACh attachment. This blocks transmission of the excitatory stimulus to the muscle fibers, preventing contraction.
2. Acetylcholinesterase is the enzyme that breaks down acetylcholine. Thus it turns off the signal telling the muscle cells to contract and allows them to relax. When organophosphates interfere with the action of the enzyme, the muscle cells are continually stimulated to contract and cannot return to their resting state.
3. The symptoms indicate myasthenia gravis. If a blood test shows antibodies to ACh receptors, the disease is confirmed.
4. When a body part increases in bulk very quickly, the dermis of the skin may tear, resulting in striae (stretch marks). The young man needs to reduce the emphasis on his thighs and add exercises to work on his calves (jogging or resistance exercises) and his arms.
5. Considering Charlie's age, low-impact aerobic exercise would be better for his joints; his articular cartilage probably is not very resilient, and he could suffer from arthritis if he put too much stress on his joints. His bones would benefit from weight-bearing exercise such as walking. He needs to progress from a slow pace to a faster one, giving his cardiovascular and muscular systems time to accommodate to the increased demands.
6. Dehydration and loss of electrolytes, which can occur with profuse sweating, can trigger muscle cramps. The man should make sure to replace salts and fluids the next time he works in the heat for a prolonged period.
7. RICE: rest, ice, compression, and elevation (of her leg).
8. Tragically, the chances are good that the boy is afflicted with muscular dystrophy, probably Duchenne muscular dystrophy. There is currently no confirmed treatment for this condition, although transplants of myoblasts are being used experimentally. The condition is fatal when it progresses to the muscles of respiration.
9. Charlie is probably suffering from intermittent claudication, caused when his poorly vascularized out-of-shape muscles contract and squeeze the blood vessels to the point of collapse. This condition is less of a problem in people with a healthy, vigorous cardiovascular system. Charlie should continue his workout to improve his heart function, but he might have to proceed more slowly than initially planned.

Stop and Think

1. All three types of muscle produce movement of some sort. Skeletal muscles move bones or skin; cardiac muscle propels blood out of the heart and into blood vessels; smooth muscle raises hairs on cold skin, pushes food along the digestive tract, and dilates the pupil of the eye, just to name a few of its functions. However, only skeletal muscle maintains posture and stabilizes joints and is important in body heat production.
2. Extensibility is the ability to lengthen. If a muscle on one side of a bone didn't lengthen, the muscle on the other side wouldn't be able to shorten. Each muscle has to stretch when its opposing muscle contracts, or else the bone will not move. For stabilizing joints, muscles on both sides of the bone contract to make sure the bone does not move.
3. Ducks are "better" flyers; although their flight muscles do not contract as fast as those of chickens, they are much more fatigue-resistant (red fibers) and can function for longer periods of time.
4. A cross section through a muscle fiber at the T tubules would reveal that this tubular system branches to encircle each myofibril.

5. Cross bridge attachment is asynchronous; some cross bridges attach and pull while others detach and recock. The holding cross bridges detach just as the others start to pull. Thus the thin myofilaments are always held in place before being pulled closer in, just as a person pulling a rope grabs hold with one hand before letting go with the other. In a precision rowing team, all oars lift in unison. If all cross bridges detached in unison, nothing would hold the thin myofilaments in place. They would slide back apart after each power stroke, and no gain would be made.

6. Calcium is moved into the central, tubular network of the SR by active transport involving calsequestrin. The resulting decrease in calcium concentration in the sarcoplasm increases the rate of detachment of calcium from troponin, leading to relaxation.

7. The elasticity of the muscle fiber, coupled with the internal tension built up when the fiber shortened, causes the myofilaments to slide back apart. Each sarcomere returns to its resting length. The H zone reappears and the I band lengthens.

8. Synaptic vesicles fusing with the axonal membrane create a high surface area for release of ACh, increasing its rate of diffusion through the synaptic cleft to its receptors. The junctional folds of the sarcolemma at the synapse increase the membrane surface for ACh receptors, increasing the speed and strength of response of the muscle cell to ACh stimulation. The elaborate T system increases the communication between the cell's interior and the sarcolemma, relative to the electrical events of the sarcolemma. In the SR, the tubular arrangement between the terminal cisternae increases the surface area for active transport of calcium, speeding the onset of relaxation. Thus, high surface area increases the number of membrane transport opportunities for ACh, sodium, potassium, and calcium and increases the sensitivity of the cell to ACh and electrical activity.

9. If the muscle cell remained refractory, it could not be restimulated for additional contractions at short intervals. Hence, tetanic contraction would not be possible.

10. Since the motor units contract in turn, each has time to recover before its next turn. If the same motor unit was stimulated continually, its muscle fibers would fatigue. Even keeping your eyelids open all day would be tiring. Muscle tone is accomplished by asynchronous motor unit summation.

11. As a muscle contracts, the blood vessels within it are squeezed and can collapse. In moderate activity, collapse is unlikely or of short duration. As activity becomes more vigorous, contraction is stronger and more prolonged and blood supply can actually decrease. Active muscles rely on oxygen stored in myoglobin. Training encourages the growth of more capillaries and increases myoglobin stores. Activity also increases blood pressure, making vessel collapse less likely and resulting in the delivery of more blood with each heartbeat.

12. Muscles are extensible, but the connective tissue that replaces atrophied muscle is not. Unless the joints are moved through their full range routinely and/or braced in an extended position, the fibrous tissue will be shorter than the original muscle and the joint will be contorted.

13. Calcium binds to troponin in skeletal muscle, triggering the attachment of myosin cross bridges and the power stroke. In smooth muscle, which does not have troponin, calcium binds to calmodulin, which activates a kinase enzyme. This triggers cross bridge attachment.

14. Without an elaborate SR, calcium flux into the cell is relatively slow and requires that Ca^{2+} enter from the cell exterior. This slows and prolongs both the contractile period and the relaxation period, since expulsion of calcium is also a membrane-limited phenomenon.

15. The heart has its own "in-house" (intrinsic) pacemaker that will continue to set the pace of the heartbeat. However, without neural controls, the heart rate could not be increased during vigorous activity.

COVERING ALL YOUR BASES

Multiple Choice

1. C, D   2. A, B, C   3. B   4. B, C   5. A, B, C, D, E   6. C, E   7. C   8. D   9. A, C, D   10. A, C   11. A, D
12. C   13. C   14. A, C, D   15. A, C   16. C   17. B   18. C   19. A, D   20. A   21. B   22. A, B, C, D   23. B
24. B, D   25. B, D   26. A, B, D   27. A, B, C   28. A, C, D   29. A, C   30. A, B, C, D   31. A, D   32. C, D
33. A, B, C

Word Dissection

Word root	Translation	Example	Word root	Translation	Example
1. fasci	bundle	fascicle	7. sarco	flesh	sarcoplasm
2. lemma	sheath	sarcolemma	8. stalsis	constriction	peristalsis
3. metric	measure	isometric	9. stria	streak	striation
4. myo	mouse	myofibril	10. synap	union	synapse
5. penna	feather, wing	unipennate	11. tetan	rigid, tense	tetanus
6. raph	seam	raphe			

Chapter 10 The Muscular System

BUILDING THE FRAMEWORK

Muscle Mechanics: The Importance of Leverage and Fascicle Arrangement

1.   1. muscles   2. bones   3. the muscle inserts   4. joint
2.   Circle: fast, moves a small load over a large distance, requires minimal shortening, and muscle force greater than the load.
3.

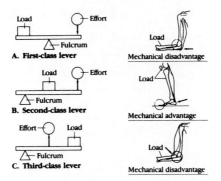

Figure 10.1

4.   The brachialis, which is inserted *closer* to the fulcrum (elbow) than the load, is structured for mechanical disadvantage; the brachioradialis, which is inserted far from the elbow, works by mechanical advantage but is a weak elbow flexor at best.
5.   1. Bipennate; C   2. Circular; D   3. Parallel; A   4. Convergent; B.   Examples of A include the biceps and triceps brachii of the arm, nearly all forearm muscles, and the rectus abdominis. Examples of B are the pectoralis major and minor. Examples of C include the rectus femoris and the flexor hallucis longus. Examples of D are the orbicularis oris and oculi muscles. Consult muscle tables for more examples.

Interactions of Skeletal Muscles in the Body

1.   1. B   2. C   3. A   4. D

Naming Skeletal Muscles

1.   1. A, F   2. D   3. G   4. A, F   5. B, C   6. A   7. B, C   8. C, F   9. C, E   10. D

Major Skeletal Muscles of the Body

1.   1. E   2. G   3. A   4. F   5. C   6. B   7. H

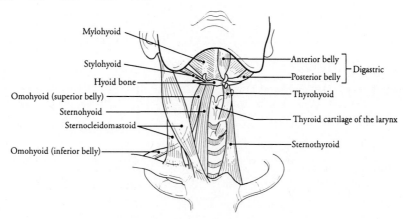

Figure 10.3

2.   1. E   2. C   3. H   4. F   5. B   6. D   7. G
3.   1. zygomatic   2. buccinator   3. orbicularis oculi   4. frontalis   5. orbicularis oris   6. masseter   7. temporalis
   8. splenius   9. sternocleidomastoid   10. pterygoid

4. They work as a group during swallowing to propel a food bolus to the esophagus.

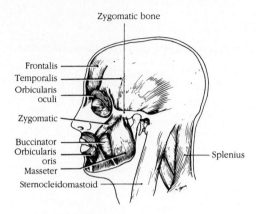

Figure 10.4

5. 1. rectus abdominis   2. pectoralis major   3. deltoid   4. external oblique   5. sternocleidomastoid   6. serratus anterior   7. rectus abdominis, external oblique, internal oblique, transversus abdominis   8. linea alba   9. transversus abdominis   10. external intercostals   11. diaphragm   12. pectoralis minor

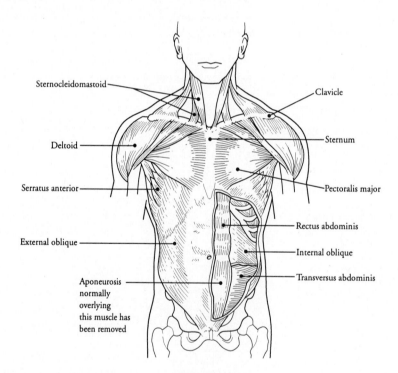

Figure 10.5

6. 1. trapezius   2. latissimus dorsi   3. deltoid   4. erector spinae   5. quadratus lumborum   6. rhomboids   7. levator scapulae   8. teres major   9. supraspinatus   10. infraspinatus   11. teres minor

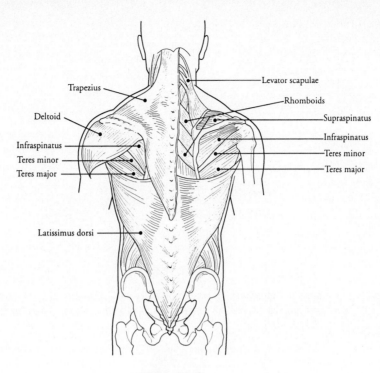

Figure 10.6

7. 1. C   2. D   3. B   4. A
8. 1. B, E   2. C, F   3. A, D, G   4. F   5. A   6. E   7. D
9. 1. C   2. A   3. B   4. D   5. F   6. E

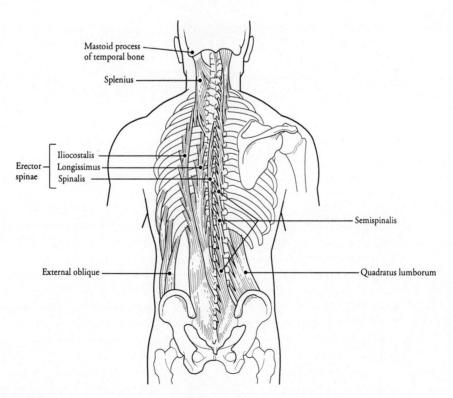

Figure 10.7

10. 1. H   2. J   3. G   4. E, F   5. I   6. A, C   7. C   8. D   9. K

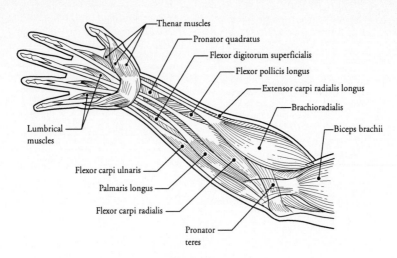

Figure 10.8

11. 1. flexor carpi ulnaris   2. extensor digitorum   3. biceps brachii   4. brachioradialis   5. triceps brachii   6. anconeus   7. brachialis   8. extensor carpi radialis brevis   9. extensor carpi radialis longus   10. abductor pollicis longus   11. extensor pollicis brevis   12. deltoid

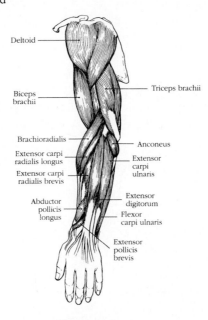

Figure 10.9

12. 1. iliopsoas   2. gluteus maximus   3. triceps surae   4. tibialis anterior   5. adductors   6. quadriceps   7. hamstrings
8. gluteus medius   9. gracilis   10. tensor fasciae latae   11. extensor digitorum longus   12. peroneus longus
13. tibialis posterior

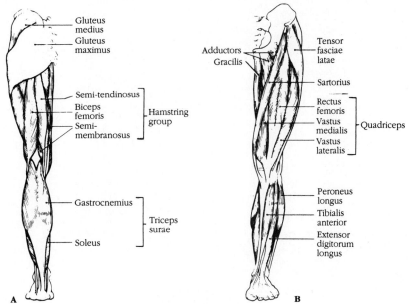

Figure 10.10

13. 1. deltoid   2. gluteus maximus or medius   3. vasti   4. quadriceps   5. calcaneal or Achilles   6. proximal
7. forearm   8. anterior   9. posteriorly   10. knee   11. flex
14. 1. iliopsoas, rectus femoris (adductors longus and brevis)   2. quadriceps   3. tibialis anterior (extensor digitorum
longus, peroneus tertius, extensor hallucis longus)
15. 1. E   2. D   3. F   4. A   5. G   6. C   7. B
16. 1. Biceps femoris   2. Antagonists   3. Gluteus minimus   4. Vastus medialis   5. Teres major   6. Peroneus longus
7. Teres minor
17. 1. 4   2. 5   3. 17   4. 16   5. 7   6. 6   7. 20   8. 14   9. 19   10. 12   11. 11   12. 10   13. 22   14. 1   15. 2
16. 3   17. 15   18. 21   19. 13   20. 9   21. 18   22. 8
18. 1. 1   2. 2   3. 5   4. 9   5. 7   6. 4   7. 6   8. 3   9. 8   10. 10   11. 11
19.

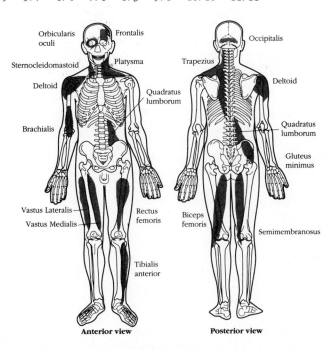

Figure 10.13

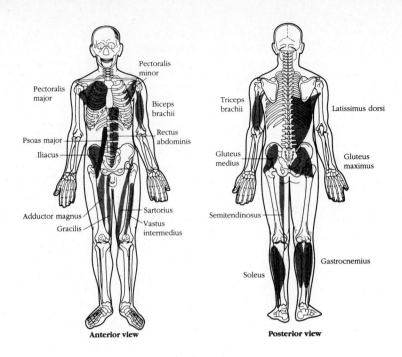

Figure 10.14

CHALLENGING YOURSELF

At the Clinic

1. The facial nerve, which innervates the muscles of the face.
2. When we are in the fully bent-over (hip flexed) position, the erector spinae are relaxed. When that movement is reversed, they are totally inactive, leaving the gluteus maximus and hamstring muscles to initiate the action. Thus sudden or improper lifting techniques are likely to injure both back ligaments and the erector spinae, causing the latter to go into painful spasms.
3. The phrenic nerve innervates the diaphragm; the intercostal nerves supply the intercostal muscles. Paralysis of these essential breathing muscles is fatal.
4. The unilateral weakening of the abdominal wall will result in scoliosis because the stronger contraction of the muscles of the opposite side will pull the spine out of alignment.
5. The rectus abdominis is a narrow, medially placed muscle that does not extend completely across the iliac regions. No, if the incision was made as described, the rectus abdominus was not cut.
6. The muscle of the urogenital diaphragm, which surrounds the vaginal opening, through which uterine prolapse would occur, is the deep transverse perineus. Lateral to the vagina is the pubococcygeus of the levator ani (part of the pelvic diaphragm).
7. The deltoid and the pectoralis major have broad origins on the inferior border of the clavicle. The sternocleidomastoid inserts on the medial superior border of the clavicle; the trapezius inserts on the lateral third of the clavicle. An arm sling inhibits use of these clavicle-moving muscles.
8. Tendons attaching at the anterior wrist are involved in wrist and finger flexion. Malcolm will lose his ability to make a fist and grasp a baseball.
9. The hamstrings can be strained (pulled) when the hip is flexed and the knee is vigorously extended at the same time.
10. The latissimus dorsi and the trapezius, which together cover most of the superficial surface of the back, are receiving most of the massage therapist's attention.
11. Gluteus maximus—action prevented is lateral rotation of the thigh; gluteus medius and minimus—actions prevented are thigh abduction and medial rotation.

Stop and Think

1. Very few muscles don't attach to bone at least at one end. The muscles in the tongue (intrinsic tongue muscles), which change the shape of the tongue, are among those few. Sphincter muscles, such as the orbicularis muscles of the face, the sphincter ani, and the sphincter urethrae, cause constriction of an opening. Muscles of facial expression mostly originate on bone but insert on fascia or skin; consequently, they alter the position of the soft tissues of the face.
2. With the help of fixators, an insertion can be the immovable end and the origin can be moved in selected cases. For example, the trapezius originates on the occipital bone and inserts on the scapula, but fixing the scapula will allow the trapezius to assist in extension of the head. Chin-ups give another example: the forearm is fixed when grasping a chin-up bar, so the bones of origin (the shoulder girdle)—and the body along with them—are moved.

3.   Muscles pull bones toward their origin as they contract: muscles originating on the front of the body pull bones anteriorly (usually flexion); muscles on the back pull bones posteriorly (usually extension); laterally placed muscles abduct (or, with the foot, evert); medially located muscles adduct (or invert the foot). Muscles *usually* move the bone beyond (distal to) their bellies: the muscles on the thorax move the arm; the muscles of the arm move the forearm; the muscles of the forearm move the wrist and fingers. Likewise, the muscles of the hip move the thigh, the muscles of the thigh move the leg, and so on.

4.   The latissimus dorsi inserts on the anterior surface of the humerus, getting there via its medial side. In rotation, the latissimus pulls the anterior surface of the humerus toward the body—this is medial rotation.

5.   The tendon of the tibialis anterior is extensive, starting at the distal third of the tibia and continuing around the medial side of the ankle, finally inserting on the plantar surface of the foot. Because it inserts on the bottom of the foot, it can pull the sole of the foot medially (inversion). Likewise, on the lateral side, the peroneus longus inserts on the plantar surface, so it can pull the sole laterally (eversion).

6.   The dorsal muscles of the lower limb are extensor muscles that act against the pull of gravity. Flexion at lower limb joints is assisted by gravity.

Closer Connections: Checking the Systems—Covering, Support, and Movement

1.   The "tummy tuck," by reducing the abdominal musculature's girth, forces the abdominal contents into a smaller space. This increases abdominal pressure, which puts pressure on the vertebral column, forcing the vertebrae farther apart. As a result, compression of the intervertebral discs is lessened, which in turn reduces pressure on the nerves.

2.   The need of the muscles for calcium is most crucial. Calcium plays an active role in muscle contraction but is fairly inert in bone tissue. Bones will give up calcium to the point of being excessively soft in order to maintain proper calcium levels for muscle contraction and other vital body functions (secretion of hormones and neurotransmitters, blood clotting, etc.).

3.   Collagen is a component of connective tissue. In the integumentary system, it is in the dermis. In the skeletal system, which is composed almost entirely of connective tissue, collagen is found in the extracellular matrix of bone, as well as in hyaline cartilage, fibrocartilage, and the dense fibrous periosteum. In the muscular system, collagen fibers are found in the deep fascia and the deeper layers of connective tissue within and around individual muscles.

4.   The greatest rate of mitosis in the systems in this unit is found in the epidermis of the skin. Least mitotically active in this group are mature skeletal muscle cells, which do not divide once full growth is achieved.

5.   Homeostasis of body temperature is maintained by the skin (its thermoreceptors monitoring external environmental temperature, as well as sweating, flushing, turning blue) and by the activity of the muscular system (shivering).

6.   The integumentary system has several lines of defense that protect the body not only from infection but also from excessive fluid and ion loss. The skeletal system contains bones that encase and protect vital organs.

COVERING ALL YOUR BASES

Multiple Choice

1. A   2. B   3. B, D   4. A, B, C, D   5. A, B, C   6. B, C, D   7. A, D   8. B, C   9. A, C, D   10. A, C, D
11. A, B   12. A, B   13. C   14. D   15. A, B, C, E   16. A, B, C, D   17. B, D   18. B   19. A, B, D   20. C
21. B, C   22. B, D   23. A, B   24. B, C, D   25. A, B, C, D   26. D   27. D   28. B   29. A, B, C, D

Word Dissection

Word root	Translation	Example	Word root	Translation	Example
1. agon	contest	agonist, antagonist	6. glossus	tongue	genioglossus
2. brevis	short	peroneus brevis	7. pectus	chest, breast	pectoralis major
3. ceps	head, origin	biceps brachii	8. perone	fibula	peroneus longus
4. cleido	clavicle	sternocleidomastoid	9. rectus	straight	rectus abdominis
5. gaster	belly	gastrocnemius			

Chapter 11   Fundamentals of the Nervous System and Nervous Tissue

BUILDING THE FRAMEWORK

1.   1. It monitors all information about changes occurring both inside and outside the body.   2. It processes and interprets the information received and integrates it in order to make decisions.   3. It commands responses by activating muscles, glands, and other parts of the nervous system.

Organization of the Nervous System

1.

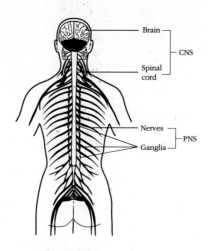

Figure 11.1

2.   1. B   2. D   3. C   4. A   5. B   6. C

Histology of Nervous Tissue

1.   1. T   2. neuroglia   3. PNS   4. neurons   5. astrocytes   6. microglia   7. ependymal cells   8. oligodendrocytes
         9. astrocytes
2.   1. B   2. D   3. E   4. C   5. A   6. C   7. D   8. A   9. E   10. F   11. E   12. C
3.   1. Centrioles   2. Mitochondria   3. Glycogen   4. Output   5. Axon collateral
4.

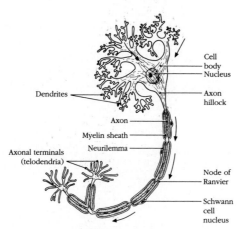

Figure 11.2

5.   1. D   2. B   3. A   4. C
6.   Neurons *have extreme longevity,* they are *amitotic,* and they have a *high metabolic rate.*
7.   1. G   2. C   3. E   4. I   5. A   6. F   7. H   8. D
8.   1. Cell body   2. Centrioles   3. Clusters of glial cells   4. Dendrites   5. Gray matter
9.   A. Bipolar, sensory   B. Unipolar, sensory   C. Multipolar, motor or association
10.   1. motor   2. multipolar   3. association   4. multipolar   5. bipolar   6. multipolar   7. sensory   8. unipolar
11.   1. SS   2. VS   3. VM   4. SM   5. VS   6. SS   7. VM
12.   A. Unmyelinated   B. Myelinated

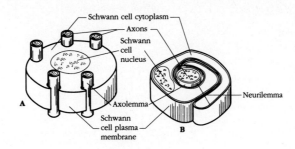

Figure 11.4

Neurophysiology

1. 1. Ohm's 2. I = V/R 3. ions 4. voltage 5. plasma membrane 6. proteins 7. passive or leakage 8. gated
9. chemical 10. electrical

2. Check 1, 4, 5, 6, 9

3.

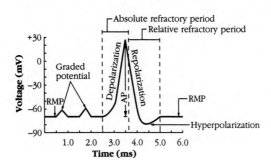

Figure 11.5

4. 1. H 2. C 3. G 4. I 5. B 6. A 7. J 8. L 9. K 10. B 11. D 12. L 13. F 14. E 15. B 16. C
17. B 18. E 19. B

5. 1. High Na$^+$ 2. High Cl$^-$ 3. 3Na$^+$ in / 2K$^+$ out 4. Nerve impulse 5. Hyperpolarization 6. Cell bodies

6. 1. faster 2. sodium 3. T 4. permeability to sodium ions 5. presynaptic 6. a neurotransmitter binds
7. a potential difference

7. 1. calcium 2. presynaptic axonal membrane 3. neurotransmitter 4. synaptic cleft 5. receptors 6. membrane
potential 7. excitatory 8. sodium and potassium 9. positive 10. action potentials 11. action potential
12. hillock 13. inhibitory 14. negative 15. hyperpolarization 16. summation 17. summate 18. axoaxonic
19. neuromodulators

8. 1. A 2. B 3. B 4. A 5. B 6. A 7. B 8. B 9. A

9. 1. B, F, G 2. D, F, H, J, K 3. A, E, G 4. C, E, F, H, I, L

10. 1. Slow response 2. Short-lived effect 3. Weak stimuli 4. Pain receptors 5. ACh

11.

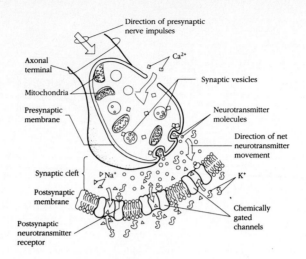

Figure 11.6

12.

Figure 11.7

Basic Concepts of Neural Integration

1.   1. PNS   2. All-or-nothing response   3. Variety of stimuli
2.   1. pin-prick pain   2. skeletal muscle   3. two (third with muscle)

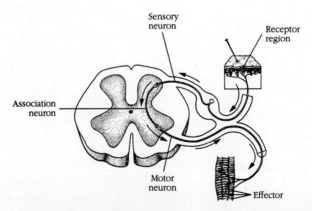

Figure 11.8

3.   1. B   2. A   3. B   4. D   5. A   6. D   7. C   8. C

Developmental Aspects of Neurons

1.   1. surface ectoderm (neuroectoderm)   2. mesoderm   3. neuroblasts   4. proliferation   5. mitosis (division)   6. migration   7. differentiation   8. neuron specialization   9. synthesis of specific neurotransmitters   10. formation of synapses   11. unipolar   12. electrical   13. glia   14. nerve cell adhesion molecule (N-CAM)   15. growth cones   16. die

CHALLENGING YOURSELF

At the Clinic

1.   Diuretics often cause a potassium deficiency, which would lead to hyperpolarization of neurons' receptive areas. This would inhibit impulse generation and reduce neuronal output.

2.   Many viruses, including some herpesviruses, travel up axons by a retrograde type of transport used for recycling of cellular materials in the cell body.

3.   Since nerve cells are amitotic, the tumor must be a glioma, developing from one of the types of neuroglia.

4.   Myelination is essential for saltatory conduction. Multiple sclerosis is characterized by loss of myelination, interrupting the rapid node-to-node propagation of action potentials.

5.   GABA is an inhibitory neurotransmitter that counterbalances the excitation of motor and integrative function. Seizures result from hyperactivity of neurons that have escaped normal GABA-induced inhibition.

6.   Caffeine is a stimulant; it lowers the threshold of neurons and makes them more irritable, decreasing the amount of stimulation required to trigger an action potential.

7.   The continuous pressure of the bannister on his axillary blood vessels impaired the delivery of O_2 and other nutrients to the nerve fibers in his right upper limb, leading to their inability to conduct nerve impulses.

Stop and Think

1.   Microglia are involved in phagocytosis of infected and damaged cells. Hence, trauma (due to stroke or a physical blow) and infection can induce an increase in the concentration of microglia.

2.   A high degree of branching increases the surface area of a cellular process, as is seen in dendritic spines and in telodendria. The increase in surface area increases the space available for synapses.

3.   Increased axonal diameter at the nodes of Ranvier increases the surface area of the axonal membrane there. This makes more room for sodium channels and increases the speed of impulse propagation and recovery.

4.   Neurons without axons can come into close contact with dendrites of other neurons and pass along their excitation by direct contact.

5.   Excitatory stimulation (EPSPs) opens channels that allow simultaneous passage of sodium and potassium. Inhibition (IPSPs) opens potassium and/or chloride gates. The net sodium flux (occurring at excitatory synapses) generates an electrical charge opposite that of the inhibitory potassium and chloride fluxes. If the ion fluxes exactly balance, the membrane voltage will not differ from the resting membrane potential and will graph as a flat line.

6.   The pathway with longer axons will have fewer synapses. Since synaptic delay significantly slows impulse transmission, the pathway with longer fibers will be faster. In a parallel after-discharge circuit, the parallel pathways have different numbers of synapses. This arrangement ensures that the final postsynaptic cell will be stimulated sequentially instead of simultaneously by the presynaptic neurons.

7.   If chloride gates open (due to inhibitory stimulation), more negative chloride ions will diffuse into the cell from the extracellular space (following their concentration gradient). Hence, the chloride flux hyperpolarizes the cell (inside becomes more negative).

8.   The effect of neurotransmitter is to create a *graded* potential, not an *action* potential. Depolarization and refractory period are characteristics of the action potential. Persistence of neurotransmitter can increase the frequency of action potentials but will not affect the character of the action potential.

9.   False. Most neurons make and release more than one neurotransmitter, simultaneously or at different times. It is quite likely that these neurotransmitters have very different effects. For example, release of an amino acid will most likely open ion channels and exert its effect only briefly, whereas a peptide neurotransmitter is more likely to produce a more prolonged response (mediated by a G protein second messenger system).

10.   Formation of new synapses can increase the influence of a particular presynaptic neuron over a postsynaptic neuron. Neurons in the facilitated zone have fewer synapses and require more impulses to move the postsynaptic cell to threshold. Increasing the number of synaptic connections can change a neuron's zone designation to "discharge zone." Fewer impulses may then be sufficient to push the postsynaptic cell to threshold.

11.   Yes, a neuron can be the postsynaptic cell of a convergent circuit and the presynaptic cell of a divergent circuit, just to name one possibility. Formation of axon collaterals increases the likelihood of a single neuron being part of multiple circuits.

12.   It's an axon because (1) dendrites do not generate and conduct action potentials (peripheral processes do generate and conduct APs) and (2) dendrites are never myelinated (many, if not most, peripheral processes of sensory neurons—particularly somatic sensory neurons—are myelinated). It's a dendrite because the electrical current is moving *toward* the cell body. (This is the older definition of dendrite/axon, i.e., dendrites conduct electrical currents toward the cell body; axons conduct an impulse away from the cell body.) Function follows structure; if the peripheral process looks and acts like an axon functionally, then it must be an axon.

13. After birth, the number of neurons actually decreases. In a case of a lost ability, the increased dependence on other senses can slow the loss of neurons involved in the "enhanced sense" pathways and facilitate those pathways. It is neuron retention and increased efficiency of neurotransmission, not an increase in neurons, that causes the enhancement.

14. In the PNS, Schwann cells reconstruct the channel to guide new axonal growth. In the CNS, the myelinating cells (oligodendrocytes) do *not* aid axonal regeneration. Furthermore, growth inhibitors are present in the CNS.

COVERING ALL YOUR BASES

Multiple Choice

1. D　**2.** D　**3.** C　**4.** B　**5.** B, D　**6.** A　**7.** B, D　**8.** B, C　**9.** D　**10.** B　**11.** C　**12.** A, B　**13.** A, C, D
14. B, C　**15.** A　**16.** B　**17.** B, D　**18.** A　**19.** D　**20.** B, D　**21.** A, B　**22.** C, D　**23.** C　**24.** D　**25.** A
26. A, B, C　**27.** A, B　**28.** B, D　**29.** C　**30.** C　**31.** D

Word Dissection

Word root	Translation	Example	Word root	Translation	Example
1. dendr	branch	dendrite	**6.** nom	law	autonomic
2. ependy	tunic	ependymal	**7.** oligo	few	oligondendrocyte
3. gangli	knot, swelling	ganglion	**8.** salta	leap, dance	saltatory conduction
4. glia	glue	neuroglia	**9.** syn	clasp, join	synapse
5. neur	cord	neuron			

Chapter 12　The Central Nervous System

BUILDING THE FRAMEWORK

The Brain

1. 　1. D　2. L　3. F　4. C　5. K　6. B　7. E　8. A　9. I　10. H　11. J　12. G　Areas B and C should be striped.

2.

Secondary brain vesicle	Adult brain structures	Neural canal regions
Telencephalon	Cerebral hemispheres (cortex, white matter, basal nuclei)	Lateral ventricles; superior portion of third ventricle
Diencephalon	Diencephalon (thalamus, hypothalamus, epithalamus)	Most of third ventricle
Mesencephalon	Brain stem: midbrain	Cerebral aqueduct
Metencephalon	Brain stem: pons; cerebellum	Fourth ventricle
Myelencephalon	Brain stem: medulla oblongata	Fourth ventricle

3.

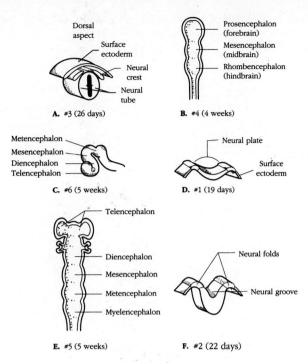

A. #3 (26 days)   B. #4 (4 weeks)

C. #6 (5 weeks)   D. #1 (19 days)

E. #5 (5 weeks)   F. #2 (22 days)

Figure 12.2

4.

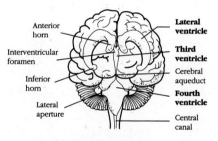

Figure 12.3

5.

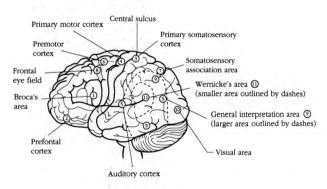

Figure 12.4

6. 1. G 2. W 3. W 4. W 5. W 6. G 7. G 8. W

7. 1. postcentral 2. temporal 3. frontal 4. Broca's 5. left 6. T 7. precentral 8. premotor 9. fingertips 10. somatosensory association area 11. occipital 12. T 13. prefontal 14. T 15. affective 16. language 17. right

8. 1. B, C, D, F 2. H 3. A 4. I 5. B 6. G 7. E 8. D 9. F

9. 1. L 2. N 3. Q 4. O 5. B 6. F 7. A 8. D 9. M 10. I 11. K 12. G 13. P 14. H 15. J 16. C 17. E Figure 12.5: Brain stem areas O, J, and K should be colored blue. The gray dotted areas contain cerebrospinal fluid and should be colored yellow.

10. 1. hypothalamus 2. pons 3. cerebellum 4. thalamus 5. medulla oblongata 6. cerebral peduncle
7. pineal body 8. pons 9. corpora quadrigemina

11.

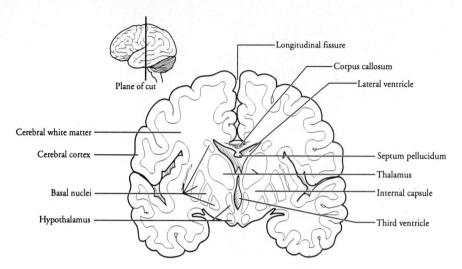

Longitudinal fissure

Corpus callosum

Lateral ventricle

Plane of cut

Cerebral white matter

Cerebral cortex

Septum pellucidum

Thalamus

Basal nuclei

Internal capsule

Hypothalamus

Third ventricle

Figure 12.6

12.

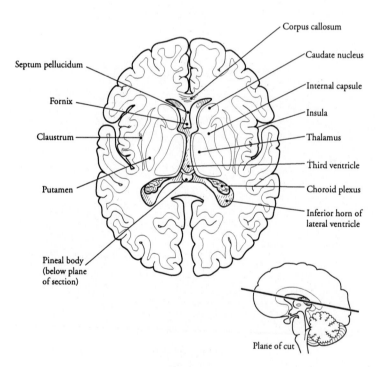

Septum pellucidum

Fornix

Claustrum

Putamen

Pineal body
(below plane
of section)

Corpus callosum

Caudate nucleus

Internal capsule

Insula

Thalamus

Third ventricle

Choroid plexus

Inferior horn of
lateral ventricle

Plane of cut

Figure 12.7

13. The cranial nerves include all named nerves (except those designated as spinal nerves). The midbrain is the area including C, I, and K. The diencephalon is the region from which the mammillary bodies and the infundibulum project. The pons is the enlarged region between K and A; the medulla is the region from A to just below D where the spinal cord begins.

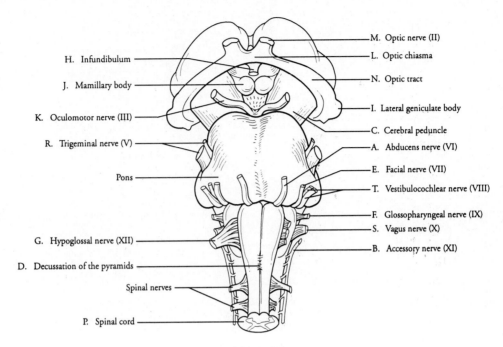

H. Infundibulum

J. Mamillary body

K. Oculomotor nerve (III)

R. Trigeminal nerve (V)

Pons

G. Hypoglossal nerve (XII)

D. Decussation of the pyramids

Spinal nerves

P. Spinal cord

M. Optic nerve (II)

L. Optic chiasma

N. Optic tract

I. Lateral geniculate body

C. Cerebral peduncle

A. Abducens nerve (VI)

E. Facial nerve (VII)

T. Vestibulocochlear nerve (VIII)

F. Glossopharyngeal nerve (IX)

S. Vagus nerve (X)

B. Accessory nerve (XI)

Figure 12.8

14. 1. Cerebral peduncles   2. Pineal gland   3. Pons   4. Paralysis   5. Speech disorder   6. Motor impairment
7. Association fibers   8. Internal capsule   9. Caudate nucleus   10. Parietal lobe

15. 1. 5   2. 2   3. 5   4. 1   5. 3   6. 4

16. 1. R   2. R   3. L   4. L   5. L   6. R   7. L   8. R   9. R   10. L   11. R   12. L   13. R   14. L

17.

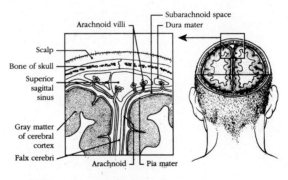

Arachnoid villi

Subarachnoid space
Dura mater

Scalp

Bone of skull

Superior
sagittal
sinus

Gray matter
of cerebral
cortex

Falx cerebri

Arachnoid   Pia mater

Figure 12.9

18. 1. pia mater   2. arachnoid villi   3. dura mater   4. subarachnoid space   5. subarachnoid space   6. superior sagittal sinus   7. falx cerebri

19. 1. blood plasma   2. hydrogen ions (H^+)   3. choroid plexuses   4. ventricle   5. ependymal   6. cerebral aqueduct
7. central canal   8. subarachnoid space   9. fourth ventricle   10. arachnoid villi   11. hydrocephalus   12. brain damage

20. 1. tight junctions   2. oxygen   3. potassium   4. hypothalamus   5. lipid-soluble   6. T

21. 1. E   2. F   3. D   4. G   5. B   6. C   7. I   8. A   9. H

The Spinal Cord

1. 1. D   2. B   3. A   4. H   5. G   6. J   7. I   8. C   9. F

2. 1. foramen magnum   2. lumbar   3. lumbar tap or puncture   4. 31   5. 8   6. 12   7. 5   8. 5   9. cauda equina

3.

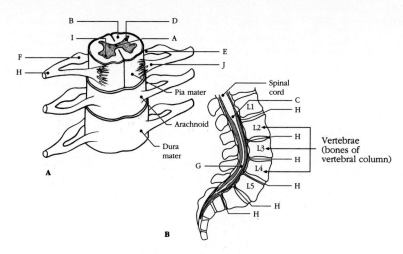

Figure 12.10

4.   1. meningeal   2. epidural   3. cerebrospinal fluid   4. anterior median fissure   5. white   6. gray commissure
7. anterior   8. lateral   9. sensory   10. motor   11. neural crest
5.   1. C   2. B   3. A
6.   1. E   2. G   3. C   4. I   5. D   6. H   7. F   8. B

Diagnostic Procedures for Assessing CNS Dysfunction

1.   1. pneumoencephalogram   2. T   3. cerebral angiogram   4. PET scan   5. above

Developmental Aspects of the Central Nervous System

1.   1. deafness   2. oxygen   3. cerebral palsy   4. cephalocaudal   5. gross   6. hypothalamus   7. anencephaly
8. spina bifida   9. myelomeningocele   10. myelination   11. synaptic   12. senility   13. stroke (CVA)
14. reduction

The Incredible Journey

1.   1. cerebellum   2. medulla oblongata   3. hypothalamus   4. memories   5. temporal
6. a motor speech area (e.g., Broca's area)   7. reasoning   8. frontal   9. vagus   10. dura mater
11. subarachnoid space   12. fourth

CHALLENGING YOURSELF

At the Clinic

1.   The cerebellum.
2.   Meningitis; by examining a sample of CSF for the presence of microbes.
3.   The RAS of the reticular formation of the brain stem.
4.   Somatosensory cortex.
5.   Broca's area and adjacent regions of the motor cortex.
6.   Left motor cortex.
7.   Somatosensory association area.
8.   Prefrontal cortex.
9.   Dyslexia.
10.   Cerebral palsy—it will not get worse.
11.   Intracranial hemorrhage.
12.   Alzheimer's disease is diagnosed with certainty only by histological examination, not by clinical signs. Many other factors, such as drug interaction and malnourishment, can cause similar behavior.
13.   Tabes dorsalis.
14.   Upper motor neuron damage will leave spinal reflexes intact (spastic paralysis), which will maintain the muscle mass.
15.   Amyotrophic lateral sclerosis.
16.   The Trevors' baby obviously has spina bifida cystica—a condition of incomplete formation of the vertebral arches, which is often accompanied by hydrocephalus and can cause severe neural defects. The most severe cases (where the cyst contains parts of the spinal cord and spinal roots) render the lower spinal cord functionless. The lower limbs, bowel, and bladder muscles are paralyzed. In the case of the Trevors' baby, the outlook appears pretty good. The shunt will drain off the excess CSF responsible for the hydrocephalus. The "weak in the ankles" prognosis probably means the child will have some degree of skeletal muscle weakness in the lower limbs but not paralysis.

Stop and Think

1. Humans rely primarily on vision for assessment of the external environment; consequently, much of the cerebral cortex is devoted to this sense.
2. The close correlation of sensory and motor neurons that innervate the same body part is a reflection of the growth of nerve processes that connect the cortex to each body part. The sensory and motor axons grow together to each part of the body.
3. Yes. All tracts from the cerebrum pass as projection fibers alongside (or into) the thalamus, and from there to the midbrain, pons, and medulla—all the brain stem regions.
4. The primary cortex receives input directly from the eye; damage here will cause blindness. Damage to the association cortex will cause inability to process visual images and make sense of them, but visual sensations will still occur.
5. Yes. If blockage of CSF flow is internal, the ventricular spaces above the blockage will enlarge, making the brain larger and more hollow and putting pressure on neural tissue from the inside. If the blockage is in the arachnoid villi, CSF will accumulate in the subarachnoid space, putting external pressure on the brain. The skull will enlarge, but the brain will not.
6. Since mannitol causes crenation and fluid loss from cells, it must be increasing the osmotic pressure of the solution bathing (outside) the cells; that is, it increases blood's solute osmotic concentration in comparison to that of the cells of the capillary wall. Osmosis hinges on differential permeability. If the cells' plasma membranes were permeable to mannitol, it would diffuse into the cells until it reached equilibrium, and osmosis would not occur. Thus, the cells must be impermeable to mannitol.
7. The correct choice is (c); the injury was at spinal cord level L_3. Recall that the spinal cord levels are several segments superior to vertebral levels (L_3 cf. T_{12}).
8. The spinal cord is so narrow that any damage severe enough to cause paralysis will most likely affect both sides of the cord. In the brain, motor control of the two sides of the body is sufficiently separated so that unilateral damage and paralysis are more likely.
9. The posterior side of the spinal cord carries predominantly ascending sensory tracts, so damage here would likely cause paresthesia.
10. Early in gestation the nervous system contains fewer total cells, each destined to give rise to large areas of the brain, spinal cord, or PNS. Hence, death of a single cell can have wide-ranging effects. Late in gestation, most areas of the nervous system have developed fairly completely, and most cell division is complete. Death of a single cell will not have much effect.
11. Each time the head (and brain) is subjected to physical trauma, neurons are injured, and the damage is cumulative and nonreversible. "Punch drunk" refers to the neural and motor deficits often expressed by former professional boxers.
12. The primary somatosensory cortex is being stimulated, so (c) is the most appropriate response.

COVERING ALL YOUR BASES

Multiple Choice

1. C 2. B 3. A 4. C 5. B 6. C, D 7. A, C 8. B 9. A 10. A, B, C 11. C 12. B, C 13. B, C, D
14. C 15. B, C 16. C 17. A, C 18. A, C 19. B, C, D 20. A, C, D 21. A, B, C 22. A, B, C, D 23. D
24. A, B, D 25. B, D 26. A, B, C 27. C 28. A, C 29. C 30. B 31. D 32. A, B, D 33. D 34. B
35. A, D 36. B 37. D

Word Dissection

Word root	Translation	Example	Word root	Translation	Example
1. campo	sea animal	hippocampus	9. infundib	funnel	infundibulum
2. collicul	little hill	superior colliculi	10. isch	suppress	ischemia
3. commis	united	commissure	11. nigr	black	substantia nigra
4. enceph	within the head	mesencephalon	12. rhin	nose	rhinencephalon
5. falx	sickle	falx cerebri	13. rostr	snout	rostral
6. forn	arch	fornix	14. uncul	little	homunculus
7. gyro	turn, twist	gyrus	15. uncus	hook	uncus (olfactory)
8. hippo	horse	hippocampus			

Chapter 13 The Peripheral Nervous System and Reflex Activity

BUILDING THE FRAMEWORK

Overview of the Peripheral Nervous System

1. 1. sensory 2. afferent 3. motor 4. efferent 5. mixed 6. somatic 7. autonomic 8. somatic motor
   9. autonomic motor

2. 1. receptor → 2. sensory nerve fiber → 3. ventral ramus of spinal nerve → 4. spinal nerve →
5. dorsal root of spinal nerve → 6. motor nerve fiber → 7. spinal nerve → 8. ventral ramus of spinal nerve →
9. effector

3. 1. C, 2   2. B, 1   3. A, 5 (3)   4. A, 4   5. A, 2   6. A, 3   7. B, 2   8. B, 2   9. A, 1

4. 1. Tendon stretch   2. Numerous in muscle   3. Light touch   4. Thermoreceptor   5. Ruffini's corpuscle
6. Nociceptors   7. Merkel discs   8. Interoceptor   9. Free dendritic endings

5.

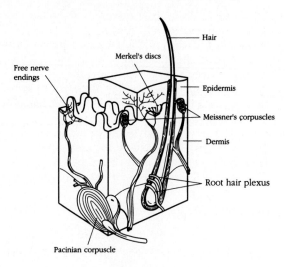

Figure 13.1

6. 1. A, G   2. A, G   3. B, F, G, H   4. B, F, H   5. D, E, I   6. A

7. 1. energy   2. permeability   3. local graded potential   4. receptor potential   5. generator potential
6. nerve impulse   7. transduction

8. Check 1, 3, 4, 8

9. somatic afferent, somatic efferent, visceral afferent, visceral efferent

10.

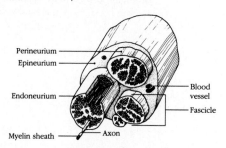

Figure 13.2

11. 1. divide   2. peripheral   3. cell body   4. degenerate   5. Schwann cells   6. macrophages   7. Schwann cells
8. endoneurium   9. neuroma   10. central   11. microglial   12. oligodendrocytes   13. die   14. astrocytes

Cranial Nerves

1. 1. H   2. K   3. I   4. L   5. B   6. C   7. G, H, L   8. D   9. A   10. F   11. J   12. D, E   13. G   14. C

2. 1. Accessory (XI)   2. Olfactory (I)   3. Oculomotor (III)   4. Vagus (X)   5. Facial (VII)   6. Trigeminal (V)
7. Vestibulocochlear (VIII)   8. Glossopharyngeal (IX)   9. III, IV, VI   10. Trigeminal (V)   11. Optic (II)
12. Vestibulocochlear (VIII)   13. Hypoglossal (XII)   14. Facial (VII)   15. Olfactory (I)   16. Trochlear (IV)

3. 1. A   2. F   3. D   4. C   5. E   6. B

4.

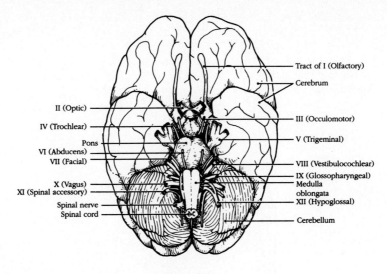

Figure 13.3

5. Temporal, zygomatic, buccal, mandibular, and cervical.

Spinal Nerves

1. 1. dorsal 2. ventral roots 3. rami 4. plexuses 5. limbs 6. thorax 7. posterior trunk

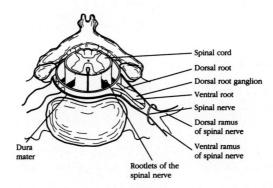

Figure 13.4

2. 1. Two roots 2. Thoracic roots 3. Afferent fibers 4. Ventral root 5. Autonomic nervous system 6. Two spinal roots

3. 1. cervical plexus 2. phrenic nerve 3. sciatic nerve (tibial division) 4. peroneal, tibial 5. median 6. musculocutaneous 7. lumbar plexus 8. femoral 9. ulnar 10. brachial plexus 11. sciatic nerve 12. sacral plexus 13. radial nerve 14. femoral nerve 15. obturator nerve 16. inferior gluteal

4.

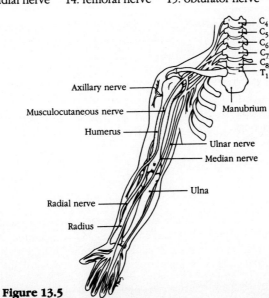

Figure 13.5

5.

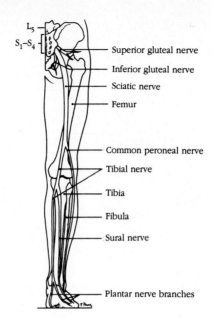

L₅

S₁–S₄

— Superior gluteal nerve

— Inferior gluteal nerve

— Sciatic nerve

— Femur

— Common peroneal nerve

— Tibial nerve

— Tibia

— Fibula

— Sural nerve

— Plantar nerve branches

Figure 13.6

6.   1. radial   2. T   3. median   4. ulnar   5. T

7.   1. free (bare)   2. noxious   3. bradykinin   4. inflammation   5. somatic pain   6. visceral pain   7. pricking
8. burning   9. aching   10. substance P   11. threshold   12. tolerance   13. substantia gelatinosa   14. touch
15. pain impulses

Reflex Activity

1.   1. A rapid, involuntary, predictable response to a stimulus.   2. No; some reflexes are learned or acquired.
3. receptor, sensory neuron, integration center, motor neuron, effector   4. somatic reflexes and autonomic reflexes
5. spinal reflex

2.   1. A   2. B   3. A   4. B   5. A   6. B   7. B   8. A

3.   1. muscle spindles   2. intrafusal fibers   3. nuclear bag fibers   4. nuclear chain fibers   5. primary   6. type Ia sensory
fibers   7. secondary   8. type II sensory fibers   9. gamma efferent fibers   10. extrafusal fibers
11. alpha efferent fibers   12. alpha motor neurons   13. reciprocal inhibition

4.   1. primary sensory afferent   2. alpha motor neuron   3. synapse   4. spinal cord   5. muscle spindle   6. nuclear bag
fiber   7. nuclear chain fiber   8. ascending fibers   9. skeletal muscle (extrafusal fibers)   10. motor ending to
extrafusal fiber

5.   Stretch: 1, 3, 5, 6, 8, 9   Deep tendon: 2, 3, 4, 7, 9, 10

6.   1. E   2. D   3. B   4. A, B, C, D, F   5. C, E   6. E   7. B   8. A, F   9. A, C, D, F   10. A, F   11. D   12. E   13. F
14. A

Developmental Aspects of the Peripheral Nervous System

1.   1. somites   2. sensory and motor (afferent and efferent)   3. maturation   4. cutaneous   5. C₁   6. dermatomes
7. trigeminal   8. migration   9. decreases   10. neurons

CHALLENGING YOURSELF

At the Clinic

1.   Abducens.

2.   Trigeminal nerve; trigeminal neuralgia (tic douloureux).

3.   Damage to the facial nerve. The symptoms are facial paralysis, loss of taste, drooping of lower eyelid; the mouth sags; the
eye tears and can't completely close.

4.   Glossopharyngeal.

5.   Vagus; complete vagal paralysis results in death.

6.   Trapezius, sternocleidomastoid.

7.   Brachial plexus.

8.   Median nerve.

9.   In suturing the nerve back together, there is no guide to ensure that each nerve fiber continues across the transection into
the same neurilemma in which it started. Nerve fibers can grow into pathways different from their original ones and
establish new synapses. The brain cannot keep track of which nerve fibers have grown into different pathways and
projects sensations back to the original point of origin.

10. The olfactory nerve (I) fibers, which are located in the roof of the nasal cavity.
11. The right hypoglossal nerve (cranial XII).
12. Spinal cord transection at L_5 (c). The posterior femoral cutaneous nerve, which arises from ventral rami S_1–S_3, provides the sensory supply in the areas described. The motor nerves for those areas also receive fibers from the ventral rami of L_4 and L_5, explaining his lack of motor problems.
13. Check the gag and swallowing reflexes (IX and X) and for weakness in the sternocleidomastoid and trapezius muscles (XI).
14. Most somatic reflexes involve fast A fibers, while pain sensations travel over slow C fibers. Pain perception that occurs *after* the somatic reflex is triggered helps reinforce the proper behavioral response (learning not to touch hot objects).

Stop and Think

1. Nissl bodies are composed of rough ER and ribosomes. Since most of the protein synthesis will be devoted to synthesizing the internal proteins needed for repair and replacement of the damaged axon, the ribosomes will predominantly be free, rather than attached. The ribosomes of the Nissl bodies will separate from the ER membrane, and the characteristic appearance of the Nissl body will be lost until repair is nearly complete. The nucleolus functions in ribosome assembly, so as the demand for ribosomes increases, nucleolar function and appearance will become more significant. The accumulation of raw materials for protein synthesis may increase the osmotic pressure of the cell, causing it to attract water and take on a swollen appearance.
2. Exteroceptors that are not cutaneous receptors include the chemoreceptors of the tongue and nasal mucosa, the photoreceptors of the eyes, and the mechanoreceptors of the inner ear. These all monitor changes in the external environment, so they are classified as exteroceptors.
3. An inability to feel the parts of your body is the result of accommodation by the proprioceptors. Although these adapt to only a minimal extent, they can slowly accommodate if you don't change your body position for a very long time.
4. Starting from twelve o'clock, for the right eye: lateral movement (abducens), inferior movement, medial movement (oculomotor), superior movement (oculomotor, trochlear). For the left eye: oculomotor for medial and inferior movements, abducens for lateral movement, oculomotor and trochlear for superior movement.
5. The superior colliculi of the midbrain's corpora quadrigemina contain nuclei for initiation of visual reflexes. This would necessarily involve connection to motor nerves of the eye.
6. No. The bundle of "nerves" in the cauda equina really consists of dorsal and ventral roots, which don't merge into the spinal nerve until after exiting the vertebral column.
7. The origin of the phrenic nerve so high up the spinal cord means that the distance between brain and phrenic is short. This decreases the likelihood of spinal cord damage above the phrenic nerve, since there isn't much spinal cord above that nerve's origin.
8. Initially, as muscle spindles are stretched, the reflex sends impulses back to contract the muscle. With prolonged stretching, accommodation decreases the vigor of the stretch reflex somewhat, and the muscle can relax and stretch more, reducing the risk of tearing muscle tissue during exercise.
9. The nerves serving the ankle joint are those serving the muscles that cause movement at the ankle joint—the posterior femoral cutaneous, common peroneal, and tibial.
10. Connective tissue.
11. Use word roots to help you remember. Glossopharyngeal means tongue-pharynx, and that nerve serves those regions. Hypoglossal means "below the tongue," and this nerve serves the small muscles that move the tongue.
12. Abdul was more correct. The facial nerve is almost entirely a motor nerve.

COVERING ALL YOUR BASES

Multiple Choice

1. C 2. A, C, D 3. D 4. C, D 5. C, D 6. A, C, D 7. B 8. B 9. A, B, C 10. D 11. A, B, C, D
12. A, C 13. A, C, D 14. B, D 15. C 16. C 17. A, B 18. B 19. A 20. D 21. A 22. C 23. D
24. A 25. D 26. A, B, C

Word Dissection

Word root	Translation	Example	Word root	Translation	Example
1. esthesi	sensation	kinesthetic	5. propri	one's own	proprioceptors
2. glosso	tongue	glossopharyngeal	6. puden	shameful	pudendal nerve
3. kines	movement	kinesthetic	7. vagus	wanderer	vagus nerve
4. noci	harmful	nociceptors			

Chapter 14 The Autonomic Nervous System

BUILDING THE FRAMEWORK

Overview of the Autonomic Nervous System

1.

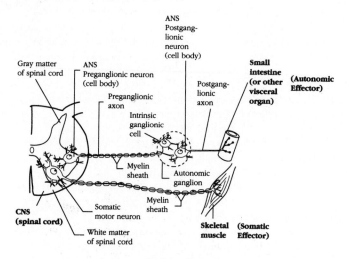

Figure 14.1

2. 1. S, A 2. S 3. A 4. S, A 5. S 6. S 7. A 8. A 9. S, A 10. A 11. A 12. S 13. S 14. S 15. A
16. S, A

3. Sympathetic: 2, 4, 7, 9   Parasympathetic: 1, 3, 5, 6, 8

Anatomy of the Autonomic Nervous System

1.

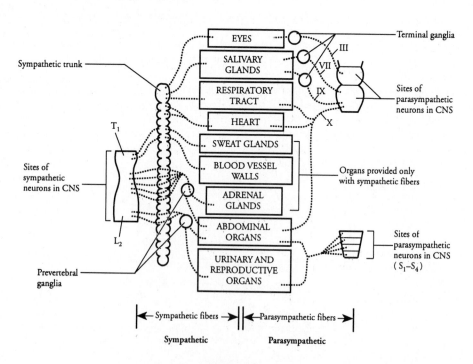

Figure 14.2

2. 1. lateral 2. T_1–L_2 3. spinal cord 4. white (myelinated) 5. paravertebral ganglia 6. sympathetic
7. postganglionic 8. rami communicantes 9. unmyelinated 10. smooth muscle 11. cervical 12. pupil (iris)
13. salivary 14. skin 15. prevertebral (collateral) 16. thoracic splanchnic 17. celiac and superior mesenteric
18. abdominal 19. adrenal 20. lumbar 21. urinary 22. brain stem 23. sacral 24. terminal 25. oculomotor

(III)   26. VII (facial)   27. IX (glossopharyngeal)   28. vagus (X)   29. plexuses   30. thorax   31. abdomen   32. sacral region

3.   1. S   2. P   3. S   4. P   5. P   6. P   7. S   8. P   9. P   10. S   11. S   12. S, P

Physiology of the Autonomic Nervous System

1.   1. Fibers that release acetylcholine   2. Nicotinic and muscarinic receptors
2.   1. Fibers that release norepinephrine   2. Alpha and beta adrenergic receptors
3.   1. C   2. A   3. C   4. D   5. B   6. A   7. B   8. B
4.   Sympathetic: 2, 3, 4, 6   Parasympathetic: 1, 5, 7
5.   1. NE   2. Nicotinic receptor   3. Heart rate slows   4. Parasympathetic tone   5. Rapid heart rate   6. Sympathetic tone

Homeostatic Imbalances of the Autonomic Nervous System

1.   1. smooth muscle   2. T   3. sympathectomy   4. adrenergic

Developmental Aspects of the Autonomic Nervous System

1.   1. neural tube   2. neural crest   3. migrations   4. spinal cord   5. effector   6. nerve growth factor   7. parasympathetic   8. mobility (tone)   9. gastrointestinal tract   10. tear   11. vasomotor   12. sympathetic

CHALLENGING YOURSELF

At the Clinic

1.   Parasympathetic.

2.   The sympathetic division causes vasoconstriction of vessels supplying the digestive organs. Stress-induced overstimulation of sympathetic nerves can cause an almost total shutdown of blood supply to the stomach. Lack of blood leads to tissue death and necrosis, which causes the ulcers.

3.   Achalasia.

4.   The condition is the result of poorly developed parasympathetic nerve supply to the lower colon. As a result, the muscle of the colon fails to contract, and fecal matter cannot be expelled. Since the nerve deficiency cannot be corrected, surgical removal of the distended area is the only course of treatment. After removal of the underdeveloped region, fecal material will be moved along healthy portions of the colon normally.

5.   A sympathectomy will be performed. It will result in dilation of the distal blood vessels as their vasoconstrictor fibers are cut.

6.   Sweating is stimulated by sympathetic nerves; after they are cut, anhidrosis will result.

7.   The condition is mass reflex reaction, or autonomic hyperreflexia. The staff will watch for hypertension-induced cerebrovascular accident.

8.   The fainting episodes are the result of orthostatic hypotension, due to slowed response of aging sympathetic vasomotor centers. The condition is exaggerated when blood supply to the skin is increased (such as when the room is warm), since the shunting of blood to the skin reduces blood flow to other body parts.

9.   Bertha probably has a urinary tract infection that has spread to her kidneys. Kidney inflammation causes pain to be referred to the lower back.

10.   Antiadrenergic drugs (such as phentolamine) that interfere with the activity of the sympathetic vasomotor fibers.

11.   Epigastric region.

12.   These are parasympathetic ganglia that cause the pupils to constrict and the lens to bulge for close vision (ciliary ganglion) and that stimulate secretory activity of the lacrimal and nasal glands (sphenopalatine ganglion) and the salivary glands (submandibular ganglion). Hence, you might have dry eyes, nasal passages, and mouth, and have a headache because your pupils are dilated in bright light but otherwise you are OK. Back to the lab!

Stop and Think

1.   No. Without sensory feedback to inform the CNS of the physical and chemical status of the viscera, the motor output of the autonomic nerves can be inappropriate and even life-threatening.

2.   Parasympathetic.

3.   Spinal reflex, autonomic reflex.

4.   Pelvic splanchnic nerves to the inferior hypogastric plexus.

5.   Cerebral input to the preganglionic neuron in the sacral spinal cord is inhibitory; cerebral stimulation of motor neurons controlling the more inferior skeletal muscle (voluntary) sphincter can shut off urine flow.

6.   Transection above the level of the reflex would cut off the possibility for cerebral inhibition. Voiding would become entirely involuntary and totally reflexive.

7.   Shunting of blood to the vessels of the hands reduces blood flow to the brain, thus reducing the vessel distension that causes the pain of a migraine headache.

8.   The cholinergic preganglionic axon runs from a lateral horn of the gray matter in the upper thoracic spinal cord to the ventral root of the spinal nerve, through the white ramus communicans to the paravertebral ganglion of the thoracic region of the sympathetic chain, and up the chain (without synapsing) to the cervical region to synapse in the superior cervical ganglion. There it releases ACh to excite the postganglionic neuron. The adrenergic postganglionic axon runs from the ganglion to join with fibers of the oculomotor nerve to reach the iris of the eye. Release of NE at the neuroeffector junctions stimulates the radial smooth muscle layer of the iris, which results in dilation of the pupil.

9. From the cardiac center in the medulla to the nucleus of the vagus nerve, where the cholinergic preganglionic axon leaves in the vagus nerve and runs to the cardiac plexuses. From there it continues on to the heart itself, ending at an intramural ganglion, where ACh is released. ACh excites the postganglionic neurons, whose cholinergic postganglionic axons travel to the target cells, where they release ACh to inhibit cardiac function.

COVERING ALL YOUR BASES

Multiple Choice

1. D   2. D   3. D   4. B, C   5. A, B, C   6. B   7. A   8. A   9. C   10. D   11. C   12. C   13. B   14. A   15. B   16. A   17. C   18. C   19. A   20. A   21. A   22. A   23. A   24. B

Word Dissection

Word root	Translation	Example	Word root	Translation	Example
1. adren	toward the kidney	adrenaline	5. ortho	straight	orthostatic
2. chales	relaxed	achalasia	6. para	beside	parasympathetic
3. epinephr	upon the kidney	epinephrine	7. pathos	feeling	sympathetic
4. mural	wall	intramural ganglia	8. splanchn	organ	splanchnic nerve

Chapter 15 Neural Integration

BUILDING THE FRAMEWORK

Sensory Integration: From Reception to Perception

1. 1. sensation   2. perception   3. receptor   4. circuit   5. ascending (sensory) pathways   6. perceptual
   7. cerebral cortex   8. action potentials   9. permeability   10. graded   11. threshold   12. action   13. anterolateral
   14. lemniscal   15. anterolateral (nonspecific)   16. module   17. thalamus

2.

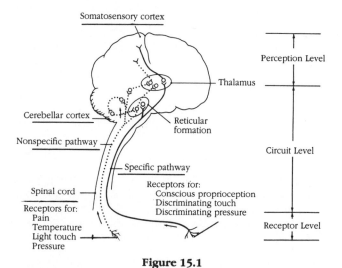

Figure 15.1

3. 1. primary sensory cortex   2. frequency   3. nonspecific   4. T   5. thalamus   6. specific   7. first-order   8. T
   9. parallel processing

4. The pathway could be accurately identified by the process of elimination. It continues all the way to the somatosensory cortex, so it is not the spinocerebellar pathway. It ascends in the *lateral* white column, so it is not either the fasciculus cuneatus or gracilis, which are dorsal column pathways (and constitute the *specific* system). Circle the junctions (synapses) between the first- and second-order neurons and between the second- and third-order neurons.

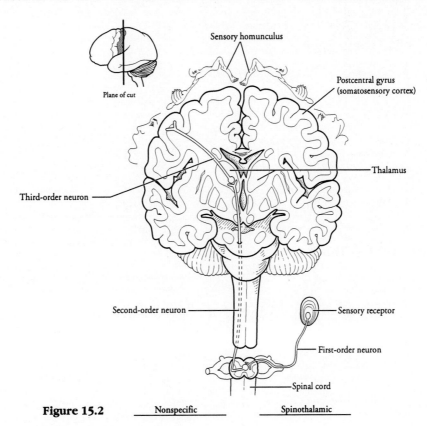

Figure 15.2 _____ Nonspecific _____ Spinothalamic _____

5.   1. F   2. C   3. D   4. B   5. B   6. A   7. E   8. A   9. E

Motor Integration: From Intention to Effect

1.   1. C   2. B, G   3. E   4. D   5. H   6. C   7. A   8. A, D

2.

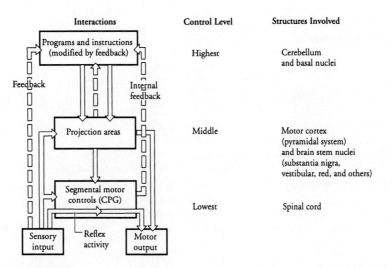

Figure 15.3

3. 1. Consciously considered actions   2. Command neurons   3. Cerebellum   4. Precommand areas   5. Basal nuclei   6. Multineuronal tracts   7. Perception disorders   8. Basal nuclei unaffected   9. Ataxia

4.

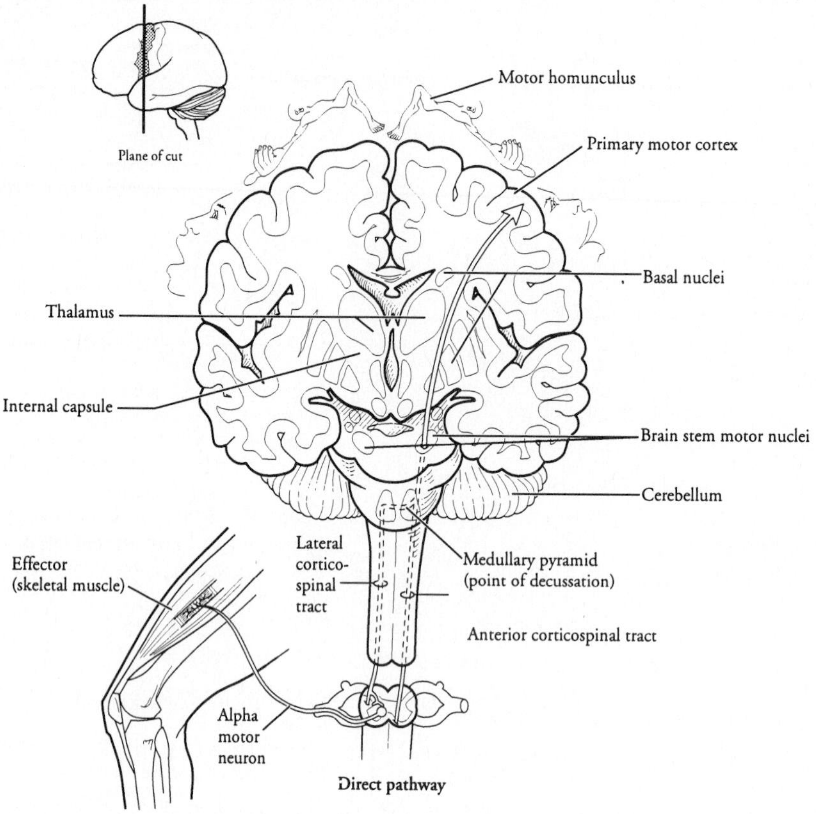

Figure labels: Motor homunculus; Primary motor cortex; Plane of cut; Thalamus; Basal nuclei; Internal capsule; Brain stem motor nuclei; Cerebellum; Lateral cortico-spinal tract; Medullary pyramid (point of decussation); Effector (skeletal muscle); Anterior corticospinal tract; Alpha motor neuron; Direct pathway

Figure 15.4

5. 1. A, F   2. A, I   3. B, J   4. C, G   5. D, K   6. E, H

Higher Mental Functions

1. 1. A   2. D   3. A   4. C   5. C   6. B
2. 1. electroencephalogram (EEG)   2. frequencies   3. Hz (cycles per second)   4. amplitude   5. sleep   6. unconsciousness   7. epilepsy   8. petit mal   9. grand mal   10. brain death
3. 1. 24   2. coma   3. brain stem   4. hypothalamus   5. increase   6. T   7. NREM   8. acetylcholine   9. REM   10. T   11. narcolepsy
4. Consciousness is evidence of holistic information-processing by the brain. Consciousness involves large areas of simultaneous cerebral cortical activity, is superimposed on other types of neural activity, and is totally interconnected throughout the cerebrum.
5. alertness, drowsiness, stupor, coma
6. 1. S   2. S   3. L   4. L   5. S   6. L
7. 1. An alert, aroused state of consciousness.   2. Repetition or rehearsal of facts or skills.   3. Association of new information with stored information.   4. Time for chemical or structural changes to occur.
8. 1. Fact memory: the learning of precise, detailed information; related to conscious thoughts when consolidation into long term memory is associated with already learned facts.   2. Skill memory: the learning of specific motor activities; not usually related to conscious thoughts; best acquired by practice and remembered in repeated performance.
9. 1. subcortical structures   2. basal forebrain   3. a more lasting memory   4. amygdala   5. anterograde amnesia   6. yes   7. no   8. throughout the cerebral cortex   9. corpus striatum
10. 1. cerebral cortex   2. words   3. grammar   4. speech/articulation   5. Wernicke's area   6. the frontal and midfrontal cortex   7. sensory   8. Broca's   9. motor

CHALLENGING YOURSELF

At the Clinic

1. Pattern recognition.
2. The lesion could be in the pyramidal tract in the cerebral cortex or in the medullary pyramid, in the left side of the brain. The term for weak muscle tone is hypotonia.

3. Internal capsule or spinal cord.
4. Marie has ataxia; cerebellum.
5. Athetosis; basal nuclei.
6. Parkinson's disease; levodopa or perhaps transplant of catecholamine (or dopamine) secreting tissue.
7. Parkinson's disease.
8. Decreases.
9. Petit mal epilepsy; the condition will probably resolve itself without treatment.
10. Psychomotor epilepsy.
11. Anticonvulsive drugs.
12. Narcolepsy; the reticular formation.
13. Four hours of sleep is not abnormal for an elderly person. A sleeping aid will reduce slow-wave (restorative) sleep, which compounds the problem.
14. Retrograde amnesia; this does not affect skill memory.
15. Left and/or midfrontal cortex (anterior cingulate region).
16. The nurse has forgotten the phenomenon of adaptation, a condition in which sensory receptor transmission declines or stops when the stimulus is unchanging. She has had her hand in the water while filling the tub, and the thermoreceptors in *her* skin have accommodated to the hot water.

Stop and Think

1. No. Perception refers to integrative activity, and this includes integration at subconscious levels. For example, the hypothalamus integrates visceral input.
2. Pricking the finger excites the nonspecific pathway for pain, beginning with the dendritic endings of a first-order neuron, a nociceptor, in the finger. This neuron synapses with the second-order neuron in the spinal cord, and the axon of the second-order neuron (after crossing over in the spinal cord) ascends in the anterolateral pathway through the lateral spinothalamic tract. Synapse with third-order neurons occurs in thalamic nuclei, but there are many synapses with nuclei of the reticular formation as well. Pain is perceived, and the cortex is aroused as processing at the perceptual level occurs. The somatosensory cortex pinpoints the sensation; the sensory association area perceives the sensation as undesirable. Integrative input to the prefrontal area sends signals to the motor speech area, which directs the primary motor cortex to activate the muscles of speech to say "ouch."
3. Nociceptors in the skin send signals via the trigeminal nerve along the corticobulbar tracts into the thalamus. Pain is perceived (in a crude sense) as impulses are sent to the somatosensory cortex, where sensation localization and pain awareness occur. Impulses are sent along association fibers to the prefrontal cortex for further integration and decision making. Integration in the prefrontal cortex leads to transmission of intention impulses to the cerebellum and basal nuclei, which prepare for motor activity. Impulses to the premotor area coordinate the primary motor output of the pyramidal cells, along the direct pathway to the pyramidal tracts of the medulla. Proprioceptive feedback to the cerebellum fine-tunes the direction, speed, and check of muscle contraction to smack the mosquito. Impulses travel to the cervical region of the spinal cord, stimulating motor neurons in the ventral gray horn. Impulses are conveyed along the brachial plexus to the muscles of the upper limb, which carry out the "execution" of the mosquito.
4. Dopamine is a "feel good" neurotransmitter. People suffering from Parkinson's disease and dopamine deficiency usually report feeling "blah"—not really having any pleasurable feelings. Tourette's victims typically feel euphoric during a seizure.
5. Short-term memory is affected by ECT. Long-term memory does not rely solely on the continuation of electrical output; rather, it is based on physical or chemical changes to the cells in the memory pathway.
6. The precommand areas *ready* the primary motor cortex for action, but the actual *command* (nerve impulse) to the muscles is issued by the motor cortex.

COVERING ALL YOUR BASES

Multiple Choice

1. C   2. A, B, D   3. B   4. A, C, D   5. C   6. A, B, C, D   7. D   8. C   9. B, C, D   10. A   11. C, D   12. D
13. C   14. B   15. C   16. A, B, C, D   17. A, B, C, D   18. B, C   19. D   20. A, C   21. A, B, D   22. B, C, D
23. A, B, C, D   24. A, B, C, D   25. C, D   26. A, B, C, D   27. A, D   28. A

Word Dissection

Word root	Translation	Example	Word root	Translation	Example
1. alge	pain	analgesic	7. mnem	memory	amnesia
2. ceps	head	perception	8. per	by means of	perception
3. cope	cut	syncope	9. prax	action	apraxia
4. duc	lead	transduction	10. sens	feeling	sensation
5. epilep	laying hold of	epilepsy	11. somnus	sleep	insomnia
6. lemnisc	ribbon	medial lemniscus	12. trans	across, through	transduction

Chapter 16 The Special Senses

BUILDING THE FRAMEWORK

The Chemical Senses: Taste and Smell

1. 1. chemoreception 2. bitter 3. olfactory 4. roof 5. limbic 6. thalamus

2.

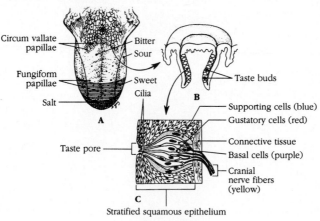

Figure 16.1

3. 1. They play a role as local integrator cells. 2. granule cells 3. olfactory adaptation 4. amygdala, hypothalamus, and other limbic system structures

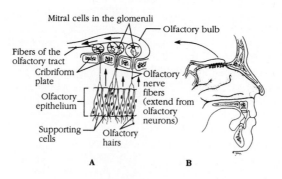

Figure 16.2

4. 1. Musky 2. Epithelial cell 3. Yellow-tinged epithelium 4. Olfactory nerve 5. Four receptor types 6. Metal ions 7. H⁺ 8. Anosmia

The Eye and Vision

1. 1. extrinsic 2. eyelids 3. tarsal or Meibomian 4. chalazion

2.

Accessory eye structures	Secretory product
Conjunctiva	Mucus
Tarsal glands	Oily secretion
Lacrimal glands	Saline solution with mucus, lysozyme, antibodies (IgA), and enkephalins

3. 1. 2 2. 4 3. 3 4. 1

4. 1. Superior rectus turns eye superiorly. 2. Inferior rectus turns eye inferiorly. 3. Superior oblique turns eye inferiorly and laterally. 4. Lateral rectus turns eye laterally. 5. Medial rectus turns eye medially. 6. Inferior oblique turns eye superiorly and laterally.

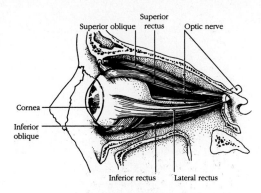

Superior oblique   Superior rectus   Optic nerve

Cornea

Inferior oblique

Inferior rectus   Lateral rectus

Figure 16.3

5. 1. E   2. G   3. B   4. A   5. D   6. F
6. 1. Vitreous humor   2. Iris   3. Far vision   4. Mechanoreceptors   5. Iris   6. Pigmented retina   7. Pigmented layer
8. Macula lutea   9. Richly vascular   10. Inner segment   11. Lens placode
7. 1. E   2. C   3. A   4. D   5. B   6. F
8. 1. In distant vision the ciliary muscle is relaxed, the lens convexity is decreased, and the degree of light refraction is decreased.   2. In close vision the ciliary muscle is contracted, the lens convexity is increased, and the degree of light refraction is increased.
9. 1. L   2. A   3. K   4. I   5. D   6. C   7. B   8. J   9. M   10. C   11. D   12. G   13. F   14.–17. E, H, M, A
18. E   19. G
10.

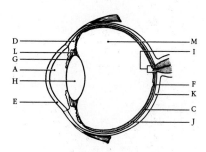

D   M
L   I
G
A
H   F
K
E
C
J

Figure 16.4

11. During reading, all intrinsic muscles and the extrinsic medial recti are contracted. Distant vision does not require accommodation and relaxes these muscles.
12. 1. E   2. F   3. D   4. A   5. C
13. 1. L   2. A   3. F   4. H   5. K   6. D   7. I   8. C   9. G   10. E   11. B   12. J
14. 1. 3   2. blue   3. green   4. red   5. at the same time   6. total color blindness   7. males   8. rods   9. fovea centralis   10. retinal periphery
15. 1. opsin; rhodopsin   2. all-*trans;* bleaching of the pigment   3. vitamin A; liver   4. Na$^+$ leaks into outer segments, causing depolarization and local currents that lead to neurotransmitter release.   5. Membrane becomes impermeable to Na$^+$; hyperpolarization occurs; neurotransmitter release stops.
16. Retina → optic nerve → optic chiasma → optic tract → synapse in thalamus → optic radiation → optic cortex.
17. Check 1, 2, 3, 4, 11, 12
18. The visual fields partially overlap, and inputs from both eyes reach each visual cortex, allowing for depth perception.
19. 1. ganglion cells   2. bipolar   3. photoreceptors   4. on-center   5. off-center   6. horizontal   7. amacrine
8. contrasts   9. lateral geniculate nuclei of thalamus   10. simple   11. complex   12. prestriate cortex
13. motion

20.

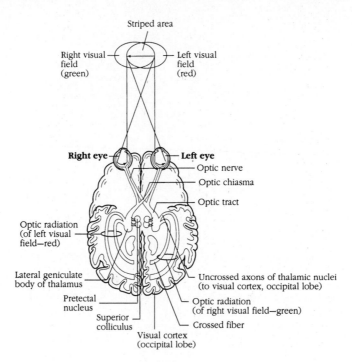

Figure 16.5

The Ear: Hearing and Balance

1. 1.–3. E, I, M   4.–6. C, K, N   7.–9. A, F, L   10. K   11. B   12. M   13. C   14. B   15.–16. K, N   17. G   18. D
     19. H   20. L

2.

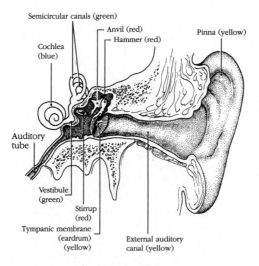

Figure 16.6

3. Eardrum → hammer → anvil → stirrup → oval window → perilymph → membrane → endolymph → hair cells.
4. 1. C   2. F   3. E   4. D   5. B   6. G   7. I   8. J   9. A
5. Stapedius attached to stapes; tensor tympani attached to malleus. When loud sounds assault the ears, these muscles contract and limit the movements of the stapes and malleus against their respective membranes.

6.

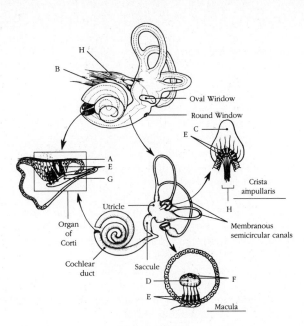

Figure 16.7

7. The ossicles deliver the total force impinging upon the eardrum to the oval window, which is about one-twentieth the size of the eardrum, in effect amplifying the sound.

8. 1. vibrating object   2. elastic   3. compression   4. rarefaction   5. wavelength   6. short   7. long   8. pitch   9. loudness   10. decibels

9. Short basilar membrane fibers resonate with higher frequency sounds, while larger fibers resonate with lower frequency sounds. At points of maximal basilar membrane vibration, the hair cells are excited.

10. 1. C   2. S   3. C   4. S   5. C, S   6. C   7. S

11. 1. C   2. I   3. A   4. D   5. B   6. J   7. H   8. K   9. E   10. G   11. L

12. 1. nausea   2. dizziness   3. balance problems

13. 1. Pinna   2. Tectorial membrane   3. Sound waves   4. Pharyngotympanic tube   5. Optic nerve   6. Retina

Developmental Aspects of the Special Senses

1. 1. optic vesicles   2. diencephalon   3. optic cups   4. retina   5. optic stalks   6. ectoderm   7. lens placode
2. 1. B   2. C   3. A
3. 1. nervous system   2. measles (rubella)   3. blindness   4. vision   5. hyperopic   6. elastic   7. presbyopia   8. presbycusis

The Incredible Journey

1. 1. bony labyrinth   2. perilymph   3. saccule   4. utricle   5. (gel) otolithic membrane   6. otoliths   7. macula   8. static   9. cochlear duct   10. organ of Corti   11. hearing   12. cochlear division of cranial nerve (VIII)   13. semicircular canals   14. cupula   15. crista ampullaris   16. dynamic

CHALLENGING YOURSELF

At the Clinic

1. Uncinate fit; irritation of the olfactory pathway sometimes follows brain surgery.
2. Anosmia; zinc deficiency.
3. Ciliary; sty.
4. Pinkeye (infectious conjunctivitis); yes, because infectious conjunctivitis is caused by a bacterial infection.
5. Patching the strong eye to force the weaker eye muscles to become stronger.
6. Cataract; UV radiation, smoking.
7. Concave lenses are needed to diverge the light before it enters the eye. Janie's lens is too strong (overconverges) for clear distance vision.
8. Nyctalopia; vitamin A; photoreceptors of retina.
9. The proximal end; sensorineural.
10. Vestibular apparatus (including the semicircular canals).
11. Glaucoma; inadequate drainage of aqueous humor; blindness due to compression of the retina and optic disc.
12. Most likely the stapedius and tensor tympani muscles became fatigued after being activated continuously for over an hour and were unable to continue to play their protective roles.

13. If Biff hasn't been suffering earaches, he probably does *not* have otitis media. This fact, along with his swimming activity, points to otitis externa.
14. Mr. Cruz has high intraocular pressure; glaucoma.
15. The right side of his visual field.

Stop and Think

1. An = without; osm = smell.
2. Uncus = region of cerebral cortex involved in olfaction.
3. Lack of melanin production results in an absence of pigment in the eye. Light rays "bounce around" in the eye because there is no melanin to absorb stray rays. Consequently, vision is poor.
4. Vitreous humor is not produced after eye development is complete. Eye size in children is close to that of adults. Since a child's head is smaller, the eyes look relatively larger.
5. Neural layer of the retina.
6. Angled light rays reflected from near objects can enter the eye; the only light rays that reach the eye from far objects are rays that are basically pointed straight at the eye (perpendicular to its surface) and thus are parallel.
7. Starlight is too dim to excite cones. Light from objects in the center of the visual field strikes the fovea centralis, which contains only cones, not rods.
8. Retinal is shuttled from the photoreceptors to the pigmented epithelial cells and restored to its 11-*cis* configuration. It then returns to the outer segments of the photoreceptors and is rejoined to opsin to make rhodopsin in the rods. The accumulation of rhodopsin increases the sensitivity of the retina, and retinal adaptation switches from cones to rods.
9. Plants reflect green, so they must be absorbing red (625 nm) and blue (455 nm).
10. In elephants, the spiral organ of Corti is elongated at the helicotrema end to increase the sensitivity to low-frequency sound. In dogs, the proximal end of the spiral organ is stiffer than it is in humans, so higher frequencies are required to excite the hair cells at that end.
11. Gustatory hairs—microvilli; olfactory hairs—cilia; spiral organ of Corti hairs—stereocilia (microvilli); hair cells in the crista ampullaris and macular hairs in the saccule and utricle—stereocilia and kinocilium (a true cilium). Cilia in the photoreceptor cells support the connection between the inner and outer segments.
12. The taste buds and olfactory epithelium can replace their receptor cells; the retina of the eye and the cochlea of the ear cannot.

COVERING ALL YOUR BASES

Multiple Choice

1. D 2. B, C 3. B, D 4. B, C, D 5. A, B, D 6. C 7. A 8. B 9. C 10. A, D 11. A 12. C
13. A, B, C 14. D 15. B 16. B, D 17. B 18. A, B, D 19. A, C 20 A, C 21. B 22. B, C, D
23. A, B 24. A, C, D 25. A, C, D 26. A 27. A, C 28. A, B, D 29. B 30. A, D 31. A, C, D 32. C
33. A, C, D 34. B, C, D 35. A, B, C, D 36. C 37. B, C 38. C 39. C 40. A, B, C, D 41. C
42. A, B, D 43. C 44. A, C 45. A, B, C 46. A, C, D 47. B 48. A, B

Word Dissection

Word root	Translation	Example	Word root	Translation	Example
1. ampulla	flask	crista ampullaris	13. olfact	smell	olfactory cell
2. branchi	gill	branchial groove	14. palpebra	eyelid	superior palpebra
3. caruncl	bit of flesh	caruncle of eye	15. papill	nipple	fungiform papilla
4. cer	wax	ceruminous gland	16. pinn	wing	pinna of ear
5. cochlea	snail shell	cochlear duct	17. presby	old	presbycusis
6. fove	small pit	fovea centralis	18. sacc	sack	saccule of inner ear
7. glauc	gray	glaucoma	19. scler	hard	sclera of eye
8. gust	taste	gustatory hair	20. tars	flat	tarsal plate
9. lut	yellow	macula lutea	21. trema	hole	helicotrema
10. macula	spot	macula lutea	22. tympan	drum	tympanic membrane
11. metr	measure	emmetropic			
12. modiol	water wheel bucket	modiolus of cochlea			

Chapter 17   The Endocrine System

BUILDING THE FRAMEWORK

The Endocrine System and Hormone Function: An Overview

1. 1. K   2. H   3. D   4. G   5. A   6.–8. B, I, L   9. E   10. J

2. Figure 17.1: A. Pineal   B. Anterior pituitary   C. Posterior pituitary   D. Thyroid   E. Thymus   F. Adrenal gland
G. Pancreas   H. Ovary   I. Testis   J. Parathyroids   K. Placenta   L. Hypothalamus

Hormones

1. 1. I   2. N   3. A   4. L   5. K   6. G   7. C   8. D   9. F   10. B   11. J   12. E   13. H

2. 1. A   2. B   3. B   4. A   5. C   6. C   7. A   8. B

3. 1. first   2. a G protein   3. adenylate cyclase   4. ATP   5. second   6. protein kinases   7. phosphate
8. target cell type   9.–11. diacylglycerol, inositol triphosphate, cyclic GMP   12. Ca^{2+}

4. Hormones are inactivated by enzymes in their target cells as well as by liver and kidney enzymes.

5. 1.–3. A, G, I   4. C   5.–8. B, E, H, K   9. L   10. D   11. F

Major Endocrine Organs of the Body

1. 1. D   2. B   3. A   4. C

2.

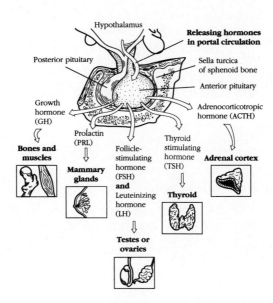

Figure 17.2

3. 1. C   2. D   3. A   4. A   5. B   6. E, I (A)   7. C   8. F   9. C   10. F   11. C   12. H   13. C   14. D   15. E, I (A)
16. C   17. G   18. J (A)   19. K   20. L   21. L   22. C

4. 1. thyroxine/T_3 2. thymosin   3. PTH   4. cortisone (glucocorticoids)   5. epinephrine   6. insulin
7.–10. TSH, FSH, ACTH, LH   11. glucagon   12. ADH   13. FSH   14. LH   15. estrogens   16. progesterone
17. aldosterone   18. prolactin   19. oxytocin

5. 1. Anterior lobe   2. Posterior lobe   3. Cortisol   4. Hypophyseal portal system   5. Oxytocin   6. Increases blood
Ca^{2+} levels   7. Depresses glucose uptake   8. Thyroxine

6. 1. Kidneys: Causes retention of Ca^{2+} and enhances excretion of PO_4^{3-}. Promotes activation of vitamin D.
2. Intestine: Causes increased absorption of Ca^{2+} from foodstuffs.   3. Bones: Causes enhanced release of Ca^{2+} from
bone matrix.

7. In high amounts, it promotes blood vessel constriction, increasing blood pressure (it has a pressor effect).

8. Check 1, 4, 6, 7, 10, 11. Circle 2, 3, 5, 8, 9.

9. 1. 2   2. 7   3. 8   4. 5   5. 9   6. 1   7. 4   8. 3   9. 6

10. 1. Polyuria—high sugar content in kidney filtrate (acts as an osmotic diuretic and) causes large amounts of water to be
lost in the urine.   2. Polydipsia—thirst due to large volumes of urine excreted.   3. Polyphagia—hunger because
blood sugar cannot be used as a body fuel even though its levels are high.

11. 1. B   2. A   3. A   4. B   5. B   6. A   7. A   8. B   9. B   10. B   11. B   12. A

12. 1. A, E   2. B   3. C   4. E   5. A   6. E   7. A   8. B, E   9. A, E   10. D

13. 1. B   2. A   3. B   4. A   5. A

14. 1. B   2. D   3. E   4. C   5. A
15. 1. estrogens/testosterone   2. PTH   3. ADH   4. thyroxine   5. thyroxine   6. insulin   7. growth hormone
8. estrogens/progesterone   9. thyroxine
16. 1. growth hormone   2. thyroxine   3. PTH   4. glucocorticoids   5. growth hormone   6. androgens

Other Endocrine Structures

1.

Hormone	Chemical makeup	Source	Effects
Gastrin	Peptide	Stomach	Stimulates stomach glands to secrete HCl
Secretin	Peptide	Duodenum	Stimulates the pancreas to secrete HCO_3^--rich juice and the liver to release more bile; inhibits stomach glands
Cholecystokinin	Peptide	Duodenum	Stimulates the pancreas to secrete enzyme-rich juice and the gallbladder to contract; relaxes sphincter of Oddi
Erythropoietin	Glycoprotein	Kidney in response to hypoxia	Stimulates production of red blood cells by bone marrow
Cholecalciferol (vitamin D_3)	Steroid	Skin; activated by kidneys	Enhances intestinal absorption of calcium
Atrial natriuretic factor (ANF)	Peptide	Heart atrial cells	Inhibits Na^+ reabsorption by kidneys; inhibits renin and aldosterone release

Developmental Aspects of the Endocrine System

1. 1. all three   2. neoplasm   3. iodine   4. estrogens   5. menopause   6. bear children   7. insulin

The Incredible Journey

1. 1. insulin   2. pancreas   3. hypothalamus   4. ADH   5. parathyroid   6. calcium   7. adrenal medulla
8. epinephrine   9. T_3/T_4

CHALLENGING YOURSELF

At the Clinic

1. Pituitary dwarfs (deficient in GH) are short in stature but have fairly normal proportions; cretins (deficient in thyroxine) retain childlike body proportions and have thick necks and protruding tongues.
2. Hypothyroidism; iodine deficiency (treated by dietary iodine supplement) or thyroid cell burnout (treated by hormonal supplement).
3. Hypersecretion of ADH causes diabetes insipidus, and insufficiency of insulin (or lack of response to insulin) causes diabetes mellitus. Both involve polyuria and consequent polydipsia. Glucose in the urine will indicate diabetes mellitus. (More complicated tests are used to diagnose diabetes insipidus.)
4. Adrenal cortex.
5. Hyperparathyroidism (resulting in hypercalcemia and undesirable calcium salt deposit), probably from a parathyroid tumor.
6. Cushing's syndrome, most likely caused by tumor; anterior pituitary (hypersecretion of ACTH) or adrenal cortex (hypersecretion of cortisol).
7. Hyperthyroidism; anterior pituitary tumor (hypersecretion of TSH) or thyroid tumor (hypersecretion of thyroxine).
8. Epinephrine and norepinephrine; pheochromocytoma, a tumor of the chromaffin cells of the adrenal medulla; exophthalmos and goiter would not be present.
9. Hypoglycemia; overdose of insulin.
10. Prolactin; tumor of the adenohypophysis.
11. Lying between the pinealocytes in the gland are grains of calcium salts. This "pineal sand," like the calcium salts of bone, is radiopaque.

Stop and Think

1. *Protein hormones*   *Steroid hormones*
   (a) rough ER, Golgi apparatus, secretory vesicles   smooth ER
   (b) storage in secretory vesicles   no storage
   (c) manufactured constantly   manufactured only as needed
   (d) secretion by exocytosis   leaves cell by diffusion through lipid bilayer
   (e) most (but not all) transported dissolved in plasma transport may require transport protein transport molecule
   (f) receptors on plasma membrane   receptors in cytoplasm or nucleus
   (g) second messenger used   no second messenger
   (h) immediate activation   lag time for protein synthesis to occur
   (i) effects end as soon as hormone is metabolized   effects prolonged after hormone is metabolized

2. (a) Growth hormone, TSH, ACTH, prolactin, FSH, LH.
   (b) Calcium, glucose, sodium, and potassium blood levels.
3. Hormone-producing cells located in organs of the digestive tract. They are called paraneurons because many of their hormones are identical to neurotransmitters released by neurons.
4. Down-regulation.
5. These hormones are lipid soluble; they must be "tied down" to prevent their escape through the plasma membrane.
6. Corticotropin-releasing hormone secreted by hypothalamus → capillaries of the hypophyseal portal system → anterior pituitary gland → release of ACTH by anterior pituitary → general circulation to zone fasciculata of the adrenal cortex → glucocorticoids (cortisol, etc.) released.
7. The thyroid gland contains thyroid follicles surrounded by soft connective tissue. The follicular cells of the follicles produce the thyroglobulin colloid from which T_3 and T_4 are split. At the external aspects of the follicles are *parafollicular* (C) cells that produce calcitonin, a completely different hormone. Usually structurally associated with the thyroid gland are the tiny parathyroid glands, which look completely different histologically. The parathyroid cells produce PTH, a calcitonin antagonist.
8. For the giant, GH is being secreted in excess by the anterior pituitary, resulting in extraordinary height. For the dwarf, GH is deficient, resulting in very small stature but normal body proportions. For the fat man, T_3 and T_4 are not being adequately produced, resulting in depressed metabolism and leading to obesity (myxedema). The bearded lady has a tumor of her adrenal cortex (androgen-secreting area) leading to hirsutism.

Closer Connections: Checking the Systems—Regulation and Integration of the Body

1. The CNS influences hormonal fluctuations through the hypothalamus, which not only synthesizes its own hormones (ADH and oxytocin) but also regulates release of anterior pituitary hormones.
2. The endocrine system includes hormones, such as thyroxine, that maintain the general health of nerve tissue and mineral-regulating hormones that maintain the proper balance of ions of sodium, calcium, and potassium required for optimal neural function. Additionally, certain hormones (T_3 and T_4) are required for normal growth and maturation of the nervous system.
3. Nervous system.
4. Endocrine system on both counts.
5. The nervous system elicits rapid responses from skeletal muscles; the endocrine system maintains the metabolism and general health of the muscle tissue and enhances its growth in mass during puberty. Both systems (sympathetic nervous system and catecholamines of the adrenal medulla) help ensure the muscles have an adequate blood supply during their activity when demands for O_2 and nutrients increase.

COVERING ALL YOUR BASES

Multiple Choice

1. B, C   **2.** A, C   **3.** C   **4.** A, C, D   **5.** C   **6.** A, B, D   **7.** C   **8.** C   **9.** B   **10.** D   **11.** B   **12.** A, B, C, D
13. A, B, C, D   **14.** A, B, C, D   **15.** D   **16.** A, B, C   **17.** A, B, C, D   **18.** D   **19.** A, C   **20.** C
21. A, B, C, D   **22.** A, B, C   **23.** B   **24.** B   **25.** C   **26.** B   **27.** D   **28.** A   **29.** A

Word Dissection

Word root	Translation	Example	Word root	Translation	Example
1. adeno	gland	adenohypophysis	**6.** hormon	excite	hormonal
2. crine	separate	endocrine	**7.** humor	fluid	humoral control
3. dips	thirst, dry	polydipsia	**8.** mell	honey	diabetes mellitus
4. diure	urinate	diuretic	**9.** toci	birth	oxytocin
5. gon	seed, offspring	gonad	**10.** trop	turn, change	tropic hormone

Chapter 18 Blood

BUILDING THE FRAMEWORK

Overview: Composition and Functions of Blood

1.   1. connective tissue   2. formed elements   3. plasma   4. blood clotting   5. red blood cells   6. hematocrit
7. plasma   8. white blood cells   9. platelets   10. 1   11. oxygen

2.   1. Distributes (1) O_2 from lungs to tissue cells and CO_2 from tissues to lungs; (2) metabolic wastes to elimination sites, such as kidneys; (3) hormones; (4) heat throughout the body.   2. Regulatory: (1) blood buffers maintain normal blood pH; (2) blood solutes maintain blood volume for normal circulation.   3. Protection: (1) clotting proteins and platelets prevent blood loss; (2) certain plasma proteins and white blood cells defend the body against foreign substances that have gained entry.

Blood Plasma

1.   1. nutrients   2. electrolytes   3. plasma proteins

2.   Proteins found in plasma:

Constituent	Description/importance
Albumin	60% of plasma proteins; important for osmotic balance
Fibrinogen	4% of plasma proteins; important in blood clotting
Globulins • alpha and beta • gamma	36% of plasma proteins: • transport proteins • antibodies
Nonprotein nitrogenous substances	Lactic acid, urea, uric acid, creatinine, ammonium salts
Nutrients	Organic chemicals absorbed from the digestive tract
Respiratory gases	O_2, CO_2

Formed Elements

1.   1. 6   2. 1   3. 4   4. 5   5. 7   6. 3   7. 2   Circle reticulocyte, underline normoblast.

2.   1. 100–120 days   2. iron, vitamin B_{12}, and folic acid   3. They are engulfed by macrophages of the liver, spleen, or bone marrow.   4. Iron is salvaged and stored, and amino acids from the breakdown of globin are conserved. The balance of hemoglobin is degraded to billirubin, which is secreted in bile by the liver, and leaves the body in feces.

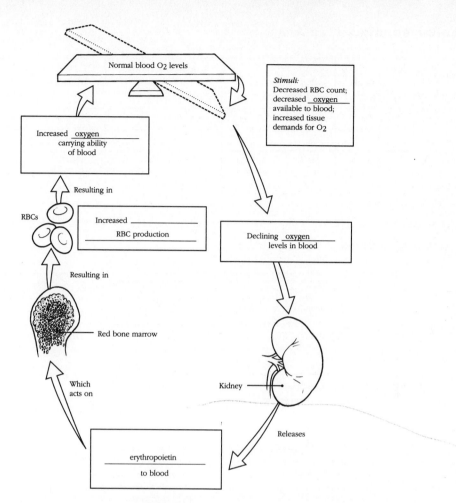

Figure 18.1

3. Check 1, 2, 3

4. 1. biconcave disk   2. anucleate   3. virtually none   4. hemoglobin molecules   5. spectrin, a protein attached to the cytoplasmic face of the plasma membrane   6. anaerobic   7. no mitochondria   8. globin protein (2 alpha and 2 beta chains) and 4 heme groups   9. heme (iron)   10. globin (amino acids)   11. lungs   12. tissues   13. oxyhemoglobin   14. reduced hemoglobin or deoxyhemoglobin

5. 1. F   2. B   3. A   4. C   5. B   6. D   7. E

6. 1. diapedesis   2. differential   3. kidneys   4. globin parts   5. 5.5   6. T   7. T   8. hematocrit   9. less   10. lymphocytes   11. T   12. transferrin

7. 1. G   2.–4. A, B, G   5. I   6. F   7. G   8. D   9. F   10. E   11. C   12. B   13. A   14. D   15. I   16. H, I   17. F   18. H   19. I   20.–24. A, B, D, F, G   25. B   26. G   27. D   28. F   29. G   30. B

8. Figure 18.2: A. neutrophil   B. monocyte   C. eosinophil   D. lymphocyte

9. 1. Erythrocytes   2. Monocytes   3. Antibodies   4. Lymphocyte   5. Platelets   6. Aneurysm   7. Increased hemoglobin   8. Hemoglobin   9. Lymphocyte

10. They are chemicals that promote the proliferation of white blood cells. Macrophages, T lymphocytes.

11. 1. macrophage-monocyte colony-stimulating factor   2. granulocyte colony-stimulating factor   3. granulocyte-macrophage colony-stimulating factor   4. interleukin 3   5. multi-colony-stimulating factor

Hemostasis

1. 1. A   2. F   3. H   4. J   5. E   6. G   7. I   8. D   9. C   10. B

2. 1. I   2. I   3. B   4. B   5. E   6. I   7. E   8. I   9. E   10. B   11. B   12. B

3. 1. A   2. C   3. H   4. E   5. G   6. I   7. D   8. B   9. J   10. L   11. K   12. F

Transfusion and Blood Replacement

1.

Blood type	Agglutinogens or antigens	Agglutinins or antibodies in plasma	Can donate blood to type	Can receive blood from type
Type A	A	Anti-B	A, AB	A, O
Type B	B	Anti-A	B, AB	B, O
Type AB	A, B	None	AB	A, B, AB, O
Type O	None	Anti-A, anti-B	A, B, AB, O	O

2.   No; A+

3.   A situation in which plasma antibodies attach to and lyse red blood cells different from your own.

Diagnostic Blood Tests

1.

Characteristic	Normal value or normal range
% of body weight	8%
Blood volume	4–5 L (female), 5–6 L (male)
Arterial pH	7.4 (7.35–7.45)
Blood temperature	38°C
RBC count	4.3–5.2/mm^3 μl (f), 5.1–5.8 (m)
Hematocrit	45% 42 ± 5% (f) 47 ± 7% (m)
Hemoglobin	12–16 g/100 ml (f), 13–18 (m)
WBC count	4,000–11,000/mm^3 (μl)
Differential WBC count #5 neutrophils	40–70%
#2 eosinophils	1–4%
#1 basophils	< 1%
#4 lymphocytes	20–45%
#3 monocytes	4–8%
Platelet count	250,000–500,000/mm^3 (μl)

2.

Tests for certain nutrients	Tests for clotting	Tests for mineral levels	Tests for anemia
1. Glucose level (glucose tolerance test)	1. Prothrombin time	1. Calcium assay	1. Hematocrit
	2. Platelet count	2. Potassium assay	2. Total RBC count
2. Fat content		3. Sodium (NaCl) assay	3. Hb determination

Developmental Aspects of Blood

1.   1. blood islands   2. F   3. jaundiced   4. iron   5. pernicious   6. B$_{12}$ 7. thrombi   8. leukemia

The Incredible Journey

1.   1. hematopoiesis   2. hemostasis   3. hemocytoblasts   4. neutrophil   5. phagocyte   6. erythropoietin   7. red blood cells   8. hemoglobin   9. oxygen   10. lymphocytes   11. antibodies   12.–15. basophils, eosinophils, monocytes, platelets   16. endothelium   17. platelets   18. seratonin   19. fibrin   20. clot   21. prothrombin activator   22. prothrombin   23. thrombin   24. fibrinogen   25. embolus

CHALLENGING YOURSELF

At the Clinic

1. Erythropoietin.
2. Aplastic anemia; short-term: transfusion; long-term: bone marrow transplant; packed red cells.
3. Hemorrhagic anemia.
4. Polycythemia vera.
5. Acute leukemia.
6. Vitamin K.
7. Other antigens; for long-term transfusions, typing and cross-matching of blood for these other antigens is also necessary.
8. 1. Hemolytic disease of the newborn (erythroblastosis fetalis).   2. Red blood cells have been destroyed by the mother's antibodies, so the baby's blood is carrying insufficient oxygen.   3. She must have received mismatched Rh+ blood in a previous transfusion.   4. Give the mother RhoGAM to prevent her from becoming sensitized to the Rh+ antigen. 5. Fetal progress will be followed in expectation of hemolytic disease of the newborn; intrauterine transfusions will be given if necessary, as well as complete blood transfusion to the newborn.
9. Although red marrow is found in the cavities of most bones of young children, red marrow and active hematopoietic sites are found only in flat bones of adults. The bones most often chosen for obtaining a marrow sample in adults are those flat bones that are most easily accessible, the sternum and ilium.
10. It seems that Rooter has "gifted" the Jones family members (at least the daughters) with pinworms. Eosinophils are the body's most effective defense against infestations by fairly large parasites (such as pinworms, tapeworms, and the like), which explains the girls' elevated eosinophil counts.
11. The stomach cells are the source of gastric intrinsic factor, which is essential as part of the carrier system for the absorption of vitamin B_{12} by the intestinal cells. Apparently, insufficient intrinsic factor–producing cells remained after the surgery, requiring that this patient receive vitamin B_{12} injections. There would be no point in giving the vitamin orally because it is not absorbable in the absence of intrinsic factor. If he refuses the shots, he will develop pernicious anemia.
12. Infectious mononucleosis: total WBC count; differential WBC count.
Bleeding problems: prothrombin time, clotting time, platelet count (tests for certain procoagulants, vitamin K levels, etc.).
13. Tests to determine whether fat absorption from the intestine is normal will indicate if vitamin K is being absorbed normally, and a history of a recent course of antibiotics may reveal that enteric bacteria, which synthesize vitamin K, have been destroyed. A low platelet count is diagnostic for thrombocytopenia.
14. The diagnosis is thalassemia. It can be treated with a blood transfusion.

Stop and Think

1. Crushing of the artery elicits the release of more chemical clotting mediators and a stronger vascular spasm, which is more effective in reducing blood loss.
2. Only hemocytoblasts form blood elements. If most hemocytoblasts are diverted to WBC production, fewer are available to form RBCs and platelets.
3. poly = many; cyt = cells; emia = blood (many cells in the blood).
dia = through; ped = foot; sis = act of (putting a foot through).
4. The extrinsic "shortcut" provides for faster clotting; the multiple steps of the intrinsic pathway result in an amplifying cascade.
5. Sodium EDTA is an anticoagulant; removal of calcium ions *prevents* clotting because ionic calcium is required for nearly every step of hemostasis.
6. Fetal hemoglobin F must receive oxygen from maternal hemoglobin A; the higher affinity of hemoglobin F ensures that it can load oxygen from hemoglobin A.
7. When an athlete is exercising vigorously and body temperature rises, the warmed blood increases in volume. A blood test done at this time to determine iron content would indicate iron-deficiency anemia. This "illusion" of anemia is called athlete's anemia.
8. RBC antigens that occur only in specific families; that is, they are highly restricted in the general population.
9. It will rise. Since CO competes with oxygen for binding sites on heme, the oxygen carrying capacity of John's blood will decline, thereby stimulating erythropoiesis.

COVERING ALL YOUR BASES

Multiple Choice

1. A, D   2. B   3. B, D   4. A, D   5. A, B, D   6. D   7. A, B, D   8. A, B, C   9. C   10. B   11. D
12. B, C   13. D   14. B   15. B   16. C   17. B, C, D   18. A, B, C   19. C   20. C   21. B, C   22. C
23. B, C   24. C   25. B, C, D   26. A   27. B   28. A, B, C, D   29. A, D   30. D   31. B   32. B, D

Word Dissection

Word root	Translation	Example		Word root	Translation	Example
1. agglutin	glued together	agglutination		**9.** karyo	nucleus	megakaryocyte
2. album	egg white	albumin		**10.** leuko	white	leukocyte
3. bili	bile	bilirubin		**11.** lymph	water	lymphatic system
4. embol	wedge	embolus		**12.** phil	love	neutrophil
5. emia	blood	anemia		**13.** poiesis	make	hemopoiesis
6. erythro	red	erythrocyte		**14.** rhage	break out	hemorrhage
7. ferr	iron	ferritin		**15.** thromb	clot	thrombocyte
8. hem	blood	hemoglobin				

Chapter 19 The Cardiovascular System: The Heart

BUILDING THE FRAMEWORK

Heart Anatomy

1.   1. mediastinum   2. diaphragm   3. second   4. midsternal line   5. fibrous   6. visceral   7. epicardium   8. friction   9. myocardium   10. cardiac muscle   11. fibrous skeleton   12. endocardium   13. endothelial   14. 4   15. atria   16. ventricles

2.

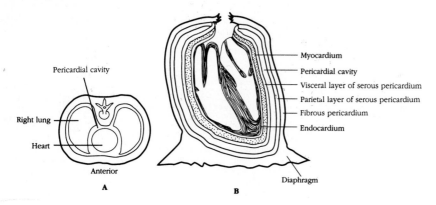

Figure 19.1

3.   1. right ventricle   2. pulmonary semilunar valve   3. pulmonary arteries   4. lungs   5. right and left pulmonary veins   6. left atrium   7. mitral (bicuspid)   8. left ventricle   9. aortic   10. aorta   11. capillary beds   12. superior vena cava   13. inferior vena cava
In Figure 19.2, the white areas represent regions transporting O_2-rich blood. The gray vessels transport O_2-poor blood.

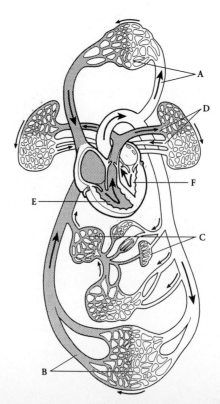

Figure 19.2

4.   Figure 19.3: 1. right atrium   2. left atrium   3. right ventricle   4. left ventricle   5. superior vena cava   6. inferior vena cava   7. aorta   8. pulmonary trunk   9. left pulmonary artery   10. right pulmonary artery   11. right pulmonary veins   12. left pulmonary veins   13. vessels of coronary circulation   14. apex of heart   15. ligamentum arteriosum

Blood Supply to the Heart: Coronary Circulation

1.

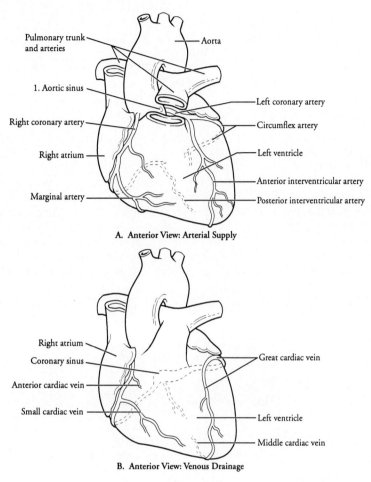

Pulmonary trunk and arteries
Aorta
1. Aortic sinus
Left coronary artery
Right coronary artery
Circumflex artery
Right atrium
Left ventricle
Anterior interventricular artery
Marginal artery
Posterior interventricular artery

A. Anterior View: Arterial Supply

Right atrium
Coronary sinus
Great cardiac vein
Anterior cardiac vein
Small cardiac vein
Left ventricle
Middle cardiac vein

B. Anterior View: Venous Drainage

Figure 19.4

Properties of Cardiac Muscle Fibers

1.   1. C   2. C   3. C   4. C   5. S   6. S   7. C   8. S   9. C   10. S   11. C   12. C
2.

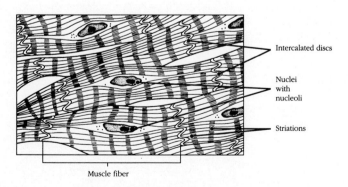

Intercalated discs
Nuclei with nucleoli
Striations
Muscle fiber

Figure 19.5

1. endomysium   2. prevent separation of adjacent cells   3. allow impulse (ions) to pass from cell to cell
4. functional syncytium   5. gap junctions

3.

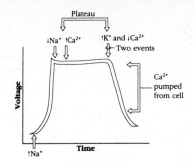

Figure 19.6

Heart Physiology

1. 1. spontaneously   2. resting membrane potential   3. action potential generation   4. pacemaker potentials   5. K$^+$
6. Na$^+$ 7. into   8. voltage-gated sodium channels   9. Na$^+$

2. A. 6   B. 7   C. 8   D. 9   E. 9   F. 8   G. 1   H. 2
1. SA node   2. AV node   3. AV bundle or bundle of His   4. bundle branches   5. Purkinje fibers   6. pulmonary valve
7. aortic valve   8. mitral (bicuspid) valve   9. tricuspid valve
Figure 19.7: Red arrows should be drawn from the left atrium to the left ventricle and out the aorta. Blue arrows should be drawn from the superior and inferior vena cavae into the right atrium, then into the right ventricle and out the pulmonary trunk. Green arrows should be drawn from 1 to 5 in numerical order. The cords, called chordae tendineae, should run from the edges of the flaps of the AV valves to the inferior ventricular walls.

3. 1. sinus rhythm   2. SA node 70–80 beats/min; AV node 50/min; AV bundle 30/min; Purkinje fibers 30/min   3. 0.3–0.5 m/s   4. 0.22

4.

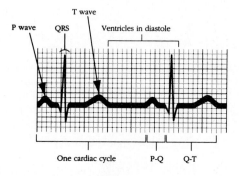

Figure 19.8

5. Figure 19.9: 1. B has extra P waves   2. C shows tachycardia   3. A has an abnormal QRS complex
6. 1. systole   2. diastole   3. lub-dup   4. atrioventricular   5. semilunar   6. ventricles   7. atria   8. atria
9. ventricles   10. murmurs
7.

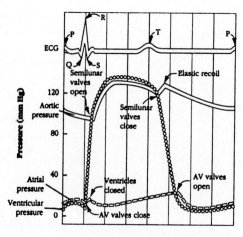

Figure 19.10

8. 1. AV valves open   2. SA node is pacemaker   3. Ventricular systole   4. Semilunar valves open   5. Heart sound after valve closes   6. AV valve open

9. 1. cardiac output   2. heart rate   3. stroke volume   4. 75 beats per minute   5. 70 ml per beat   6. 5250 ml per minute   7. minute   8. 120 ml   9. 50 ml   10. cardiac reserve   11. running, climbing, or any aerobic activity   12. end diastolic volume (EDV)   13. end systolic volume (ESV)   14. preload   15. cross–bridge   16. force   17. venous blood   18. ventricles

10. 1. Increased force of heart beat that is independent of EDV.   2. Sympathetic nervous system activation; thyroxine; epinephrine.   3. Exercise enhances the effectiveness of the respiratory and skeletal muscle "pumps" and activates the sympathetic nervous system.

11. Check 1, 2, 4, 5, 6, 8, 9, 10, 13, 15, 19, 22.

12. 1. sympathetic   2. T   3. increases   4. high   5. fetal   6. resting heart rate   7. systolic   8. left   9. T   10. T   11. T

13. 1. H   2. D   3. G   4. C   5. B   6. E   7. J   8. F   9. A   10. I

14. 1. D   2. A   3. C   4. B   5. E

Developmental Aspects of the Heart

1. 1. fourth   2. 140–160 beats per minute   3. foramen ovale   4. ductus arteriosus   5. pulmonary trunk   6. occluded   7. cyanosis   8. exercise   9. valve flaps   10. nodes of the intrinsic conduction system   11. coronary arteries   12.–14. animal fat, salt, cholesterol

CHALLENGING YOURSELF

At the Clinic

1. Cardiac tamponade; compression of the heart by excess pericardial fluid reduces the space for ventricular activity and impairs ventricular filling.
2. Cor pulmonale; pulmonary embolism.
3. Mitral valve prolapse; valve replacement.
4. Myocarditis caused by the strep infection.
5. Valvular stenosis; heart murmur will be high-pitched during systole.
6. Complete heart block; conduction pathway between SA node and AV node is damaged.
7. Angina pectoris.
8. Zero; myocardial infarction. The posterior interventricular artery supplies much of the left ventricle, the systemic pump.
9. Bradycardia, which results from excessive vagal stimulation of the heart, can be determined by taking pulse.
10. Peripheral congestion due to right heart failure.
11. BP measurement: A high blood pressure may hint that the patient has hypertensive heart disease; when BP is chronically elevated, the heart has to work harder. Blood lipid and cholesterol levels: High levels of triglycerides and cholesterol are risk factors for coronary heart disease. Electrocardiogram: Will indicate if the pacing and electrical events of the heart are abnormal. Chest X ray: Will reveal if the heart is enlarged or abnormally located in the thorax.
12. A defective valve would be detected during auscultation. Since the valve flaps are not electrically excitable, an ECG does not reveal a valvular problem.
13. Ductus arteriosus is a canal-like connection between the aorta and pulmonary trunk that allows blood to bypass the immature fetal lungs. Since this shunt is still open (patent) in the girl, there is some mixing of oxygenated and deoxygenated blood, accounting for the breathlessness.

Stop and Think

1. They will not be compressed when the ventricles contract, as are the vessels within the myocardium.
2. Contraction from the top down is like squeezing a tube of toothpaste from the bottom up. The blood is moved in the direction of the outflow at the AV valves. Ventricles contract from the bottom up, due to the arrangement of the conduction system, from apex up the lateral walls. This also moves blood toward the valves but, in this case, against the pull of gravity.
3. Referred pain from damage to the heart is felt in the left chest and radiates down the medial side of the left arm to the fifth digit in severe cases.
4. Prolonged contraction of the myocardium allows sufficient tension to build to overcome the inertia of the blood and actually move the fluid out of the chambers.
5. It takes about one minute, the time it takes for the entire blood volume to circulate through the heart/body.
6. With exercise, blood flow to the myocardium is maintained and expanded, and the heart becomes stronger and more efficient. With CHF, myocardial blood flow is diminished, and the heart weakens. A healthy heart attains sufficient ventricular pressure to provide sufficient stroke volume with a reduced heart rate. The weakened, congested heart cannot exert sufficient force to eject much blood during ventricular systole, hence heart rate often increases.
7. Pulmonary stenosis and coarctation of the aorta increase the workload; a septal defect allows mixing of oxygenated and unoxygenated blood. Tetralogy of Fallot involves both types of defects.
8. Hypothyroidism leads to reduced heart rate.

9.   The flap over the foramen ovale acts as a valve, preventing backflow from the left atrium to the right atrium. The groove channels blood from the inferior vena cava to the left atrium (via the foramen ovale), providing the systemic circulation with most of the freshly oxygenated blood.

10. Exercise increases heart rate only during the period of increased activity; during rest (which is most of the time), the heart rate decreases as a result of regular exercise. If the resting heart rate drops from 80 to 60 bpm, and one hour of exercise a day causes heart rate to be elevated to 180 bpm, the net change in number of heart beats per day can be calculated as follows:

Sedentary: 80 bpm × 60 min/hr × 24 hr/day = 115,200 beats/day

Exerciser: 60 bpm × 60 min/hr × 23 hr/day +

180 bpm × 60 min/hr × 1 hr/day = 93,600 beats/day

As you can see, an "exerciser" gains, not loses, time.

COVERING ALL YOUR BASES

Multiple Choice

1. D   **2.** A, D   **3.** A, D   **4.** A, B, C, D   **5.** A, B, C, D   **6.** C   **7.** A, D   **8.** A, C   **9.** A   **10.** C, D   **11.** A, C, D   **12.** A, D   **13.** C   **14.** C   **15.** C   **16.** C   **17.** B, D   **18.** C   **19.** A, C   **20.** D   **21.** B   **22.** B   **23.** A, D   **24.** C   **25.** A, B, C   **26.** A, B, C, D   **27.** B   **28.** B   **29.** A, C, D   **30.** D   **31.** B   **32.** A, B, C

Word Dissection

Word root	Translation	Example		Word root	Translation	Example
1. angina	choked	angina pectoris		**8.** ectop	displaced	ectopic focus
2. baro	pressure	baroreceptor		**9.** intercal	insert	intercalated disc
3. brady	slow	bradycardia		**10.** pectin	comb	pectinate muscle
4. carneo	flesh	trabeculae carneae		**11.** sino	hollow	sinoatrial node
5. cusp	pointed	tricuspid valve		**12.** stenos	narrow	mitral stenosis
6. diastol	stand apart	diastole		**13.** systol	contract	systole
7. dicro	forked	dicrotic notch		**14.** tachy	fast	tachycardia

Chapter 20   The Cardiovascular System: Blood Vessels

BUILDING THE FRAMEWORK

Overview of Blood Vessel Structure and Function

1.   1. femoral artery   2. brachial artery   3. popliteal artery   4. facial artery   5. radial artery   6. posterior tibial artery   7. temporal artery   8. dorsalis pedis

2.   1. A   2. B   3. A   4. A   5. C   6. B   7. C

A. artery; thick media; small round lumen   B. vein; thin media; relatively larger lumen; valves

C. capillary; single layer of endothelium

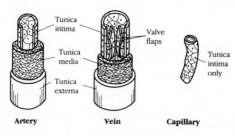

Figure 20.1

3.   1. A   2. E   3. G   4. B   5. C   6. D   7. F   8. H   9. K   10. I   11. J

4.   Arterial anastomoses provide alternate pathways for blood to reach a given organ. If one branch is blocked, an alternate branch can still supply the organ.

5.

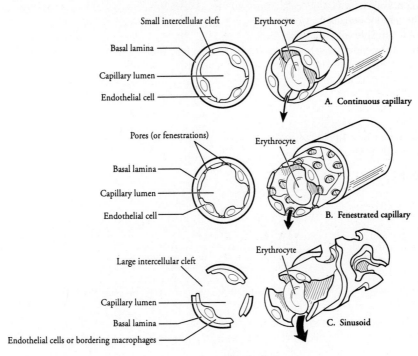

Figure 20.2

6. 1. Muscular   2. Pressure points   3. Heart   4. Gap junctions   5. Kidney   6. High pressure   7. End arteries
8. Thick media   9. True capillaries

7. The venous valves prevent backflow of blood.

8. Veins have large lumens and thin walls and can hold large volumes of blood. At any time, up to 65% of total blood volume can be contained in veins. These reservoirs are most abundant in the skin and visceral organs (particularly the digestive viscera).

Physiology of Circulation

1. 1. Blood Flow = difference in blood pressure ÷ resistance

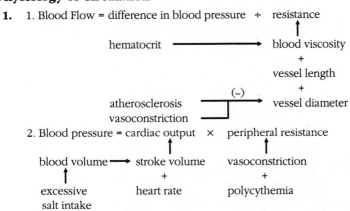

2. Blood pressure = cardiac output × peripheral resistance

2. When the heart contracts, blood is forced into the large arteries, stretching the elastic tissue in the artery walls. During diastole, the artery walls recoil against the blood, maintaining continuous pressure and blood flow.

3. 1. H   2. B   3. C   4. F   5. D   6. J   7. E   8. A   9. G   10. I

4. 1. D   2. I   3. I   4. I   5. I   6. D   7. D   8. I   9. D   10. D   11. D   12. I   13. I   14. D   15. I   16. I
17. D   18. I   19. I   20. D   21. I   22. I

5. 1. Cardiac output   2. Blood flow   3. Low viscosity   4. Blood pressure   5. High blood pressure   6. Vasodilation
7. Vein   8. 120 mm Hg   9. Cardiac cycle   10. Inactivity   11. Sympathetic activity

6. 1. increase   2. orthostatic   3. brain   4. stethoscope   5. low   6. vasoconstricting   7. hypertension   8. arterioles
9. medulla/brain   10. T   11. reduction   12. vasoconstriction   13. blood vessel length   14. capillaries
15. arterial system   16. autoregulation   17. T

7. 1. interstitial fluid   2. concentration gradient (via diffusion)   3. fat-soluble substances like fats and gases   4. water and water-soluble substances like sugars and amino acids   5. through the metarteriole-thoroughfare channels 6. A   7. capillary blood   8. capillary hydrostatic (blood) pressure   9. blood pressure   10. capillary osmotic pressure   11. albumin   12. at the arterial end   13. It is picked up by lymphatic vessels for return to the bloodstream.   14. capillary hydrostatic (blood) pressure

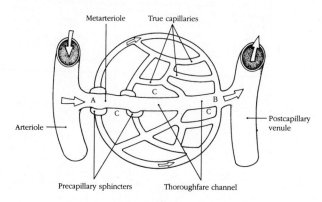

Figure 20.3

8. Hypovolemic shock is a decrease in total blood volume, as from acute hemorrhage. Increased heart rate and vasoconstriction result. In vascular shock, blood volume is normal, but pronounced vasodilation expands the vascular bed, resistance decreases, and blood pressure falls. A common cause of vascular shock is bacterial infection.

9. 1. capillaries   2. arteries   3. capillaries   4. arteries   5. veins

10. 1. F   2. A   3. D   4. B   5. A   6. D   7. C   8. E   9. B   10. D   11. C   12. A   13. A   14. A 15. F   16. E   17. E

Circulatory Pathways: Blood Vessels of the Body

1. The right atrium and ventricle and all vessels with "pulmonary" in their name should be colored blue; the left atrium and ventricle and the aortic arch and lobar arteries should be colored red.

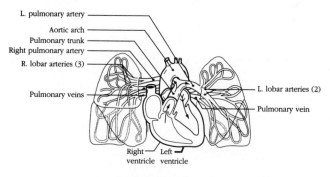

Figure 20.4

2.

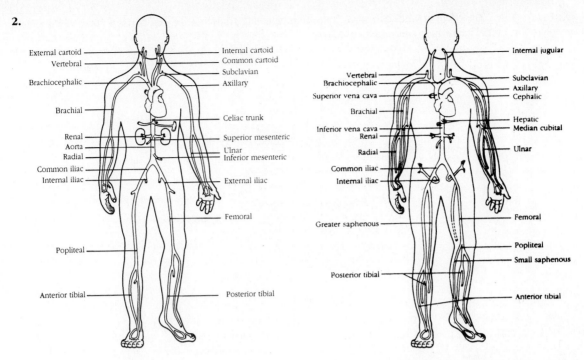

Figure 20.5 Arteries

Figure 20.6 Veins

3.

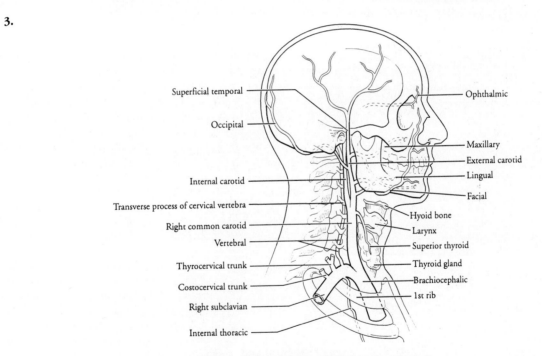

Figure 20.7 Arteries of the head and neck

4.   1. F   2. C, D   3. A   The circle of Willis consists of the communicating arteries and those parts of the cerebral arteries that complete the arterial anastomosis around the pituitary.

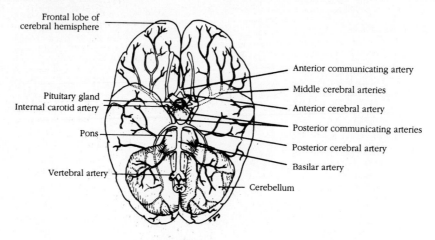

Figure 20.8

5.   1. F   2. W   3. H   4. P   5. Y   6. B   7. J   8. I   9. S   10. C   11. C   12. N   13. Q   14. L   15. C   16. X   17. G   18. E   19. K   20.–22. A, R, T   23. U

6.

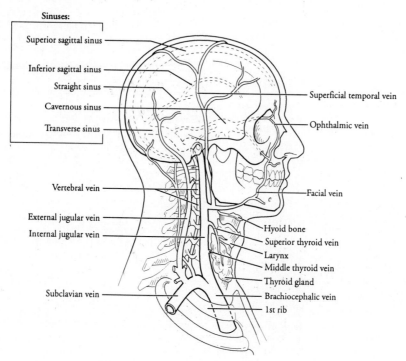

Figure 20.9 Veins of the head and neck

7.

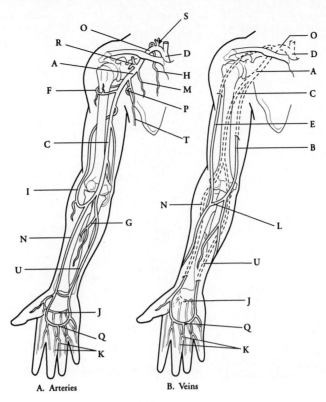

A. Arteries   B. Veins

Figure 20.10

8.

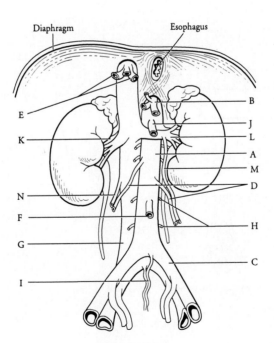

Figure 20.11

9.

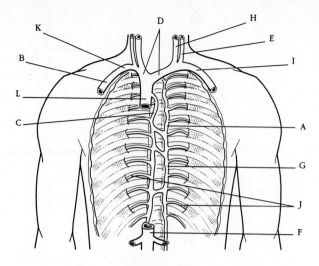

Figure 20.12

10.

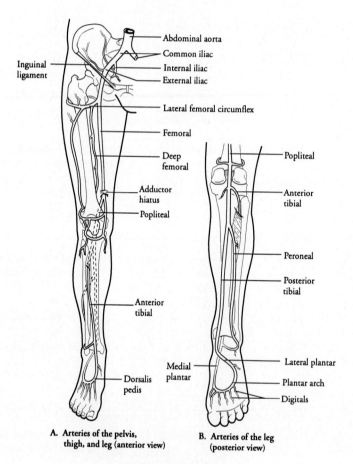

Inguinal ligament

Abdominal aorta
Common iliac
Internal iliac
External iliac

Lateral femoral circumflex

Femoral

Deep femoral

Adductor hiatus

Popliteal

Anterior tibial

Dorsalis pedis

Popliteal

Anterior tibial

Peroneal

Posterior tibial

Medial plantar

Lateral plantar

Plantar arch

Digitals

A. Arteries of the pelvis, thigh, and leg (anterior view)

B. Arteries of the leg (posterior view)

Figure 20.13

11.

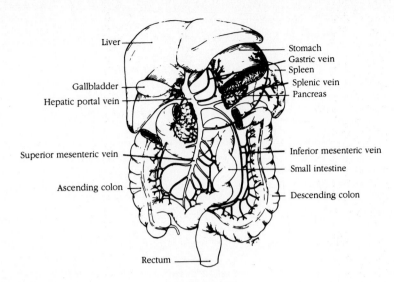

Liver
Stomach
Gastric vein
Spleen
Splenic vein
Pancreas
Gallbladder
Hepatic portal vein
Superior mesenteric vein
Inferior mesenteric vein
Small intestine
Ascending colon
Descending colon
Rectum

Figure 20.14

12.   1. S   2. X   3. U   4. E   5. T   6. Q   7. D   8. A   9. R   10. M   11. F   12. J   13. B   14. O   15. L
16.–18. I, N, V   19. K   20. G   21. H

Developmental Aspects of the Blood Vessels

1.   1. mesoderm   2. blood islands   3. mesenchymal   4. lungs   5. ductus venosus   6. umbilical vein   7. placenta
8. fetal liver   9. umbilical arteries   10. lower   11. atherosclerosis   12. varicose veins   13. legs and feet
14. hypertension   15.–17. good diet, exercise, not smoking

The Incredible Journey

1.   1. left atrium   2. left ventricle   3. mitral (bicuspid)   4. chordae tendineae   5. diastole   6. systole/contraction
7. aortic semilunar   8. aorta   9. superior mesenteric   10. endothelial   11. superior mesenteric vein   12. splenic
13. nutrients   14. phagocytic (van Kupffer)   15. hepatic   16. inferior vena cava   17. right atrium
18. pulmonary   19. lungs   20. capillaries   21. gas (O_2 and CO_2)   22. subclavian

CHALLENGING YOURSELF

At the Clinic

1. The veins, particularly the superficial saphenous veins, become very prominent and tortuous. Typically it is due to failure of the venous valves. The problem here is the restriction of blood flow by her enlarged uterus during her many pregnancies. Elevate the legs whenever possible; avoid standing still.
2. Thrombosis, atherosclerosis; arterial anastomosis (circle of Willis).
3. High; polycythemia increases blood viscosity (thus peripheral resistance), which increases blood pressure.
4. Hypovolemic shock due to blood loss.
5. Loss of vasomotor tone due to damage to the vasomotor center in the medulla could cause extreme vasodilation, resulting in vascular shock.
6. Chronically elevated; chronic hypersecretion of ADH will trigger vasoconstriction (due to the vasopressin effect) and excessive water retention, both of which increase peripheral resistance.
7. Occlusion of the renal blood supply will reduce blood pressure in the kidneys, triggering the release of renin, which in turn precipitates activation of angiotensin, a powerful vasoconstrictor.
8. Beta blockers prevent the effect of epinephrine, which constricts the arterioles and raises the blood pressure. Diuretics increase urine output, which reduces blood volume. Hence, the treatment will result in lower blood pressure.
9. Balloon (or another type of) angioplasty.
10. Septicemia; blood-borne bacterial toxins trigger widespread vasodilation, which causes blood pressure to plummet (vascular shock).
11. Cardiogenic shock.
12. The dorsalis pedis pulse is the most distal palpable pulse. Chances are if his popliteal and pedal pulses are strong, his right limb is getting adequate circulation. Also, the skin will be warm and *non*cyanotic if the blood supply is adequate.
13. Inflammation of the veins; thrombophlebitis, which can lead to pulmonary embolism.
14. Pulse pressure = 20 mm Hg; low; loss of elasticity.
15. Nicotine is a vasoconstrictor; a high dose of nicotine in a body not accustomed to it can reduce blood flow to the brain and cause dizziness.

Stop and Think

1. A thin layer is sufficient to withstand the frictional forces of the blood. A thicker layer would reduce the size of the lumen and require more nutrients to maintain it.
2. Decrease; vasoconstriction increases resistance, which decreases blood flow.
3. Open; all true capillaries will be flushed with blood. In the face exposed to cold, the capillary networks are closed off. The cells deplete oxygen and nutrients, and waste products build up. Precapillary sphincters relax, but no blood flows to the capillaries since the supplying arteriole is closed off. In warm air, the arterioles vasodilate, and blood flows into all the capillaries.
4. In a warm room, the skin capillaries are flushed with blood, which helps lose heat from the body. Rerouting blood to the head takes longer in this circumstance than in a cold room, in which many skin capillary beds are being bypassed by the metarteriole-thoroughfare channel shunts.
5. Pressure of the fetus on the inferior vena cava will reduce the return of blood from the lower torso and lower limbs.
6. Pericardial sac.
7. Organs supplied by end arteries have no collateral circulation. If the supplying end artery is blocked, no other circulatory route is available.
8. Formation of new tissues (adipose, muscle).
9. Yes.
10. Vertebral arteries to basilar artery to posterior cerebral artery to posterior communicating artery to R. internal carotid artery beyond the blockage (hopefully).
11. Reduced plasma proteins (particularly albumin) will reduce the osmotic pressure of the blood. Consequently, less fluid will be drawn back at the venous end of capillaries, and edema will result.
12. Decreased cardiac output would reduce blood pressure to the brain. If compensatory mechanisms are inadequate to maintain blood supply to the vasomotor center, vasomotor tone will be reduced. The resulting vasodilation will cause vascular shock, which will reduce venous return to the heart. This will lead to even lower cardiac output, which will further diminish blood delivery to the vasomotor center.
13. Hepatic circulation in an adult is important in processing nutrients from the digestive tract. In the fetus, preprocessed nutrients are supplied via the umbilical vein. Bypassing the liver capillaries gives a more direct return of oxygenated, nutrified blood from the placenta to the general fetal circulation.
14. An aneurysm is a ballooned-out and weakened area in a blood vessel. The primary problem in all cases is that it might rupture and cause a fatal stroke. The second problem in the patient discussed here is that the enlarged vessel is pressing on brain tissue and nerves. Since neural tissue is very fragile, it is susceptible to irreversible damage from physical trauma as well as from deprivation of a blood supply (another possible consequence of the compression of nervous tissue). Surgery was done to replace the weakened region of the vessel with inert plastic tubing.
15. There is no "great choice" to make here. Try to compress the subclavian artery that runs just deep to the clavicle by forcing your fingers inferiorly just posterior to the clavicular midline and lateral to the sternocleidomastoid muscle. Blockage of the subclavian artery would prevent the blood from reaching the axilla.
16. Erythrocytes are 7–8 μm in diameter, thus five of them side by side would measure a lumen diameter of 35–40 μm (the approximate size of the observed vessel). Since the average capillary is 8–10 μm, one erythrocyte would just about fill a capillary's entire lumen. Hence, the vessel is most likely a postcapillary venule.

COVERING ALL YOUR BASES

Multiple Choice

1. A, B, C   2. A, B, C, D   3. A, C   4. A, B   5. C   6. B   7. A, B, C, D   8. B, C   9. B, D   10. C   11. A, B, C
12. B, C, D   13. C, D   14. A, B, D   15. A, B, C, D   16. D   17. B, C, D   18. B   19. A, B, D   20. A, C
21. A, B, C, D   22. C   23. A, C, D   24. B   25. A   26. A, B, C   27. A, B, C   28. A, B   29. C   30. A, B, C
31. C   32. D   33. A   34. C   35. B

Word Dissection

Word root	Translation	Example	Word root	Translation	Example
1. anastomos	coming together	anastomoses	11. epiplo	membrane	epiploic artery
2. angio	vessel	angiogram	12. fenestr	window	fenestrated capillary
3. aort	great artery	aorta	13. jugul	throat	jugular vein
4. athera	gruel	atherosclerosis	14. ortho	straight	orthostatic hypotension
5. auscult	listen	auscultation	15. phleb	vein	phlebitis
6. azyg	unpaired	azygos vein	16. saphen	clear, apparent	great saphenous vein
7. capill	hair	capillary	17. septi	rotten	septicemia
8. carot	stupor	carotid artery	18. tunic	covering	tunica intima
9. celia	abdominal	celiac artery	19. vaso	vessel	vasodilation
10. entero	intestine	mesenteric arteries	20. viscos	sticky	viscosity

Chapter 21 The Lymphatic System

BUILDING THE FRAMEWORK

Lymphatic Vessels

1. 1. pump   2. arteries   3. veins   4. valves   5. lymph   6. 3 liters

2. Lymphatic vessels pick up fluid leaked from the cardiovascular system. The overlapping edges of the endothelial cells form minivalves that allow fluid to enter, but not leave, the lymph capillaries. When the interstitial pressure rises, gaps are exposed between the endothelial flaps, and fluid enters. Collapse of the vessel is prevented by fine filaments that anchor the lymph capillaries to surrounding tissues.

3.

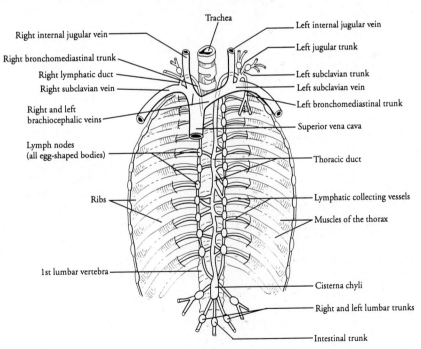

Figure 21.1

4. 1. Fat-free lymph   2. Blood capillary   3. Abundant supply of lymphatics   4. High-pressure gradient   5. Impermeable

5.

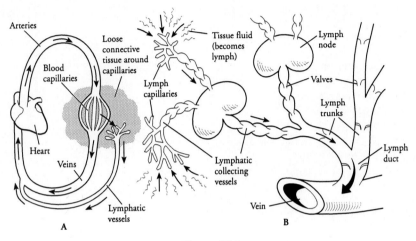

Figure 21.2

Lymphoid Cells, Tissues, and Organs: An Overview

1. 1. lymphocytes   2. antigens   3. T cells   4. B cells   5. macrophages   6. phagocytizing   7. reticular cells   8. reticular   9. diffuse   10. nodular

2.

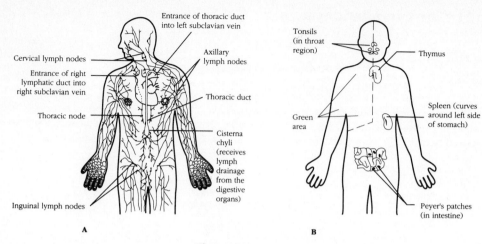

Figure 21.3

Lymph Nodes

1. 1. B lymphocytes 2. They produce and release antibodies. 3. T lymphocytes 4. macrophages 5. phagocytes 6. reticular 7. This slows the flow of lymph through the node, allowing time for immune cells and macrophages to respond to foreign substances present in the lymph. 8. valves in the afferent and efferent lymphatics 9. cervical, axillary, inguinal 10. They act to protect the body by removing bacteria or other debris from the lymphatic stream.

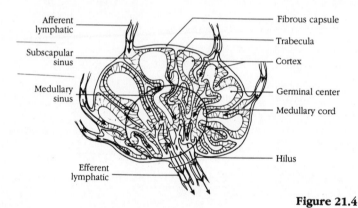

Figure 21.4

Other Lymphoid Organs

1. 1. C 2. A 3. D 4. B, E 5. C 6. C 7. D 8. E 9. B 10. E
2. 1. Iron 2. Platelets 3. RBCs 4. Erythropoiesis 5. Red 6. Sinuses 7. RBCs (erythrocytes) 8. White 9. Immune
3.

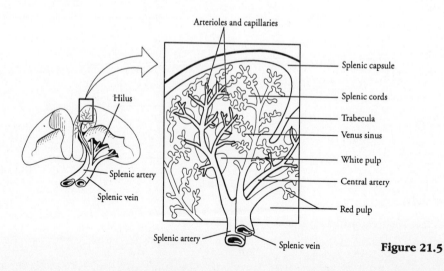

Figure 21.5

4.   MALT, being located just deep to the mucosa, acts as a sentinel to protect the upper respiratory and digestive tracts from entry of foreign matter.

5.   palatine tonsils, pharyngeal tonsils, and lingual tonsils

Developmental Aspects of the Lymphatic System

1.   1. veins   2. jugular   3. thorax   4. upper extremities   5. head   6. inferior vena cava   7. iliac veins   8. mesoderm   9. thymus   10. pharynx   11. spleen   12. lymphocytes   13. hormones

CHALLENGING YOURSELF

At the Clinic

1.   Lymphangitis (inflammation of the lymphatic vessels) due to infection.

2.   Lymphedema; no, the lymphatic vessels will be replaced by budding.

3.   Bubo; bubonic plague.

4.   Her lymph nodes are enlarged but firm, and not tender to the touch.

5.   Hemorrhage; the spleen is a blood reservoir, and the circulatory pattern in the spleen makes it impossible to close off blood vessels and prevent bleeding. No; the liver, bone marrow, and other tissues can take over the spleen's functions.

6.   Pharyngeal tonsils, which lie close to the orifices of the auditory tubes in the pharynx.

7.   Elephantiasis due to parasitic worm infection.

8.   Enlarged spleen; septicemia, mononucleosis, leukemia.

Stop and Think

1.   Only lymph nodes filter lymph.

2.   Cancer cells that have entered lymphatic vessels can get trapped in the lymph node sinuses.

3.   Macrophages are found in the endothelial lining of organs like the liver and are associated with reticular connective tissue in lymph nodes.

4.   If proteins accumulated in the interstitial fluid, the osmotic gradient of the capillaries would be lost. Fluid would be retained in the tissue spaces, resulting in edema.

Closer Connections: Checking the Systems—Circulation and Protection

1.   Maintaining homeostasis of the body tissues, particularly concerning the delivery of needed substances and the carrying away of waste.

2.   The circulatory system includes the cardiovascular and lymphatic systems.

3.   Cardiac muscle: provides the CO to maintain systolic arterial blood pressure; smooth muscle: vasoconstriction maintains mean arterial pressure and propels blood in vessels with valves; skeletal muscle: squeezes low-pressure, valved vessels to help propel blood (or lymph) back to the heart.

4.   No; filtration of fluid from the plasma to the tissue spaces involves movement of fluid out of the closed vessels. This escaped fluid flows through the tissue spaces either to return to the cardiovascular system or to pass into the vessels of the lymphatic system.

5.   The lymphatic vessels pick up leaked fluid, proteins, and various types of debris (including pathogenic microorganisms and cancer cells) from tissue spaces over virtually the entire body. Although the lymph is cleansed (debris is phagocytized) in the lymph nodes, some foreign substances do get delivered to the bloodstream. The cardiovascular system, in turn, distributes blood and whatever it contains throughout the body.

6.   Lymphoid tissue contains immunocompetent lymphocytes, which elicit the immune response when exposed to specific pathogens.

COVERING ALL YOUR BASES

Multiple Choice

1. C, D   **2.** A, B, C, D   **3.** A, B   **4.** C   **5.** A   **6.** B   **7.** A, B   **8.** A, B, C   **9.** A, B, C, D   **10.** A, B, D   **11.** A, B, D   **12.** B, D   **13.** A, B, C, D   **14.** D   **15.** A, C, D   **16.** A   **17.** A   **18.** A, B   **19.** B, C, D   **20.** A, B, C,D   **21.** A, C   **22.** A, B, C, D   **23.** A, C   **24.** D

Word Dissection

Word root	Translation	Example	Word root	Translation	Example
1. adeno	a gland	lymphadenopathy	**4.** lact	milk	lacteals
2. angi	vessel	lymphangiography	**5.** lymph	water; clear water	lymphatic vessels
3. chyle	juice	cisterna chyli			

Chapter 22   Nonspecific Body Defenses and Immunity

BUILDING THE FRAMEWORK

Part I: Nonspecific Body Defenses

1.   1. surface membrane barriers, mucosae   2. natural killer cells   3. chemicals

Surface Membrane Barriers

1.   1. tears, saliva   2. stomach, vagina   3. sebaceous glands; skin   4. goblet cells; digestive
2.   1. A, B, E, F   2. C, G   3. A, B, D, E, F   4. D   5. A–G
3.   They propel mucus laden with trapped debris superiorly away from the lungs to the throat, where it can be spit out.

Nonspecific Cellular and Chemical Defenses

1.   Phagocytosis is ingestion and destruction of particulate material. To occur, the phagocyte must adhere to the particle; this is difficult if the particle is smooth. The rougher the particle, the more easily it is ingested.
2.   Neutrophils release potent oxidizing chemicals to the interstitial fluid, which, besides killing pathogens, also damages them. Macrophages rely only on intracellular killing and do not exhibit this behavior.
3.   1. F   2. D   3. E   4. A   5. C   6. G   7. B   8. I   9. H
4.   1. Nausea   2. Natural killer cells   3. Neutrophils   4. Interferon   5. Antibacterial   6. Macrophages   7. Defensins
5.

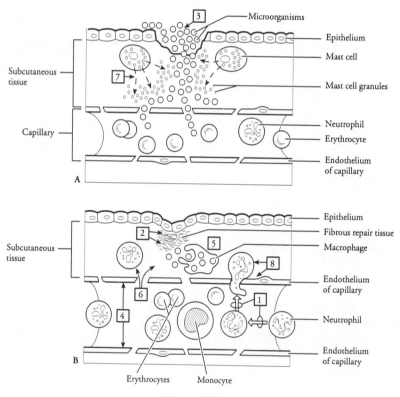

Figure 22.1

6.   1. Dilutes harmful chemicals in the area.   2. Brings in oxygen and nutrients needed for repair.   3. Allows entry of clotting proteins.   4. Also forms a structural basis for permanent repair.
7.   1. 20   2. classical   3. antigen-antibody   4. alternative   5. P   6. polysaccharide   7. C3b
       8. membrane attack complex (MAC)   9. lysis   10. opsonization   11. C3a
8.   Interferon is synthesized in response to viral infection of a cell. The cell produces and releases interferon proteins, which diffuse to nearby cells and attach to their surface receptors. This binding event prevents viruses from multiplying within those cells.
9.   Check 1, 3, 4.

Part II: Specific Body Defenses: Immunity

Antigens

1. 1. immune system   2. immunogenicity   3. reactivity   4. proteins   5. haptens   6. size   7. antigenic determinants
    8. simple   9. rejected

2. 1. MHC proteins   2. Major histocompatibility complex   3. immune system

Cells of the Immune System: An Overview

1. antigen-specific; systemic; has memory

2. The ability to recognize foreign substances in the body by binding to them.

3. 1. The appearance of antigen-specific receptors on the membrane of the lymphocyte.   2. fetal stage   3. its genes
    4. binding to "its" antigen   5. self

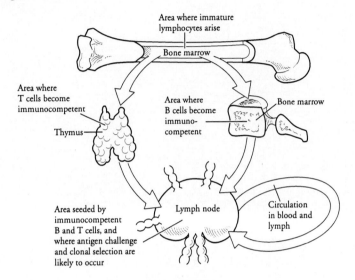

Figure 22.2

4. 1. A   2. E   3. D   4. B   5. I   6. H   7. C   8. F   9. G

5. T cell: 1, 3, 4, 5, 7, 8, 9   B cell: 1, 2, 4, 6, 7, 8

Humoral Immune Response

1. 1. The V part   2. The C part

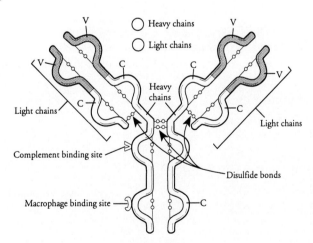

Figure 22.3

2. 1. B, M   2. D   3. E   4. D, E   5. D   6. C   7. A   8. E

3. 1. A   2. P   3. P   4. P   5. A   6. A

4. 1. P   2. P   3. S   4. P   5. S

5. 1. antigen   2. complement fixation and activation   3. neutralization   4. agglutination   5. M   6. precipitation
    7. phagocytes

Cell-Mediated Immune Response

1.   1. C   2. C   3. D   4. A   5. B   6. C   7. A, D
2.   1. F   2. C   3. B   4. K   5. E   6. G   7. D   8. I   9. A   10. J
3.   1. Lymphokines   2. Liver   3. T cell activation   4. Xenograft   5. Natural killer cells   6. Na$^+$-induced polymerization
         7. Class I MHC proteins
4.

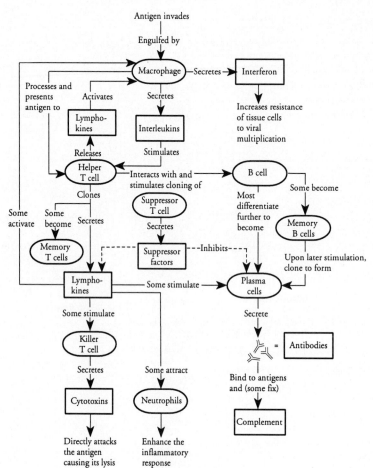

Figure 22.4

5.   1. cell-surface receptors   2. chemicals   3. tolerates   4. destruction
6.   1. allografts   2. an unrelated person   3. cytotoxic T cells and macrophages   4. To prevent rejection, the recipient's
         immune system must be suppressed.   5. The patient is unprotected from foreign antigens, and bacterial or viral
         infection is a common cause of death.

Homeostatic Imbalances of Immunity

1.   1. C   2. A   3. A   4. C   5. B   6. C   7. B   8. C   9. B
2.   1. A, B, C, E   2. A, B   3. C   4. E   5. D   6. A   7. A, B   8. A, B   9. D   10. E   11. A, B   12. D   13. D
         14. C, E

Developmental Aspects of the Immune System

1.   1. liver/spleen   2. lymphoid tissue   3. thymus   4. birth   5. genetically determined   6. depressed   7. nervous
         8. opiate   9. epinephrine   10. corticosteroids   11. declines   12. autoimmune disease   13. immunodeficiencies

The Incredible Journey

1.   1. protein   2. lymph node   3. B cells   4. plasma cell   5. antibodies   6. macrophage   7. antigens
         8. antigen-presenters   9. T   10. clone   11. immunological memory

CHALLENGING YOURSELF

At the Clinic

1. Active artificial immunity.
2. Mother: Rh negative; father: Rh positive; IgG antibodies.
3. Valve replacement surgery with pig valves.
4. Graft-versus-host reaction.
5. Anaphylactic shock; epinephrine (adrenaline).
6. Contact dermatitis (delayed hypersensitivity), probably caused by a reaction to the chemicals in the detergent used to launder the diapers.
7. Thrombocytopenia; autoimmune disorder.
8. Hashimoto's thyroiditis (an autoimmune response to the formerly "hidden" antigens in the thyroid hormone colloid).
9. Digestive complaints often indicate a food allergy.
10. Sterility associated with mumps occurs only after production of sperm antigens begins, after the onset of puberty.

Stop and Think

1. The acidity of the vaginal tract inhibits bacterial growth. Bacterial establishment would trigger an inflammatory response, resulting in vaginitis.
2. No; salmonella and other pathogens such as intestinal viruses survive passage through the stomach. Parasitic worms' eggs or cysts provide protection.
3. Swelling of the nasal and pharyngeal mucosa or glands inhibits drainage from the sinuses and middle ear cavity. Without a drainage route, pathogens in the surface secretions accumulate instead of being flushed out.
4. Chewing anything stimulates saliva secretion. Saliva washes over the gums and teeth, helping to flush out pathogens; lysozyme in saliva can kill bacteria.
5. Histamine: hist, tissue; amine, amino acid; diapedesis: dia, through; ped, foot; interleukin: inter, between; leuk, white (blood cells).
6. Olfactory receptors.
7. Lipid-soluble; it is absorbed through the skin, which is characteristic of lipid-soluble but not water-soluble molecules.
8. The constant recycling of lymphocytes through blood, lymph nodes, and lymph. Its role is to provide for the broadest and quickest sampling of the body for antigen recognition and interception.

COVERING ALL YOUR BASES

Multiple Choice

1. A, B, C 2. B, D 3. A, B, C, D 4. D 5. A 6. C 7. B 8. A, C 9. A, B, C, D 10. A, C 11. B, C, D 12. C 13. B 14. A, B, C, D 15. B, D 16. B, C, D 17. A, C 18. A, B, C, D 19. B 20. C, D 21. B, C, D 22. A, C, D 23. D 24. A, B, C, D 25. A

Word Dissection

Word root	Translation	Example	Word root	Translation	Example
1. hapt	fastened	hapten	6. phylax	preserve, guard	anaphylaxis
2. humor	fluid	humoral immunity	7. phago	eat	phagocyte
3. macro	big/large	macrophage	8. pyro	fire	pyrogen
4. opso	delicacy	opsonization	9. vacc	cow*	vaccine
5. penta	five	pentamer			

The first vaccine was produced using the cowpox virus.

Chapter 23 The Respiratory System

BUILDING THE FRAMEWORK

Functional Anatomy of the Respiratory System

1. pulmonary ventilation, external respiration, transport of respiratory gases, and internal respiration
2. 1. respiratory bronchioles, alveolar ducts, and alveoli 2. gas exchange 3. nasal cavity, pharynx, larynx, trachea, bronchi, and all of their branches except those of the respiratory zone

3. Nose: external nares → middle meatus → internal nares

   ↓

   Pharynx: nasopharynx → adenoids → oropharynx → laryngopharynx

   ↓

   Larynx: epiglottis → vocal folds

   ↓

   Trachea: carina→primary bronchi

4.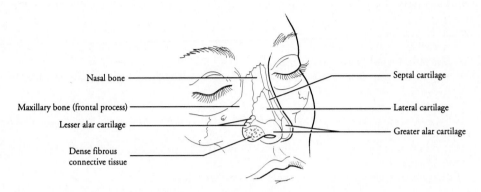

Nasal bone

Maxillary bone (frontal process)

Lesser alar cartilage

Dense fibrous
connective tissue

Septal cartilage

Lateral cartilage

Greater alar cartilage

Figure 23.1

5. Note that the frontal and sphenoidal sinuses should be colored with the same color. Likewise, the subdivisions of the pharynx (the nasopharynx, oropharynx, and laryngopharynx) should be identified visually with a single color.

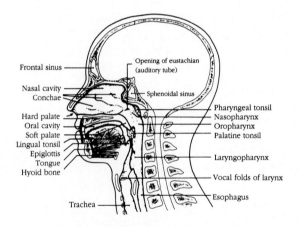

Frontal sinus

Nasal cavity
Conchae

Hard palate
Oral cavity
Soft palate
Lingual tonsil
Epiglottis
Tongue
Hyoid bone

Trachea

Opening of eustachian
(auditory tube)

Sphenoidal sinus

Pharyngeal tonsil
Nasopharynx
Oropharynx
Palatine tonsil

Laryngopharynx

Vocal folds of larynx

Esophagus

Figure 23.2

6. 1. External nares   2. Nasal septum   3.–5. (in any order): Warm; Moisten; Trap debris in   6. Sinuses   7. Speech
   8. Pharynx   9. Larynx   10. Tonsils

7. 1. Mandibular   2. Alveolus   3. Larynx   4. Peritonitis   5. Nasal septum   6. Choanae   7. Tracheal cartilage
   8. Nasopharynx

8. 1. Provides a patent airway; serves as a switching mechanism to route food into the posterior esophagus; voice production location (contains vocal cords).   2. arytenoid cartilages   3. elastic   4. hyaline   5. The epiglottis has to be flexible to be able to flap over the glottis during swallowing. The more rigid hyaline cartilages support the walls of the larynx.
   6. Adam's apple

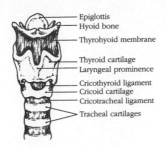

Figure 23.3

9.   1. B   2. G   3. I   4. E   5. D   6. K   7. A   8. H   9. L   10. F   11. C   12. O   13. N   14. K
10.   1. vocal folds (true vocal cords)   2. speak   3. glottis   4. intrinsic laryngeal   5. arytenoid   6. higher   7. wide
8. louder   9. laryngitis
11.

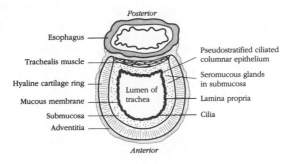

Figure 23.4

1. The C-shaped cartilage rings prevent the trachea from collapsing during the pressure changes occurring during
breathing.   2. Allows the esophagus to expand anteriorly when we swallow food or fluids.   3. It constricts the
tracheal lumen, causing air to be expelled with more force—coughing, yelling.
12.   1. B   2. B   3. A   4. B   5. E   6. C
13.   1. C   2. D   3. E   4. A   5. B
14.

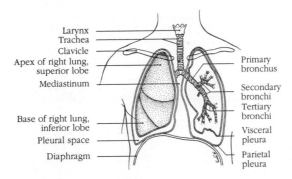

Figure 23.5

15.

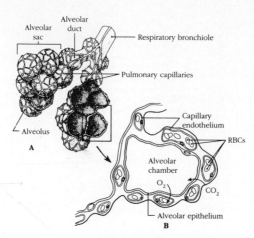

Figure 23.6

16. 1. elastic connective   2. type I   3. gas exchange   4. type II   5. surfactant   6. decrease the surface tension
7. bronchial

Mechanics of Breathing

1. 1. C   2. A   3. B   4. B   5. C   6. B   7. B
2. When the diaphragm is contracted, the internal volume of the thorax increases, the internal pressure in the thorax decreases, the size of the lungs increases, and the direction of air flow is into the lungs. When the diaphragm is relaxed, the internal volume of the thorax decreases, the internal pressure in the thorax increases, the size of the lungs decreases, and the direction of air flow is out of the lungs.
3. 1. D   2. A   3. C   4. B   5. D   6. C   7. B   8. A   9. A
4. 1. scalenes, sternocleidomastoids, pectorals   2. obliques and transversus muscles   3. internal intercostals and latissimus dorsi
5.

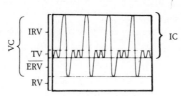

Figure 23.7

6. Check 1, 3, 5.
7. 1. E   2. A   3. G   4. D   5. B   6. F
8. 1. spirometer   2. obstructive pulmonary disease   3. restrictive disorders   4. FEV or forced expiratory volume
5. FVC or forced vital capacity
9. 1. hiccup   2. cough   3. sneeze

Gas Exchanges in the Body

1. 1. B   2. A   3. C
2. Hyperbaric oxygen; Henry's law.
3. Mechanism: generation of free radicals; consequences: CNS disturbances, coma, death.
4. 1. C   2. B   3. E   4. A   5. D
5. 1. 4   2. 5   3. 3   4. 2   5. 1   6. 2   7. 1   8. 3   9. 4   10. 5   11. 2   12. 1   13. 2   14. 1
6. 1. F   2. G   3. H   4. B   5. E   6. J   7. D   8. C   9. I
7. 1. low   2. constrict   3. high   4. dilate

Transport of Respiratory Gases by Blood

1. 1. hemoglobin   2. plasma   3. bicarbonate ion   4. carbonic anhydrase   5. chloride ions   6. chloride shift
7. carbamino Hb   8. Haldane effect   9. more   10. hemoglobin   11. Bohr effect   12. oxygen
2. high PO_2 of blood; low temperature; alkalosis; low levels of 2,3-BPG
3. 1. Hb can still be nearly completely saturated at lower atmospheric PO_2 or in those with respiratory disease.   2. Much more oxygen can be unloaded to the tissues without requiring cardiovascular or respiratory system adjustments to meet increased tissue demands.
4. 1. oxygen-hemoglobin dissociation curve   2. lower pH; higher temperature, increase in BPG

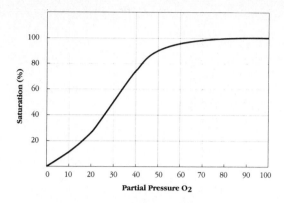

Figure 23.8

Control of Respiration

1.   1. A, E   2. C, D   3. F   4. B   5. D   6. C   7. D   8. E

2.   In healthy people, rising CO_2 levels provide the respiratory drive by stimulating medullary centers. In people who retain CO_2, declining O_2 levels provide the respiratory drive; therefore, administering pure oxygen would result in apnea.

3.   1. peripheral   2. peripheral   3. central

4.   1. ↑ pH   2. Hyperventilation   3. ↑ Oxygen   4. ↑ CO_2 in blood   5. N toxicity   6. ↑ PCO_2

Respiratory Adjustments During Exercise and at High Altitudes

1.   1. Hyperpnea increases depth of ventilation but not necessarily the rate of ventilation; hyperventilation is deep, and typically rapid, ventilation. Hyperpnea does not lead to significant changes in blood levels of O_2 and CO_2, whereas hyperventilation may result in hypocapnia.   2. Psychic stimuli, cortical motor activation, and proprioceptors.

2.   1. Hyperventilation.   2. Hemoglobin's affinity for oxygen declines so that more oxygen is unloaded to the tissues. 3. Enhanced erythropoiesis.

Homeostatic Imbalances of the Respiratory System

1.   1. A   2. F   3. D   4. G   5. E   6. C   7. B   8. C, E   9. H   10. E   11. J   12. I

Developmental Aspects of the Respiratory System

1.   1. olfactory placode   2. endoderm   3. 40   4. 12–15   5. asthma   6. chronic bronchitis   7. emphysema or tuberculosis   8. elasticity   9. vital capacity   10. respiratory infections, particularly pneumonia

The Incredible Journey

1.   1. nasal conchae   2. pharyngeal tonsil   3. nasopharynx   4. mucous   5. vocal fold   6. larynx   7. digestive   8. epiglottis   9. trachea   10. cilia   11. oral cavity   12. primary bronchi   13. left   14. bronchiole   15. alveolus   16. red blood cells   17. red   18. oxygen   19. carbon dioxide   20. cough

CHALLENGING YOURSELF

At the Clinic

1. Atelectasis; the lungs are contained in separate pleural sacs, so only the left lung will collapse.

2. The mucus secreted in the conducting zone will be abnormally thick and difficult to clear. Consequently, the respiratory passageways tend to become blocked and infection is more easily established.

3. The lower oxygen pressure of high altitudes prompts renal secretion of erythropoietin, leading to accelerated RBC production. Len will notice that he will begin to hyperventilate. His minute ventilation will increase by about 2–3 L/min. His arterial PCO_2 will be lower than the normal 40 mm Hg, and his hemoglobin saturation will be only about 67%.

4. Chest surgery causes painful breathing, and many patients try not to cough because the pain can be intense. However, coughing is necessary to clear mucus; if allowed to accumulate, mucus can cause blockage and increase the risk of infection.

5. Inflammation in the alveoli causes fluid to accumulate in the air spaces, which increases the apparent thickness of the respiratory membrane and reduces the lungs' ability to oxygenate the blood.

6. Pleurisy.

7. Stagnant hypoxia.

8. Michael most likely is suffering from carbon monoxide poisoning.

9. Sudden infant death syndrome.

10. Small cell (oat cell) carcinoma.

11. Emphysema; although ventilation is difficult, oxygenation is sufficient, and cyanosis does not occur until late in the disease progress.
12. Chronic bronchitis; smoking inhibits ciliary action.
13. Failure of the epiglottis and soft palate to close the respiratory channels completely during swallowing. The former will place the patient at risk for aspiration pneumonia.
14. The baby most likely swallowed the safety pin (babies put everything in their mouths). It is probably in the R. primary bronchus, which is larger in diameter and runs more vertically.
15. Pharyngeal tonsils.

Stop and Think

1. When air is inhaled through the nose, it is warmed by the nasal mucosa. Exhaling through the mouth expels the warmed air, resulting in heat loss.
2. Deep-sea divers risk nitrogen narcosis, the bends, and air emboli. Effective treatment requires a hyperbaric chamber.
3. Fetal hemoglobin receives oxygen from maternal hemoglobin; its higher affinity means more oxygen can be carried by fetal than by maternal hemoglobin at the same oxygen partial pressure.
4. If chloride channels are impaired, chloride ions cannot leave the cell to enter the respiratory tract lumen. Retention of the negatively charged chloride ions leads to retention of positive sodium ions to maintain electrical balance. Without secretion of NaCl, no osmotic gradient exists to draw water into the lumen. Insufficient water secretion causes the mucus to be very thick and viscous.
5. Expired air has a higher partial pressure of oxygen than does alveolar air. Expired air is a mixture of (fresh) air from the dead air space and oxygen-depleted air from the alveoli.
6. 1. When we laugh the mechanism for closing off the nasopharynx from the oropharynx (upward movement of the soft palate and uvula) is not operative.   2. Peristalsis carries foods *toward* the stomach regardless of the body's position.
7. Because each bronchopulmonary segment is isolated by connective tissue septa and has its own vascular supply.
8. Those in the upper passageways move mucus toward the *esophagus* to be swallowed; those in the lower passageways do *likewise* but in the opposite direction, which prevents mucus from "pooling" in the lungs.
9. The elastin is responsible for the natural elasticity of the lungs, which allows them to recoil passively during expiration.
10. It could be if he was hitting the right spot (just inferior to the rib cage) with a vigorous upward thrust, which would rapidly propel the air out of his lungs upward through the respiratory passageways. (Of course, he might also break a few ribs or rupture his spleen or liver in the process.)
11. The choanae constitute the funnel-shaped junction between the nasal cavities and nasopharynx. The conchae are the mucosa-covered projections protruding medially from the lateral walls of the nasal cavities. The carina is the shield-shaped cartilage at the anteroinferior aspect of the trachea (junction with primary bronchi).

COVERING ALL YOUR BASES

Multiple Choice

1. B, D   **2.** C   **3.** B, C, D   **4.** A   **5.** B   **6.** D   **7.** D   **8.** B   **9.** B   **10.** A   **11.** A, C, D   **12.** C   **13.** B, C, D
14. B, C   **15.** B   **16.** B, D   **17.** C   **18.** B, C, D   **19.** A, B, C   **20.** D   **21.** C   **22.** C   **23.** A, D   **24.** C
25. A, D   **26.** A, C   **27.** A, B, D   **28.** A, D   **29.** A   **30.** C   **31.** A, B   **32.** B, C   **33.** C

Word Dissection

Word root	Translation	Example	Word root	Translation	Example
1. alveol	cavity	alveolus	**12.** pleur	side; rib	pleura
2. bronch	windpipe	bronchus	**13.** pne	breath	eupnea, apnea
3. capn	smoke	hypercapnia	**14.** pneum	air, lungs	pneumothorax
4. carin	keel	carina	**15.** pulmo	lung	pulmonary
5. choan	funnel	choanae	**16.** respir	breathe	respiration
6. crico	ring	cricoid cartilage	**17.** spire	breathe	inspiration
7. ectasis	dilation	atelectasis	**18.** trach	rough	trachea
8. emphys	inflate	emphysema	**19.** ventus	wind	ventilation
9. flat	blow, blown	inflation	**20.** vestibul	porch	vestibule
10. nari	nostril	nares	**21.** vibr	shake, vibrate	vibrissae
11. nas	nose	nasal			

Chapter 24 The Digestive System

BUILDING THE FRAMEWORK

Overview of the Digestive System

1.   1. pharynx   2. esophagus   3. stomach   4. duodenum   5. jejunum   6. ileum   7. ascending colon   8. transverse colon   9. descending colon   10. sigmoid colon   11. rectum   12. anal canal

2.   Mouth: (I) teeth, tongue; (O) salivary glands and ducts.
Duodenum: (O) liver, gall bladder, bile ducts, pancreas.

3.   The salivary glands include the sublingual, submandibular, and parotid glands.

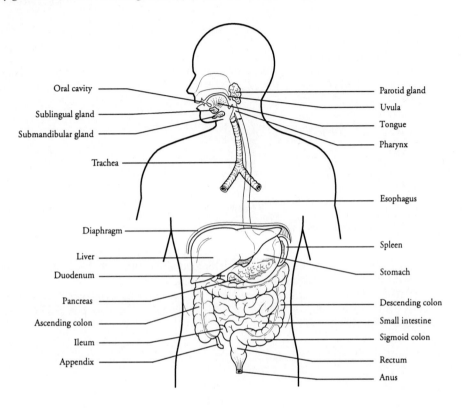

Figure 24.1

4.   1. D   2. G, H   3. E, F, H   4. B   5. A   6. C

5.   1. stretch, osmolarity/pH, and presence of substrates and end products of digestion   2. submucosal plexus and myenteric plexus   3. long reflexes involve CNS centers and extrinsic autonomic nerves and can involve long regions of the tract; short reflexes involve only local (enteric) plexuses, and are very limited in their range

6.   1. peritoneum   2. mesentery   3. retroperitoneal organs   4. intraperitoneal or peritoneal organs

7.   1. A   2. D, E   3. D   4. C   5. B

8.   1. J   2. X   3. N   4. P   5. L, U   6. V   7. O   8. E, I, J   9. D   10. R   11. G   12. K   13. H   14. S   15. C   16. W   17. B   18. U   19. I   20. S   21. Q   22. T   23. S   24. M   25. C   26. A   27. F   28. E   29. D

9.   1. mucosa   2. muscularis externa   3. submucosa   4. serosa

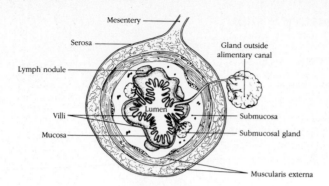

Figure 24.2

Functional Anatomy of the Digestive System

1.

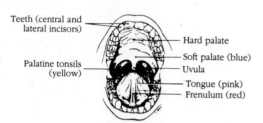

Figure 24.3

2. 1. Cleans the mouth, dissolves chemicals for tasting, moistens and compacts food, contains starch-digesting enzymes that begin the chemical digestion of starch.   2. Parotid glands, open next to second upper molar; submandibular glands, open at base of lingual frenulum; sublingual glands, open along floor of mouth   3. Serous cells secrete watery, enzyme-containing fluid; mucous cells secrete mucus   4. Lysozyme

3. Circle: 1. parasympathetic   2. salivatory; glossopharyngeal   3. sour   4. conditioned

4. 1. deciduous   2. 6 mouths   3. 6 years   4. permanent   5. 32   6. 20   7. incisors   8. canine (eyetooth)   9. premolars (bicuspids)   10. molars   11. wisdom

5. 1. A   2. C   3. D   4. B   5. E   6. C

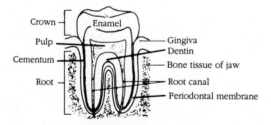

Figure 24.4

6. 1. oropharynx   2. laryngopharynx   3. stratified squamous   4. pharyngeal constrictor muscles   5. esophagus   6. gastroesophageal sphincter (valve)   7. deglutition   8. buccal   9. pharyngeal-esophageal   10. tongue   11. uvula   12. larynx   13. epiglottis   14. peristalsis   15. gastroesophageal

7. On part B, the HCl-secreting parietal cells should be colored red, the mucus neck cells yellow, and the cells identified as chief cells (which produce protein-digesting enzymes) blue.

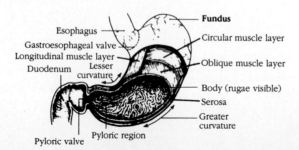

Figure 24.5

8.   Check 2, 3, 4, 5, 7.

9.

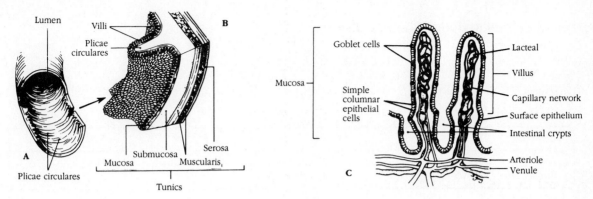

Figure 24.6

10.   1. D   2. A   3. B   4. C

11.   1. Secretin   2. Gastric emptying   3. Intestinal phase   4. HCl   5. Sympathetic   6. Esophagus   7. Salts

12.

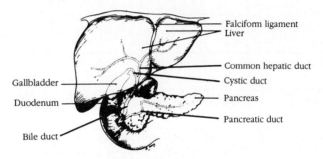

Figure 24.7

13.   Branches of hepatic artery, hepatic vein, and bile duct. Located at the lobule periphery in the surrounding connective tissue.

14.   1. A, C   2. B   3. A   4. C   5. A, C, D

15.   1. B   2. H   3. F   4. G   5. C   6. A

16.

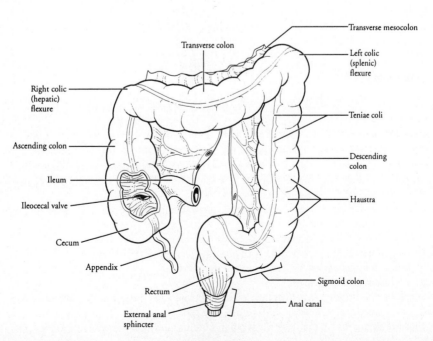

Figure 24.8

17. 1. peristalsis   2. segmentation   3. segmentation   4. mass movements   5. rectum   6. defecation   7. emetic
8. vomiting

18. 1. I   2. G   3. H   4. J   5. D   6. E   7. C   8. F   9. A

Physiology of Chemical Digestion and Absorption

1. 1. B, D, F, G, H, I, J, L, M   2. F, G, H   3. H   4. I, J, M   5. L   6. D   7. C, K   8. A   9. E

2. 1. P   2. A   3. A   4. P   5. A.   Circle 4, Fatty acids.

3. 1. Q   2. N   3. P   4. C   5. G   6. B   7. O   8. I   9. M   10. C   11. R   12. A, F, K

4. 1. Trypsin   2. Chylomicrons   3. Lost during hemorrhage   4. Nuclease   5. Removal of water   6. Pepsin
7. Vitamin B_{12}

5.

Enzyme	Substrate	Product	Secreted by	Active in
Salivary amylase	Starch	Maltose and oligosaccharides	Salivary glands	Mouth
Pepsin	Protein	Peptides	Gastric glands	Stomach
Trypsin	Protein, peptides	Dipeptides, amino acids	Pancreas	Small intestine
Dipeptidase	Dipeptides	Amino acids	Not secreted; brush border enzyme	Small intestine
Pancreatic amylase	Starch, oligosaccharides	Maltose	Pancreas	Small intestine
Lipase	Triglycerides	Monoglycerides, fatty acids	Pancreas	Small intestine
Maltase	Maltose	Glucose	Not secreted; brush border enzyme	Small intestine
Carboxy-peptidase	Polypeptide	Amino acids	Not secreted; brush border enzyme	Small intestine
Nuclease	Nucleic acids	Nucleotides	Pancreas	Small intestine

Developmental Aspects of the Digestive System

1. 1. B   2. A   3. D   4. D, N   5. E   6. F, O   7. K   8. M   9. C   10. G   11. I   12. J

The Incredible Journey

1. 1. mucosa   2. vestibule   3. tongue   4. salivary amylase   5. peristalsis   6. esophagus   7. larynx   8. epiglottis
9. stomach   10. mucus   11. pepsin   12. hydrochloric acid   13. pyloric   14. lipase   15. pancreas   16. villi
17. ileocecal

CHALLENGING YOURSELF

At the Clinic

1. Mumps.

2. Gastroesophageal sphincter.

3. Heartburn due to a hiatal hernia; esophagitis and esophageal ulcers.

4. Histamine is one of the chemical stimuli for HCl secretion; thus an antihistamine drug will inhibit HCl secretion; perforation, peritonitis, and massive hemorrhage. She was told not to take aspirin because it can cause stomach bleeding.

5. Leakage of HCl and pepsin from aich perforating gastric ulcer will literally erode and digest away any other tissues with which these chemicals come into contact.

6. Probably alcoholic cirrhosis; ascites (distended abdomen); jaundice (yellow skin).

7. Appendicitis; surgical removal; burst appendix with life-threatening peritonitis.

8. Lack of lactase (lactose intolerance); addition of lactase to milk.

9. Gluten enteropathy (adult celiac disease).

10. Tracheoesophageal fistula; surgery to correct the defect and close the connection between the baby's trachea and esophagus.

11. Palpation of other pelvic organs (in this case the woman's uterus) through the wall of the rectum.

Stop and Think

1. *Gap junctions* provide a continuous contraction along the digestive tract (peristalsis) because each muscle cell stimulates the next. The *stress-relaxation reflex* is important when mass peristalsis delivers fecal material to the rectum at an inconvenient moment. After stretching, the rectal smooth muscle will relax to accommodate to the increased mass. Because smooth muscle is *nonstriated*, it can stretch considerably without losing its ability to contract.

2. Salivary amylase is denatured by the acidity of the stomach; since it is a protein, it is digested along with dietary proteins. In the duodenum, pepsin is inactivated by the relatively high pH there and will be digested along with other proteins in the intestinal lumen.

3. Since histamine increases blood flow and capillary permeability, it might enhance the blood's ability to pick up nutrients from the digestive epithelium.

4. Since virtually all calcium is absorbed in the duodenum, a massive load of calcium can saturate the calcium transport mechanism. Smaller, more frequent doses will decrease likelihood of saturation and result in better absorption of the entire daily dose.

5. Fats are *emulsified* by bile salts. Digestion of triglycerides typically involves hydrolysis of two of the three fatty acids by *pancreatic lipase,* resulting in monoglycerides and free fatty acids. Bile salts form *micelles* that transport fat end products to the intestinal epithelium. *Absorption* occurs passively through the phospholipid bilayer. Triglycerides are re-formed in the epithelial cells and packaged with other lipids into protein-coated *chylomicrons,* which enter the *lacteals* to circulate from the lymphatic system to the general circulation. Entrance to the liver is via the hepatic artery of the systemic arterial system.

6. Examination of the plasma would quickly reveal the presence of chylomicrons, which give the plasma a milky-white appearance.

7. The liver manufactures albumin, which is essential in maintaining the osmotic balance of the blood. An albumin deficiency reduces the blood's osmotic pressure, and fluid is retained in the tissue spaces.

8. (a) In the stomach, these cells are found at the junctions of the gastric pits with the gastric glands. In the small intestine, they are at the base of the intestinal glands (crypts). (b) Their common function is to replace the epithelial cells (exposed to the harsh conditions of the digestive tract) as they die and slough off.

9. (a) Serous (gland) cells produce a watery enzyme-rich secretion; serous membranes produce a lubricating fluid within the ventral body cavity. (b) Caries are cavities (decayed areas) in teeth; (bile) canaliculi are tiny canals in the liver that carry bile. (c) Anal canal is the last portion of the tubular alimentary canal; runs from the rectum to the anus (external opening). (d) Diverticulosis is a condition in which the (weakened) walls of the large intestine pouch out; diverticulitis is inflammation of these diverticuli. (e) Hepatic vein drains venous blood *from* the liver; hepatic portal vein brings nutrient-rich venous blood *to* the liver from the digestive viscera.

10. Rough ER and Golgi apparatus: Liver makes tremendous amounts of proteins for export. Smooth ER: Liver is an important site of fat metabolism and cholesterol synthesis and breakdown. Peroxisomes: Liver is an important detoxifying organ.

COVERING ALL YOUR BASES

Multiple Choice

1. A, C, D   **2.** B   **3.** C   **4.** C   **5.** A   **6.** B, C, D   **7.** A, B, C, D   **8.** D   **9.** A, B, C, D   **10.** B   **11.** C   **12.** B, C
13. C   **14.** A, D   **15.** D   **16.** A, B, C   **17.** A, D   **18.** A, C, D   **19.** A, C   **20.** D   **21.** A, B   **22.** B, C, D
23. B   **24.** B, C, D   **25.** A, B, C   **26.** A, C, D   **27.** A, C   **28.** A, C   **29.** C, D   **30.** A, B, D   **31.** D   **32.** A, D
33. A   **34.** B   **35.** A, C, D   **36.** A, C, D   **37.** B, C   **38.** A, B, C   **39.** C   **40.** A, D   **41.** B, C, D   **42.** A
43. A   **44.** B   **45.** A   **46.** C   **47.** D   **48.** D   **49.** B   **50.** A

Word Dissection

Word root	Translation	Example	Word root	Translation	Example
1. aliment	nourish	alimentary canal	**17.** hiat	gap	hiatal hernia
2. cec	blind	cecum	**18.** ile	intestine	ileum
3. chole	bile	cholecystokinin	**19.** jejun	hungry	jejunum
4. chyme	juice	chyme	**20.** micell	a little crumb	micelle
5. decid	falling off	deciduous teeth	**21.** oligo	few	oligosaccharides
6. duoden	twelve each	duodenum	**22.** oment	fat skin	greater omentum
7. enter	intestine	mesentery	**23.** otid	ear	parotid gland
8. epiplo	thin membrane	epiploic appendages	**24.** pep	digest	pepsin
9. eso	within, inward	esophagus	**25.** plic	fold	plicae circulares
10. falci	sickle	falciform ligament	**26.** proct	anus, rectum	proctodeum
11. fec	dregs	feces	**27.** pylor	gatekeeper	pylorus
12. fren	bridle	lingual frenulum	**28.** ruga	wrinkle	rugae
13. gaster	stomach	gastric juice	**29.** sorb	suck in	absorb
14. gest	carry	digestion	**30.** splanch	the viscera	splanchnic circ.
15. glut	swallow	deglutition	**31.** stalsis	constriction	peristalsis
16. haustr	draw up	haustra	**32.** teni	ribbon	tenia coli

Chapter 25 Nutrition, Metabolism, and Body Temperature Regulation

BUILDING THE FRAMEWORK

Nutrition

1. 1. carbohydrates, lipids, proteins, vitamins, minerals, water   2. Molecules the body cannot make and must ingest.
    3. 1 kcal = the heat necessary to raise the temperature of 1 kg of water by 1° centigrade.   4. glucose   5. brain cells
    and red blood cells   6. Carbohydrate foods that provide energy sources but no other nutrients.   7. egg yolk
    8. linoleic acid (ingested as lecithin) and (possibly) linolenic acid   9. A protein containing all the different kinds of amino
    acids required by the body.   10. The amount of nitrogen ingested in proteins equals the amount of nitrogen excreted
    in urine or feces (rate of protein synthesis equals rate of protein degradation and loss).   11. structural and functional
    uses   12. Most vitamins function as coenzymes.   13. liver   14. calcium and phosphorous

2. 1. I   2. E   3. D   4. B   5. H   6. G   7.–10. A, G, H, I   11. M   12. L   13. F   14. A   15. C   16. K

3. 1. J   2. C   3. I   4. F   5. G   6. H   7. B   8. D   9. E

4. 1. Trisaccharides   2. Nuts   3. Fats   4. Vitamin C   5. K^+ 6. Amino acids   7. Must be ingested   8. Zinc
    9. About 16 oz   10. Catabolic hormones

5. 1. A   2. B   3. C   4. B   5. B   6. C   7. C   8. B

Metabolism

1. 1. metabolism   2. catabolism   3. anabolism   4. cellular respiration   5. ATP   6. GI tract   7. mitochondria
    8. oxygen   9. water   10. carbon dioxide   11. hydrogen atoms   12. electrons   13. reduction   14. coenzymes
    15. NAD^+ 16. FAD

2. 1. glycolysis   2. cytoplasm   3. not used   4. NAD^+ 5. two   6. sugar activation   7. sugar cleavage   8. sugar
    oxidation   9. pyruvic acid   10. lactic acid   11. skeletal muscle   12. brain   13. pH   14. red blood   15. carbon
    dioxide   16. NAD^+ 17. coenzyme A   18. oxaloacetic acid   19. citric acid   20. Krebs   21. mitochondrion
    22. keto   23. two   24. three   25. one   26. fatty acids   27. amino acids   28. electron transport   29. protons
    30. water   31. ADP   32. oxidative phosphorylation   33. 38 or 36

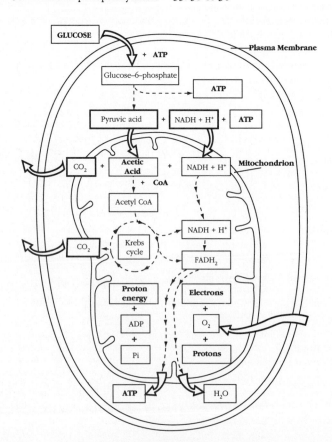

Figure 25.1

3. 1. hydrogen atoms   2. electrons   3. protons   4. metal   5. inner   6. EC_1, EC_2, and EC_3 7. oxygen   8. protons
9. mitochondrial matrix   10. intermembrane space   11. lower   12. protons   13. ATP synthase   14. ADP +
$P_i \rightarrow$ ATP

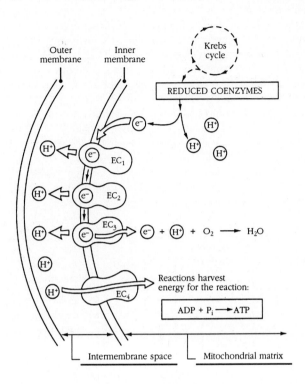

Figure 25.2

4. 1. C   2. D   3. B   4. A
5. 1. D   2. F   3. B   4. G   5. C   6. A   7. E   8. C   9. F   10. Ammonia is too toxic; blood pH may rise.   11. Keto
acids   12. There are no storage depots for amino acids in the body. Proteins continually degrade and newly ingested
amino acids are required. If an essential amino acid is not consumed, the remaining ones cannot be used for protein
synthesis and are oxidized for energy.

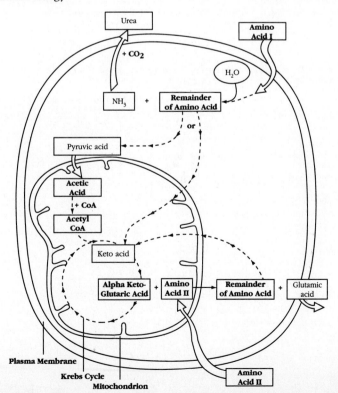

Figure 25.3

6.   1. F   2. C   3. E   4. D   5. A   6. B   7. G   8. B

7.   1. 40% or less   2. ATP than ADP   3. into   4. glycogenesis   5. liver and skeletal muscle

8.   1. Glucose   2. Hydrolyzed in mitochondria   3. Cytoplasm   4. ATP deficit   5. Low blood sugar level

9.   1. anabolic   2. T   3. glycogen   4. triglycerides   5. protein synthesis   6. T   7. rising blood glucose levels   8. facilitated diffusion   9. glycogenolysis and gluconeogenesis   10. other body cells   11. T

10.   1. 90–100 mg glucose per 100 ml blood   2. The brain prefers glucose as its energy source.   3. The liver, by glycogenolysis; skeletal muscle, by glycogenolysis; adipose tissue and liver, by lipolysis; tissue cells, by protein catabolism.   4. Skeletal muscle cells cannot dephosphorylate glucose, but they can produce pyruvic acid or lactic acid. The liver reconverts these to glucose, which is released to the blood.   5. Lipolysis produces fatty acids and glycerol. Only the glycerol can be converted directly to glucose by the liver. (Acetyl CoA, from fatty acid oxidation, is produced beyond the reversible steps of glycolysis and cannot be converted to glucose.)   6. The liver deaminates amino acids and converts the residues to glucose, which is released to the blood.   7. Fats   8. Ketone bodies from the oxidation of fats by the liver.   9. Epinephrine and glucagon   10. Alpha cells of pancreatic islets produce glucagon, which targets the liver and adipose tissue.   11. Falling glucose levels and rising amino acid levels.   12. High amino acid levels in blood stimulate release of insulin. Glucagon counteracts effects of insulin and prevents abrupt hypoglycemia.   13. Hypothalamic receptors sense declines in blood glucose levels. Stimulation of the adrenal medulla causes release of epinephrine, which produces the same effects as glucagon.

11.

Hormone	Blood glucose	Blood amino acids	Glycogen-olysis	Lipogenesis	Protein synthesis
Insulin	↓	↓	↓	↑	↑
Glucagon	↑	XXX	↑	↓	XXX
Epinephrine	↑	XXX	↑	↓	XXX
Growth hormone	↑ (over the long term)	Å	↑	↓	↑
Thyroxine	↓	XXX	↑	↓	↑

12.   1. albumin   2. clotting proteins   3. cholesterol   4. hyperglycemia   5. glycogen   6. hypoglycemia   7. glycogenolysis   8. gluconeogenesis   9. detoxification   10. phagocytic   11. lipoproteins   12. insoluble   13. lower   14. high   15. chylomicrons   16. triglycerides (fats)   17. cholesterol   18. membranes   19. steroid hormones   20. liver   21. bile salts   22. atherosclerosis   23. unsaturated   24. H (high)   25. A, D, and B_{12} 26. iron   27. red blood

Body Energy Balance

1.   1. TMR   2. ↓ Metabolic rate   3. Child   4. Fats   5. Fasting   6. Body's core   7. Vasoconstriction

2.   1. C ↓   2. A ↓   3. B ↓   4. C ↑   5. D ↓↑   6. B ↑   7. C ↓

3.   1. The intake of energy by the catabolism of food equals the total energy output of heat plus work plus energy storage.   2. heat   3. Body weight remains stable when energy intake equals energy outflow. Weight change accompanies energy inequality.

4.   1. D   2. B, K   3. A   4. F   5. J   6. C   7. H, I   8. G   9. E   10. J

5.   1. skeletal muscle   2. 37.8°C (100°F)   3. 10%   4. proteins   5. T   6. heat stroke

Developmental Aspects of Nutrition and Metabolism

1.   1. protein   2. brain   3. three   4. cystic fibrosis   5. phenylketonuria (PKU)   6. phenylalanine   7. brain   8. phenylalanine   9. tyrosine   10. galactosemia   11. glycogen storage disease   12. liver   13. skeletal muscles   14. diabetes mellitus   15. declines   16. hypokalemia   17. diuretics   18. absorption   19. liver

CHALLENGING YOURSELF

At the Clinic

1.   Many vegetables contain incomplete proteins. Unless complete proteins are ingested the value of the dietary protein for anabolism is lost, because the amino acids will be oxidized for energy. Beans and grains.

2.   "Empty calories" means simple carbohydrates with no other nutrients, such as candy, cookies, and soft drinks.

3.   Glucagon; hypoglycemia.

4. Dieting triggers catabolism of proteins as well as fat, resulting in weight loss due to loss of muscle and adipose tissue. Subsequent weight gain without exercise usually means regaining only fat, not muscle. Continual cycling of muscle/adipose loss followed by adipose gain can significantly alter body fat composition and "rev up" metabolic systems that increase the efficiency of fat storage and retention.
5. Rickets; deficiency of vitamin D and calcium.
6. Ketosis; her self-starvation has resulted in deficiency of carbohydrate fuels (and oxaloacetic acid). This deficiency promotes conversion of acetyl CoA to ketone bodies such as acetone.
7. Heat exhaustion; they should drink a "sports drink" containing electrolytes or lemonade to replace lost fluids.
8. Bert has heat stroke. Heavy work in an environment that restricts heat loss results in a spiraling upward of body temperature and cessation of thermoregulation. Bert should be immersed in cool water immediately to bring his temperature down and avert brain damage.
9. Children have a greater requirement for fat than adults, particularly up to age 2 or 3 when myelination of the nervous system is still a major consideration.
10. Hypercholesterolemia; <200 mg/100 ml blood; atherosclerosis, strokes, and heart attacks.
11. Iron. She has hemorrhagic anemia compounded by iron loss.
12. Mr. Hodges has scurvy due to vitamin C deficiency. Recommend citrus fruits and plenty of tomatoes.

Stop and Think

1. Osmotic effect would be retention of water in cells maintaining a pool of amino acids; pH would decrease.
2. A higher concentration of glucose-phosphate in a cell would attract water, and the cell would swell.
3. Oxygen is required as the final electron acceptor in the electron transport chain. Lack of oxygen brings electron transport to a halt and backs up the Krebs cycle as well, since NAD^+ and FAD cannot be recycled. ATP production grinds to a halt in cells that cannot rely on anaerobic respiration (such as brain cells), and cell death occurs.
4. Urea production increases. Protein synthesis will not proceed unless all necessary amino acids are available. Hence, any amino acids that are not utilized are deaminated and oxidized for energy or converted to fat and stored. Excess amino groups will be converted to urea.
5. Absorption of simple sugars is quite rapid compared to absorption of complex carbohydrates, which must be digested first. The result is a much more rapid rise in blood glucose level when simple sugars are ingested.
6. No; the ability of brain cells to take in glucose is not regulated by insulin.
7. Since omega-3 fatty acids reduce the stickiness of platelets, excessive bleeding might result from ingesting an excess of these lipids.
8. Males tend to have a higher ratio of muscle to fat than females. Since muscle tissue has a higher BMR than fat, a higher proportion of muscle requires a higher calculation factor.
9. Since PKU sufferers cannot manufacture melanin, pigment in the interior of the eye could be deficient, resulting in poor vision. (Albinos likewise have poor vision.)
10. Depending on the diuretic, different electrolytes have an increased rate of excretion. A common type of diuretic promotes potassium excretion, requiring potassium supplementation.
11. The excess of glucose can form abnormal cross-bridges between protein fibers in the vessel walls, resulting in hardening of the arteries.
12. Hypothyroidism would result in hypothermia, since thyroid hormones are thermogenic (heat generating).
13. When muscles are "at work" they are generating and using large amounts of ATP to power the sliding of their myofilaments. Since some heat is "lost" in every chemical reaction, a large amount of heat is generated at the same time.

COVERING ALL YOUR BASES

Multiple Choice

1. B, D   2. A, B, D   3. C   4. D   5. A, B, C, D   6. B   7. B, C   8. A, C   9. C, D   10. A, B, C, D   11. A
12. C, D   13. A, B   14. A   15. B, D   16. A   17. B, C, D   18. B, D   19. B, D   20. A   21. C   22. A, B, C
23. C   24. A, B, ,C   25. D   26. A, B, D   27. D   28. A

Word Dissection

Word root	Translation	Example	Word root	Translation	Example
1. acet	vinegar	acetyl CoA	6. lecith	egg yolk	lecithin
2. calor	heat	calorie	7. linol	flax oil	linoleic acid
3. flav	yellow	riboflavin	8. nutri	feed, nourish	nutrient
4. gluco	sweet	glucose	9. pyro	fire	pyrogen
5. kilo	thousand	kilocalorie			

Chapter 26 The Urinary System

BUILDING THE FRAMEWORK

1. formation of renin (involved in blood pressure regulation) and erythropoietin (which stimulates erythropoiesis); activation of vitamin D

2. Ptosis is the dropping of the kidneys to a more inferior position, usually caused by loss of support by the adipose capsule. Hydronephrosis is a backing up of urine in the kidney, leading to pressure on the renal tissue. Its usual cause is kinking of a ureter due to ptosis.

3.

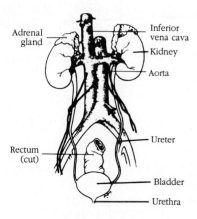

Figure 26.1

Kidney Anatomy

1. The fibrous membrane surrounding the kidney is the renal capsule; the basinlike area is the pelvis; the cuplike extension of the pelvis is the calyx; the cortexlike tissue running through the medulla is the renal column. The cortex contains the bulk of the nephron structures; the medullary pyramids are primarily formed by collecting ducts.

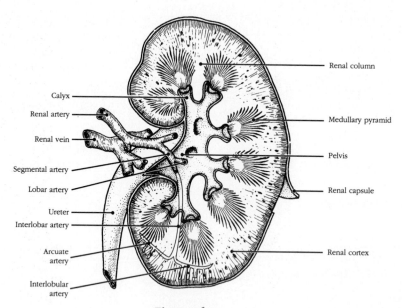

Figure 26.2

2.

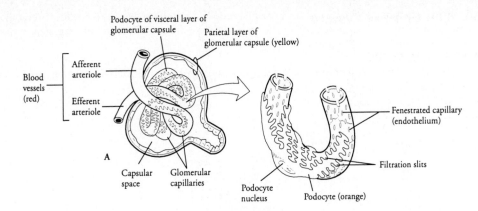

Figure 26.3

B1 is a collecting tubule; few microvilli, large diameter. B2 is a DCT; has microvilli but less abundant than those in the PCTs. B3 is a PCT; very dense and long microvilli.

3. 1. glomerular capsule   2. afferent arteriole   3. efferent arteriole   4. interlobular artery   5. interlobular vein
6. arcuate artery   7. arcuate vein   8. interlobar artery   9. interlobar vein   10. loop of Henle   11. collecting duct
12. DCT   13. PCT   14. peritubular capillaries   15. glomerulus   Number 1 is green, 15 is red, 14 is blue, 11 is
yellow, 13 is orange.

4. 1. Renal pyramid   2. Renal corpuscle   3. Glomerulus   4. Collecting duct   5. Cortical nephrons   6. Collecting duct
7. Glomeruli

5. 1. 4   2. 5   3. 1   4. 3   5. 2

Kidney Physiology: Mechanisms of Urine Formation

1. 1. afferent   2. efferent   3. blood plasma   4. diffusion   5. active transport   6. microvilli   7. secretion   8.–10. diet,
cellular metabolism, urine output   11. 1–1.8   12. urochrome   13.–15. urea, uric acid, creatinine   16. lungs
17. evaporation   18. decreases   19. dialysis (artificial kidney)

2. 1. B   2. A   3. C   4. B

3. Black arrows are at the site of filtrate formation, the glomerulus. Arrows leave the glomerulus and enter the glomerular
capsule. Red arrows are at the site of amino acid and glucose reabsorption. They go from the PCT interior and pass
through the PCT walls to the capillary bed surrounding the PCT. Nutrients leave the filtrate. Green arrows are at the
site of ADH action. Arrows indicating water movement leave the interior of the collecting duct. Water passes through
its walls to enter the peritubular capillary bed. Water leaves the filtrate. Yellow arrows are at the site of aldosterone
action. Arrows indicating Na$^+$ movement leave the DCT and pass through their walls into the surrounding capillary
bed. Na$^+$ leaves the filtrate. Blue arrows at the site of tubular secretion enter the PCT and cortical part of the collecting
duct to enter the filtrate.

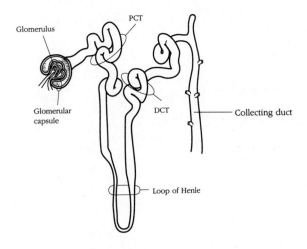

Figure 26.5

4. 1. (Glomerular osmotic pressure + Capsular hydrostatic pressure)   2. 17   3. Glomerular hydrostatic pressure
4. Glomerular osmotic pressure

5. Myogenic mechanism: smooth muscle cells in afferent arterioles respond to degree of stretch. Tubuloglomerular feedback mechanism: macula densa of distal convoluted tubules. (Indirect effect: Renin-angiotensin mechanism: juxtaglomerular cells of the JG apparatus.)

6. Check 1, 2, 4, 5.

7. 1. active   2. passive, electrical gradient   3. passive, electrical gradient   4. solvent drag   5. active   6. active   7. active   8. active   9. solvent drag   10. active   11. passive, osmotic

8.

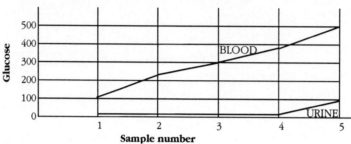

1. 375 mg/100 ml   2. yes

9. H^+, K^+, creatinine, ammonia (as ammonium ion)

10. In the absence of ADH, it permits very dilute urine to be excreted, thus decreasing blood volume. In the presence of ADH, it allows very small amounts of very concentrated urine to be excreted, thus increasing blood volume.

11. 1. descending   2. ascending   3. thick   4. urea   5. vasa recta   6. papilla (apex)   7. 1200   8. ADH

12.

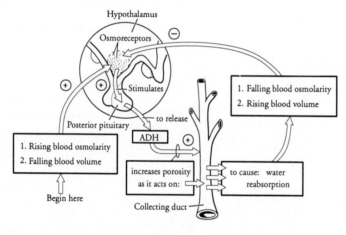

Figure 26.6

13. 1. Aldosterone   2. Secretion   3. ↑ K^+ reabsorption   4. ↓ BP   5. ↓ K^+ retention

14. 1. 180 ml/min   2. high   3. secreted   4., 5. reabsorbed   6. secreted

15. 1. D   2. D   3. I   4. D   5. I   6. I

16. 1. A   2. B   3. A   4. A   5. B

17. 1. L   2. G   3. G   4. A   5. L   6. A   7. G   8. G   9. G   10. A   11. G   12. L

18. 1. Hematuria is caused by bleeding in the urinary tract.   2. Ketonuria is caused by diabetes mellitus or starvation. 3. Albuminuria is caused by glomerulonephritis or pregnancy.   4. Pyuria is caused by urinary tract infection. 5. Bilirubinuria is caused by liver disease.   6. The presence of "sand" has no official terminology and is caused by kidney stones.   7. Glycosuria is caused by diabetes mellitus.

19. 1. Glucose is usually totally reabsorbed by tubule cells.   2. Albumin usually does not pass through the glomerular filter.

Ureters, Urinary Bladder, and Urethra

1. 1. kidneys   2. ureters   3. peristalsis   4. bladder   5. urethra   6. 8   7. sperm/semen   8. 1.5

2. 1. Kidney   2. Forms urine   3. Continuous with renal pelvis   4. Blocks urethra   5. Female

Micturition

1. 1. micturition   2. stretch receptors   3. contract   4. internal urethral sphincter   5. external urethral sphincter 6. voluntarily   7. approximately 600   8. incontinence   9. infants, toddlers   10. emotional or neural problems 11. pressure (pregnancy)   12. urinary retention   13. prostate

Developmental Aspects of the Urinary System

1. 1. placenta   2. mesoderm   3. metanephros   4. pronephros   5. mesonephros   6. polycystic kidney   7. hypospadias   8. male babies   9. bladder   10. 18 to 20   11. arteriosclerosis   12. renal tubule   13. frequency   14. nocturia
2. 1. E   2. H   3. J   4. I   5. F   6. G   7. C   8. D   9. B   10. A

The Incredible Journey

1. 1. tubule   2. renal   3. afferent   4. glomerulus   5. Glomerular capsule   6. blood plasma   7. proteins   8. loop of Henle   9. microvilli   10. reabsorption   11. glucose   12. amino acids   13. 7.35–7.45   14. nitrogenous   15. sodium   16. potassium   17. urochrome   18. antidiuretic hormone   19. collecting duct   20. pelvis   21. peristalsis   22. urine   23. micturition   24. urethra

CHALLENGING YOURSELF

At the Clinic

1. Renal ptosis and subsequent hydronephrosis.
2. Anuria; renal dialysis.
3. Urinary tract infection (urethritis and cystitis), probably by *E. coli.*
4. Acute glomerulonephritis due to a reaction to *Streptococcus* bacteria.
5. Most likely bladder cancer; both pesticides and some of the chemicals in cigarette smoke are carcinogenic.
6. Test for glucose in the urine; diabetes insipidus; 1.001 sg with no sugar in the urine.
7. Cystocele.
8. Nocturnal enuresis; perhaps Eddie is a very heavy sleeper and is, thus, unresponsive to the "urge" to urinate.
9. Renal infarct; blockage of one or more interlobar arteries.
10. The kidneys lie retroperitoneal in the superior lumbar region and receive some protection from the rib cage. If the parietal pleura (adjacent) is penetrated during surgery, the lung will collapse.

Stop and Think

1. The glomerular capillaries are a high-pressure bed, considerably higher than typical capillaries. There is no low-pressure end on the glomerulus; rather, the peritubular capillaries provide the low-pressure bed for reabsorption of filtrate. Compared to a typical capillary bed, the combination of the glomerulus and the peritubular capillaries provides an enormous surface area for filtration and reabsorption.
2. A decrease in plasma proteins will reduce the glomerular colloid osmotic pressure. The change in the net filtration pressure will increase GFR, decrease blood volume (and therefore flow rate) in the peritubular capillaries, and increase flow rate in the renal tubule. Also, the colloid osmotic pressure in the peritubular capillaries will be low, so reabsorption of water into the peritubular capillaries will be reduced. The high flow rate in the proximal convoluted tubule will decrease the reabsorption of solutes and thus reduce the obligatory reabsorption of water. Urine output will increase.
3. The mitochondria congregate along the basal side of the cells, indicating that active transport is occurring on that side.
4. Secretion of ADH increases reabsorption of water from the collecting tubules, thereby briefly diluting the interstitial fluid of the medullary pyramids. But equilibration with the blood in the vasa recta then dilutes the blood, carrying off the excess water and reestablishing the osmotic gradient.
5. Alcohol inhibits ADH secretion, resulting in excess water loss by the formation of dilute urine. The resulting dehydration leads to decreased secretion by body glands, including the salivary glands, making the mouth dry.
6. Caffeine increases GFR by causing vasodilation of the glomerular capillaries. The increase in flow rate through the renal tubule reduces the time available for reabsorption of solutes resulting in dilute urine.
7. A dialysis patient has a deficiency of erythropoietin, resulting in a low RBC count. Erythropoietin supplementation will counteract the problem.
8. Transection above the pelvic splanchnic nerves does not alter the reflex, but it prevents conscious inhibition of micturition. Urinary incontinence results as voiding becomes totally reflexive.
9. Damage to the sacral region of the spinal cord destroys the micturition reflex. Contraction of the detrusor muscles can no longer be stimulated by the parasympathetic NS, and the internal urethral sphincter is paralyzed. Urine dribbles out but cannot be forcefully drained. Catheterization is likely in such cases.

COVERING ALL YOUR BASES

Multiple Choice

1. C, D   2. A   3. A, B   4. D   5. B, C, D   6. C, D   7. A, B, C   8. B, C   9. C, D   10. A, B   11. A, D   12. A, C
13. C   14. A, D   15. C, D   16. C   17. B, C, D   18. A, B, C, D   19. A   20. A, B, D   21. A, B, D
22. A, C, D   23. A, C   24. A, B, D   25. C, D   26. A, C, D   27. B   28. D   29. A   30. C   31. D   32. B

Word Dissection

Word root	Translation	Example		Word root	Translation	Example
1. azot	nitrogen	azotemia		**7.** nephr	kidney	nephron
2. calyx	cup-shaped	minor calyx		**8.** pyel	pelvis	pyelitis
3. diure	urinate	diuretic		**9.** ren	kidney	renal
4. glom	ball	glomerulus		**10.** spado	castrated	hypospadias
5. gon	angle	trigone		**11.** stroph	twist, turn	exstrophy
6. mictur	urinate	micturition		**12.** trus	thrust, push	detrusor

Chapter 27 Fluid, Electrolyte, and Acid-Base Balance

BUILDING THE FRAMEWORK

Body Fluids

1. 1. Male adult 2. Lean adult 3. Claude Bernard's "internal environment" 4. ICF 5. Electric charge

2. 1. N 2. E 3. E 4. E 5. N 6. E

3. mEq/L in plasma: bicarbonate, 26; calcium, 5; chloride, 105; magnesium, 3; hydrogen, 2; potassium, 4; sodium, 142.
1. proteins 2. Na^+ 3. Cl^- and proteins 4. ↓ are HCO_3^-, Ca^{2+}, Cl^-, and Na^+; the other are ↑.

4. 1. T 2. hydrostatic pressure 3. T 4. lymphatic vessels 5. tissue cell 6. plasma 7. equal 8. D

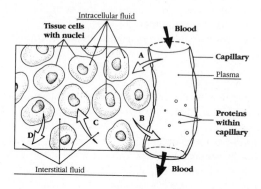

Figure 27.1

Water Balance

1. 1. Most water (60%) comes from ingested fluids. Other sources are moist foods and cellular metabolism. 2. The greatest water loss (60%) is from excretion of urine. Other routes are as water vapor in air expired from lungs, through the skin in perspiration, and in feces. 3. Insensible water loss is water loss of which we are unaware. This type continually occurs via evaporation from skin and in water vapor that is expired from the lungs. It is uncontrollable.

2. 1. hypothalamus 2. decrease 3. increase 4. osmoreceptors 5. stretch 6. stomach and intestines 7. dilution 8. decrease 9. ADH 10. dilute 11. Na^+ 12. obligatory 13. 500

3. 1. B 2. C 3. A, B 4. A 5. C 6. A 7. A 8. C 9. C 10. B

Electrolyte Balance

1. 1. salts 2. foods 3.–5. in feces, in perspiration, in urine 6. kidneys 7. Na^+ 8. osmotic 9. volume 10. water

2. 1. juxtaglomerular 2. macula densa 3. retention/reabsorption 4. water 5. increase 6. hours to days 7. sympathetic nervous 8. baroreceptors 9. medulla 10. arterioles 11. constrict 12. juxtaglomerular cells 13. renin 14. rise

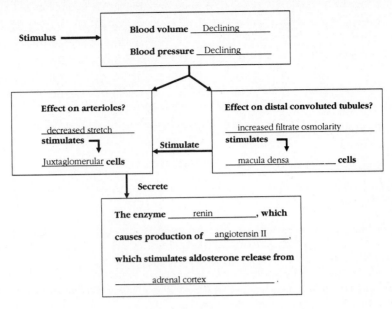

Figure 27.2

3.   1. K^+, Mg^{2+} 2. Ca^{2+} 3. Cl^- 4. K^+ 5. Mg^{2+} 6. HCO_3^- 7. K^+ 8. Ca^{2+} 9. K^+ 10. Ca^{2+}

4.   1. $\downarrow H^+$ secretion   2. $\uparrow$ ADH   3. $\uparrow Na^+$ retention   4. Progesterone   5. $\uparrow K^+$ reabsorption   6. $\downarrow$ BP

Acid-Base Balance

1.   1. E   2. B   3. D   4. F   5. C   6. B   7. A

2.   1. anaerobic respiration   2. combination of carbon dioxide (from cellular respiration) and water   3. stomach secretions   4. fat metabolism   5. oxidation of phosphorus-containing substances, such as proteins and nucleic acids   6. events of the Kreb's cycle

3.   1. Chemical buffering; response in less than 1 second.   2. Adjustment in respiratory rate and depth; response in minutes.   3. Regulation by kidneys; response in hours or days.

4.   1. D   2. E   3. B   4. E   5. D   6. A   7. B   8. D   9. C   10. E

5.   1. Weak acids only partially dissociate (free some H^+) in solution, but can be forced to dissociate more by an alkaline environment.   2. Strong acids dissociate completely and free all H^+ regardless of pH.   3. Weak bases partially dissociate and bind relatively few H^+.   4. Strong bases dissociate completely and bind all available H^+.   5. A buffer system includes molecular species that quickly resist large changes in pH by releasing or binding H^+.

6.

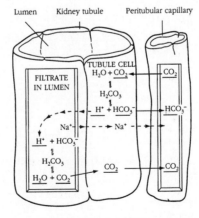

Figure 27.3

1. rises   2. H_2CO_3 3. filtrate   4. within them   5. peritubular capillary blood   6. tubule lumen (filtrate)   7. HCO_3^-
8. H_2CO_3 9. H_2O and CO_2 10. H_2O 11. one   12. peritubular capillary blood   13. active   14. excreted
15. phosphate   16. ammonia-ammonium ion

7.   1. H^+ in H_2O 2. Plasma   3. Na_2CO_3 4. NH_4HCO_3 5. $\downarrow H^+$ in filtrate   6. $\uparrow HCO_3^-$ in urine   7. Plasma
8. $\downarrow$ plasma H^+ 9. $\downarrow$ in blood CO_2 10. $\uparrow$ pH

8. 1. incompletely   2. together   3. T   4. NaHCO$_3$ 5. weak acid, H$_2$PO$_4^-$.   6. proteins   7. R-NH$_2$ 8. T
9. bicarbonate   10. kidneys

9. 1. B   2. C   3. E   4. D   5. A   6. B

Developmental Aspects of Fluid, Electrolyte, and Acid-Base Balance

1. 1. higher (70% to 80%)   2. ECF   3. electrolytes   4. lower   5. ICF   6. fat   7. high   8. skin   9. metabolic rate
10. residual volume   11. kidneys   12. decreases   13. ICF   14. few or no   15. becomes slower   16. thirst
17. congestive heart failure   18. diabetes mellitus   19. body water content

CHALLENGING YOURSELF

At the Clinic

1. Pituitary; hyposecretion of ADH; diabetes insipidus.
2. Hypotonic hydration; hyponatremia.
3. Pica; an electrolyte deficiency other than sodium.
4. Edema.
5. High sodium content and copious urine volume (although the glucocorticoids can partially take over the role of aldosterone).
6. Increased due to suppression of the renin-angiotensin-aldosterone mechanism.
7. Yes; estrogen has aldosteronelike effects.
8. Metabolic acidosis with partial respiratory compensation; the high potassium level has triggered increased renal secretion of potassium and a resulting retention of hydrogen ions.
9. Hypercalcemia.
10. Sodium, potassium, chloride, and bicarbonate; hyperventilation indicates acidosis, which can be corrected with bicarbonate. Since infants are both more susceptible to and more at risk from dehydration and acidosis, treatment should be sought immediately. Sunken fontanels in an infant are an obvious sign of severe dehydration.

Stop and Think

1. Cells are impermeable to mannitol. It remains in the capillaries, where it can draw the excess water into the bloodstream for removal by the kidneys. It will appear in filtrate but won't be reabsorbed, so its osmotic activity will extend to the urine.
2. Decreased blood volume leads to a lower stroke volume. Consequently, homeostatic regulatory mechanisms will increase the heart rate in an attempt to maintain normal cardiac output.
3. Because high levels of glucocorticoids have aldosteronelike effects, blood volume will increase.
4. Protein synthesis increases cellular uptake of potassium, reducing its concentration in the ECF and causing hypokalemia.
5. Release of ICF will increase the potassium concentration in the plasma; rapid or massive transfusion can cause transient hyperkalemia.
6. Since calcitonin's sole activity results in increased deposition of calcium in bone, its role is more important in children, in whom bone growth is more rapid.
7. Compensation for alkalosis by plasma protein buffer components will mean release of hydrogen ions that have been bound to the proteins. This will simultaneously increase binding of calcium to the proteins, causing hypocalcemia. Hypoalbuminemia means fewer plasma proteins on which calcium can attach; this will result in hypocalcemia.
8. A high ratio of phosphate to calcium would increase the formation of insoluble calcium phosphate. The kidneys maintain the required high level of dissociated calcium ions by concurrently keeping phosphate levels low.

COVERING ALL YOUR BASES

Multiple Choice

1. D   **2.** C   **3.** B   **4.** B, D   **5.** A   **6.** A, B, C   **7.** B, D   **8.** A, B, D   **9.** A, C, D   **10.** A, B, D   **11.** C
12. A, B, C, D   **13.** A, D   **14.** A, B, C, D   **15.** A, B, D   **16.** B, D   **17.** B   **18.** C   **19.** C, D   **20.** A
21. A, B, C, D   **22.** A, C, D   **23.** A, B, C, D   **24.** A, C   **25.** B, C, D   **26.** A, C, D   **27.** C

Word Dissection

Word root	Translation	Example
1. kal	potassium	hyperkalemia
2. natri	sodium	hyponatremia
3. osmo	pushing	osmosis

Chapter 28   The Reproductive System

BUILDING THE FRAMEWORK

Anatomy of the Male Reproductive System

1.   1. When external temperature is low, they contract, drawing the testes closer to the warmth of the body wall. When the external temperature is high, they are relaxed, and the testes hang away from the body wall.   2. This venous plexus absorbs heat from the arterial blood, cooling the blood before it enters the testes.

2.   The spongy tissue that becomes engorged with blood during erection is the corpus cavernosum and spongiosum of the penis. The portion of the duct system that also serves the urinary system is the urethra. The structure that provides ideal temperature conditions for sperm formation is the scrotum. The structure that is cut or cauterized during a vasectomy is the ductus deferens. The prepuce is removed by circumcision. The glands producing a secretion that contains sugar are the seminal vesicles.

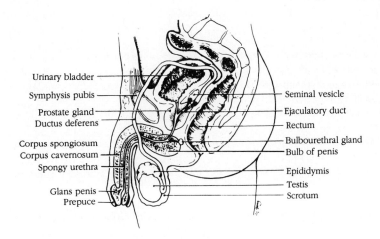

Figure 28.1

3.   1. E   2. K   3. C   4. L   5. A, G, H, K   6. I   7. B   8. F   9. G   10. H   11. A   12. J
4.   seminiferous tubule → rete testis → epididymis → ductus deferens
5.   The site of spermatogenesis is the seminiferous tubule. Sperm mature in the epididymis. The fibrous coat is the tunica albuginia.

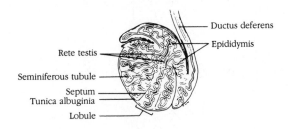

Figure 28.2

Physiology of the Male Reproductive System

1.   1. A   2. B   3. C   4. A   5. B   6. A   7. A   8. B   9. C   10. B   11. B   12. B
2.

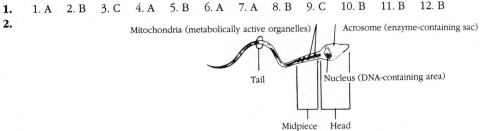

Figure 28.3

3.   1. D   2. C, E, F   3. C   4. F   5. E   6. A, G

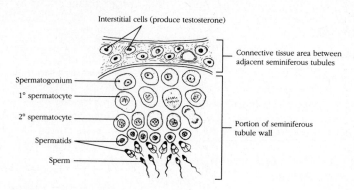

Figure 28.4

4.   1. A   2. E   3. C   4. D, F

5.   1. Spermiogenesis involves sloughing off of superfluous cytoplasm, fashioning a penetrating device (acrosome) over the nuclear "head," and forming a tail for propulsion.   2. sustentacular (Sertoli) cells   3. The blood-testis barrier is an unbroken barrier, formed by junctions between sustentacular cells, that divides the seminiferous tubule wall into two compartments. It is important because it prevents the antigens of sperm (not previously recognized as self by the immune system) from escaping through the basement membrane into the blood, where they might promote an autoimmune response.

6.   1. deepening voice   2. enlargement of skeletal muscles   3. increased density of skeleton   4. increased hair growth, particularly in axillary and genital regions

7.   1. Acidic   2. Type B cell   3. Spermatogenic   4. Parasympathetic nervous system   5. Inhibin   6. Vasoconstriction

Anatomy of the Female Reproductive System

1.   The clitoris should be colored blue, the hymen should be colored yellow, and the recess enclosed by the labia minora, red.

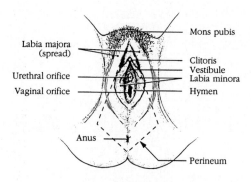

Figure 28.5

2. The egg travels along the uterine tube to the uterus after it is released from the ovary. Broad ligaments suspend the reproductive organs. The ovary forms hormones and gametes, and the labia majora are the homologue of the male scrotum.

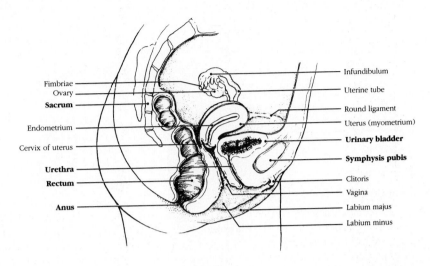

Figure 28.6

3. Because of this structural condition, many "eggs" (oocytes) are lost in the peritoneal cavity; therefore, they are unavailable for fertilization. The discontinuity also provides infectious microorganisms with access to the peritoneal cavity, possibly leading to PID.

4. 1. uterus   2. vagina   3. uterine tube   4. clitoris   5. uterine tube   6. hymen   7. ovary   8. fimbriae

5. The alveolar glands should be colored blue, and the rest of the internal breast, excluding the duct system, should be colored yellow.

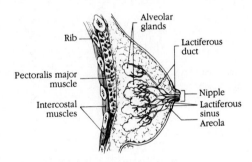

Figure 28.7

Physiology of the Female Reproductive System

1. 1. FSH   2. LH   3. estrogen, progesterone   4. estrogen   5. LH   6. LH

2. The granulosa cells produce estrogen. The corpus luteum produces progesterone. Oocytes are the central cells in all immature follicles and the larger cells pushed to one side in the mature vesicular follicles.
1. ovulation   2. peritoneal cavity   3. when sperm penetration occurs   4. ruptured (ovulated) follicle, which first becomes a corpus hemorrhagicum   5. one ovum, three polar bodies   6. They (the polar bodies) deteriorate because they lack nutrient-containing cytoplasm.

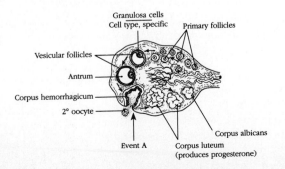

Figure 28.8

3. 1. C 2. D 3. D 4. B
4. From left to right on part C, the structures are: the primordial follicle, the secondary follicle, the vesicular follicle, the ovulating follicle, and the corpus luteum. In part D, menses is from day 0 to day 5, the proliferative stage is from day 6 to day 14, and the secretory stage is from day 14 to day 28.

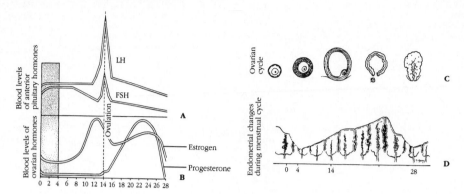

Figure 28.9

5. 1. A, B 2. B 3. A 4. B 5. A 6. A 7. B
6. 1. appearance of axillary and pubic hair 2. development of breasts 3. widening of pelvis 4. onset of menses
7. The erection phase is similar in both males and females. Blood pressure rises, the genitals (and breasts for females) become engorged with blood, and reproductive secretions are released. The male penis and the female clitoris and nipples become erect. The second phase, referred to as orgasm, involves pleasurable sensations and rhythmic muscle contractions. Ejaculation occurs in males but not in females.

Sexually Transmitted Diseases

1. 1. B 2. D 3. C 4. A 5. B 6. A 7. B 8. A 9. C, D 10. E

Developmental Aspects of the Reproductive System: Chronology of Sexual Development

1. 1. gonadal ridges 2. sexually indifferent 3. XY (X and Y) 4. XX 5. mesonephric 6. testosterone 7. female ducts 8. genital tubercle 9. gubernaculum 10. spermatic cord 11. cryptorchidism 12. puberty 13. penis 14. scrotum 15. budding breasts 16. fifth 17. menopause 18. hot flashes 19. declines 20. rise 21. estrogen 22. eighth

The Incredible Journey

1. 1. uterus 2. ovary 3. fimbriae 4. ovulation 5. 2° oocyte 6. granulosa cells (corona radiata) 7. peristalsis 8. cilia 9. sperm 10. acrosomes 11. meiotic 12. ovum 13. second polar body 14. dead 15. fertilization 16. zygote 17. endometrium 18. vagina

CHALLENGING YOURSELF

At the Clinic

1. Testicular cancer.
2. Cryptorchidism; surgery to reposition the right testis in the scrotum.
3. He developed an inguinal hernia from the heavy lifting.
4. Enlarged prostate gland; prostatitis, prostate cancer.
5. Possibility of cervical cancer.
6. Endometriosis.
7. Pelvic inflammatory disease; the causative organism must be identified to determine the proper antibiotic.
8. Ovarian cancer, an extremely serious condition.
9. Uterine prolapse.
10. STDs (sexually transmitted diseases); these infections can be asymptomatic in females; lack of treatment can lead to sterility and PID. In addition, these STDs often go hand in hand with infections caused by HIV, the AIDS virus.

Stop and Think

1. Testosterone and estrogens contribute to maintenance of bone density and muscle mass as well as metabolic rate and (in females) low cholesterol level.
2. The woman appears to be a partial hermaphrodite: external genitalia that do not match the gonads. Presence of testosterone indicates that testes, not ovaries, are present; an explanation of this condition would be faulty testosterone receptors during (her) early development in utero. Inability to respond to testosterone would result in default development of female external genitalia. Males have no homologue of the uterus.

3.   Mumps would increase scrotal temperature transiently. This would have no effect before spermatogenesis begins, but would impair sperm production in an adult male. If inflammation led to a breakdown in the blood-testis barrier, antibodies to sperm could be produced (see next answer).

4.   The blood-testis barrier is maintained by the tight junctions between adjacent sustentacular cells. This prevents exposure of the immune system to sperm antigens, which are not produced during the period of development of immune (self) tolerance and would be recognized as foreign by the immune system. A breakdown in this barrier, by trauma, poisoning, or inflammation, would be necessary to allow sperm antigens to escape into the interstitial spaces. Once exposed to sperm antigens, B lymphocytes would be activated and mount the immune attack.

5.   *Estrus* is the period of receptivity in a female animal; this has been characterized as a "frenzy" in other animals.

COVERING ALL YOUR BASES

Multiple Choice

1. D   2. A, C   3. A, D   4. A, C   5. A, B   6. B, C   7. C   8. C   9. D   10. A, D   11. B, C, D   12. B
13. B, C, D   14. C, D   15. B, C, D   16. C   17. B   18. A, B   19. A, C   20. C   21. A, B   22. C
23. A, B, C, D   24. C   25. D   26. D   27. B   28. B   29. B   30. D   31. A   32. A   33. A   34. B

Word Dissection

Word root	Translation	Example	Word root	Translation	Example
1. acro	tip	acrosome	20. mamma	teat	mammary gland
2. alb	white	tunica albuginea	21. mei	less	meiosis
3. apsi	juncture	synapsis	22. men	month	menopause
4. arche	first	menarche	23. menstru	monthly	menstrual cycle
5. cremaster	suspenders	cremaster muscle	24. metr	womb, mother	endometrium
6. didym	testis	epididymis	25. oo	egg	oogenesis
7. estro	frenzy	estrogen	26. orchi	testis	cryptorchidism
8. forn	arch	vaginal fornix	27. ov	egg	ovary
9. gamet	spouse	gametes	28. pampin	tendril	pampiniform plexus
10. gen	bear, produce	genitalia	29. penis	tail	glans penis
11. gest	carried	progesterone	30. pub	of the pubis	puberty
12. gono	seed, offspring	gonads	31. pudend	shameful	pudendal arteries
13. gubern	a rudder; govern	gubernaculum	32. salpin	trumpet	mesosalpinx
14. gyneco	woman	gynecology	33. scrot	pouch	scrotum
15. herp	creeping	herpes	34. semen	seed, sperm	seminal vesicle
16. hymen	membrane	hymen of the vagina	35. uter	womb	uterus
17. hyster	uterus	hysterectomy	36. vagin	sheath	tunica vaginalis
18. inguin	the groin	inguinal canal	37. vener	of Venus	venereal disease
19. lut	yellow	corpus luteum	38. vulv	covering, wrapper	vulva

Chapter 29   Pregnancy and Human Development

BUILDING THE FRAMEWORK

From Egg to Embryo

1.   This allows the digestive enzymes to be released where they are needed for penetration of the oocyte. If they were released in the male tract, they might cause autolysis of the male reproductive organs.

2.   Meiosis II, to form the ovum and second polar body.

3.   1. Only its nucleus.   2. Clears the path through the corona radiata to the oocyte membrane by breaking down the intercellular cement holding the granulosa cells together.   3. beta protein; alpha protein

4.   1. N   2. M   3. C   4. A   5. L (E)   6. I   7. B   8. G   9. D   10. K   11. F, G   12. J (E)   13. F   14. H (E)

5.   Part I: 1. fertilization (sperm penetration)   2. 2° oocyte; zygote after fertilization   3. cleavage   4. blastocyst (chorionic vesicle)   5. implanting embryo undergoes gastrulation
     Part II: 1. secondary oocyte   2. ovum   3. second polar body   4. pronuclei   5. zygote (fertilized ovum)   6. cleavage   7. morula   8. blastocyst   9. trophoblast   10. gastrulation   11. embryo (embryonic disc)   12. chorion

6.   The syncytiotrophoblast cells of the trophoblast, a multinuclear cytoplasmic mass, projects invasively into the endometrium and rapidly erodes it away (by digesting the uterine cells) until it has become buried in the uterine wall. The process takes about one week from its start to its finish when it has been sealed off from the uterine cavity by the proliferation of the endometrial cells.

7.

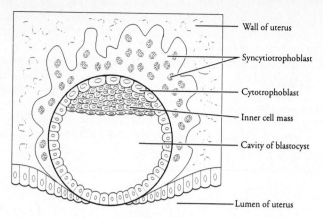

Wall of uterus

Syncytiotrophoblast

Cytotrophoblast

Inner cell mass

Cavity of blastocyst

Lumen of uterus

Figure 29.2

8. 1. D 2. B, C 3. E 4. F 5. B 6. E 7. E 8. A

9. The conceptus and then the placenta release hCG, which is like LH and sustains the function of the corpus luteum temporarily.

10. 1. and 2. fetal and maternal 3. trophoblast 4. chorionic villi 5. intervillus spaces 6. functionalis 7. chorionic villi 8. decidua basalis

Events of Embryonic Development

1. 1. D 2. B 3. A, C

2. The surface layer of the embryonic disc is ectoderm and should be colored blue. The lower layer is endoderm, which should be colored yellow.

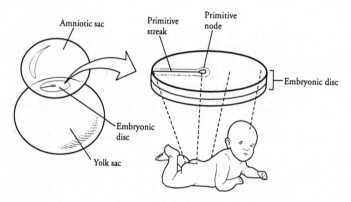

Amniotic sac

Primitive streak

Primitive node

Embryonic disc

Embryonic disc

Yolk sac

Figure 29.3

3.

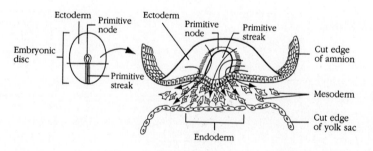

Ectoderm

Primitive node

Embryonic disc

Primitive streak

Ectoderm

Primitive node

Primitive streak

Cut edge of amnion

Mesoderm

Cut edge of yolk sac

Endoderm

Figure 29.4

1. primary germ layers 2. amnion 3. embryonic disc 4. ectoderm 5. primitive groove 6. mesoderm 7. yolk sac 8. endoderm

4. 1. Helps form the dermis of the skin. 2. Forms skeletal muscles of the neck and trunk. 3. Forms vertebrae and ribs

5. 1. G 2. F 3. H 4. E 5. A 6. B 7. C 8. D 9. I 10. D 11. C

Events of Fetal Development

1. The fetal lungs are not functioning in oxygen exchange, and they are collapsed. The placenta makes the gas exchanges with the fetal blood. Two lung bypass shunts exist, the foramen ovale and the ductus arteriosus.

2.

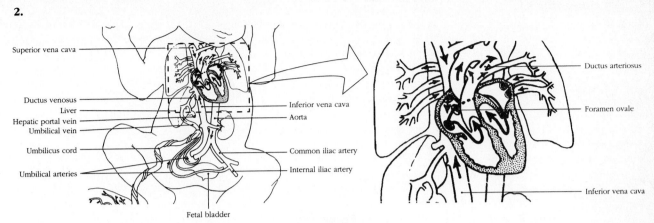

Figure 29.5

3.

System	Ectoderm	Mesoderm	Endoderm
Nervous	All nervous tissue; cornea and lens of the eye	Fibrous and vascular tunics of the eye	X
Circulatory	X	Blood; heart, and blood vessels, in their entirety	X
Digestive	Oral and anal epithelium; tooth enamel	Smooth muscle, connective tissue, and serosa of alimentary canal	Epithelium of digestive tract; liver and pancreas
Respiratory	Epithelium of nasal cavities and sinuses	Connective tissue of lungs; smooth muscle and serosa of respiratory tubes	Epithelium of respiratory tract
Endocrine	Pineal, pituitary, adrenal medulla	Gonads, adrenal cortex	Thyroid, parathyroid, thymus, pancreas
Muscular	X	Skeletal muscle and associated connective tissue sheaths	X
Skeletal	Some cranial bones and cartilages	Bone, cartilage, ligaments, synovial membranes, marrow	X
Integumentary	Epidermis and derivatives, melanocytes	Dermis and other connective tissues	X
Urogenital	X	Kidneys; gonads; smooth muscle and serosa of the ureters, bladder, and reproductive ducts	Epithelium of the urethra, bladder, reproductive ducts, and glands

4.   1. Tissue specialization   2. growth

Effects of Pregnancy on the Mother

1.   Check 1, 2, 4, 6, 10, 11, 12, 13, 15.
2.   1. Maintains corpus luteum (LH-like)   2. Maintains endometrium, stimulates mammary glands   3. Inhibits uterine smooth muscle contraction, stimulates differentiation of mammary glands, maintains endometrium   4. Increases flexibility of pelvic ligaments and pubic symphysis   5. Stimulates maturation of mammary glands, promotes growth of fetus, causes glucose sparing in mother   6. Increases metabolic rate of mother; increases her blood calcium level   7. Stimulates release of prostaglandins in uterus during labor; causes ejection of milk during lactation   8. Stimulate powerful uterine contractions during labor   9. Stimulates milk production before birth and for the duration of lactation

Parturition (Birth)

1.

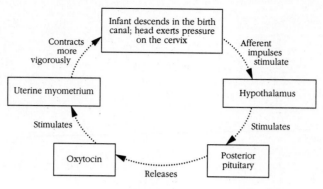

Figure 29.6

2. Each pass forces the baby farther in the birth passage. The cycle ends with the birth of the baby.
3. 1. Dilation stage: The period from the beginning of labor until full dilation (approx. 10 cm diameter) of the cervix; the longest phase.   2. Expulsion stage: The period from full dilation to birth (delivery).   3. Placental stage: Delivery of the placenta, which follows delivery of the infant.
4. False labor (irregular, ineffective uterine contractions). They occur because rising estrogen levels promote enhanced response of the uterus to oxytocin and antagonize progesterone's quieting influence on the myometrium.

Adjustments of the Infant to Extrauterine Life

1. The foramen ovale and ductus arteriosus, which allow blood returning to the heart to bypass the lungs, are occluded after birth. In order for the infant to make the crucially important gas (O_2 and CO_2) exchanges, the pulmonary circuit must become active and functional. The ductus venosus, which bypasses the fetal liver, is also occluded. The fetal liver must now carry out its normal metabolic roles, roles the mother's placenta had been performing before birth.
2. Rising blood levels of CO_2.

Lactation

1. The response to the stimulus enhances the stimulus. For example, the more a baby suckles at its mother's breast, the more prolactin and oxytocin are produced.
2. Both are products of the mammary glands. Colostrum is "premilk" produced for two or three days after birth; then true milk production begins. Colostrum contains less lactose and fat but more protein, vitamin A, and minerals than true milk.

CHALLENGING YOURSELF

At the Clinic

1. No; tests are for the presence of hCG, which first appears one week after fertilization (which typically occurs within a day of ovulation).
2. Fetal alcohol syndrome (FAS).
3. Mrs. Duclos' blood will be tested for estrogens and progesterone. hCG levels, which have been maintaining the corpus luteum's production of estrogens and progesterone, will begin to wane during the third month. By the fourth month, luteal hormone levels fall. If placental production of estrogens and progesterone is insufficient, then the endometrium will be sloughed off, ending the pregnancy.
4. hPL, human placental lactogen, has a glucose-sparing effect, which liberates glucose for the fetus.
5. No; morning sickness usually ends after the first few months.
6. The increased pressure in the pelvic cavity from carrying the fetus as well as from the birthing process can cause varicosities in the vessels of the anal canal.
7. Ibuprofen is a prostaglandin inhibitor; such drugs can inhibit labor in the early stages.
8. Passing meconium after labor has been ongoing might indicate a breech birth.
9. Cesarian section.
10. Perhaps nonclosure of the ductus arteriosus and the foramen ovale.
11. Oxytocin, secreted as part of the let-down reflex of nursing, stimulates uterine contractions, which (while painful) help the uterus shrink back to a smaller size.
12. Physiological jaundice; no; as the liver matures, the jaundice should disappear.
13. Eclampsia.

Stop and Think

1. The size of normal body cells.
2. As the preembryo gets to the blastocyst stage, accumulation of fluid causes its size to increase; it is at this time that the conceptus "hatches" from the zona.
3. Anything that delays implantation can prevent the proper signaling of the corpus luteum by hCG. Without that signal, the corpus luteum will fail and the endometrium will deteriorate, ending the chance of successful implantation and pregnancy.
4. More, since organogenesis is just beginning in the second week and each primitive tissue type is at an earlier stage of differentiation. Hence, the earlier the interference, the more serious the deficits.
5. Progesterone inhibits smooth muscle contraction, leading to constipation.
6. Relaxin loosens ligaments, including foot ligaments. The added weight of the fetus and fluid and new maternal tissue strains the arches of the feet, causing fallen arches and elongated feet.
7. Amniotic fluid volume is increased by fetal urine output. A deficiency of fluid volume could indicate fetal renal failure or blockage of urine flow.
8. No; there are 40 somite pairs and only about 33 vertebrae (some of which later fuse to form the composite sacrum and coccyx).
9. Endothelium; mesoderm.
10. Only the umbilical vein.
11. The brain (within its bony skull) cannot enlarge more than will be permitted to pass through the pelvic opening.
12. Cyanosis due to inadequate oxygenation of the blood.
13. No. The weight of the uterus would press on the inferior vena cava, significantly reducing venous return from the lower torso and legs.
14. Lysosomes.
15. Cesarian section.
16. Closure of the ductus arteriosus will increase venous return to the left atrium as more blood is allowed to flow to the lungs; ligation of the umbilical vein and arteries will shut down the significant flow of blood to the right atrium. Consequently, right atrial pressure declines and left atrial pressure increases. The flap over the foramen ovale will press against the interatrial septum rather than being pushed open.
17. No. A little research on the incidences of congenital diseases reveals that all of these numbers are within normal limits.

COVERING ALL YOUR BASES

Multiple Choice

1. B, C, D   2. A, B   3. A, C, D   4. B   5. A, B, C, D   6. A, D   7. A, B, D   8. A, C   9. A, B, C   10. B   11. A, D
12. A, B, D   13. B. C   14. C, D   15. B, D   16. B, C, D   17. B, C, D   18. C   19. C   20. C   21. A, B, C
22. A, B   23. A, B, C   24. B, C   25. A, B, C   26. B   27. C   28. A, B, C, D   29. A, D   30. A, C   31. D
32. A, C, D   33. A, D   34. A

Word Dissection

Word root	Translation	Example	Word root	Translation	Example
1. bryo	swollen	embryo	9. noto	back	notochord
2. chori	membrane	chorion	10. partur	give birth	parturition
3. coel	hollow	coelom	11. placent	flat cake	placenta
4. decid	falling off	decidua basalis	12. terato	a monster	teratogen
5. episio	vulva	episiotomy	13. toc	birth	dystocia
6. fetus	fruitful	fetus, fetal	14. troph	nourish	trophoblast
7. gest	carry	gestation	15. ula	little	gastrula
8. nata	birth	neonatal	16. zyg	yoke	zygote

Chapter 30   Heredity

BUILDING THE FRAMEWORK

The Vocabulary of Genetics

1. 1. Freckled female   2. Identical chromosomes   3. Heterozygous   4. *Ff*   5. Recessive is expressed   6. Diploid
   7. Haploid   8. Code for different traits
2. 1. T   2. Alleles   3. genotype   4. T   5. Linked   6. incomplete dominance or intermediate inheritance   7. T

Sexual Sources of Genetic Variation

1. 1. G   2. D, E, G   3. 2 and 3   4. one

2. 1. Independent assortment of homologous chromosomes during meiosis I. 2. Recombination of genes during crossing over of homologues. 3. Random fertilization of any egg by any sperm.

Types of Inheritance

1. 1. F 2. F 3. E 4. D 5. C 6. B 7. F 8. B 9. E

2. 1. Straight hairline, thin lips, free earlobes, freckles, dimples. 2. *hh, tt, EE, Ff, Dd* 3. no; yes 4. Figure 30.3: *htEXDf, htEXdF, htEXdf, htEXDF*. Recessive traits are straight hairline and thin lips. Dominant traits are free earlobes, freckles, and dimples.

3.

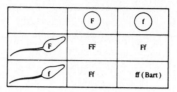

Figure 30.3

1. 25% or 1 in 4 2. FF

4. 1. $X^B X^B$ or $X^B X^b$ 2. $X^B X^b$ 3. $X^B Y$ 4. $X^b Y$

5. $I^A I^A$, $I^B I^B$, or $I^A I^B$

6. 1. *aa* 2. *Aa* 3. *AA* 4. *Aa*

7. 1. *aabbcc* 2. *AaBbCc* 3. *AaBbCc* 4. No, because albinism is a recessive Mendelian trait unrelated to these genes with genotype *aa* (homozygous recessive). 5. Yes, the child could inherit *AABBCC* or other multiple dominant combinations.

The Influence of Environmental Factors on Gene Expression

1. 1. the genes 2. a normal 3. T 4. nutrition 5. hormones

Nontraditional Inheritance

1. 1. A 2. A 3. B 4. A 5. B 6. B 7. A

Genetic Screening, Counseling, and Therapy

1. 1. karyotype 2. chromosome 21 3. 35 4. carrier 5. 25% or 1 in 4 6. pedigree analysis 7. blood tests 8. amniocentesis 9. genetic disorder 10. fourteenth 11. cultured (grown) 12. chorionic villi 13. eight

2. 1. yes 2. recessive 3. 2, 5, 7, 18 4. 9, 14, 16, 23, 25 5. no 6. consanguineous 7. no 8. 100% 9. 0% 10. 100% 11. 50% 12. 25%

CHALLENGING YOURSELF

At the Clinic

1. The homozygous condition for the achondroplasia gene is lethal; this most likely explains the miscarriages. Since people with achondroplasia are necessarily heterozygous, their children will have a one in four chance of being normal.

2. Huntington's disease is another example of a condition in which the homozygous genotype is lethal; therefore, the young man must be heterozygous or homozygous recessive. Having the condition in members of both his mother's and his father's families does increase his risk of inheriting the condition, though, if his mother has Huntington's disease. If she is not stricken, then she does not carry the gene, and he could have received it only from his father's gamete. His chance of having the gene is 50% (if his mother does not carry it); the chance of his passing it to his children if he has it is also 50%. Thus, the chance of his children getting the gene is 50% times 50%, or 25%.

3. Since albinism is recessive and her husband is probably homozygous dominant, all of the children would be heterozygous and have normal skin coloration. Hence, the chance of an albino offspring in this generation is zero. (However, all of the children would be carriers of the gene.)

4. Since cystic fibrosis is a recessive condition, both parents must be heterozygous (carriers) of the defective gene. The chance of any single child being homozygous recessive and suffering from the disorder is 25% in this mating.

5. Because of the earlier offspring's death, a genetic disorder is suspect. If the parents trace their origin to Eastern European Jews, then Tay-Sachs disease is a likely diagnosis.

6. Because the younger boy has a 50% chance of having the sickle-cell trait, he needs to avoid prolonged exposure to low oxygen levels, such as living at high altitude and excessively vigorous playing activity.

7. None. Since the boy has inherited his hemophilia from his mother and this is an X-linked condition, the second child, being a girl, has inherited a normal gene on the paternal X chromosome. This is a recessive condition, so the girl will be normal.

8. PKU is a genetic disorder, but the condition can be managed with dietary restrictions that limit the amount of phenylalanine-containing foods.

9. Down syndrome is a chromosomal nondisjunction defect, not a genetic defect. It is the result of an "accident" during meiosis. Thus, it cannot be predicted as precisely as genetic disorders in cases where parental genotypes are known. Predictions are typically made based on statistical evidence (number of cases per 1000 births). The current approach is to perform an amniocentesis and do a karyotype (which will show three chromosome 21s if the condition is present).

10. Chorionic villi sampling; she must wait at least until 8 weeks gestational age (10 weeks is safer).

Stop and Think

1. Increasing maternal age reduces the ability of the reproductive system to discriminate and reject defective embryos.

2. All mitochondria are inherited from the mother, in the ovum. This allows tracing of matrilineal heritage by examining genes on the mitochondrial DNA. In fact, many genetic conditions are now being traced to mitochondrial DNA (extrachromosomal inheritance).

3. If testosterone is secreted in substantial enough amounts (e.g., if the mother has an adrenal cortex tumor) when the indifferent reproductive structures are in the process of differentiating to male or female structures, a genetic female (XX) will develop male structures. By the same token, if testosterone is *not* present at the appropriate developmental time (e.g., the male fetus's testes fail to secrete testosterone), a genetic male (XY) will form a female reproductive tract and external genitalia.

4. The closer two gene loci are to each other, the less likely they are to be separated at crossover.

5. Since the degeneration associated with Huntington's disease does not appear until after typical child-bearing years, a genetic marker would allow people to know whether they carry the gene before they decide to have children. A genetic counselor would inform anyone found to carry the gene that the chance of passing it on is 50%. The test would not find a homozygous dominant individual, since that genotype is lethal.

6. She inherited this X-linked condition from both her mother (who is a carrier) and her father, who is color-blind.

COVERING ALL YOUR BASES

Multiple Choice

1. C   2. A   3. D   4. A, B, C, D   5. A, B, D   6. B   7. B   8. A   9. A, B, D   10. B, C, D   11. A, B   12. D
13. A, C, D   14. C   15. C   16. A, C   17. B C, D   18. A, B, C   19. C   20. A, B, C, D

Word Dissection

Word root	Translation	Example		Word root	Translation	Example
1. allel	of one another	allele		5. karyo	kernel, nucleus	karyotype
2. centesis	pricking	amniocentesis		6. pheno	to show or see	phenotype
3. chiasm	cross	chiasma		7. sanguin	blood	consanguineous
4. gen	born, become	gene, genome				

Appendix B
How to Use A.D.A.M. with This *Study Guide*

This *Study Guide* is a valuable tool to enhance your knowledge of anatomy and physiology. You will notice that some illustrations (figures) include a "walking man" icon. For these figures, you can use A.D.A.M. software, Standard version, to enhance your learning of the associated anatomy.

When you see the A.D.A.M. icon, use the table that follows to see which view and layer in A.D.A.M. best illustrates the structures shown in the figure. For example, an icon is included with Figure 7.8 in the *Study Guide*. Using the table that follows, you will find that the best view to see the scapula in A.D.A.M. is "posterior" and that the layer is "scapula." Some figures are best seen in A.D.A.M. using "View" on the Menu Bar and choosing the appropriate system instead of selecting a specific view and layer using the "Full Anatomy" option. Also listed in the table that follows are the corresponding figure number in *Human Anatomy and Physiology*, Third Edition, by Elaine Marieb ("Text Figure"), and the key structures shown in the figures and in A.D.A.M.

After you have opened A.D.A.M., you will see three windows. The large one that covers the left half of the screen is the Primary Window. The top part of the right half of the screen is the Library Window, and the bottom part of the right half of the screen is the Structure List Window. The Primary Window contains the content material you are studying and is where you can identify anatomical structures on the image. Electronic publications are located in the Library Window, and the Structure List Window contains the layer lists and helps you navigate through A.D.A.M.

When you are studying a specific figure in the *Study Guide*, you can choose the corresponding A.D.A.M. view in two ways. One is to click once on the selected view (A—anterior, L—lateral, P—posterior, or M—medial) in the Anatomy View Panel located in the Primary Window. The left arrow on the panel moves back to the previous view, and the right arrow moves forward to the next view.

Another way to choose the appropriate view is to select "View" on the Menu Bar located above the Primary Window. After choosing "Full Anatomy" from the list, select "Options" on the Menu Bar and click on the recommended view.

The A.D.A.M. layer that corresponds to the *Study Guide* figure can be located by choosing the "List Options" button in the Structure List Window. When the "List Options" screen appears, choose "Alphabetic" and close the window by clicking on "OK." The list of layers is now sorted alphabetically in the Structure List Window, and you can look through the list using the scroll bars on the right side of the box. When you have located the layer you want to see, simply click on it. The layer will be highlighted, and you will be able to view that layer in the Primary Window.

In addition to using the "List Options" in the Structure List Window, you can choose the A.D.A.M. layer you want to review by clicking and holding down the mouse button on the Title Bar indicator located in the Primary Window. All structures in your selected view will be shown in alphabetical order. When you have chosen the layer you want to view by scrolling down the list, release the mouse button and the layer will appear.

You can also locate a layer quickly by clicking the "Search List" button in the Structure List Window. When the dialog box appears, type the name of the layer you are searching for. Close the dialog box by clicking on "OK." The layer in the Structure List Window will now be highlighted. Click on the layer name, and it will appear in the Primary Window.

Once you have chosen the recommended view and layer in A.D.A.M., you are ready to identify the structures that can be seen in the image. Click the "Identify" tool located on the left side of the Primary Window. After the pointer changes to an arrow, place it on the structure you want to identify and hold the mouse button down. The name of the system, structure, and substructure (if present) will be shown. Use the "Isolate" button to highlight selected structures.

There are many additional options that can be chosen when using A.D.A.M. Once you have learned how to navigate and use all of its available tools, you will find that there is a wealth of information contained in A.D.A.M., and you will really enjoy learning the anatomy of the human body.

Study Guide Figure Number	Text Figure Number	Key Structures	A.D.A.M. View and Layer
1.1	1.2f	Cardiovascular system	Using "View" on the Menu Bar, choose Cardiovascular system
1.2	1.2i	Respiratory system	Using "View" on the Menu Bar, choose Respiratory system
1.3	1.2d	Nervous system	Using "View" on the Menu Bar, choose Nervous system
1.4	1.2k	Female urinary system	Using "Options," choose Female; using "View" on the Menu Bar, choose Urinary system
1.5	1.2j	Digestive system	Using "View" on the Menu Bar, choose Digestive System

Study Guide Figure Number	Text Figure Number	Key Structures	A.D.A.M. View and Layer
1.6	1.2m	Female reproductive system	Using "Options," choose female; using "View" on the Menu Bar, choose Reproductive system
4.4	4.9c	Pericardium	Go to "Full Anatomy" View: Anterior Layer: Heart—cut section
4.4	4.9c	Pleura	Go to "Full Anatomy" View: Anterior Layer: Lungs
4.4	4.9c	Peritoneum	Go to "Full Anatomy" View: Medial Layer: Mesentery
6.1	6.3	Femur—coronal section	Go to "Full Anatomy" View: Anterior Layer: Bones—coronal section
7.1	7.1	Skeletal system	Using "View" on the Menu Bar, choose Skeletal system
7.1	7.1	Skull bones	Go to "Full Anatomy" View: Lateral Layer: Skull
7.2A	7.3a	Skull bones	Go to "Full Anatomy" View: Lateral Layer: Skull
7.2C	7.2a	Skull bones	Go to "Full Anatomy" View: Anterior Layer: Skull
7.3A	7.11a	Paranasal sinuses	Go to "Full Anatomy" View: Anterior Layer: Anterior jugular vein
7.3B	7.11b	Paranasal sinuses	Go to "Full Anatomy" View: Medial Layer: Nerves of the adductor canal
7.4	7.13	Vertebral column	Go to "Full Anatomy" View: Medial Layer: Subcutaneous fat
7.7	7.19a	Thorax	Using "View" on the Menu Bar, choose Skeletal system
7.8	7.21e	Scapula	Go to "Full Anatomy" View: Posterior Layer: Scapula

Study Guide Figure Number	Text Figure Number	Key Structures	A.D.A.M. View and Layer
7.9A	7.22a	Humerus	Go to "Full Anatomy" View: Anterior Layer: Humerus
7.9B	7.23a	Radius	Go to "Full Anatomy" View: Anterior Layer: Radius
7.9C	7.23a	Ulna	Go to "Full Anatomy" View: Anterior Layer: Ulna
7.10	7.24a	Bones of the hand	Go to "Full Anatomy" View: Anterior Layer: Bones—coronal section
7.11	7.25a	Pelvic girdle	Go to "Full Anatomy" View: Anterior Layer: Femur
7.12	7.27b	Femur	Go to "Full Anatomy" View: Anterior Layer: Femur
7.12	7.28a	Tibia	Go to "Full Anatomy" View: Anterior Layer: Tibia
7.12	7.28a	Fibula	Go to "Full Anatomy" View: Anterior Layer: Fibula
8.1	8.3a	Knee joint	Go to "Full Anatomy" View: Anterior Layer: Bones—coronal section
8.3A	8.10a	Hip joint	Go to "Full Anatomy" View: Anterior Layer: Coronary ligaments of the knee
10.3	10.8a	Muscles of the anterior neck	Go to "Full Anatomy" View: Anterior Layer: Sternohyoid muscle
10.4	10.6	Muscles of the scalp, face, and neck	Go to "Full Anatomy" View: Lateral Layer: Orbicularis oculi muscle
10.5	10.4b	Muscles of the anterior trunk	Using "View" on the Menu Bar, choose Muscular system
10.6	10.13b	Superficial muscles of the back; muscles of the scapular region	Go to "Full Anatomy" View: Posterior Layers: Trapezius muscle; Rhomboideus muscle

Study Guide Figure Number	Text Figure Number	Key Structures	A.D.A.M. View and Layer
10.7	10.9d	Deep muscles of the back	Go to "Full Anatomy" View: Posterior Layers: Iliocostalis lumborum muscle; Semispinalis capitis muscle
10.8	10.15a	Superficial muscles of the anterior forearm and hand	Using "View" on the Menu Bar, choose Muscular system
10.9	10.16a	Superficial muscles of the posterior arm and forearm	Go to "Full Anatomy" View: Posterior Layer: Deltoid muscle
10.10A	10.19a 10.22a	Superficial muscles of the posterior hip, thigh, and leg	Go to "Full Anatomy" View: Posterior Layer: Gluteus maximus muscle
10.10B	10.18a 10.20a	Superficial muscles of the anterior and medial thigh and anterior leg	Go to "Full Anatomy" View: Anterior Layer: Tensor fasciae latae muscle
10.11	10.4b	Superficial muscles of the anterior body	Using "View" on the Menu Bar, choose Muscular system
10.12	10.5b	Superficial muscles of the posterior body	Go to "Full Anatomy" View: Posterior Layer: Deltoid muscle
10.13 Anterior View 10.14 Anterior View	10.4b	Superficial muscles of the anterior body	Using "View" on the Menu Bar, choose Muscular system
10.13 Posterior View 10.14 Posterior View	10.5b	Superficial muscles of the posterior body	Go to "Full Anatomy" View: Posterior Layer: Deltoid muscle
11.1	11.1b	Spinal cord	Go to "Full Anatomy" View: Posterior Layer: Spinal cord
12.1	12.8	Surface of the brain	Go to "Full Anatomy" View: Lateral Layer: Brain
12.5	12.14a	Midsagittal brain	Go to "Full Anatomy" View: Medial Layer: Nerves of the adductor canal
12.6	12.9	Regions of the cerebral hemisphere	Go to "Full Anatomy" View: Anterior Layer: Bones—coronal section

Study Guide Figure Number	Text Figure Number	Key Structures	A.D.A.M. View and Layer
12.9	12.21	Meninges of the brain	Go to "Full Anatomy" View: Anterior Layer: Bones—coronal section
13.5	13.8c	Peripheral nerves of the brachial plexus in the upper limb	Using "View" on the Menu Bar, choose Nervous system
13.6	13.10b	Peripheral nerves of the sacral plexus in the lower limb	Go to "Full Anatomy" View: Posterior Layer: Sciatic nerve and branches
13.6	13.10b	Peripheral nerves of the sacral plexus in the lower limb	Go to "Full Anatomy" View: Posterior Layer: Tibial nerve and branches
16.3	16.6a	Extrinsic muscles of the eye	Go to "Full Anatomy" View: Lateral Layer: Eye muscles—lateral
17.1	17.1	Endocrine system	Using "View" on the Menu Bar, choose Endocrine system
19.1	19.2	Pericardium	Go to "Full Anatomy" View: Anterior Layer: Heart—cut section
19.3	19.4a	Heart and great vessels	Go to "Full Anatomy" View: Anterior Layer: Cardiac veins
19.4A 19.4B	19.4a	Coronary vessels	Go to "Full Anatomy" View: Anterior Layer: Cardiac veins
19.7	19.4c	Interior of the heart	Go to "Full Anatomy" View: Anterior Layer: Heart—cut section
20.4	20.15b	Pulmonary circulation	Go to "Full Anatomy" View: Anterior Layer: Pulmonary arteries
20.5	20.17b	Arteries of the systemic circulation	Using "View" on the Menu Bar, choose Cardiovascular system
20.6	20.22b	Veins of the systemic circulation	Using "View" on the Menu Bar, choose Cardiovascular system
20.7	20.18b	Arteries of the head and neck	Go to "Full Anatomy" View: Lateral Layer: Common carotid artery and branches

Study Guide Figure Number	Text Figure Number	Key Structures	A.D.A.M. View and Layer
20.9	20.23b	Veins of the head and neck	Go to "Full Anatomy" View: Lateral Layer: External jugular vein and tributaries
20.9	20.23c	Dural sinuses of the brain	Go to "Full Anatomy" View: Lateral Layer: Cardiac veins
20.10A	20.19b	Arteries of the upper limb	Using "View" on the Menu Bar, choose Cardiovascular system
20.10A	20.19b	Branches of the axillary artery	Go to "Full Anatomy" View: Anterior Layer: Axillary artery and branches
20.10B	20.24b	Superficial veins of the upper limb	Go to "Full Anatomy" View: Anterior Layer: Superficial veins
20.11	20.20c	Arteries of the abdomen	Go to "Full Anatomy" View: Anterior Layer: Abdominal aorta and branches
20.11	20.25b	Veins of the abdomen	Go to "Full Anatomy" View: Anterior Layer: Inferior vena cava
20.12	20.24b	Veins of the deep thorax	Go to "Full Anatomy" View: Anterior Layer: Azygos vein and tributaries
20.13A	20.21b	Arteries of the pelvis, thigh, and leg	Go to "Full Anatomy" View: Anterior Layer: Abdominal aorta and branches
20.14	20.25c	Hepatic portal circulation	Go to "Full Anatomy" View: Anterior Layer: Portal vein and tributaries
21.2	21.2b	Thoracic duct	Go to "Full Anatomy" View: Anterior Layer: Thoracic duct
21.3A	21.2a	Lymphatic vessels and nodes	Using "View" on the Menu Bar, choose Lymphatic system
21.3B	21.4	Remnant of thymus gland	Go to "Full Anatomy" View: Anterior Layer: Remnant of thymus gland
21.3B	21.4	Spleen	Go to "Full Anatomy" View: Anterior Layer: Spleen

Study Guide Figure Number	Text Figure Number	Key Structures	A.D.A.M. View and Layer
23.1	23.2a	External nose	Go to "Full Anatomy" View: Anterior Layer: Nasal cartilage
23.2	23.3a	Upper respiratory tract	Go to "Full Anatomy" View: Medial Layer: Nasal conchae
23.3	23.4a	Trachea	Go to "Full Anatomy" View: Anterior Layer: Tracheal lymph nodes
23.5	23.7a	Lower respiratory tract	Go to "Full Anatomy" View: Anterior Layer: Tracheal lymph nodes
24.1	24.1	Digestive system	Using "View" on the Menu Bar, choose Digestive system
24.3	24.7b	Oral cavity	Go to "Full Anatomy" View: Anterior Layer: Palatoglossus muscle
24.4	24.11	Tooth	Go to "Full Anatomy" View: Lateral Layer: Brain
24.5A	24.14a	Stomach	Go to "Full Anatomy" View: Anterior Layer: Stomach—coronal section
24.7	24.20a	Liver, gallbladder, and associated ducts	Go to "Full Anatomy" View: Anterior Layer: Gallbladder
24.7	24.20a	Pancreas and duodenum	Go to "Full Anatomy" View: Anterior Layer: Pancreas
24.8	24.20a	Large intestine	Go to "Full Anatomy" View: Anterior Layer: Transverse colon
26.1	26.1a	Female urinary system Male urinary system	Using "View" on the Menu Bar, choose Urinary system; using "Options," choose Female; using "Options," choose Male
26.2	26.3b	Internal kidney	Go to "Full Anatomy" View: Anterior Layer: Ureter
28.1	28.1	Male reproductive system	Using "Options," choose Male View: Medial Layer: Celiac trunk

Study Guide Figure Number	Text Figure Number	Key Structures	A.D.A.M. View and Layer
28.6	28.11b	Female reproductive system	Using "Options," choose Female View: Medial Layer: Celiac trunk and branches
28.7	28.17	Breast	Using "Options," choose Female View: Anterior Layer: Breast

Study Guide Figure Number	Text Figure Number	Key Structures	A.D.A.M. View and Layer